Information Theory

Coding Theorems for Discrete Memoryless Systems

This book is widely regarded as a classic in the field of information theory, providing deep insights and expert treatment of the key theoretical issues. It includes in-depth coverage of the mathematics of reliable information transmission, both in two-terminal and multi-terminal network scenarios. Updated and considerably expanded, this new edition presents unique discussions of information-theoretic secrecy and of zero-error information theory, including substantial connections of the latter with extremal combinatorics. The presentations of all core subjects are self-contained, even the advanced topics, which helps readers to understand the important connections between seemingly different problems. Finally, 320 end-of-chapter problems, together with helpful solving hints, allow readers to develop a full command of the mathematical techniques. This is an ideal resource for graduate students and researchers in electrical and electronic engineering, computer science, and applied mathematics.

Imre Csiszár is a Research Professor at the Alfréd Rényi Institute of Mathematics of the Hungarian Academy of Sciences, where he has worked since 1961. He is also Professor Emeritus of the University of Technology and Economics, Budapest, a Fellow of the IEEE, and former President of the Hungarian Mathematical Society. He has received numerous awards, including the Shannon Award of the IEEE Information Theory Society (1996).

János Körner is a Professor of Computer Science at the Sapienza University of Rome, Italy, where he has worked since 1992. Prior to this, he was a member of the Institute of Mathematics of the Hungarian Academy of Sciences for over 20 years, and he also worked at AT&T Bell Laboratories, Murray Hill, New Jersey, for two years.

The field of applied mathematics known as Information Theory owes its origins and early development to three pioneers: Shannon (USA), Kolmogorov (Russia) and Rényi (Hungary). This book, authored by two of Rényi's leading disciples, represents the elegant and precise development of the subject by the Hungarian School. This second edition contains new research of the authors on applications to secrecy theory and zero-error capacity with connections to combinatorial mathematics.

Andrew Viterbi, USC

Information Theory: Coding Theorems for Discrete Memoryless Systems, by Imre Csiszár and János Körner, is a classic of modern information theory. "Classic" since its first edition appeared in 1979. "Modern" since the mathematical techniques and the results treated are still fundamentally up to date today. This new edition was long overdue. Beyond the original material, it contains two new chapters on zero-error information theory and connections to extremal combinatorics, and on information theoretic security, a topic that has garnered very significant attention in the last few years. This book is an indispensable reference for researchers and graduate students working in the exciting and ever-growing area of information theory.

Giuseppe Caire, USC

The first edition of the Csiszár and Körner book on information theory is a classic, in constant use by most mathematically-oriented information theorists. The second edition expands the first with two new chapters, one on zero-error information theory and one on information theoretic security. These use the same consistent set of tools as edition 1 to organize and prove the central results of these currently important areas. In addition, there are many new problems added to the original chapters, placing many newer research results into a consistent formulation.

Robert Gallager, MIT

The classic treatise on the fundamental limits of discrete memoryless sources and channels –an indispensable tool for every information theorist.

Sergio Verdu, Princeton

Information Theory

Coding Theorems for Discrete Memoryless Systems

IMRE CSISZÁR

Alfréd Rényi Institute of Mathematics, Hungarian Academy of Sciences, Hungary

JÁNOS KÖRNER

Sapienza University of Rome, Italy

CAMBRIDGE
UNIVERSITY PRESS

CAMBRIDGE
UNIVERSITY PRESS

University Printing House, Cambridge CB2 8BS, United Kingdom

One Liberty Plaza, 20th Floor, New York, NY 10006, USA

477 Williamstown Road, Port Melbourne, VIC 3207, Australia

314-321, 3rd Floor, Plot 3, Splendor Forum, Jasola District Centre, New Delhi - 110025, India

103 Penang Road, #05-06/07, Visioncrest Commercial, Singapore 238467

Cambridge University Press is part of the University of Cambridge.

It furthers the University's mission by disseminating knowledge in the pursuit of education, learning and research at the highest international levels of excellence.

www.cambridge.org
Information on this title: www.cambridge.org/9781107565043

First published 1981
Second edition 2011
Paperback edition first published 2016

A catalogue record for this publication is available from the British Library

ISBN 978-0-521-19681-9 Hardback
ISBN 978-1-107-56504-3 Paperback

To the memory of Alfréd Rényi,
the outstanding mathematician
who established information theory in Hungary

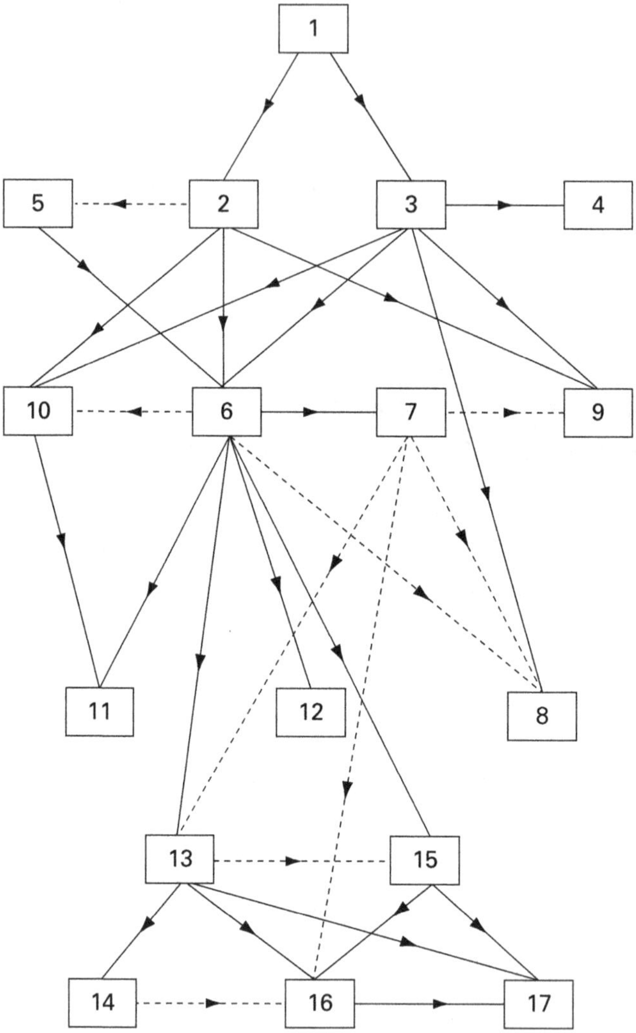

Dependence graph of the text; numbers refer to chapters

Contents

Preface to the first edition

Information theory was created by Claude E. Shannon for the study of certain quantitative aspects of information, primarily as an analysis of the impact of coding on information transmission. Research in this field has resulted in several mathematical theories. Our subject is the stochastic theory, often referred to as the Shannon theory, which directly descends from Shannon's pioneering work.

This book is intended for graduate students and research workers in mathematics (probability and statistics), electrical engineering and computer science. It aims to present a well-integrated mathematical discipline, including substantial new developments of the 1970s. Although applications in engineering and science are not covered, we hope to have presented the subject so that a sound basis for applications has also been provided. A heuristic discussion of mathematical models of communication systems is given in the Introduction, which also offers a general outline of the intuitive background for the mathematical problems treated in the book.

As the title indicates, this book deals with discrete memoryless systems. In other words, our mathematical models involve independent random variables with finite range. Idealized as these models are from the point of view of most applications, their study reveals the characteristic phenomena of information theory without burdening the reader with the technicalities needed in the more complex cases. In fact, the reader needs no other prerequisites than elementary probability and a reasonable mathematical maturity. By limiting our scope to the discrete memoryless case, it was possible to use a unified, basically combinatorial approach. Compared with other methods, this often led to stronger results and yet simpler proofs. The combinatorial approach also seems to lead to a deeper understanding of the subject.

The dependence graph of the text is shown on p. vi.

There are several ways to build up a course using this book. A one-semester graduate course can be made up of Chapters 1, 2, 6, 7 and the first half of Chapter 13. A challenging short course is provided by Chapters 2, 9 and 10. In both cases, the technicalities from Chapter 3 should be used when necessary. For students with some information theory background, a course on multi-terminal Shannon theory can be based on Part III, using Chapters 2 and 6 as preliminaries. The problems offer a lot of opportunities for creative work for the students. It should be noted, however, that illustrative examples are scarce; thus the teacher is also supposed to do some homework of his own by supplying such examples.

Every chapter consists of text followed by a Problems section. The text covers the main ideas and proof techniques, with a sample of the results they yield. The selection of the latter was influenced both by didactic considerations and the authors' research interests. Many results of equal importance are given in the Problem sections. While the text is self-contained, there are several points at which the reader is advised to supplement formal understanding by consulting specific problems. This suggestion is indicated by the Problem number in the margin of the text. For all but a few problems sufficient hints are given to enable a serious student familiar with the corresponding text to give a solution. The exceptions, marked by an asterisk, serve mainly for supplementary information; these problems are not necessarily more difficult than the others, but their solution requires methods not treated in the text.

In the text the origins of the results are not mentioned, but credits to authors are given at the end of each chapter. Concerning the Problems, an appropriate attribution accompanies each Problem. An absence of references indicates that the assertion is either folklore or else an unpublished result of the authors. Results were attributed on the basis of publications in journals or books with complete proofs. The number after the author's name indicates the year of appearance of the publication. Conference talks, theses and technical reports are quoted only if – to our knowledge – their authors have never published their result in another form. In such cases, the word "unpublished" is attached to the reference year, to indicate that the latter does not include the usual delay of "regular" publications.

We are indebted to our friends Rudy Ahlswede, Péter Gács and Katalin Marton for fruitful discussions which contributed to many of our ideas. Our thanks are due to R. Ahlswede, P. Bártfai, J. Beck, S. Csibi, P. Gács, S. I. Gelfand, J. Komlós, G. Longo, K. Marton, A. Sgarro and G. Tusnády for reading various parts of the manuscript. Some of them have saved us from vicious errors.

The patience of Mrs. Éva Várnai in typing and retyping the ever-changing manuscript should be remembered, as well as the spectacular pace of her doing it.

Special mention should be made of the friendly assistance of Sándor Csibi who helped us to overcome technical difficulties with the preparation of the manuscript. Last but not least, we are grateful to Eugene Lukács for his constant encouragement without which this project might not have been completed.

Preface to the second edition

When the first edition of this book went to print, information theory was only 30 years old. At that time we covered a large part of the topic indicated in the title, a goal that is no longer realistic. An additional 30 years have passed, the Internet revolution occurred, and information theory has grown in breadth, volume and impact. Nevertheless, we feel that, despite many new developments, our original book has not lost its relevance since the material therein is still central to the field.

The main novelty of this second edition is the addition of two new chapters. These cover zero-error problems and their connections to combinatorics (Chapter 11) and information-theoretic security (Chapter 17). Of several new research directions that emerged in the 30 years between the two editions, we chose to highlight these two because of personal research interests. As a matter of fact, these topics started to intrigue us when writing the first edition; back then, this led us to a last-minute addition of problems on secrecy.

Except for the new chapters, new results are presented only in the form of problems. These either directly complete the original material or, occasionally, illustrate a new research area. We made only minor changes, mainly corrections, to the text of the original chapters. (Hence the words *recent* and *new* refer to the time of the first edition, unless the context indicates otherwise.) We have updated the history part of each chapter and, in particular, we have included pointers to new developments. We have not broadened the original scope of the book. Readers interested in a wider perspective may consult Cover and Thomas (2006).

In the preface to the first edition we suggested several ways in which to construct courses using this book. In addition, either of the new Chapters 11 or 17 can be used for a short graduate course.

As in the first edition, this book is dedicated to the memory of Alfréd Rényi, whose mathematical heritage continues to influence information theory and to inspire us.

Special thanks are due to Miklós Simonovits, who, sacrificing his precious research time, assisted us to overcome TeX-nical difficulties as only the most selfless friend would do. We are indebted to our friends Prakash Narayan and Gábor Simonyi, as well as to the Ph.D. students Lóránt Farkas, Tamás Kói, Sirin Nitinawarat and Himanshu Tyagi for a careful reading of parts of the manuscript.

Basic notation and conventions

$\triangleq$	equal by definition
iff	if and only if
○	end of a theorem, definition, remark, etc.
□	end of a proof
$\mathsf{A}, \mathsf{B}, \ldots, \mathsf{X}, \mathsf{Y}, \mathsf{Z}$	sets (finite unless stated otherwise; infinite sets will be usually denoted by script capitals)
$\emptyset$	void set
$x \in \mathsf{X}$	x is an element of the set X; as a rule, elements of a set will be denoted by the same letter as the set
$\mathsf{X} \triangleq \{x_1, \ldots, x_k\}$	X is a set having elements $x_1, \ldots, x_k$
$\lvert\mathsf{X}\rvert$	number of elements of the set X
$\mathbf{x} = (x_1, \ldots, x_n)$, $\mathbf{x} = x_1 \ldots x_n$	vector (finite sequence) of elements of a set X
$\mathsf{X} \times \mathsf{Y}$	Cartesian product of the sets X and Y
X^n	nth Cartesian power of X, i.e., the set of n-length sequences of elements of X
X^*	set of all finite sequences of elements of X
$\mathsf{A} \subset \mathsf{X}$	A is a (not necessarily proper) subset of X
$\mathsf{A} - \mathsf{B}$	the set of those elements $x \in \mathsf{A}$ which are not in B
$\bar{\mathsf{A}}$	complement of a set $\mathsf{A} \subset \mathsf{X}$, i.e., $\bar{\mathsf{A}} \triangleq \mathsf{X} - \mathsf{A}$ (will be used only if a finite ground set X is specified)
$\mathsf{A} \circ \mathsf{B}$	symmetric difference: $\mathsf{A} \circ \mathsf{B} \triangleq (\mathsf{A} - \mathsf{B}) \cup (\mathsf{B} - \mathsf{A})$
$f : \mathsf{X} \to \mathsf{Y}$	mapping of X into Y
$f^{-1}(y)$	the inverse image of $y \in \mathsf{Y}$, i.e., $f^{-1}(y) \triangleq \{x : f(x) = y\}$
$\lVert f \rVert$	number of elements of the range of the mapping f
PD	abbreviation of "probability distribution"
$P \triangleq \{P(x) : x \in \mathsf{X}\}$	PD on X
$P(\mathsf{A})$	probability of the set $\mathsf{A} \subset \mathsf{X}$ for the PD P, i.e., $P(\mathsf{A}) \triangleq \sum_{x \in \mathsf{A}} P(x)$
$P \times Q$	direct product of the PDs P on X and Q on Y, i.e., $P \times Q \triangleq \{P(x)Q(y) : x \in \mathsf{X},\ y \in \mathsf{Y}\}$
P^n	nth power of the PD P, i.e., $P^n(\mathbf{x}) \triangleq \prod_{i=1}^{n} P(x_i)$
support of P	the set $\{x : P(x) > 0\}$

$W : \mathsf{X} \to \mathsf{Y}$ $W = \{W(y\|x) : x \in \mathsf{X}, y \in \mathsf{Y}\}$	stochastic matrix with rows indexed by elements of X and columns indexed by elements of Y; i.e., $W(\cdot\|x)$ is a PD on Y for every $x \in \mathsf{X}$
$W(\mathsf{B}\|x)$	probability of the set $\mathsf{B} \subset Y$ for the PD $W(\cdot\|x)$
$W^n : \mathsf{X}^n \to \mathsf{Y}^n$	nth direct power of W, i.e., $W^n(\mathbf{y}\|\mathbf{x}) \triangleq \prod_{i=1}^n W(y_i\|x_i)$
RV	abbreviation for "random variable"
X, Y, Z	RVs ranging over finite sets
$X^n = (X_1, \ldots, X_n)$ $X^n = X_1 \ldots X_n$	alternative notations for the vector-valued RV with components $X_1, \ldots, X_n$
$\Pr\{X \in \mathsf{A}\}$	probability of the event that the RV X takes a value in the set A
P_X	distribution of the RV X, defined by $P_X(x) \triangleq \Pr\{X = x\}$
$P_{Y\|X=x}$	conditional distribution of Y given $X = x$, i.e., $P_{Y\|X=x}(y) \triangleq \Pr\{Y = y\|X = x\}$; not defined if $P_X(x) = 0$
$P_{Y\|X}$	the stochastic matrix with rows $P_{Y\|X=x}$, called the conditional distribution of Y given X; here x ranges over the support of P_X
$P_{Y\|X} = W$	means that $P_{Y\|X=x} = W(\cdot\|x)$ if $P_X(x) > 0$, involving no assumption on the remaining rows of W
EX	expectation of the real-valued RV X
$\mathrm{var}(X)$	variance of the real-valued RV X
$X \multimap Y \multimap Z$	means that these RVs form a Markov chain in this order
$(a, b),\ [a, b],\ [a, b)$	open, closed resp. left-closed interval with endpoints $a < b$
$\|r\|^+$	positive part of the real number r, i.e., $\|r\|^+ \triangleq \max(r, 0)$
$\lfloor r \rfloor$	largest integer not exceeding r
$\lceil r \rceil$	smallest integer not less than r
$\min[a, b],\ \max[a, b]$	the smaller resp. larger of the numbers a and b
$\mathbf{r} \geqq \mathbf{s}$	means for vectors $\mathbf{r} = (r_1, \ldots, r_n)$, $\mathbf{s} = (s_1, \ldots, s_n)$ of the n-dimensional Euclidean space that $r_i \geqq s_i,\ i = 1, \ldots, n$
$\overline{\mathcal{A}}$	convex closure of a subset $\mathcal{A}$ of a Euclidean space, i.e., the smallest closed convex set containing $\mathcal{A}$
exp, log	are understood to the base 2
ln	natural logarithm
$a \log(a/b)$	equals zero if $a = 0$ and $+\infty$ if $a > b = 0$
$h(r)$	the binary entropy function $h(r) \triangleq -r \log r - (1 - r) \log(1 - r),\ r \in [0, 1]$

Most asymptotic results in this book are established with uniform convergence. Our way of specifying the extent of uniformity is to indicate in the statement of results all those parameters involved in the problem upon which threshold indices depend. In this context, e.g., $n_0 = n_0(|\mathsf{X}|, \varepsilon, \delta)$ means some threshold index which could be explicitly given as a function of $|\mathsf{X}|, \varepsilon, \delta$ alone.

Preliminaries on random variables and probability distributions

As we shall deal with RVs ranging over finite sets, the measure-theoretic foundations of probability theory will never really be needed. Still, in a formal sense, when speaking of RVs it is understood that a Kolmogorov probability space $(\Omega, \mathcal{F}, \mu)$ is given (i.e., Ω is some set, $\mathcal{F}$ is a σ-algebra of its subsets, and μ is a probability measure on $\mathcal{F}$). Then a RV with values in a finite set X is a mapping $X : \Omega \to \mathsf{X}$ such that $X^{-1}(x) \in \mathcal{F}$ for every $x \in \mathsf{X}$. The probability of an event defined in terms of RVs means the μ-measure of the corresponding subset of Ω, e.g.,

$$\Pr\{X \in \mathsf{A}\} \triangleq \mu(\{\omega : X(\omega) \in \mathsf{A}\}).$$

Throughout this book, it will be assumed that the underlying probability space $(\Omega, \mathcal{F}, \mu)$ is "rich enough" in the following sense. To any pair of finite sets X, Y, any RV X with values in X and any distribution P on $\mathsf{X} \times \mathsf{Y}$ whose marginal on X coincides with P_X, there exists a RV Y with values in Y such that $P_{XY} = P$. This assumption is certainly fulfilled, e.g., if Ω is the unit interval, $\mathcal{F}$ is the family of its Borel subsets, and μ is the Lebesgue measure.

The set of all PDs on a finite set X will be identified with the subset of the $|\mathsf{X}|$-dimensional Euclidean space, consisting of all vectors with non-negative components summing up to unity. Linear combinations of PDs and convexity are understood accordingly. For example, the convexity of a real-valued function $f(P)$ of PDs on X means that

$$f(\alpha P_1 + (1-\alpha)P_2) \leqq \alpha f(P_1) + (1-\alpha) f(P_2)$$

for every P_1, P_2 and $\alpha \in (0, 1)$. Similarly, topological terms for PDs on X refer to the metric topology defined by Euclidean distance. In particular, the convergence $P_n \to P$ means that $P_n(x) \to P(x)$ for every $x \in \mathsf{X}$.

The set of all stochastic matrices $W : \mathsf{X} \to \mathsf{Y}$ is identified with a subset of the $|\mathsf{X}||\mathsf{Y}|$-dimensional Euclidean space in an analogous manner. Convexity and topological concepts for stochastic matrices are understood accordingly.

Finally, for any distribution P on X and any stochastic matrix $W : \mathsf{X} \to \mathsf{Y}$ we denote by PW the distribution on Y defined as the matrix product of the (row) vector P and the matrix W, i.e.,

$$(PW)(y) \triangleq \sum_{x \in \mathsf{X}} P(x) W(y|x) \quad \text{for every } y \in \mathsf{Y}.$$

Introduction

Information is a fashionable concept with many facets, among which the quantitative one–our subject–is perhaps less striking than fundamental. At the intuitive level, for our purposes, it suffices to say that *information* is some knowledge of predetermined type contained in certain data or pattern and wanted at some destination. Actually, this concept will not explicitly enter the mathematical theory. However, throughout the book certain functionals of random variables will be conveniently interpreted as measures of the amount of information provided by the phenomena modeled by these variables. Such information measures are characteristic tools of the analysis of optimal performance of codes, and they have turned out to be useful in other branches of mathematics as well.

Intuitive background

The mathematical discipline of information theory, created by C. E. Shannon (1948) on an engineering background, still has a special relation to communication engineering, the latter being its major field of application and the source of its problems and motivation. We believe that some familiarity with the intuitive communication background is necessary for a more than formal understanding of the theory, let alone for doing further research. The heuristics, underlying most of the material in this book, can be best explained on Shannon's idealized model of a communication system (which can also be regarded as a model of an information storage system). The important question of how far the models treated are related to, and the results obtained are relevant for, real systems will not be addressed. In this respect we note that although satisfactory mathematical modeling of real systems is often very difficult, it is widely recognized that significant insight into their capabilities is given by phenomena discovered on apparently overidealized models. Familiarity with the mathematical methods and techniques of proof is a valuable tool for system designers in judging how these phenomena apply in concrete cases.

Shannon's famous block diagram of a (two-terminal) communication system is shown in Fig. I.1. Before turning to the mathematical aspects of Shannon's model, let us take a glance at the objects to be modeled.

The *source* of information may be nature, a human being, a computer, etc. The data or pattern containing the information at the source is called the *message*; it may consist of observations on a natural phenomenon, a spoken or written sentence, a sequence of

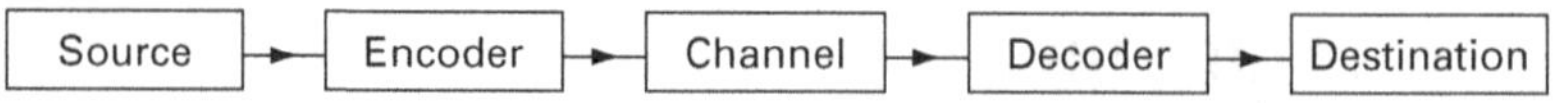

Figure I.1

binary digits, etc. Part of the information contained in the message (e.g., the shape of characters of a handwritten text) may be immaterial to the particular *destination.* Small distortions of the relevant information might be tolerated as well. These two aspects are jointly reflected in a *fidelity criterion* for the reproduction of the message at the destination. For example, for a person watching a color TV program on a black-and-white set, the information contained in the colors must be considered immaterial and the fidelity criterion is met if the picture is not perceivably worse than it would be by a good black-and-white transmission. Clearly, the fidelity criterion of a person watching the program in color would be different.

The source and destination are separated in space or time. The communication or storing device available for bridging over this separation is called the *channel.* As a rule, the channel does not work perfectly and thus its output may significantly differ from the input. This phenomenon is referred to as *channel noise.* While the properties of the source and channel are considered unalterable, characteristic to Shannon's model is the liberty of transforming the message before it enters the channel. Such a transformation, called *encoding,* is always necessary if the message is not a possible input of the channel (e.g., a written sentence cannot be directly radioed). More importantly, encoding is an effective tool of reducing the *cost of transmission* and of combating channel noise (trivial examples are abbreviations such as cable addresses in telegrams on the one hand, and spelling names on telephone on the other). Of course, these two goals are conflicting and a compromise must be found. If the message has been encoded before entering the channel – and often even if not – a suitable processing of the channel output is necessary in order to retrieve the information in a form needed at the destination; this processing is called *decoding.* The devices performing encoding and decoding are the *encoder* and *decoder* of Fig. I.1. The rules determining their operation constitute the *code.* A code accomplishes *reliable transmission* if the joint operation of encoder, channel and decoder results in reproducing the source messages at the destination within the prescribed fidelity criterion.

Informal description of the basic mathematical model

Shannon developed information theory as a mathematical study of the problem of *reliable transmission* at a possibly low cost (for a given source, channel and fidelity criteria). For this purpose mathematical models of the objects in Fig. I.1 had to be introduced. The terminology of the following models reflects the point of view of communication between terminals separated in space. Appropriately interchanging the roles of time and space, these models are equally suitable for describing *data storage.*

Having in mind a source which keeps producing information, its output is visualized as an infinite sequence of *symbols* (e.g., Latin characters, binary digits, etc.). For an observer, the successive symbols cannot be predicted. Rather, they seem to appear randomly according to probabilistic laws representing potentially available prior knowledge about the nature of the source (e.g., in the case of an English text we may think of language statistics, such as letter or word frequencies, etc.). For this reason the source is identified with a discrete-time stochastic process. The first k random variables of the source process represent a *random message* of length k; realizations thereof are called *messages of length k*. The theory is largely of asymptotic character: we are interested in the transmission of long messages. This justifies restricting our attention to messages of equal length, although, e.g., in an English text, the first k letters need not represent a meaningful piece of information; the point is that a sentence cut at the tail is of negligible length compared to a large k. In non-asymptotic investigations, however, the structure of messages is of secondary importance. Then it is mathematically more convenient to regard them as realizations of an arbitrary random variable, the so-called random message (which may be identified with a finite segment of the source process or even with the whole process, etc.). Hence we shall often speak of messages (and their transformation) without specifying a source.

An obvious way of taking advantage of a stochastic model is to disregard undesirable events of small probability. The simplest fidelity criterion of this kind is that the *probability of error,* i.e., the overall probability of not receiving the message accurately at the destination, should not exceed a given small number. More generally, viewing the message and its reproduction at the destination as realizations of stochastically dependent random variables, a *fidelity criterion* is formulated as a global requirement involving their joint distribution. Usually, one introduces a numerical measure of the loss resulting from a particular reproduction of a message. In information theory this is called a *distortion measure.* A typical fidelity criterion is that the expected distortion be less than a threshold, or that the probability of a distortion transgressing this threshold be small.

The channel is supposed to be capable of successively transmitting symbols from a given set, the *input alphabet.* There is a starting point of the transmission and each of the successive uses of the channel consists of putting in one symbol and observing the corresponding symbol at the output. In the ideal case of a *noiseless channel* the output is identical to the input; in general, however, they may differ and the output need not be uniquely determined by the input. Also, the *output alphabet* may differ from the input alphabet. Following the stochastic approach, it is assumed that for every finite sequence of input symbols there exists a probability distribution on output sequences of the same length. This distribution governs the successive outputs if the elements of the given sequence are successively transmitted from the start of transmission on, as the beginning of a potentially infinite sequence. This assumption implies that no output symbol is affected by possible later inputs, and it amounts to certain consistency requirements among the mentioned distributions. The family of these distributions represents all possible knowledge about the channel noise, prior to transmission. This family defines the *channel* as a mathematical object.

The encoder maps messages into sequences of channel input symbols in a not necessarily one-to-one way. Mathematically, this very mapping is the *encoder.* The images of messages are referred to as *codewords.* For convenience, attention is usually restricted to encoders with fixed codeword length, mapping the messages into channel input sequences of length n, say. Similarly, from a purely mathematical point of view, a *decoder* is a mapping of output sequences of the channel into reproductions of messages. By a *code* we shall mean, as a rule, an encoder–decoder pair or, in specific problems, a mathematical object effectively determining this pair.

A random message, an encoder, a channel and a decoder define a joint probability distribution over messages, channel input and output sequences, and reproductions of the messages at the destination. In particular, it can be decided whether a given fidelity criterion is met. If it is, we speak of *reliable transmission* of the random message. The cost of transmission is not explicitly included in the above mathematical model. As a rule, one implicitly assumes that its main factor is the cost of channel use, the latter being proportional to the length of the input sequence. (In the case of telecommunication this length determines the channel's operation time and, in the case of data storage, the occupied space, provided that each symbol requires the same time or space, respectively.) Hence, for a given random message, channel and fidelity criterion, the problem consists in finding the smallest codeword length n for which reliable transmission can be achieved.

We are basically interested in the reliable transmission of long messages of a given source using *fixed-length-to-fixed-length codes,* i.e., encoders mapping messages of length k into channel input sequences of length n and decoders mapping channel output sequences of length n into reproduction sequences of length k. The average number n/k of channel symbols used for the transmission of one source symbol is a measure of the performance of the code, and it will be called the *transmission ratio.* The goal is to determine the limit of the minimum transmission ratio (LMTR) needed for reliable transmission, as the message length k tends to infinity. Implicit in this problem statement is that fidelity criteria are given for all sufficiently large k. Of course, for the existence of a finite LMTR, let alone for its computability, proper conditions on source, channel and fidelity criteria are needed.

The intuitive problem of transmission of long messages can also be approached in another – more ambitious – manner, incorporating into the model certain constraints on the complexity of encoder and decoder, along with the requirement that the transmission be indefinitely continuable. Any fixed-length-to-fixed-length code, designed for transmitting messages of length k by n channel symbols, say, may be used for *non-terminating transmission* as follows. The infinite source output sequence is partitioned into consecutive blocks of length k. The encoder mapping is applied to each block separately and the channel input sequence is the succession of the obtained blocks of length n. The channel output sequence is partitioned accordingly and is decoded blockwise by the given decoder. This method defines a code for non-terminating transmission. The transmission ratio is n/k; the block lengths k and n constitute a rough measure of complexity of the code. If the channel has no "input memory," i.e., the transmission of the individual blocks is not affected by previous inputs, and if the source and channel are time-invariant, then each source block will be reproduced within the same fidelity criterion as the first one. Suppose, in addition, that the fidelity criteria for messages of different length have the following property: if successive blocks and their reproductions individually meet the fidelity criterion, then so does their juxtaposition. Then, by this very coding, messages of potentially infinite length are reliably transmitted, and one can speak of *reliable non-terminating transmission.* Needless to say, this blockwise coding is a very special way of realizing non-terminating transmission. Still, within a very general class of codes for reliable non-terminating transmission, in order to minimize the transmission ratio[1] under conditions such as above, it suffices to restrict attention to blockwise codes. In such cases the present minimum equals the previous LMTR and the two approaches to the intuitive problem of transmission of long messages are equivalent.

While in this book we basically adopt the first approach, a major reason of considering mainly fixed-length-to-fixed-length codes consists in their appropriateness also for non-terminating transmission. These codes themselves are often called *block codes* without specifically referring to non-terminating transmission.

Measuring information

A remarkable feature of the LMTR problem, discovered by Shannon and established in great generality by further research, is a phenomenon suggesting the heuristic interpretation that information, like liquids, "has volume but no shape," i.e., the amount of information is measurable by a scalar. Just as the time necessary for conveying the liquid content of a large container through a pipe (at a given flow velocity) is determined by the ratio of the volume of the liquid to the cross-sectional area of the pipe, the LMTR equals the ratio of two numbers, one depending on the source and fidelity criterion, the other depending on the channel. The first number is interpreted as a measure of the *amount of information* needed, on average, for the reproduction of one source symbol, whereas the second is a measure of the *channel's capacity,* i.e., of how much information is transmissible on average by one channel use. It is customary to take as a standard the simplest channel that can be used for transmitting information, namely the noiseless channel with two input symbols, 0 and 1, say. The capacity of this *binary noiseless channel,* i.e., the amount of information transmissible by one binary digit, is considered the unit of the amount of information, called 1 *bit.* Accordingly, the amount of information needed on average for the reproduction of one symbol of a given source (relative to a given fidelity criterion) is measured by the LMTR for this source and the binary noiseless channel. In particular, if the most demanding fidelity criterion is imposed, which within a stochastic theory is that of a small probability of error, the corresponding LMTR provides a measure of the total amount of information carried, on average, by one source symbol.

[1] The relevance of this minimization problem to data storage is obvious. In typical communication situations, however, the transmission ratio of non-terminating transmission cannot be chosen freely. Rather, it is determined by the rates at which the source produces and the channel transmits symbols. Then one question is whether a given transmission ratio admits reliable transmission, but this is mathematically equivalent to the above minimization problem.

The above ideas naturally suggest the need for a measure of the amount of information individually contained in a single source output. In view of our source model, this means to associate some information content with an arbitrary random variable. One relies on the intuitive postulate that the observation of a collection of independent random variables yields an amount of information equal to the sum of the information contents of the individual variables. Accordingly, one defines the *entropy* (information content) of a random variable as the amount of information carried, on average, by one symbol of a source which consists of a sequence of independent copies of the random variable in question. This very entropy is also a measure of the amount of *uncertainty* concerning this random variable before its observation.

We have sketched a way of assigning information measures to sources and channels in connection with the LMTR problem and arrived, in particular, at the concept of entropy of a single variable. There is also an opposite way: starting from entropy, which can be expressed by a simple formula, one can build up more complex functionals of probability distributions. On the basis of heuristic considerations (quite independent of the above communication model), these functionals can be interpreted as information measures corresponding to different connections of random variables. The operational significance of these information measures is not a-priori evident. Still, under general conditions the solution of the LMTR problem can be given in terms of these quantities. More precisely, the corresponding theorems assert that the operationally defined information measures for source and channel can be given by such functionals, just as intuition suggests. This consistency underlines the importance of entropy-based information measures, both from a formal and a heuristic point of view.

The relevance of these functionals, corresponding to their heuristic meaning, is not restricted to communication or storage problems. Still, there are also other functionals which can be interpreted as information measures with an operational significance not related to coding.

Multi-terminal systems

Shannon's block diagram (Fig. I.1) models one-way communication between two terminals. The communication link it describes can be considered as an artificially isolated elementary part of a large communication system involving exchange of information among many participants. Such an isolation is motivated by the implicit assumptions that

(i) the source and channel are in some sense independent of the remainder of the system, the effects of the environment being taken into account only as channel noise,

(ii) if exchange of information takes place in both directions, they do not affect each other.

Note that dropping assumption (ii) is meaningful even in the case of communication between two terminals. Then the new phenomenon arises that transmission in one direction has the byproduct of feeding back information on the result of transmission in the opposite direction. This *feedback* can conceivably be exploited for improving the performance of the code; this, however, will necessitate a modification of the mathematical concept of the encoder.

Problems involving feedback will be discussed in this book only casually. On the other hand, the whole of Part III will be devoted to problems arising from dropping assumption (i). This leads to models of *multi-terminal systems* with several sources, channels and destinations, such that the stochastic interdependence of individual sources and channels is taken into account. A heuristic description of such mathematical models at this point would lead too far. However, we feel that readers familiar with the mathematics of two-terminal systems treated in Parts I and II will have no difficulty in understanding the motivation for the multi-terminal models of Part III.

Part I

Information measures in simple coding problems

1 Source coding and hypothesis testing; information measures

A (discrete) *source* is a sequence $\{X_i\}_{i=1}^{\infty}$ of random variables (RVs) taking values in a finite set X called the *source alphabet*. If the X_i's are independent and have the same distribution P, we speak of a *discrete memoryless source* (DMS) with *generic distribution* P.

A k-to-n binary *block code* is a pair of mappings

$$f : \mathsf{X}^k \to \{0, 1\}^n, \quad \varphi : \{0, 1\}^n \to \mathsf{X}^k.$$

For a given source, the *probability of error* of the code (f, φ) is

$$e(f, \varphi) \triangleq \Pr\{\varphi(f(X^k)) \neq X^k\},$$

where X^k stands for the k-length initial string of the sequence $\{X_i\}_{i=1}^{\infty}$. We are interested in finding codes with small ratio n/k and small probability of error. → 1.1

More exactly, for every k let $n(k, \varepsilon)$ be the smallest n for which there exists a k-to-n binary block code satisfying $e(f, \varphi) \leqq \varepsilon$; we want to determine $\lim_{k\to\infty} \frac{n(k,\varepsilon)}{k}$. → 1.2

THEOREM 1.1 For a DMS with generic distribution $P = \{P(x) : x \in \mathsf{X}\}$

$$\lim_{k\to\infty} \frac{n(k, \varepsilon)}{k} = H(P) \quad \text{for every} \quad \varepsilon \in (0, 1), \tag{1.1}$$

where $H(P) \triangleq -\sum_{x\in\mathsf{X}} P(x) \log P(x)$. ○

COROLLARY 1.1

$$0 \leqq H(P) \leqq \log |\mathsf{X}|. \tag{1.2}$$

○

Proof The existence of a k-to-n binary block code with $e(f, \varphi) \leqq \varepsilon$ is equivalent to the existence of a set $\mathsf{A} \subset \mathsf{X}^k$ with $P^k(\mathsf{A}) \geqq 1 - \varepsilon$, $|\mathsf{A}| \leqq 2^n$ (let A be the set of those sequences $\mathbf{x} \in \mathsf{X}^k$ which are reproduced correctly, i.e., $\varphi(f(\mathbf{x})) = \mathbf{x}$). Denote by $s(k, \varepsilon)$ the minimum cardinality of sets $\mathsf{A} \subset \mathsf{X}^k$ with $P^k(\mathsf{A}) \geqq 1 - \varepsilon$. It suffices to show that

$$\lim_{k\to\infty} \frac{1}{k} \log s(k, \varepsilon) = H(P) \quad (\varepsilon \in (0, 1)). \tag{1.3}$$

To this end, let $\mathsf{B}(k, \delta)$ be the set of those sequences $\mathbf{x} \in \mathsf{X}^k$ which have probability

$$\exp\{-k(H(P) + \delta)\} \leqq P^k(\mathbf{x}) \leqq \exp\{-k(H(P) - \delta)\}.$$

We first show that $P^k(\mathsf{B}(k,\delta)) \to 1$ as $k \to \infty$, for every $\delta > 0$. In fact, consider the real-valued RVs

$$Y_i \triangleq -\log P(X_i);$$

these are well defined with probability 1 even if $P(x) = 0$ for some $x \in \mathsf{X}$. The Y_i's are independent, identically distributed and have expectation $H(P)$. Thus by the weak law of large numbers

$$\lim_{k\to\infty} \Pr\left\{\left|\frac{1}{k}\sum_{i=1}^{k} Y_i - H(P)\right| \leqq \delta\right\} = 1 \quad \text{for every} \quad \delta > 0.$$

As $X^k \in \mathsf{B}(k,\delta)$ iff $|\frac{1}{k}\sum_{i=1}^{k} Y_i - H(P)| \leqq \delta$, the convergence relation means that

$$\lim_{k\to\infty} P^k(\mathsf{B}(k,\delta)) = 1 \quad \text{for every} \quad \delta > 0, \tag{1.4}$$

as claimed. The definition of $\mathsf{B}(k,\delta)$ implies that

$$|\mathsf{B}(k,\delta)| \leqq \exp\{k(H(P)) + \delta)\}.$$

Thus (1.4) gives for every $\delta > 0$

$$\overline{\lim_{k\to\infty}} \frac{1}{k} \log s(k,\varepsilon) \leqq \overline{\lim_{k\to\infty}} \frac{1}{k} \log |\mathsf{B}(k,\delta)| \leqq H(P) + \delta. \tag{1.5}$$

On the other hand, for every set $\mathsf{A} \subset \mathsf{X}^k$ with $P^k(\mathsf{A}) \geqq 1 - \varepsilon$, (1.4) implies

$$P^k(\mathsf{A} \cap \mathsf{B}(k,\delta)) \geqq \frac{1-\varepsilon}{2}$$

for sufficiently large k. Hence, by the definition of $\mathsf{B}(k,\delta)$,

$$\begin{aligned}|\mathsf{A}| \geqq |\mathsf{A} \cap \mathsf{B}(k,\delta)| &\geqq \sum_{\mathbf{x}\in \mathsf{A}\cap\mathsf{B}(k,\delta)} P^k(\mathbf{x}) \exp\{k(H(P) - \delta)\} \\ &\geqq \frac{1-\varepsilon}{2} \exp\{k(H(P) - \delta)\},\end{aligned}$$

proving that for every $\delta > 0$

$$\underline{\lim_{k\to\infty}} \frac{1}{k} \log s(k,\varepsilon) \geqq H(P) - \delta.$$

This and (1.5) establish (1.3). The corollary is immediate. □

For intuitive reasons expounded in the Introduction, the limit $H(P)$ in Theorem 1.1 is interpreted as a measure of the information content of (or the uncertainty about) a RV X with distribution $P_X = P$. It is called the *entropy* of the RV X or of the distribution P:

$$H(X) = H(P) \triangleq -\sum_{x\in\mathsf{X}} P(x) \log P(x).$$

This definition is often referred to as *Shannon's formula*.

The mathematical essence of Theorem 1.1 is formula (1.3). It gives the asymptotics for the minimum size of sets of large probability in X^k. We now generalize (1.3) for the case when the elements of X^k have unequal weights and the size of subsets is measured by total weight rather than cardinality.

Let us be given a sequence of positive-valued "mass functions" $M_1(x), M_2(x), \ldots$ on X and set

$$M(\mathbf{x}) \triangleq \prod_{i=1}^{k} M_i(x_i) \quad \text{for} \quad \mathbf{x} = x_1 \cdots x_k \in \mathsf{X}^k.$$

For an arbitrary sequence of X-valued RVs $\{X_i\}_{i=1}^{\infty}$ consider the minimum of the M-mass

$$M(\mathsf{A}) \triangleq \sum_{\mathbf{x} \in \mathsf{A}} M(\mathbf{x})$$

of those sets $\mathsf{A} \subset \mathsf{X}^k$ which contain X^k with high probability: let $s(k, \varepsilon)$ denote the minimum of $M(\mathsf{A})$ for sets $\mathsf{A} \subset \mathsf{X}^k$ of probability

$$P_{X^k}(\mathsf{A}) \geqq 1 - \varepsilon.$$

The previous $s(k, \varepsilon)$ is a special case obtained if all the functions $M_i(x)$ are identically equal to 1.

THEOREM 1.2 If the X_i's are independent with distributions $P_i \triangleq P_{X_i}$ and $|\log M_i(x)| \leqq c$ for every i and $x \in \mathsf{X}$ then, setting

$$E_k \triangleq \frac{1}{k} \sum_{i=1}^{k} \sum_{x \in \mathsf{X}} P_i(x) \log \frac{M_i(x)}{P_i(x)},$$

we have for every $0 < \varepsilon < 1$

$$\lim_{k \to \infty} \left(\frac{1}{k} \log s(k, \varepsilon) - E_k \right) = 0.$$

More precisely, for every $\delta, \varepsilon \in (0, 1)$,

$$\left| \frac{1}{k} \log\ s(k, \varepsilon) - E_k \right| \leqq \delta \quad \text{if} \quad k \geqq k_0 = k_0(|\mathsf{X}|, c, \varepsilon, \delta). \tag{1.6}$$

○

Proof Consider the real-valued RVs

$$Y_i \triangleq \log \frac{M_i(X_i)}{P_i(X_i)}.$$

Since the Y_i's are independent and $E\left(\frac{1}{k} \sum_{i=1}^{k} Y_i\right) = E_k$, Chebyshev's inequality gives for any $\delta' > 0$

$$\Pr \left\{ \left| \frac{1}{k} \sum_{i=1}^{k} Y_i - E_k \right| \geqq \delta' \right\} \leqq \frac{1}{k^2 \delta'^2} \sum_{i=1}^{k} \text{var}\,(Y_i) \leqq \frac{1}{k \delta'^2} \max_i \text{var}\,(Y_i).$$

This means that for the set

$$\mathsf{B}(k,\delta') \triangleq \left\{\mathbf{x} : \mathbf{x} \in \mathsf{X}^k,\, E_k - \delta' \leqq \frac{1}{k}\log\frac{M(\mathbf{x})}{P_{X^k}(\mathbf{x})} \leqq E_k + \delta'\right\}$$

we have

$$P_{X^k}(\mathsf{B}(k,\delta')) \geqq 1 - \eta_k, \quad \text{where} \quad \eta_k \triangleq \frac{1}{k\delta'^2}\max_i \operatorname{var}(Y_i).$$

Since by the definition of $\mathsf{B}(k,\delta')$

$$M(\mathsf{B}(k,\delta')) = \sum_{\mathbf{x}\in\mathsf{B}(k,\delta')} M(\mathbf{x}) \leqq \sum_{\mathbf{x}\in\mathsf{B}(k,\delta')} P_{X^k}(\mathbf{x})\exp[k(E_k+\delta')] \leqq \exp[k(E_k+\delta')],$$

it follows that

$$\frac{1}{k}\log s(k,\varepsilon) \leqq \frac{1}{k}\log M(\mathsf{B}(k,\delta')) \leqq E_k + \delta' \quad \text{if} \quad \eta_k \leqq \varepsilon.$$

On the other hand, we have $P_{X^k}(\mathsf{A}\cap\mathsf{B}(k,\delta')) \geqq 1-\varepsilon-\eta_k$ for any set $\mathsf{A}\subset\mathsf{X}^k$ with $P_{X^k}(\mathsf{A}) \geqq 1-\varepsilon$. Thus for every such A, again by the definition of $\mathsf{B}(k,\delta')$,

$$\begin{aligned} M(\mathsf{A}) \geqq M(\mathsf{A}\cap\mathsf{B}(k,\delta')) &\geqq \sum_{\mathbf{x}\in\mathsf{A}\cap\mathsf{B}(k,\delta')} P_{X^k}(\mathbf{x})\exp\{k(E_k-\delta')\} \\ &\geqq (1-\varepsilon-\eta_k)\exp[(E_k-\delta')], \end{aligned}$$

implying

$$\frac{1}{k}\log s(k,\varepsilon) \geqq \frac{1}{k}\log(1-\varepsilon-\eta_k) + E_k + \delta'.$$

Setting $\delta' \triangleq \delta/2$, these results imply (1.6) provided that

$$\eta_k = \frac{4}{k\delta^2}\max_i \operatorname{var}(Y_i) \leqq \varepsilon \quad \text{and} \quad \frac{1}{k}\log(1-\varepsilon-\eta_k) \geqq -\frac{\delta}{2}.$$

By the assumption $|\log M_i(x)| \leqq c$, the last relations hold if $k \geqq k_0(|\mathsf{X}|, c, \varepsilon, \delta)$. □

An important corollary of Theorem 1.2 relates to *testing statistical hypotheses*. Suppose that a probability distribution of interest for the statistician is given by either $P = \{P(x) : x \in \mathsf{X}\}$ or $Q = \{Q(x) : x \in \mathsf{X}\}$. She or he has to decide between P and Q on the basis of a *sample* of size k, i.e., the result of k independent drawings from → 1.3 the unknown distribution. A (non-randomized) test is characterized by a set $\mathsf{A}\subset\mathsf{X}^k$, in the sense that if the sample $X_1 \ldots X_k$ belongs to A, the statistician accepts P and else accepts Q. In most practical situations of this kind, the role of the two hypotheses is not symmetric. It is customary to prescribe a bound ε for the tolerated probability of wrong decision if P is the true distribution. Then the task is to minimize the probability of a → 1.4 wrong decision if hypothesis Q is true. The latter minimum is

$$\beta(k,\varepsilon) \triangleq \min_{\substack{\mathsf{A}\subset\mathsf{X}^k \\ P^k(\mathsf{A})\geqq 1-\varepsilon}} Q^k(\mathsf{A}).$$

COROLLARY 1.2 For any $0 < \varepsilon < 1$,

$$\lim_{k\to\infty} \frac{1}{k} \log \beta(k, \varepsilon) = -\sum_{x\in\mathsf{X}} P(x) \log \frac{P(x)}{Q(x)}. \qquad \bigcirc$$

Proof If $Q(x) > 0$ for each $x \in \mathsf{X}$, set $P_i \triangleq P$, $M_i \triangleq Q$ in Theorem 1.2. If $P(x) > Q(x) = 0$ for some $x \in \mathsf{X}$, the P-probability of the set of all k-length sequences containing this x tends to 1. This means that $\beta(k, \varepsilon) = 0$ for sufficiently large k, so that both sides of the asserted equality are $-\infty$. □

It follows from Corollary 1.2 that the sum on the right-hand side is non-negative. It measures how much the distribution Q differs from P in the sense of statistical distinguishability, and is called *informational divergence* or *I-divergence*:

$$D(P||Q) \triangleq \sum_{x\in\mathsf{X}} P(x) \log \frac{P(x)}{Q(x)}.$$

Another common name given to this quantity is *relative entropy*. Intuitively, one can say that the larger $D(P||Q)$ is, the more information for discriminating between the hypotheses P and Q can be obtained from one observation. Hence $D(P||Q)$ is also called the *information for discrimination*. The amount of information measured by $D(P||Q)$ is, however, conceptually different from entropy, since it has no immediate coding interpretation.

On the space of infinite sequences of elements of X one can build up product measures both from P and Q. If $P \neq Q$, the two product measures are mutually orthogonal; $D(P||Q)$ is a (non-symmetric) measure of how fast their restrictions to k-length strings approach orthogonality.

REMARK Both entropy and informational divergence have a form of expectation:

$$H(X) = E(-\log P(X)), \qquad D(P||Q) = E \log \frac{P(X)}{Q(X)},$$

where X is a RV with distribution P. It is convenient to interpret $-\log P(x)$, resp. $\log P(x)/Q(x)$, as a measure of the amount of information, resp. the weight of evidence in favor of P against Q provided by a particular value x of X. These quantities are important ingredients of the mathematical framework of information theory, but have less direct operational meaning than their expectations. ○

The entropy of a pair of RVs (X, Y) with finite ranges X and Y needs no new definition, since the pair can be considered a single RV with range $\mathsf{X} \times \mathsf{Y}$. For brevity, instead of $H((X, Y))$ we shall write $H(X, Y)$; similar notation will be used for any finite collection of RVs.

The intuitive interpretation of entropy suggests to consider as further information measures certain expressions built up from entropies. The difference $H(X, Y) - H(X)$ measures the additional amount of information provided by Y if X is already known.

It is called the *conditional entropy* of Y given X:

$$H(Y|X) \triangleq H(X,Y) - H(X).$$

Expressing the entropy difference by Shannon's formula we obtain

$$H(Y|X) = -\sum_{x \in \mathsf{X}} \sum_{y \in \mathsf{Y}} P_{XY}(x,y) \log \frac{P_{XY}(x,y)}{P_X(x)} = \sum_{x \in \mathsf{X}} P_X(x) H(Y|X=x), \quad (1.7)$$

where

$$H(Y|X=x) \triangleq -\sum_{y \in \mathsf{Y}} P_{Y|X}(y|x) \log P_{Y|X}(y|x).$$

Thus $H(Y|X)$ is the expectation of the entropy of the conditional distribution of Y given $X = x$. This gives further support to the above intuitive interpretation of conditional entropy. Intuition also suggests that the conditional entropy cannot exceed the unconditional one.

→ 1.5

LEMMA 1.3

$$H(Y|X) \leqq H(Y). \qquad \bigcirc$$

Proof

$$H(Y) - H(Y|X) = H(Y) - H(X,Y) + H(X)$$

$$= \sum_{x \in \mathsf{X}} \sum_{y \in \mathsf{Y}} P_{XY}(x,y) \log \frac{P_{XY}(x,y)}{P_X(x) P_Y(y)} = D(P_{XY} \| P_X \times P_Y) \geqq 0. \qquad \square$$

REMARK For certain values of x, $H(Y|X=x)$ may be larger than $H(Y)$. ○

The entropy difference in the preceding proof measures the decrease of uncertainty about Y caused by the knowledge of X. In other words, it is a measure of the amount of information about Y contained in X. Note the remarkable fact that this difference is symmetric in X and Y. It is called *mutual information*:

$$I(X \wedge Y) \triangleq H(Y) - H(Y|X) = H(X) - H(X|Y) = D(P_{XY} \| P_X \times P_Y). \quad (1.8)$$

Of course, the amount of information contained in X about itself is just the entropy:

$$I(X \wedge X) = H(X).$$

Mutual information is a measure of stochastic dependence of the RVs X and Y. The fact that $I(X \wedge Y)$ equals the informational divergence of the joint distribution of X and Y from what it would be if X and Y were independent reinforces this interpretation. There is no compelling reason other than tradition to denote mutual information by a different symbol than entropy. We keep this tradition, although our notation $I(X \wedge Y)$ differs slightly from the more common $I(X; Y)$.

Discussion

Theorem 1.1 says that the minimum number of binary digits needed – on average – to represent one symbol of a DMS with generic distribution P equals the entropy $H(P)$. This fact – and similar ones discussed later on – are our basis for interpreting $H(X)$ as a measure of the amount of information contained in the RV X, resp. of the uncertainty about this RV. In other words, in this book we adopt an operational or *pragmatic approach* to the concept of information. Alternatively, one could start from the intuitive concept of information and set up certain postulates which an information measure should fulfil. Some representative results of this *axiomatic approach* are treated in Problems 1.11–1.14.

Our starting point, Theorem 1.1, has been proved here in the conceptually simplest way. The key idea is that, for large k, all sequences in a subset of X^k with probability close to 1, namely $\mathsf{B}(k, \delta)$, have "nearly equal" probabilities in an exponential sense. This proof easily extends also to non-DM cases (not in the scope of this book).

On the other hand, in order to treat DM models at depth, another – purely combinatorial – approach will be more suitable. The preliminaries to this approach will be given in Chapter 2.

Theorem 1.2 demonstrates the intrinsic relationship of the basic source coding and hypothesis testing problems. The interplay of information theory and mathematical statistics goes much further; its more substantial examples are beyond the scope of this book. ○

Problems

1.1. (a) Check that the problem of determining $\lim_{k\to\infty} \frac{1}{k} n(k, \varepsilon)$ for a discrete source is just the formal statement of the LMTR problem (see the Introduction) for the given source and the binary noiseless channel, with the probability of error fidelity criterion.

(b) Show that for a DMS and a noiseless channel with arbitrary alphabet size m the LMTR is $H(P)/\log m$, where P is the generic distribution of the source.

1.2. Given an encoder $f : \mathsf{X}^k \to \{0, 1\}^n$, show that the probability of error $e(f, \varphi)$ is minimized iff the decoder $\varphi : \{0, 1\}^n \to \mathsf{X}^k$ has the property that $\varphi(\mathbf{y})$ is a sequence of maximum probability among those $\mathbf{x} \in \mathsf{X}^k$ for which $f(\mathbf{x}) = \mathbf{y}$.

1.3. A *randomized test* introduces a chance element into the decision between the hypotheses P and Q in the sense that if the result of k successive drawings is $\mathbf{x} \in \mathsf{X}^k$, one accepts the hypothesis P with probability $\pi(\mathbf{x})$, say. Define the analog of $\beta(k, \varepsilon)$ for randomized tests and show that it still satisfies Corollary 1.2.

1.4. (*Neyman–Pearson lemma*) Show that for any given bound $0 < \varepsilon < 1$ on the probability of wrong decision if P is true, the best randomized test is given by

$$\pi(\mathbf{x}) = \begin{cases} 1 & \text{if } P^k(\mathbf{x}) > c_k Q^k(\mathbf{x}) \\ \gamma_k & \text{if } P^k(\mathbf{x}) = c_k Q^k(\mathbf{x}) \\ 0 & \text{if } P^k(\mathbf{x}) < c_k Q^k(\mathbf{x}), \end{cases}$$

where c_k and γ_k are appropriate constants. Observe that the case $k = 1$ contains the general one, and there is no need to restrict attention to independent drawings.

1.5. (a) Let $\{X_i\}_{i=1}^{\infty}$ be a sequence of independent RVs with common range X but with arbitrary distributions. As in Theorem 1.1, denote by $n(k, \varepsilon)$ the smallest n for which there exists a k-to-n binary block code having probability of error $\leqq \varepsilon$ for the source $\{X_i\}_{i=1}^{\infty}$. Show that for every $\varepsilon \in (0, 1)$ and $\delta > 0$

$$\left| \frac{n(k,\varepsilon)}{k} - \frac{1}{k}\sum_{i=1}^{k} H(X_i) \right| \leqq \delta \quad \text{if} \quad k \geqq k_0(|\mathsf{X}|, \varepsilon, \delta).$$

Hint Use Theorem 1.2 with $M_i(x) = 1$.

(b) Let $\{(X_i, Y_i)\}_{i=1}^{\infty}$ be a sequence of independent replicas of a pair of RVs (X, Y) and suppose that X^k should be encoded and decoded in the knowledge of Y^k. Let $\tilde{n}(k, \varepsilon)$ be the smallest n for which there exists an encoder $f : \mathsf{X}^k \times \mathsf{Y}^k \to \{0, 1\}^n$ and a decoder $\varphi : \{0, 1\}^n \times \mathsf{Y}^k \to \mathsf{X}^k$ such that the probability of error is $\Pr\{\varphi(f(X^k, Y^k), Y^k) \neq X^k\} \leqq \varepsilon$.

Show that

$$\lim_{k\to\infty} \frac{\tilde{n}(k,\varepsilon)}{k} = H(X|Y) \quad \text{for every} \quad \varepsilon \in (0, 1).$$

Hint Use part (a) for the conditional distributions of the X_i's given various realizations $\mathbf{y}$ of Y^k.

1.6. (*Random selection of codes*) Let $\mathcal{F}(k, n)$ be the class of all mappings $f : \mathsf{X}^k \to \{0, 1\}^n$. Given a source $\{X_i\}_{i=1}^{\infty}$, consider the class of codes (f, φ_f), where f ranges over $\mathcal{F}(k, n)$ and $\varphi_f : \{0, 1\}^n \to \mathsf{X}^k$ is defined so as to minimize $e(f, \varphi)$; see Problem 1.2. Show that for a DMS with generic distribution P we have

$$\frac{1}{|\mathcal{F}(k,n)|} \sum_{f\in\mathcal{F}(k,n)} e(f, \varphi_f) \to 0,$$

if k and n tend to infinity, so that

$$\inf \frac{n}{k} > H(P).$$

Hint Consider a random mapping F of X^k into $\{0, 1\}^n$, assigning to each $\mathbf{x} \in \mathsf{X}^k$ one of the 2^n binary sequences of length n with equal probabilities 2^{-n}, independently of each other and of the source RVs. Let $\Phi : \{0, 1\}^n \to \mathsf{X}^k$ be the random mapping taking the value φ_f if $F = f$. Then

$$\begin{aligned}\frac{1}{|\mathcal{F}(k,n)|} \sum_{f\in\mathcal{F}(k,n)} e(f, \varphi_f) &= \Pr\{\Phi(F(X^k)) \neq X^k\} \\ &= \sum_{\mathbf{x}\in\mathsf{X}^k} P^k(\mathbf{x}) \Pr\{\Phi(F(\mathbf{x})) \neq \mathbf{x}\}.\end{aligned}$$

Here

$$\Pr\{\Phi(F(\mathbf{x})) \neq \mathbf{x}\} \leqq 2^{-n}|\{\mathbf{x}' : P^k(\mathbf{x}') \geqq P^k(\mathbf{x})\}|$$

and this is less than $2^{-n+k(H(P)+\delta)}$ if $P^k(\mathbf{x}) \geqq 2^{-k(H(P)+\delta)}$.

1.7. (a) (*Linear source codes*) Let X be a Galois field (i.e., any finite field) and consider X^k as a vector space over this field. A *linear source code* is a pair of mappings $f : \mathsf{X}^k \to \mathsf{X}^n$ and $\varphi : \mathsf{X}^n \to \mathsf{X}^k$ such that f is a linear mapping (φ is arbitrary). Show that for a DMS with generic distribution P there exist linear source codes with $n/k \to H(P)/\log|\mathsf{X}|$ and $e(f, \varphi) \to 0$. Compare this result with Problem 1.1(b). (Implicit in Elias (1955), cf. Wyner (1974).)

Hint Verify that the class of all linear mappings $f : \mathsf{X}^k \to \mathsf{X}^n$ satisfies the condition in (b) below.

(b) Extend the result of Problem 1.6 to the case when the role of {0, 1} is played by any finite set Y, and $\mathcal{F}(k, n)$ is *any class* of mappings $f : \mathsf{X}^k \to \mathsf{Y}^n$ satisfying

$$\frac{1}{|\mathcal{F}(k,n)|}\left|\{f \in \mathcal{F}(k,n) \colon f(\mathbf{x}) = f(\mathbf{x}')\}\right| \leqq |\mathsf{Y}|^{-n} \qquad \text{for} \qquad \mathbf{x} \neq \mathbf{x}'.$$

(Such a class of mappings is called a *universal hash family*; see Carter and Wegman (1979).)

Hint If $|\mathsf{Y}| = 2$, the hint to Problem 1.6 applies verbatim for the random mapping F selected from the present $\mathcal{F}(k, n)$, by the uniform distribution. If $|\mathsf{Y}| > 2$, the crucial bound on $\Pr\{\Phi(F(\mathbf{x})) \neq \mathbf{x}\}$ will hold with $|\mathsf{Y}|^{-n}$ instead of 2^{-n}; accordingly, the assertion follows if in the hypothesis $H(P)$ is replaced by $H(P)/\log|\mathsf{Y}|$.

1.8.* Show that the $s(k, \varepsilon)$ of Theorem 1.2 satisfies

$$\left|\log s(k,\varepsilon) - E_k - \sqrt{k}\lambda S_k + \frac{1}{2}\log k\right| \leqq \frac{140}{\delta^8}$$

whenever

$$\delta \leqq \min\left(S_k, \frac{1}{R_k}\right), \delta \leqq \varepsilon \leqq 1-\delta, \sqrt{k} \geqq \frac{140}{\delta^8}.$$

Here $S_k \triangleq \left(\frac{1}{k}\sum_{i=1}^k \operatorname{var}(Y_i)\right)^{1/2}$, $R_k \triangleq \left(\frac{1}{k}\sum_{i=1}^k E|Y_i - EY_i|^3\right)^{1/3}$ and λ is determined by $\Phi(\lambda) = 1 - \varepsilon$, where Φ denotes the distribution function of the standard normal distribution; E_k and Y_i are the same as in the text. (See Strassen (1964).)

1.9. In hypothesis testing problems it sometimes makes sense to speak of "prior probabilities" $\Pr\{P \text{ is true}\} = p_0$ and $\Pr\{Q \text{ is true}\} = q_0 = 1 - p_0$. On the basis of a sample $\mathbf{x} \in \mathsf{X}^k$, the posterior probabilities are then calculated as

$$\Pr\{P \text{ is true } | X^k = \mathbf{x}\} \triangleq p_k(\mathbf{x}) = \frac{p_0 P^k(\mathbf{x})}{p_0 P^k(\mathbf{x}) + q_0 Q^k(\mathbf{x})},$$
$$\Pr\{Q \text{ is true } | X^k = \mathbf{x}\} \triangleq q_k(\mathbf{x}) = 1 - p_k(\mathbf{x}).$$

Show that if P is true then $p_k(X^k) \to 1$ and $\frac{1}{k} \log q_k(X^k) \to -D(P\|Q)$ with probability 1, no matter what was $p_0 \in (0, 1)$.

1.10. The interpretation of entropy as a measure of uncertainty suggests that "more uniform" distributions have larger entropy. For two distributions P and Q on X we call P more uniform than Q, in symbols $P > Q$, if for the non-increasing ordering $p_1 \geqq p_2 \geqq \cdots \geqq p_n$, $q_1 \geqq q_2 \geqq \cdots \geqq q_n$ $(n = |\mathsf{X}|)$ of their probabilities, $\sum_{i=1}^k p_i \leqq \sum_{i=1}^k q_i$ for every $1 \leqq k \leqq n$. Show that $P > Q$ implies $H(P) \geqq H(Q)$; compare this result with (1.2).

(More generally, $P > Q$ implies $\sum_{i=1}^k \psi(p_i) \leqq \sum_{i=1}^k \psi(q_i)$ for every convex function ψ; see Karamata (1932).)

Postulational characterizations of entropy (Problems 1.11–1.14)

In the following problems, $H_m(p_1, \ldots, p_m)$, $m = 2, 3, \ldots$, designates a sequence of real-valued functions defined for non-negative p_i's with sum 1 such that H_m is invariant under permutations of the p_i's. Some simple postulates on H_m will be formulated which ensure that

$$H_m(p_1, \ldots, p_m) = -\sum_{i=1}^{m} p_i \log p_i, \quad m = 2, 3 \ldots \tag{*}$$

In particular, we shall say that $\{H_m\}$ is

(i) *expansible* if $H_{m+1}(p_1, \ldots, p_m, 0) = H_m(p_1, \ldots, p_m)$;

(ii) *additive* if

$$H_m(p_1, \ldots, p_m) + H_n(q_1, \ldots, q_n)$$
$$= H_{mn}(p_1 q_1, \ldots, p_1 q_n, \ldots, p_m q_1, \ldots, p_m q_n);$$

(iii) *subadditive* if $H_m(p_1, \ldots, p_n) + H_n(q_1, \ldots, q_n) \geqq H_{mn}(r_{11}, \ldots, r_{mn})$ whenever $\sum_{j=1}^{n} r_{ij} = p_i$, $\sum_{i=1}^{m} r_{ij} = q_j$;

(iv) *branching* if there exist functions $J_m(x, y)$ (with $x, y \geqq 0$, $x + y \leqq 1$, $m = 3, 4, \ldots$) such that

$$H_m(p_1, \ldots, p_m) - H_{m-1}(p_1 + p_2, \ldots, p_m) = J_m(p_1, p_2);$$

(v) *recursive* if it is branching with

$$J_m(p_1, p_2) = (p_1 + p_2) H_2\left(\frac{p_1}{p_1 + p_2}, \frac{p_2}{p_1 + p_2}\right), \quad m = 3, 4, \ldots;$$

(vi) *normalized* if $H_2\left(\frac{1}{2}, \frac{1}{2}\right) = 1$.

For a complete exposition of this subject, we refer to Aczél and Daróczy (1975).

1.11. Show that if $\{H_m\}$ is recursive, normalized and $H_2(p, 1-p)$ is a continuous function of p then (*) holds. (See Faddeev (1956); the first "axiomatic" characterization of entropy, using somewhat stronger postulates, was given by Shannon (1948).)

Hint The key step is to prove $H_m\left(\frac{1}{m}, \ldots, \frac{1}{m}\right) = \log m$. To this end, check that $f(m) \triangleq H_m\left(\frac{1}{m}, \ldots, \frac{1}{m}\right)$ is additive, i.e., $f(mn) = f(m) + f(n)$, and that $f(m+1) - f(m) \to 0$ as $m \to \infty$. Show that these properties and $f(2) = 1$ imply $f(m) = \log m$. (The last implication is a result of Erdös (1946); for a simple proof, see Rényi (1961).)

1.12.* (a) Show that if $H_m(p_1, \ldots, p_m) = \sum_{i=1}^m g(p_i)$ with a continuous function $g(p)$, and $\{H_m\}$ is additive and normalized, then (*) holds. (Chaundy and McLeod (1960)).

(b) Show that if $\{H_m\}$ is expansible and branching then $H_m(p_1, \ldots, p_m) = \sum_{i=1}^m g(p_i)$, with $g(0) = 0$ (Ng, (1974).)

1.13.* (a) Show that if $\{H_m\}$ is expansible, additive, subadditive, normalized and $H_2(p, 1-p) \to 0$ as $p \to 0$ then (*) holds.

(b) If $\{H_m\}$ is expansible, additive and subadditive, show that there exist constants $A \geqq 0$, $B \geqq 0$ such that

$$H_m(p_1, \ldots, p_m) = A\left(-\sum_{i=1}^m p_i \log p_i\right) + B \log |\{i : p_i > 0\}|.$$

(Forte (1975), Aczél–Forte–Ng (1974).)

1.14.* Suppose that $H_m(p_1, \ldots, p_m) = -\log \Phi^{-1}\left[\sum_{i=1}^m p_i \Phi(p_i)\right]$ with some strictly monotonic continuous function Φ on (0,1] such that $t\Phi(t) \to 0\Phi(0) \triangleq 0$ as $t \to 0$. Show that if $\{H_m\}$ is additive and normalized then either (*) holds or

$$H_m(p_1, \ldots, p_m) = \frac{1}{1-\alpha} \log \sum_{i=1}^m p_i^\alpha \quad \text{with some } \alpha > 0,\ \alpha \neq 1.$$

(Conjectured by Rényi (1961) and proved by Daróczy (1964). The preceding expression is called *Rényi's entropy of order* α. A similar expression was used earlier by Schützenberger (1954) as "pseudo information.")

1.15. For $P = (p_1, \ldots, p_m)$, denote by $H_\alpha(P)$ the Rényi entropy of order α if $\alpha \neq 1$, $\alpha > 0$, and the Shannon entropy $H(P)$ if $\alpha = 1$. Show that $H_\alpha(P)$ is a continuous, non-increasing function of α, whose limits as $\alpha \to 0$, resp. $\alpha \to +\infty$, are

$$H_0(P) \triangleq \log |\{i : p_i > 0\}|, \quad H_\infty(P) \triangleq \min(-\log p_i),$$

called the *maxentropy*, resp. *minentropy*, of P.

Hint Check that $\log \sum_{i=1}^m p_i^\alpha$ is a convex function of α.

1.16. (*Fisher's information*) Let $\{P_\vartheta\}$ be a family of distributions on a finite set X, where ϑ is a real parameter ranging over an open interval. Suppose that the probabilities $P_\vartheta(x)$ are positive and that they are continuously differentiable functions of ϑ. Write

$$I(\vartheta) \triangleq \sum_{x\in\mathsf{X}} \frac{1}{P_\vartheta(x)} \left(\frac{\partial}{\partial\vartheta} P_\vartheta(x)\right)^2 .$$

(a) Show that for every ϑ

$$\lim_{\vartheta'\to\vartheta} \frac{1}{(\vartheta'-\vartheta)^2} D(P_{\vartheta'} \| P_\vartheta) = \frac{1}{\ln 4} I(\vartheta)$$

(Kullback and Leibler, 1951).

(b) Show that every unbiased estimator f of ϑ from a sample of size n, i.e., every real-valued function f on X^n such that $E_\vartheta f(X^n) = \vartheta$ for each ϑ, satisfies

$$\mathrm{var}_\vartheta(f(X^n)) \geqq \frac{1}{nI(\vartheta)}.$$

Here E_ϑ and var_ϑ denote expectation, resp. variance, in the case when X^n has distribution P^n_ϑ.

(Fisher (1925) introduced $I(\vartheta)$ as a measure of the information contained in one observation from P_ϑ for estimating ϑ. His motivation was that the maximum likelihood estimator of ϑ from a sample of size n has asymptotic variance $1/(nI(\vartheta_0))$ if $\vartheta = \vartheta_0$. The assertion of (b) is a special case of the *Cramér–Rao inequality,* see e.g., Schmetterer (1974).)

Hint (a) directly follows by L'Hospital's rule. For (b), it suffices to consider the case $n = 1$. But

$$\sum_{x\in\mathsf{X}} P_\vartheta(x) \left(\frac{1}{P_\vartheta(x)} \frac{\partial}{\partial\vartheta} P_\vartheta(x)\right)^2 \cdot \sum_{x\in\mathsf{X}} P_\vartheta(x)(f(x)-\vartheta)^2 \geqq 1$$

follows from Cauchy's inequality, since

$$\sum_{x\in\mathsf{X}} \frac{\partial}{\partial\vartheta} P_\vartheta(x) \cdot (f(x)-\vartheta) = \frac{\vartheta}{\partial\vartheta} \sum_{x\in\mathsf{X}} P_\vartheta(x) f(x) = 1.$$

Story of the results

The basic concepts of information theory are due to Shannon (1948). In particular, he proved Theorem 1.1, introduced the information measures entropy, conditional entropy and mutual information, and established their basic properties. The name entropy has been borrowed from physics, as entropy in the sense of statistical physics is expressed by a similar formula, due to Boltzmann (1877). The very idea of measuring information regardless of its content dates back to Hartley (1928), who assigned to a symbol out of m alternatives the amount of information $\log m$. An information measure in a specific context was used by Fisher (1925), as in Problem 1.16. Informational divergence was introduced by Kullback and Leibler (1951) (under the name information for discrimination; they used the term divergence for its symmetrized version). Corollary 1.2 is known as Stein's lemma; it appears in Chernoff (1956), attributed to C. Stein. Theorem 1.2 is a common generalization of Theorem 1.1 and Corollary 1.2; a stronger result of this kind was given by Strassen (1964), see Problem 1.8. For a nice discussion of the pragmatic and axiomatic approaches to information measures, see Rényi (1965).

Addition. For more on the interplay of information theory and statistics, see Kullback (1959), Rissanen (1989), and Csiszár and Shields (2004).

2 Types and typical sequences

Most of the proof techniques used in this book will be based on a few simple combinatorial lemmas, summarized below.

Drawing k times independently with distribution Q from a finite set X, the probability of obtaining the sequence $\mathbf{x} \in \mathsf{X}^k$ depends only on how often the various elements of X occur in $\mathbf{x}$. In fact, denoting by $N(a|\mathbf{x})$ the number of occurrences of $a \in \mathsf{X}$ in $\mathbf{x}$, we have

$$Q^k(\mathbf{x}) = \prod_{a \in \mathsf{X}} Q(a)^{N(a|\mathbf{x})}. \tag{2.1}$$

DEFINITION 2.1 The *type* of a sequence $\mathbf{x} \in \mathsf{X}^k$ is the distribution $P_{\mathbf{x}}$ on X defined by

$$P_{\mathbf{x}}(a) \triangleq \frac{1}{k} N(a|\mathbf{x}) \quad \text{for every} \quad a \in \mathsf{X}.$$

For any distribution P on X, the set of sequences of type P in X^k is denoted by T_P^k or simply T_P. A distribution P on X is called a *type of sequences in* X^k if $\mathsf{T}_P^k \neq \emptyset$. ○

Sometimes the term "type" will also be used for the sets $\mathsf{T}_P^k \neq \emptyset$ when this does not lead to ambiguity. These sets are also called *type classes* or *composition classes*.

REMARK In mathematical statistics, if $\mathbf{x} \in \mathsf{X}^k$ is a sample of size k consisting of the results of k observations, the type of $\mathbf{x}$ is called the *empirical distribution* of the sample $\mathbf{x}$. ○

By (2.1), the Q^k-probability of a subset of T_P is determined by its cardinality. Hence the Q^k-probability of any subset A of X^k can be calculated by combinatorial counting arguments, looking at the intersections of A with the various sets T_P separately. In doing so, it will be relevant that the number of different types in X^k is much smaller than the number of sequences $\mathbf{x} \in \mathsf{X}^k$.

→ 2.1

LEMMA 2.2 (*Type counting*) The number of different types of sequences in X^k is less than $(k+1)^{|\mathsf{X}|}$. ○

Proof For every $a \in \mathsf{X}$, $N(a|\mathbf{x})$ can take $k+1$ different values. □

The next lemma explains the role of entropy from a combinatorial point of view, via the asymptotics of a multinomial coefficient.

LEMMA 2.3 For any type P of sequences in X^k → 2.2

$$(k+1)^{-|\mathsf{X}|} \exp[kH(P)] \leqq |\mathsf{T}_P| \leqq \exp[kH(P)]. \qquad \bigcirc$$

Proof Since (2.1) implies

$$P^k(\mathbf{x}) = \exp[-kH(P)] \quad \text{if} \quad \mathbf{x} \in \mathsf{T}_P$$

we have

$$|\mathsf{T}_P| = P^k(\mathsf{T}_P) \exp[kH(P)].$$

Hence it is enough to prove that

$$P^k(\mathsf{T}_P) \geqq (k+1)^{-|\mathsf{X}|}.$$

This will follow by the Type counting lemma if we show that the P^k-probability of $\mathsf{T}_{\widehat{P}}$ is maximized for $\widehat{P} = P$.

By (2.1) we have

$$P^k(\mathsf{T}_{\widehat{P}}) = |\mathsf{T}_{\widehat{P}}| \cdot \prod_{a \in \mathsf{X}} P(a)^{k\widehat{P}(a)} = \frac{k!}{\prod_{a \in \mathsf{X}} (k\widehat{P}(a))!} \prod_{a \in \mathsf{X}} P(a)^{k\widehat{P}(a)}$$

for every type $\widehat{P}$ of sequences in X^k.

It follows that

$$\frac{P^k(\mathsf{T}_{\widehat{P}})}{P^k(\mathsf{T}_P)} = \prod_{a \in \mathsf{X}} \frac{(kP(a))!}{(k\widehat{P}(a))!} P(a)^{k(\widehat{P}(a) - P(a))}.$$

Applying the obvious inequality $\frac{n!}{m!} \leqq n^{n-m}$, this gives

$$\frac{P^k(\mathsf{T}_{\widehat{P}})}{P^k(\mathsf{T}_P)} \leqq \prod_{a \in \mathsf{X}} k^{k(P(a) - \widehat{P}(a))} = 1. \qquad \square$$

If X and Y are two finite sets, the *joint type* of a pair of sequences $\mathbf{x} \in \mathsf{X}^k$ and $\mathbf{y} \in \mathsf{Y}^k$ is defined as the type of the sequence $\{(x_i, y_i)\}_{i=1}^k \in (\mathsf{X} \times \mathsf{Y})^k$. In other words, it is the distribution $P_{\mathbf{x},\mathbf{y}}$ on $\mathsf{X} \times \mathsf{Y}$ defined by

$$P_{\mathbf{x},\mathbf{y}}(a, b) \triangleq \frac{1}{k} N(a, b|\mathbf{x}, \mathbf{y}) \quad \text{for every} \quad a \in \mathsf{X}, b \in \mathsf{Y}.$$

Joint types will often be given in terms of the type of $\mathbf{x}$ and a stochastic matrix $V : \mathsf{X} \to \mathsf{Y}$ such that

$$P_{\mathbf{x},\mathbf{y}}(a, b) = P_{\mathbf{x}}(a) V(b|a) \quad \text{for every} \quad a \in \mathsf{X}, b \in \mathsf{Y}. \tag{2.2}$$

Note that the joint type $P_{\mathbf{x},\mathbf{y}}$ uniquely determines $V(b|a)$ for those $a \in \mathsf{X}$ which do occur in the sequence $\mathbf{x}$. For conditional probabilities of sequences $\mathbf{y} \in \mathsf{Y}^k$, given a sequence $\mathbf{x} \in \mathsf{Y}^k$, the matrix V of (2.2) will play the same role as the type of $\mathbf{y}$ does for unconditional probabilities.

DEFINITION 2.4 We say that $\mathbf{y} \in \mathsf{Y}^k$ has *conditional type* V given $\mathbf{x} \in \mathsf{X}^k$ if

$$N(a, b|\mathbf{x}, \mathbf{y}) = N(a|\mathbf{x})V(b|a) \quad \text{for every} \quad a \in \mathsf{X}, b \in \mathsf{Y}.$$

For any given $\mathbf{x} \in \mathsf{Y}^k$ and stochastic matrix $V : \mathsf{X} \to \mathsf{Y}$, the set of sequences $\mathbf{y} \in \mathsf{Y}^k$ having conditional type V given $\mathbf{x}$ will be called the *V-shell* of $\mathbf{x}$, denoted by $\mathsf{T}^k_V(\mathbf{x})$ or simply $\mathsf{T}_V(\mathbf{x})$. ○

REMARK The conditional type of $\mathbf{y}$ given $\mathbf{x}$ is not uniquely determined if some $a \in \mathsf{X}$ do not occur in $\mathbf{x}$. Still, the set $\mathsf{T}_V(\mathbf{x})$ containing $\mathbf{y}$ is unique. ○

→ 2.3

Note that conditional type is a generalization of types. In fact, if all the components of the sequence $\mathbf{x}$ are equal (say x) then the V-shell of $\mathbf{x}$ coincides with the set of sequences of type $V(\cdot|x)$ in Y^k.

In order to formulate the basic size and probability estimates for V-shells, it will be convenient to introduce some notations. The average of the entropies of the rows of a stochastic matrix $V : \mathsf{X} \to \mathsf{Y}$ with respect to a distribution P on X will be denoted by

$$H(V|P) \triangleq \sum_{x \in \mathsf{X}} P(x)H(V(\cdot|x)). \tag{2.3}$$

The analogous average of the informational divergences of the corresponding rows of two stochastic matrices $V : \mathsf{X} \to \mathsf{Y}$ and $W : \mathsf{X} \to \mathsf{Y}$ will be denoted by

$$D(V\|W|P) \triangleq \sum_{x \in \mathsf{X}} P(x)D(V(\cdot|x)\|W(\cdot|x)). \tag{2.4}$$

Note that $H(V|P)$ is the conditional entropy $H(Y|X)$ of RVs X and Y such that X has distribution P and Y has conditional distribution V given X. The quantity $D(V\|W|P)$ is called the *conditional informational divergence.* A counterpart of Lemma 2.3 for V-shells is

LEMMA 2.5 For every $\mathbf{x} \in \mathsf{X}^k$ and stochastic matrix $V : \mathsf{X} \to \mathsf{Y}$ such that $\mathsf{T}_V(\mathbf{x})$ is non-void, we have

$$(k+1)^{-|\mathsf{X}||\mathsf{Y}|} \exp[kH(V|P_{\mathbf{x}})] \leqq |\mathsf{T}_V(\mathbf{x})| \leqq \exp[kH(V|P_{\mathbf{x}})].$$ ○

Proof This is an easy consequence of Lemma 2.2. In fact, $|\mathsf{T}_V(\mathbf{x})|$ depends on $\mathbf{x}$ only through the type of $\mathbf{x}$. Hence we may assume that $\mathbf{x}$ is the juxtaposition of sequences $\mathbf{x}_a$, $a \in \mathsf{X}$, where $\mathbf{x}_a$ consists of $N(a|\mathbf{x})$ identical elements a. In this case $\mathsf{T}_V(\mathbf{x})$ is the Cartesian product of the sets of sequences of type $V(\cdot|a)$ in $\mathsf{Y}^{N(a|\mathbf{x})}$, with a running over those elements of X which occur in $\mathbf{x}$.

Thus Lemma 2.3 gives

$$\prod_{a \in \mathsf{X}} (N(a|\mathbf{x}) + 1)^{-|\mathsf{Y}|} \exp[N(a|\mathbf{x})H(V(\cdot|a))]$$
$$\leqq |\mathsf{T}_V(\mathbf{x})| \leqq \prod_{a \in \mathsf{X}} \exp[N(a|\mathbf{x})H(V(\cdot|a))],$$

whence the assertion follows by (2.3). □

LEMMA 2.6 For every type P of sequences in X^k and distribution Q on X

$$Q^k(\mathbf{x}) = \exp[-k(D(P\|Q) + H(P))] \quad \text{if} \quad \mathbf{x} \in \mathsf{T}_P, \tag{2.5}$$

$$(k+1)^{-|\mathsf{X}|}\exp[-kD(P\|Q)] \leqq Q^k(\mathsf{T}_P) \leqq \exp[-kD(P\|Q)]. \tag{2.6}$$

Similarly, for every $\mathbf{x} \in \mathsf{X}^k$ and stochastic matrices $V : \mathsf{X} \to \mathsf{Y}$, $W : \mathsf{X} \to \mathsf{Y}$ such that $\mathsf{T}_V(\mathbf{x})$ is non-void,

$$W^k(\mathbf{y}|\mathbf{x}) = \exp[-k(D(V\|W|P_\mathbf{x}) + H(V|P_\mathbf{x}))] \quad \text{if} \quad \mathbf{y} \in \mathsf{T}_V(\mathbf{x}), \tag{2.7}$$

$$(k+1)^{-|\mathsf{X}||\mathsf{Y}|}\exp[-kD(V\|W|P_\mathbf{x})] \leqq W^k(\mathsf{T}_V(\mathbf{x})|\mathbf{x})$$

$$\leqq \exp[-kD(V\|W|P_\mathbf{x})]. \tag{2.8}$$

○

Proof Equation (2.5) is just a rewriting of (2.1). Similarly, (2.7) is a rewriting of the identity

$$W^k(\mathbf{y}|\mathbf{x}) = \prod_{a\in\mathsf{X},\, b\in\mathsf{Y}} W(b|a)^{N(a,b|\mathbf{x},\mathbf{y})}.$$

The remaining assertions now follow from Lemmas 2.3 and 2.5. □

The quantity $D(P\|Q) + H(P) = -\sum_{x\in\mathsf{X}} P(x)\log Q(x)$ appearing in (2.5) is sometimes called *inaccuracy.*

For $Q \neq P$, the Q^k-probability of the set T_P^k is exponentially small (for large k); cf. Lemma 2.6. It can be seen that even $P^k(\mathsf{T}_P^k) \to 0$ as $k \to \infty$. Thus sets of large probability must contain sequences of different types. Dealing with such sets, the continuity of the entropy function plays a relevant role. The next lemma gives more precise information on this continuity. → 2.2

The *variation distance* of two distributions P and Q on X is

$$d(P, Q) \triangleq \sum_{x\in\mathsf{X}} |P(x) - Q(x)|.$$

(Some authors use the term for the half of this.)

LEMMA 2.7 If $d(P, Q) = \Theta \leqq 1/2$ then

$$|H(P) - H(Q)| \leqq -\Theta \log \frac{\Theta}{|\mathsf{X}|}.$$

○

For a sharpening of this lemma, see Problem 3.10.

Proof Write $\vartheta(x) \triangleq |P(x) - Q(x)|$. Since $f(t) \triangleq -t\log t$ is concave and $f(0) = f(1) = 0$, we have for every $0 \leqq t \leqq 1 - \tau$, $0 \leqq \tau \leqq 1/2$,

$$|f(t) - f(t+\tau)| \leqq \max(f(\tau), f(1-\tau)) = -\tau\log\tau.$$

Hence for $0 \leqq \Theta \leqq 1/2$

$$|H(P) - H(Q)| \leqq \sum_{x \in \mathsf{X}} |f(P(x)) - f(Q(x))| \leqq -\sum_{x \in \mathsf{X}} \vartheta(x) \log \vartheta(x)$$

$$= \Theta \left(-\sum_{x \in \mathsf{X}} \frac{\vartheta(x)}{\Theta} \log \frac{\vartheta(x)}{\Theta} - \log \Theta \right) \leqq \Theta \log |\mathsf{X}| - \Theta \log \Theta,$$

where the last step follows from Corollary 1.1. □

DEFINITION 2.8 For any distribution P on X, a sequence $\mathbf{x} \in \mathsf{X}^k$ is called *P-typical* with constant δ if

$$\left| \frac{1}{k} N(a|\mathbf{x}) - P(a) \right| \leqq \delta \quad \text{for every} \quad a \in \mathsf{X}$$

and, in addition, no $a \in \mathsf{X}$ with $P(a) = 0$ occurs in $\mathbf{x}$. The set of such sequences will be denoted by $\mathsf{T}^k_{[P]_\delta}$ or simply $\mathsf{T}_{[P]_\delta}$. If X is a RV with values in X, we refer to P_X-typical sequences as *X-typical,* and write $\mathsf{T}^k_{[X]_\delta}$ or $\mathsf{T}_{[X]_\delta}$ for $\mathsf{T}^k_{[P_X]_\delta}$. ○

REMARK $\mathsf{T}^k_{[P]_\delta}$ is the union of the sets $\mathsf{T}^k_{\widehat{P}}$ for those types $\widehat{P}$ of sequences in X^k which satisfy

$$|\widehat{P}(a) - P(a)| \leqq \delta \quad \text{for every} \quad a \in \mathsf{X}$$

and $\widehat{P}(a) = 0$ whenever $P(a) = 0$. ○

DEFINITION 2.9 For a stochastic matrix $W : \mathsf{X} \to \mathsf{Y}$, a sequence $\mathbf{y} \in \mathsf{Y}^k$ is *W-typical* under the condition $\mathbf{x} \in \mathsf{X}^k$ (or *W-generated* by the sequence $\mathbf{x} \in \mathsf{X}^k$) with constant δ if

$$\left| \frac{1}{k} N(a, b|\mathbf{x}, \mathbf{y}) - \frac{1}{k} N(a|\mathbf{x}) W(b|a) \right| \leqq \delta \quad \text{for every} \quad a \in \mathsf{X}, b \in \mathsf{Y},$$

and, in addition, $N(a, b|\mathbf{x}, \mathbf{y}) = 0$ whenever $W(b|a) = 0$. The set of such sequences $\mathbf{y}$ will be denoted by $\mathsf{T}^k_{[W]_\delta}(\mathbf{x})$ or simply by $\mathsf{T}_{[W]_\delta}(\mathbf{x})$. Further, if X and Y are RVs with values in X resp. Y and $P_{Y|X} = W$, then we shall speak of *$Y|X$-typical* or *$Y|X$-generated* sequences and write $\mathsf{T}^k_{[Y|X]_\delta}(\mathbf{x})$ or $\mathsf{T}_{[Y|X]_\delta}(\mathbf{x})$ for $\mathsf{T}^k_{[W]_\delta}(\mathbf{x})$. ○

Sequences $Y|X$-generated by an $\mathbf{x} \in \mathsf{X}^k$ are defined only if the condition $P_{Y|X} = W$ uniquely determines $W(\cdot|a)$ for $a \in \mathsf{X}$ with $N(a|\mathbf{x}) > 0$, that is, if no $a \in \mathsf{X}$ with $P_X(a) = 0$ occurs in the sequence $\mathbf{x}$; this automatically holds if $\mathbf{x}$ is X-typical.

The set $\mathsf{T}^k_{[XY]_\delta}$ of (X, Y)-typical pairs $(\mathbf{x}, \mathbf{y}) \in \mathsf{X}^k \times \mathsf{Y}^k$ is defined applying Definition 2.8 to (X, Y) in the role of X. When the pair $(\mathbf{x}, \mathbf{y})$ is typical, we say that $\mathbf{x}$ and $\mathbf{y}$ are *jointly typical.*

LEMMA 2.10 If $\mathbf{x} \in \mathsf{T}^k_{[X]_\delta}$ and $\mathbf{y} \in \mathsf{T}^k_{[Y|X]_{\delta'}}(\mathbf{x})$ then $(\mathbf{x}, \mathbf{y}) \in \mathsf{T}^k_{[XY]_{\delta+\delta'}}$ and, consequently, $\mathbf{y} \in \mathsf{T}^k_{[Y]_{\delta''}}$ for $\delta'' \triangleq (\delta + \delta')|\mathsf{X}|$. ○

For reasons which will be obvious from Lemmas 2.12 and 2.13, typical sequences will be used with δ depending on k such that → 2.4

$$\delta_k \to 0, \quad \sqrt{k} \cdot \delta_k \to \infty \quad \text{as} \quad k \to \infty. \tag{2.9}$$

Throughout this book, we adopt the following convention.

CONVENTION 2.11 *(Delta-convention)* To every set X resp. ordered pair of sets (X, Y) there is given a sequence $\{\delta_k\}_{k=1}^{\infty}$ satisfying (2.9). Typical sequences are understood with these δ_k. The sequences $\{\delta_k\}$ are considered as fixed, and in all assertions dependence on them will be suppressed. Accordingly, the constant δ will be omitted from the notation, i.e., we shall write $\mathsf{T}^k_{[P]}$, $\mathsf{T}^k_{[W]}(\mathbf{x})$, etc. In most applications, some simple relations between these sequences $\{\delta_k\}$ will also be needed. In particular, whenever we need that typical sequences should generate typical ones, we assume that the corresponding δ_k are chosen according to Lemma 2.10. ○

LEMMA 2.12 There exists a sequence $\varepsilon_k \to 0$ depending only on $|\mathsf{X}|$ and $|\mathsf{Y}|$ (see the delta-convention) so that for every distribution P on X and stochastic matrix $W : \mathsf{X} \to \mathsf{Y}$

$$P^k(\mathsf{T}^k_{[P]}) \geqq 1 - \varepsilon_k,$$

$$W^k(\mathsf{T}^k_{[W]}(\mathbf{x})|\mathbf{x}) \geqq 1 - \varepsilon_k \quad \text{for every} \quad \mathbf{x} \in \mathsf{X}^k.$$ ○

REMARK More explicitly,

$$P^k(\mathsf{T}^k_{[P]_\delta}) \geqq 1 - \frac{|\mathsf{X}|}{4k\delta^2}, \quad W^k(\mathsf{T}^k_{[W]_\delta}(\mathbf{x})|\mathbf{x}) \geqq 1 - \frac{|\mathsf{X}||\mathsf{Y}|}{4k\delta^2},$$

for every $\delta > 0$, and here the terms subtracted from 1 could be replaced even by $2|\mathsf{X}|e^{-2k\delta^2}$ resp. $2|\mathsf{X}||\mathsf{Y}|e^{-2k\delta^2}$. ○

Proof It suffices to prove the inequalities of the Remark. Clearly, the second inequality implies the first one as a special case (choose in the second inequality a one-point set for X). Now if $\mathbf{x} = x_1 \dots x_k$, let $Y_1, Y_2, \dots, Y_k$ be independent RVs with distributions $P_{Y_i} = W(\cdot|x_i)$. Then the RV $N(a, b|\mathbf{x}, Y^k)$ has binominal distribution with expectation $N(a|\mathbf{x})W(b|a)$ and variance

$$N(a|\mathbf{x})W(b|a)(1 - W(b|a)) \leqq \frac{1}{4}N(a|\mathbf{x}) \leqq \frac{k}{4}.$$

Thus by Chebyshev's inequality

$$\Pr\{|N(a, b|\mathbf{x}, Y^k) - N(a|\mathbf{x})W(b|a)| > k\delta\} \leqq \frac{1}{4k\delta^2}$$

for every $a \in \mathsf{X}$, $b \in \mathsf{Y}$. Hence the inequality with $1 - \frac{|\mathsf{X}||\mathsf{Y}|}{4k\delta^2}$ follows. The claimed sharper bound is obtained similarly, employing Hoeffding's inequality (see Problem 3.18 (b)) instead of Chebyshev's. □

LEMMA 2.13 There exists a sequence $\varepsilon_k \to 0$ depending only on $|\mathsf{X}|$ and $|\mathsf{Y}|$ (see the delta-convention) so that for every distribution P on X and stochastic matrix $W : \mathsf{X} \to \mathsf{Y}$

$$\left| \frac{1}{k} \log |\mathsf{T}^k_{[P]}| - H(P) \right| \leqq \varepsilon_k$$

and

$$\left| \frac{1}{k} \log |\mathsf{T}^k_{[W]}(\mathbf{x})| - H(W|P) \right| \leqq \varepsilon_k \quad \text{for every} \quad \mathbf{x} \in \mathsf{T}^k_{[P]}. \qquad \bigcirc$$

Proof The first assertion immediately follows from Lemma 2.3 and the uniform continuity of the entropy function (Lemma 2.7). The second assertion, containing the first one as a special case, follows similarly from Lemmas 2.5 and 2.7. To be formal, observe that, by the type counting lemma, $\mathsf{T}^k_{[W]}(\mathbf{x})$ is the union of at most $(k+1)^{|\mathsf{X}||\mathsf{Y}|}$ disjoint V-shells $\mathsf{T}_V(\mathbf{x})$. By Definitions 2.4 and 2.9, all the underlying V satisfy

$$|P_{\mathbf{x}}(a)V(b|a) - P_{\mathbf{x}}(a)W(b|a)| \leqq \delta'_k \quad \text{for every} \quad a \in \mathsf{X}, b \in \mathsf{Y}, \tag{2.10}$$

where $\{\delta'_k\}$ is the sequence corresponding to the pair of sets X, Y by the delta-convention. By (2.10) and Lemma 2.7, the entropies of the joint distributions on $\mathsf{X} \times \mathsf{Y}$ determined by $P_{\mathbf{x}}$ and V resp. by $P_{\mathbf{x}}$ and W differ by at most $-|\mathsf{X}||\mathsf{Y}|\delta'_k \log \delta'_k$ (if $|\mathsf{X}||\mathsf{Y}|\delta'_k \leqq 1/2$) and thus also

$$|H(V|P_{\mathbf{x}}) - H(W|P_{\mathbf{x}})| \leqq -|\mathsf{X}||\mathsf{Y}|\delta'_k \log \delta'_k.$$

On account of Lemma 2.5, it follows that

$$(k+1)^{-|\mathsf{X}||\mathsf{Y}|} \exp\left[k(H(W|P_{\mathbf{x}}) + |\mathsf{X}||\mathsf{Y}|\delta'_k \log \delta'_k)\right]$$
$$\leqq |\mathsf{T}^k_{[W]}(\mathbf{x})| \leqq (k+1)^{|\mathsf{X}||\mathsf{Y}|} \exp\left[k(H(W|P_{\mathbf{x}}) - |\mathsf{X}||\mathsf{Y}|\delta'_k \log \delta'_k)\right]. \tag{2.11}$$

Finally, since $\mathbf{x}$ is P-typical, i.e.,

$$|P_{\mathbf{x}}(a) - P(a)| \leqq \delta_k \quad \text{for every} \quad a \in \mathsf{X},$$

we have by Corollary 1.1

$$|H(W|P_{\mathbf{x}}) - H(W|P)| \leqq \delta_k \log|\mathsf{Y}|.$$

Substituting this into (2.11), the assertion follows. □

The last basic lemma of this chapter asserts that no "large probability set" can be substantially smaller than $\mathsf{T}_{[P]}$ resp. $\mathsf{T}_{[W]}(\mathbf{x})$.

LEMMA 2.14 Given $0 < \eta < 1$, there exists a sequence $\varepsilon_k \to 0$ depending only on η, $|\mathsf{X}|$ and $|\mathsf{Y}|$ such that

(i) if $\mathsf{A} \subset \mathsf{X}^k$, $P^k(\mathsf{A}) \geqq \eta$ then $\frac{1}{k}\log|\mathsf{A}| \geqq H(P) - \varepsilon_k$;

(ii) if $\mathsf{B} \subset \mathsf{Y}^k$, $W^k(\mathsf{B}|\mathbf{x}) \geqq \eta$ then $\frac{1}{k}\log|\mathsf{B}| \geqq H(W|P_{\mathbf{x}}) - \varepsilon_k$. ○

COROLLARY 2.14 There exists a sequence $\varepsilon'_k \to 0$ depending only on η, $|\mathsf{X}|$, $|\mathsf{Y}|$ (see the delta-convention) such that if $\mathsf{B} \subset \mathsf{Y}^k$ and $W^k(\mathsf{B}|\mathbf{x}) \geqq \eta$ for some $\mathbf{x} \in \mathsf{T}_{[P]}$ then

$$\frac{1}{k}\log|\mathsf{B}| \geqq H(W|P) - \varepsilon'_k. \qquad \bigcirc$$

Proof It is sufficient to prove (ii). By Lemma 2.12, the condition $W^k(\mathsf{B}|\mathbf{x}) \geqq \eta$ implies

$$W^k(\mathsf{B} \cap \mathsf{T}_{[W]}(\mathbf{x})|\mathbf{x}) \geqq \frac{\eta}{2}$$

for $k \geqq k_0(\eta, |\mathsf{X}|, |\mathsf{Y}|)$. Recall that $\mathsf{T}_{[W]}(\mathbf{x})$ is the union of disjoint V-shells $\mathsf{T}_V(\mathbf{x})$ satisfying (2.10); see the proof of Lemma 2.13. Since $W^k(\mathbf{y}|\mathbf{x})$ is constant within a V-shell of $\mathbf{x}$, it follows that

$$|\mathsf{B} \cap \mathsf{T}_V(\mathbf{x})| \geqq \frac{\eta}{2}|\mathsf{T}_V(\mathbf{x})|$$

for at least one $V : \mathsf{X} \to \mathsf{Y}$ satisfying (2.10). Now the proof can be completed using Lemmas 2.5 and 2.7 just as in the proof of the previous lemma. □

Observe that the preceding three lemmas contain a proof of Theorem 1.1. Namely, the fact that about $kH(P)$ binary digits are sufficient for encoding k-length messages of a DMS with generic distribution P, is a consequence of Lemmas 2.12 and 2.13, while the necessity of this many binary digits follows from Lemma 2.14. Most coding theorems in this book will be proved using typical sequences in a similar manner. The merging of several nearby types has the advantage of facilitating computations. When dealing with the more refined questions of the speed of convergence of error probabilities, however, the method of typical sequences will become inappropriate. In such problems, we shall have to consider each type separately, relying on the first part of this chapter. Although this will not occur until Chapter 9, as an immediate illustration of the more subtle method we now refine the basic source coding result, Theorem 1.1. → 2.5

THEOREM 2.15 For any finite set X and $R > 0$ there exists a sequence of k-to-n_k binary block codes (f_k, φ_k) with

$$\frac{n_k}{k} \to R$$

such that for every DMS with alphabet X and arbitrary generic distribution P, the probability of error satisfies

$$e(f_k, \varphi_k) \leqq \exp\left\{-k\left[\inf_{Q:H(Q)>R} D(Q||P) - \eta_k\right]\right\} \tag{2.12}$$

with

$$\eta_k \triangleq \frac{\log(k+1)}{k}|\mathsf{X}|.$$

This result is asymptotically sharp for every particular DMS, in the sense that for any sequence of k-to-n_k binary block codes, $n_k/k \to R$ implies

$$\underline{\lim}_{k\to\infty} \frac{1}{k} \log e(f_k, \varphi_k) \geqq - \inf_{Q:H(Q)>R} D(Q||P). \tag{2.13}$$

The infimum in (2.12) and (2.13) is finite iff $R < \log s(P)$, and then it equals the minimum subject to $H(Q) \geqq R$. ○

Here $s(P)$ denotes the size of the support of P, that is, the number of those $a \in \mathsf{X}$ for which $P(a) > 0$.

REMARK This result sharpens Theorem 1.1 in two ways. First, for a DMS with generic distribution P, and $R > H(P)$, it gives the precise asymptotics, in the exponential sense, of the probability of error of the best codes with $n_k/k \to R$ (the result is also true, but uninteresting, for $R \leqq H(P)$). Second, it shows that this optimal performance can be achieved by codes not depending on the generic distribution of the source. The remaining assertion of Theorem 1.1, namely that for $n_k/k \to R < H(P)$
→ 2.6 the probability of error tends to 1, can be sharpened similarly. ○

Proof of Theorem 2.15. Write

$$\mathsf{A}_k \triangleq \bigcup_{Q:H(Q)\leqq R} \mathsf{T}_Q.$$

Then, by Lemmas 2.2 and 2.3,

$$|\mathsf{A}_k| \leqq (k+1)^{|\mathsf{X}|} \exp(kR); \tag{2.14}$$

further, by Lemmas 2.2 and 2.6,

$$P^k(\mathsf{X}^k - \mathsf{A}_k) \leqq (k+1)^{|\mathsf{X}|} \exp\left\{ -k \min_{Q:H(Q)>R} D(Q||P) \right\}. \tag{2.15}$$

Let us encode the sequences in A_k in a one-to-one way and all others by a fixed codeword, say. Equation (2.14) shows that this can be done with binary codewords of length n_k satisfying $n_k/k \to R$. For the resulting code, (2.15) gives (2.12), with

$$\eta_k \triangleq \frac{\log(k+1)}{k} |\mathsf{X}|.$$

The last assertion of Theorem 2.15 is obvious, and implies that it suffices to prove (2.13) for $R < \log s(P)$. The number of sequences in X^k correctly reproduced by a k-to-n_k binary block code is at most 2^{n_k}. Thus, by Lemma 2.3, for every type Q of sequences in X^k satisfying

$$(k+1)^{-|\mathsf{X}|} \exp[kH(Q)] \geqq 2^{n_k+1}, \tag{2.16}$$

at least half of the sequences in T_Q will not be reproduced correctly. On account of Lemma 2.6, it follows that

$$e(f_k, \varphi_k) \geqq \frac{1}{2}(k+1)^{-|\mathsf{X}|} \exp[-kD(Q||P)]$$

for every type Q satisfying (2.16). Hence

$$e(f_k, \varphi_k) \geqq \frac{1}{2}(k+1)^{-|\mathsf{X}|} \exp\left\{-k \min_{Q: H(Q) \geqq R+\varepsilon_k} D(Q||P)\right\},$$

where Q runs over types of sequences in X^k and

$$\varepsilon_k \triangleq \frac{n_k}{k} - R + \frac{1}{k} + \frac{\log(k+1)}{k}|\mathsf{X}|.$$

Using that $R < \log s(P)$, for large k the last minimum changes little if Q is not restricted to types, and ε_k is omitted. □ → 2.7

Discussion

The simple combinatorial lemmas concerning types are the basis of the proof of most coding theorems treated in this book. Merging "nearby" types, i.e., the formalism of typical sequences, has the advantage of shortening computations. In the literature, there are several concepts of typical sequences. Often one merges more types than we have done in Definition 2.8; in particular, the entropy-typical sequences of Problem 2.5 are widely used. The latter kind of typicality has the advantage that it easily generalizes to models with memory and with abstract alphabets. For the discrete memoryless systems treated in this book, the adopted concept of typicality often leads to stronger results. Still, the formalism of typical sequences has a limited scope, for it does not allow evaluation of convergence rates of error probabilities. This is illustrated by the fact that typical sequences led to a simple proof of Theorem 1.1 while to prove Theorem 2.15 types had to be considered individually.

The technique of estimating probabilities without merging types is also more appropriate for the purpose of deriving *universal coding* theorems. Intuitively, universal coding means that codes have to be constructed in complete ignorance of the probability distributions governing the system; then the performance of the code is evaluated by the whole spectrum of its performance indices for the various possible distributions. Theorem 2.15 is the first universal coding result in this book. It is clear that two codes are not necessarily comparable from the point of view of universal coding. In view of this, it is somewhat surprising that for the class of DMSs with a fixed alphabet X there exist codes *universally optimal* in the sense that for every DMS they have asymptotically the same probability of error as the best code designed for that particular DMS. ○

Problems

2.1. Show that the exact number of types of sequences in X^k equals

$$\binom{k+|\mathsf{X}|-1}{|\mathsf{X}|-1}.$$

Draw the conclusion that the lower bounds in Lemmas 2.3, 2.5 and 2.6 can be sharpened replacing the power of $(k+1)$ by this number.

2.2. Prove that the size of T^k_P is of order of magnitude $k^{-(s(P)-1)/2}\exp\{kH(P)\}$, where $s(P)$ is the number of elements $a \in \mathsf{X}$ with $P(a) > 0$. More precisely, show that

$$\log|\mathsf{T}^k_P| = kH(P) - \frac{s(P)-1}{2}\log(2\pi k) - \frac{1}{2}\sum_{a:P(a)>0}\log P(a) - \frac{\vartheta(k,P)}{12\ln 2}s(P),$$

where $0 \leqq \vartheta(k,P) \leqq 1$.

Hint Use Robbins' sharpening of Stirling's formula:

$$\sqrt{2\pi}\,n^{n+\frac{1}{2}}e^{-n+\frac{1}{12(n+1)}} \leqq n! \leqq \sqrt{2\pi}\,n^{n+\frac{1}{2}}e^{-n+\frac{1}{12n}}$$

(see e.g. Feller (1968), p. 54), noting that $P(a) \geqq 1/k$ whenever $P(a) > 0$.

2.3. Clearly, every $\mathbf{y} \in \mathsf{Y}^k$ in the V-shell of an $\mathbf{x} \in \mathsf{X}^k$ has the same type Q where

$$Q(b) \triangleq \sum_{a\in\mathsf{X}} P_{\mathbf{x}}(a)V(b|a).$$

(a) Show that $\mathsf{T}_V(\mathbf{x}) \neq \mathsf{T}_Q$ even if all the rows of the matrix V are equal to Q (unless $\mathbf{x}$ consists of identical elements).

(b) Show that if $P_{\mathbf{x}} = P$ then

$$(k+1)^{-|\mathsf{X}||\mathsf{Y}|}\exp[-kI(P,V)] \leqq \frac{|\mathsf{T}_V(\mathbf{x})|}{|\mathsf{T}_Q|} \leqq (k+1)^{|\mathsf{Y}|}\exp[-kI(P,V)],$$

where $I(P,V) \triangleq H(Q) - H(V|P)$ is the mutual information of RVs X and Y such that $P_X = P$ and $P_{Y|X} = V$. In particular, if all rows of V are equal to Q then the size of $\mathsf{T}_V(\mathbf{x})$ is not "exponentially smaller" than that of T_Q.

2.4. Prove that the first resp. second condition of (2.9) is necessary for Lemmas 2.13 resp. 2.12 to hold.

2.5. (*Entropy-typical sequences*) Let us say that a sequence $\mathbf{x} \in \mathsf{X}^k$ is entropy-P-typical with constant δ if

$$\left| -\frac{1}{k}\log P^k(\mathbf{x}) - H(P) \right| \leqq \delta;$$

further, $\mathbf{y} \in \mathsf{Y}^k$ is entropy-W-typical under the condition $\mathbf{x}$ if

$$\left| -\frac{1}{k}\log W^k(\mathbf{y}|\mathbf{x}) - H(W|P_{\mathbf{x}}) \right| \leqq \delta.$$

(a) Check that entropy-typical sequences also satisfy the assertions of Lemmas 2.12 and 2.13 (if $\delta = \delta_k$ is chosen as in the delta-convention).

Hint These properties were implicitly used in the proofs of Theorems 1.1 and 1.2.

(b) Show that typical sequences – with constants chosen according to the delta-convention – are also entropy-typical, with some constants $\delta'_k = c_P \cdot \delta_k$ resp. $\delta'_k = c_W \cdot \delta_k$. On the other hand, entropy-typical sequences are not necessarily typical with constants of the same order of magnitude.

(c) Show that the analog of Lemma 2.10 for entropy-typical sequences does not hold.

2.6. (*Codes with rate below entropy*) Prove the following counterpart of Theorem 2.15.

(a) For every DMS with generic distribution P, the probability of error of k-to-n_k binary block codes with $n_k/k \to R < H(P)$ tends to 1 exponentially; more exactly,

$$\overline{\lim_{k\to\infty}} \log(1 - e(f_k, \varphi_k)) \leqq - \min_{Q:H(Q)\leqq R} D(Q\|P).$$

(b) The bound in (a) is exponentially tight. More exactly, for every $R > 0$ there exist k-to-n_k binary block codes with $n_k/k \to R$ such that for every DMS with an arbitrary generic distribution P we have

$$\underline{\lim_{k\to\infty}} \frac{1}{k} \log(1 - e(f_k, \varphi_k)) \geqq - \min_{Q:H(Q)\leqq R} D(Q\|P).$$

(The limit given by (a) and (b) has been determined in Csiszár and Longo (1971), in a different algebraic form.)

Hint (a) The ratio of correctly decoded sequences within a T_Q is at most $(k+1)^{|\mathsf{X}|} \exp[-|kH(Q) - n_k|^+]$, by Lemma 2.3. Hence by Lemma 2.6 and the type counting lemma

$$\overline{\lim_{k\to\infty}} \frac{1}{k} \log(1 - e(f_k, \varphi_k)) \leqq -\min_{Q}(D(Q\|P) + |H(Q) - R|^+).$$

In order to prove that the last minimum is achieved when $H(Q) \leqq R$, it suffices to consider the case $R = 0$. Then, however, we have the identity

$$\min_{Q}(D(Q\|P) + H(Q)) = \min_{x\in\mathsf{X}}(-\log P(x)) = \min_{Q:H(Q)=0} D(Q\|P).$$

(b) Let the encoder be a one-to-one mapping on the union of the sets T_Q with $H(Q) \leqq R$.

2.7. (*Non-typewise upper bounds*)

(a) For any set $\mathsf{F} \subset \mathsf{X}^k$, show that

$$|\mathsf{F}| \leqq \exp[kH(P_\mathsf{F})], \quad \text{where} \quad P_\mathsf{F} \triangleq \frac{1}{|\mathsf{F}|}\sum_{\mathbf{x}\in\mathsf{F}} P_\mathbf{x}.$$

(See Massey (1974), citing T. A. Kriz.)

(b) For any set $\mathsf{F} \subset \mathsf{X}^k$ and distribution Q on X, show that

$$Q^k(\mathsf{F}) \leqq \exp[-kD(P\|Q)], \quad \text{where} \quad P(a) \triangleq \sum_{\mathbf{x}\in\mathsf{F}} \frac{Q^k(\mathbf{x})}{Q^k(\mathsf{F})} P_\mathbf{x}(a).$$

Note that these upper bounds generalize those of Lemmas 2.3 and 2.6.

Hint Consider RVs $X_1, \dots, X_k$ such that the vector $(X_1, \dots, X_k)$ is uniformly distributed on F and let J be a RV uniformly distributed on $\{1, \dots, k\}$ and independent of $X_1, \dots, X_k$. Then

$$\log |\mathsf{F}| = H(X_1, \ldots, X_k) \leqq \sum_{i=1}^{k} H(X_i) = kH(X_J \,|\, J) \leqq kH(X_J) = kH(P_{\mathsf{F}}).$$

This proves (a). Part (b) follows similarly, defining now the distribution of $(X_1, \ldots, X_k)$ by

$$\Pr\{X_1 \ldots X_k = \mathbf{x}\} \triangleq \frac{Q^k(\mathbf{x})}{Q^k(\mathsf{F})} \quad \text{if} \quad \mathbf{x} \in \mathsf{F}, \quad \text{and 0 otherwise.}$$

(c) Conclude from (b) that the upper bound in Theorem 2.15, though asymptotically sharp, can be significantly improved for small k. Namely, show that in the proof of Theorem 2.15 the polynomial factor in (2.15) may be omitted.

2.8. Show that

(a) $\sum_{i=1}^{d} \binom{k}{i} \leqq \exp\left[kh\,(d/k)\right]$ if $d < k/2$;

(b) the probability that in k independent trials an event of probability q occurs d times or less/more, according to whether d is less or greater than kq, is bounded above by $\exp\left[-kD(d/k \,||\, q)\right]$, where

$$D(p \,||\, q) \triangleq p \log \frac{p}{q} + (1-p) \log \frac{1-p}{1-q}.$$

Moreover, show in both cases that the upper bound divided by $k+1$ is a lower bound.

Hint Apply Problem 2.7, letting F consist of those binary sequences of length k in which the number of 1s is at most/at least d. For the lower bound use Problem 2.1.

2.9. Show that for independent RVs X and Y most pairs of typical sequences are jointly typical in the following sense: to any sequence $\{\delta_k\}$ satisfying (2.9) there exists a sequence $\{\delta'_k\}$ also satisfying (2.9) such that

$$\lim_{k\to\infty} \frac{|\mathsf{T}^k_{[XY]_{\delta'_k}} \cap (\mathsf{T}^k_{[X]_{\delta_k}} \times \mathsf{T}^k_{[Y]_{\delta_k}})|}{|\mathsf{T}^k_{[X]_{\delta_k}} \times \mathsf{T}^k_{[Y]_{\delta_k}}|} = 1.$$

(Compare this with Problem 2.3.)

2.10. Determine the asymptotic cardinality of the set of sequences in Y^k which are in the intersection of two different V-shells. Specifically, show that for any stochastic matrices $V : \mathsf{X} \to \mathsf{Y}$, $V' : \mathsf{X} \to \mathsf{Y}$ and every $\mathbf{x}, \mathbf{x}'$ the relation $\mathsf{T}_V(\mathbf{x}) \cap \mathsf{T}_{V'}(\mathbf{x}') \neq \emptyset$ implies that

$$\left| \frac{1}{k} \log |\mathsf{T}_V(\mathbf{x}) \cap \mathsf{T}_{V'}(\mathbf{x}')| - \max H(Y|X, X') \right| \leqq \varepsilon_k,$$

where $\varepsilon_k \to 0$ and the maximum refers to RVs X, X', Y such that $P_{X,X'} = P_{\mathbf{x},\mathbf{x}'}$ and $P_{Y|X} = V$, $P_{Y|X'} = V'$.

Hint Represent the intersection as the union of W-shells $\mathsf{T}_W(\mathbf{x}, \mathbf{x}')$ for stochastic matrices $W : \mathsf{X} \times \mathsf{X} \to \mathsf{Y}$ equal to $P_{Y|XX'}$ for RVs X, X', Y as in the assertion.

2.11. Prove that the assertions of Lemma 2.14 remain true if the constant $\eta > 0$ is replaced by a sequence $\{\eta_k\}$ which tends to zero slower than exponentially, i.e., $(1/k)\log \eta_k \to 0$.

2.12. (*Large deviation probabilities for empirical distributions*)

(a) Let $\mathcal{P}$ be any set of PDs on X and let $\mathcal{P}_k$ be the set of those PDs in $\mathcal{P}$ which are types of sequences in X^k. Show that for every distribution Q on X

$$\left|\frac{1}{k}\log Q^k(\{\mathbf{x} : P_{\mathbf{x}} \in \mathcal{P}\}) + \min_{P\in\mathcal{P}_k} D(P\|Q)\right| \leqq \frac{\log(k+1)}{k}|\mathsf{X}|.$$

(b) Let $\mathcal{P}$ be a set of PDs on X such that the closure of the interior of $\mathcal{P}$ equals $\mathcal{P}$. Show that for k independent drawings from a distribution Q the probability of a sample with empirical distribution belonging to $\mathcal{P}$ has the asymptotics

$$\lim_{k\to\infty}\frac{1}{k}\log Q^k(\{\mathbf{x} : P_{\mathbf{x}} \in \mathcal{P}\}) = -\min_{P\in\mathcal{P}} D(P\|Q).$$

(c) Show that if $\mathcal{P}$ is a convex set of distributions on X then

$$\frac{1}{k}\log Q^k(\{\mathbf{x} : P_{\mathbf{x}} \in \mathcal{P}\}) \leqq -\inf_{P\in\mathcal{P}} D(P\|Q),$$

for every k and every distribution Q on X.

(See Sanov (1957) and Hoeffding (1965).)

Hint Part (a) follows from Lemma 2.6 and the type counting lemma; (b) is an easy consequence of (a). Part (c) follows from the result of Problem 2.7(b).

2.13. (*Hypothesis testing*) Strengthen the result of Corollary 1.2 as follows.

(a) For a given P there exist tests which are asymptotically optimal simultaneously against all alternatives Q, i.e., there exist sets $\mathsf{A}_k \subset \mathsf{X}^k$ such that $P^k(\mathsf{A}_k) \to 1$ while

$$\lim_{k\to\infty}\frac{1}{k}\log Q^k(\mathsf{A}_k) = -D(P\|Q) \quad \text{for every} \quad Q.$$

Hint Set $\mathsf{A}_k \triangleq \mathsf{T}^k_{[P]}$, and apply (a) of Problem 2.12.

(b) For any given P and $a > 0$ there exist sets $\mathsf{A}_k \subset \mathsf{X}^k$ such that

$$\lim_{k\to\infty}\frac{1}{k}\log(1 - P^k(\mathsf{A}_k)) \leqq -a \qquad (*)$$

and for every Q

$$\frac{1}{k}\log Q^k(\mathsf{A}_k) \leqq -b(a, P, Q),$$

where

$$b(a, P, Q) \triangleq \min_{\tilde{P}:D(\tilde{P}\|P)\leqq a} D(\tilde{P}\|Q).$$

This result is optimal in the sense that if any sets A_k satisfy (*) then for every Q

$$\lim_{k\to\infty} \frac{1}{k} \log Q^k(\mathsf{A}_k) \geqq -b(a, P, Q).$$

(See Hoeffding (1965).)

Hint The sets $\mathsf{A}_k \triangleq \bigcup_{\tilde{P}:\ D(\tilde{P}\|P) \leqq a} \mathsf{T}^k_{\tilde{P}}$ do have the claimed properties by Problem 2.12. On the other hand, for every $\varepsilon > 0$, any set A_k with $1 - P^k(\mathsf{A}_k) \leqq \exp[-k(a-\varepsilon)]$ must contain at least half of the sequences of type $\tilde{P}$ whenever $D(\tilde{P}\|P) \leqq a - 2\varepsilon$, by Lemma 2.6. Hence the last assertion follows by another application of Lemma 2.6 and a continuity argument.

2.14. (*Evaluation of error exponents*) The error exponents of source coding resp. hypothesis testing (see Theorem 2.15 and Problem 2.13(b)) have been given as divergence minima. Determine the minimizing distributions.

(a) For an arbitrary distribution P on X and any $0 < \alpha \leqq 1$, define the distribution P_α by

$$P_\alpha(x) \triangleq P^\alpha(x)\Big(\sum_{a\in\mathsf{X}} P^\alpha(a)\Big)^{-1}.$$

Show that $H(P_\alpha)$ is a continuous function of α, that its limit as $\alpha \to 0$ is $\log s(P)$, and that this function is strictly decreasing unless $H(P) = \log s(P)$. Draw the conclusion that if $H(P) \leqq R < s(P)$, there exists a unique α^* with

$$H(P_{\alpha^*}) = R, \quad 0 < \alpha^* \leqq 1.$$

(b) Show that for $H(P) \leqq R < \log s(P)$ the divergence minimum figuring in Theorem 2.15 is equal to

$$F(R) \triangleq D(P_{\alpha^*}\|P),$$

with $\alpha^* = \alpha^*(R)$ defined in (a).

Hint Verify that for every Q and $0 < \alpha \leqq 1$

$$D(Q\|P) = \frac{1-\alpha}{\alpha} H(Q) + \frac{1}{\alpha}(D(Q\|P_\alpha) - \log \sum_{x\in\mathsf{X}} P^\alpha(x)).$$

Hence for $H(Q) \geqq R$, and α^* defined in (a),

$$D(Q\|P) - D(P_{\alpha^*}\|P) \geqq \frac{1}{\alpha^*} D(Q\|P_{\alpha^*}).$$

(c) For given distributions $P \neq Q$ on X, supposing for convenience that Q is strictly positive, and for $0 < \alpha \leqq 1$, define the distribution $\tilde{P}_\alpha$ by

$$\tilde{P}_\alpha(x) \triangleq P^\alpha(x) Q^{1-\alpha}(x)\Big(\sum_{a\in\mathsf{X}} P^\alpha(a) Q^{1-\alpha}(a)\Big)^{-1}.$$

Show that $D(\tilde{P}_\alpha \| P)$ is a continuous function of α whose limit as $\alpha \to 0$ is $D(Q' \| P)$, where $Q'(x)$ equals a constant times $Q(x)$ if $P(x) > 0$ and zero if $P(x) = 0$. Moreover, this function is strictly decreasing unless $Q' = P$.

(d) Show that for $0 \leqq a < D(Q' \| P)$ the divergence minimum defining the exponent $b(a, P, Q)$ of Problem 2.13(b) is achieved for $\tilde{P} = \tilde{P}_{\alpha^*}$, where α^* is the unique $0 < \alpha \leqq 1$ with

$$D(\tilde{P}_\alpha \| P) = a,$$

whereas for $a \geqq D(Q' \| P)$ the divergence minimum is achieved by $\tilde{P} = Q'$.

Hint For an arbitrary $\tilde{P}$, express $D(\tilde{P} \| Q)$ by $D(\tilde{P} \| P)$ and $D(\tilde{P} \| \tilde{P}_\alpha)$ in a way analogous to the hint to part (b).

2.15. (*An operational meaning of Rényi's information measures*)

(a) Show that the error exponent for source coding, evaluated in Problem 2.14(b), also equals the supremum of $\frac{1-\alpha}{\alpha}(R - H_\alpha(P))$ for $0 < \alpha \leqq 1$, where $H_\alpha(P)$ denotes entropy of order α; see Problem 1.15. Check that for $H(P) \leqq R < \log s(P)$ this supremum is a maximum, attained by $\alpha^* = \alpha^*(R)$ defined in Problem 2.14(a).

Hint Referring to Problem 2.14(b), for $H(P) \leqq R < \log s(P)$ verify that

$$F(R) = \frac{1-\alpha^*}{\alpha^*}(R - H_{\alpha^*}(P)), \qquad \frac{dF}{dR} = \frac{1-\alpha^*}{\alpha^*},$$

and that this implies the assertion when $R < \log s(P)$. Finally, check that the supremum in the assertion equals $+\infty$ if $R \geqq \log s(P)$.

(b) Given a DMS with generic distribution P and any $\beta > 0$, show that for each $R > H_\alpha(P)$, where $\alpha \triangleq 1/(1+\beta)$, there exist k-to-n_k binary block codes with $n_k/k \to R$ whose probability of error is at most $\exp[-\beta k(R - H_\alpha(P)) + o(k)]$; moreover, here $H_\alpha(P)$ can not be replaced by any smaller number.

(c) Establish the analog of (a) for the hypothesis testing error exponent, evaluated in Problem 2.14(d). Use the result to give an operational meaning to *Rényi's divergence of order* α

$$D_\alpha(P \| Q) \triangleq \frac{1}{\alpha - 1} \log \sum_{x \in \mathsf{X}} P^\alpha(x) Q^{1-\alpha}(x),$$

similar to that of entropy of order α in (b).

(See Csiszár (1995); for divergence of order α, see Rényi (1961).)

2.16. (*Exact asymptotics of error probability*)

(a) Prove directly that for a DMS with generic distribution P the best k-to-n_k binary block codes with $n_k/k \to R < \log s(P)$ yield

$$\lim_{k\to\infty} \frac{1}{k} \log e(f_k, \varphi_k) = -F(R),$$

where $F(R)$ has been defined in Problem 2.14(b). More exactly, show that if $\mathsf{A}_k \subset \mathsf{X}^k$ has maximum P^k-probability under the condition $|\mathsf{A}_k| = \lceil \exp kR \rceil$ then

$$\lim_{k\to\infty} \frac{1}{k} \log (1 - P^k(\mathsf{A}_k)) = -F(R).$$

(b) Show that, more precisely,

$$\left| \log(1 - P^k(\mathsf{A}_k)) + kF(R) + \frac{1}{2\alpha^*} \log k \right| \leqq K(P)$$

for every k, where $K(P)$ is a suitable constant.

Hint Let α_k be determined by the condition $P^k_{\alpha_k}(\mathsf{A}_k) = 1/2$, where P_α is the same as in Problem 2.14(a). Then $\alpha_k \to \alpha^*$ by Theorem 1.1. Now (a) follows from the Neyman–Pearson lemma (Problem 1.4) and Corollary 1.2. For (b), use the asymptotic formula of Problem 1.8 rather than Theorem 1.1 and Corollary 1.2.

(See Dobrušin (1962a) and Jelinek (1968b); the proof hinted above is of Csiszár and Longo (1971) who extended the same approach to the hypothesis testing problem.)

Story of the results

The asymptotics of the number of sequences of type P in terms of $H(P)$ plays a basic role in statistical physics, cf. Boltzmann (1877). The idea of using typical sequences in information-theoretic arguments (in fact, even the word) emerges in Shannon (1948) in a heuristic manner. A unified approach to information theory based on an elaboration of this concept was given by Wolfowitz (1964).

By now, typical sequences have become a standard tool; however, several different definitions are in use. We have adopted a definition similar to that of Wolfowitz (1964). "Type" was not an established name in the literature prior to this book. It has been chosen here in order to stress the importance of the proof technique based on types directly, rather than through typical sequences.

The material of this chapter is essentially folklore. Lemma 2.7 goes back to Fannes (1973). Lemmas 2.10–2.14 paraphrase Wolfowitz (1964). Theorem 2.15 comprises results of several authors. The exponentially tight bound of error probability for a given DMS was established in the form of Problem 2.16 by Jelinek (1968b) and earlier, in another context, by Dobrušin (1962a). The present form of the exponent appears in Blahut (1974) and Marton (1974). The universal attainability of this exponential bound is pointed out in Kričevskiĭ and Trofimov (1977). The simple derivation of

Theorem 2.15 given in the text was proposed independently (but relying on the first part of this chapter) by Longo and Sgarro (1979).

Addition. The method of types, introduced in the first edition of this book, has become a standard technique in the theory of DM systems. A tutorial based on this book is Csiszár (1998).

3 Formal properties of Shannon's information measures

The information measures introduced in Chapter 1 are important formal tools of information theory, often used in quite complex computations. Familiarity with a few identities and inequalities will make such computations perspicuous. Also, these formal properties of information measures have intuitive interpretations that enable one to remember them and use them properly.

Let $X, Y, Z, \dots$ be RVs with finite ranges $\mathsf{X}, \mathsf{Y}, \mathsf{Z}, \dots$. We shall consistently use the notational convention introduced in Chapter 1 that information quantities featuring a collection of RVs in the role of a single RV will be written without putting this collection into brackets. We shall often use a notation explicitly indicating that information measures associated with RVs are actually determined by their (joint) distribution. Let P be a distribution on X and let $W = \{W(y|x) : x \in \mathsf{X}, y \in \mathsf{Y}\}$ be a stochastic matrix, i.e., $W(\cdot|x) \triangleq \{W(y|x) : y \in \mathsf{Y}\}$ is a distribution on Y for every fixed $x \in \mathsf{X}$. Then for a pair of RVs (X, Y) with $P_X = P,\ P_{Y|X} = W$, we shall write $H(W|P)$ for $H(Y|X)$ as we did in Chapter 2, and similarly we shall write $I(P, W)$ for $I(X \wedge Y)$.

Then, cf. (1.7), (1.8), we have

$$H(W|P) = \sum_{x \in \mathsf{X}} P(x) H(W(\cdot|x)), \tag{3.1}$$

$$I(P, W) = H(PW) - H(W|P) = \sum_{x \in \mathsf{X}} P(x) D(W(\cdot|x) || PW). \tag{3.2}$$

Here PW designates the distribution of Y if $P_X = P,\ P_{Y|X} = W$, i.e.,

$$PW(y) \triangleq \sum_{x \in \mathsf{X}} P(x) W(y|x) \quad (y \in \mathsf{Y}).$$

Since information measures of RVs are functionals of their (joint) distribution, they are automatically defined also under the condition that some other RVs take some fixed values (provided that the conditioning event has positive probability). For entropy and mutual information determined by conditional distributions we shall use a self-explanatory notation like $H(X|Y = y, Z = z)$, $I(X \wedge Y|Z = z)$. Averages of such quantities by the (conditional) distribution of some of the conditioning RVs given the values of the remaining ones (if any) will be denoted similarly, omitting the specification of values of those RVs which were averaged out. For example,

$$H(X|Y, Z = z) \triangleq \sum_{y \in \mathsf{Y}} \Pr\{Y = y|Z = z\} H(X|Y = y, Z = z),$$

$$I(X \wedge Y|Z) \triangleq \sum_{z \in \mathsf{Z}} \Pr\{Z = z\} I(X \wedge Y|Z = z),$$

with the understanding that an undefined term multiplied by zero is zero. These conventions are consistent with the notation introduced in Chapter 1 for conditional entropy.

Unless stated otherwise, the terms *conditional entropy* (of X given Y) and *conditional mutual information* (of X and Y given Z) will always stand for quantities averaged with respect to the conditioning variable(s).

Sometimes information measures are associated also with individual (non-random) sequences $\mathbf{x} \in \mathsf{X}^n$, $\mathbf{y} \in \mathsf{Y}^n$, etc. These are defined as the entropy, mutual information, etc. determined by the (joint) types of the sequences in question. Thus, e.g.,

$$H(\mathbf{x}) \triangleq H(P_{\mathbf{x}}), \quad I(\mathbf{x} \wedge \mathbf{y}) \triangleq H(\mathbf{x}) + H(\mathbf{y}) - H(\mathbf{x}, \mathbf{y}).$$

We send forward an elementary lemma which is equivalent to the fact that $D(P||Q) \geqq 0$, with equality iff $P = Q$; the simple proof will not rely on Corollary 1.2. Most inequalities in this chapter are consequences of this lemma.

LEMMA 3.1 (*Log-sum inequality*) For arbitrary non-negative numbers a_i, b_i, $i = 1, \ldots, n$ we have

$$\sum_{i=1}^{n} a_i \log \frac{a_i}{b_i} \geqq a \log \frac{a}{b},$$

where

$$a \triangleq \sum_{i=1}^{n} a_i, \quad b \triangleq \sum_{i=1}^{n} b_i.$$

The equality holds iff $a_i b = b_i a$ for $i = 1, \ldots, n$. ○

Proof We may assume that the a_i are positive since by deleting the pairs (a_i, b_i) with $a_i = 0$ (if any), the left-hand side remains unchanged while the right-hand side does not decrease. Next, the b_i may also be assumed to be positive, else the inequality is trivial. Further, it suffices to prove the lemma for $a = b$, since multiplying the b_i by a constant does not affect the inequality. For this case, however, the statement follows from the inequality

$$\log x \leqq \frac{x - 1}{\ln 2}$$

substituting $x = b_i/a_i$. □

The following four lemmas summarize some immediate consequences of the definition of information measures. They will be used throughout the book, usually without reference.

LEMMA 3.2 (*Non-negativity*)

$$\text{(a) } H(X) \geqq 0, \text{ (b) } H(Y|X) \geqq 0, \quad \text{(c) } D(P||Q) \geqq 0,$$
$$\text{(d)} I(X \wedge Y) \geqq 0, \quad \text{(e) } I(X \wedge Y|Z) \geqq 0.$$

The equality holds iff (a) X is constant with probability 1, (b) there is a function $f : \mathsf{X} \to \mathsf{Y}$ such that $Y = f(X)$ with probability 1, (c) $P = Q$, (d) X and Y are independent, (e) X and Y are conditionally independent given Z (i.e., under the condition $Z = z$, whenever $\Pr\{Z = z\} > 0$). ○

Proof (a) is trivial, (c) follows from Lemma 3.1 and (d) follows from (c) since $I(X \wedge Y) = D(P_{XY}||P_X \times P_Y)$. (b) and (e) follow from (a) and (d), respectively. □

LEMMA 3.3

$$H(X) = E(-\log P_X(X)), \quad H(Y|X) = E(-\log P_{Y|X}(Y|X)),$$
$$I(X \wedge Y) = E\left(\log \frac{P_{XY}(X, Y)}{P_X(X)P_Y(Y)}\right),$$
$$I(X \wedge Y|Z) = E\left(\log \frac{P_{XY|Z}(X, Y|Z)}{P_{X|Z}(X|Z)P_{Y|Z}(Y|Z)}\right). \quad ○$$

→ 3.1

LEMMA 3.4 (*Additivity*)

$$H(X, Y) = H(X) + H(Y|X), \qquad H(X, Y|Z) = H(X|Z) + H(Y|X, Z),$$
$$H(X) = H(X|Y) + I(X \wedge Y), \qquad H(X|Z) = H(X|Y, Z) + I(X \wedge Y|Z),$$
$$I(X, Y \wedge Z) = I(X \wedge Z) + I(Y \wedge Z|X),$$
$$I(X, Y \wedge Z|U) = I(X \wedge Z|U) + I(Y \wedge Z|X, U). \quad ○$$

Proof The first identities in the first two rows hold by definition and imply those to their right by averaging. The fifth identity follows from

$$\frac{P_{XYZ}(x, y, z)}{P_{XY}(x, y)P_Z(z)} = \frac{P_{XZ}(x, z)}{P_X(x)P_Z(z)} \cdot \frac{P_{YZ|X}(y, z|x)}{P_{Y|X}(y|x)P_{Z|X}(z|x)}$$

by Lemma 3.3, and it implies the last one again by averaging. □

COROLLARY 3.4 (*Chain rules*)

$$H(X_1, \ldots, X_k) = \sum_{i=1}^{k} H(X_i|X_1, \ldots, X_{i-1}),$$
$$I(X_1, \ldots, X_k \wedge Y) = \sum_{i=1}^{k} I(X_i \wedge Y|X_1, \ldots, X_{i-1});$$

→ 3.2 similar identities hold for conditional entropy and conditional mutual information. ○

It is worth emphasizing that the content of Lemmas 3.2 and 3.4 completely conforms with the intuitive interpretation of information measures. For example, the identity $I(X, Y \wedge Z) = I(X \wedge Z) + I(Y \wedge Z|X)$ means that the information contained in (X, Y) about Z consists of the information provided by X about Z plus the information Y provides about Z in the knowledge of X. Further, the additivity relations of Lemma 3.4 combined with Lemma 3.2 give rise to a number of inequalities with equally obvious intuitive meaning. We thus have

$$H(X|Y, Z) \leqq H(X|Z), \quad I(Y \wedge Z|X) \leqq I(X, Y \wedge Z),$$

etc. Such inequalities will be used without reference.

LEMMA 3.5 (*Convexity*)
(a) $H(P)$ is a concave function of P; (b) $H(W|P)$ is a concave function of W and a linear function of P; (c) $D(P||Q)$ is a convex function of the pair (P, Q); (d) $I(P, W)$ is a concave function of P and a convex function of W. ○ → 3.3

Proof Suppose that $P = \alpha P_1 + (1-\alpha)P_2$, $Q = \alpha Q_1 + (1-\alpha)Q_2$, i.e., $P(x) = \alpha P_1(x) + (1-\alpha)P_2(x)$ for every $x \in \mathsf{X}$ and similarly for Q, where $0 < \alpha < 1$. Then, by the convexity of the function $f(t) = t \log t$ resp. by the log-sum inequality, we have

$$\alpha P_1(x) \log P_1(x) + (1-\alpha)P_2(x) \log P_2(x) \geqq P(x) \log P(x)$$

resp.

$$\alpha P_1(x) \log \frac{P_1(x)}{Q_1(x)} + (1-\alpha)P_2(x) \log \frac{P_2(x)}{Q_2(x)}$$
$$= \alpha P_1(x) \log \frac{\alpha P_1(x)}{\alpha Q_1(x)} + (1-\alpha)P_2(x) \log \frac{(1-\alpha)P_2(x)}{(1-\alpha)Q_2(x)} \geqq P(x) \log \frac{P(x)}{Q(x)}.$$

Summing for $x \in \mathsf{X}$, it follows that

$$\alpha H(P_1) + (1-\alpha)H(P_2) \leqq H(P),$$
$$\alpha D(P_1 \| Q_1) + (1-\alpha)D(P_2 \| Q_2) \geqq D(P \| Q),$$

proving (a) and (c). Now (b) follows from (a) and (3.1) while (d) follows from (a), (c) and (3.2). □

The additivity properties of information measures can be viewed as formal identities for RVs as free variables. There is an interesting correspondence between these identities and those valid for an arbitrary additive set-function μ. To establish this correspondence, let us replace RVs $X, Y, \ldots$ by set variables $\mathsf{A}, \mathsf{B}, \ldots$ and use the following substitutions of symbols:

$$, \leftrightarrow \cup;$$
$$| \leftrightarrow -;$$
$$\wedge \leftrightarrow \cap.$$

Thereby we associate a set-theoretic expression with every formal expression of RVs occurring in the various information quantities. Putting these set-theoretic expressions into the argument of μ, we associate a real-valued function of several set variables with each information quantity (the latter being conceived as a function of the RVs therein). In other words, we make the following correspondence:

$$\begin{aligned} H(X) &\leftrightarrow \mu(\mathsf{A}), & H(X,Y) &\leftrightarrow \mu(\mathsf{A}\cup\mathsf{B}), \\ H(X \mid Y) &\leftrightarrow \mu(\mathsf{A}-\mathsf{B}), & H(X \mid Y,Z) &\leftrightarrow \mu(\mathsf{A}-(\mathsf{B}\cup\mathsf{C})), \\ I(X \wedge Y) &\leftrightarrow \mu(\mathsf{A}\cap\mathsf{B}), & I(X\wedge Y \mid Z) &\leftrightarrow \mu((\mathsf{A}\cap\mathsf{B})-\mathsf{C}), \quad \text{etc.} \end{aligned}$$

In this way, every information measure corresponds to a set-theoretic expression of form $\mu((\mathsf{A}\cap\mathsf{B})-\mathsf{C})$, where A, B, C stand for finite unions of set variables (A and B non-void, C possibly void). Conversely, every expression of this form corresponds to an information measure.

THEOREM 3.6 A linear equation for entropies and mutual informations is an identity iff the corresponding equation for additive set functions is an identity. ○

Proof The identities

$$I(X\wedge Y) = H(X) - H(X|Y), \quad I(X\wedge Y|Z) = H(X|Z) - H(X|Y,Z),$$
$$H(X|Y) = H(X,Y) - H(X)$$

have the trivial analogs

$$\begin{aligned} \mu(\mathsf{A}\cap\mathsf{B}) &= \mu(\mathsf{A}) - \mu(\mathsf{A}-\mathsf{B}), \\ \mu((\mathsf{A}\cap\mathsf{B})-\mathsf{C}) &= \mu(\mathsf{A}-\mathsf{C}) - \mu(\mathsf{A}-(\mathsf{B}\cup\mathsf{C})), \\ \mu(\mathsf{A}-\mathsf{B}) &= \mu(\mathsf{A}\cup\mathsf{B}) - \mu(\mathsf{B}). \end{aligned}$$

Using them, one can transform linear equations for information measures resp. their analogs for set functions into linear equations involving solely unconditional entropies resp. set-function analogs of the latter. Thus it suffices to prove the assertion for such equations. To this end, we shall show that a linear expression of form

$$\sum_\sigma c_\sigma H(\{X_i\}_{i\in\sigma}) \quad \text{or} \quad \sum_\sigma c_\sigma \mu\left(\bigcup_{i\in\sigma} \mathsf{A}_i\right)$$

with σ ranging over the subsets of $\{1,2,\ldots,k\}$ vanishes identically iff all coefficients c_σ are zero. Both statements are proved in the same way, hence we give the proof only for entropies. We show by induction that if

$$\sum_\sigma c_\sigma H(\{X_i\}_{i\in\sigma}) = 0 \quad \text{for every choice of} \quad (X_1,\ldots,X_k) \tag{3.3}$$

then $c_\sigma = 0$ for each $\sigma \subset \{1,\ldots,k\}$. Clearly, this statement is true for $k=1$. For an arbitrary k, setting $X_i = \text{const}$ for i $< k$, (3.3) yields $\sum_{\sigma:k\in\sigma} c_\sigma = 0$. This implies for arbitrary $X_1,\ldots,X_{k-1}$ that the choice $X_k \triangleq (X_1,\ldots,X_{k-1})$ makes the contribution of the terms containing X_k vanish. Hence

$$\sum_{\sigma\subset\{1,\ldots,k-1\}} c_\sigma H(\{X_i\}_{i\in\sigma}) = 0,$$

and the induction hypothesis implies $c_\sigma = 0$ whenever $k \notin \sigma$. It follows, by symmetry, that $c_\sigma = 0$ whenever $\sigma \neq \{1, \ldots, k\}$. Then (3.3) gives $c_\sigma = 0$ also for $\sigma = \{1, \ldots, k\}$. □

REMARK The set-function analogy might suggest that we introduce further information quantities corresponding to arbitrary Boolean expressions of sets. For example, the "information quantity" corresponding to $\mu(\mathsf{A} \cap \mathsf{B} \cap \mathsf{C}) = \mu(\mathsf{A} \cap \mathsf{B}) - \mu((\mathsf{A} \cap \mathsf{B}) - \mathsf{C})$ would be $I(X \wedge Y) - I(X \wedge Y|Z)$; this quantity has, however, no natural intuitive meaning. ○ → 3.4

For non-negative additive set functions, $\mu(\mathsf{A} \circ \mathsf{B})$ is a pseudo-metric on the subsets of a given set, where $\mathsf{A} \circ \mathsf{B} \triangleq (\mathsf{A} - \mathsf{B}) \cup (\mathsf{B} - \mathsf{A})$ is the symmetric difference of the sets A and B. Although $\mu(\mathsf{A} \circ \mathsf{B})$ has no direct information-theoretic analog, the identity $\mu(\mathsf{A} \circ \mathsf{B}) = \mu(\mathsf{A} - \mathsf{B}) + \mu(\mathsf{B} - \mathsf{A})$ suggests that we consider $H(X|Y) + H(Y|X)$. This turns out to be a pseudo-metric.

LEMMA 3.7 $\Delta(X, Y) \triangleq H(X|Y) + H(Y|X)$ is a pseudo-metric among RVs, i.e., → 3.5

(i) $\Delta(X, Y) \geqq 0, \quad \Delta(X, X) = 0$;
(ii) $\Delta(X, Y) = \Delta(Y, X)$;
(iii) $\Delta(X, Y) + \Delta(Y, Z) \geqq \Delta(X, Z)$. ○

Proof It suffices to prove the triangle inequality (iii). To this end we show that

$$H(X|Y) + H(Y|Z) \geqq H(X|Z).$$

In fact, on account of Lemmas 3.4 and 3.2,

$$\begin{aligned} H(X|Z) \leqq H(X, Y|Z) &= H(X|Y, Z) + H(Y|Z) \\ &\leqq H(X|Y) + H(Y|Z). \end{aligned}$$ □

The *entropy metric* $\Delta(X, Y)$ is continuous with respect to the simple metric $\Pr\{X \neq Y\}$, as shown by the next lemma, frequently used in information theory.

LEMMA 3.8 (*Fano's inequality*) For RVs X and Y with the same range X,

$$H(X|Y) \leqq \Pr\{X \neq Y\} \log(|\mathsf{X}| - 1) + h(\Pr\{X \neq Y\}).$$ ○

Proof Introduce a new RV Z, setting $Z = 0$ if $X = Y$ and $Z = 1$ otherwise. Then

$$H(X|Y) = H(X, Z|Y) = H(X|Y, Z) + H(Z|Y) \leqq H(X|Y, Z) + H(Z).$$

Clearly, $H(Z) = h(\Pr\{X \neq Y\})$. Further, for every $y \in \mathsf{Y}$,

$$H(X|Y = y, Z = 0) = 0, \quad H(X|Y = y, Z = 1) \leqq \log(|\mathsf{X}| - 1),$$

where the second inequality follows from (1.2). Hence

$$H(X|Y, Z) \leqq \Pr\{X \neq Y\} \log(|\mathsf{X}| - 1).$$ □

REMARK The same proof shows that if Y can also take values outside the range X of X, a slightly weaker bound still holds, namely with $\log(|\mathsf{X}| - 1)$ replaced by $\log |\mathsf{X}|$. ○

In the space of k-length sequences of elements of a given set X, a useful metric is provided by the *Hamming distance* $d_{\mathrm{H}}(\mathbf{x}, \mathbf{y})$, $\mathbf{x}, \mathbf{y} \in \mathsf{X}^k$, defined as the number of positions where the sequences **x** and **y** differ. The following corollary relates the entropy metric to expected Hamming distance.

COROLLARY 3.8 For arbitrary sequences of X-valued RVs $X^k \triangleq X_1 \dots X_k$, $Y^k \triangleq Y_1 \dots Y_k$ we have

$$H(X^k|Y^k) \leqq Ed_{\mathrm{H}}(X^k, Y^k) \log (|\mathsf{X}| - 1) + kh\left(\frac{1}{k} Ed_{\mathrm{H}}(X^k, Y^k)\right). \qquad ○$$

Proof Using the chain rule (Corollary 3.4), Lemma 3.8 yields

$$\begin{aligned} H(X^k|Y^k) &= \sum_{i=1}^{k} H(X_i|X^{i-1}, Y^k) \leqq \sum_{i=1}^{k} H(X_i|Y_i) \\ &\leqq \sum_{i=1}^{k} \Pr\{X_i \neq Y_i\} \log(|\mathsf{X}| - 1) + \sum_{i=1}^{k} h\Big(\Pr\{X_i \neq Y_i\}\Big). \end{aligned}$$

Since

$$\sum_{i=1}^{k} \Pr\{X_i \neq Y_i\} = Ed_{\mathrm{H}}(X^k, Y^k)$$

and $h(t)$ is a concave function, Corollary 3.8 follows. □

The following lemmas show how information quantities reflect Markov dependence of RVs.

DEFINITION 3.9 A finite or infinite sequence $X_1, X_2, \dots$ of RVs is a *Markov chain*, denoted by $X_1 \multimap X_2 \multimap \dots$, if for every i the RV X_{i+1} is conditionally independent of $(X_1, \dots, X_{i-1})$ given X_i. We shall say that $X_1, X_2, \dots$ is a conditional Markov chain given Y if, for every i, X_{i+1} is conditionally independent of $(X_1, \dots, X_{i-1})$ given (X_i, Y). ○

LEMMA 3.10 $X_1 \multimap X_2 \multimap \dots$ iff $I(X_1, \dots, X_{i-1} \wedge X_{i+1}|X_i) = 0$ for every i. Moreover, the RVs $X_1, X_2, \dots$ form a conditional Markov chain given a RV Y iff $I(X_1, \dots, X_{i-1} \wedge X_{i+1}|X_i, Y) = 0$ for every i. ○

Proof See Lemma 3.2. □

→ 3.7 COROLLARY 3.10 If $X_1 \multimap X_2 \multimap \dots$ and $1 \leqq k_1 \leqq n_1 < k_2 \leqq n_2 < \dots$ then the blocks $Y_j \triangleq (X_{k_j}, X_{k_j+1}, \dots, X_{n_j})$ also form a Markov chain. The same holds for conditional Markov chains. ○

Proof It suffices to prove the first assertion. We show that the assumption $I(X_1, \dots, X_{i-1} \wedge X_{i+1}|X_i) = 0$ for every i implies the same for the Y_j:

$$
\begin{aligned}
&I(Y_1, \dots, Y_{j-1} \wedge Y_{j+1}|Y_j) \\
&\leqq I(X_1, \dots, X_{k_j-1} \wedge X_{n_j+1}, \dots, X_{n_{j+1}}|X_{k_j}, \dots, X_{n_j}) \\
&= \sum_{l=n_j}^{n_{j+1}-1} I(X_1, \dots, X_{k_j-1} \wedge X_{l+1}|X_{k_j}, \dots, X_l) \\
&\leqq \sum_{l=n_j}^{n_{j+1}-1} I(X_1, \dots, X_{l-1} \wedge X_{l+1}|X_l) = 0.
\end{aligned}
$$

□

In a Markov chain, any two RVs are dependent on one another only through the intermediate ones, hence intuition suggests that their mutual information cannot exceed that of two intermediate RVs. A related phenomenon for hypothesis testing is that one cannot gain more information for discriminating between the hypotheses P and Q when observing the outcome of the experiment with less accuracy, possibly subject to random errors. These simple but remarkably useful facts are the assertions of the following lemma.

LEMMA 3.11 (*Data processing*)

(i) If $X_1 \multimap X_2 \multimap X_3 \multimap X_4$ then $I(X_1 \wedge X_4) \leqq I(X_2 \wedge X_3)$.
(ii) For any distributions P and Q on X and stochastic matrix $W : \mathsf{X} \to \mathsf{Y}$,

$$D(PW\|QW) \leqq D(P\|Q).$$

○

Proof (i) By Corollary 3.10, $I(X_1 \wedge X_4|X_2) = I(X_2 \wedge X_4|X_3) = 0$. Thus, by the additivity lemma

$$
\begin{aligned}
I(X_1 \wedge X_4) &\leqq I(X_1, X_2 \wedge X_4) = I(X_2 \wedge X_4) \\
&\leqq I(X_2 \wedge X_3, X_4) = I(X_2 \wedge X_3).
\end{aligned}
$$

(ii) Using the log-sum inequality, → 3.8

$$
\begin{aligned}
D(P\|Q) &= \sum_{x \in \mathsf{X}} P(x) \log \frac{P(x)}{Q(x)} = \sum_{y \in \mathsf{Y}} \sum_{x \in \mathsf{X}} P(x)W(y|x) \log \frac{P(x)W(y|x)}{Q(x)W(y|x)} \\
&\geqq \sum_{y \in \mathsf{Y}} PW(y) \log \frac{PW(y)}{QW(y)} = D(PW\|QW).
\end{aligned}
$$

□

Finally, the log-sum inequality implies various upper bounds on the entropy of a RV. Such bounds are given by Lemma 3.12.

LEMMA 3.12 Let $f(x)$ be a real-valued function on the range X of the RV X, and let α be an arbitrary real number. Then

$$H(X) \leqq \alpha E f(X) + \log \sum_{x \in \mathsf{X}} \exp(-\alpha f(x)).$$

The equality holds iff

$$P_X(x) = \frac{1}{A}\exp(-\alpha f(x)); \quad A \triangleq \sum_{x \in \mathsf{X}} \exp(-\alpha f(x)). \qquad \bigcirc$$

Proof Apply the log-sum inequality with $P_X(x)$ and $\exp(-\alpha f(x))$ in the role of a_i and b_i, respectively. □

COROLLARY 3.12 For a positive integer-valued RV N we have

$$H(N) < \log EN + \log e. \qquad \bigcirc$$

Proof With $f(N) \triangleq N$, $\alpha \triangleq \log EN/(EN-1)$ the lemma yields

$$H(N) \leqq \alpha EN + \log \frac{\exp(-\alpha)}{1-\exp(-\alpha)} = (EN-1)\log\left(1 + \frac{1}{EN-1}\right) + \log EN.$$

□

Problems

3.1. (a) Check that for an arbitrary function f with domain X

$$H(X, f(X)) = H(X), \quad H(Y|X, f(X)) = H(Y|X),$$
$$I(X, f(X) \wedge Y) = I(X \wedge Y).$$

(b) Deduce from (a) the inequalities

$$H(f(X)) \leqq H(X), \quad H(Y|f(X)) \geqq H(Y|X), \quad I(f(X) \wedge Y) \leqq I(X \wedge Y),$$

and determine in each case the condition of equality.

(c) State the analogs of the above inequalities with a conditioning RV Z and a function f with domain $\mathsf{X} \times \mathsf{Z}$, e.g., $H(f(X, Z)|Z) \leqq H(X|Z)$.

3.2. Show that if $X_1, \ldots, X_n$ are mutually independent then

$$I(X_1, \ldots, X_n \wedge Y_1, \ldots, Y_n) \geqq \sum_{i=1}^{n} I(X_i \wedge Y_i),$$

and that the reverse inequality holds if, given X_i, the RV Y_i is conditionally independent of the remaining RVs, for $i = 1, \ldots, n$.

3.3. Show that the concavity of $H(P)$ resp. $I(P, W)$ as a function of P is equivalent to $H(X|Y) \leqq H(X)$ resp. to $I(X \wedge Y|Z) \leqq I(X \wedge Y)$ for Y and Z conditionally independent given X.

3.4. Show that the last inequality of Problem 3.3 holds if X, Y, Z form a Markov chain in any order, but does not hold in general.

3.5. Prove the following continuity properties of information measures with respect to the entropy metric:

$$|H(X_1) - H(X_2)| \leqq \Delta(X_1, X_2),$$
$$|H(X_1 \mid Y_1) - H(X_2 \mid Y_2)| \leqq \Delta(X_1, X_2) + \Delta(Y_1, Y_2),$$
$$|I(X_1 \wedge Y_1) - I(X_2 \wedge Y_2)| \leqq \Delta(X_1, X_2) + \Delta(Y_1, Y_2).$$

3.6. Show that $X_1 -\!\!\circ\!\!- X_2 -\!\!\circ\!\!- \ldots -\!\!\circ\!\!- X_n$ iff $X_n -\!\!\circ\!\!- X_{n-1} -\!\!\circ\!\!- \ldots -\!\!\circ\!\!- X_1$.

3.7. Is it true that if $X_1 -\!\!\circ\!\!- X_2 -\!\!\circ\!\!- \ldots -\!\!\circ\!\!- X_n$ and f is an arbitrary function on the common range of the X_i then $f(X_1) -\!\!\circ\!\!- f(X_2) -\!\!\circ\!\!- \ldots -\!\!\circ\!\!- f(X_n)$? Give a counterexample.

3.8. Deduce assertion (i) of Lemma 3.11 from its assertion (ii).
Hint Set $P \triangleq P_{X_2X_3}$, $Q \triangleq P_{X_2} \times P_{X_3}$, $W \triangleq P_{X_1X_4|X_2X_3}$.

3.9. A natural candidate for the mutual information of three RVs is the divergence $D(P_{XYZ} \| P_X \times P_Y \times P_Z)$. Show that it equals $H(X) + H(Y) + H(Z) - H(X, Y, Z) = I(X, Y \wedge Z) + I(X \wedge Y)$. More generally, show that

$$\sum_{i=1}^{n} H(X_i) - H(X_1, \ldots, X_n) = D(P_{X_1,\ldots,X_n} \| P_{X_1} \times \ldots \times P_{X_n}),$$

and give decompositions of the latter into sums of mutual informations. (This divergence is frequently called *multi-information.*)

3.10. Prove the following improvement to Lemma 2.7:

$$|H(P) - H(Q)| \leqq \frac{1}{2} d(P, Q) \log(|\mathsf{X}| - 1) + h\left(\frac{1}{2} d(P, Q)\right).$$

(See Audenaert (2007) and Zhang (2007).)
Hint Show that there exist RVs X, Y with distributions P and Q such that $d(P, Q) = 2\Pr\{X \neq Y\}$. Then use the fact that $H(X) - H(Y) = H(X|Y) - H(Y|X)$ and apply Fano's inequality.

3.11. Determine the condition of equality in Lemmas 3.7 and 3.8.

3.12. Show that if the common range of X and Y consists of two elements then $H(X|Y) + H(Y|X) \geqq h(\Pr\{X \neq Y\})$. Is this true in general? Give a counterexample.

3.13. Show that

$$\Delta'(X, Y) \triangleq \begin{cases} \dfrac{\Delta(X, Y)}{H(X, Y)} = 1 - \dfrac{I(X \wedge Y)}{H(X, Y)} & \text{if} \quad H(X, Y) \neq 0 \\ 0 & \text{if} \quad H(X, Y) = 0 \end{cases}$$

is also a pseudo-metric among RVs. (See Rajski (1961); for a simple proof see Horibe (1973).)

3.14. Show that if $X_1 -\!\!\circ\!\!- X_2 -\!\!\circ\!\!- X_3 -\!\!\circ\!\!- X_4 -\!\!\circ\!\!- X_5$ then $X_3 -\!\!\circ\!\!- (X_2, X_4) -\!\!\circ\!\!- (X_1, X_5)$.

3.15. Let W be *doubly stochastic*, i.e., a square matrix of non-negative elements with row and column sums equal to 1. Show that then

$$H(PW) \geqq H(P)$$

for every distribution P on X.
Hint Use the data processing lemma with the uniform distribution as Q.

3.16. Show that $P > Q$ iff there exists a doubly stochastic matrix W such that $P = QW$, so that the previous problem is equivalent to Problem 1.10. (Hardy, Littlewood and Pólya (1934), Th. 46).

Properties of informational divergence (Problems 3.17–3.20)

3.17. $D(P\|Q)$ is not a distance among PDs, for it is not symmetric. Show that the symmetrized divergence $J(P\|Q) \triangleq D(P\|Q) + D(Q\|P)$ is not a distance, either; moreover, show that there exist PDs P_1, P_2, P_3 such that

$$D(P_1\|P_2) + D(P_2\|P_3) < D(P_1\|P_3)$$

and

$$D(P_3\|P_2) + D(P_2\|P_1) < D(P_3\|P_1).$$

($J(P\|Q)$ was introduced earlier than $D(P\|Q)$ by Jeffreys (1946).)

3.18. (*Divergence and variational distance*)

(a) Prove the *Pinsker inequality*

$$D(P\|Q) \geqq \frac{1}{2\ln 2} d^2(P, Q).$$

Show that this bound is tight in the sense that the ratio of $D(P\|Q)$ and $d^2(P, Q)$ can be arbitrarily close to $1/2\ln 2$. (See Csiszár (1967), Kemperman (1969) and Kullback (1967); the bound $D(P\|Q) \geqq cd^2(P, Q)$ – with a worse constant c – was first given by Pinsker (1960). Several authors gave sharper bounds involving higher powers of $d(P, Q)$; strong results of this kind, as well as a parametric representation of the minimum of $D(P\|Q)$ when $d(P, Q)$ is fixed, appear in Fedotov, Harremoës and Topsøe (2003).)

Hint With $\mathsf{A} \triangleq \{x : P(x) \geqq Q(x)\}$, $\hat{P} \triangleq (P(\mathsf{A}), P(\bar{\mathsf{A}}))$, $\hat{Q} \triangleq (Q(\mathsf{A}), Q(\bar{\mathsf{A}}))$, we have $D(P\|Q) \geqq D(\hat{P}\|\hat{Q})$, $d(P, Q) = d(\hat{P}, \hat{Q})$. Hence it suffices to consider the case $\mathsf{X} = \{0, 1\}$, i.e., to determine the largest c such that

$$p \log \frac{p}{q} + (1 - p) \log \frac{1 - p}{1 - q} - 4c(p - q)^2 \geqq 0$$

for every

$$0 \leqq q \leqq p \leqq 1.$$

For $q = p$ the equality holds; further, the derivative of the left-hand side with respect to q is negative for $q < p$ if $c \leqq 1/2\ln 2$ while for $c > 1/2\ln 2$ and $p = 1/2$ it is positive in the neighborhood of p.

(b) For the number N of occurrences in k independent trials of an event of probability q, show that $\Pr\{|N - kq| > \delta\} < 2e^{-2\delta^2 k}$. (This special case of Hoeffding's inequality, Hoeffding (1963), was used in Chapter 2, Remark to Lemma 2.12.)

Hint Use Problem 2.8 and part (a).

3.19. (*Strong data processing*) Show that if the stochastic matrix W is such that for some $y_0 \in \mathsf{Y}$

$$W(y_0|x) \geqq c > 0 \quad \text{for every } x \in \mathsf{X},$$

then assertion (ii) of the data processing lemma can be strengthened to

$$D(PW\|QW) \leqq (1-c)D(P\|Q).$$

Hint Write $W = (1-c)W_1 + cW_2$, where $W_2(y_0|x) = 1$ for every $x \in \mathsf{X}$, and use the convexity of informational divergence.

3.20. (*Divergence geometry*) Prove that $D(P\|Q)$ is an analog of squared Euclidean distance in the following sense.

(a) (*Parallelogram identity*) For any three PDs P, Q, R on X,

$$D(P\|R) + D(Q\|R) = D\left(P\left\|\frac{P+Q}{2}\right.\right) + D\left(Q\left\|\frac{P+Q}{2}\right.\right) + 2D\left(\frac{P+Q}{2}\right\|R\right).$$

(b) (*Projection*) Let $\mathcal{P}$ be a closed convex set of PDs on X. Show that any PD R such that $D(P\|R) < \infty$, for some $P \in \mathcal{P}$, has a unique "projection" onto $\mathcal{P}$, i.e., there exists a unique minimizer of $D(P\|R)$ subject to $P \in \mathcal{P}$.

(c) (*Pythagoras theorem*) Let $\mathcal{P}$ be a set of PDs on X determined by linear constraints, i.e., the set of all P such that for some fixed matrix M (with arbitrary real entries) PM equals some fixed vector. Show that the projection Q of a PD R onto $\mathcal{P}$ satisfies the identity

$$D(P\|Q) + D(Q\|R) = D(P\|R) \quad \text{for every} \quad P \in \mathcal{P}.$$

Hint If

$$\mathcal{P} = \left\{P : \sum_{x \in \mathsf{X}} P(x)M(y|x) = a(y) \text{ for every } y \in \mathsf{Y}\right\},$$

let A be the set of those $x \in \mathsf{X}$ which have positive probability for at least one $P \in \mathcal{P}$ with $D(P\|R) < \infty$. Show that the projection Q of R onto $\mathcal{P}$ has the following form:

$$Q(x) = \begin{cases} cR(x)\exp\left[\sum_{y \in \mathsf{Y}} b(y)M(y|x)\right] & \text{if } x \in \mathsf{A} \\ 0 & \text{if } x \notin \mathsf{A}. \end{cases}$$

(d) (*Maximum entropy*) Conclude that for $\mathcal{P}$, as in part (b), a unique maximizer of $H(P)$ subject to $P \in \mathcal{P}$ exists, the projection to $\mathcal{P}$ of the uniform distribution on X. Find the implication of (c) for entropy maximization subject to linear constraints, thus for $\mathcal{P}$ as in (c).

(e) (*Iterated projection*) Show that if $\mathcal{P}$ is a set of PDs determined by linear constraints, and $\mathcal{P}_1$ is any closed convex subset of $\mathcal{P}$, then the projection of any R onto $\mathcal{P}_1$ can be obtained by first projecting R onto $\mathcal{P}$ and then projecting the resulting distribution onto $\mathcal{P}_1$.

(See Csiszár (1975).)

Structural results on entropy (Problems 3.21–3.22)

3.21. (a) Sharpening Theorem 3.6, assign to RVs $X_1, \ldots, X_k$ subsets $\mathsf{A}_1, \ldots, \mathsf{A}_k$ of a suitable set Ω, and an additive set function μ on the subsets of Ω, such that the information-theoretic expressions involving the RVs X_i are actually equal to their set-theoretic counterparts in Theorem 3.6. (Yeung (1991).)

Hint Take $\Omega \triangleq \{0,1\}^k - \{0\}^k$ and $\mathsf{A}_i \triangleq \{\omega = a_1 \ldots a_k \colon a_i = 1\}$. Then specify a (unique) μ as required.

(b)* Show that if $X_1 - \circ - \ldots - \circ - X_k$ is a Markov chain then the μ above is non-negative-valued. (Kawabata-Yeung (1992).)

3.22. A set function h on the subsets of $\{1, \ldots, k\}$ is an *entropy function* if $h(\sigma) = H(\{X_i\}_{i \in \sigma})$, $\sigma \subset \{1, \ldots, k\}$, for some RVs $X_1, \ldots, X_k$. A set function h is a *polymatroid* if it satisfies the axioms

$$h(\emptyset) = 0,\ h(\sigma) \leqq h(\xi) \text{ if } \sigma \subset \xi,\ h(\sigma \cup \xi) + h(\sigma \cap \xi) \leqq h(\sigma) + h(\xi).$$

Denote the set of all polymatroids resp. all entropy functions, for a given k, by Γ_k and Γ_k^*.

(a) Show that all entropy functions are polymatroids: $\Gamma_k^* \subset \Gamma_k$ (Fujishige (1978).)

Hint Immediate from Lemma 3.2.

(b) Show that Γ_k^* is not convex if $k \geq 3$, but its closure $cl(\Gamma_k^*)$ is a convex cone (a cone contains all positive multiples of each of its elements). (Zhang and Yeung (1997).)

Hint Show that the set function taking value a on the one-point subsets of $\{1,2,3\}$ and value $2a$ on the other non-empty subsets is in Γ_3^* iff a equals the log of a positive integer. To prove the second assertion, check first that if two set functions are in Γ_k^* then so is their sum.

(c) A *linear information inequality* is any inequality of form

$$\sum_{\sigma \subset \{1,\ldots,k\}} c_\sigma h(\sigma) \geqq 0$$

that holds for all entropy functions $h \in \Gamma_k^*$. N. Pippenger asked in 1986 (in a conference talk cited in Yeung (2008)) whether all such inequalities are implied by the basic inequalities of Lemma 3.2 (a), (b), (d), (e); in other words, whether they hold for all $h \in \Gamma_k$. Show that an equivalent question is whether $cl(\Gamma_k^*) = \Gamma_k$.

(d)* Show that the answer to Pippenger's question is yes if $k \leqq 3$ but no if $k > 3$. (The case $k = 3$ was settled, in effect, by Han (1981). For $k = 4$, Zhang and Yeung (1998) explicitly constructed a non-Shannon-type linear information inequality, that is, one not implied by the basic inequalities.)

(e)* Strengthening the negative result of (d), show that for $k > 3$ no finite number of linear information inequalities can imply all such inequalities. (Matúš (2007, unpublished).)

Story of the results

The standard introductory material in this chapter is essentially due to Shannon (1948). Some formal properties of entropy emerged earlier in physics; entropy inequalities of Gibbs (1902) are reviewed in Falk (1970).

Since in the early days of information theory many results appeared only in internal publications of the MIT, where most of the research was concentrated, by now it is hard to trace individual contributions. In particular, Lemma 3.8 is unanimously attributed to Fano (1952, unpublished), see e.g. Gallager (1968). Corollary 3.8 appears in Gallager (1964). Theorem 3.6 was proved by Hu Guo Ding (1962); the analogy between the algebraic properties of information quantities and those of additive set functions was noted also by Reza (1961). For the pseudo-metric $\Delta(X, Y)$ of Lemma 3.7 see Shannon (1953).

Addition. Since the first edition of this book, two research directions emerged addressing formal properties of information measures. *Information geometry* deals with geometric properties of I-divergence and related divergence measures, similar to those of (squared) Euclidean distance. It dates back at least to Čencov (1972) but has flourished since the 1990s. Problem 3.20 illustrates some basic results which do not require differential geometry, otherwise a main tool of the subject; see Amari and Nagaoka (2000). Another, more recent direction studies structural properties of information functions, along the lines of which a first example is Theorem 3.6. See Problems 3.21 and 3.22 for some typical results, and Yeung (2008) for details.

4 Non-block source coding

In this section we revisit the source coding problem of Chapter 1 that has motivated the introduction of entropy. First we consider more general codes for a DMS, and then turn to sources with memory and codes with variable symbol costs, giving an outlook also on subjects beyond the scope of this book. The solutions will still be given in terms of entropy, providing additional support to its intuitive interpretation as a measure of the amount of information.

Let $\{X_i\}_{i=1}^{\infty}$ be a (discrete) source with (finite) alphabet X, and let Y be another finite set, the *code alphabet.* A *code for k-length messages* is a pair of mappings $f : \mathsf{X}^k \to \mathsf{Y}^*$, $\varphi : \mathsf{Y}^* \to \mathsf{X}^k$. The sequences in the range of f are called *codewords*. Here Y^* denotes the set of all finite sequences of elements of Y. Thus, contrary to Chapter 1, the codewords may be of different length, i.e., we are dealing with *variable-length codes* (more precisely, *fixed-to-variable-length* codes). We shall assume that the void sequence is not a possible codeword.

As a fidelity criterion, it is reasonable to impose one of the following:

$$\begin{aligned} &\text{(i)} \quad \Pr\{\varphi(f(X^k)) = X^k\} = 1, \\ &\text{(ii)} \quad \Pr\{\varphi(f(X^k)) = X^k\} \geqq 1 - \varepsilon, \\ &\text{(iii)} \quad Ed(X^k, \varphi(f(X^k))) \quad \leqq \varepsilon, \end{aligned} \tag{4.1}$$

where $d(\mathbf{x}, \mathbf{x}') \triangleq \frac{1}{k} d_{\mathrm{H}}(\mathbf{x}, \mathbf{x}')$ is the fraction of positions where the sequences $\mathbf{x}$ and $\mathbf{x}'$ of length k differ. Clearly, these criteria are of decreasing strength if $0 < \varepsilon < 1$, and all are equivalent if $\varepsilon = 0$. If f is a one-to-one mapping of X^k into Y^*, we always choose $\varphi = f^{-1}$ ensuring that (i) is satisfied. With some abuse of terminology, we shall then speak of a code f (suppressing $\varphi \triangleq f^{-1}$).

Having non-terminating transmission in mind, i.e., successive application of the encoder f to consecutive k-length blocks of source outputs, it is desirable that f has *separable range* in the sense that finite sequences obtained by juxtaposition of codewords be unambiguously decomposable into these codewords. A sufficient condition for this is the *prefix property* of the range of f, which means that no codeword is the prefix of another one. → 4.1 If $f : \mathsf{X}^k \to \mathsf{Y}^*$ is a one-to-one mapping and its range is separable resp. it has the prefix property, we shall speak of a *separable code* resp. *prefix code f.* → 4.2 Separable codes are often called *uniquely decodable* or *uniquely decipherable* codes.

One often visualizes codeword sets on an infinite (rooted) tree, with $|\mathsf{Y}|$ (directed) edges – labeled by the different elements of Y – starting from each vertex. To this end,

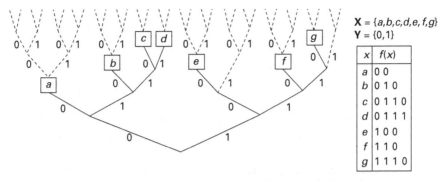

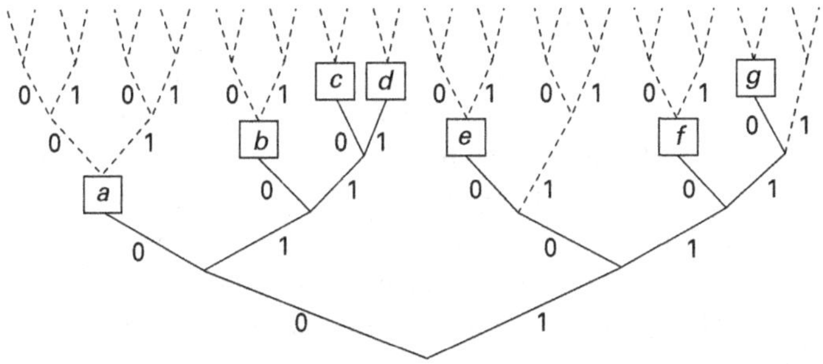

$X = \{a,b,c,d,e,f,g\}$
$Y = \{0,1\}$

x	$f(x)$
a	0 0
b	0 1 0
c	0 1 1 0
d	0 1 1 1
e	1 0 0
f	1 1 0
g	1 1 1 0

Figure 4.1 Representation of a prefix code on an infinite tree. The solid subtree is the code tree.

assign to each vertex the sequence of labels on the path from the root to that vertex. This defines a one-to-one correspondence between the sequences $\mathbf{y} \in \mathsf{Y}^*$ and the vertices of the infinite tree. Clearly, a subset of Y^* has the prefix property iff the corresponding set of vertices is the set of terminal vertices of a finite subtree. In particular, prefix → 4.3 codes $f : \mathsf{X} \to \mathsf{Y}^*$ are in a one-to-one correspondence with finite subtrees having $|\mathsf{X}|$ terminal vertices labeled by the different elements of X (the vertex corresponding to the codeword $f(x)$ is labeled by x). Such a finite tree is called a *code tree.*

A natural performance index of a code is the per letter *average length* of the codewords:

$$\bar{l}(f) \triangleq E\frac{1}{k}\, l(f(X^k)),$$

where $l(\mathbf{y})$ designates the length of the sequence $\mathbf{y} \in \mathsf{Y}^*$.

THEOREM 4.1 (*Average length*) If $\{X_i\}_{i=1}^{\infty}$ is a DMS with generic distribution P then every separable code $f : \mathsf{X}^k \to \mathsf{Y}^*$ has per letter average codeword length

$$\bar{l}(f) \geqq \frac{H(P)}{\log|\mathsf{Y}|}, \tag{4.2}$$

and there exists a prefix code satisfying

$$\bar{l}(f) < \frac{H(P)}{\log|\mathsf{Y}|} + \frac{1}{k}. \tag{4.3}$$

The bound (4.2) "almost holds" even for non-separable codes whenever they meet the weakest fidelity criterion of (4.1), with a "small" ε. More exactly, if (iii) of (4.1) holds with $\varepsilon < 1/2$ then

$$\bar{l}(f) \geqq \frac{H(P)}{\log|\mathsf{Y}|} - \frac{1}{\log|\mathsf{Y}|}(\varepsilon\log(|\mathsf{X}|-1) + h(\varepsilon)) - \frac{|\log(dk)|^+}{k\log|\mathsf{Y}|}, \tag{4.4}$$

where

$$d \triangleq e\frac{\log|\mathsf{X}|}{\log|\mathsf{Y}|},$$

e being the base of natural logarithms. If the range of f is separable, the last term in (4.4) may be omitted. ○

COMMENT In order to compare Theorem 4.1 with the simple block-coding result of Theorem 1.1, it is convenient to reformulate the latter so that for binary block codes meeting criterion (ii) of (4.1), the minimum of $\bar{l}(f)$ converges to $H(P)$ as $k \to \infty$. Theorem 4.1 says that this asymptotic performance bound cannot be significantly improved even if the weaker fidelity criterion (iii) is imposed and also variable-length codes are permitted, provided that one insists on a small average error frequency. On the other hand, the same limiting performance can also be achieved with prefix codes, admitting correct decoding with certainty, rather than with probability close to 1. ○

Proof First we prove (4.4). To this end we bound $H(f(X^k))$ from above in terms of $\bar{l}(f)$. Consider any sequence $Y_1 \dots Y_N$ of Y-valued RVs of random length N. Then

$$\begin{aligned} H(Y_1 \dots Y_N) &= H(Y_1 \dots Y_N|N) + H(N) \\ &= \sum_n \Pr\{N = n\} H(Y_1 \dots Y_n|N = n) + H(N). \end{aligned} \tag{4.5}$$

Here

$$H(Y_1 \dots Y_n|N = n) \leqq \sum_{i=1}^{n} H(Y_i|N = n) \leqq n \log|\mathsf{Y}| \tag{4.6}$$

and by Corollary 3.12

$$H(N) < \log(eEN). \tag{4.7}$$

When applying (4.5) to $Y_1 \dots Y_N \triangleq f(X^k)$, one may assume that $EN < dk/e$ since otherwise $\bar{l}(f) = \frac{1}{k}EN \geqq d/e = \log|\mathsf{X}|/\log|\mathsf{Y}|$ and (4.4) would automatically hold. Thus (4.5), (4.6) and (4.7) yield

$$H(f(X^k)) \leqq EN \cdot \log|\mathsf{Y}| + \log(dk) = k\bar{l}(f) \cdot \log|\mathsf{Y}| + \log(dk). \tag{4.8}$$

Note that condition (iii) of (4.1) implies by Fano's inequality (Corollary 3.8)

$$H(X^k|\varphi(f(X^k))) \leqq k(\varepsilon \log(|\mathsf{X}| - 1) + h(\varepsilon)).$$

This yields

$$\begin{aligned} kH(P) = H(X^k) &= H(X^k|\varphi(f(X^k))) + H(\varphi(f(X^k))) \\ &\leqq k(\varepsilon \log(|\mathsf{X}| - 1) + h(\varepsilon)) + H(f(X^k)). \end{aligned}$$

Thus from (4.8) we obtain (4.4).

We now show that if the range of f is separable then the last term in (4.4) may be omitted; this will also prove (4.2), setting $\varepsilon = 0$. To this end, extend the code (f, φ) to a code (f_m, φ_m) for messages of length mk as follows: decompose each $\mathbf{x} \in \mathsf{X}^{mk}$ into consecutive blocks $\mathbf{x}_i \in \mathsf{X}^k$, $i = 1, \dots, m$, and let $f_m(\mathbf{x})$ be the juxtaposition of the codewords $f(\mathbf{x}_i)$. By assumption, sequences of form $f_m(\mathbf{x})$, $\mathbf{x} \in \mathsf{X}^{mk}$ can be uniquely decomposed into f-codewords, thus $\varphi_m : \mathsf{Y}^* \to \mathsf{X}^{mk}$ can be naturally defined so that $\varphi_m(f_m(\mathbf{x}))$ be the juxtaposition of the blocks $\varphi(f(\mathbf{x}_i))$, $i = 1, \dots, m$. Clearly, $\bar{l}(f_m) = \bar{l}(f)$, and the code (f_m, φ_m) also meets criterion (iii) of (4.1) if (f, φ) does. Applying

(4.4) to (f_m, φ_m) instead of (f, φ), we see that (4.4) is true also if in the last term k is replaced by km. Since m can be arbitrarily large, this term may actually be omitted.

To complete the proof of Theorem 4.1, we have to construct a prefix code satisfying (4.3). Clearly, it suffices to give the construction for $k = 1$.

Without restricting generality, suppose that Y is the set of non-negative integers less than $q \triangleq |\mathsf{Y}|$. Assign to each $\mathbf{y} = y_1 \dots y_l \in \mathsf{Y}^*$ the number $a(\mathbf{y}) \in [0, 1)$ represented in the q-ary number system as $0.y_1 \dots y_l$. Now order the elements of X by decreasing probability: $P(x_1) \geqq P(x_2) \geqq \dots \geqq P(x_r)$, $r \triangleq |\mathsf{X}|$, and write

$$a_i \triangleq \sum_{j=1}^{i-1} P(x_j).$$

Define $f(x_i)$ as the sequence $\mathbf{y} \in \mathsf{Y}^*$ of smallest length l for which the interval $[a(\mathbf{y}), a(\mathbf{y}) + q^{-l})$ contains a_i but does not contain any other a_k. Clearly, this is a prefix code. Also, by definition, if $f(x_i) = y_1 \dots y_l$ then the interval $[a(y_1 \dots y_{l-1}), a(y_1 \dots y_{l-1}) + q^{-(l-1)})$ contains at least one of a_{i+1} and a_{i-1}. Hence, putting $l_i \triangleq l(f(x_i))$, we have

$$q^{-(l_i - 1)} > \min(P(x_{i-1}), P(x_i)) = P(x_i).$$

Thus

$$-\log P(x_i) > (l_i - 1) \log q$$

and

$$H(P) = -\sum_{i=1}^{r} P(x_i) \log P(x_i) > \log q \left(\sum_{i=1}^{r} P(x_i) l_i - 1 \right),$$

proving that

$$\bar{l}(f) = \sum_{i=1}^{r} P(x_i) l_i < \frac{H(P)}{\log |\mathsf{Y}|} + 1. \qquad \square$$

REMARK A prefix code $f: X \to \{0, 1\}^*$ is called a *Shannon code* for a distribution P on X if its codeword lengths are $\lceil -\log P(x) \rceil$, $x \in \mathsf{X}$. The above proof (with $q = 2$) contains a construction of a prefix code with codeword lengths at most $\lceil -\log P(x) \rceil$, thus also of a Shannon code (as extra bits can always be added if necessary). ○ → 4.4

Now let $X^\infty \triangleq \{X_i\}_{i=1}^\infty$ be an arbitrary discrete source with alphabet X. If the limit

$$\bar{H}(X^\infty) \triangleq \lim_{k \to \infty} \frac{1}{k} H(X^k)$$

exists, it will be called the *entropy rate* of the source X^∞. The entropy rate of a DMS is just the entropy of its generic distribution.

A source X^∞ is *stationary* if the joint distribution of $X_i, X_{i+1}, \dots, X_{i+k-1}$ does not depend on i $(k = 1, 2, \dots)$.

→ 4.5

LEMMA 4.2 For a stationary source, $(1/k)\,H(X^k)$, $k = 1, 2, \ldots$, is a non-increasing sequence and, consequently, $\bar{H}(X^\infty)$ exists. Moreover,

$$\bar{H}(X^\infty) = \lim_{k\to\infty} H(X_k|X_1, \ldots, X_{k-1}). \qquad \bigcirc$$

Proof For any positive integer k,

$$H(X_k|X_1, X_2, \ldots, X_{k-1}) \leqq H(X_k|X_2, \ldots, X_{k-1}) = H(X_{k-1}|X_1, \ldots, X_{k-2}),$$

where the last equality holds by stationarity. In view of the chain rule (Corollary 3.4) this implies both

$$\frac{1}{k} H(X_1, \ldots, X_k) \leqq \frac{1}{k-1} H(X_1, \ldots, X_{k-1})$$

and the last assertion of the lemma. □

We shall generalize the average length Theorem 4.1 to sources with memory, considering at the same time more general performance indices than average codeword length. Suppose that each $y \in \mathsf{Y}$ has a *cost* $c(y) > 0$ (measuring, e.g., the time or space needed for the transmission or storage of the symbol y) and let the cost of a sequence $\mathbf{y} = y_1 \ldots y_n$ be

$$c(\mathbf{y}) \triangleq \sum_{i=1}^{n} c(y_i).$$

Given a source $\{X_i\}_{i=1}^\infty$, the per letter *average cost* of a code for k-length messages is

$$\bar{c}(f) \triangleq \frac{1}{k} Ec(f(X^k)).$$

LEMMA 4.3 The maximum of $H(Y)/Ec(Y)$ for Y-valued RVs Y is the positive root α_0 of the equation

$$\sum_{y\in\mathsf{Y}} \exp[-\alpha c(y)] = 1. \tag{4.9}$$

$\bigcirc$

Proof The left-hand side of (4.9) is a strictly decreasing function of α taking the value $|\mathsf{Y}|$ at $\alpha = 0$ and approaching zero as $\alpha \to \infty$. Hence the equation has a unique positive root α_0. Now the assertion follows from Lemma 3.12. □

The promised generalization of the average length Theorem 4.1 is as follows.

THEOREM 4.4 (*Average cost*) For an arbitrary discrete source $\{X_i\}_{i=1}^\infty$, any code (f, φ) for k-length messages meeting the error frequency criterion (iii) of (4.1) with $\varepsilon < 1/2$, satisfies

$$\bar{c}(f) \geqq \frac{H(X^k)}{k\alpha_0} - \frac{1}{\alpha_0}(\varepsilon \log(|\mathsf{X}| - 1) + h(\varepsilon)) - \frac{|\log(dk)|^+}{\alpha_0 k}, \tag{4.10}$$

where α_0 is the positive root of (4.9) and

$$d \triangleq \frac{e \log |\mathsf{X}|}{\alpha_0 \min_{y \in \mathsf{Y}} c(y)}.$$

If the range of f is separable, the last term in (4.10) may be omitted. In particular, every separable code has per letter average cost

$$\bar{c}(f) \geqq \frac{H(X^k)}{k\alpha_0}. \tag{4.11}$$

Moreover, for every k there exists a prefix code such that

$$\bar{c}(f) < \frac{H(X^k)}{k\alpha_0} + \frac{c^*}{k}, \qquad c^* \triangleq \max_{y \in \mathsf{Y}} c(y). \tag{4.12}$$

○

COROLLARY 4.4 If the given source has entropy rate $\bar{H}(X^\infty)$ then to any $\delta > 0$ there exist $\varepsilon > 0$ and k_0 such that every code for messages of length $k \geqq k_0$ meeting (iii) of (4.1) with this ε satisfies

$$\bar{c}(f) > \frac{\bar{H}(X^\infty)}{\alpha_0} - \delta.$$

Further, for every $\delta > 0$ and sufficiently large k there exist prefix codes satisfying

$$\bar{c}(f) < \frac{\bar{H}(X^\infty)}{\alpha_0} + \delta.$$

For stationary sources and separable codes, the lower bound can be sharpened to

$$\bar{c}(f) \geqq \frac{\bar{H}(X^\infty)}{\alpha_0}.$$

○

COMMENTS It is instructive to send forward the following interpretation: suppose that a noiseless channel is capable of transmitting sequences of symbols $y \in \mathsf{Y}$, and $c(y)$ is the cost of transmission of symbol y. Corollary 4.4 says that the minimum average per letter cost at which long messages of a given source are transmissible over the given channel (allowing variable-length codes), is asymptotically equal to $\frac{1}{\alpha_0}\bar{H}(X^\infty)$. This → 4.6 fact has two consequences for the intuition. On the one hand, it justifies the interpretation of the entropy rate as the measure of the amount of information carried, on average, by one source symbol. On the other hand, it suggests that we interpret α_0 as the *capacity per unit cost* of the given channel; this capacity can be effectively exploited for a wide class of sources, applying suitable codes. For stationary sources, the latter result holds also in the stronger sense of non-terminating transmission; in fact, by successively encoding consecutive blocks, one can achieve non-terminating transmission with average per letter cost equal to that for the first block. It is intuitively significant that the capacity per unit cost of a noiseless channel in the above operational sense equals the maximum of $H(Y)/Ec(Y)$, as one would heuristically expect based on the intuitive meaning of entropy.

○

Proof of Theorem 4.4. The proof of (4.10) is like that of (4.4), but now in (4.6) $H(Y_i|N=n)$ should be upper-bounded by $\alpha_0 E(c(Y_i)|N=n)$, see Lemma 4.3, rather than by $\log|\mathsf{Y}|$. If the range of f is separable, in order to get rid of the last term in (4.10) we introduce a new source $\{\tilde{X}_i\}_{i=1}^{\infty}$ such that the consecutive blocks $(\tilde{X}_{lk+1},\ldots,\tilde{X}_{(l+1)k})$, $l=0,1,\ldots,$ are independent copies of X^k. We construct the codes (f_m,φ_m) as in the proof of Theorem 4.1. Applying inequality (4.10) to these codes and the source $\{\tilde{X}_i\}_{i=1}^{\infty}$ we obtain

$$\bar{c}(f)=\bar{c}(f_m)\geqq\frac{H(\tilde{X}^{km})}{km\alpha_0}-\frac{1}{\alpha_0}(\varepsilon\log(|\mathsf{X}|-1)+h(\varepsilon))-\frac{|\log(dkm)|^{+}}{\alpha_0 km}.$$

Since here $H(\tilde{X}^{km})=mH(X^k)$, on letting $m\to\infty$ the desired result follows. Inequality (4.11) is just the particular case $\varepsilon=0$.

To establish the existence part (4.12), the construction in the proof of Theorem 4.1 has to be modified as follows (again, it suffices to consider the case $k=1$). Identifying the set Y with the integers from 0 to $q-1$ as there, to every $\mathbf{y}=y_1\ldots y_l\in\mathsf{Y}^*$ we now assign the real number

$$\tilde{a}(\mathbf{y})\triangleq\sum_{\mathbf{y}'\in\mathsf{D}(\mathbf{y})}\exp\left[-\alpha_0 c(\mathbf{y}')\right],$$

where $\mathsf{D}(\mathbf{y})$ is the set of those $\mathbf{y}'\in\mathsf{Y}^*$ with $l(\mathbf{y}')=l$ for which $a(\mathbf{y}')<a(\mathbf{y})$, where $a(\mathbf{y})$ denotes the q-ary number $0.y_1\ldots y_l$. It follows from the definition of α_0 that the intervals

$$\mathcal{I}(\mathbf{y})\triangleq[\tilde{a}(\mathbf{y}),\tilde{a}(\mathbf{y})+\exp\left[-\alpha_0 c(\mathbf{y})\right]),\quad \mathbf{y}\in\mathsf{Y}^l$$

form a partition of [0,1) for every fixed l. Define $f(x_i)$ as the sequence $\mathbf{y}\in\mathsf{Y}^*$ of smallest length l for which $\mathcal{I}(\mathbf{y})$ contains a_i but does not contain any other a_k, where a_i is the same as in the proof of Theorem 4.1. Then

$$\exp\left[-\alpha_0\sum_{j=1}^{l_i-1}c(y_j)\right]>P(x_i)$$

if $f(x_i)=y_1\ldots y_{l_i}$. This implies

$$-\log P(x_i)>\alpha_0[c(f(x_i))-c^*],$$

whence the assertion follows.

The corollary is immediate, using for the last assertion that, by Lemma 4.2,

$$\frac{H(X^k)}{k}\geqq\bar{H}(X^{\infty}).$$ □

The following lemma highlights the concept of capacity of a noiseless channel from a different point of view.

Let $\mathsf{A}(t)\subset\mathsf{Y}^*$ be the set of those sequences of cost not exceeding t which cannot be prolonged without violating this property. Formally, put

$$\mathsf{A}(t)\triangleq\left\{\mathbf{y}:\mathbf{y}\in\mathsf{Y}^*,\quad t-c_0<c(\mathbf{y})\leqq t\right\},\tag{4.13}$$

where $t > 0$ is arbitrary and $c_0 \triangleq \min_{y \in \mathsf{Y}} c(y)$. Then the largest l for which the l-length binary sequences can be encoded in a one-to-one manner into sequences $\mathbf{y} \in \mathsf{A}(t)$ equals $\lfloor \log |\mathsf{A}(t)| \rfloor$ Thus, intuitively,

$$\lim_{t \to \infty} \frac{1}{t} \log |\mathsf{A}(t)|$$

is the average number of binary digits transmissible (by suitable coding) with unit cost over the given channel.

LEMMA 4.5

$$\lim_{t \to \infty} \frac{1}{t} \log |\mathsf{A}(t)| = \alpha_0 ;$$

more exactly → 4.7

$$\exp [\alpha_0 (t - c^*)] \leqq |\mathsf{A}(t)| \leqq \exp [\alpha_0 t] \quad (t > 0),$$

where $c^* \triangleq \max_{y \in \mathsf{Y}} c(y)$. ○

Proof Consider a DMS $\{Y_i\}_{i=1}^{\infty}$ with alphabet Y and generic distribution defined by

$$P(y) \triangleq \exp [-\alpha_0 c(y)] ;$$

by the definition of α_0 this is, indeed, a distribution on Y, cf. (4.9).

For any $\tau > 0$, let N_τ be the smallest integer n for which $\sum_{i=1}^{n} c(Y_i) > \tau$, and let $\mathsf{B}(\tau)$ be the set of possible values of the sequence of random length $Z_\tau \triangleq Y_1 \dots Y_{N_\tau}$. In other words, $\mathsf{B}(\tau)$ is the set of those sequences $\mathbf{y} \in \mathsf{Y}^*$ for which $c(\mathbf{y}) > \tau$, but after deleting the last symbol of $\mathbf{y}$ this inequality is no longer valid. Then

$$1 = \sum_{\mathbf{y} \in \mathsf{B}(\tau)} \Pr \{Z_\tau = \mathbf{y}\} = \sum_{\mathbf{y} \in \mathsf{B}(\tau)} \exp [-\alpha_0 c(\mathbf{y})] \leqq |\mathsf{B}(\tau)| \exp [-\alpha_0 \tau].$$

Since every sequence $y \in \mathsf{Y}^*$ of cost exceeding $t - c^*$ has a unique prefix in $\mathsf{B}(t - c^*)$, we have

$$|\mathsf{A}(t)| \geqq |\mathsf{B}(t - c^*)|.$$

This and the previous inequality establish the lower bound of the lemma. Moreover, since $\mathsf{A}(t) \subset \mathsf{B}(t - c_0)$, one also gets

$$|\mathsf{A}(t)| \exp [-\alpha_0 t] \leqq \sum_{\mathbf{y} \in \mathsf{A}(t)} \exp [-\alpha_0 c(\mathbf{y})]$$

$$\leqq \sum_{\mathbf{y} \in \mathsf{B}(t - c_0)} \exp [-\alpha_0 c(\mathbf{y})] = 1.$$ □

COROLLARY 4.5 For every separable code $f : \mathsf{X} \to \mathsf{Y}^*$, → 4.8

$$\sum_{x \in \mathsf{X}} \exp [-\alpha_0 c(f(x))] \leqq 1.$$ ○

Proof Apply Lemma 4.5 to the set of codewords $\{f(x) : x \in \mathsf{X}\}$ in the role of Y, and let $\mathsf{A}_1(t)$ be the set playing the role of $\mathsf{A}(t)$ with this choice. Then

$$\lim_{t\to\infty} \frac{1}{t} \log |\mathsf{A}_1(t)| = \alpha_1,$$

where

$$\sum_{x\in\mathsf{X}} \exp[-\alpha_1 c(f(x))] = 1. \tag{4.14}$$

By the separability assumption, different elements of $\mathsf{A}_1(t)$ are represented by different sequences $\mathbf{y} \in \mathsf{Y}^*$. Further, by the definition of $\mathsf{A}_1(t)$, every such $\mathbf{y}$ has cost $c(\mathbf{y}) \leqq t$, while $c(\mathbf{y}) + c(f(x)) > t$ for each $x \in \mathsf{X}$. This implies that every such sequence $\mathbf{y}$ may be extended to a $\mathbf{y}' \in \mathsf{A}(t)$ by adding a suffix of length less than $n_0 \triangleq \min_{x\in\mathsf{X}} l(f(x))$, and thus

$$|\mathsf{A}_1(t)| \leqq n_0|\mathsf{A}(t)| \quad \text{for every} \quad t > 0.$$

It follows that

$$\alpha_1 = \lim_{t\to\infty} \frac{1}{t} \log |\mathsf{A}_1(t)| \leqq \lim_{t\to\infty} \frac{1}{t} \log |\mathsf{A}_1(t)| = \alpha_0,$$

which together with (4.14) gives the assertion. □

Though Theorems 4.1 and 4.4 are of rather general nature, their existence parts involve an assumption seldom met in practice, namely that the pertinent distributions are exactly known at the encoder. One way to deal with the source coding problem without this assumption is to adopt the point of view of universal coding, see the Discussion of Chapter 2. Theorem 2.15 illustrated the phenomenon that for certain classes of sources it is possible to construct codes having asymptotically optimal performance for each source in the class. There we were concerned with block codes and the performance index was the probability of error. We conclude this chapter with a theorem exhibiting universally optimal codes within the framework of variable-length codes with probability of error equal to zero, and with the average cost as the performance index to be optimized.

THEOREM 4.6 Given a cost function c on Y, for every k there exists a prefix code $f : \mathsf{X}^k \to \mathsf{Y}^*$ such that, for every distribution P on X, the application of this code to k-length messages of a DMS with generic distribution P yields per letter average cost

$$\bar{c}(f) \leqq \frac{H(P)}{\alpha_0} + a\frac{\log(k+1)}{k}.$$

Here a is a constant depending only on $|\mathsf{X}|$, $|\mathsf{Y}|$ and the cost function c. ○

Proof Let the codeword $f(\mathbf{x})$ associated with an $\mathbf{x} \in \mathsf{X}^k$ consist of two parts, the first one determining the type of $\mathbf{x}$ and the second one specifying $\mathbf{x}$ within the set of k-length sequences having the same type. More exactly, let $f_1 : \mathsf{X}^k \to \mathsf{Y}^l$ be defined as

$f_1(\mathbf{x}) \triangleq \hat{f}(P_{\mathbf{x}})$, where $\hat{f}$ is a one-to-one mapping of types (of sequences in X^k) into Y^l. By the type counting lemma, one may choose

$$l \triangleq \left\lceil \frac{|\mathsf{X}| \log(k+1)}{\log |\mathsf{Y}|} \right\rceil. \tag{4.15}$$

Further, for any type Q of sequences in X^k, set

$$t(Q) \triangleq \frac{1}{\alpha_0} \log |\mathsf{T}^k_Q| + c^*. \tag{4.16}$$

Let $f_2 : \mathsf{X}^k \to \mathsf{Y}^*$ map each type class T^k_Q in a one-to-one way into $\mathsf{A}(t(Q))$, cf. (4.13). Such an f_2 exists by Lemma 4.5, and it yields, by definition,

$$c(f_2(\mathbf{x})) \leqq t(P_{\mathbf{x}}) \quad \text{for every} \quad \mathbf{x} \in \mathsf{X}^k. \tag{4.17}$$

Let $f(\mathbf{x})$ be the juxtaposition of $f_1(\mathbf{x})$ and $f_2(\mathbf{x})$. As the set $\mathsf{A}(t)$ has the prefix property for every $t > 0$, $f : \mathsf{X}^k \to \mathsf{Y}^*$ is a prefix code.

Since for any DMS sequences $\mathbf{x} \in \mathsf{X}^k$ of the same type have equal probability, the conditional distribution of X^k given that $X^k \in \mathsf{T}^k_Q$ is uniform on T^k_Q. Using (4.16) and (4.17), it follows that

$$\begin{aligned} kH(P) = H(X^k) &\geqq H(X^k | f_1(X^k)) = \sum_Q P^k(\mathsf{T}^k_Q) \log |\mathsf{T}^k_Q| \\ &= \sum_Q P^k(\mathsf{T}^k_Q) \alpha_0 (t(Q) - c^*) = \sum_{\mathbf{x} \in \mathsf{X}^k} P_{X^k}(\mathbf{x}) \alpha_0 (t(P_{\mathbf{x}}) - c^*) \\ &\geqq \alpha_0 E c(f_2(X^k)) - \alpha_0 c^*. \end{aligned}$$

Hence, taking into account the obvious inequality $Ec(f_1(X^k)) \leqq lc^*$ and (4.15), we get

$$\begin{aligned} \bar{c}(f) &= \frac{1}{k} E\left(c(f_1(X^k)) + c(f_2(X^k))\right) \\ &\leqq \frac{c^*}{k} \left\lceil \frac{|\mathsf{X}| \log(k+1)}{\log |\mathsf{Y}|} \right\rceil + \frac{H(P)}{\alpha_0} + \frac{c^*}{k}. \end{aligned}$$ □

Problems

4.1. (*Instantaneous codes*) Every mapping $f : \mathsf{X} \to \mathsf{Y}^*$ has an extension $f : \mathsf{X}^* \to \mathsf{Y}^*$, defined by letting $f(\mathbf{x})$ be the juxtaposition of the codewords $f(x_1), \ldots, f(x_k)$ if $\mathbf{x} = x_1 \ldots x_k$. This mapping is called an *instantaneous code* if, upon receiving an initial string of form $\mathbf{y} = f(\mathbf{x})$ of any sequence of code symbols, one can immediately conclude that $\mathbf{x}$ is an initial string of the encoded message. Show that this holds iff $f : \mathsf{X} \to \mathsf{Y}^*$ is a prefix code.

4.2. (a) Show that the prefix property is not necessary for separability. Find separable codes which are neither prefix nor are obtained from a prefix code by reversing the codewords. (For a deeper result, see Problem 4.10.)

(b) Give an example of a separable code $f : \mathsf{X} \to \mathsf{Y}^*$ such that for two different infinite sequences $x_1 x_2 \ldots$ and $x'_1 x'_2 \ldots$ the juxtapositions of the codewords $f(x_i)$ resp. $f(x'_i)$ give the same infinite sequence.

4.3. (*Kraft inequality*)

(a) Show that a prefix code $f : \mathsf{X} \to \mathsf{Y}^*$ with given codeword lengths $l(f(x)) = n(x)$ $(x \in \mathsf{X})$ exists iff $\sum_{x \in \mathsf{X}} |\mathsf{Y}|^{-n(x)} \leqq 1$ (see also Problem 4.8).

(b) Show that for a prefix code, the Kraft inequality holds with equality iff the code tree is *saturated*, i.e., if exactly $|\mathsf{Y}|$ edges start from every non-terminal vertex.

Hint Count at the nth level of the infinite tree the vertices which can be reached from the terminal vertices of the code tree, where $n \triangleq \max_{x \in \mathsf{X}} n(x)$. (Kraft (1949, unpublished).)

4.4. (*Huffman code*) Give an algorithm for constructing to a RV X a prefix code $f : \mathsf{X} \to \mathsf{Y}^*$ minimizing $\bar{l}(f) \triangleq El(f(X))$.

Hint One may assume that the code tree is saturated, except possibly for one vertex which is the starting point of d edges, where $d \equiv |\mathsf{X}| \pmod{(|\mathsf{Y}| - 1)}$, $2 \leqq d \leqq |\mathsf{Y}|$. Replace the d least probable values of X by a single new value x_0. Show that any optimal code for the so obtained RV X' gives rise to an optimal code for X when adding a single symbol to the codeword of x_0 in d different ways. (Huffman (1952).)

4.5. Show that for a stationary source $\bar{H}(X^\infty) = H(X_1)$ iff the source is a DMS, and $\bar{H}(X^\infty) = H(X_2|X_1)$ iff the source is a Markov chain.

4.6. (a) Show that for a DMS the second inequality of Corollary 4.4 can be achieved also with block codes, meeting fidelity criterion (ii) of (4.1), for arbitrary $0 < \varepsilon < 1$.

(b) The entropy rate of a stationary source is not necessarily a relevant quantity from the point of view of block coding, with fidelity criterion (ii) of (4.1). Let $\{X_i\}_{i=1}^\infty$ and $\{Y_i\}_{i=1}^\infty$ be two DMSs with entropy rates $H_1 > H_2$. Let U be a RV independent of both sources, $\Pr\{U = 1\} = \alpha$, $\Pr\{U = 2\} = 1 - \alpha$, and set

$$Z_i \triangleq \begin{cases} X_i & \text{if} \quad U = 1 \\ Y_i & \text{if} \quad U = 2 \end{cases} \qquad i = 1, 2, \ldots$$

Show for this mixed source $\{Z_i\}_{i=1}^\infty$ that – with the notation of Chapter 1 –

$$\lim_{k \to \infty} \frac{n(k, \varepsilon)}{k} = \begin{cases} H_1 & \text{if} \quad \varepsilon < \alpha \\ H_2 & \text{if} \quad \varepsilon > \alpha, \end{cases}$$

while $\bar{H}(Z^\infty) = \alpha H_1 + (1 - \alpha) H_2$.

(c) Show that $\frac{1}{k} n(k, \varepsilon)$ converges to the entropy rate $\bar{H}$ for those sources that have the following *asymptotic equipartition property* (AEP): for k sufficiently large there exists a "typical" subset of X^k whose P_{X^k}-probability is arbitrarily close to 1, and the normalized amount of information $-\frac{1}{k} \log P_{X^k}(\mathbf{x})$, provided by a particular realization $\mathbf{x}$ of X^k, is arbitrarily close to $\bar{H}$ if $\mathbf{x}$ is in this "typical" set. Show that "good" linear codes also exist for any source having the AEP, in the sense that the result of Problem 1.7(a) holds replacing $H(P)$ by the entropy rate $\bar{H}$.

(The significance of the AEP has been discovered by Shannon (1948). The *Shannon–McMillan–Breiman theorem*, not in the scope of this book, says that all stationary ergodic sources have the AEP, and even the stronger property that the normalized amount of information converges to $\bar{H}$ with probability 1 as $k \to \infty$; see, e.g., Cover and Thomas (2006). A stationary source is *ergodic* if it cannot be represented as a mixture of other stationary sources.)

4.7. Consider the sets $\mathsf{A}_0(t) \triangleq \{\mathbf{y} : c(\mathbf{y}) = t\}$ and $\hat{\mathsf{A}}(t) \triangleq \{\mathbf{y} : c(\mathbf{y}) \leqq t\}$, in addition to $\mathsf{A}(t)$ defined by (4.13). Then $\mathsf{A}_0(t) \subset \mathsf{A}(t) \subset \hat{\mathsf{A}}(t)$. Show that

$$\overline{\lim_{t\to\infty}} \frac{1}{t} \log |\mathsf{A}_0(t)| = \lim_{t\to\infty} \frac{1}{t} \log |\hat{\mathsf{A}}(t)|.$$

4.8. (*Generalized Kraft inequality*) The inequality

$$\sum_{x\in\mathsf{X}} \exp\left[-\alpha_0 c(f(x))\right] \leqq 1 \qquad (*)$$

of Corollary 4.5 is a generalization of Kraft's inequality (Problem 4.3) to code symbols of different costs and to separable (rather than prefix) codes.

(a) Conclude that to any separable code there exists a prefix code with the same set of codeword lengths. (However, Shor (1985) showed that not every separable code can be obtained from a prefix code by permuting the letters of each codeword.)

(b) Show that, in general, the inequality $\sum_{x\in\mathsf{X}} \exp\left[-\alpha_0 \tilde{c}(x)\right] \leqq 1$ is not sufficient for the existence of a prefix (or separable) code with codeword costs $c(f(x)) = \tilde{c}(x)$.

(c) Give a direct proof of the generalized Kraft inequality (*).

Hint Expand $\left(\sum_{x\in\mathsf{X}} \exp\{-\alpha_0 c(f(x))\}\right)^n$, where n is an arbitrary positive integer. Grouping terms corresponding to Y-sequences of the same length, check that this expression is bounded by a constant multiple of n.

(Karush (1961), Csiszár (1969).)

(d) Show that the inequality (*) implies assertion (4.11) of Theorem 4.4. (McMillan (1956), Krause (1962).)

Hint Use the log-sum inequality.

4.9. Find an algorithm for deciding whether a given code $f : \mathsf{X} \to \mathsf{Y}^*$ has separable range. (Sardinas and Patterson (1953).)

Hint Denote by S_1 the set of codewords, i.e., $\mathsf{S}_1 \triangleq \{f(x) : x \in \mathsf{X}\}$. Define the sets $\mathsf{S}_i \subset \mathsf{Y}^*$ $(i = 2, 3, \ldots)$ successively, so that $\mathbf{y} \in \mathsf{S}_n$ iff there is a codeword $\mathbf{y}^* \in \mathsf{S}_1$ such that $\mathbf{y}^*\mathbf{y} \in \mathsf{S}_{n-1}$. Show that f has separable range iff none of the S_i with $i > 1$ contains a codeword.

4.10. (*Composed codes*)

(a) Let $g : \mathsf{X} \to \mathsf{Y}^*$ and $h : \mathsf{Y} \to \mathsf{Z}^*$ be separable codes and consider the *composed code* $f : \mathsf{X} \to \mathsf{Z}^*$ defined as follows: if $g(x) = y_1 \ldots y_l$ then let $f(x)$ be the juxtaposition of $h(y_1), \ldots, h(y_l)$. Show that this composed code is again a separable code.

(b)* A *suffix code* is one obtainable from a prefix code by reversing the codewords. Show that not every separable code is a result of successive composition of prefix and suffix codes.

(Césari (1974).)

Hint Consider the binary code having codeword set $\mathsf{B} \triangleq \mathsf{B}_1 \cup \mathsf{B}_2$, where $\mathsf{B}_1 \triangleq \{1, 01, 100, 0000\}$ and B_2 is obtained from $\hat{\mathsf{B}}_2 \triangleq \{01, 10, 11, 0000, 0100, 1000, 1100\}$ by prefixing to each of its elements the sequence 0100.

(Boë (1978) proved by algebraic methods that this binary code belongs to a class of indecomposable codes.)

4.11. (*Synchronizing codes*) A separable code $f : \mathsf{X} \to \mathsf{Y}^*$ is *synchronizing* if there exists a sequence of codewords $\mathbf{s}$ such that an arbitrary sequence in Y^* that has $\mathbf{s}$ as a suffix is a sequence of codewords. (Synchronizing codes are very useful in practice. In fact, a long sequence of code symbols can be cut into shorter ones, each delimited by the synchronizing sequence $\mathbf{s}$, in such a manner that the shorter sequences are known to be sequences of codewords. Thus these sequences can be decoded into elements of X^* independently of each other.)

(a) Show that if a mapping $f : \mathsf{X} \to \mathsf{Y}^*$ is a synchronizing separable code then (i) the codeword lengths $n(x) \triangleq l(f(x))$ satisfy Kraft's inequality with equality (see Problem 4.3), (ii) the greatest common divisor of these lengths is 1.

(b)* Show that to every collection of positive integers $n(x)$, $x \in \mathsf{X}$, satisfying (i) and (ii) there exists a synchronizing prefix code $f : \mathsf{X} \to \mathsf{Y}^*$ with codeword lengths $l(f(x)) = n(x),\ x \in \mathsf{X}$.

(Schützenberger (1967).)

4.12. (*Codeword length and information of an event*) The negative logarithm of the probability of an event, measuring the amount of information provided by its occurrence (see Chapter 1) is often interpreted as the *ideal codelength*, supported by the near optimality of Shannon codes in the sense of average length. This problem provides further support to that interpretation, not relying upon averaging, and also points to a limitation.

(a) Given any source $\{X_i\}_{i=1}^{\infty}$ with alphabet X, and arbitrary separable codes $f_k \colon \mathsf{X}^k \to \{0, 1\}^*$, $k = 1, 2, \ldots$, show that for almost all infinite source output sequences $\mathbf{x}^\infty = x_1 x_2 \ldots$ the codeword lengths satisfy

$$l(f_k(x_1 \ldots x_k)) > -\log P_{X^k}(x_1 \ldots x_k) - 2\log k \quad \text{if} \quad k > k_0(\mathbf{x}^\infty).$$

Hint Bound the probability of $l(f_k(x_1 \ldots x_k)) \leqq -\log P_{X^k}(x_1 \ldots x_k) - 2\log k$ applying Kraft's inequality to the separable code f_k. Then use the Borel–Cantelli lemma.

(b) In the setting of (a), show that if the given source is ergodic, with entropy rate $\bar{H}$, then $\underline{\lim}_{k\to\infty} \frac{1}{k} l(f_k(x_1 \ldots x_k)) \geqq \bar{H}$, and there exist prefix codes f_k for which the limit exists and equals $\bar{H}$, with probability 1.

Hint Use (a) and the Shannon–McMillan–Breiman theorem, Problem 4.6(c). (Barron (1985, unpublished).)

(c)* For a DMS with generic distribution P, show that prefix codes $f : \mathsf{X}^k \to \{0,1\}^*$ of minimum average length satisfy, for some sequence $\varepsilon_k \to 0$,

$$\left|\frac{1}{k} l(f(\mathbf{x})) + \log P^k(\mathbf{x})\right| \leqq \varepsilon_k \quad \text{for every} \quad \mathbf{x} \in \mathsf{X}^k.$$

(Nemetz and Simon (1977).)

(d) Show that for any $p \in (0,1)$, and distribution P on any set X such that $P(x) = p$ for some $x \in \mathsf{X}$, each prefix code $f : \mathsf{X} \to \{0,1\}^*$ of minimum average length with respect to P satisfies

$$l(f(x)) \leqq l_p - 1,$$

and that this bound is best possible. Here l_p denotes the largest integer l with $f_l < 1/p$, where $\{f_l\}_{l=1}^{\infty}$ is the *Fibonacci sequence* defined recursively by

$$f_1 = f_2 \triangleq 1, \quad f_l \triangleq f_{l-1} + f_{l-2} \quad \text{if} \quad l \geqq 3.$$

Show also that

$$\lim_{p \to 0} \frac{l_p}{-\log p} = \left[\log \frac{1+\sqrt{5}}{2}\right]^{-1} > 1.$$

(Katona and Nemetz (1976). This result shows that a prefix code optimal with respect to some distribution P on a set that has an element x with $P(x) = p$, may assign to this x a codeword substantially longer than the "ideal" $-\log p$.)

Hint Let P be any distribution on X with $P(x) = p$ for some $x \in \mathsf{X}$. Consider the code tree of an optimal prefix code $f : \mathsf{X} \to \{0,1\}^*$. Denote by $A_0, A_1, \ldots, A_l$ the vertices on the path from the root A_0 to the terminal vertex A_l corresponding to x, and by B_i the vertex connected with A_{i-1} which is not on this path. The optimality of the code implies that the probability of the set of terminal vertices reachable from B_i cannot be smaller than that of those reachable from A_{i+1}. This proves $f_{l+1} p < 1$, i.e., $l \leqq l_p - 1$. Further, a distribution P on a suitable set X which achieves this bound can be easily constructed so that the only vertices of the optimal code tree be $A_0, A_1, \ldots, A_l, B_1, \ldots, B_l$.

4.13. (*Search strategies and codes*) Suppose that an unknown element x^* of a set X is to be located on the basis of successive experiments of the following kind. The possible states of knowledge about x^* are that x^* belongs to some subset X' of X. Given such a state of knowledge, the next experiment partitions X' into at most q subsets, the result specifying the subset containing x^*. A q-ary *search strategy* gives successively – starting from the state of ignorance, i.e. $\mathsf{X}' \triangleq \mathsf{X}$ – the possible states of knowledge and the corresponding partitions; each atom of the latter is a possible state of knowledge for the next step.

(a) Show that q-ary search strategies are equivalent to prefix codes $f : \mathsf{X} \to \mathsf{Y}^*$ with $|\mathsf{Y}| = q$, i.e., to code trees. Each possible state of knowledge is represented by a vertex of the tree; the edges starting from this vertex represent the possible results of the experiment performed at this stage. Reaching a terminal vertex of the code tree means that the unknown x^* has been located. (Sobel (1960) used this equivalence of search strategies and codes to solve a search problem.)

(b) When searching for the unknown element x^* of an ordered set X, the possible search strategies are often restricted by allowing only partitions into intervals. Show that these search strategies correspond to *alphabetic prefix codes*, i.e., to order-preserving mappings $f : \mathsf{X} \to \mathsf{Y}^*$, where Y is an ordered set and Y^* is endowed with the lexicographic order.

(c) To an arbitrary distribution P on X, construct alphabetic prefix codes $f : \mathsf{X} \to \mathsf{Y}^*$ with

$$\bar{l}(f) \leqq \frac{H(P)+1}{\log|\mathsf{Y}|} + 1 \quad \text{resp.} \quad \bar{c}(f) \leqq \frac{H(P)+1}{\alpha_0} + c^*.$$

Hint Use the construction in the proof of Theorem 4.1, resp. 4.4, without reordering the source symbols, and giving the role of a_i to

$$\tilde{a}_i \triangleq \sum_{j=1}^{i-1} P(x_j) + \frac{1}{2} P(x_i).$$

(Gilbert and Moore (1959).)

4.14. (a) Given a code tree with terminal vertices labeled by the elements of X having probabilities $P(x)$, for each vertex A let $P(A)$ be the sum of probabilities of terminal vertices reachable from A. If the edges starting from A lead to vertices $B_1, \ldots, B_q$, set $P_A \triangleq \{P(B_i|A) : i = 1, \ldots, q\}$, where $P(B_i|A) \triangleq P(B_i)/P(A)$. Show that

$$\sum_A P(A) H(P_A) = H(P);$$

here summation refers to the non-terminal vertices. Interpret this identity in terms of the search model of the previous problem, supposing that x^* is a RV with distribution P.

(b) Deduce from the above identity the bounds (4.2) and (4.11) for the average codeword length resp. cost of prefix codes, including the condition of equality.

Hint Use the bound $H(P_A) \leqq \log|\mathsf{Y}|$ resp. $H(P_A) \leqq \alpha_0 \bar{c}_A$, where $\bar{c}_A$ is the expectation of $c(y)$ with respect to the distribution P_A.

4.15. Let $X^\infty = \{X_i\}_{i=1}^\infty$ be a discrete source and let N be a positive integer-valued RV with finite range such that the values $X_1, \ldots, X_n$ uniquely determine whether or not $N = n$ (i.e., N is a *stopping time*). Represent the sequence $\{X_i\}_{i=1}^\infty$ by an infinite rooted tree and show that stopping times are equivalent to saturated (finite) subtrees (see Problem 4.3(b)). Deduce from the result of Problem 4.14 (a) that for a DMS with generic distribution P

$$H(X_1, \ldots, X_N) = EN \cdot H(P).$$

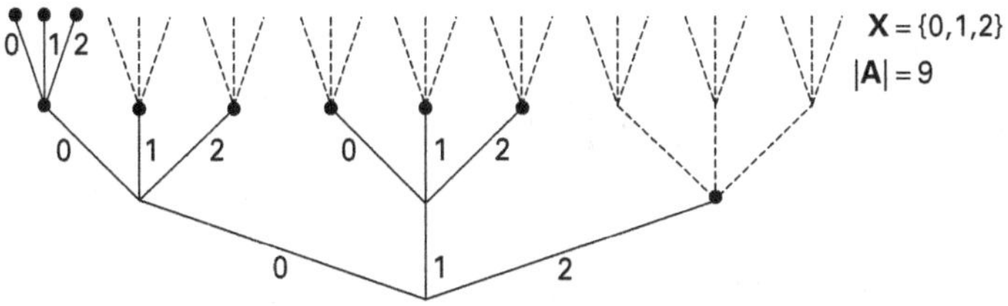

Figure P4.16 Code tree of a variable-to-fixed-length code (solid). In a successive construction of optimal code trees, the next tree is obtained by extending a path of maximum probability, say path 10.

4.16. (*Variable-to-fixed-length codes*) Let X^∞ be a discrete source and let N be a stopping time as in Problem 4.15. Let $\mathsf{A} \subset \mathsf{X}^*$ be the range of $X^N = (X_1, \dots, X_N)$. A one-to-one mapping $f : \mathsf{A} \to \mathsf{Y}^l$ is a *variable-to-fixed-length code* for X^∞. Set

$$\bar{l}(f) \triangleq \frac{l}{EN}.$$

(a) Show that if X^∞ is a DMS with generic distribution P, then $\bar{l}(f) \geqq H(P)/\log|\mathsf{Y}|$ and, moreover, to any $\delta > 0$ there exist variable-to-fixed-length codes with $\bar{l}(f) < H(P)/\log|\mathsf{Y}| + \delta$.

Hint To prove the existence part, construct in a recursive manner code trees such that

$$\min_{\mathbf{x}\in\mathsf{A}} P(\mathbf{x}) \geqq \min_{x\in\mathsf{X}} P(x) \cdot \max_{\mathbf{x}\in\mathsf{A}} P(\mathbf{x}),$$

where $P(\mathbf{x}) \triangleq P^k(\mathbf{x})$ if $\mathbf{x} \in \mathsf{X}^k$. (Jelinek and Schneider (1972); they attribute the optimal algorithm to Tunstall (1968, unpublished).)

(b) Show by counterexample that for stationary sources

$$\bar{l}(f) \geqq \frac{\bar{H}(X^\infty)}{\log|\mathsf{Y}|}$$

need not hold.

Hint Consider the mixed source of Problem 4.6 (b).

4.17. (*Conservation of entropy*)

(a) Let X^∞ be any discrete source whose entropy rate exists and let $\{N_k\}_{k=1}^\infty$ be a sequence of positive integer-valued RVs (not necessarily stopping times) such that $\lim_{k\to\infty} E\left|\frac{N_k}{k} - c\right| = 0$ for some constant $c \geqq 0$. Show that

$$\lim_{k\to\infty} \frac{1}{k} H(X_1 \dots X_{N_k}) = c\bar{H}(X^\infty)$$

and hence deduce the result of Problem 4.15.

(b) Show that if $\{N_k^{(1)}\}_{k=1}^\infty$ and $\{N_k^{(2)}\}_{k=1}^\infty$ are two sequences of positive integer-valued RVs such that

$$\lim_{k\to\infty} \frac{1}{k} E\left|N_k^{(1)} - N_k^{(2)}\right| = 0,$$

then

$$\lim_{k\to\infty} \frac{1}{k} \left| H\left(X_1, \ldots, X_{N_k^{(1)}}\right) - H\left(X_1, \ldots, X_{N_k^{(2)}}\right) \right| = 0.$$

Hint

$$\begin{aligned} &H\left(X_1, \ldots, X_{N_k^{(1)}} \mid X_1, \ldots, X_{N_k^{(2)}}\right) \\ &\leqq H\left(X_1, \ldots, X_{N_k^{(1)}} \mid X_1, \ldots, X_{N_k^{(2)}}, N_k^{(1)}\right) + H\left(N_k^{(1)} \mid N_k^{(2)}\right) \\ &\leqq E|N_k^{(1)} - N_k^{(2)}|^+ \log |\mathsf{X}| + H\left(N_k^{(1)} - N_k^{(2)}\right). \end{aligned}$$

Hence the assertion follows by Problem 3.5 and Corollary 3.12.

(c) Let the mapping $f : \mathsf{X}^* \to \mathsf{Y}^*$ be a *sequential encoder*, i.e., if $\mathbf{x}'$ is a prefix of $\mathbf{x}$ then so is $f(\mathbf{x}')$ for $f(\mathbf{x})$. Let X^∞ be a discrete source with alphabet X, having an entropy rate, and suppose that $l(f(X^k)) \to \infty$ as $k \to \infty$, with probability 1. Then encoding X^∞ by f results in a well-defined source Y^∞ with alphabet Y. Suppose that there exists a constant m such that for every k at most m different sequences $\mathbf{x} \in \mathsf{X}^k$ can have the same codeword. Show that if for some $\bar{l} > 0$ one has $E\left|l(f(X^k))/k - \bar{l}\right| \to 0$ then $\bar{H}(Y^\infty)$ exists and equals $\bar{H}(X^\infty) / \bar{l}$.

Hint Apply (a) to Y^∞, setting $N_k \triangleq l(f(X^k))$. (Csiszár, Katona and Tusnády (1969).)

4.18. (*Conservation of entropy and ergodic theory*) In this problem, a source with alphabet X means a doubly infinite sequence of RVs with values in X. A *sliding block code* for such a source is a mapping f of doubly infinite sequences $\ldots x_{-1}x_0x_1 \ldots$ into sequences $\ldots y_{-1}y_0y_1 \ldots$ determined by a mapping $f_0 : \mathsf{X}^{2m+1} \to \mathsf{Y}$, setting $y_i \triangleq f_0(x_{i-m}, \ldots, x_{i+m}), i = 0, \pm 1, \ldots$. Unlike the codes considered so far, sliding block codes have the property that on applying them to a stationary source $\{X_i\}_{i=-\infty}^{\infty}$, the resulting source $\{Y_i\}_{i=-\infty}^{\infty}$ (where $Y_i \triangleq f_0(X_{i-m}, \ldots, X_{i+m})$) is again stationary. An *infinite code* f for the source $\{X_i\}_{i=-\infty}^{\infty}$ is a stochastic limit of a sequence of sliding block codes determined by mappings $f_0^{(n)} : \mathsf{X}^{2m_n+1} \to \mathsf{Y}$, i.e., f maps $\{X_i\}_{i=-\infty}^{\infty}$ into $\{Y_i\}_{i=-\infty}^{\infty}$, where

$$\Pr\{Y_i = f_0^{(n)}(X_{i-m_n}, \ldots, X_{i+m_n})\} \to 1 \quad \text{as} \quad n \to \infty.$$

Two stationary sources are called *isomorphic* if there exists an infinite code f for the first source which is invertible with probability 1 and its inverse is also an infinite code, such that it maps the first source into a source having the same joint distributions as the second one.

(a) Show that the application of a sliding block code to a stationary source $\{X_i\}_{i=-\infty}^{\infty}$ cannot increase the entropy rate, i.e.,

$$\lim_{k\to\infty} \frac{1}{k} H(Y_1 \ldots Y_k) \leqq \lim_{k\to\infty} \frac{1}{k} H(X_1, \ldots, X_k).$$

If the mapping f is one-to-one, the equality holds.

(b) Prove the inequality of (a) if $\{Y_i\}_{i=-\infty}^{\infty}$ is obtained from $\{X_i\}_{i=-\infty}^{\infty}$ by any infinite code.

Hint Writing $Y_i^{(n)} \triangleq f_0^{(n)}(X_{i-m_n}, \dots, X_{i+m_n})$, note that

$$H(Y_1, \dots, Y_k) \leqq H(Y_1^{(n)}, \dots, Y_k^{(n)}) + H(Y_1, \dots, Y_k | Y_1^{(n)}, \dots, Y_k^{(n)})$$
$$\leqq H(Y_1^{(n)}, \dots, Y_k^{(n)}) + kH(Y_1 | Y_1^{(n)}).$$

Now use (a) and Fano's inequality.

(c) Conclude from (b) that sources with different entropy rates cannot be isomorphic.

(The question of when two stationary sources are isomorphic is an equivalent formulation of the isomorphy problem of dynamical systems in ergodic theory. The discovery that the latter problem is inherently information-theoretic is due to Kolmogorov (1958). He established (c) and thereby proved – settling a long-standing problem – that two DMSs with generic distributions of different entropy cannot be isomorphic. The result of (b) is known as the *Kolmogorov–Sinaĭ theorem* (Kolmogorov (1958), Sinaĭ (1959)). Ornstein (1970) proved that for DMSs the equality of the entropies of the generic distributions already implies isomorphism. For this and further related results see Ornstein (1973).)

4.19. (a) Give a non-probabilistic proof for Lemma 4.5.

Hint Write a difference equation for $|\mathsf{A}(t)|$, splitting $\mathsf{A}(t)$ into sets of sequences with the same initial symbol, and use induction on the multiples of the minimum cost c_0.

(Shannon (1948), Csiszár (1969).)

(b) Show that $\lim_{t\to\infty} |\mathsf{A}(t)| \exp(-\alpha_0 t)$ exists whenever the costs $c(y)$ are not all integral multiples of the same real number d. In the opposite case, prove that the limit exists if $t \to \infty$ running over the integral multiples of the largest possible d. Evaluate these limits.

Hint Proceed as in the text, and apply the renewal theorem, see Feller (1966), p. 347. (Smorodinsky (1968).)

General noiseless channels (Problems 4.20–4.22)

So far we have considered a noiseless channel as a device capable of transmitting any sequence in Y^*, the cost of transmission being the sum of the costs of the individual symbols. In the following three problems we drop the assumption that every sequence in Y^* is transmissible. Also, we shall consider more general cost functions.

4.20. A general noiseless channel with alphabet Y is a pair (V, c), where $\mathsf{V} \subset \mathsf{Y}^*$ is the set of admissible input sequences and $c(\mathbf{y})$ is a cost function on V such that if $\mathbf{y}'$ is a prefix of $\mathbf{y}$ then $c(\mathbf{y}') \leqq c(\mathbf{y})$. Let $\mathsf{A}_1(t) \subset \mathsf{V}$ consist of those sequences $\mathbf{y} \in \mathsf{V}$ for which $c(\mathbf{y}) \leqq t$ and let $\mathsf{A}(t) \subset \mathsf{A}_1(t)$ consist of those elements of $\mathsf{A}_1(t)$ which are not proper prefixes of any sequence in $\mathsf{A}_1(t)$.

Consider two kinds of capacity of (V, c):

$$C_1 \triangleq \lim_{t\to\infty} \frac{1}{t} \log |\mathsf{A}_1(t)|, \quad C \triangleq \lim_{t\to\infty} \frac{1}{t} \log |\mathsf{A}(t)|,$$

provided that the limits exist.

(a) Show that if $c(\mathbf{y})/\log l(\mathbf{y}) \to \infty$ uniformly as $l(\mathbf{y}) \to \infty$ then $C_1 = C$ (provided that either limit exists).

(b) Show that for encoders $f : \mathsf{X}^k \to \mathsf{V}$, the first assertion of Corollary 4.4 holds if α_0 is replaced by C_1.

(c) Denote by $\mathsf{A}_{\mathbf{y}}(t)$ the subset of $\mathsf{A}(c(\mathbf{y}) + t)$ consisting of those sequences of which $\mathbf{y}$ is a prefix. Suppose that there exists a constant $a > 0$ such that

$$|\mathsf{A}_{\mathbf{y}}(t)| \geqq a|\mathsf{A}(t)| \quad \text{for} \quad t \geqq t_0, \quad \mathbf{y} \in \mathsf{V}.$$

Show that then C exists and the second assertion of Corollary 4.4 is valid (with prefix codes having codewords in V) if α_0 is replaced by C.

(Csiszár (1970).)

4.21. (*Finite state noiseless channels*)

(a) Suppose that at any time instant the admissible channel inputs are determined by the "state of the channel," and that this input and state determine the state at the next time instant. Formally, let Y and S be finite sets, and let g be a mapping of a subset of $\mathsf{Y} \times \mathsf{S}$ into S. Note that $y \in \mathsf{Y}$ is an admissible input at state $s \in \mathsf{S}$ if $g(y, s)$ is defined, which then gives the next state (this model is also called a *Moore automaton* with restricted input). Fixing an initial state $s_1 \in \mathsf{S}$, the set V of admissible input sequences consists of those $\mathbf{y} \in \mathsf{Y}^*$ for which $s_{i+1} = g(y_i, s_i)$, $i = 1, 2, \ldots, l(\mathbf{y})$, are defined.

Suppose that for any choice of s_1, every $s \in \mathsf{S}$ can be reached by some admissible input sequence, i.e., there exists $\mathbf{y} \in \mathsf{V}$ yielding $s_{l(\mathbf{y})+1} = s$. Show that if $c(\mathbf{y}) \triangleq l(\mathbf{y})$ for all $\mathbf{y} \in \mathsf{V}$ then the capacity C (see Problem 4.20) equals the logarithm of the greatest positive eigenvalue of the matrix $\{n_{ss'}\}_{s,s' \in \mathsf{S}}$, where $n_{ss'} \triangleq |\{y : g(y, s) = s'\}|$. More generally, set $c(\mathbf{y}) \triangleq \sum_{i=1}^{l} c(y_i, s_i)$ for $\mathbf{y} = y_1 \ldots y_l$, where $c(y, s)$ is a given positive-valued function on $\mathsf{Y} \times \mathsf{S}$. Denoting $d_{ss'}(\alpha) \triangleq \sum_{y: g(y,s)=s'} \exp[-\alpha c(y, s)]$, prove that C equals the largest positive root α_0 of the equation $\mathrm{Det}\{d_{ss'}(\alpha) - \delta_{ss'}\}_{s,s' \in \mathsf{S}} = 0$, where δ is the Kronecker symbol; more exactly, prove the analog of Lemma 4.5 with this α_0. (Shannon (1948), Ljubič (1962), Csiszár (1969).)

(b) In a more general type of finite state noiseless channels the output is not necessarily identical to the input but is a given function of the input and the state. Let $g : \mathsf{Y} \times \mathsf{S} \to \mathsf{S}$ and $h : \mathsf{Y} \times \mathsf{S} \to \mathsf{Z}$ be the mappings determining the state transition and the output (for the sake of simplicity, input restrictions are not imposed; this model is called a *Mealy automaton),* and let $\mathsf{W} \subset \mathsf{Z}^*$ be the set of all possible output sequences. Construct a channel such as in (a), having alphabet Z, for which the set of admissible input sequences equals W.

(Csiszár and Komlós (1968).)

(c) For channel (b), the admissible separable codes are those for which the output sequences corresponding to different codewords are different. Prove the analog of Corollary 4.4 for the case $c(\mathbf{y}) \triangleq l(\mathbf{y})$, with the capacity of the finite state channel constructed in (b).

4.22. (*Conservation of entropy*) Generalize Problem 4.17(c) to codes for transmission over an arbitrary noiseless channel (V, c), cf. Problem 4.20. Let X^∞, f and Y^∞ be as in Problem 4.17 (c), with the additional assumption that f maps X^* into V.

(a) Defining the RV N_k as the largest n for which $c(Y_1 ... Y_n) \leqq k$, show that if

$$E\left|\frac{l(f(X^k))}{k} - \bar{c}\frac{N_k}{k}\right| \to 0 \quad \text{for some} \quad \bar{c} > 0$$

then

$$\bar{c} \lim_{k\to\infty} \frac{1}{k} H(Y_1, \ldots, Y_{N_k}) = \bar{H}(X^\infty).$$

Hint Use the results of Problem 4.17.

(b) Give regularity conditions under which the convergence in probability of $c(f(X^k))/k$ to $\bar{c}$ implies the assumption of (a).

(c) How are these results related to Problem 4.20 (b)?

(See Csiszár, Katona and Tusnády (1969).)

Universal variable-length codes (Problems 4.23–4.26)

4.23. (a) Show that to any separable code $f : \mathsf{X}^k \to \{0, 1\}^*$ there exists a PD Q on X^k such that a Shannon code for Q is no worse than f, namely $\lceil -\log Q(\mathbf{x})\rceil \leqq l(f(\mathbf{x}))$ for all $\mathbf{x} \in \mathsf{X}^k$.

Hint Take $Q(\mathbf{x}) = c \exp[-l(f(\mathbf{x})]$. Use Kraft's inequality to show that $c \geqq 1$.

(b) The *redundancy* of a code $f : \mathsf{X}^k \to \{0, 1\}^*$ for a source $\{X_i\}_{i=1}^\infty$ is

$$r(f, X^k) \triangleq \frac{1}{k} El(f(X^k)) - \frac{1}{k} H(X^k).$$

Show that if f is a Shannon code for a PD Q on X^k then

$$0 \leqq r(f, X^k) - \frac{1}{k} D(P_{X^k} \| Q) \leqq \frac{1}{k}.$$

(c) A sequence of separable codes $f_k : \mathsf{X}^k \to \{0, 1\}^*$, $k = 1, 2, \ldots,$ is *weakly* resp. *strongly universal* for a class of sources with alphabet X if $r(f_k, X^k) \to 0$, resp. this holds uniformly, for the sources in the given class. Show that the existence of weakly resp. strongly universal codes is equivalent to the existence of PDs Q_k on the sets X^k such that $\frac{1}{k} D(P_{X^k} \| Q_k) \to 0$, resp. this holds uniformly, for the sources in the given class. Moreover, if this condition is satisfied, Shannon codes for these PDs Q_k are weakly resp. strongly universal.

(It is desirable if the PDs Q_k can be chosen as the k-dimensional distributions of an auxiliary source. In practice, instead of Shannon codes, an arithmetic code is used, see Problem 4.24, not increasing redundancy by more than $1/k$. Arithmetic coding with respect to a suitable auxiliary source is a key technique of data compression, see Problem 4.26.)

(d) For a given class of sources with alphabet X, the *minimax redundancy* $r(k)$ is the minimum for all separable codes $f\colon \mathsf{X}^k \to \{0,1\}^*$ of the supremum of $r(f, X^k)$ taken for all sources in the class. Show that

$$r(k) - \frac{1}{k} < \inf_{Q} \sup_{P \in \mathcal{P}_k} D(P\|Q) \leqq r(k),$$

where $\mathcal{P}_k$ is the family of the k-dimensional distributions P_{X^k} of the sources in the class, and the infimum is for all PDs Q on X^k.

(Davisson (1973). See also Problem 8.10.)

4.24. (*Arithmetic codes*) Given a source $\{X_i\}_{i=1}^{\infty}$ with alphabet X and an ordering of X, denote for $\mathbf{x} = x_1 \ldots x_k \in \mathsf{X}^k$

$$b(\mathbf{x}) \triangleq \sum_{i=1}^{k-1} \sum_{a < x_{i+1}} Q(\mathbf{x}^i a), \quad \mathbf{x}^i = x_1 \ldots x_i,$$

where $Q(\mathbf{x}^i) \triangleq \Pr\{X^i = \mathbf{x}^i\}$.

(a) Let the mapping $f : \mathsf{X}^* \to \{0,1\}^*$ assign to $\mathbf{x} = x_1 \ldots x_k$ the shortest binary sequence $y_1 \ldots y_l$ such that

$$0.y_1 \ldots y_l \leqq b(\mathbf{x}) + \frac{1}{2} Q(\mathbf{x}) < 0.y_1 \ldots y_l + 2^{-l}.$$

Check that the restriction of this mapping to X^k is the same as the alphabetic code in Problem 4.13 (c), the role of X and P there played by X^k with the lexicographic order, and by Q. In particular, $f\colon \mathsf{X}^k \to \{0,1\}^*$ is a prefix code with codeword lengths $l(f(\mathbf{x})) < -\log Q(\mathbf{x}) + 2$.

(b) Let $g : \mathsf{X}^* \to \{0,1\}^*$ assign to $\mathbf{x} = x_1 \ldots x_k$ the longest binary sequence $y_1 \ldots y_l$ for which the interval $[0.y_1 \ldots y_l,\ 0.y_1 \ldots y_l + 2^{-l})$ contains $[b(\mathbf{x}), b(\mathbf{x}) + Q(\mathbf{x}))$. Show that $l(g(\mathbf{x})) \leqq -\log Q(\mathbf{x})$, and that the restriction of g to X^k, though in general not a prefix code (even the void sequence may be a codeword), meets the fidelity criterion (i) in (4.1), with a suitable decoder. Further, g admits sequential encoding: if $\mathbf{x}'$ is a prefix of $\mathbf{x}$ then $g(\mathbf{x}')$ is a prefix of $g(\mathbf{x})$.

(Arithmetic codes as above, and variants, became basic tools after Rissanen (1976) and Pasco (1976, unpublished) showed – overcoming an apparent difficulty – that they can be efficiently implemented using fixed-precision arithmetic. Yielding effectively as good compression as Shannon codes, their computational complexity is low if that of the probabilities defining $b(\mathbf{x})$ is low. In most applications, the latter are not the (typically unknown) probabilities of the actual source, but those of a suitable auxiliary source, see Problems 4.23 and 4.26.)

4.25. (*Asymptotics of minimax redundancy*) Show that for the class of DMSs with alphabet X the minimax redundancy $r(k)$, see Problem 4.23 (d), is asymptotically

$$\frac{|\mathsf{X}|-1}{2}\frac{\log k}{k},$$

in the sense that the ratio of the two quantities tends to 1 as $k \to \infty$. Comparing this with Theorem 4.1, interpret the slower convergence of order $\log k/k$ rather than $1/k$ as the price that must be paid in redundancy for universal asymptotic optimality. (Kričevskiĭ (1968), (1970).)

Hint Note that Theorem 4.6 gives $r(k) \leqq |\mathsf{X}|[\log(k+1)/k] + 1/k$. To get a sharper bound, consider prefix codes $f : \mathsf{X}^k \to \{0,1\}^*$ defined by juxtaposing two mappings f_1 and f_2 – as in the proof of Theorem 4.6 – where f_1 is a prefix code of types of sequences in X^k, and f_2 maps each type class T^k_P into binary sequences of length $\lceil \log|\mathsf{T}^k_P|\rceil$. In this case

$$\frac{1}{k}(El(f_1(X^k)) - H(f_1(X^k))) \leqq r(f, X^k)$$
$$\leqq \frac{1}{k}(El(f_1(X^k)) - H(f_1(X^k)) + 1) \qquad (*)$$

for every DMS. Now, using Problem 2.2, one can see that for a suitable constant b the numbers

$$n(P) \triangleq \left\lceil -\log P^k(\mathsf{T}^k_P) + \frac{|\mathsf{X}|-1}{2}\log k + b\right\rceil$$

satisfy Kraft's inequality. If f_1 has these codeword lengths, we obtain from (*)

$$r(f, X^k) \leqq \frac{|\mathsf{X}|-1}{2}\frac{\log k}{k} + \frac{b+2}{k}.$$

A lower bound on $r(k)$ is the minimum over separable codes $f : \mathsf{X}^k \to \{0,1\}^*$ of the average of $r(f, X^k)$ for all possible generic distributions P. Take this average with respect to the Lebesgue measure on the set of PDs on X (considering this set as a simplex in the $|\mathsf{X}|$-dimensional Euclidean space). Note that the resulting averaged distribution Q on X^k assigns equal probabilities to each type class T^k_P; further, $Q(\mathbf{x})$ is constant within each T^k_P. For any f, the average of $El(f(X^k))$ equals $\sum_{\mathbf{x}\in\mathsf{X}^k} Q(\mathbf{x})l(f(\mathbf{x}))$. Thus by Theorem 4.1, the minimum of this average is approximated within constant 2 by a code f obtained by suffixing the same f_2 as above to an f_1 which assigns a fixed-length binary sequence to each type. Now the desired lower bound follows from (*) and Problem 2.2.

4.26. (*Practical universal code*) Consider an auxiliary source whose k-dimensional distributions Q_k are defined recursively by

$$Q_1(a) \triangleq \frac{1}{|\mathsf{X}|}, \quad Q_{k+1}(\mathbf{x}a) \triangleq Q_k(\mathbf{x})\frac{N(a|\mathbf{x}) + 1/2}{k + |\mathsf{X}|/2}, \qquad a \in \mathsf{X},\ \mathbf{x} \in \mathsf{X}^k.$$

Show that arithmetic coding with respect to this auxiliary source is strongly universal for the class of DMSs with alphabet X, with worst case redundancy asymptotically equal to the minimax redundancy determined in Problem 4.25. Prove the stronger result that for no DMS does the codelength of any $\mathbf{x} \in \mathsf{X}^k$ exceed the ideal codelength for that source (see Problem 4.12) by more than $(1/2)(|\mathsf{X}| - 1) \log k$ plus a constant.

(See Davisson *et al.* (1981). The above auxiliary source is a key building block for constructing auxiliary sources adequate for wide classes of sources with memory, underlying the powerful *context tree weighting* data compression algorithm of Willems, Shtarkov and Tjalkens (1995).)

Hint Represent $Q_k(\mathbf{x})$ and the largest possible probability $\max_P P^k(\mathbf{x})$ of $\mathbf{x}$ in explicit form, and apply the sharp version of Stirling's formula, see Problem 2.2.

Story of the results

The results of this chapter, except for Theorem 4.6, are essentially due to Shannon (1948). He credited R. M. Fano with the independent discovery of the code construction in the text achieving (4.3); it is known as the *Shannon–Fano code*. Its extension to the case of unequal symbol costs is due to Krause (1962); a similar construction was suggested by Bloh (1960).

For the lower bound of the average codeword length resp. cost (see (4.2) and (4.11)) Shannon gave a heuristic argument. The first rigorous proof of (4.2) was apparently that of McMillan (1956), based on his generalization of Kraft's inequality (see Problem 4.3) to separable codes. Inequality (4.11) was proved similarly by Krause (1962), starting with Corollary 4.5. We stated the lower bounds under somewhat weaker conditions than usual. The proofs are elaborations of Shannon's original heuristic argument, following Csiszár, Katona and Tusnády (1969) and Csiszár (1969); we preferred this information-theoretic approach to the rather ad hoc proofs mentioned above. Lemma 4.5 is a variant of the capacity formula of Shannon (1948). Corollary 4.5 was proved by Schützenberger and Marcus (1959) and Krause (1962). Universally optimal codes were first constructed by Davisson (1966), Fitingof (1966) and Lynch (1966). Theorem 4.6 uses the construction of Davisson (1966), extended to unequal symbol costs.

Addition. A glimpse of concepts that have become fundamental in data compression since the first edition of this book is given in Problems 4.24 and 4.26. Progress concerning noiseless channels has also been substantial (but no problems were added to illustrate it), and is characterized by efficient use of tools of *symbolic dynamics*, see Lind and Marcus (1995).

5 Blowing up lemma: a combinatorial digression

Given a finite set Y, the set Y^n of all n-length sequences of elements from Y is sometimes considered as a metric space with the Hamming metric. Recall that the Hamming distance of two n-length sequences is the number of positions in which these two sequences differ. The Hamming metric can be extended to measure the distance of subsets of Y^n, setting

$$d_{\mathrm{H}}(\mathsf{B}, \mathsf{C}) \triangleq \min_{\mathbf{y}\in\mathsf{B}, \hat{\mathbf{y}}\in\mathsf{C}} d_{\mathrm{H}}(\mathbf{y}, \hat{\mathbf{y}}).$$

Some classical problems in geometry have exciting combinatorial analogs in this setup. One of them is the *isoperimetric problem,* which will turn out to be relevant for information theory. The results of this chapter will be used mainly in Part III.

Given a set $\mathsf{B} \subset \mathsf{Y}^n$, the *Hamming l-neighborhood* of B is defined as the set

$$\Gamma^l \mathsf{B} \triangleq \{\mathbf{y} : \mathbf{y} \in \mathsf{Y}^n, d_{\mathrm{H}}(\{\mathbf{y}\}, \mathsf{B}) \leqq l\}.$$

We shall write Γ for Γ^1. The *Hamming boundary* $\partial\mathsf{B}$ of $\mathsf{B} \subset \mathsf{Y}^n$ is defined by $\partial\mathsf{B} \triangleq \mathsf{B} \cap \Gamma\bar{\mathsf{B}}$.

Considering the boundary $\partial\mathsf{B}$ as a discrete analog of the surface, one can ask how small the "size" $|\partial\mathsf{B}|$ of the "surface" of a set $\mathsf{B} \subset \mathsf{Y}^n$ can be if the "volume" $|\mathsf{B}|$ is fixed. Theorem 5.3 below answers (a generalized form of) this question in an asymptotic sense. Afterwards, the result will be used to see how the probability of a set is changed by adding or deleting relatively few sequences close to its boundary.

One easily sees that if $l_n/n \to 0$ then the cardinality of B and of its l_n-neighborhood have the same exponential order of magnitude, and the same holds also for their P^n-probabilities, for every distribution P on Y. More precisely, one has the following lemma.

LEMMA 5.1 Given a sequence of positive integers $\{l_n\}$ with $l_n/n \to 0$ and a distribution P on Y with positive probabilities, there exists a sequence $\varepsilon_n \to 0$ depending only on $\{l_n\}$, $|\mathsf{Y}|$ and $m_P \triangleq \min_{y\in\mathsf{Y}} P(y)$ such that for every $\mathsf{B} \subset \mathsf{Y}^n$

$$0 \leqq \frac{1}{n}\log|\Gamma^{l_n}\mathsf{B}| - \frac{1}{n}\log|\mathsf{B}| \leqq \varepsilon_n,$$

$$0 \leqq \frac{1}{n}\log P^n(\Gamma^{l_n}\mathsf{B}) - \frac{1}{n}\log P^n(\mathsf{B}) \leqq \varepsilon_n.$$

○

Proof Since $\mathsf{B} \subset \Gamma^l \mathsf{B} = \bigcup_{\mathbf{y}\in\mathsf{B}} \Gamma^l\{\mathbf{y}\}$, it suffices to prove both assertions for one-point sets. Clearly,

$$|\Gamma^{l_n}\{\mathbf{y}\}| \leqq \binom{n}{l_n} |\mathsf{Y}|^{l_n}.$$

As $P^n(\mathbf{y}') \leqq m_P^{-l_n} P^n(\mathbf{y})$ for every $\mathbf{y}' \in \Gamma^{l_n}\{\mathbf{y}\}$, this implies

$$P^n(\Gamma^{l_n}\{\mathbf{y}\}) \leqq \binom{n}{l_n} |\mathsf{Y}|^{l_n} m_P^{-l_n} P^n(\mathbf{y}).$$

Since $\binom{n}{l_n}$ equals the number of length-n binary sequences of type $(l_n/n, 1 - l_n/n)$, by Lemma 2.3 we have

$$\binom{n}{l_n} \leqq \exp\left[nH\left(\frac{l_n}{n}, 1 - \frac{l_n}{n}\right)\right] = \exp\left[nh\left(\frac{l_n}{n}\right)\right].$$

Thus the assertions follow with

$$\varepsilon_n \triangleq h\left(\frac{l_n}{n}\right) + \frac{l_n}{n}(\log|\mathsf{Y}| - \log m_P). \qquad \square$$

Knowing that the probability of the l_n-neighborhood of a set has the same exponential order of magnitude as the probability of the set itself (if $l_n/n \to 0$), we would like to have a deeper insight into the question of how passing from a set to its l_n-neighborhood increases probability. The answer will involve the function

$$f(s) \triangleq \begin{cases} \varphi(\Phi^{-1}(s)) & \text{if} \quad s \in (0, 1) \\ 0 & \text{if} \quad s = 0 \quad \text{or} \quad s = 1, \end{cases}$$

where $\varphi(t) \triangleq (2\pi)^{-\frac{1}{2}} e^{-t^2/2}$ and $\Phi(t) \triangleq \int_{-\infty}^{t} \varphi(u)d(u)$ are the density resp. distribution function of the standard normal distribution, and $\Phi^{-1}(s)$ denotes the inverse function of $\Phi(t)$. Some properties of $f(s)$ are summarized in Lemma 5.2.

LEMMA 5.2 The function $f(s)$, defined on $[0, 1]$, is symmetric around the point $s = 1/2$; it is non-negative, concave and satisfies

$$\text{(i)} \qquad \lim_{s\to 0} \frac{f(s)}{s\sqrt{-2\ln s}} = \lim_{s\to 0} \frac{-\Phi^{-1}(s)}{\sqrt{-2\ln s}} = 1,$$

$$\text{(ii)} \qquad f'(s) = -\Phi^{-1}(s) \qquad (s \in (0, 1)),$$

$$\text{(iii)} \qquad f''(s) = -\frac{1}{f(s)} \qquad (s \in (0, 1)). \qquad \bigcirc$$

COROLLARY 5.2 There exists a constant $K_0 > 0$ such that $f(s) \leqq K_0 s\sqrt{-\ln s}$ for all $s \in [0, 1/2]$. $\bigcirc$

Proof The obvious relation (ii) implies (iii), establishing the concavity of the non-negative function $f(s)$. The symmetry is also clear, so that it remains to prove (i). Observe first that because of (ii)

$$\lim_{s\to 0}\frac{f(s)}{\sqrt{-2\ln s}}=\lim_{s\to 0}\frac{f'(s)}{\sqrt{-2\ln s}}\frac{1}{1+\dfrac{1}{2\ln s}}=\lim_{s\to 0}\frac{-\Phi^{-1}(s)}{\sqrt{-2\ln s}}.$$

Hence, applying the substitution $s=\Phi(t)$ and using the well-known fact

$$\lim_{t\to-\infty}\frac{-t\Phi(t)}{\varphi(t)}=1$$

(see Feller (1968), p. 175), it follows that the above limit further equals

$$\lim_{t\to-\infty}\frac{-t}{\sqrt{-2\ln\Phi(t)}}=\lim_{t\to-\infty}\frac{-t}{\sqrt{-2\ln\dfrac{\varphi(t)}{-t}}}=\lim_{t\to\infty}\frac{t}{\sqrt{-2\ln\varphi(t)+2\ln t}}=1. \quad \square$$

Now we are ready to give an asymptotically rather sharp lower bound on the probability of the boundary of an arbitrary set in terms of its own probability. This and the following results will be stated somewhat more generally than Lemma 5.1, for conditional probabilities, because this is the form needed in the subsequent parts of the book.

THEOREM 5.3 For every stochastic matrix $W:\mathsf{X}\to\mathsf{Y}$, integer n, set $\mathsf{B}\subset\mathsf{Y}^n$ and each $\mathbf{x}\in\mathsf{X}^n$ one has

$$W^n(\partial\mathsf{B}|\mathbf{x})\geqq\frac{a}{\sqrt{n}}f(W^n(\mathsf{B}|\mathbf{x})),\quad a=a_W\triangleq K\frac{m_W}{\sqrt{-\ln m_W}},$$

where m_W denotes the smallest positive entry of W and K is an absolute constant. ○

Proof The statement is trivial if all the positive entries of W equal 1. In the remaining cases the smallest positive entry m_W of W does not exceed 1/2.

The proof goes by induction. The case $n=1$ is simple. In fact, then $\partial\mathsf{B}=\mathsf{B}$ for every $\mathsf{B}\subset\mathsf{Y}$. Hence one has to prove that for some absolute constant K

$$t\geqq K\frac{m_W}{\sqrt{-\ln m_W}}f(t)\quad\text{for}\quad t\geqq m_W.$$

As $m_W\leqq 1/2$ and by Lemma 5.2 $f(t)\leqq f(1/2)=1/\sqrt{2\pi}$, this inequality obviously holds if

$$K\leqq\sqrt{2\pi\ln 2}. \tag{5.1}$$

Suppose now that the statement of the theorem is true for $n-1$. Fix some set $\mathsf{B}\subset\mathsf{Y}^n$ and sequence $\mathbf{x}=x_1\ldots x_{n-1}x_n\in\mathsf{X}^n$. Write $\mathbf{x}^*\triangleq x_1\ldots x_{n-1}$; further, for every $y\in\mathsf{Y}$ denote by B_y the set of those sequences $y_1\ldots y_{n-1}\in\mathsf{Y}^{n-1}$ for which $y_1\ldots y_{n-1}y\in\mathsf{B}$. Then, obviously,

$$W^n(\mathsf{B}|\mathbf{x})=\sum_{y\in\mathsf{Y}}W(y|x_n)\,W^{n-1}(\mathsf{B}_y|\mathbf{x}^*), \tag{5.2}$$

and since

$$\partial\mathsf{B}\supset\bigcup_{y\in\mathsf{Y}}[\partial\mathsf{B}_y\times\{y\}],$$

also

$$W^n(\partial \mathsf{B}|\mathbf{x}) \geqq \sum_{y \in \mathsf{Y}} W(y|x_n)\, W^{n-1}(\partial \mathsf{B}_y|\mathbf{x}^*). \tag{5.3}$$

Put $\mathsf{S} \triangleq \{y : W(y|x_n) > 0\}$ and

$$d \triangleq \max_{y \in \mathsf{S}} W^{n-1}(\mathsf{B}_y|\mathbf{x}^*) - \min_{y \in \mathsf{S}} W^{n-1}(\mathsf{B}_y|\mathbf{x}^*).$$

Since $\partial \mathsf{B} \supset (\mathsf{B}_{y'} - \mathsf{B}_{y''}) \times \{y'\}$ for any y', y'' in Y, one gets

$$W^n(\partial \mathsf{B}|\mathbf{x}) \geqq m_W \cdot d. \tag{5.4}$$

If

$$d \geqq \frac{a}{m_W \sqrt{n}} f(W^n(\mathsf{B}|\mathbf{x})),$$

the statement of the theorem for n immediately follows from (5.4). Let us turn therefore to the contrary case of

$$d < \frac{a}{m_W \sqrt{n}} f(W^n(\mathsf{B}|\mathbf{x})). \tag{5.5}$$

Combining (5.3) and the induction hypothesis we see that

$$\begin{aligned} W^n(\partial \mathsf{B}|\mathbf{x}) &\geqq \sum_{y \in \mathsf{Y}} W(y|x_n) \cdot W^{n-1}(\partial \mathsf{B}_y|\mathbf{x}^*) \\ &\geqq \frac{a}{\sqrt{n-1}} \sum_{y \in \mathsf{Y}} W(y|x_n) \cdot f(W^{n-1}(\mathsf{B}_y|\mathbf{x}^*)). \end{aligned} \tag{5.6}$$

Write

$$s \triangleq W^n(\mathsf{B}|\mathbf{x}), \quad s_y \triangleq W^{n-1}(\mathsf{B}_y|\mathbf{x}^*),$$

and consider the interval of length d

$$\Delta \triangleq [\min_{y \in \mathsf{S}} s_y, \max_{y \in \mathsf{S}} s_y].$$

Since $\sum_{y \in \mathsf{Y}} W(y|x_n) s_y = s$ by (5.2), it follows from Taylor's formula,

$$f(s_y) = f(s) + (s_y - s) f'(s) + \frac{1}{2}(s_y - s)^2 f''(\sigma_y)$$

(where $\sigma_y \in \Delta$ if $y \in \mathsf{S}$), that if $\sigma \in \Delta$ satisfies

$$|f''(\sigma)| = \max_{s \in \Delta} |f''(s)| \tag{5.7}$$

then

$$\sum_{y \in \mathsf{Y}} W(y|x_n) \cdot f(s_y) \geqq f(s) - \frac{1}{2} d^2 |f''(\sigma)|.$$

This, (5.5) and Lemma 5.2 yield, by substitution into (5.6), the estimate

$$W^n(\partial\mathsf{B}|\mathbf{x}) \geqq \frac{a}{\sqrt{n-1}}\left[f(s) - \frac{a^2}{2m_W^2 n}\frac{f^2(s)}{f(\sigma)}\right].$$

Rearranging this we get

$$W^n(\partial\mathsf{B}|\mathbf{x}) \geqq \frac{a}{\sqrt{n}} \cdot f(s)\left[\sqrt{\frac{n}{n-1}} - \frac{a^2 f(s)}{2m_W^2\sqrt{n(n-1)}f(\sigma)}\right].$$

To complete the proof, we show that

$$\sqrt{\frac{n}{n-1}} - \frac{a^2 f(s)}{2m_W^2\sqrt{n(n-1)}\cdot f(\sigma)} \geqq 1,$$

or, equivalently, that

$$\frac{f(\sigma)}{f(s)} \geqq \frac{a^2}{m_W^2} \cdot \frac{\sqrt{n}+\sqrt{n-1}}{2\sqrt{n}}.$$

It is sufficient to prove

$$\frac{f(\sigma)}{f(s)} \geqq \frac{a^2}{m_W^2}. \tag{5.8}$$

Note that on account of Lemma 5.2 and (5.7), σ is an endpoint of the interval Δ. Thus, with the notation $\bar{r} \triangleq \min(r, 1-r)$, one sees that

$$\bar{s} - d \leqq \bar{\sigma} \leqq \bar{s},$$

and therefore, by the symmetry and the concavity of $f(s)$,

$$\frac{f(\sigma)}{f(s)} = \frac{f(\bar{\sigma})}{f(\bar{s})} \geqq \frac{\bar{\sigma}}{\bar{s}} \geqq 1 - \frac{d}{s}.$$

Thus, using (5.5) and Corollary 5.2,

$$\frac{f(\sigma)}{f(s)} \geqq 1 - \frac{a}{m_W\sqrt{n}}\frac{f(\bar{s})}{\bar{s}} \geqq 1 - \frac{a}{m_W\sqrt{n}}K_0\sqrt{-\ln\bar{s}}.$$

Hence, substituting a and using the fact that $\bar{s} \geqq m_W^n$, we get

$$\frac{f(\sigma)}{f(s)} \geqq 1 - \frac{KK_0}{\sqrt{-n\ln m_W}}\sqrt{-\ln m_W^n} = 1 - KK_0.$$

Choosing K satisfying (5.1) and $1 - KK_0 \geqq K^2/\ln 2$, (5.8) will follow since

$$\frac{a}{m_W K} = \frac{1}{\sqrt{\ln\frac{1}{m_W}}} < \frac{1}{\sqrt{\ln 2}}.$$ □

For our purpose, the importance of this theorem lies in the following corollary establishing a lower bound on the probability of the l-neighborhood of a set in terms of its own probability.

COROLLARY 5.3 For every $n, l, \mathsf{B} \subset \mathsf{Y}^n$ and $\mathbf{x} \in \mathsf{X}^n$

$$W^n(\Gamma^l \mathsf{B}|\mathbf{x}) \geqq \Phi\left[\Phi^{-1}(W^n(\mathsf{B}|\mathbf{x})) + \frac{l-1}{\sqrt{n}} a_W\right].$$

Proof We shall use the following two obvious relations giving rise to estimates of the probability of 1-neighborhoods by that of boundaries:

$$\Gamma\mathsf{B} - \mathsf{B} \supset \partial(\Gamma\mathsf{B}), \quad \Gamma\mathsf{B} - \mathsf{B} = \partial\bar{\mathsf{B}}.$$

Denoting $t_k \triangleq \Phi^{-1}(W^n(\Gamma^k \mathsf{B}|\mathbf{x}))$, one has

$$\Phi(t_{k+1}) - \Phi(t_k) = W^n(\Gamma^{k+1}\mathsf{B} - \Gamma^k \mathsf{B}|\mathbf{x}),$$

and hence the above relations yield by Theorem 5.3 that

$$\begin{aligned} \Phi(t_{k+1}) - \Phi(t_k) &\geqq \max\{W^n(\partial(\Gamma^{k+1}\mathsf{B})|\mathbf{x}), W^n(\partial(\overline{\Gamma^k \mathsf{B}})|\mathbf{x})\} \\ &\geqq \frac{a}{\sqrt{n}} \max\{\varphi(t_{k+1}), \varphi(t_k)\}. \end{aligned} \tag{5.9}$$

As φ is monotone on both $(-\infty, 0)$ and $(0, \infty)$, we have, unless $t_k < 0 < t_{k+1}$,

$$\max_{t_k \leqq u \leqq t_{k+1}} \varphi(u) = \max\{\varphi(t_k), \varphi(t_{k+1})\}.$$

This, substituted into (5.9), yields by Lagrange's theorem

$$t_{k+1} - t_k \geqq [\Phi(t_{k+1}) - \Phi(t_k)] \cdot \left(\max_{t_k \leqq u \leqq t_{k+1}} \varphi(u)\right)^{-1} \geqq \frac{a}{\sqrt{n}}$$

unless $t_k < 0 < t_{k+1}$. Hence

$$t_l - t_0 = \sum_{k=0}^{l-1} (t_{k+1} - t_k) \geqq \frac{l-1}{\sqrt{n}} a. \qquad \square$$

We conclude this series of estimates by a counterpart of Lemma 5.1. In fact, Lemma 5.1 and the next Lemma 5.4 are those results of this chapter that will be often used in the following chapters.

LEMMA 5.4 (*Blowing up*) To any finite sets X and Y and sequence $\varepsilon_n \to 0$ there exist a sequence of positive integers l_n with $l_n/n \to 0$ and a sequence $\eta_n \to 1$ such that for every stochastic matrix $W : \mathsf{X} \to \mathsf{Y}$ and every $n, \mathbf{x} \in \mathsf{X}^n, \mathsf{B} \subset \mathsf{Y}^n$

$$W^n(\mathsf{B}|\mathbf{x}) \geqq \exp(-n\varepsilon_n) \quad \text{implies} \quad W^n(\Gamma^{l_n}\mathsf{B}|\mathbf{x}) \geqq \eta_n. \tag{5.10}$$

○

Proof For a fixed W, the existence of sequences $\{l_n\}_{n=1}^{\infty}$ and $\{\eta_n\}_{n=1}^{\infty}$ satisfying (5.10) is an easy consequence of Corollary 5.3. The bound of Corollary 5.3 depends on W through m_W, as $a_W = K m_W / \sqrt{-\ln m_W}$. Thus, in order to get such sequences which are good for every W, for matrices with small m_W an approximation argument is needed.

Let X, Y and the sequence $\varepsilon_n \to 0$ be given. We first claim that for a suitable sequence of positive integers k_n with $k_n/n \to 0$ the following statement is true. Setting

$$\delta_n \triangleq \frac{k_n}{2n|\mathsf{X}||\mathsf{Y}|} \tag{5.11}$$

for every pair of stochastic matrices $W : \mathsf{X} \to \mathsf{Y}$, $\widetilde{W} : \mathsf{X} \to \mathsf{Y}$ such that

$$|W(b|a) - \widetilde{W}(b|a)| \leqq \delta_n \text{ for every } \quad a \in \mathsf{X}, b \in \mathsf{Y} \tag{5.12}$$

and for every $\mathbf{x} \in \mathsf{X}^n$, $\mathsf{B} \subset \mathsf{Y}^n$, we have

$$\widetilde{W}(\Gamma^{k_n}\mathsf{B}|\mathbf{x}) \geqq W^n(\mathsf{B}|\mathbf{x}) - \frac{1}{2}\exp(-n\varepsilon_n). \tag{5.13}$$

To prove this, note that inequality (5.12) implies the existence of a stochastic matrix $\widehat{W} : \mathsf{X} \to \mathsf{Y} \times \mathsf{Y}$ having W and $\widetilde{W}$ as marginals, i.e.,

$$\sum_{\tilde{b}\in\mathsf{Y}} \widehat{W}(b, \tilde{b}|a) = W(b|a), \quad \sum_{b\in\mathsf{Y}} \widehat{W}(b, \hat{b}|a) = \widetilde{W}(\tilde{b}|a),$$

such that

$$\sum_{b\in\mathsf{Y}} \widehat{W}(b, b|a) \geqq 1 - \delta_n|\mathsf{Y}| \quad \text{for every } a \in \mathsf{X}.$$

By the last property of $\widehat{W}$ we have for every $(\mathbf{y}, \tilde{\mathbf{y}}) \in \mathsf{T}^n_{[\widehat{W}]_{\delta_n}}(\mathbf{x})$

$$d_{\mathrm{H}}(\mathbf{y}, \tilde{\mathbf{y}}) = n - \sum_{a\in\mathsf{X}}\sum_{b\in\mathsf{Y}} N(a, b, b|\mathbf{x}, \mathbf{y}, \tilde{\mathbf{y}}) \leqq 2n\delta_n|\mathsf{X}||\mathsf{Y}| = k_n,$$

so that

$$(\mathsf{B} \times \mathsf{Y}^n) \cap \mathsf{T}^n_{[\widehat{W}]_{\delta_n}}(\mathbf{x}) \subset \mathsf{B} \times \Gamma^{k_n}\mathsf{B}.$$

Hence

$$\begin{aligned}\widetilde{W}^n(\Gamma^{k_n}\mathsf{B}|\mathbf{x}) &\geqq \widehat{W}^n((\mathsf{B} \times \mathsf{Y}^n) \cap \mathsf{T}^n_{[\widehat{W}]_{\delta_n}}(\mathbf{x})|\mathbf{x}) \\ &\geqq W^n(\mathsf{B}|\mathbf{x}) - (1 - \widehat{W}^n(\mathsf{T}^n_{[\widehat{W}]_{\delta_n}}(\mathbf{x})|\mathbf{x})).\end{aligned}$$

Here, by the Remark to Lemma 2.12,

$$1 - \widehat{W}^n(\mathsf{T}^n_{[\widehat{W}]_{\delta_n}}(\mathbf{x})|\mathbf{x}) \leqq 2|\mathsf{X}||\mathsf{Y}|^2 e^{-2n\delta_n^2}. \tag{5.14}$$

Thus, if δ_n in (5.11) satisfies $\delta_n^2/\varepsilon_n \to +\infty$, the claim (5.13) follows.

We shall henceforth assume that

$$\lim_{n\to\infty} \frac{\delta_n}{\sqrt{-\varepsilon_n \ln \delta_n}} = +\infty. \tag{5.15}$$

Consider an arbitrary $W : \mathsf{X} \to \mathsf{Y}$ and $\mathbf{x} \in \mathsf{X}^n$, $\mathsf{B} \subset \mathsf{Y}^n$ for which

$$W^n(\mathsf{B}|\mathbf{x}) \geqq \exp(-n\varepsilon_n).$$

Approximate the matrix W in the sense of (5.12) by a $\widetilde{W} : \mathsf{X} \to \mathsf{Y}$ satisfying

$$m_{\widetilde{W}} \geqq \frac{\delta_n}{|\mathsf{Y}|},$$

and apply Corollary 5.3 to the matrix $\widetilde{W}$ and the set $\tilde{\mathsf{B}} \triangleq \Gamma^{k_n}\mathsf{B}$. Since by (5.13) we have

$$\widetilde{W}^n(\Gamma^{k_n}\mathsf{B}|\mathbf{x}) \geqq \exp(-n\varepsilon_n - 1),$$

it follows, using also Lemma 5.2(i), that for every positive integer l

$$\begin{aligned}\widetilde{W}^n(\Gamma^{k_n+l}\mathsf{B}|\mathbf{x}) &\geqq \Phi\left[\Phi^{-1}(\exp\,(-n\varepsilon_n - 1)) + \frac{l-1}{\sqrt{n}}K\frac{m_{\widetilde{W}}}{\sqrt{-\ln m_{\widetilde{W}}}}\right] \\ &\geqq \Phi\left[-b\sqrt{n\varepsilon_n + 1} + \frac{l-1}{\sqrt{n}}K\frac{\delta_n}{|\mathsf{Y}|\sqrt{\ln\frac{|\mathsf{Y}|}{\delta_n}}}\right]. \qquad (5.16)\end{aligned}$$

Here K is the constant from Theorem 5.3 and $b > 0$ is another absolute constant. By (5.15), there exists a sequence of positive integers $\tilde{l}_n$ such that $\tilde{l}_n/n \to 0$ and

$$-b\sqrt{n\varepsilon_n + 1} + \frac{\tilde{l}_n - 1}{\sqrt{n}}K\frac{\delta_n}{|\mathsf{Y}|\sqrt{\ln\frac{|\mathsf{Y}|}{\delta_n}}} \to \infty.$$

Denoting by $\tilde{\eta}_n$ the lower bound resulting from (5.16) for $l \triangleq \tilde{l}_n$, we have $\tilde{\eta}_n \to 1$ as $n \to \infty$. Applying (5.13) once more (interchanging the roles of W and $\widetilde{W}$), the assertion (5.10) follows with $l_n \triangleq 2k_n + \tilde{l}_n$, $\eta_n \triangleq \tilde{\eta}_n - \frac{1}{2}\exp\,(-n\varepsilon_n)$. □

Problems

5.1. (*Isoperimetric problem*) The n-dimensional *Hamming space* is $\{0, 1\}^n$ with the Hamming metric. In this space, a *sphere* with center $\mathbf{y}$ and radius m is the set $\Gamma^m\{\mathbf{y}\}$.

(a) Check that Theorem 5.3 implies

$$|\partial\mathsf{B}| \geqq L\frac{2^n}{\sqrt{n}}f\left(\frac{|\mathsf{B}|}{2^n}\right)$$

for every $\mathsf{B} \subset \{0, 1\}^n$, where L is an absolute constant. Prove that this bound is tight (up to the constant factor L) if B is a sphere.

(b)* Show that if B is a sphere in the Hamming space then every set $\mathsf{B}' \subset \{0, 1\}^n$ with $|\mathsf{B}'| = |\mathsf{B}|$ has boundary of size $|\partial\mathsf{B}'| \geqq |\partial\mathsf{B}|$. (Harper (1966); for further geometric properties of the Hamming space, see Ahlswede and Katona (1977).)

5.2. Show that if $\mathsf{A}_n \subset \mathsf{Y}^n$, $\mathsf{B}_n \subset \mathsf{Y}^n$, $\lim_{n\to\infty} \frac{1}{n} d_{\mathrm{H}}(\mathsf{A}_n, \mathsf{B}_n) > 0$, then for every distribution P on Y

$$P^n(\mathsf{A}_n) \cdot P^n(\mathsf{B}_n) \to 0.$$

5.3. For a set $\mathsf{B} \subset \{0,1\}^n$ denote by $d(\mathsf{B})$ the largest integer d for which there exists a d-dimensional Hamming subspace of $\{0,1\}^n$ with the property that the projection of B onto this space is the whole $\{0,1\}^d$. Show that if

$$\varliminf_{n\to\infty} \frac{1}{n}\log|\mathsf{B}_n| > 0, \text{ then } \varliminf_{n\to\infty} \frac{d(\mathsf{B}_n)}{n} > 0.$$

Hint Show that for every $\mathsf{B} \subset \{0,1\}^n$ the relation

$$|\mathsf{B}| > \sum_{i=0}^{k} \binom{n}{i}$$

implies $d(\mathsf{B}) > k$.

Use induction. For $n = 1$ the statement is obvious. Suppose that it is true for every $n' < n$. Consider any set $\mathsf{B} \subset \{0,1\}^n$ for which $|\mathsf{B}| > \sum_{i=0}^{k}\binom{n}{i}$. Denote by B^+ the set of those $\mathbf{x} \in \{0,1\}^{n-1}$ for which both $1\mathbf{x}$ and $0\mathbf{x}$ are in B, and denote by B^- the set of those $\mathbf{x} \in \{0,1\}^{n-1}$ for which precisely one of $\mathrm{l}\mathbf{x}$ and $0\mathbf{x}$ is in B. Clearly,

$$|\mathsf{B}| = 2|\mathsf{B}^+| + |\mathsf{B}^-| = |\mathsf{B}^+| + |\mathsf{B}^+ \cup \mathsf{B}^-|.$$

Since

$$\sum_{i=0}^{k}\binom{n}{i} = \sum_{i=0}^{k-1}\binom{n-1}{i} + \sum_{i=0}^{k}\binom{n-1}{i},$$

one of the following inequalities has to hold:

$$|\mathsf{B}^+| > \sum_{i=0}^{k-1}\binom{n-1}{i} \quad \text{or} \quad |\mathsf{B}^+ \cup \mathsf{B}^-| > \sum_{i=0}^{k}\binom{n-1}{i}.$$

Apply the induction hypothesis to both sets and note that

$$d(\mathsf{B}) \geqq d(\mathsf{B}^+) + 1.$$

(Vapnik and Červonenkis (1971), Sauer (1972) and S. Shelah and M. Perles, cited in Shelah (1972). This property of the Hamming space has a variety of applications; in particular, it is crucial for statistical learning theory.)

5.4. (a) Show that the n-dimensional Hamming space can be partitioned into spheres of radius 1 iff $n+1$ is a power of 2. (Hamming (1950).)

Hint Necessity is obvious. For sufficiency, suppose $n + 1 = 2^r$, and let M be a matrix whose rows are all vectors in $\{0,1\}^r$ of which two or more components equal 1. Let C be the set of all $\mathbf{x} \in \{0,1\}^n$ whose last r bits $\mathbf{x}'' = x_{n-r+1}\dots x_n$ are related to the first $n-r$ bits $\mathbf{x}' = x_1 \dots x_{n-r}$ by the parity check equations $\mathbf{x}'' = \mathbf{x}'M$, in mod 2 arithmetic. Verify that to each $\mathbf{y} \in \{0,1\}^n$ there exists a unique $\mathbf{x} \in \mathsf{C}$ with $d_{\mathrm{H}}(\mathbf{x},\mathbf{y}) \leqq 1$.

(b) Let $g(n)$ denote the maximum cardinality of sets $\mathsf{B} \subset \{0,1\}^n$ with $\partial\mathsf{B} = \mathsf{B}$. Show that

$$\lim_{n\to\infty}\frac{g(n)}{2^n}=1.$$

Hint Verify that if a set $\mathsf{D}\subset\{0,1\}^n$ satisfies $\Gamma\mathsf{D}=\{0,1\}^n$ then $g(n)\geqq 2^n-|\mathsf{D}|$. Deduce from (a) the existence of such D of size less than $\delta 2^n$, for any $\delta>0$, if n is sufficiently large.

5.5. (*Sparse sets in Hamming space*) For a set $\mathsf{C}\subset\{0,1\}^n$ denote by $d_{\min}$ the minimum Hamming distance of distinct sequences in C.

(a) Show that always

$$d_{\min}\leqq\frac{n}{2}\cdot\frac{|\mathsf{C}|}{|\mathsf{C}|-1}.$$

(Proved by M. Plotkin in 1951; see Plotkin (1960).)

Hint For $\mathsf{C}=\{\mathbf{x}_1,\dots,\mathbf{x}_N\}$ with $\mathbf{x}_i=x_{i1}\dots x_{in}$,

$$N(N-1)d_{\min}\leqq\sum_{i=1}^{N}\sum_{j=1}^{N}d_{\mathrm{H}}(\mathbf{x}_i,\mathbf{x}_j)=\sum_{l=1}^{n}|\{(i,j):\ x_{il}\neq x_{jl}\}|.$$

Denoting $\mathbf{y}_l\triangleq x_{1l}\dots x_{Nl}\in\{0,1\}^N$, the lth term of the preceding sum equals

$$|\{(i,j):\ x_{il}=1,\ x_{jl}=0\}|+|\{(i,j):\ x_{il}=0,\ x_{jl}=1\}|$$
$$=2N^2P_{\mathbf{y}_l}(1)P_{\mathbf{y}_l}(0)\leqq\frac{N^2}{2}.$$

(b) Show that for each n and $d\leqq n/2$ there exists a set $\mathsf{C}\subset\{0,1\}^n$ with

$$\frac{1}{n}\log|\mathsf{C}|\geqq 1-h\left(\frac{d}{n}\right),\qquad d_{\min}>d.$$

(Gilbert (1952).)

Hint Construct C by the greedy algorithm, using Problem 2.8.

Remark. The results in Problems 5.4 (a) and 5.5 were among the earliest ones of coding theory. It is the longest standing open problem of coding theory whether Gilbert's bound is asymptotically tight when $n\to\infty$ and d/n approaches a constant in $(0,1/2)$. In that context, the best available upper bound is the so-called JPL bound of McEliece *et al.* (1977).

Story of the results

This chapter is based on Ahlswede, Gács and Körner (1976). The key result, Theorem 5.3, is their generalization of a lemma of Margulis (1974). The uniform version of their blowing up lemma is new.

Addition. The blowing up lemma is an instance of the *measure concentration phenomenon*, whose first substantial applications may have been those in this book. For an inherently information-theoretic proof of this lemma, and its extensions, see Marton (1986). Measure concentration is now a major reasearch area in probability theory and beyond; see Talagrand (1995) and Ledoux (2001).

Part II

Two-terminal systems

6 The noisy channel coding problem

Let X and Y be finite sets. A (discrete) *channel* with *input set* X and *output set* Y is defined as a stochastic matrix $W : \mathsf{X} \to \mathsf{Y}$. The entry $W(y|x)$ is interpreted as the probability that if x is the channel input then the output will be y. We shall say that two RVs X and Y are *connected by the channel* W if $P_{Y|X} = W$, more exactly if $P_{Y|X}(y|x) = W(y|x)$ whenever $P_X(x) > 0$.

A *code* for channels with input set X and output set Y is a pair of mappings (f, φ), where f maps some finite set $\mathsf{M} = \mathsf{M}_f$ into X and φ maps Y into a finite set M'. The elements of M are called *messages*, the mapping f is the *encoder* and φ is the *decoder*. The images of the messages under f are called *codewords*. One reason for allowing the range of φ to differ from M is mathematical convenience; more substantial reasons will be apparent in Chapter 10. However, unless stated otherwise, we shall always assume $\mathsf{M}' \supset \mathsf{M}$.

Given a channel $W : \mathsf{X} \to \mathsf{Y}$, a *code for channel* W is any code (f, φ) as above. Such a code and the channel W define a new channel $T : \mathsf{M} \to \mathsf{M}'$, where

$$T(m'|m) \triangleq W(\varphi^{-1}(m')|f(m)) \qquad (m \in \mathsf{M},\ m' \in \mathsf{M}') \tag{6.1}$$

is the probability that using the code (f, φ) over the channel W the decoding results in m', provided that m was actually transmitted. The probability of erroneous transmission of message m is

$$e_m = e_m(W, f, \varphi) \triangleq 1 - T(m|m).$$

The *maximum probability of error* of the code (f, φ) is → 6.1

$$e = e(W, f, \varphi) \triangleq \max_{m \in \mathsf{M}} e_m.$$

If a probability distribution is given on the message set M, the performance of the code can be evaluated by the corresponding overall probability of error. In particular, the overall probability of error corresponding to equiprobable messages is called the *average probability of error* of the code (f, φ): → 6.2

$$\bar{e} = \bar{e}(W, f, \varphi) \triangleq \frac{1}{|\mathsf{M}|} \sum_{m \in \mathsf{M}} e_m.$$

The channel coding problem consists in making the message set M as large as possible while keeping the maximum probability of error e as low as possible. We shall solve this problem in an asymptotic sense, for channels $W^n : \mathsf{X}^n \to \mathsf{Y}^n$.

Intuitively speaking, the "channel" of the communication engineer operates by successively transmitting symbols, one at each time unit, say. Its operation through n time units or "n uses of the channel" can be modeled by a stochastic matrix $W_n : \mathsf{X}^n \to \mathsf{Y}^n$, where $W_n(\mathbf{y}|\mathbf{x})$ is the probability that the input $\mathbf{x} = x_1 \dots x_n$ results in the output $\mathbf{y} = y_1 \dots y_n$. Thus the n-length operation of the physical channel is described by the mathematical channel W_n, and the mathematical model of a physical channel is a sequence $\{W_n : \mathsf{X}^n \to \mathsf{Y}^n\}_{n=1}^{\infty}$ of channels in the mathematical sense. Of course, not every sequence of stochastic matrices $W_n : \mathsf{X}^n \to \mathsf{Y}^n$ is a proper model of a physical channel.

With some abuse of terminology, justified by the intuitive background, a sequence $\{W_n : \mathsf{X}^n \to \mathsf{Y}^n\}_{n=1}^{\infty}$ will also be called a (discrete) *channel*. (Later in this book, the term channel will be used also for some other families of individual channels.) The finite sets X and Y are called the *input alphabet* and the *output alphabet* of the channel $\{W_n : \mathsf{X}^n \to \mathsf{Y}^n\}_{n=1}^{\infty}$. Within this framework, noiseless channels are characterized by the condition that W_n is the identity matrix, for every n.

Given $\{W_n : \mathsf{X}^n \to \mathsf{Y}^n\}_{n=1}^{\infty}$, one is interested in the trade-off between the size of the message set and the probability of error for the channel W_n, as $n \to \infty$. In this book the discussion will be centered around the special case $W_n = W^n$, where

$$W^n(\mathbf{y}|\mathbf{x}) \triangleq \prod_{i=1}^{n} W(y_i|x_i).$$

A sequence of channels $\{W^n : \mathsf{X}^n \to \mathsf{Y}^n\}_{n=1}^{\infty}$ is called a *discrete memoryless channel* (DMC) with transition probability matrix W. This DMC is denoted by $\{W : \mathsf{X} \to \mathsf{Y}\}$ or simply $\{W\}$.

An *n-length block code* for a channel $\{W_n : \mathsf{X}^n \to \mathsf{Y}^n\}_{n=1}^{\infty}$ is a code (f, φ) for the channel W_n. The *rate* of such a code is $\frac{1}{n} \log |\mathsf{M}_f|$. Note that for an error-free transmission of messages $m \in \mathsf{M}_f$ over a binary noiseless channel $\log |\mathsf{M}_f|$ binary digits are needed. Thus transmitting messages over the given channel by means of the code (f, φ), one channel use corresponds to $\frac{1}{n} \log |\mathsf{M}_f|$ uses of the "standard" binary noiseless channel. Of course, such a comparison makes sense only if the error probability of the code is reasonably small.

An n-length block code for the channel $\{W_n\colon \mathsf{X}^n \to \mathsf{Y}^n\}_{n=1}^{\infty}$ with maximum probability of error $e(W_n, f, \varphi) \leqq \varepsilon$ will be called an *(n, ε)-code*.

DEFINITION 6.1 For $0 \leqq \varepsilon < 1$, a non-negative number R is an *ε-achievable rate* for the channel $\{W_n : \mathsf{X}^n \to \mathsf{Y}^n\}_{n=1}^{\infty}$ if for every $\delta > 0$ and every sufficiently large n there exist (n, ε)-codes of rate exceeding $R - \delta$. A number R is an *achievable rate* if it is
→ 6.3 ε-achievable for all $0 < \varepsilon < 1$. The supremum of ε-achievable resp. achievable rates is called the *ε-capacity* C_ε resp. the *capacity* C of the channel. ○

REMARK $\lim_{\varepsilon \to 0} C_\varepsilon = C$. ○ → 6.4

In Definition 6.1 each supremum is actually a maximum, thus the (ε-) capacity equals the largest (ε-) achievable rate. Some authors admit only codes of rates not less than R in the definition of (ε-) achievable rates. This does not affect their supremum but causes this supremum to be not necessarily an achievable rate. An equivalent version of Definition 6.1 is that R is an ε-achievable or achievable rate if there exist (n, ε)-codes, or (n, ε_n)-codes with $\varepsilon_n \to 0$, whose rates approach R as $n \to \infty$.

In this chapter we shall determine the capacity of a DMC, showing, in particular, that $C > 0$ except for trivial cases. This fact, i.e., the possibility that messages may be transmitted over a noisy channel at a fixed positive rate with as small probability of error as desired, is by no means obvious at first sight; rather, it has been considered a major result of information theory. It will also turn out that for a DMC $C_\varepsilon = C$ for every $0 < \varepsilon < 1$.

More generally, by using properties of typical sequences established in Chapter 2, we shall obtain asymptotic results on the maximum rate of (n, ε)-codes with codewords belonging to some prescribed sets $\mathsf{A}_n \subset \mathsf{X}^n$. This maximum rate will turn out to be asymptotically independent of ε and related to the minimum size of sets $\mathsf{B} \subset \mathsf{Y}^n$, which have "large" $W^n(\cdot|\mathbf{x})$-probability for every $\mathbf{x} \in \mathsf{A}_n$.

DEFINITION 6.2 A set $\mathsf{B} \subset \mathsf{Y}$ is an *η-image* ($0 < \eta \leqq 1$) of a set $\mathsf{A} \subset \mathsf{X}$ over a channel $W : \mathsf{X} \to \mathsf{Y}$ if $W(\mathsf{B}|x) \geqq \eta$ for every $x \in \mathsf{A}$. The minimum cardinality of η-images of A will be denoted by $g_W(\mathsf{A}, \eta)$. ○

If (f, φ) and $(\tilde{f}, \tilde{\varphi})$ are codes for a channel $W : \mathsf{X} \to \mathsf{Y}$, we shall say that $(\tilde{f}, \tilde{\varphi})$ is an *extension* of (f, φ), or (f, φ) is a *subcode* of $(\tilde{f}, \tilde{\varphi})$, if $\mathsf{M}_f \subset \mathsf{M}_{\tilde{f}}$ and $f(m) = \tilde{f}(m)$ for $m \in \mathsf{M}_f$. Note that no assumption is made about the relationship of φ and $\tilde{\varphi}$.

LEMMA 6.3 (*Maximal code*) For every DMC $\{W: \mathsf{X} \to \mathsf{Y}\}$, distribution P on X, set $\mathsf{A} \subset \mathsf{T}^n_{[P]}$ and $0 < \tau < \varepsilon < 1$ there exists an (n, ε)-code (f, φ) such that

(i) $f(m) \in \mathsf{A}$,
(ii) $\varphi^{-1}(m) \subset \mathsf{T}^n_{[W]}(f(m))$

for every $m \in \mathsf{M}_f$ and the rate of the code satisfies

$$\frac{1}{n} \log |\mathsf{M}_f| \geqq \frac{1}{n} \log g_{W^n}(\mathsf{A}, \varepsilon - \tau) - H(W|P) - \tau, \tag{6.2}$$

provided that[1] $n \geqq n_0(|\mathsf{X}|, |\mathsf{Y}|, \tau)$. ○

COROLLARY 6.3 For every DMC $\{W : \mathsf{X} \to \mathsf{Y}\}$, distribution P on X and τ, ε in (0,1) there exists an (n, ε)-code (f, φ) such that for every $m \in \mathsf{M}_f$

[1] Actually, n_0 depends also on the sequences $\{\delta_n\}$ occurring in the definition of $\mathsf{T}^n_{[P]}$ and $\mathsf{T}^n_{[W]}(\mathbf{x})$; following the delta-convention (Convention 2.11), this dependence is suppressed.

(i) $f(m) \in \mathsf{T}^n_{[P]}$,
(ii) $\varphi^{-1}(m) \subset \mathsf{T}^n_{[W]}(f(m))$,

and the rate is at least $I(P, W) - 2\tau$, provided that $n \geqq n_0(|\mathsf{X}|, |\mathsf{Y}|, \tau, \varepsilon)$. ○

REMARKS We shall actually prove that (6.2) holds for every (n, ε)-code satisfying (i), (ii) of the lemma that has no extension with the same properties. The name "maximal code lemma" refers to this. Condition (ii) is a technical one; the main assertion of the lemma is the existence of (n, ε)-codes with codewords in A and of a rate satisfying (6.2). ○

Proof Let (f, φ) be an (n, ε)-code satisfying (i), (ii) that has no extension with the same properties, and set

$$\mathsf{B} \triangleq \bigcup_{m \in \mathsf{M}} \varphi^{-1}(m), \quad \mathsf{M} \triangleq \mathsf{M}_f;$$

if no (n, ε)-code satisfies (i), (ii), we set $\mathsf{M} = \mathsf{B} \triangleq \emptyset$. Note first that if we had

$$W^n(\mathsf{T}_{[W]}(\tilde{\mathbf{x}}) - \mathsf{B}|\tilde{\mathbf{x}}) \geqq 1 - \varepsilon$$

for some $\tilde{\mathbf{x}} \in \mathsf{A}$, then the code (f, φ) would have an extension satisfying (i), (ii), contradicting our assumption. This can be seen by adding a new message $\tilde{m}$ to the message set and setting $f(\tilde{m}) \triangleq \tilde{\mathbf{x}}$. Now, if for $\mathbf{y} \in \mathsf{T}_{[W]}(\tilde{\mathbf{x}}) - \mathsf{B}$ we modify the decoder accordingly, requiring $\varphi(\mathbf{y}) \triangleq \tilde{m}$, we obtain the promised extension. This contradiction proves that

$$W^n(\mathsf{T}_{[W]}(\mathbf{x}) - \mathsf{B}|\mathbf{x}) < 1 - \varepsilon \quad \text{for every} \quad \mathbf{x} \in \mathsf{A}.$$

Since for sufficiently large n (depending on $|\mathsf{X}|, |\mathsf{Y}|, \tau$)

$$W^n(\mathsf{T}_{[W]}(\mathbf{x})|\mathbf{x}) \geqq 1 - \tau \quad \text{for every} \quad \mathbf{x} \in \mathsf{X}^n$$

(cf. Lemma 2.12), it follows that

$$W^n(\mathsf{B}|\mathbf{x}) > \varepsilon - \tau \quad \text{for every} \quad \mathbf{x} \in \mathsf{A}. \tag{6.3}$$

Thus, by Definition 6.2,

$$|\mathsf{B}| \geqq g_{W^n}(\mathsf{A}, \varepsilon - \tau).$$

On the other hand, by (ii) and Lemma 2.13, for sufficiently large n (depending on $|\mathsf{X}|, |\mathsf{Y}|, \tau$)

$$|\mathsf{B}| = \sum_{m \in \mathsf{M}} |\varphi^{-1}(m)| \leqq \sum_{m \in \mathsf{M}} |\mathsf{T}_{[W]}(f(m))| \leqq |\mathsf{M}| \exp[n(H(W|P) + \tau)].$$

Comparing this with the previous inequality we obtain

$$|\mathsf{M}| \geqq g_{W^n}(\mathsf{A}, \varepsilon - \tau) \exp[-n(H(W|P) + \tau)]$$

proving (6.2).

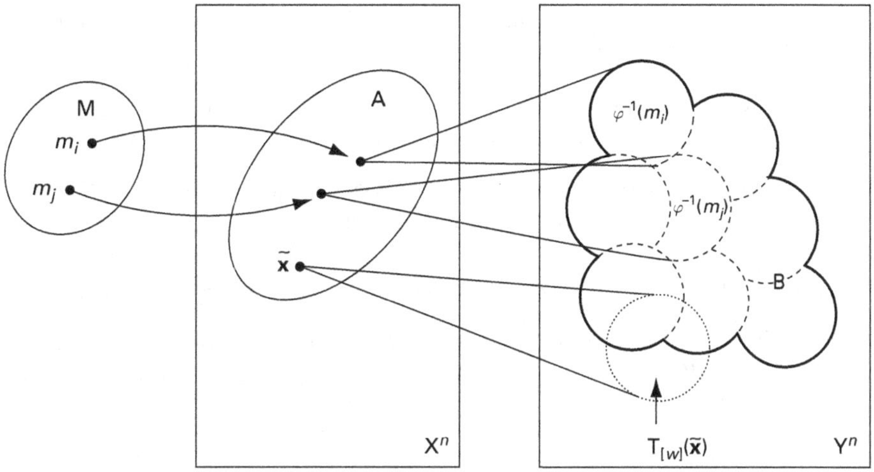

Figure 6.1 Maximal code.

The corollary is immediate as, by Lemma 2.14,

$$\frac{1}{n}\log g_{W^n}(\mathsf{T}_{[P]}, \varepsilon - \tau) \geqq H(Y) - \tau$$

if n is sufficiently large (depending on $|\mathsf{X}|$, $|\mathsf{Y}|$, τ, ε). □

A counterpart of the existence result Lemma 6.3 is given by

LEMMA 6.4 For arbitrary $\varepsilon, \tau \in (0, 1)$, if (f, φ) is an (n, ε)-code for the DMC $\{W : \mathsf{X} \to \mathsf{Y}\}$ such that all the codewords belong to the set $\mathsf{A} \subset \mathsf{T}^n_{[P]}$, then

$$\frac{1}{n}\log|\mathsf{M}_f| < \frac{1}{n}\log g_{W^n}(\mathsf{A}, \varepsilon + \tau) - H(W|P) + \tau,$$

whenever $n \geqq n_0$ ($|\mathsf{X}|$, $|\mathsf{Y}|$, τ). ○

COROLLARY 6.4 For arbitrary $\varepsilon, \tau \in (0, 1)$, if (f, φ) is an (n, ε)-code for the DMC $\{W : \mathsf{X} \to \mathsf{Y}\}$ such that all the codewords belong to $\mathsf{T}^n_{[P]}$, then

$$\frac{1}{n}\log|\mathsf{M}_f| < I(P, W) + 2\tau,$$

whenever $n \geqq n_0$ ($|\mathsf{X}|$, $|\mathsf{Y}|$, τ, ε). ○

Proof Let (f, φ) be any (n, ε)-code for $\{W\}$ such that $\mathsf{A}' \triangleq \{f(m) : m \in \mathsf{M}_f\} \subset \mathsf{A}$. Further, let $\mathsf{B} \subset \mathsf{Y}^n$ be an $(\varepsilon + \tau)$-image of A' for which $|\mathsf{B}| = g_{W^n}(\mathsf{A}', \varepsilon + \tau)$. Then, as $e_m = 1 - W^n(\varphi^{-1}(m)|f(m)) \leqq \varepsilon$, we have $W^n(\mathsf{B} \cap \varphi^{-1}(m)|f(m)) \geqq \tau$ for every $m \in \mathsf{M}_f$. Hence, applying Lemma 2.14, it follows for n large enough (depending on $|\mathsf{X}|$, $|\mathsf{Y}|$ and τ) that

$$|\mathsf{B} \cap \varphi^{-1}(m)| \geqq \exp[n(H(W|P) - \tau)],$$

and, since the sets $\varphi^{-1}(m)$ are disjoint,

$$g_{W^n}(\mathsf{A}', \varepsilon + \tau) = |\mathsf{B}| \geqq \sum_{m \in \mathsf{M}_f} |\mathsf{B} \cap \varphi^{-1}(m)| \geqq |\mathsf{M}_f| \exp[n(H(W|P) - \tau)].$$

As $\mathsf{A}' \subset \mathsf{A}$ implies $g_{W^n}(\mathsf{A}', \varepsilon + \tau) \leqq g_{W^n}(\mathsf{A}, \varepsilon + \tau)$, the lemma follows.

In order to prove the corollary, note that by Lemma 2.10 $\mathsf{T}^n_{[PW]_{\delta''_n}}$ is an $(\varepsilon + \tau)$-image of every subset of $\mathsf{T}^n_{[P]_{\delta_n}}$, provided that $\delta''_n = 2\delta_n|\mathsf{X}|$. Consequently, the corollary follows from the first statement of Lemma 2.13. □

The capacity of a DMC can be determined by combining Corollaries 6.3 and 6.4.

THEOREM 6.5 (*Noisy channel coding theorem*) For any $0 < \varepsilon < 1$, the ε-capacity of the DMC $\{W\}$ is

$$C_\varepsilon = C = \max_P I(P, W) = \max I(X \wedge Y),$$

where the last maximum refers to RVs connected by channel W. ○

Proof Since $I(P, W)$ is a continuous function of the distribution P, the maximum is, in fact, attained.

By Corollary 6.3, for every $0 < \varepsilon < 1$ and every distribution P on X, the mutual information $I(P, W)$ is an ε-achievable rate for the DMC $\{W\}$. Hence

$$C_\varepsilon \geqq \max_P I(P, W).$$

In order to prove the opposite inequality, consider an arbitrary (n, ε)-code for the DMC $\{W\}$ and denote by M the corresponding message set M_f. For any type P of sequences in X^n, let $\mathsf{M}(P)$ be the set of those messages which are encoded into sequences of type P, i.e.,

$$\mathsf{M}(P) \triangleq \{m : f(m) \in \mathsf{T}_P\}.$$

By Corollary 6.4, for every $\tau > 0$, every type P and sufficiently large n,

$$|\mathsf{M}(P)| \leqq \exp[n(I(P, W) + 2\tau)]. \tag{6.4}$$

Thus, by the type counting lemma,

$$\begin{aligned}|\mathsf{M}| = \sum_P |\mathsf{M}(P)| &\leqq (n+1)^{|\mathsf{X}|} \max_P |\mathsf{M}(P)| \\ &\leqq (n+1)^{|\mathsf{X}|} \exp[n(\max_P I(P, W) + 2\tau)],\end{aligned}$$

where P runs over the types of sequences in X^n. Clearly, we further increase the last bound when taking the maximum for all PDs on X. Hence

$$C_\varepsilon \leqq \max_P I(P, W). \tag{6.5}$$

□

COMMENTS The mathematical significance of Theorem 6.5 is that it makes possible the computation of channel capacity to any desired degree of accuracy, at least in principle (the question of actual computation will be considered in Chapter 8). The fact that capacity equals the maximum mutual information of RVs connected by the channel reinforces the intuitive interpretation of mutual information. It should be pointed out that, although the proof of Theorem 6.5 suggests that (n, ε)-codes with rate close to capacity can be constructed in an arbitrary manner by successive extension, such a construction would involve an insurmountable amount of computation, even for moderate block lengths. Thus, Theorem 6.5 should be considered an existence result. Its practical significance consists in clarifying the theoretical capabilities of communication systems. This provides an objective basis for evaluating the efficiency of actual codes by comparing them with the theoretical optimum. Similar comments apply to all the other coding theorems treated in the following. ○

Theorem 6.5 was obtained as a consequence of the asymptotically coinciding lower and upper bounds of Corollaries 6.3 and 6.4. It is less obvious but easily follows from the results of Chapter 5 that the lower and upper bounds of Lemmas 6.3 and 6.4 also coincide asymptotically. The proof of this fact relies on Lemma 6.6.

LEMMA 6.6 For every $\tau, \varepsilon', \varepsilon'' \in (0, 1)$, DMC $\{W : \mathsf{X} \to \mathsf{Y}\}$ and set $\mathsf{A} \subset \mathsf{X}^n$ we have

$$\left| \frac{1}{n} \log g_{W^n}(\mathsf{A}, \varepsilon') - \frac{1}{n} \log g_{W^n}(\mathsf{A}, \varepsilon'') \right| < \tau$$

whenever $n \geqq n_0(|\mathsf{X}|, |\mathsf{Y}|, \tau, \varepsilon', \varepsilon'')$. ○

Proof Let $\varepsilon' > \varepsilon''$, say. Then clearly

$$\frac{1}{n} \log g_{W^n}(\mathsf{A}, \varepsilon') - \frac{1}{n} \log g_{W^n}(\mathsf{A}, \varepsilon'') \geqq 0.$$

Further, by the blowing up Lemma 5.4 and Lemma 5.1, there exists a sequence $\{l_n\}$ such that for sufficiently large n (depending on ε', ε'', $|\mathsf{X}|$ and $|\mathsf{Y}|$)

$$\frac{1}{n} \log |\Gamma^{l_n} \mathsf{B}| - \frac{1}{n} \log |\mathsf{B}| < \tau \quad \text{for every} \quad \mathsf{B} \subset \mathsf{Y}^n \tag{6.6}$$

and

$$W^n(\Gamma^{l_n} \mathsf{B} | \mathbf{x}) \geqq \varepsilon' \quad \text{if} \quad W^n(\mathsf{B} | \mathbf{x}) \geqq \varepsilon''.$$

Now, if B is an ε''-image of A with $|\mathsf{B}| = g_{W^n}(\mathsf{A}, \varepsilon'')$, the preceding relation means that $\Gamma^{l_n} \mathsf{B}$ is an ε'-image of A. Thus

$$|\Gamma^{l_n} \mathsf{B}| \geqq g_{W^n}(\mathsf{A}, \varepsilon').$$

This and (6.6) complete the proof. □

DEFINITION 6.7 The *ε-capacity* $C(\mathsf{A}, \varepsilon) = C_W(\mathsf{A}, \varepsilon)$ of a set $\mathsf{A} \subset \mathsf{X}^n$ is the maximum rate of those (n, ε)-codes for the DMC $\{W\}$, all codewords of which belong to A. ○

LEMMA 6.8 For any $\tau, \varepsilon', \varepsilon'' \in (0, 1)$, DMC $\{W : \mathsf{X} \to \mathsf{Y}\}$, distribution P on X and set $\mathsf{A} \subset \mathsf{T}^n_{[P]}$ we have

$$|C_W(\mathsf{A}, \varepsilon') - \frac{1}{n} \log g_{W^n}(\mathsf{A}, \varepsilon'') + H(W|P)| < \tau,$$

provided that $n \geqq n_0(|\mathsf{X}|, |\mathsf{Y}|, \tau, \varepsilon', \varepsilon'')$. ○

Proof The statement is a combination of Lemmas 6.3, 6.4 and 6.6. □

THEOREM 6.9 For any $\tau, \varepsilon', \varepsilon'' \in (0, 1)$, every DMC $\{W : \mathsf{X} \to \mathsf{Y}\}$ and set $\mathsf{A} \subset \mathsf{X}^n$

$$|C(\mathsf{A}, \varepsilon') - C(\mathsf{A}, \varepsilon'')| < \tau,$$

whenever $n \geqq n_0(|\mathsf{X}|, |\mathsf{Y}|, \tau, \varepsilon', \varepsilon'')$. ○

COROLLARY 6.9 If $\varepsilon' > \varepsilon''$ and $n \geqq n_0(|\mathsf{X}|, |\mathsf{Y}|, \tau, \varepsilon', \varepsilon'')$ then every (n, ε')- code (f', φ') for the DMC $\{W\}$ has a subcode (f'', φ'') with maximum probability of error less than ε'' and with rate $(1/n) \log |\mathsf{M}_{f''}| > (1/n) \log |\mathsf{M}_{f'}| - \tau$. ○

Proof Suppose $\varepsilon' > \varepsilon''$. Then $C(\mathsf{A}, \varepsilon') - C(\mathsf{A}, \varepsilon'') \geqq 0$. To prove the other inequality, note that the sets

$$\mathsf{A}(P) \triangleq \mathsf{A} \cap \mathsf{T}_P$$

partition A as P runs over the types of sequences in X^n. By the type counting lemma, for sufficiently large n (depending on τ) there exists a type P' such that $\mathsf{A}' \triangleq \mathsf{A}(P')$ yields

$$C(\mathsf{A}', \varepsilon') \geqq C(\mathsf{A}, \varepsilon') - \frac{\tau}{2}. \tag{6.7}$$

By Lemma 6.8, if n is large enough (depending on $|\mathsf{X}|, |\mathsf{Y}|, \varepsilon', \varepsilon'', \tau$), then both for $\varepsilon = \varepsilon'$ and $\varepsilon = \varepsilon''$

$$\left|C(\mathsf{A}', \varepsilon) - \frac{1}{n} \log g_{W^n}(\mathsf{A}', \varepsilon') + H(W|P')\right| < \frac{\tau}{4}. \tag{6.8}$$

Equations (6.7) and (6.8) imply

$$C(\mathsf{A}, \varepsilon'') \geqq C(\mathsf{A}', \varepsilon'') \geqq C(\mathsf{A}', \varepsilon') - \frac{\tau}{2} \geqq C(\mathsf{A}, \varepsilon') - \tau.$$

This proves Theorem 6.9. The corollary follows by taking for A the codeword set of f'. □

Finally, we determine the limiting ε-capacity in a particular case. More than an interesting example, the next result will be useful in later chapters.

THEOREM 6.10 For every $\eta, \varepsilon, \tau \in (0, 1)$ and every DMC $\{W : \mathsf{X} \to \mathsf{Y}\}$

$$P^n(\mathsf{A}) \geqq \eta \quad \text{implies} \quad C_W(\mathsf{A}, \varepsilon) \geqq I(P, W) - 3\tau,$$

whenever $n \geqq n_0(|\mathsf{X}|, |\mathsf{Y}|, \tau, \varepsilon, \eta)$. ○

Proof Consider the set $\tilde{\mathsf{A}} \triangleq \mathsf{A} \cap \mathsf{T}^n_{[P]}$. By Lemma 2.12, for $n \geqq n_1$ we have $P^n(\tilde{\mathsf{A}}) \geqq \eta/2$. In virtue of Lemma 6.3 it suffices to prove for $n \geqq n_2$ that

$$\frac{1}{n} \log g_{W^n}(\tilde{\mathsf{A}}, \varepsilon - \tau) \geqq H(Y) - 2\tau.$$

Let $\mathsf{B} \subset \mathsf{Y}^n$ be any $(\varepsilon - \tau)$-image of $\tilde{\mathsf{A}}$. Then $(PW)^n(\mathsf{B}) \geqq (\eta/2)(\varepsilon - \tau)$, by definition, and hence the above inequality follows by Lemma 2.14. □

In order to model certain engineering devices appropriately it is necessary to incorporate into our model the possibility that not every combination of the elements of the input alphabet of the channel can be used for transmission. The physical restriction on possible channel input sequences can often be given in terms of a non-negative function $c(x)$ defined on the input alphabet X. This function is extended to X^n by

$$c(\mathbf{x}) \triangleq \frac{1}{n} \sum_{i=1}^{n} c(x_i) \quad \text{for} \quad \mathbf{x} = x_1 x_2 \ldots x_n,$$

and the *constraint upon the codewords* is $c(f(m)) \leqq \Gamma$ for some number Γ.

Definition 6.1 can be modified incorporating this additional constraint on possible codes so that one arrives at the notion of the *ε-capacity* resp. *capacity of a channel under input constraint* (c, Γ), denoted by $C_\varepsilon(\Gamma)$ resp. $C(\Gamma)$. These capacities are defined if

$$\Gamma \geqq \Gamma_0 \triangleq \min_{x \in \mathsf{X}} c(x).$$

For a DMC $\{W\}$, setting

$$\mathsf{A}_n(\Gamma) \triangleq \{\mathbf{x} : \mathbf{x} \in \mathsf{X}^n,\ c(\mathbf{x}) \leqq \Gamma\},$$

we have

$$C_\varepsilon(\Gamma) = \varliminf_{n \to \infty} C_W(\mathsf{A}_n(\Gamma), \varepsilon), \quad C(\Gamma) = \lim_{\varepsilon \to 0} C_\varepsilon(\Gamma) \quad (\Gamma \geqq \Gamma_0).$$

Clearly, $C(\Gamma) \leqq C$, and for sufficiently large Γ the equality holds.

For distributions P on X write

$$c(P) \triangleq \sum_{x \in \mathsf{X}} P(x) c(x).$$

The results obtained hitherto easily imply the following.

THEOREM 6.11 For any $\varepsilon \in (0, 1)$ and $\Gamma \geqq \Gamma_0$ the ε-capacity of the DMC $\{W\}$ under the input constraint (c, Γ) is

$$C_\varepsilon(\Gamma) = C(\Gamma) = \max_{P : c(P) \leqq \Gamma} I(P, W) = \max I(X \wedge Y),$$

where the last maximum refers to RVs connected by the channel W and such that $Ec(X) \leqq \Gamma$. The capacity $C(\Gamma)$ is a non-decreasing concave function of $\Gamma \geqq \Gamma_0$. ○

Proof Write, for a moment,

$$\tilde{C}(\Gamma) \triangleq \max_{P:c(P) \leqq \Gamma} I(P, W) \qquad (\Gamma \geqq \Gamma_0).$$

The maximum is attained as $I(P, W)$ is a continuous function of P. Clearly, $\tilde{C}(\Gamma)$ is a non-decreasing function of Γ; to check its concavity, suppose that P_1 resp. P_2 maximize $I(P, W)$ under the constraint $c(P) \leqq \Gamma_1$ resp. $c(P) \leqq \Gamma_2$. Then, by the concavity of $I(P, W)$ as a function of P (Lemma 3.5), we have for arbitrary $\alpha \in (0, 1)$

$$\alpha\tilde{C}(\Gamma_1) + (1-\alpha)\tilde{C}(\Gamma_2) = \alpha I(P_1, W) + (1-\alpha) I(P_2, W) \leqq I(P, W), \tag{6.9}$$

where $P \triangleq \alpha P_1 + (1-\alpha)P_2$, and consequently $c(P) = \alpha c(P_1) + (1-\alpha)c(P_2) \leqq \alpha\Gamma_1 + (1-\alpha)\Gamma_2$. Thus (6.9) establishes the claim

$$\alpha\tilde{C}(\Gamma_1) + (1-\alpha)\tilde{C}(\Gamma_2) \leqq \tilde{C}(\alpha\Gamma_1 + (1-\alpha)\Gamma_2).$$

For $\Gamma = \Gamma_0$, the assertion $C_\varepsilon(\Gamma) = \tilde{C}(\Gamma)$ follows from Theorem 6.5, applying it to the DMC with input alphabet $\mathsf{X}_0 \triangleq \{x : c(x) = \Gamma_0\} \subset \mathsf{X}$. Thus suppose that $\Gamma > \Gamma_0$.

If P is any PD with $c(P) < \Gamma$, we have for every $\mathbf{x} \in \mathsf{T}_{[P]} = \mathsf{T}^n_{[P]\delta_n}$

$$c(\mathbf{x}) = \frac{1}{n}\sum_{a\in\mathsf{X}} N(a|\mathbf{x})c(a) \leqq c(P) + \delta_n \sum_{a\in\mathsf{X}} c(a) < \Gamma$$

if n is sufficiently large. Hence, by Corollary 6.3, $I(P, W)$ is an achievable rate for the DMC $\{W\}$ under input constraint (c, Γ). On account of the continuity of $\tilde{C}(\Gamma)$ (implied by its concavity), it follows that

$$C(\Gamma) \geqq \sup_{\Gamma'<\Gamma} \tilde{C}(\Gamma') = \tilde{C}(\Gamma).$$

To complete the proof, it suffices to show that

$$C_\varepsilon(\Gamma) \leqq \tilde{C}(\Gamma) \quad \text{for every} \quad 0 < \varepsilon < 1. \tag{6.10}$$

We proceed analogously to the proof of (6.5). If (f, φ) is an arbitrary (n, ε)-code for $\{W\}$, meeting the input constraint (c, Γ), let $\mathsf{M}(P)$ be the set of those messages $m \in \mathsf{M}_f$ for which $f(m)$ is of type P; by assumption, $\mathsf{M}(P)$ is non-void only if $c(P) \leqq \Gamma$. Thus, by Corollary 6.4 and the type counting lemma

$$\begin{aligned} |\mathsf{M}_f| &= \sum_{P:c(P)\leqq\Gamma} |\mathsf{M}(P)| \leqq \sum_{P:c(P)\leqq\Gamma} \exp[n(I(P, W) + 2\tau)] \\ &\leqq (n+1)^{|\mathsf{X}|} \cdot \exp\left[n\left(\max_{P:c(P)\leqq\Gamma} I(P, W) + 2\tau\right)\right], \end{aligned}$$

establishing (6.10). □

When proving Theorems 6.5 and 6.11, we have actually proved slightly more than asserted. In fact, (6.5) and (6.10) only mean that it is impossible to construct (n, ε)-codes for *every* sufficiently large n with rates exceeding the claimed value of the capacity by some fixed $\delta > 0$. We have proved that such codes do not exist for *any* sufficiently large n. For its conceptual importance we note this fact as follows.

THEOREM 6.12 For any sequence (f_n, φ_n) of (n, ε)-codes for the DMC $\{W\}$ satisfying the input constraint (c, Γ),

$$\overline{\lim_{n\to\infty}} \frac{1}{n} \log |\mathsf{M}_{f_n}| \leqq C(\Gamma). \qquad \bigcirc$$

COROLLARY 6.12 The functions → 6.6

$$C_k(\Gamma) \triangleq \max_{P: c(P) \leqq \Gamma} \frac{1}{k} I(P, W^k),$$

where the maximum refers to PDs P on X^k, satisfy

$$C_k(\Gamma) = C_1(\Gamma) = C(\Gamma) \quad \text{for} \quad k = 2, 3, \ldots. \qquad \bigcirc$$

Proof By Theorem 1.11, $kC_k(\Gamma)$ is the capacity under input constraint (c, Γ) of the DMC $\{\widehat{W}\}$ with input alphabet X^k, and output alphabet Y^k, where $\widehat{W} \triangleq W^k$. Every (nk, ε)-code for the DMC $\{W\}$ is an (n, ε)-code for $\{\widehat{W}\}$ and conversely. Thus on account of Theorem 1.12 the capacity of the DMC $\{\widehat{W}\}$ under input constraint (c, Γ) equals $kC(\Gamma)$. □

Discussion

The typical results of information theory are of asymptotic character and relate to the existence of codes with certain properties. Theorems asserting the existence of codes are often called *direct results,* while those asserting non-existence are called *converse results.* A combination of such results giving a complete asymptotic solution of a code existence problem is called a *coding theorem* (such as Theorems 6.5 and 6.11 or the earlier Theorems 1.1, 4.1, etc.). In particular, a result stating that for every $\varepsilon \in (0,1)$ the ε-achievable rates are achievable rates as well, is called a *strong converse.* Note that for → 6.7 a DMC more than this is true, namely that every "bad" code has a "good" subcode of "almost" the same rate (Corollary 6.9). A converse result not asserting $C_\varepsilon = C$ for any $\varepsilon > 0$ is often called a *weak converse.*

Definition 6.1 reflects a "pessimistic point of view" inasmuch as achievable rates are defined by the existence of "good" codes for *every* sufficiently large n. An optimist might be satisfied if "good" codes existed for some sequence of blocklengths $n_k \to \infty$. Similar alternatives also arise in other problems treated later.

In general, the pessimist–optimist alternatives are genuinely different, but they are not → 6.8 for DMCs, nor for other problems involving DM systems addressed in this book. Indeed, the converse proofs of all coding theorems will show (even if not stated explicitly) that codes of blocklengths $n_k \to \infty$ whose rates have a limit which is not an achievable rate, cannot have error probabilities approaching zero. Moreover, all strong converse proofs will show that these error probabilities actually go to unity. In the literature, the term "strong converse" often means this stronger statement. In the theory of general channels, this stronger form of the strong converse is sometimes called the *strong converse property*. By Theorem 6.12, all DMCs are included in the class of channels $\{W_n : \mathsf{X}^n \to \mathsf{Y}^n\}_{n=1}^{\infty}$ that have this property.

At the time this book was first published, the noisy channel coding theorem was merely a theoretical result, see the comments to Theorem 6.5. Since then, there has been a breakthrough: *turbo codes* have been discovered that can be actually implemented and, in a practical sense, achieve channel capacity. Then it was recognized that *low density parity check (LDPC) codes* (known previously) are also suitable for this purpose. A common feature of these efficient codes is that an iterative algorithm called a *sum-product* or *belief propagation* algorithm is employed for the computationally hard task of decoding. The study of turbo and LDPC codes, including the mathematical analysis of the belief propagation algorithm, became a major research direction in information theory. It requires mathematical techniques different from those in this book, and will not be discussed. In the following chapters, we always concentrate on determining *fundamental limits*, that is, the best rates theoretically achievable, not asking the question how (if at all) they can be achieved in practice. However, for at least part of the problems treated in the rest of this book, it is no longer out of reach to achieve fundamental limits in practice, by clever application of the belief propagation algorithm. ○

Problems

6.1. (a) Let (f, φ) be a code for a channel $W : \mathsf{X} \to \mathsf{Y}$ with message set $\mathsf{M}_f = \mathsf{M}$. Show that $\bar{e}(W, f, \varphi) \leqq e(W, f, \varphi)$, and that, for some $\hat{\mathsf{M}} \subset \mathsf{M}$ with $|\hat{\mathsf{M}}| \geqq \frac{1}{2}|\mathsf{M}|$, the restriction $\hat{f}$ of f to $\hat{\mathsf{M}}$ yields $e(W, \hat{f}, \varphi) \leqq 2\bar{e}(W, f, \varphi)$.

(b) When a PD Q is given on the message set M_f, the *overall probability of error* is the weighted average of the individual error probabilities $e_m(W, f, \varphi)$ with weights $Q(m)$. Consider a DMS $\{S_i\}_{i=1}^{\infty}$ with alphabet S and generic distribution P, and any code (f, φ) for a channel $W : \mathsf{X} \to \mathsf{Y}$ with message set of size

$$|\mathsf{M}_f| \geqq \exp[k(H(P) + \delta)], \quad k \geqq k_0(|\mathsf{S}|, \varepsilon, \delta).$$

Construct a code (f', φ') for the same channel which has message set $\mathsf{M}_{f'} = \mathsf{S}^k$ and overall probability of error less than $e(W, f, \varphi) + \varepsilon$.

6.2. Given an encoder f for a channel $W : \mathsf{X} \to \mathsf{Y}$, show that the average probability of error is minimized iff $\varphi : \mathsf{Y} \to \mathsf{M}_f$ is a *maximum likelihood decoder*, i.e., iff φ satisfies

$$W(y|f(\varphi(y))) = \max_{m \in \mathsf{M}_f} W(y|f(m)) \quad \text{for every } y \in \mathsf{Y}.$$

Find a decoder minimizing the overall probability of error corresponding to an arbitrary fixed distribution on M.

6.3. (a) Show that the capacity of any channel $\{W_n : \mathsf{X}^n \to \mathsf{Y}^n\}$ remains the same if achievable rates are defined using the condition $\bar{e} \leqq \varepsilon$ rather than $e \leqq \varepsilon$.

Hint See Problem 6.1(a).

(b) Check that the capacity of a DMC equals the reciprocal of the LMTR (see the Introduction) for the transmission over the given DMC of a DMS with binary alphabet and uniform generic distribution, provided that the probability of error fidelity criterion is used.

(c) Show that the capacity of a DMC $\{W\}$ is positive unless all rows of W are identical.

Hint Use Theorem 6.5.

6.4. (*Zero-error capacity*)

(a) Check that in general $C_0 \neq \lim_{\varepsilon \to 0} C_\varepsilon \; (= C)$.

(b) Show that C_0 is positive iff there exist $x_1 \in \mathsf{X}$, $x_2 \in \mathsf{X}$ with $W(y|x_1)W(y|x_2) = 0$ for every $y \in \mathsf{Y}$.

(Shannon (1956).)

6.5. For a DMC $\{W: \mathsf{X} \times \mathsf{S} \to \mathsf{Y}\}$ with input alphabet $\mathsf{X} \times \mathsf{S}$, determine the best asymptotic rate of such (n, ε)-codes whose codewords $(\mathbf{x}, \mathbf{s})$ have a prescribed second component $\mathbf{s} \in \mathsf{S}^n$. Show that for any $\varepsilon \in (0, 1)$, for large n the best rate of such codes is arbitrarily close to the maximum with respect to X of $I(X \wedge Y|S)$, where S is a dummy RV whose distribution equals the type of $\mathbf{s}$ and $P_{Y|X,S} = W$.

Hint Given X, Y, S as above, consider the subset $\mathsf{A} \triangleq \mathsf{T}_{[X|S]}(\mathbf{s}) \times \{\mathbf{s}\}$ of $\mathsf{T}_{[X,S]}$. To obtain a lower bound to the best rate, apply Lemma 6.3, bounding $\frac{1}{n} \log g_{W^n}(\mathsf{A}, \varepsilon - \tau)$ from below via $H(Y|S)$. To obtain the upper bound, use Lemma 6.4 and proceed similarly as in the proof of Theorem 6.5.

6.6. (*Weak converse*)

(a) Give a direct proof of Corollary 6.12 using only the properties of mutual information established in Chapter 3.

(b) Use this result to show that $C(\Gamma) \leqq \max_{c(P) \leqq \Gamma} I(P, W)$.

(c) When defining $C(\Gamma)$, the input constraint $c(f(m)) \leqq \Gamma$ has been imposed individually for every $m \in \mathsf{M}_f$. Show that the result of (b) holds also under the weaker *average input constraint*

$$|\mathsf{M}_f|^{-1} \sum_{m \in \mathsf{M}_f} c(f(m)) \leqq \Gamma.$$

(Fano (1952, unpublished); see also Fano (1961).)

Hint Any code (f, φ) defines a Markov chain $M \multimap X^n \multimap Y^n \multimap M'$, where M is uniformly distributed over M_f, $X^n \triangleq f(M)$, X^n and Y^n are connected by channel W^n and $M' \triangleq \varphi(Y^n)$. Then

$$\log |\mathsf{M}_f| = H(M|M') + I(M \wedge M').$$

Bound the first term by Fano's inequality. Further, show via the data processing lemma, Problem 3.2 and the convexity Lemma 3.5 that

$$I(M \wedge M') \leqq \sum_{i=1}^{n} I(X_i \wedge Y_i) \leqq n I(P, W),$$

where

$$P \triangleq \frac{1}{n}\sum_{i=1}^{n} P_{X_i}.$$

6.7. (*Invalidity of strong converse*) Let $\{W_n : \mathsf{X}^n \to \mathsf{Y}^n\}$ be the mixture of two DMCs, i.e., for some $\widetilde{W} : \mathsf{X} \to \mathsf{Y}$, $\widehat{W} : \mathsf{X} \to \mathsf{Y}$ and $0 < \alpha < 1$ set $W_n(\mathbf{y}|\mathbf{x}) \triangleq \alpha \widetilde{W}^n(\mathbf{y}|\mathbf{x}) + (1-\alpha)\widehat{W}^n(\mathbf{y}|\mathbf{x})$. Show that the ε-capacity of this channel depends on ε, unless $\{\widetilde{W}\}$ and $\{\widehat{W}\}$ have the same capacity. (Wolfowitz (1963).)

6.8. Consider the channel $\{W_n : \mathsf{X}^n \to \mathsf{Y}^n\}_{n=1}^{\infty}$ obtained from the DMC in Problem 6.5 by fixing an infinite sequence $s_1 s_2 \ldots$ of elements of S, that is, let $W_n(\mathbf{y}|\mathbf{x}) \triangleq \prod_{i=1}^{n} W(y_i|x_i, s_i)$.

(a) Show that for any $\varepsilon \in (0, 1)$ the maximum rate of (n, ε)-codes for this channel is arbitrarily close to

$$C_n \triangleq \frac{1}{n}\sum_{i=1}^{n} C(W_{s_i}),$$

where $C(W_s)$ denotes the capacity of the DMC with matrix $W_s = W(.\,|\,.,s)$.

Hint Show that the maximum of $I(X \wedge Y|S)$ in Problem 6.5 is equal to C_n.

(b) Draw the conclusion that the ε-capacity of this channel equals the liminf of C_n for $n \to \infty$, while the "optimistic" definition would yield ε-capacity equal to the limsup, which may be strictly larger. In that case, although the strong converse holds for this channel, the "strong converse property" does not; see the last passage of the Discussion.

6.9. (*Random selection of channel codes*) Given a DMC $\{W : \mathsf{X} \to \mathsf{Y}\}$, a distribution P on X and a message set M of size $|\mathsf{M}| \lesssim \exp\{n(I(P, W) - \delta)\}$, select an encoder $f : \mathsf{M} \to \mathsf{T}^n_{[P]}$ at random, so that each codeword $f(m)$ is selected from $\mathsf{T}^n_{[P]}$ with uniform distribution, independently of all the others.

(a) Associate with any encoder $f : \mathsf{M} \to \mathsf{T}^n_{[P]}$ a decoder $\varphi : \mathsf{Y}^n \to \mathsf{M}'$, called a *joint typicality decoder*, as follows: (i) if there is a unique element of M for which the pair $(f(m), \mathbf{y})$ is typical with respect to the joint distribution Q on $\mathsf{X} \times \mathsf{Y}$ determined by P and W, then put $\varphi(\mathbf{y}) \triangleq m$; (ii) if there is no such m, or there are several, then put $\varphi(\mathbf{y}) \triangleq m' \in \mathsf{M}' - \mathsf{M}$. Show that for the randomly selected f and the associated φ as above, the expectation of $\bar{e}(W^n, f, \varphi)$ is arbitrarily small if n is sufficiently large. Conclude that for any $\varepsilon > 0$ we have $\bar{e}(W^n, f, \varphi) < \varepsilon$ with probability tending to 1.

Hint The expectation of $\bar{e}$ equals that of e_m, for any fixed $m \in \mathsf{M}$. Clearly,

$$Ee_m = \sum_{\mathbf{x}\in\mathsf{T}^n_{[P]}} \frac{1}{|\mathsf{T}^n_{[P]}|}\left[W^n\left(\{\mathbf{y} : (\mathbf{x},\mathbf{y}) \notin \mathsf{T}^n_{[Q]}\}|\mathbf{x}\right) + \sum_{\mathbf{y}:(\mathbf{x},\mathbf{y})\in\mathsf{T}^n_{[Q]}} W^n(\mathbf{y}|\mathbf{x})\Pr\left\{(f(\tilde{m}), \mathbf{y}) \in \mathsf{T}^n_{[Q]} \quad \text{for some} \quad \tilde{m} \neq m\right\}\right].$$

Denoting by $\widetilde{W} : \mathsf{Y} \to \mathsf{X}$ the backward channel, i.e.,

$$\widetilde{W}(x|y) \triangleq \frac{P(x)W(y|x)}{(PW)(y)},$$

the relation $(\mathbf{x}, \mathbf{y}) \in \mathsf{T}^n_{[Q]}$ implies $\mathbf{x} \in \mathsf{T}^n_{[\widetilde{W}]}(\mathbf{y})$. Thus

$$\Pr\left\{(f(\tilde{m}), \mathbf{y}) \in \mathsf{T}^n_{[Q]} \quad \text{for some} \quad \tilde{m} \neq m\right\} \leqq |\mathsf{M}| \frac{|\mathsf{T}^n_{[\widetilde{W}]}(\mathbf{y})|}{|\mathsf{T}^n_{[P]}|},$$

whence by Lemma 2.13 the assertion follows. (This proof was sketched by Shannon (1948).)

(b) Show that the result of (a) holds also if the following decoder is associated with the randomly selected f: (i) put $\varphi(\mathbf{y}) \triangleq m$ if m is the unique element of M with $W^n(\mathbf{y}|f(m)) > \exp\{-n[H(W|P) + \delta/2]\}$; (ii) if no such m exists, or several do, then put $\varphi(\mathbf{y}) \triangleq m' \in \mathsf{M}' - \mathsf{M}$.

Hint Proceeding as in (a), the event $(f(\tilde{m}), \mathbf{y}) \in \mathsf{T}^n_{[Q]}$ is now replaced by $W^n(\mathbf{y}|f(\tilde{m})) > \exp\{-n[H(W|P) + \delta/2]\}$ (a received sequence $\mathbf{y}$ jointly typical with $\mathbf{x} = f(m)$ is not decoded correctly iff the latter occurs for some $\tilde{m} \neq m$). Accordingly, in the last bound in the hint of (a) the set $\mathsf{T}^n_{[\widetilde{W}]}(\mathbf{y})$ is now replaced by $\mathsf{T}^n_{[P]} \cap \{\tilde{\mathbf{x}} \colon W^n(\mathbf{y}|\tilde{\mathbf{x}}) > \exp\{-n[H(W|P) + \delta/2]\}\}$.

6.10. (*Capacity of simple channels*)

(a) The DMC with $\mathsf{X} = \mathsf{Y} = \{0, 1\}$ and $W(1|0) = W(0|1) = p$ is called a *binary symmetric channel* (BSC) with *crossover probability* p. Show that the capacity of this BSC is $C = 1 - h(p)$.

(b) Given an encoder f, a decoder φ is called a *minimum distance decoder* if

$$d_{\mathrm{H}}(\mathbf{y}, f(\varphi(\mathbf{y}))) = \min_{m \in \mathsf{M}} d_{\mathrm{H}}(\mathbf{y}, f(m))$$

for every $\mathbf{y} \in \mathsf{Y}^n$. Show that for a BSC with crossover probability $p < 1/2$ the maximum likelihood decoders are just the minimum distance decoders.

(c) The DMC with $\mathsf{X} = \{0, 1\}$, $\mathsf{Y} = \{0, 1, 2\}$ and

$$W(1|0) = W(0|1) = 0, \;\; W(2|0) = W(2|1) = p$$

is called a *binary erasure channel* (BEC). Show that the capacity of this BEC is $C = 1 - p$.

6.11. Let $W : \mathsf{X} \to \mathsf{Y}$ be a stochastic matrix and let $\tilde{W} : \tilde{\mathsf{X}} \to \mathsf{Y}$ be a maximal submatrix of W consisting of linearly independent rows. Show that the DMCs $\{W\}$ and $\{\tilde{W}\}$ have the same capacity. (Shannon (1957b).)

6.12. (*Symmetric channels*) Let the rows of W be permutations of the same distribution P and let also the columns of W be permutations of each other. Show that in this case the capacity of the DMC $\{W\}$ equals $\log|\mathsf{Y}| - H(P)$. (Shannon (1948).)

6.13. (*Linear channel codes*) Let X be a Galois field. An (n, k) *linear code* for a channel with input alphabet X is an n-length block code (f, φ) with message set $\mathsf{M}_f \triangleq \mathsf{X}^k$ such that $f(\mathbf{m}) \triangleq \mathbf{m}F$, where $\mathbf{m} \in \mathsf{X}^k$ denotes the message and F

is some $k \times n$ matrix over the field X. Further, (f, φ) is a *shifted linear code* if $f(\mathbf{m}) \triangleq \mathbf{m}F + \mathbf{x}_0$ for some fixed $\mathbf{x}_0 \in \mathsf{X}^n$.

(a) Show that if $W : \mathsf{X} \to \mathsf{X}$ is a channel with *additive noise*, i.e., $W(y|x) \triangleq P(y - x)$ for some fixed distribution P on X, then the capacity of the DMC $\{W\}$ can be attained by linear codes. Namely, for every $\varepsilon \in (0, 1)$, $\delta > 0$ and sufficiently large n there exist (n, k) linear codes with $\frac{k}{n} \log |\mathsf{X}| > C - \delta$ and maximum probability of error less than ε. Here $C = \log |\mathsf{X}| - H(P)$, by Problem 6.12.

(Elias (1955). A simple example of channels with additive noise is the BSC of Problem 6.10(a).)

Hint Let $g : \mathsf{X}^n \to \mathsf{X}^{\hat{k}}$ be the encoder of a linear source code with probability of error less than ε for a DMS with generic distribution P. By Problem 1.7, $\hat{k}/n$ can be arbitrarily close to $H(P)/\log |\mathsf{X}|$. Now set $k \triangleq n - \hat{k}$, and consider a linear channel code whose codewords are exactly those $\mathbf{x} \in \mathsf{X}^n$ for which $g(\mathbf{x}) = \mathbf{0}$.

(b) For any DMC $\{W : \mathsf{X} \to \mathsf{Y}\}$, determine the largest R for which there exist shifted linear codes with rates converging to R and with average probability of error tending to zero. Show that this largest R equals $I(P_0, W)$, where P_0 is the uniform distribution on X. (Gabidulin (1967).)

Hint Select the shifted linear encoder f at random, choosing the entries of the matrix F and of the vector $\mathbf{x}_0$ independently and with uniform distribution from X. For each particular selection of f consider the corresponding maximum likelihood decoder. Show that if $k \log |\mathsf{X}| < n(I(P_0, W) - \delta)$ then the expectation of the average probability of error of this randomly selected code (f, φ) tends to zero. The converse follows from Fano's inequality, see the hint to Problem 6.6.

(c) A general channel $\{W_n : \mathsf{X}^n \to \mathsf{X}^n\}_{n=1}^{\infty}$ is a channel with additive noise if X is a Galois field and $W_n(\mathbf{y}|\mathbf{x}) \triangleq P_{Z^n}(\mathbf{y} - \mathbf{x})$, where $\{Z_i\}_{i=1}^{\infty}$ is any noise source with alphabet X. Extend the result of (a) to such channels when the noise source has the AEP; see Problem 4.6(c).

6.14. (*Product of channels*) The product of two channels $W_1 : \mathsf{X}_1 \to \mathsf{Y}_1$ and $W_2 : \mathsf{X}_2 \to \mathsf{Y}_2$ is the channel $W_1 \times W_2 : \mathsf{X}_1 \times \mathsf{X}_2 \to \mathsf{Y}_1 \times \mathsf{Y}_2$ defined by

$$(W_1 \times W_2)(y_1, y_2|x_1, x_2) \triangleq W_1(y_1|x_1)W_2(y_2|x_2).$$

Show that the capacity of the DMC $\{W_1 \times W_2\}$ equals the sum of the capacities of the DMCs $\{W_1\}$ and $\{W_2\}$.

6.15. (*Sum of channels*) The sum of two channels $W_1 : \mathsf{X}_1 \to \mathsf{Y}_1$ and $W_2 : \mathsf{X}_2 \to \mathsf{Y}_2$ with $\mathsf{X}_1 \cap \mathsf{X}_2 = \mathsf{Y}_1 \cap \mathsf{Y}_2 = \emptyset$ is the channel

$$W_1 \oplus W_2 : (\mathsf{X}_1 \cup \mathsf{X}_2) \to (\mathsf{Y}_1 \cup \mathsf{Y}_2)$$

defined by

$$(W_1 \oplus W_2)(y|x) \triangleq \begin{cases} W_1(y|x) & \text{if} \quad x \in \mathsf{X}_1, y \in \mathsf{Y}_1 \\ W_2(y|x) & \text{if} \quad x \in \mathsf{X}_2, y \in \mathsf{Y}_2 \\ 0 & \quad\text{else.} \end{cases}$$

Show that the capacity of the DMC $\{W_1 \oplus W_2\}$ equals $\log(\exp C_1 + \exp C_2)$, where C_i is the capacity of the DMC $\{W_i\}$. (Shannon (1948).)

Comparison of channels (Problems 6.16–6.18)

6.16. A channel $\widetilde{W} : \mathsf{X} \to \mathsf{Z}$ is a *degraded version* of a channel $W : \mathsf{X} \to \mathsf{Y}$ if there exists a channel $V : \mathsf{Y} \to \mathsf{Z}$ such that $\widetilde{W}$ equals the product matrix WV, i.e.,

$$\widetilde{W}(z|x) = \sum_{y \in \mathsf{Y}} W(y|x)V(z|y).$$

Show that in this case to every n-length block code $(f, \tilde{\varphi})$ for the DMC $\{\widetilde{W}\}$ there exists a block code (f, φ) for the DMC $\{W\}$ (with the same encoder f) such that $\bar{e}(W^n, f, \varphi) \leqq \bar{e}(\widetilde{W}^n, f, \tilde{\varphi})$. (Intuitively, the channel $\widetilde{W}^n$ is interpreted as a noisy observation of the output $\mathbf{y}$ of the channel W^n.)

6.17. A DMC $\{W : \mathsf{X} \to \mathsf{Y}\}$ is *better in the Shannon sense* than a DMC $\{\widetilde{W} : \tilde{\mathsf{X}} \to \tilde{\mathsf{Y}}\}$ if to each code for $\widetilde{W}^n$ there exists a code for W^n with the same message set and not larger average probability of error, $n = 1, 2, \ldots$.

(a) Show that for any channels $U : \tilde{\mathsf{X}} \to \mathsf{X}$, $W : \mathsf{X} \to \mathsf{Y}$, $V : \mathsf{Y} \to \tilde{\mathsf{Y}}$ the DMC $\{W\}$ is better in the Shannon sense than $\{\widetilde{W}\}$ if $\widetilde{W} = UWV$. Further, if a DMC $\{W : \mathsf{X} \to \mathsf{Y}\}$ is better in the Shannon sense than both of the channels $\{\widetilde{W}_i : \tilde{\mathsf{X}} \to \tilde{\mathsf{Y}}\}$, $i = 1, 2$, then $\{W\}$ is better than $\{\alpha \widetilde{W}_1 + (1 - \alpha)\widetilde{W}_2\}$, for every $0 \leqq \alpha \leqq 1$. (Shannon (1958a).)

(b) Give an example of DMCs where $\{W\}$ is better in the Shannon sense than $\{\widetilde{W}\}$ but $\widetilde{W}$ cannot be represented as a convex combination of channels obtained from W as in (a). (Karmažin (1964).)

Hint Take $W \triangleq \begin{pmatrix} 1-2\delta & \delta & \delta \\ \delta & 1-2\delta & \delta \\ \delta & \delta & 1-2\delta \end{pmatrix}$, $\quad \widetilde{W} \triangleq \begin{pmatrix} 1 & 0 & 0 \\ 1-\varepsilon & \varepsilon & 0 \\ 1 & 0 & 0 \end{pmatrix}$.

6.18. The DMC $\{W : \mathsf{X} \to \mathsf{Y}\}$ is *more capable* than a DMC $\{\widetilde{W} : \mathsf{X} \to \mathsf{Z}\}$ if $C_W(\mathsf{A}, \varepsilon) \geqq C_{\widetilde{W}}(\mathsf{A}, \varepsilon) - \delta$ for every $0 < \varepsilon < 1$, $\delta > 0$, $n \geqq n_0(\varepsilon, \delta)$ and $\mathsf{A} \subset \mathsf{X}^n$. Show that this holds iff

$$I(P, W) \geqq I(P, \widetilde{W}) \quad \text{for every distribution } P \text{ on } \mathsf{X}. \tag{*}$$

Hint Show by induction that $(*)$ implies $I(X^n \wedge Y^n) \geqq I(X^n \wedge Z^n)$ whenever $P_{Y^n|X^n} = W^n$, $P_{Z^n|X^n} = \widetilde{W}^n$, for which it may be assumed that $Y^n \multimap X^n \multimap Z^n$. Check first that

$$I(X^n \wedge Y^n) - I(X^n \wedge Z^n) = [I(X_n \wedge Y_n|Y^{n-1}) - I(X_n \wedge Z_n|Y^{n-1})] + [I(X^{n-1} \wedge Y^{n-1}|Z_n) - I(X^{n-1} \wedge Z^{n-1}|Z_n)].$$

To prove the relation "more capable" one may suppose $\mathsf{A} \subset \mathsf{T}^n_P$. Show that for some $\eta = \eta(|\mathsf{Y}|, \tau) > 0$

$$\frac{1}{n} \log g_{W^n}(\mathsf{A}, \eta) \geqq \frac{1}{n} H(Y^n) - \tau = \frac{1}{n} I(X^n \wedge Y^n) + H(W|P) - \tau$$

if $\Pr\{X^n \in \mathsf{A}\} = 1$. Letting $(1/n)I(X^n \wedge Z^n) > C_{\tilde{W}}(\mathsf{A}, \varepsilon) - \tau$, apply Lemma 6.8. (Körner and Marton (1977a). For more on this, see Problem 15.11.)

6.19. (*Constant composition codes*) Show that for a DMC $\{W : \mathsf{X} \to \mathsf{Y}\}$ Corollary 6.3 remains true if the codewords are required to have type exactly P, provided that P is a possible type of sequences in X^n. Conclude that there exist n-length block codes (f_n, φ_n) with rate tending to C and such that the codewords $f_n(m)$, $m \in \mathsf{M}_{f_n}$ have the same type P_n.
Hint Use Problem 2.11 for lower bounding $g_{W^n}(\mathsf{T}^n_P, \varepsilon)$.

6.20. (*Maximum mutual information decoding*) For sequences $\mathbf{x} \in \mathsf{X}^n$, $\mathbf{y} \in \mathsf{Y}^n$ define $I(\mathbf{x} \wedge \mathbf{y})$ as the mutual information of RVs with joint distribution $P_{\mathbf{x},\mathbf{y}}$. Given an n-length block encoder $f : \mathsf{M} \to \mathsf{X}^n$ for a DMC $\{W : \mathsf{X} \to \mathsf{Y}\}$, a MMI decoder is a mapping $\varphi : \mathsf{Y}^n \to \mathsf{M}$ satisfying

$$I(f(\varphi(\mathbf{y})) \wedge \mathbf{y}) = \max_{m \in \mathsf{M}} I(f(m) \wedge \mathbf{y}).$$

The significance of the MMI decoder is that, unlike the maximum likelihood or joint typicality decoder, it depends only on the encoder and not on W.

(a) Show that for any DMC there exist encoders f_n with rate converging to C such that if φ_n is a MMI decoder then $e(W^n, f_n, \varphi_n) \to 0$. (Goppa (1975).)

Hint A more general result will be proved in Chapter 10.

(b) For a binary channel, show that for a given $\mathbf{y} \in \{0, 1\}^n$ an $\mathbf{x} \in \{0, 1\}^n$ maximizes $I(\mathbf{x} \wedge \mathbf{y})$ iff $d_H(\mathbf{x}, \mathbf{y}) = 0$ or n. Further, if the types of $\mathbf{x}$ and $\mathbf{y}$ are fixed, then $I(\mathbf{x} \wedge \mathbf{y})$ is a function of $d_H(\mathbf{x}, \mathbf{y})$; show that this function is convex. Draw the conclusion that if all codewords $f(m)$ are of the same type, a MMI decoder assigns to a received sequence $\mathbf{y}$ a message $m \in \mathsf{M}$ that either minimizes or maximizes the Hamming distance $d_H(f(m), \mathbf{y})$.

6.21. (*Several input constraints*) Consider a DMC $\{W : \mathsf{X} \to \mathsf{Y}\}$ with several input constraints (c_j, Γ_j), $j = 1, \dots, r$ on the admissible codewords, and define the capacities $C_\varepsilon(\Gamma_1, \dots, \Gamma_r)$, $C(\Gamma_1, \dots, \Gamma_r)$. Show that they are equal to the maximum of $I(P, W)$ for PDs P on X such that $c_j(P) \leqq \Gamma_j$, $j = 1, \dots, r$.

6.22. (*Output constraint*) Consider a DMC $\{W : \mathsf{X} \to \mathsf{Y}\}$ and let $c(y) \geqq 0$ be a given function on Y extended to Y^n by

$$c(\mathbf{y}) \triangleq \frac{1}{n} \sum_{i=1}^{n} c(y_i) \quad (\mathbf{y} = y_1 \dots y_n).$$

Suppose that, for some physical reason, the decoder is unable to accept any $m \in \mathsf{M}$ if, for the received sequence, $c(\mathbf{y}) > \Gamma$. Define the ε-capacity resp. capacity of the DMC $\{W\}$ under this output constraint, i.e., admitting only such codes (f, φ) for which $c(\mathbf{y}) > \Gamma$ implies $\varphi(\mathbf{y}) \notin \mathsf{M}$. Show that these capacities are equal to max $I(X \wedge Y)$ for RVs connected by the channel W and such that $Ec(Y) \leqq \Gamma$.

Zero-error capacity and graphs (Problems 6.23–6.25)

6.23. With any channel $W : \mathsf{X} \to \mathsf{Y}$ we associate a graph $G = G(W)$ with vertex set X such that two vertices $x' \in \mathsf{X}$ and $x'' \in \mathsf{X}$ are adjacent iff $W(y|x')W(y|x'') = 0$ for every $y \in \mathsf{Y}$. For any graph G, let $\omega(G)$ be the maximum number of mutually adjacent vertices.

(a) Show that the zero-error capacity C_0 of $\{W\}$ depends only on the graph $G = G(W)$ and that $C_0 \geqq \log \omega(G)$.

(b) Define the product of two graphs G_1 and G_2 in such a way that if $G_1 = G(W_1)$, $G_2 = G(W_2)$ then $G_1 \times G_2 = G(W_1 \times W_2)$. Show that $\omega(G_1 \times G_2) \geqq \omega(G_1)\omega(G_2)$, and give an example for the strict inequality.

Hint Let $G_1 = G_2$ be a pentagon.

(c) Show that $C_0 = \lim_{n\to\infty} \frac{1}{n} \log \omega(G^n)$, where $G = G(W)$.

Hint The existence of the limit follows from (b); see Lemma 11.2. (Shannon (1956) assigned to the channel the complementary graph of $G(W)$ here, which we are using for the sake of consistency with later developments, see Chapter 11.)

(d) Show that for every (finite, simple) graph G there exists a stochastic matrix W such that $G = G(W)$.

Remark Determining the limit in (c) for a general G is an open problem. Even the simplest non-trivial case, the pentagon, withstood attacks for more than 20 years, until Lovász (1979) proved that for this case $C_0 = \frac{1}{2} \log 5$. It is interesting to note that the zero-error capacity of the product of two channels (see Problem 6.14) may be larger than the sum of their zero-error capacities, as shown by Haemers (1979).

6.24. For a graph G with vertex set X, a function $f : \mathsf{X} \to \mathsf{C}$, where C is an arbitrary set, is called a *coloring* (or vertex coloring) of G if f takes different values on adjacent vertices. The *chromatic number* of the graph G, denoted by $\chi(G)$, is the minimum cardinality of C for colorings as above.

(a) Show that for any G and positive integer n

$$[\chi(G)]^n \geqq \chi(G^n) \geqq \omega(G^n) \geqq [\omega(G)]^n.$$

Conclude that if $\chi(G) = \omega(G)$ for $G = G(W)$ then $C_0 = \log \omega(G)$.

(b)* Let $\{W : \mathsf{X} \to \mathsf{Y}\}$ be a DMC with output alphabet $\mathsf{Y} = \{1, \ldots, |\mathsf{Y}|\}$ and such that $\{y : W(y|x) > 0\}$ is an interval in Y for every $x \in \mathsf{X}$. Show that in this case $\chi(G) = \omega(G)$ and thus $C_0 = \log \omega(G)$. (From a result of T. Gallai; see Hajnal and Surányi (1958).)

6.25. (*Perfect graphs*) The zero-error capacity problem stimulated interest in the class of graphs all induced subgraphs of which satisfy $\omega(G') = \chi(G')$ (G' is an *induced subgraph* of G if its vertex set is a subset of that of G, and vertices in G' are adjacent iff they are in G). Such graphs, called *perfect graphs*, have various interesting properties, e.g., the complement of a perfect graph is also perfect, as conjectured by Berge (1962) and proved by Lovász (1972). Show, however, that the product of two perfect graphs need not be perfect.

Hint Prove that G^n is perfect for every $n \geqq 1$ iff the adjacency of vertices is an equivalence relation in G. To this end, note that if L is a graph with three vertices and two edges, then L^3 has a pentagon subgraph. (Körner (1973c).)

6.26. (*Remainder terms in the coding theorem*) For a DMC $\{W : \mathsf{X} \to \mathsf{Y}\}$, denote by $N(n, \varepsilon)$ the maximum message set size of (n, ε)-codes.

(a) Show that to every $\varepsilon > 0$ there exists a constant $K = K(|\mathsf{X}|, |\mathsf{Y}|, \varepsilon)$ such that $\exp\{nC - K\sqrt{n}\} \leqq N(n, \varepsilon) \leqq \exp\{nC + K\sqrt{n}\}$ for every n. (Wolfowitz (1957).)

(b)* Prove that

$$\log N(n, \varepsilon) = \begin{cases} nC - \sqrt{n}\lambda T_{\min} + K_{n,\varepsilon} \log n & \text{if } \ 0 < \varepsilon \leqq \dfrac{1}{2} \\ nC - \sqrt{n}\lambda T_{\max} + K_{n,\varepsilon} \log n & \text{if } \ \dfrac{1}{2} < \varepsilon < 1, \end{cases}$$

where $|K_{n,\varepsilon}|$ is bounded by a constant depending on W and ε while $\lambda = \lambda(\varepsilon)$ is defined as in Problem 1.8; $T_{\min}$ resp. $T_{\max}$ is the minimum resp. maximum standard deviation of the "information density" $\log(W(Y|X)/(P_X W)(Y))$ for RVs X and Y connected by the channel W and achieving $I(X \wedge Y) = C$. (Strassen (1964); the fact that $\log N(n, \varepsilon) < nC - K\sqrt{n}$ if $\varepsilon < 1/2$ was proved earlier for a BSC by Weiss (1960).)

6.27. (*Variable-length channel codes*) A variable-length channel code for a channel $\{W_n : \mathsf{X}^n \to \mathsf{Y}^n\}_{n=1}^{\infty}$ is any code (f, φ) for the sum of the channels W_n (this sum channel, see Problem 6.15, has countable input and output sets X^* and Y^*). The *rate* of a variable-length code is defined as $(1/l(f)) \log |\mathsf{M}_f|$, where

$$l(f) \triangleq \frac{1}{|\mathsf{M}_f|} \sum_{m \in \mathsf{M}_f} l(f(m))$$

is the average codeword length. (If $l(f(m)) = n$ for every $m \in \mathsf{M}_f$, this gives the usual rate of an n-length block code.)

(a) Prove that for a DMC with capacity C every variable-length code of average probability of error $\bar{e}$ has rate not exceeding $(1 - \bar{e})^{-1} (C + \delta(l(f)))$, where $\delta(t) \to 0$ as $t \to \infty$. More exactly, $\delta(t) = t^{-1} \log(2et)$, where e is the base of natural logarithms.

(b) Prove the analogous result in case of an input constraint.

Hint Show that for any X^N and Y^N connected by the sum of the channels W^n, $n = 1, 2, \ldots$ (where $X^N = X_1 \ldots X_N$, $Y^N = Y_1 \ldots Y_N$ are sequences of RVs of random length N), the condition $E \sum_{i=1}^{N} c(X_i) \leqq \Gamma \cdot EN$ implies

$$I(X^N \wedge Y^N) \leqq EN \cdot C(\Gamma) + \log(eEN).$$

Then proceed as in Problem 6.6. (A more precise result appears in Ahlswede and Gács (1977).)

6.28. (*Capacity per unit cost*) So far we have tacitly assumed that the cost of transmitting a codeword is proportional to its length. Suppose, more generally, that the cost of transmitting $\mathbf{x} = x_1 \ldots x_n \in \mathsf{X}^n$ equals $\sum_{i=1}^{n} c(x_i)$, where c is some given positive-valued function on X. For an encoder $f : \mathsf{M}_f \to \mathsf{X}^*$, let $c(f)$ be the arithmetic mean of the costs of the codewords. The *capacity per unit cost* of a channel $\{W_n : \mathsf{X}^n \to \mathsf{Y}^n\}$ is defined as the supremum of

$$\varliminf_{k \to \infty} \frac{1}{c(f_k)} \log |\mathsf{M}_{f_k}|$$

for sequences of codes (f_k, φ_k) with $|\mathsf{M}_{f_k}| \to \infty$ and average probability of error tending to zero. Show that for a DMC $\{W\}$, this capacity equals $\max_P (I(P, W)/c(P))$; moreover, it can be attained by constant composition block codes. (See also Problem 8.9.)

Hint See also Problems 6.27 and 6.19.

6.29. (*Feedback does not increase the capacity of a DMC*) The previous code concepts have disregarded the possibility that at every time instant some information may be available at the channel input about the previous channel outputs. As an extreme case, suppose that at the encoder's end all previously received symbols are known exactly before selecting the next channel input (*complete feedback*).

(a) An n-length block code with complete feedback for a DMC $\{W : \mathsf{X} \to \mathsf{Y}\}$ is a pair (f, φ), where the encoder $f = (f_1, \ldots, f_n)$ is a sequence of mappings $f_i : \mathsf{M}_f \times \mathsf{Y}^{i-1} \to \mathsf{X}$ and the decoder φ is a mapping $\varphi : \mathsf{Y}^n \to \mathsf{M}'$ (where $\mathsf{M}' \supset \mathsf{M}_f$). Using the encoder f, the probability that a message $m \in \mathsf{M}_f$ gives rise to an output sequence $\mathbf{y} \in \mathsf{Y}^n$ equals

$$W_f(\mathbf{y}|m) \triangleq \prod_{i=1}^{n} W(y_i | f_i(m, y_1, \ldots, y_{i-1})).$$

The error probabilities associated with the code (f, φ) are defined as in the non-feedback case, except that instead of (6.1) we now write $T(m'|m) \triangleq W_f(\varphi^{-1}(m')|m)$. Prove that admitting codes with complete feedback does not increase the capacity of a DMC. (Shannon (1956), Dobrušin (1958).)

(b) Generalize the above model and result for variable-length codes where the length of transmission may depend both on the message sent and the sequence received. As a further generalization, show that feedback does not increase the capacity per unit cost. (Csiszár (1973).)

Hint Let M be a message RV and X^N, Y^N the corresponding input and output sequences (of random length) when using a variable-length encoder f with complete feedback. Though $I(X^N \wedge Y^N)$ now cannot be bounded as in the hint to Problem 6.27, show that $I(M \wedge Y^N)$ can, by looking at the decomposition

$$I(M \wedge Y^N) = \sum_{j=1}^{\infty} I(M \wedge Y_j | Y^{j-1}) \leq \sum_{j=1}^{\infty} I(X_j \wedge Y_j | Y^{j-1}),$$

where for $j > N$ we set $X_j = Y_j \triangleq x^* \notin \mathsf{X} \cup Y$.

6.30. (*Zero-error capacity with feedback*) Let $C_{0,\mathrm{f}}$ be the zero-error capacity of a given DMC $\{W\}$ when admitting block codes with complete feedback.

(a) Show that $C_{0,\mathrm{f}} = 0$ iff $C_0 = 0$. Further, if C_0 is positive, $C_{0,\mathrm{f}}$ may be larger than C_0.

(b)* Show that if $C_0 > 0$ then

$$C_{0,\mathrm{f}} = \max_P \min_{y \in \mathsf{Y}} \{-\log \sum_{x : W(y|x) > 0} P(x)\}.$$

(Shannon (1956), who attributes the observation $C_0 \neq C_{0,\mathrm{f}}$ to P. Elias.)

(c) Show that whenever W has at least one zero entry in a non-zero column, one can attain zero probability of error at any rate below capacity if variable-length codes with complete feedback are admitted. (Burnašev (1976).)

Hint One can build from any (n, ε)-code (f, φ) a variable-length code with feedback that has almost the same rate (if n is large and ε is small) and zero probability of error. Pick $x' \in \mathsf{X}$, $x'' \in \mathsf{X}$, $y' \in \mathsf{Y}$ such that $W(y'|x') > W(y'|x'') = 0$. To transmit message m, first send the codeword $f(m)$. Then, depending on whether the received sequence $\mathbf{y}$ was such that $\varphi(\mathbf{y}) = m$ or not, send k times x' or k times x''. If at least one of the last k received symbols was y' then stop transmission and decode m. Otherwise retransmit $f(m)$ and continue as above.

6.31. (*Channels with random states*) Consider a DMC $\{W : \mathsf{X} \times \mathsf{S} \to \mathsf{Y}\}$ as in Problem 6.5, where the sender controls only the first components of the inputs, while the sequence of the second components, interpreted as "states of the channel," is now assumed to be random, i.i.d. with distribution Q. If the sender and receiver are ignorant of the states, they actually have a simple DMC $\{W_Q : \mathsf{X} \to \mathsf{Y}\}$ where

$$W_Q(y|x) \triangleq \sum_{s \in \mathsf{S}} Q(s) W(y|x, s).$$

Suppose the sender has *channel state information (CSI)*, knowing at each time instant i either $s^{i-1} = s_1 \dots s_{i-1}$ (Model (a)) or $s^i = s_1 \dots s_i$ (Model (b)), while the receiver has no CSI. For another model, where the sender knows all states in advance (non-causal CSI), see Problem 12.24.

(a) In Model (a), an encoder of block length n is, to the analogy of Problem 6.29 (a), a sequence $f = (f_1, \dots, f_n)$ of mappings $f_i\colon \mathsf{M}_f \times \mathsf{S}^{i-1} \to \mathsf{X}$. If message $m \in \mathsf{M}_f$ is sent via this encoder, the probability of output $\mathbf{y} \in \mathsf{Y}^n$ is

$$W_f(\mathbf{y}|m) = \sum_{\mathbf{s} \in \mathsf{S}^n} Q^n(\mathbf{s}) \prod_{i=1}^{n} W(y_i | f_i(m, s^{i-1}), s_i).$$

Show that this model admits no achievable rates above the maximum of $I(X \wedge Y)$ for RVs X, Y such that $P_{Y|X,S} = W$ for S of distribution Q, independent of X, which maximum is the capacity of the DMC $\{W_Q\}$. Thus, the CSI consisting in the sender's knowledge of past states does not increase capacity.

Hint Use the same idea as in Problem 6.29.

(b) In Model (b), encoders and the corresponding probabilities $W_f(\mathbf{y}|m)$ are defined similarly to (a), but now with mappings $f_i\colon \mathsf{M}_f \times \mathsf{S}^i \to \mathsf{X}$. Determine the largest achievable rate for this model, and show that encoders with $f_i = f_i(m, s_i)$, $i = 1, \dots, m$, are sufficient to achieve it ("past states may be forgotten").

Hint Consider an auxiliary channel $\{\widetilde{W}\colon \tilde{\mathsf{X}} \times \mathsf{S} \to \mathsf{Y}\}$ where the auxiliary alphabet $\tilde{\mathsf{X}}$ consists of all mappings $\tilde{x}\colon \mathsf{S} \to \mathsf{X}$, and $\widetilde{W}(y|\tilde{x}, s) \triangleq W(y|\tilde{x}(s), s)$. Show that encoders in Model (b) for $\{W\}$ are equivalent to encoders in Model (a) for $\{\widetilde{W}\}$, and apply the result of (a). (Shannon (1958b)).

In practice, the states may represent interference. For example, let all three alphabets be binary, let $W(y|x, s)$ equal 1 or 0 according as $y = x + s \pmod 2$ or not, and $Q = (1/2, 1/2)$. If the sender knows at each instant i the state s_i, he can achieve *interference cancellation* and transmit at full rate, using the members $\tilde{x}_0(s) = s$, $\tilde{x}_1(s) = 1 - s$ of the auxiliary alphabet $\tilde{\mathsf{X}}$. In absence of CSI, the capacity is zero.

(c) Show that no new mathematical problems arise when the receiver has CSI.

Hint Treat the receiver's CSI as part of the channel output. Formally, consider the channel $W'\colon \mathsf{X} \times \mathsf{S} \to \mathsf{Y} \times \mathsf{S}$, where $W'(y, s'|x, s)$ equals $W(y|x, s)$ if $s = s'$ and zero otherwise.

Story of the results

The noisy channel coding theorem (Theorem 6.5) was a key result of Shannon (1948). He accepted the converse on an intuitive basis, and sketched a proof for the direct part, elaborated later in Shannon (1957a). The first rigorous proof of the capacity formula of Theorem 6.5 appears in Feinstein (1954), who attributes the (weak) converse to Fano (1952, unpublished). Shannon (1948) also claimed the independence of C_ε of ε (strong converse) though this was not proved until Wolfowitz (1957). The present proof of Theorem 6.5 via Corollaries 6.3 and 6.4 follows Wolfowitz (1957).

In this chapter the noisy channel coding problem is presented in a more general framework than usual. Lemmas 6.3, 6.4 and 6.8 are streamlined versions of results in Körner and Marton (1977c). The proof technique of Lemma 6.3, i.e., the use of maximal codes, originates in Feinstein (1954); see also Thomasian (1961). Lemma 6.6 was found by Ahlswede *et al.* (1976). Theorem 6.9 is an immediate consequence of Lemma 6.8; Corollary 6.9 was, however, proved earlier by Ahlswede and Dueck (1976). Theorem 6.10 appears in Thomasian (1961). Theorems 6.11 and 6.12 are hard to trace; see Thomasian (1961).

Addition. The "strong converse property" in the Discussion has been introduced by Verdú and Han (1994) to distinguish a well behaved class of general channels; see also Han (2003).

Turbo codes were discovered by Berrou, Glavieux and Thitimajshima (1993) and LDPC codes date back to R. Gallager's Ph.D. thesis, see Gallager (1963). For the theory of these codes and the belief propagation algorithm, see Richardson and Urbanke (2008).

7 Rate-distortion trade-off in source coding and the source–channel transmission problem

Let $\{X_i\}_{i=1}^{\infty}$ be a discrete source with alphabet X and let Y be another finite set called the *reproduction alphabet*. The sequences $\mathbf{y} \in \mathsf{Y}^k$ are considered the possible distorted versions of the sequences $\mathbf{x} \in \mathsf{X}^k$. Let the degree of distortion be measured by a non-negative function d_k with domain $\mathsf{X}^k \times \mathsf{Y}^k$; the family $d \triangleq \{d_k\}_{k=1}^{\infty}$ of these functions is called the *distortion measure*.

A *k-length block code* for sources with alphabet X and reproduction alphabet Y is a pair of mappings (f, φ), where f maps X^k into some finite set and φ maps the range of f into Y^k. The mapping f is the *source encoder* and φ is the *source decoder*. The *rate* of such a code is defined as $\frac{1}{k} \log ||f||$.

Observe that a k-length block code for a source is a code in the sense of Chapter 6 for a noiseless channel, with message set $\mathsf{M}_f \triangleq \mathsf{X}^k$ and with $\mathsf{M}' \triangleq \mathsf{Y}^k$. The reason for now defining the rate in terms of the range of f (rather than its domain) will become apparent.

As a result of the application of encoder f and decoder φ, a source output $\mathbf{x} \in \mathsf{X}^k$ is reproduced as $g(\mathbf{x}) \triangleq \varphi(f(\mathbf{x})) \in \mathsf{Y}^k$. The smaller the distortion $d_k(\mathbf{x}, g(\mathbf{x}))$, the better $\mathbf{x}$ is reproduced by the code (f, φ). We shall say that the source code (f, φ) *meets the ε-fidelity criterion* (d, Δ) if

$$\Pr\{d_k(X^k, g(X^k)) \leqq \Delta\} \geqq 1 - \varepsilon.$$

Instead of this local condition, one often imposes the global one that the source code should meet the *average fidelity criterion* (d, Δ), i.e.,

$$E d_k(X^k, g(X^k)) \leqq \Delta.$$

Given a fidelity criterion, the source coding problem consists in constructing codes meeting this criterion and achieving maximum data compression, i.e., having rates as small as possible. The very first theorem in this book (Theorem 1.1) dealt with such a problem. → 7.1

DEFINITION 7.1 Given a distortion measure d, a non-negative number R is an *ε-achievable rate at distortion level* Δ for the source $\{X_i\}_{i=1}^{\infty}$ if for every $\delta > 0$ and sufficiently large k there exist k-length block codes of rate less than $R + \delta$, meeting the ε-fidelity criterion (d, Δ). Further, R is an *achievable rate at distortion level* Δ if it is ε-achievable for every $0 < \varepsilon < 1$. If R is an ε-achievable resp. achievable rate at distortion level Δ, we also say that (R, Δ) is an *ε-achievable* resp. *achievable rate-distortion pair*. The infimum of ε-achievable resp. achievable rates at distortion level Δ will be denoted by $R_\varepsilon(\Delta)$ resp. $R(\Delta)$. The latter will be called the *Δ-distortion rate* of the source $\{X_i\}_{i=1}^\infty$ with respect to the distortion measure d. ○

→ 7.2
→ 7.3
→ 7.4

REMARK $\lim_{\varepsilon\to 0} R_\varepsilon(\Delta) = R(\Delta)$. ○

The Δ-distortion rate might have been defined also by using the average fidelity criterion. For the DM model we shall treat, however, this approach leads to the same result, as will be seen below.

In the following we assume that the distortion between sequences is defined as the average of the distortion between their corresponding elements, i.e.,

$$d_k(\mathbf{x}, \mathbf{y}) = d(\mathbf{x}, \mathbf{y}) \triangleq \frac{1}{k}\sum_{i=1}^{k} d(x_i, y_i) \tag{7.1}$$

$$\text{if} \quad \mathbf{x} = x_1 \dots x_k, \quad \mathbf{y} = y_1 \dots y_k.$$

→ 7.5

In this case we shall speak of an *averaging distortion measure*. Here $d(x, y)$ is a non-negative-valued function on $\mathsf{X} \times \mathsf{Y}$. It will also be supposed that to every x in X there exists at least one $y \in \mathsf{Y}$ such that $d(x, y) = 0$. With some abuse of terminology, we shall identify an averaging distortion measure with the function $d(x, y)$. This function is assumed finite-valued, unless stated otherwise. The reader is invited to check which results below remain valid if also $d(x, y) = +\infty$ is allowed, for some pairs (x, y).

We shall show that for a DMS with generic distribution P,

$$R(\Delta) = R(P, \Delta) = \min_{\substack{P_X = P \\ Ed(X,Y) \leqq \Delta}} I(X \wedge Y).$$

Temporarily, we denote this minimum by $\tilde{R}(P, \Delta)$, i.e., we set

$$\tilde{R}(P, \Delta) \triangleq \min_{W: d(P,W) \leqq \Delta} I(P, W),$$

where W ranges over stochastic matrices $W : \mathsf{X} \to \mathsf{Y}$, and

$$d(P, W) \triangleq \sum_{x\in\mathsf{X}}\sum_{y\in\mathsf{Y}} P(x)W(y|x)d(x, y).$$

Later, after having proved Theorem 7.3, no distinction will be made between $R(P, \Delta)$ and $\tilde{R}(P, \Delta)$.

LEMMA 7.2 For P fixed, $\tilde{R}(P, \Delta)$ is a finite-valued, non-increasing convex function of $\Delta \geqq 0$. Further, $\tilde{R}(P, \Delta)$ is a continuous function of the pair (P, Δ), where P ranges over the distributions on X, and $\Delta \geqq 0$. ○

Proof The minimum in the definition of $\tilde{R}(P, \Delta)$ is achieved as $I(P, W)$ is a continuous function of W and the minimization is over a non-void compact set. The monotonicity is obvious. The convexity follows from that of $I(P, W)$ as a function of W (Lemma 3.5) since the inequalities $d(P, W_1) \leqq \Delta_1, d(P, W_2) \leqq \Delta_2$ imply $d(P, \alpha W_1 + (1-\alpha)W_2) \leqq \alpha\Delta_1 + (1-\alpha)\Delta_2$ for any $0 < \alpha < 1$.

To prove the joint continuity of $\tilde{R}(P, \Delta)$, suppose that $P_n \to P, \Delta_n \to \Delta$. If $\Delta > 0$, pick some $W : \mathsf{X} \to \mathsf{Y}$ such that $d(P, W) < \Delta, I(P, W) < \tilde{R}(P, \Delta) + \varepsilon$; this is possible for every $\varepsilon > 0$ as for fixed P the convexity of $\tilde{R}(P, \Delta)$ implies its continuity. If $\Delta = 0$, pick a $W : \mathsf{X} \to \mathsf{Y}$ such that $W(y|x) = 0$ whenever $d(x, y) > 0$ and $I(P, W) = \tilde{R}(P, 0)$. By the continuity of $I(P, W)$ and $d(P, W)$, it follows in both cases that for n sufficiently large both $I(P_n, W) < \tilde{R}(P, \Delta) + \varepsilon$ and $d(P_n, W) \leqq \Delta_n$. This proves

$$\overline{\lim_{n\to\infty}} \tilde{R}(P_n, \Delta_n) \leqq \tilde{R}(P, \Delta).$$

On the other hand, let $W_n : \mathsf{X} \to \mathsf{Y}$ achieve the minimum in the definition of $\tilde{R}(P_n, \Delta_n)$. Consider a sequence of integers $\{n_k\}$ such that

$$\tilde{R}(P_{n_k}, \Delta_{n_k}) \to \underline{\lim}_{n\to\infty} \tilde{R}(P_n, \Delta_n)$$

and $W_{n_k} \to W$, say. Since then $d(P, W) = \lim_{k\to\infty} d(P_{n_k}, W_{n_k}) = \Delta$, it follows that

$$\tilde{R}(P, \Delta) \leqq I(P, W) = \lim_{k\to\infty} I(P_{n_k}, W_{n_k}) = \underline{\lim}_{n\to\infty} \tilde{R}(P_n, \Delta_n). \qquad \square$$

THEOREM 7.3 (*Rate-distortion theorem*) For a DMS $\{X_i\}_{i=1}^{\infty}$ with generic distribution P, we have for every $0 < \varepsilon < 1$ and $\Delta \geqq 0$

$$R_\varepsilon(\Delta) = R(\Delta) = \min_{\substack{P_X = P \\ Ed(X,Y) \leqq \Delta}} I(X \wedge Y). \qquad \bigcirc$$

Proof First we prove the existence part of the theorem, i.e., that $\tilde{R}(P, \Delta)$ is an ε-achievable rate at distortion level Δ. To this end we construct a "backward" DMC $\{\widehat{W} : \mathsf{Y} \to \mathsf{X}\}$ and k-length block codes for the DMC $\{\widehat{W}\}$. Source codes meeting the ε-fidelity criterion (d, Δ) will be obtained by choosing for the source encoder the channel decoder and for the source decoder the channel encoder. The point is that for this purpose channel codes with large error probability are needed.

If $\Delta > 0$, let X and Y be RVs such that $P_X = P$, $Ed(X, Y) < \Delta$, and fix $0 < \tau < \hat{\varepsilon} < 1$. Setting $\mathsf{Y}_0 \triangleq \{y : P_Y(y) > 0\}$, consider the DMC $\{\widehat{W} : \mathsf{Y}_0 \to \mathsf{X}\}$ with $\widehat{W} \triangleq P_{X|Y}$. Let $(\hat{f}, \hat{\varphi})$ be a $(k, \hat{\varepsilon})$-code for this DMC such that for every $m \in \mathsf{M}_{\hat{f}}$

$$\hat{f}(m) \in \mathsf{T}^k_{[Y]} \subset \mathsf{Y}^k_0, \tag{7.2}$$

$$\hat{\varphi}^{-1}(m) \subset \mathsf{T}^k_{[X|Y]}(\hat{f}(m)), \tag{7.3}$$

and the code $(\hat{f}, \hat{\varphi})$ has no extension with these properties.

Then for sufficiently large k (see (6.3) in the proof of Lemma 6.3), the set

$$\mathsf{B} \triangleq \bigcup_{m\in\mathsf{M}_{\hat{f}}} \hat{\varphi}^{-1}(m) \subset \mathsf{X}^k$$

satisfies

$$\widehat{W}^k(\mathsf{B}|\mathbf{y}) \geqq \hat{\varepsilon} - \tau \quad \text{for every} \quad \mathbf{y} \in \mathsf{T}^k_{[Y]}.$$

Hence, as by Lemma 2.12 $P_{Y^k}(\mathsf{T}^k_{[Y]}) \to 1$, we obtain

$$P^k(\mathsf{B}) \geqq \sum_{\mathbf{y}\in \mathsf{T}^k_{[Y]}} P_{Y^k}(\mathbf{y})\widehat{W}^k(\mathsf{B}|\mathbf{y}) \geqq \hat{\varepsilon} - 2\tau. \tag{7.4}$$

Further, if k is large enough then

$$\frac{1}{k}\log|\mathsf{M}_{\hat{f}}| < I(X \wedge Y) + 2\tau, \tag{7.5}$$

by Corollary 6.4.

The channel decoder $\hat{\varphi}$ maps X^k into a set $\mathsf{M}' \supset \mathsf{M}_{\hat{f}}$. We may assume that $|\mathsf{M}'| = |\mathsf{M}_{\hat{f}}| + 1$, for changing $\hat{\varphi}$ outside B does not affect the conditions on $(\hat{f}, \hat{\varphi})$. Now define a source code by

$$f(\mathbf{x}) \triangleq \hat{\varphi}(\mathbf{x}); \quad \varphi(m) \triangleq \begin{cases} \hat{f}(m) & \text{if} \quad m \in \mathsf{M}_{\hat{f}} \\ \text{arbitrary otherwise.} \end{cases} \tag{7.6}$$

Put $d_{\mathsf{M}} \triangleq \max_{a\in\mathsf{X}, b\in\mathsf{Y}} d(a, b)$. By (7.2) and (7.3), for $\mathbf{x} \in \mathsf{B}$, $\mathbf{y} \triangleq \varphi(f(\mathbf{x}))$ we have $\mathbf{y} \in \mathsf{T}^k_{[Y]}$ and $\mathbf{x} \in \mathsf{T}^k_{[X|Y]}(\mathbf{y})$, thus by Lemma 2.10,

$$\begin{aligned} d(\mathbf{x}, \varphi(f(\mathbf{x}))) &= \frac{1}{k}\sum_{a\in\mathsf{X}}\sum_{b\in\mathsf{Y}} N(a, b|\mathbf{x}, \mathbf{y})d(a, b) \\ &\leqq \sum_{a\in\mathsf{X}}\sum_{b\in\mathsf{Y}} P_{XY}(a, b)d(a, b) + 2\delta_k|\mathsf{X}||\mathsf{Y}|d_{\mathsf{M}}. \end{aligned}$$

This means that for k sufficiently large, every $\mathbf{x} \in \mathsf{B}$ is reproduced with distortion less than Δ.

On account of (7.4) and (7.5), choosing $\tau > 0$ sufficiently small and $\hat{\varepsilon} \triangleq 1 - \varepsilon + 2\tau$, we arrive at a source code (f, φ) that meets the ε-fidelity criterion (d, Δ) and has rate

$$\frac{1}{k}\log\|f\| < I(X \wedge Y) + 2\tau.$$

As X and Y were arbitrary subject to the conditions $P_X = P$, $Ed(X, Y) < \Delta$, it follows that $\tilde{R}(P, \Delta')$ is an ε-achievable rate at distortion level Δ, for any $\Delta' < \Delta$. By the continuity of $\tilde{R}(P, \Delta)$, this proves the existence part of the theorem for $\Delta > 0$.

If $\Delta = 0$, we repeat the above construction starting from any RVs X and Y with $P_X = P$, $Ed(X, Y) = 0$. Note that, on account of (7.3), $\hat{\varphi}(\mathbf{x}) = m \in \mathsf{M}_f$ implies $\widehat{W}^k(\mathbf{x}|\hat{f}(m)) > 0$. Hence, by the condition $Ed(X, Y) = 0$, the resulting source code (7.6) has the property that

$$d(\mathbf{x}, \varphi(f(\mathbf{x}))) = 0 \quad \text{for} \quad \mathbf{x} \in \mathsf{B}.$$

Thus $\tilde{R}(P, 0)$ is an ε-achievable rate at distortion level zero.

Now we turn to the (strong) converse. Somewhat more ambitiously, we shall prove the following uniform estimate. Given any $\varepsilon \in (0, 1)$, $\delta > 0$ and distortion measure d on $\mathsf{X} \times \mathsf{Y}$, the rate of a k-length block code meeting the ε-fidelity criterion (d, Δ) for a DMS with generic distribution P satisfies

$$\frac{l}{k} \log ||f|| \geqq \tilde{R}(P, \Delta) - \delta \tag{7.7}$$

whenever $k \geqq k_0(\varepsilon, \delta, d)$. The dependence of k_0 on the alphabet sizes is not indicated, as we understand that the alphabets X and Y are specified when the distortion measure d is.

Let (f, φ) be such a code; i.e., setting $g(\mathbf{x}) \triangleq \varphi(f(\mathbf{x}))$, suppose that

$$\Pr\{d(X^k, g(X^k)) \leqq \Delta\} \geqq 1 - \varepsilon.$$

Fix some $\tau > 0$ to be specified later. Then, by Corollary 2.14, the set

$$\mathsf{A} \triangleq \{\mathbf{x} : \mathbf{x} \in \mathsf{T}^k_{[P]}, \quad d(\mathbf{x}, g(\mathbf{x})) \leqq \Delta\}$$

has cardinality

$$|\mathsf{A}| \geqq \exp[k(H(P) - \tau)] \tag{7.8}$$

if $k \geqq k_1(\tau, \varepsilon, |\mathsf{X}|)$.

For every fixed $\mathbf{y} \in \mathsf{Y}^k$ the number of sequences $\mathbf{x} \in \mathsf{X}^k$ with joint type $P_{\mathbf{x},\mathbf{y}} = \tilde{P}$ is upper bounded by $\exp[kH(\tilde{X}|\tilde{Y})]$, where $\tilde{X}, \tilde{Y}$ denote RVs with joint distribution $\tilde{P}$, see Lemma 2.5. Note that $d(\mathbf{x}, \mathbf{y}) \leqq \Delta$ and $\mathbf{x} \in \mathsf{T}^k_{[P]}$ imply

$$Ed(\tilde{X}, \tilde{Y}) \leqq \Delta; \quad |P_{\tilde{X}}(a) - P(a)| \leqq \delta_k \quad \text{for every} \quad a \in \mathsf{X}. \tag{7.9}$$

Consider the pair $(\tilde{X}, \tilde{Y})$ which maximizes $H(\tilde{X}, \tilde{Y})$ subject to (7.9). Denoting by C the set of those $\mathbf{y} \in \mathsf{Y}^k$ which satisfy $\mathbf{y} = g(\mathbf{x})$ for some $\mathbf{x} \in \mathsf{A}$, we have

$$|\mathsf{A}| \leqq \sum_{\mathbf{y} \in \mathsf{C}} |\{\mathbf{x} : \mathbf{x} \in \mathsf{T}^k_{[P]}, \; d(\mathbf{x}, \mathbf{y}) \leqq \Delta\}|$$

$$\leqq ||g||(k+1)^{|\mathsf{X}|\cdot|\mathsf{Y}|} \exp[kH(\tilde{X}|\tilde{Y})]. \tag{7.10}$$

As by Lemma 2.7 the relation (7.9) implies

$$|H(P) - H(\tilde{X})| < \tau$$

if k is sufficiently large, (7.8) and (7.10) yield

$$\exp[k(H(\tilde{X}) - 2\tau)] \leqq |\mathsf{A}| \leqq ||g|| \exp[k(H(\tilde{X}|\tilde{Y}) + \tau)].$$

Hence,

$$\frac{1}{k} \log ||f|| \geqq \frac{1}{k} \log ||g|| \geqq I(\tilde{X} \wedge \tilde{Y}) - 3\tau \geqq \tilde{R}(P_{\tilde{X}}, \Delta) - 3\tau. \tag{7.11}$$

On account of the uniform continuity of $\tilde{R}(P, \Delta)$ – which follows from Lemma 7.2 and the fact that $\tilde{R}(P, \Delta)$ vanishes outside a compact set – the choice $\tau = \delta/4$ in (7.11) gives (7.7). □

COMMENTS Theorem 7.3 is as fundamental for source coding as are Theorems 6.5 and 6.10 for channel coding, and the comments to Theorem 6.5 apply to it as well. These results have a common mathematical background in Corollaries 6.3 and 6.4. Note that the definition of ε-achievable rate-distortion pairs contains a slight asymmetry. However, by Theorem 7.3 and the continuity of $R(\Delta)$, one sees that a pair (R, Δ) is ε-achievable for a DMS iff for every $\delta > 0$ and sufficiently large k there exist k-length block codes having rate less than $R + \delta$ and satisfying $\Pr\{d(X^k, g(X^k)) \leqq \Delta + \delta\} \geqq 1 - \varepsilon$. Finally, as $R_\varepsilon(\Delta)$ is the infimum of

$$\overline{\lim_{k\to\infty}} \frac{1}{k} \log ||g_k|| \tag{7.12}$$

over all sequences of mappings $g_k : \mathsf{X}^k \to \mathsf{Y}^k$ satisfying $\Pr\{d(X^k, g_k(X^k)) \leqq \Delta\} \geqq 1 - \varepsilon$, Theorem 7.3 asserts that this infimum equals

$$\min_{\substack{P_X = P \\ Ed(X,Y) \leqq \Delta}} I(X \wedge Y).$$

By (7.7), the previous infimum does not decrease if in (7.12) the $\overline{\lim}$ is replaced by $\underline{\lim}$, i.e., once again, the "pessimistic" and the "optimistic" viewpoints lead to the same result. For channels, the analogous observation was stated as Theorem 6.12. ○

Corollary 6.12 has now the following analog.

→ 7.6

COROLLARY 7.3 Set

$$R_n(\Delta) \triangleq \min_{Ed(X^n, Y^n) \leqq \Delta} \frac{1}{n} I(X^n \wedge Y^n),$$

where the minimum refers to RVs X^n and Y^n with values in X^n resp. Y^n such that $P_{X^n} = P^n$. Then

$$R_n(\Delta) = R_1(\Delta) = R(\Delta) \quad \text{for} \quad n = 2, 3, \ldots. \qquad ○$$

Proof By Theorem 7.3, $nR_n(\Delta)$ is the Δ-distortion rate of a DMS with alphabet X^n and generic distribution P^n. Clearly, k-length block codes for the latter DMS meeting the ε-fidelity criterion (d, Δ) are the same as nk-length block codes for the original DMS meeting the ε-fidelity criterion (d, Δ). Hence, taking into account (7.7), the Δ-distortion rate of the new DMS equals $nR(\Delta)$. □

Let us turn now to the problem of reliable transmission of a DMS over a DMC, illustrated in Fig. 7.1. Combining the source and channel coding theorems treated so far, we can answer the LMTR problem exposed on an intuitive level in the Introduction. Namely, we shall show by composing source and channel codes that the LMTR equals the ratio of Δ-distortion rate and channel capacity.

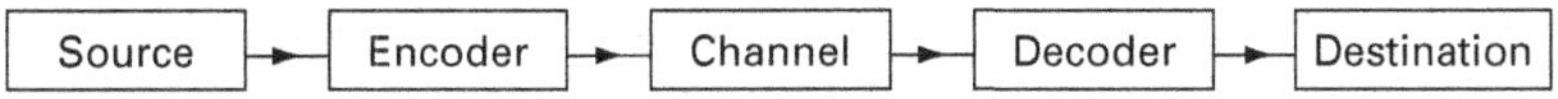

Figure 7.1

Given a source $\{S_i\}_{i=1}^{\infty}$ with alphabet S, a channel $\{W_n : \mathsf{X}^n \to \mathsf{Y}^n\}_{n=1}^{\infty}$ with input and output alphabets X and Y, and a reproduction alphabet U, a k-to-n *block code* is a pair of mappings $f : \mathsf{S}^k \to \mathsf{X}^n$, $\varphi : \mathsf{Y}^n \to \mathsf{U}^k$. When using this code, the channel input is the RV $X^n \triangleq f(S^k)$. The channel output is a RV Y^n connected with X^n by the channel W_n and depending on S^k only through X^n (i.e., $S^k \multimap X^n \multimap Y^n$). The destination receives $U^k \triangleq \varphi(Y^n)$. For a given distortion measure d on $\mathsf{S} \times \mathsf{U}$, we say that the code (f, φ) *meets the average fidelity criterion* (d, Δ) $(\Delta \geqq 0)$ if

$$Ed(S^k, U^k) = \sum_{\mathbf{s} \in \mathsf{S}^k} \sum_{\mathbf{y} \in \mathsf{Y}^n} P_{S^k}(\mathbf{s}) W_n(\mathbf{y}|f(\mathbf{s})) d(\mathbf{s}, \varphi(\mathbf{y})) \leqq \Delta. \tag{7.13}$$

Further, for a given constraint function c on X, the code (f, φ) is said to *satisfy the input constraint* (c, Γ) if $c(\mathbf{x}) \leqq \Gamma$ for every $\mathbf{x}$ in the range of f.

THEOREM 7.4 (*Source–channel transmission theorem*) If $\{S_i\}_{i=1}^{\infty}$ is a DMS and $\{W : \mathsf{X} \to \mathsf{Y}\}$ is a DMC then to every (d, Δ) and (c, Γ) with $\Delta > 0$ there exists a sequence of k-to-n_k block codes (f_k, φ_k) meeting the average fidelity criterion (d, Δ) and satisfying the input constraint (c, Γ) such that

$$\lim_{k \to \infty} \frac{n_k}{k} = \frac{R(\Delta)}{C(\Gamma)}.$$

On the other hand, if a k-to-n block code meets the average fidelity criterion (d, Δ) and satisfies the input constraint (c, Γ), or the weaker constraint $Ec(f(S^k)) \leqq \Gamma$, then

$$\frac{n}{k} \geqq \frac{R(\Delta)}{C(\Gamma)}. \tag{7.14}$$

○

Proof We shall prove that for every $0 < \delta < \Delta$ and sufficiently large k there exists a k-to-n_k block code with

$$\frac{n_k}{k} < \frac{R(\Delta - \delta)}{C(\Gamma)} + \delta \tag{7.15}$$

such that this code meets the average fidelity criterion (d, Δ) and satisfies the input constraint (c, Γ). On account of the continuity of $R(\Delta)$, this will prove the existence part.

Fix an $\varepsilon > 0$ to be specified later. Consider k-length source block codes $(\tilde{f}_k, \tilde{\varphi}_k)$ meeting the ε-fidelity criterion $(d, \Delta - \delta)$ and (n, ε)-codes $(\hat{f}_n, \hat{\varphi}_n)$ for the DMC $\{W\}$ satisfying the input constraint (c, Γ), such that

$$\lim_{k \to \infty} \frac{1}{k} \log ||\tilde{f}_k|| = R(\Delta - \delta), \tag{7.16}$$

$$\lim_{n \to \infty} \frac{1}{n} \log |\mathsf{M}_{\hat{f}_n}| = C(\Gamma). \tag{7.17}$$

Let n_k be the smallest integer for which the size of the message set of $\hat{f}_{n_k}$ is at least $||\tilde{f}_k||$. Then by (7.16) and (7.17) we have

$$\lim_{k\to\infty} \frac{n_k}{k} = \frac{R(\Delta-\delta)}{C(\Gamma)}.$$

We may suppose that the range of $\tilde{f}_k$, to be denoted by M_k, is a subset of the message set of $\hat{f}_{n_k}$. Then the source code $(\tilde{f}_k, \tilde{\varphi}_k)$ and the channel code $(\hat{f}_{n_k}, \hat{\varphi}_{n_k})$ can be combined to form a k-to-n_k block code (f_k, φ_k) setting

$$f_k(\mathbf{s}) \triangleq \hat{f}_{n_k}(\tilde{f}_k(\mathbf{s})) \qquad (\mathbf{s}\in \mathsf{S}^k),$$

$$\varphi_k(\mathbf{y}) \triangleq \begin{cases} \tilde{\varphi}_k(\hat{\varphi}_{n_k}(\mathbf{y})) & \text{if} \quad \hat{\varphi}_{n_k}(\mathbf{y})\in \mathsf{M}_k \\ \text{arbitrary otherwise.} \end{cases}$$

This code satisfies the input constraint (c, Γ) by construction. We check that it also meets the average fidelity criterion (d, Δ) provided that $\varepsilon > 0$ is sufficiently small. In fact, $d(S^k, \varphi_k(Y^{n_k})) > \Delta - \delta$ can occur only if either $d(S^k, \tilde{\varphi}_k(\tilde{f}_k(S^k))) > \Delta - \delta$ or $\hat{\varphi}_{n_k}(Y^{n_k}) \neq \tilde{f}_k(S^k)$. As $(\tilde{f}_k, \tilde{\varphi}_k)$ meets the ε-fidelity criterion (d, Δ) and $(\hat{f}_{n_k}, \hat{\varphi}_{n_k})$ has maximum error probability not exceeding ε, both events have probability at most ε. Thus writing $d_{\mathrm{M}} \triangleq \max_{s,u} d(s, u)$ we have

$$Ed(S^k, \varphi_k(Y^{n_k})) \leqq \Delta - \delta + 2\varepsilon d_{\mathrm{M}} \leqq \Delta$$

if $\varepsilon \leqq \delta(2d_{\mathrm{M}})^{-1}$.

Turning to the converse part of the theorem, suppose that (f, φ) is any k-to-n block code meeting the average fidelity criterion (d, Δ) and satisfying the constraint $Ec(f(S^k)) \leqq \Gamma$. Then

$$kR(\Delta) = kR_k(\Delta) \leqq I(S^k \wedge \varphi(Y^n)), \tag{7.18}$$

where the equality follows from Corollary 7.3 and the inequality holds by the definition of $R_k(\Delta)$. By the data processing lemma Lemma 3.11,

$$I(S^k \wedge \varphi(Y^n)) \leqq I(f(S^k) \wedge Y^n). \tag{7.19}$$

By the definition of $C_n(\Gamma)$ and Corollary 6.12 we further have

$$I(f(S^k) \wedge Y^n) \leqq nC_n(\Gamma) = nC(\Gamma).$$

Comparing this with (7.18) and (7.19) we obtain (7.14). □

Discussion

It is an interesting aspect of Theorem 7.4, both from the mathematical and the engineering points of view, that asymptotically no loss arises if the encoder and decoder of Shannon's block diagram, Fig. 7.1, are composed of two devices: one depending only on the source and the other only on the channel; see Fig. 7.2. This phenomenon is a major

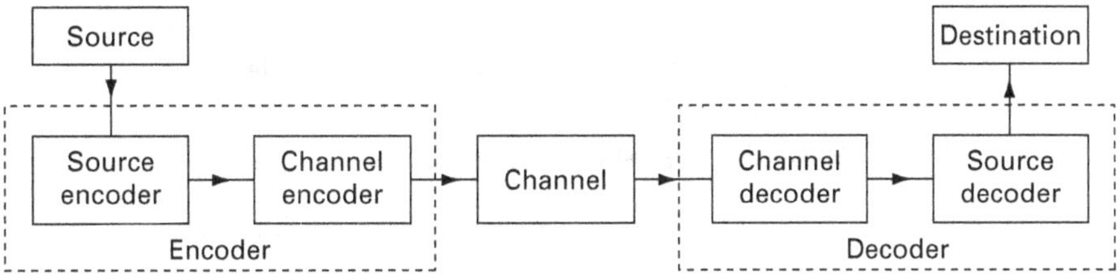

Figure 7.2

reason for studying source and channel coding problems separately. It is no wonder that source block codes, i.e., codes designed for transmission over a noiseless channel, perform well also when combined with a good channel code for a noisy channel. In fact, this is a consequence of the almost noiseless character of the new channel defined in the sense of (6.1) by such a channel code.

Theorem 7.4 is also relevant to non-terminating transmission. As explained in the Introduction, reliable non-terminating transmission can be achieved by blockwise coding whenever the fidelity criterion has the following property: if successive blocks and their reproductions individually meet the fidelity criterion then so do their juxtapositions. This property is now ensured by assumption (7.1), if the average fidelity criterion (d, Δ) is used. Note that this is not the case for the ε-fidelity criterion (d, Δ). ○

Problems

7.1. (*Probability of error and error frequency fidelity criteria*) Let $\mathsf{X} = \mathsf{Y}$, and let the distortion between k-length sequences be defined as the Hamming distance divided by k, i.e., $d_k(\mathbf{x}, \mathbf{y}) \triangleq (1/k)d_{\mathrm{H}}(\mathbf{x}, \mathbf{y})$. Check that in this case a code (f, φ) meets the ε-fidelity criterion $(d, 0)$ iff $\Pr\{\varphi(f(X^k)) \neq X^h\} \leqq \varepsilon$. Further, show that (f, φ) meets the average fidelity criterion (d, Δ) iff the expected relative frequency of the erroneously reproduced digits of X^k is at most Δ.

7.2. Suppose that $\mathsf{X} = \mathsf{Y}$ and d_k is a metric on X^k for every k. Consider the spheres $\mathsf{S}(\mathbf{y}, \Delta) \triangleq \{\mathbf{x} : d_k(\mathbf{x}, \mathbf{y}) \leqq \Delta\}$. Let $N_\varepsilon(k, \Delta)$ be the minimum number of such spheres covering X^k up to a set of probability at most ε, i.e., the smallest integer N for which there exist $\mathbf{y}_1, \mathbf{y}_2, \ldots, \mathbf{y}_N$ such that

$$P_{X^k}\left(\bigcup_{i=1}^{N} \mathsf{S}(\mathbf{y}_i, \Delta)\right) \geqq 1 - \varepsilon.$$

Verify that

$$\overline{\lim_{k\to\infty}} \frac{1}{k} \log N_\varepsilon(k, \Delta) = R_\varepsilon(\Delta)$$

and that the same is true for an arbitrary distortion measure, except for the geometric interpretation of the sets $\mathsf{S}(\mathbf{y}, \Delta)$.

Hint Let the $\mathbf{y}_i$ be the possible values of $\varphi(f(\mathbf{x}))$.

7.3. Check that $R_\varepsilon(\Delta)$ equals the LMTR (see the Introduction) for reliably transmitting the given source over a binary noiseless channel, in the sense of the ε-fidelity criterion (d, Δ).

7.4. (*Zero-error rate*) Show that in general

$$R_0(\Delta) \neq R(\Delta) \triangleq \lim_{\varepsilon \to 0} R_\varepsilon(\Delta).$$

(The zero-error problem for sources, unlike for channels, has been solved; see Theorem 9.2.)

7.5. If $\hat{d}$ is an arbitrary finite-valued function defined on $\mathsf{X} \times \mathsf{Y}$, set

$$d(x, y) \triangleq \hat{d}(x, y) - \min_{y \in \mathsf{Y}} \hat{d}(x, y).$$

Find equivalents of the results of this chapter for $\hat{d}$ by means of this transformation.

7.6. (*Weak converse*)

(a) Prove Corollary 7.3 directly, using the properties of mutual information established in Chapter 3.

(b) Use this result to show that

$$R(\Delta) \geqq \min_{\substack{P_X = P \\ Ed(X, Y) \leqq \Delta}} I(X \wedge Y).$$

7.7. (*Average fidelity criterion*)

(a) Check that Theorem 7.4 contains the counterpart of Theorem 7.3 for codes meeting the average fidelity criterion (d, Δ), provided that $\Delta > 0$.

(b) Show that for $\Delta = 0$ that counterpart does not hold; rather, the minimum achievable rate in the latter case is $R_0(0)$.

7.8. Under the conditions of Theorem 7.4, prove that the LMTR for reliable transmission in the sense of the ε-fidelity criterion (d, Δ) also equals $R(\Delta)/C(\Gamma)$. Show that, unlike Theorem 7.4, this is true also for $\Delta = 0$.

7.9. (*Error frequency fidelity criterion*) Show that if $\mathsf{X} = \mathsf{Y}$ and the distortion measure $d(x, y)$ is 0 if $x = y$ and 1 otherwise, then for a DMS with arbitrary generic distribution P

$$R(\Delta) \geqq H(P) - \Delta \log(|\mathsf{X}| - 1) - h(\Delta);$$

in case $\Delta \leqq (|\mathsf{X}| - 1) \min_{x \in \mathsf{X}} P(x)$ the equality holds.

Hint Use Fano's inequality and follow carefully its proof for the condition of equality. (For this case, Erokhin (1958) gave a formula of $R(\Delta)$ for every $\Delta \geqq 0$.)

7.10. (*Variable distortion level*)

(a) For a DMS, show that the minimum of $\overline{\lim}_{k \to \infty}(1/k) \log ||f_k||$ for sequences of codes meeting the ε-fidelity criteria (d, Δ_k) equals $R(\underline{\lim}_{k \to \infty} \Delta_k)$.

(b) Prove the same for average fidelity criteria, providing $\underline{\lim}_{k \to \infty} \Delta_k > 0$.

7.11. (*Product sources*) The *product* of two DMSs with input alphabets X_1, X_2 and generic distributions P_1, P_2 is the DMS with input alphabet $\mathsf{X}_1 \times \mathsf{X}_2$ and generic distribution $P_1 \times P_2$. Show that if the two sources have Δ-distortion rates $R_1(\Delta)$ resp. $R_2(\Delta)$ for given distortion measures d_1 resp. d_2, then the Δ-distortion rate of the product source with respect to $d((x_1, x_2), (y_1, y_2)) \triangleq d_1(x_1, y_1) + d_2(x_2, y_2)$, is

$$R(\Delta) = \min_{\Delta_1+\Delta_2=\Delta} \{R_1(\Delta_1) + R_2(\Delta_2)\}.$$

(Shannon (1959).)

7.12. (*Peak distortion measures*) Let the distortion between sequences be defined as the largest distortion between their corresponding digits, i.e.,

$$d_k(\mathbf{x}, \mathbf{y}) \triangleq \max_{1 \leqq i \leqq k} d(x_i, y_i) \quad (\mathbf{x} \in \mathsf{X}^k, \mathbf{y} \in \mathsf{Y}^k).$$

Show that, with respect to this distortion measure, a DMS with generic distribution P has Δ-distortion rate

$$R(\Delta) = R_\varepsilon(\Delta) = \min_{\substack{P_X = P \\ \Pr\{d(X,Y) \leqq \Delta\} = 1}} I(X \wedge Y) \qquad \text{for every} \quad 0 < \varepsilon < 1.$$

Check that this $R(\Delta)$ is a staircase function and that unless Δ is a point of jump, the alternative definition of the Δ-distortion rate imposing the average fidelity criterion (d, Δ) leads to the same result.

Hint Apply Theorem 7.3 to the averaging distortion measure (see (7.1)) built from

$$d_\Delta(x, y) \triangleq \begin{cases} 0 & \text{if} \quad d(x, y) \leqq \Delta \\ 1 & \text{if} \quad d(x, y) > \Delta \end{cases}$$

with distortion level 0. Or proceed directly as in the proof of Theorem 7.3.

7.13. (*Non-finite distortion measures*) Check that Theorem 7.3 holds also if $d(x, y) = +\infty$ for some pairs $(x, y) \in \mathsf{X} \times \mathsf{Y}$. Prove a similar statement for the average fidelity criterion (d, Δ), when $\Delta > 0$, provided that there exists a $y_0 \in \mathsf{Y}$ such that $d(x, y_0) < +\infty$ for every $x \in \mathsf{X}$ with $P(x) > 0$. (Gallager (1968), Th. 9.6.2.)

7.14. (*Several distortion measures*) Let $\{d_\lambda\}_{\lambda \in \Lambda}$ be a not necessarily finite family of averaging distortion measures with common alphabets (see (7.1)) such that

(i) to every $x \in \mathsf{X}$ there is a $y \in \mathsf{Y}$ with $d_\lambda(x, y) = 0$ for every $\lambda \in \Lambda$;

(ii) $d_\lambda(x, y)$ is less than some constant $D < \infty$ unless it is infinite.

Given a source $\{X_i\}_{i=1}^\infty$, a k-length block code (f, φ) meets the ε-fidelity criterion $\{d_\lambda, \Delta_\lambda\}_{\lambda \in \Lambda}$ if

$$\Pr\{d_\lambda(X^k, \varphi(f(X^k))) \leqq \Delta_\lambda \text{ for every } \lambda \in \Lambda\} \geqq 1 - \varepsilon.$$

Define $R_\varepsilon(\{\Delta_\lambda\})$ and $R(\{\Delta_\lambda\})$ analogously to $R_\varepsilon(\Delta)$ resp. $R(\Delta)$ and show that if $\inf_{\Delta_\lambda \neq 0} \Delta_\lambda > 0$ then for a DMS with generic distribution P

$$R_\varepsilon(\{\Delta_\lambda\}) = \min I(X \wedge Y),$$

the minimum taken for RVs X and Y such that $P_X = P$ and $Ed_\lambda(X, Y) \leqq \Delta_\lambda$ for every $\lambda \in \Lambda$.

7.15. (*Variable-length codes*)

(a) Given a DMS $\{X_i\}_{i=1}^\infty$ and a distortion measure $d(x, y)$ on $\mathsf{X} \times \mathsf{Y}$, show that to every $\Delta \geqq 0$ there exist codes (f_k, φ_k) with $f_k : \mathsf{X}^k \to \{0, 1\}^*$, $\varphi_k : \{0, 1\}^* \to \mathsf{Y}^k$, such that $d(X^k, \varphi_k(f_k(X^k))) \leqq \Delta$ with probability 1 and $\lim_{k\to\infty}(1/k)\, El(f_k(X^k)) = R(\Delta)$.

Hint Use Theorem 7.3.

(b) With the notation of Theorem 7.4, show that if a variable-length code (f, φ), where $f : \mathsf{S}^k \to \mathsf{X}^*, \varphi : \mathsf{Y}^* \to \mathsf{U}^k$, meets the average fidelity criterion (d, Δ) then the average codeword length $\bar{l}(f) \triangleq (1/k)El(f(X^k))$ satisfies

$$\bar{l}(f) \geqq \frac{R(\Delta) - \dfrac{\log(ke\bar{l}(f))}{k\bar{l}(f)}}{C(\Gamma)}.$$

Hint See the hint of Problem 6.27.

(c) Prove that the assertion of (b) holds also for codes with complete feedback.

Hint Use Problem 6.29.

7.16. (*Optimal transmission without coding*) Let $\mathsf{X} = \mathsf{Y} = \{0, 1\}$, let $\{X_i\}_{i=1}^\infty$ be a DMS with generic distribution $P \triangleq (1/2, 1/2)$ and let $\{W\}$ be a binary symmetric channel with crossover probability $p < 1/2$. Then if both the encoder and the decoder are the identity mapping, the average error frequency is p. Show that no code with transmission ratio 1 can give a smaller average error frequency.

Hint See Problems 7.9 and 6.9.

(Jelinek (1968b), §11.8.)

7.17. (*Remote sources*) Supposing that the encoder has no direct access to the source outputs and the destination has no direct access to the decoder's output, the following mathematical model is of interest. Given a DMS $\{X_i\}_{i=1}^\infty$ and two DMCs $\{W_1 : \mathsf{X} \to \mathsf{X}\}$, $\{W_2 : \mathsf{Y} \to \mathsf{Y}\}$, a k-length block code is a pair of mappings as at the beginning of this chapter. However, while the source output is X^k, the encoder f is applied to $\hat{X}^k$, the output of W_1^k corresponding to input X^k. Similarly, the destination receives Y^k, the output of W_2^k corresponding to input $\hat{Y}^k \triangleq \varphi(f(\hat{X}^k))$.

(a) Show that the Δ-distortion rate corresponding to the average fidelity criterion $Ed(X^k, Y^k) \leqq \Delta$ equals $\min I(\hat{X} \wedge \hat{Y})$ for RVs $X \multimap \hat{X} \multimap \hat{Y} \multimap Y$ satisfying $Ed(X, Y) \leqq \Delta$, where X and $\hat{X}$ resp. $\hat{Y}$ and Y are connected by the channels W_1 resp. W_2, and $P_X = P_{X_1}$.

(b) Prove the corresponding source–channel transmission theorem.

(Dobrušin and Tsybakov (1962), Berger (1971).)

Story of the results

The theorems of this chapter were heuristically formulated by Shannon (1948). Theorem 7.4 and the equivalent of Theorem 7.3 for average fidelity criteria were proved by Shannon (1959); similar results were obtained in a complex general framework by Dobrušin (1959b). The present Theorem 7.3 is implicit in Wolfowitz (1966); the given proof is his.

8 Computation of channel capacity and Δ-distortion rates

We have seen that the capacity of a DMC $\{W : \mathsf{X} \to \mathsf{Y}\}$ under input constraint (c, Γ) is

$$C(\Gamma) = C(W, \Gamma) = \max_{P: c(P) \leqq \Gamma} I(P, W) \quad (\Gamma \geqq \Gamma_0), \tag{8.1}$$

where $\Gamma_0 \triangleq \min_{x \in \mathsf{X}} c(x)$. Similarly, the Δ-distortion rate of a DMS with generic distribution P equals

$$R(\Delta) = R(P, \Delta) = \min_{W: d(P,W) \leqq \Delta} I(P, W) \quad (\Delta \geqq 0). \tag{8.2}$$

An analytic solution of the extremum problems (8.1) and (8.2) is possible only in a few special cases. In this chapter, we present efficient algorithms for computing $C(\Gamma)$ and $R(\Delta)$ for arbitrary DMCs and DMSs, respectively. As a byproduct, we arrive at new characterizations of the functions $C(\Gamma)$ and $R(\Delta)$.

Fixing a DMC and a constraint function, we shall consider the capacity with input constraint (c, Γ) as a function of Γ, called the *capacity-constraint function.* Similarly, for a fixed DMS and distortion measure, the Δ-distortion rate as a function of Δ will be called the *rate-distortion function.* The latter is positive iff

$$\Delta < \Delta^* \triangleq \min_y \sum_x P(x) d(x, y), \tag{8.3}$$

since a W with identical rows $W(\cdot\,|x)$ satisfying $d(P, W) \leqq \Delta$ exists iff $\Delta \geqq \Delta^*$. Thus when studying the rate-distortion function, attention may be restricted to $\Delta < \Delta^*$. The analog of Δ^* for the capacity-constraint function is the smallest value of Γ for which $C(\Gamma)$ equals the unconstrained capacity C; however, this value Γ^* is not so simply determined as Δ^*.

As $C(\Gamma)$ is a non-decreasing concave function by Theorem 6.11, and $R(\Delta)$ is a non-increasing convex one by Lemma 7.2, we see that $C(\Gamma)$ is strictly increasing for $\Gamma_0 \leqq \Gamma \leqq \Gamma^*$ and $R(\Delta)$ is strictly decreasing for $0 \leqq \Delta \leqq \Delta^*$. In particular, in these intervals, when the extrema in (8.1) resp. (8.2) are achieved, then the constraints are satisfied with equality.

We shall need the fact that the curve $C(\Gamma)$ is the lower envelope of straight lines with vertical axis intercept

$$F(\gamma) \triangleq \max_P \left[I(P, W) - \gamma c(P) \right]$$

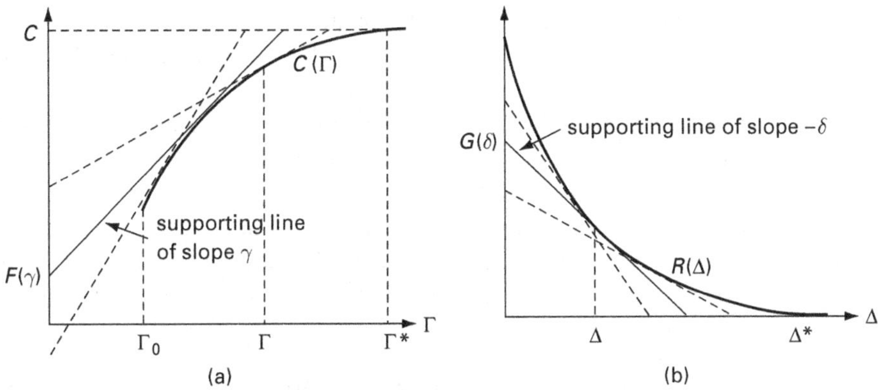

Figure 8.1 (a) Capacity-constraint function. (b) Rate-distortion function.

and slope γ ($\gamma \geqq 0$), while the curve $R(\Delta)$ is the upper envelope of straight lines of vertical axis intercept

$$G(\delta) \triangleq \min_{W} [I(P, W) + \delta d(P, W)]$$

and slope $-\delta$ ($\delta \geqq 0$):

LEMMA 8.1 We have

$$C(\Gamma) = \min_{\gamma \geqq 0} [F(\gamma) + \gamma\Gamma] \quad \text{if} \quad \Gamma > \Gamma_0, \tag{8.4}$$

$$R(\Delta) = \max_{\delta > 0} [G(\delta) - \delta\Delta] \quad \text{if} \quad 0 < \Delta < \Delta^*, \tag{8.5}$$

$$R(0) = \sup_{\delta > 0} G(\delta). \tag{8.6}$$

Moreover, if for some $\gamma \geqq 0$ a P maximizes $I(P, W) - \gamma c(P)$ and $c(P) = \Gamma$ then $I(P, W) = C(\Gamma)$. Similarly, if for some $\delta > 0$ a W minimizes $I(P, W) + \delta d(P, W)$ and $d(P, W) = \Delta$ then $I(P, W) = R(\Delta)$. ○

REMARK In the formula for the rate-distortion function, $\delta = 0$ may be excluded since $R(\Delta) > 0$ for $\Delta < \Delta^*$. ○

Proof $c(P) \leqq \Gamma$ implies $I(P, W) - \gamma c(P) \geqq I(P, W) - \gamma\Gamma$, whence the inequality $C(\Gamma) \leqq \inf_{\gamma \geqq 0}[F(\gamma) + \gamma\Gamma]$ is obvious. It remains to check that for every $\Gamma > \Gamma_0$ there exists a $\gamma \geqq 0$ with $C(\Gamma) = F(\gamma) + \gamma\Gamma$. If $\Gamma \geqq \Gamma^*$, take $\gamma = 0$. If $\Gamma_0 < \Gamma < \Gamma^*$ then, by the concavity and strict monotonicity of $C(\Gamma)$, there exists a $\gamma > 0$ such that $C(\Gamma') \leqq C(\Gamma) + \gamma(\Gamma' - \Gamma)$ for every $\Gamma' \geqq \Gamma_0$. Let P achieve the maximum in (8.1). Then $c(P) = \Gamma$, and the preceding inequality implies for every P', setting $\Gamma' \triangleq c(P')$, that

$$I(P', W) - \gamma c(P') \leqq C(\Gamma') - \gamma\Gamma' \leqq C(\Gamma) - \gamma\Gamma = I(P, W) - \gamma c(P).$$

Thus $F(\gamma) = I(P, W) - \gamma c(P) = C(\Gamma) - \gamma\Gamma$, completing the proof of (8.4).

The proof of (8.5) is similar and is omitted. Equation (8.6) follows from (8.5) since $R(\Delta)$ is continuous at $\Delta = 0$ by Lemma 7.2; see Fig. 8.1(b). The remaining assertions are obvious. □

Lemma 8.1 shows that the capacity-constraint and rate-distortion functions are easily computed if the functions $F(\gamma)$ and $G(\delta)$ are known. In particular, $C = F(0)$. We shall deal with the computation of $F(\gamma)$ resp. $G(\delta)$.

Our aim is to replace the maximum in the definition of $F(\gamma)$ by a double maximum in such a way that by fixing one variable, the maximum with respect to the other one can be found readily. Similarly, we shall express $G(\delta)$ as a double minimum.

We state the obvious identity

$$D(W\|Q|P) \triangleq \sum_{x\in\mathsf{X}} P(x)D(W(\cdot|x)\|Q) = I(P,W) + D(PW\|Q), \tag{8.7}$$

which holds for every distribution Q on Y.

Consider – for a fixed channel $W\colon \mathsf{X}\to\mathsf{Y}$ and fixed $\gamma \geqq 0$ – the following function of two distributions on X:

$$F(P,P') = F(P,P',\gamma,W) \triangleq D(W\|P'W|P) - D(P\|P') - \gamma c(P).$$

LEMMA 8.2 For fixed P, $F(P,P')$ is maximized if $P' = P$ and

$$\max_{P'} F(P,P') = I(P,W) - \gamma c(P).$$

For fixed P', $F(P,P')$ is maximized if

$$P(x) = \frac{1}{A}P'(x)\exp\left[D(W(\cdot|x)\|P'W) - \gamma c(x)\right],$$

where A is a norming constant, and

$$\max_{P} F(P,P') = \log\sum_{x\in\mathsf{X}} P'(x)\exp\left[D(W(\cdot|x)\|P'W) - \gamma c(x)\right]. \qquad \bigcirc$$

COROLLARY 8.2A *For any PDs P on X and Q on Y,*

$$\log\sum_{x\in\mathsf{X}} P(x)\exp\left[D(W(\cdot|x)\|PW) - \gamma c(x)\right]$$

$$\leqq F(\gamma) \leqq \max_{x\in\mathsf{X}}\left[D(W(\cdot|x)\|Q) - \gamma c(x)\right]. \qquad \bigcirc$$

COROLLARY 8.2B

$$F(\gamma) = \max_{P,P'} F(P,P'). \qquad \bigcirc$$

Proof By (8.7),

$$F(P,P') = I(P,W) + D(PW\|P'W) - D(P\|P') - \gamma c(P),$$

whence the first assertion follows by the data processing lemma. The second assertion is a consequence of Lemma 3.12, applying it with $\alpha = -1$ and

$$f(x) \triangleq \log P'(x) + D(W(\cdot\,|x)\|P'W) - \gamma c(x).$$

It remains to check the second inequality of Corollary 8.2A. This follows from (8.7), if applied to a PD $\tilde{P}$ maximizing $I(P, W) - \gamma c(P)$:

$$\max_{x \in \mathsf{X}} [D(W(\cdot\,|x)\|Q) - \gamma c(x)] \geqq D(W\|Q|\tilde{P}) - \gamma c(\tilde{P})$$
$$= I(\tilde{P}, W) + D(\tilde{P}W\|Q) - \gamma c(\tilde{P}) = F(\gamma) + D(\tilde{P}W\|Q). \tag{8.8}$$

□

Corollary 8.2B suggests that $F(\gamma)$ might be computed by an iteration, maximizing $F(P, P')$ with respect to P resp. P' in an alternating manner. The next theorem shows that this iteration indeed converges to $F(\gamma)$ whenever we start from a strictly positive distribution P_1.

THEOREM 8.3 (*Capacity computing algorithm*) Let P_1 be an arbitrary distribution on X such that $P_1(x) > 0$ for every $x \in \mathsf{X}$, and define the distributions P_n, $n = 2, 3, \ldots$, recursively by

$$P_n(x) \triangleq A_n^{-1} P_{n-1}(x) \exp [D(W(\cdot\,|x)\|P_{n-1}W) - \gamma c(x)], \tag{8.9}$$

where A_n is defined by the condition that P_n be a PD on X. Then

$$\log \sum_{x \in \mathsf{X}} P_n(x) \exp [D(W(\cdot\,|x)\|P_n W) - \gamma c(x)] = \log A_{n+1}$$

converges from below and $\max_{x \in \mathsf{X}}[D(W(\cdot\,|x)\|P_n W) - \gamma c(x)]$ converges from above to $F(\gamma)$. Moreover, the sequence of distributions P_n converges to a distribution P^* such that $I(P^*, W) - \gamma c(P^*) = F(\gamma)$. ○

Proof On account of Lemma 8.2,

$$F(P_1, P_1) \leqq F(P_2, P_1) \leqq F(P_2, P_2) \leqq F(P_3, P_2) \leqq \ldots$$

thus $F(P_n, P_n) = I(P_n, W) - \gamma c(P_n)$ and $F(P_n, P_{n-1}) = \log A_n$ converge increasingly to the same limit not exceeding $F(\gamma)$.

If P is any PD for which $I(P, W) - \gamma c(P) = F(\gamma)$ then, by the foregoing, (8.9) and (8.7), we get

$$0 \leqq F(\gamma) - \log A_n = I(P, W) - \gamma c(P) - \log A_n$$
$$= I(P, W) + \sum_{x \in \mathsf{X}} P(x) \log \frac{P_n(x)}{P_{n-1}(x)} - D(W\|P_{n-1}W\|P)$$
$$\leqq \sum_{x \in \mathsf{X}} P(x) \log \frac{P_n(x)}{P_{n-1}(x)} = D(P\|P_{n-1}) - D(P\|P_n). \tag{8.10}$$

It follows that the series $\sum_{n=2}^{\infty}(F(\gamma) - \log A_n)$ is convergent, thus $\log A_n \to F(\gamma)$ as asserted. To check that the sequence P_n also converges, consider a convergent subsequence, $P_{n_k} \to P^*$, say. Then, clearly, $I(P^*, W) - \gamma c(P^*) = F(\gamma)$. Substituting $P = P^*$ in (8.10), we see that the sequence of divergences $D(P^* \| P_n)$ is non-increasing. Thus $D(P^* \| P_{n_k}) \to 0$ results in $D(P^* \| P_n) \to 0$, proving $P_n \to P^*$.

Finally, by the convergence relations proved so far, the recursion defining P_n gives

$$\lim_{n\to\infty} \frac{P_n(x)}{P_{n-1}(x)} = \exp\left[D(W(\cdot|x)\| P^* W) - \gamma c(x) - F(\gamma)\right].$$

But this limit is 1 if $P^*(x) > 0$ and does not exceed 1 if $P^*(x) = 0$. Hence

$$D(W(\cdot|x)\| P^* W) - \gamma c(x) \leqq F(\gamma) \tag{8.11}$$

for every $x \in \mathsf{X}$, with equality if $P^*(x) > 0$. This proves that

$$\max_x \left[D(W(\cdot|x)\| P_n W) - \gamma c(x)\right] \to F(\gamma),$$

and the convergence is from above by Corollary 8.2A. □

THEOREM 8.4 For any $\gamma \geqq 0$,

$$F(\gamma) = \min_Q \max_{x\in\mathsf{X}} \left[D(W(\cdot|x)\| Q) - \gamma c(x)\right], \tag{8.12}$$

where Q ranges over the PDs on Y. The minimum is achieved iff $Q = PW$ for a PD P on X such that

$$I(P, W) - \gamma c(P) = F(\gamma). \tag{8.13}$$

This Q is unique. A PD P satisfies (8.13) iff $D(W(\cdot|x)\| PW) - \gamma c(x)$ is constant on the support of P and does not exceed this constant elsewhere. ○

COROLLARY 8.4 For $\Gamma > \Gamma_0$

$$C(\Gamma) = \min_Q \min_{\gamma \geqq 0} \max_{x\in\mathsf{X}} \left[D(W(\cdot|x)\| Q) + \gamma(\Gamma - c(x))\right].$$

The minimizing Q is the output distribution of channel W corresponding to any input distribution which achieves the capacity under input constraint (c, Γ). ○

→ 8.1 COMMENT As $F(0) = C$, Theorem 8.4 gives the new formula

$$C = \min_Q \max_{x\in\mathsf{X}} D(W(\cdot|x)|| Q)$$

for the capacity of the DMC $\{W\}$. This formula has an interesting "geometric" interpretation. Since informational divergence is a (non-metric) "distance" between distributions, C may be interpreted as the radius of the smallest "sphere" containing the set of distributions $W(\cdot|x), x \in \mathsf{X}$. Then the minimizing Q is the centre of this
→ 8.2 "sphere." ○

Proof of Theorem 8.4. Equation (8.12) follows from (8.8) and (8.11). If P is any PD with $I(P, W) - \gamma c(P) = F(\gamma)$, then we see from (8.8) (with $\tilde{P} \triangleq P$) that if a PD

Q on Y achieves the minimum in (8.12) then $D(PW\|Q) = 0$, i.e., $Q = PW$. Thus, even though there may be several Ps maximizing $I(P, W) - \gamma c(P)$, the corresponding output distribution PW is unique and it is the unique Q achieving the minimum in (8.12). Further, if P satisfies (8.13) then by the above we have

$$\max_{x\in\mathsf{X}} [D(W(\cdot|x)\|PW) - \gamma c(x)] = F(\gamma),$$

thus (8.8) gives that

$$D(W(\cdot|x)\|PW) - \gamma c(x) = F(\gamma) \quad \text{if} \quad P(x) > 0.$$

The preceding two formulas prove that $D(W(\cdot|x)\|PW) - \gamma c(x)$ is constant on the support of P and it nowhere exceeds this constant. Conversely, if a P has this property then

$$I(P, W) - \gamma c(P) = \max_{x\in\mathsf{X}} [D(W(\cdot|x)\|PW) - \gamma c(P)],$$

thus P satisfies (8.13) by (8.8). This completes the proof of the theorem. The corollary follows by Lemma 8.1. □

Analogous results hold for the function $G(\delta)$ and the rate-distortion function. Fixing a distribution P on X and a $\delta > 0$, consider the following function of a channel $W : \mathsf{X} \to \mathsf{Y}$ and of a PD Q on Y:

$$G(W, Q) = G(W, Q, \delta, P) \triangleq D(W\|Q|P) + \delta d(P, W).$$

LEMMA 8.5 The minimum of $G(W, Q)$ for a fixed W is attained by $Q = PW$, and is equal to $I(P, W) + \delta d(P, W)$. The minimum of $G(W, Q)$ for a fixed Q is attained by

$$W(y|x) = \frac{1}{A(x)} Q(y) \exp[-\delta d(x, y)],$$

where

$$A(x) \triangleq \sum_{y\in\mathsf{Y}} Q(y) \exp[-\delta d(x, y)]$$

and

$$\min_W G(W, Q) = -\sum_{x\in\mathsf{X}} P(x) \log A(x). \qquad \bigcirc$$

COROLLARY 8.5

$$G(\delta) = \min_{W,Q} G(W, Q) = \min_Q \left\{ -\sum_{x\in\mathsf{X}} P(x) \log \sum_{y\in\mathsf{Y}} Q(y) \exp[-\delta d(x, y)] \right\}. \qquad \bigcirc$$

Proof The first assertion follows from (8.7). To check the second one, observe that for any positive numbers $B(x)$, $x \in \mathsf{X}$,

$$\begin{aligned} G(W, Q) &= \sum_{x,y} P(x)W(y|x)\left[\log \frac{W(y|x)}{Q(y)} + \delta d(x, y)\right] \\ &= \sum_{x} P(x) \log \frac{B(x)}{P(x)} + \sum_{x,y} P(x)W(y|x) \log \frac{P(x)W(y|x)}{B(x)Q(y) \exp[-\delta d(x, y)]} \\ &\geqq \sum_{x} P(x) \log \frac{B(x)}{P(x)} - \log \sum_{x,y} B(x)Q(y) \exp[-\delta d(x, y)], \end{aligned} \tag{8.14}$$

where the last step follows from the log-sum inequality. The equality holds iff

$$P(x)W(y|x) = cB(x)Q(y) \exp[-\delta d(x, y)]$$

for some constant $c > 0$. Setting $B(x) \triangleq P(x)/A(x)$, the assertion follows.

The corollary is immediate. □

THEOREM 8.6 (*Δ-distortion rate computing algorithm*) Let Q_0 be a strictly positive distribution on Y, and define the channels $W_n : \mathsf{X} \to \mathsf{Y}$ and the PDs Q_n on Y, $n = 1, 2, \ldots$, recursively by

$$W_n(y|x) \triangleq A_n^{-1}(x)Q_{n-1}(y) \exp[-\delta d(x, y)], \qquad Q_n \triangleq PW_n,$$

where the $A_n(x)$ are norming constants. Then

$$\begin{aligned} -\sum_{x\in\mathsf{X}} P(x) \log A_n(x) - \max_{y\in\mathsf{Y}} \log \sum_{x\in\mathsf{X}} \frac{P(x)}{A_n(x)} \exp[-\delta d(x, y)] \\ \leqq G(\delta) \leqq -\sum_{x\in\mathsf{X}} P(x) \log A_n(x) \end{aligned} \tag{8.15}$$

and both the lower and upper bounds converge to $G(\delta)$ as $n \to \infty$. Moreover, the matrices W_n converge to a matrix W^* such that $I(P, W^*) + \delta d(P, W^*) = G(\delta)$. ○

Proof The first inequality in (8.15) follows by substituting $B(x) \triangleq P(x)/A_n(x)$ in (8.14) and using Corollary 8.5. On account of Lemma 8.5, one has

$$G(W_1, Q_0) \geqq G(W_1, Q_1) \geqq G(W_2, Q_1) \geqq G(W_2, Q_2) \geqq G(W_3, Q_2) \geqq \ldots$$

thus

$$G(W_n, Q_n) = I(P, W_n) + \delta d(P, W_n) \text{ and } G(W_n, Q_{n-1}) = -\sum_{x\in\mathsf{X}} P(x) \log A_n(x)$$

both converge decreasingly to a common limit which is not less than $G(\delta)$.

If $W : \mathsf{X} \to \mathsf{Y}$ is a channel with $I(P, W) + \delta d(P, W) = G(\delta)$ then, by the foregoing and (8.7),

$$\begin{aligned}
0 &\leqq -\sum_{x\in\mathsf{X}} P(x)\log A_n(x) - G(\delta)\\
&= -\sum_{x\in\mathsf{X}} P(x)\log A_n(x) - \delta d(P, W) - I(P, W)\\
&= D(W\|Q_{n-1}|P) - D(W\|W_n|P) - I(P, W)\\
&= D(PW\|Q_{n-1}) - D(W\|W_n|P) \leqq D(PW\|Q_{n-1}) - D(PW\|Q_n), \qquad (8.16)
\end{aligned}$$

where the last step follows by the convexity of informational divergence, since $PW_n = Q_n$. From (8.16) one concludes that $-\sum_{x\in\mathsf{X}} P(x)\log A_n(x) \to G(\delta)$ and that the sequence Q_n is convergent, just as the analogous assertions of Theorem 8.4 were deduced from (8.10).

Setting $Q^* \triangleq \lim_{n\to\infty} Q_n$ and $A^*(x) \triangleq \lim_{n\to\infty} A_n(x)$, the recursion implies

$$\sum_{x\in\mathsf{X}} \frac{P(x)}{A^*(x)} \exp\left[-\delta d(x, y)\right] = \lim_{n\to\infty} \frac{Q_n(y)}{Q_{n-1}(y)} \leqq 1, \qquad (8.17)$$

with equality if $Q^*(y) > 0$. This completes the proof of the theorem. □

THEOREM 8.7 For any fixed $\delta > 0$, a channel $W : \mathsf{X} \to \mathsf{Y}$ achieves → 8.3

$$I(P, W) + \delta d(P, W) = G(\delta) \qquad (8.18)$$

iff there exist non-negative numbers $B(x)$, $x \in \mathsf{X}$, such that for every $y \in \mathsf{Y}$

$$\sum_{x\in\mathsf{X}} B(x)\exp\left[-\delta d(x, y)\right] \leqq 1, \qquad (8.19)$$

with equality if $PW(y) > 0$, and

$$P(x)W(y|x) = B(x)PW(y)\exp\left[-\delta d(x, y)\right]. \qquad (8.20)$$

Further

$$G(\delta) = \max_B \sum_{x\in\mathsf{X}} P(x)\log\frac{B(x)}{P(x)}, \qquad (8.21)$$

where B ranges over the collections of non-negative numbers satisfying (8.19) for every $y \in \mathsf{Y}$. The maximum is achieved iff the numbers $B(x)$ correspond to a W satisfying (8.18); this B is unique. ○

COROLLARY 8.7 For $0 < \Delta < \Delta^*$,

$$R(\Delta) = -\min_Q \min_{\delta>0} \left\{ D(P\|Q) + \max_{y\in\mathsf{Y}} \log \sum_{x\in\mathsf{X}} Q(x)\exp\left[\delta(\Delta - d(x, y))\right] \right\},$$

where Q ranges over the PDs on X, and

$$R(0) = -\min_{Q}\left\{D(P||Q) + \max_{y\in\mathsf{Y}}\log\sum_{x:d(x,y)=0} Q(x)\right\}. \qquad \bigcirc$$

Proof For any channel $W : \mathsf{X} \to \mathsf{Y}$ and non-negative numbers $B(x)$ satisfying (8.19) for every $y \in \mathsf{Y}$, (8.14) gives

$$I(P, W) + \delta d(P, W) = G(W, PW) \geqq \sum_{x\in\mathsf{X}} P(x)\log\frac{B(x)}{P(x)}.$$

The equality holds iff (8.20) is fulfilled and in (8.19) the equality holds whenever $PW(y) > 0$. To complete the proof of the theorem, it suffices to show the existence of W and B satisfying the above conditions. (The uniqueness of B achieving the maximum in (8.21) is clear by convexity.) This follows, however, from Theorem 8.6, as $W^* \triangleq \lim_{n\to\infty} W_n$ satisfies (8.20) with $B(x) = P(x)/A^*(x)$ (where $A^*(x) \triangleq \lim_{n\to\infty} A_n(x)$). This choice satisfies (8.19), with equality if $PW^*(y) > 0$, as one sees from (8.17).

To prove the corollary, observe that by Lemma 8.1 the theorem gives

$$R(\Delta) = \max\left[\sum_{x\in\mathsf{X}} P(x)\log\frac{B(x)}{P(x)} - \delta\Delta\right],$$

the maximum being taken for positive numbers $B(x)$ and δ satisfying (8.19). There is a one-to-one correspondence between non-negative numbers $B(x)$ satisfying (8.19) with equality for at least one $y \in \mathsf{Y}$, and PDs Q on X, given by

$$B(x) = Q(x)\left(\max_{y\in\mathsf{Y}}\sum_{\bar{x}\in\mathsf{X}} Q(\bar{x})\exp\left[-\delta d(\bar{x}, y)\right]\right)^{-1}.$$

This proves the asserted formula for $0 < \Delta < \Delta^*$. The formula for $R(0)$ follows by letting $\Delta \to 0$, since clearly, for any fixed Q, the value of δ minimizing $\max_{y\in\mathsf{Y}}\log\sum_{x\in\mathsf{X}} Q(x)\exp[\delta(\Delta - d(x, y))]$ tends to $+\infty$ as $\Delta \to 0$, uniformly in Q. □

Problems

8.1. (*Information radius*)

(a) Prove directly the capacity formula

$$C = \min_{Q}\max_{x\in\mathsf{X}} D(W(\cdot\,|x)||Q).$$

Hint Clearly, $\max_{x\in\mathsf{X}} D(W(\cdot\,|x)||Q) = \max_P D(W||Q|P)$. Thus, in view of (8.7), the assertion to be proved is

$$\max_{P}\min_{Q} D(W||Q|P) = \min_{Q}\max_{P} D(W||Q|P).$$

This follows, however, from the minimax theorem (see, e.g., Karlin (1959), Th. 1.1.5). (This proof is from Csiszár (1972); the use of identity (8.7) in this context was suggested by Topsøe (1967), see also Topsøe (1972).)

(b) Give a similar proof of formula (8.12) for $F(\gamma)$.

(c) An "information radius" of the channel W might be defined also as $\min_Q \max_{x \in \mathsf{X}} D(Q||W(\cdot|x))$. This extremum is, in general, different from C. Show that it equals

$$\sup_P \min_Q D(Q||W|P) = \sup_P \left(-\log \sum_{y \in \mathsf{Y}} \prod_{x \in \mathsf{X}} [W(y|x)]^{P(x)} \right),$$

where P runs over the PDs on X with $P(x) > 0$ for every $x \in \mathsf{X}$. (Csiszár (1972).)

8.2. Conclude from Theorem 8.4 that a distribution P on X maximizes $I(P, W)$ subject to $c(P) \leqq \Gamma$ iff for some $\gamma \geqq 0$ and K we have

(i) $D(W(\cdot|x)||PW) - \gamma c(x) \leqq K$ for every $x \in \mathsf{X}$, with equality whenever $P(x) > 0$, and

(ii) $c(P) = \Gamma$ if $\gamma > 0$ resp. $c(P) \leqq \Gamma$ if $\gamma = 0$.

8.3. Given a PD P on X and a channel $W : \mathsf{X} \to \mathsf{Y}$, the "backward channel" is $\widetilde{W} : \tilde{\mathsf{Y}} \to \mathsf{X}$, where $\tilde{\mathsf{Y}}$ is the support of PW and $\widetilde{W}(x|y) \triangleq P(x)W(y|x)/PW(y)$.

(a) Conclude from Theorem 8.7 that if $0 < \Delta < \Delta^*$, a channel W minimizes $I(P, W)$ subject to $d(P, W) \leqq \Delta$ iff $d(P, W) = \Delta$ and the backward channel is of the form $\widetilde{W}(x|y) = B(x) \exp[-\delta d(x, y)]$, where $\sum_x B(x) \exp[-\delta d(x, y)] \leqq 1$ also for $y \in \mathsf{Y} - \tilde{\mathsf{Y}}$.

(b) Prove this assertion directly.

(c) Show that, although the minimizing W need not be unique, the corresponding backward channel is unique.

(The minimizing W is characterized in Gallager (1968) and Berger (1971); the latter attributes the sufficiency part to Gerrish (1963, unpublished).)

8.4. (*Differentiability of capacity-constraint function*) Show that the concave function $C(\Gamma)$ is continuous also at the point $\Gamma = \Gamma_0$ and that it is differentiable in (Γ_0, ∞), with the possible exception of $\Gamma = \Gamma^*$.

Hint Show that the minimizing γ in Corollary 8.4 is unique.

8.5. (a) Show that if P maximizes $I(P, W)$ then PW is a strictly positive distribution, though P need not be even if W is a regular square matrix.

(b) If W is a regular square matrix and $\max_P I(P, W)$ is achieved by a strictly positive P, determine this maximum by solving a system of linear equations. (Muroga (1953).)

8.6. (*Differentiability of rate-distortion function*) Show that $R(\Delta)$ is differentiable in $(0, \infty)$ with the possible exception of $\Delta = \Delta^*$.

Hint Use Problem 8.3. (Gallager (1968).)

8.7. (*Non-finite distortion measures and zero distortion*)

(a) If the distortion measure can take the value $+\infty$ then $R(\Delta)$ can be positive for every $\Delta \geqq 0$. Show that even in this case

$$R(\Delta) = \max_{\delta \geqq 0}[G(\delta) - \delta\Delta] \text{ for every } \Delta > 0.$$

(b) Extend Lemma 8.5 and Theorems 8.6 and 8.7 to such distortion measures, replacing $\exp[-\delta d(x, y)]$ by zero if $d(x, y) = \infty$ (for every $\delta \geqq 0$).

(c) For any distortion measure d, as in the passage containing (7.1), define the distortion measure $\bar{d}$ setting $\bar{d}(x, y)$ equal to zero if $d(x, y) = 0$ and $+\infty$ otherwise. Denote by $\bar{G}$ the G-function corresponding to $\bar{d}$, and show that $R(0) = \bar{G}(\delta)$ for each $\delta \geqq 0$. Using this, formulate implications of (b) for $R(0)$ and give direct proofs for them.

(d) Show by an example that the assertion of Problem 8.6 need not hold for non-finite distortion measures.

(Gallager (1968).)

8.8. Show that if $\mathsf{X} = \mathsf{Y}$ and $d(x, y) = 0$ iff $x = y$, then, for sufficiently small Δ, $R(\Delta)$ can be given a parametric representation by solving a system of linear equations.

Hint The solution $\{B(x)\}$ of $\sum_x B(x) \exp[-\delta d(x, y)] = 1$, $y \in \mathsf{X}$, is positive if δ is sufficiently large. Then with $G(\delta) = \sum_x P(x) \log(B(x)/P(x))$, one can write $R(\Delta) = G(\delta) - \delta\Delta(\delta)$, $\Delta = \Delta(\delta)$. (Jelinek (1967).)

8.9. (*Capacity per unit cost*) Extend the result of Problem 6.28 to the case when one input symbol is free, $c(x_0) = 0$ for some $x_0 \in \mathsf{X}$, while for all other $x \in \mathsf{X}$ $c(x) > 0$. Show that in this case the capacity per unit cost is equal to

$$\max_{x \neq x_0} \frac{1}{c(x)} D\left(W(\,.\,|x) \| W(\,.\,|x_0)\right).$$

(Verdú (1990), who also gave a simple code achieving this capacity.)

Hint Verify first that the capacity per unit cost equals the smallest γ with $F(\gamma) = 0$ (or $+\infty$ if no such γ exists). Then show that the minimum in (8.12) is zero iff the minimizing Q equals $W(\,.\,|x_0)$ and

$$D(W(\,.\,|x) \| W(\,.\,|x_0)) - \gamma c(x) \leqq 0 \quad \text{for all} \quad x \in \mathsf{X}.$$

8.10. (*Minimax redundancy and channel capacity*) By Problem 4.23(d), the minimax redundancy $r(k)$ of a class of sources differs by less than $1/k$ from the information radius of the family $\mathcal{P}_k$ consisting of the k-dimensional distributions of the sources in the class. Extend the result of Problem 8.1, obtained for the finite family $\{W(\,.\,|x),\ x \in \mathsf{X}\}$, to this case.

(a) Regard the family $\mathcal{P}_k$ as a channel with (typically) infinite input alphabet, assigning an input symbol to each $P \in \mathcal{P}_k$. If $\mathcal{P}_k$ is a closed set, define for each probability measure μ on $\mathcal{P}_k$ the mutual information over this channel by

$$I(\mu) \triangleq H(P_\mu) - \int H(P)\,\mu(dP), \quad \text{where } P_\mu(\mathbf{x}) \triangleq \int P(\mathbf{x})\,\mu(dP).$$

Show that the maximum of $I(\mu)$ is attained and is equal to the information radius of the family $\mathcal{P}_k$. Moreover, the PD P_μ corresponding to a maximizer of $I(\mu)$ is unique (though the maximizer need not be), and it is the "centroid" of the family $\mathcal{P}_k$, that is, the minimizer of $\sup_{P\in\mathcal{P}_k} D(P\|Q)$. (Gallager (1974b, unpublished), Ryabko (1979), Davisson and Leon-Garcia (1980).)

Hint Extend (8.7) to this setting, and apply a suitable version of the minimax theorem. Or prove the existence of a maximizer of $I(\mu)$ by a compactness argument, and show directly that the corresponding P_μ satisfies $D(P\|P_\mu) \leqq I(\mu)$ for each $P \in \mathcal{P}_k$ (would a $P \in \mathcal{P}_k$ violate this, $I((1-t)\mu + t\delta_P)$ were larger than $I(\mu)$ for sufficiently small $t > 0$, where δ_P is the unit mass at P.)

(b) Show that $\sup_{P\in\mathcal{P}_k} D(P\|Q)$ has a unique minimizer Q even if $\mathcal{P}_k$ is not a closed set. Namely, that this "centroid" of $\mathcal{P}_k$ is equal to that of its closure.

Story of the results

The capacity computing algorithm of Theorem 8.3 was suggested independently by Arimoto (1972) and Blahut (1972); the convergence proof is due to the former. (He considered the unconstrained case.) Theorem 8.4 for the unconstrained case dates back to Shannon (1948); for the general case see Blahut (1972). Its derivation via the algorithm is original here. The Δ-distortion rate computing algorithm is due to Blahut (1972); a gap in his convergence proof was filled by Csiszár (1974). Theorem 8.7 is a result of Gallager (1968); its derivation via the algorithm is new.

9 A covering lemma and the error exponent in source coding

In this chapter the rate-distortion theorem will be generalized and sharpened, by considering more general source models and by evaluating more precisely the asymptotically best code performance. We shall rely on a covering lemma. Its proof is the first example (in the text rather than in the problems sections) of an important technique widely used in information theory as well as in other branches of mathematics. This technique, *random selection*, is a simple but very efficient tool for proving the existence of some mathematical objects without actually constructing them. The principle of random selection is the following: one proves that a real-valued function w takes a value less than λ on some element of a set Z by introducing a probability distribution on Z and showing that the mean value of w is less than λ. Of course, the principle is efficient only if an appropriate distribution is used. In many cases the least sophisticated choice, the uniform distribution, is suitable. When applied to proving the existence of codes with certain properties, this technique is called *random coding*.

Let d be an arbitrary distortion measure on $\mathsf{X} \times \mathsf{Y}$. As before, dependence on alphabet sizes of thresholds depending on d will not be indicated, since the specification of the distortion measure includes that of the alphabets X and Y. Let

$$R(P, \Delta) = \min_{W : d(P,W) \leq \Delta} I(P, W)$$

be the rate-distortion function of a DMS with generic distribution P.

LEMMA 9.1 (*Type covering*) For any distortion measure d on $\mathsf{X} \times \mathsf{Y}$, any type P of sequences in X^k, and numbers $\Delta \geq 0$, $\delta > 0$, there exists a set $\mathsf{B} \subset \mathsf{Y}^k$ such that

$$d(\mathbf{x}, \mathsf{B}) \triangleq \min_{\mathbf{y} \in \mathsf{B}} d(\mathbf{x}, \mathbf{y}) \leq \Delta \quad \text{for every} \quad \mathbf{x} \in \mathsf{T}_P^k \tag{9.1}$$

and

$$\frac{1}{n} \log |\mathsf{B}| \leq R(P, \Delta) + \delta \tag{9.2}$$

→ 9.1 provided that $k \geq k_0(d, \delta)$. ○

Proof Let P be an arbitrary but fixed type of sequences in X^k. For every set $\mathsf{B} \subset \mathsf{Y}^k$ denote by $\mathsf{U}(\mathsf{B})$ the set of those $\mathbf{x} \in \mathsf{T}_P^k$ for which $d(\mathbf{x}, \mathsf{B}) > \Delta$. Fix some $\eta > 0$ and consider a pair of RVs (X, Y) such that $Ed(X, Y) \leq |\Delta - \eta|^+$ and $P_X = P$. Further, let m be an integer to be specified later.

We prove the existence of a set $\mathsf{B} \subset \mathsf{Y}^k$ with $\mathsf{U}(\mathsf{B}) = \emptyset$ and $|\mathsf{B}| \leqq m$ by the method of random selection applied to the family $\mathcal{B}_m$ of all collections of m (not necessarily distinct) elements of $\mathsf{T}^k_{[Y]}$. Let Z^m be a RV ranging over $\mathcal{B}_m$ with uniform distribution. Then $Z^m = Z_1 Z_2 \ldots Z_m$, where Z_i is the ith element of the random m-element collection Z^m; the Z_i are independent and uniformly distributed over $\mathsf{T}^k_{[Y]}$. Consider the random set $\mathsf{U}(Z^m)$, i.e., the set of those $\mathbf{x} \in \mathsf{T}^k_P$ for which $d(\mathbf{x}, Z_i) > \Delta$ for $i = 1, 2, \ldots, m$. It is enough to show that $E|\mathsf{U}(Z^m)| < 1$, for this guarantees the existence of a set $\mathsf{B} \subset \mathsf{T}^k_{[Y]}$ with $|\mathsf{B}| \leqq m$ and $|\mathsf{U}(\mathsf{B})| < 1$, i.e., $\mathsf{U}(\mathsf{B}) = \emptyset$.

Let us denote by $\chi(\mathbf{x})$ the characteristic function of the random set $\mathsf{U}(Z^m)$, i.e.,

$$\chi(\mathbf{x}) \triangleq \begin{cases} 1 & \text{if} \quad \mathbf{x} \in \mathsf{U}(Z^m) \\ 0 & \text{if} \quad \mathbf{x} \notin \mathsf{U}(Z^m). \end{cases}$$

Then $|\mathsf{U}(Z^m)| = \sum_{\mathbf{x} \in \mathsf{T}_P} \chi(\mathbf{x})$, and thus

$$E|\mathsf{U}(Z^m)| = \sum_{\mathbf{x} \in \mathsf{T}_P} E\chi(\mathbf{x}) = \sum_{\mathbf{x} \in \mathsf{T}_P} \Pr\{\mathbf{x} \in \mathsf{U}(Z^m)\}. \tag{9.3}$$

By the same argument as in the proof of Theorem 7.3, $\mathbf{x} \in \mathsf{T}^k_P$ and $\mathbf{y} \in \mathsf{T}^k_{[Y|X]}(\mathbf{x})$ imply $d(\mathbf{x}, \mathbf{y}) \leqq \Delta$ for $k \geqq k_1(d, \eta)$ (resp. for every k if $\Delta \leqq \eta$). Thus

$$\Pr\{d(\mathbf{x}, Z_i) > \Delta\} \leqq 1 - |\mathsf{T}^k_{[Y]}|^{-1} |\mathsf{T}^k_{[Y|X]}(\mathbf{x})|$$

and consequently, by the independence of the Z_i and Lemma 2.13,

$$\begin{aligned} \Pr\{\mathbf{x} \in \mathsf{U}(Z^m)\} &= \Pr\{d(\mathbf{x}, Z_i) > \Delta \quad \text{for} \quad 1 \leqq i \leqq m\} \\ = \prod_{i=1}^{m} \Pr\{d(\mathbf{x}, Z_i) > \Delta\} &\leqq \left\{1 - \exp\left[-k\left(H(Y) + \frac{\delta}{4}\right) + k\left(H(Y|X) - \frac{\delta}{4}\right)\right]\right\}^m \\ &= \left\{1 - \exp\left[-k\left(I(X \wedge Y) + \frac{\delta}{2}\right)\right]\right\}^m, \end{aligned} \tag{9.4}$$

provided that $k \geqq k_2(d, \eta, \delta)$. Applying the inequality $(1-t)^m \leqq \exp(-tm)$ to $t \triangleq \exp\left[-k\left(I(X \wedge Y) + \delta/2\right)\right]$, the right-most side of (9.4) is upper bounded by

$$\exp\left\{-m \exp\left[-k\left(I(X \wedge Y) + \frac{\delta}{2}\right)\right]\right\}.$$

Now choose $m = m(k)$ as some integer satisfying

$$\exp\left[k\left(I(X \wedge Y) + \frac{2}{3}\delta\right)\right] \leqq m \leqq \exp\left[k\left(I(X \wedge Y) + \frac{3}{4}\delta\right)\right].$$

Then, by the last upper bound, we get from (9.4) that

$$\Pr\{\mathbf{x} \in \mathsf{U}(Z^m)\} \leqq \exp\left[-\exp\frac{k\delta}{6}\right].$$

Substituting this into (9.3) results in

$$E|\mathsf{U}(Z^m)| \leqq |\mathsf{T}_P| \cdot \exp\left[-\exp\frac{k\delta}{6}\right] \leqq \exp\left[k\log|\mathsf{X}| - \exp\frac{k\delta}{6}\right] < 1$$

if $k \geqq k_3(d, \eta, \delta)$. This proves the existence of a set $\mathsf{B} \subset \mathsf{Y}^k$ with $\mathsf{U}(\mathsf{B}) = \emptyset$ and

$$|\mathsf{B}| \leqq m \leqq \exp\left[k\left(I(X \wedge Y) + \frac{3}{4}\delta\right)\right]$$

if $k \geqq k_3(d, \eta, \delta)$.

On account of the uniform continuity of the function $R(P, \Delta)$, if η is sufficiently small, the RVs X, Y can be chosen to satisfy

$$I(X \wedge Y) < R(P, \Delta) + \frac{\delta}{4}. \qquad \square$$

As a first application of Lemma 9.1, we complete Theorem 7.3 by determining $R_0(\Delta)$ (see Definition 7.1). The importance of the 0-fidelity criterion (d, Δ) for data compression is obvious: this is the proper criterion if nothing is known about the statistics of the source.

THEOREM 9.2 For every $\Delta \geqq 0$ and every DMS the generic distribution of which has support X_0, we have

$$R_0(\Delta) = \max_P R(P, \Delta),$$

where the maximization involves all distributions P vanishing outside X_0. ○

Proof The inequality

$$R_0(\Delta) \geqq \max_P R(P, \Delta)$$

is obvious. To prove the opposite inequality, consider X_0^k as the disjoint union of the sets T_P, with P ranging over the types with support in X_0. Clearly, the source output sequence belongs to X_0^k with probability 1. Let $\mathsf{B}_P \subset \mathsf{Y}^k$ be the set corresponding to the type P by Lemma 9.1 and define B as the union of the B_P. Then, using the type counting lemma, for sufficiently large k we get

$$\frac{1}{k}\log|\mathsf{B}| \leqq \frac{1}{k}|\mathsf{X}|\log(k+1) + \max_P R(P, \Delta) + \delta < \max_P R(P, \Delta) + 2\delta,$$

while $d(\mathbf{x}, \mathsf{B}) \leqq \Delta$ for every $\mathbf{x} \in \mathsf{X}_0^k$. Now let $f_k : \mathsf{X}^k \to \mathsf{B}$ be any mapping such that $d(\mathbf{x}, f_k(\mathbf{x})) = d(\mathbf{x}, \mathsf{B})$ and let φ_k be the identity mapping. $\square$

→ 9.2

The codes guaranteed by the rate-distortion theorem essentially depend on the source statistics. Thus the theorem applies only to communication situations in which both the encoder and the decoder know exactly these statistics. Dropping this assumption means that we look for codes meeting some ε-fidelity criterion for a class of sources. Lemma 9.1 is well suited for treating such problems, even if the source statistics may vary from letter to letter in an arbitrary unknown manner.

→ 9.3

Let $\mathcal{P} = \{P_s, s \in \mathsf{S}\}$ be a (not necessarily finite) family of distributions $P_s = \{P(x|s) : x \in \mathsf{X}\}$ on a finite set X, the source alphabet. An *arbitrarily varying source* (AVS) defined by $\mathcal{P}$ is a sequence of RVs $\{X_i\}_{i=1}^{\infty}$ such that the distribution of X^k is some unknown element of $\mathcal{P}^k$, the kth Cartesian power of $\mathcal{P}$. In other words, the X_i are independent and $\Pr\{X^k = \mathbf{x}\}$ can be either of $P^k(\mathbf{x}|\mathbf{s}) \triangleq \prod_{i=1}^{k} P(x_i|s_i)$, where $\mathbf{s} = s_1 s_2 \dots s_k \in \mathsf{S}^k$, $\mathbf{x} = x_1 x_2 \dots x_k \in \mathsf{X}^k$.

Given a distortion measure d on $\mathsf{X} \times \mathsf{Y}$, we say that a k-length block code (f_k, φ_k) meets the ε-fidelity criterion (d, Δ) for this AVS if the criterion is met for every possible choice of $\mathbf{s}$, i.e.,

$$\sum_{\substack{\mathbf{x} \in \mathsf{X}^k \\ d(\mathbf{x}, \varphi_k(f_k(\mathbf{x}))) \leqq \Delta}} P^k(\mathbf{x}|\mathbf{s}) \geqq 1 - \varepsilon \quad \text{for every} \quad \mathbf{s} \in \mathsf{S}^k. \tag{9.5}$$

Note that $R_\varepsilon(\Delta)$ and $R(\Delta)$ are defined for an AVS in the same way as in Chapter 7. As an easy consequence of the type covering Lemma 9.1, we obtain

THEOREM 9.3 For the AVS defined by $\mathcal{P}$, for every $\Delta \geqq 0$ and $0 < \varepsilon < 1$ we have → 9.4

$$R_\varepsilon(\Delta) = R(\Delta) = \max_{P \in \overline{\mathcal{P}}} R(P, \Delta)$$

where $\overline{\mathcal{P}}$ is the convex closure of $\mathcal{P}$. ○

Proof Implicit in the statement of the theorem is that $\max_{P \in \overline{\mathcal{P}}} R(P, \Delta)$ is attained at some distribution in $\overline{\mathcal{P}}$. This is an obvious consequence of the continuity of $R(P, \Delta)$. We start by proving

$$R(\Delta) \geqq \max_{P \in \overline{\mathcal{P}}} R(P, \Delta). \tag{9.6}$$

By definition, $\overline{\mathcal{P}}$ is the closure of the family of distributions $\hat{P}$ of form

$$\hat{P}(x) = \sum_{s \in \mathsf{S}} L(s) P(x|s) \quad \text{for every} \quad x \in \mathsf{X}, \tag{9.7}$$

where L ranges over the distributions concentrated on finite subsets of S. As (9.7) implies that

$$\hat{P}^k(\mathbf{x}) = \sum_{\mathbf{s} \in \mathsf{S}^k} L^k(\mathbf{s}) P^k(\mathbf{x}|\mathbf{s}) \quad \text{for every} \quad \mathbf{x} \in \mathsf{X}^k,$$

it follows by (9.5) that if a code (f_k, φ_k) meets the ε-fidelity criterion (d, Δ) for the AVS, it meets the same criterion for every DMS with generic distribution belonging to $\overline{\mathcal{P}}$. This proves (9.6).

To prove the opposite inequality, i.e., the existence part of the theorem, set

$$\mathsf{T} \triangleq \bigcup_{\hat{P} \in \overline{\mathcal{P}}} \mathsf{T}^k_{[\hat{P}]_{\delta_k}},$$

where $\{\delta_k\}$ is some sequence meeting the delta-convention, Convention 2.11. For every $\mathbf{s} \in \mathsf{S}^k$, the $P^k(\cdot|\mathbf{s})$-probability of the set

$$\left\{\mathbf{x} : \left|\frac{1}{k}N(a|\mathbf{x}) - \frac{1}{k}\sum_{i=1}^{k} P(a|s_i)\right| > \delta_k\right\}$$

is less than $(4k\delta_k^2)^{-1}$, by Chebyshev's inequality. Thus for $\hat{P}_\mathbf{s} \triangleq (1/k)\sum_{i=1}^k P_{s_i} \in \overline{\mathcal{P}}$ and the sequence $\varepsilon_k \triangleq |\mathsf{X}|(4k\delta_k^2)^{-1} \to 0$ we have

$$P^k(\mathsf{T}^k_{[\hat{P}_\mathbf{s}]\delta_k}|\mathbf{s}) \geqq 1 - \varepsilon_k,$$

whence

$$P^k(\mathsf{T}|\mathbf{s}) \geqq 1 - \varepsilon_k \quad \text{for every} \quad \mathbf{s} \in \mathsf{S}^k. \tag{9.8}$$

Let $\mathbf{B}_P \subset \mathsf{Y}^k$ be the set corresponding to the type P by Lemma 9.1, and set

$$\mathsf{B} \triangleq \bigcup_{P:\mathsf{T}_P \subset \mathsf{T}} \mathsf{B}_P.$$

If $\mathsf{T}_P \subset \mathsf{T}$, then, by the definition of T, there exists a $\hat{P} \in \overline{\mathcal{P}}$ such that $|P(a) - \hat{P}(a)| < \delta_k$ for every $a \in \mathsf{X}$. Thus by the type counting lemma and the continuity of $R(P, \Delta)$, Lemma 9.1 implies

$$\begin{aligned}\frac{1}{k}\log|\mathsf{B}| &\leqq \frac{1}{k}|\mathsf{X}|\log(k+1) + \max_{P:\mathsf{T}_P \subset \mathsf{T}} \frac{1}{k}\log|\mathsf{B}_P| \\ &\leqq \max_{P \in \overline{\mathcal{P}}} R(P, \Delta) + 2\delta, \end{aligned} \tag{9.9}$$

if $k \geqq k^*(d, \delta)$, while

$$d(\mathbf{x}, \mathsf{B}) \leqq \Delta \quad \text{for} \quad \mathbf{x} \in \mathsf{T}. \tag{9.10}$$

Now let $f_k : \mathsf{X}^k \to \mathsf{B}$ be any mapping such that $d(\mathbf{x}, f_k(\mathbf{x})) = d(\mathbf{x}, \mathsf{B})$ and let φ_k be the identity mapping. In virtue of (9.8), (9.10) and (9.9), the proof is complete. □

So far we have dealt with a generalization of the rate-distortion theorem, Theorem 7.3. Let us revisit now the theorem itself in the light of the type covering lemma. Theorem 7.3 says that given a DMS $\{X_i\}_{i=1}^{\infty}$, a distortion measure d and a distortion level Δ, there is a number $R(\Delta)$, the Δ-distortion rate, such that if $R > R(\Delta)$ then a sequence of k-length block codes (f_k, φ_k) exists such that

$$\frac{1}{k}\log||f_k|| \to R \tag{9.11}$$

and $\Pr\{d(X^k, \varphi_k(f_k(X^k))) > \Delta\} \to 0$. On the other hand, if $R < R(\Delta)$ then this probability goes to 1 for every sequence of k-length block codes satisfying (9.11), a stronger result than just the negation of its convergence to zero. Our next aim is to investigate the speed of these convergences, as we have done in Theorem 2.15 for the special case of the probability of error fidelity criterion.

DEFINITION 9.4 For a given distortion measure d on $\mathsf{X} \times \mathsf{Y}$ and a k-length block code (f, φ) for sources with alphabet X and reproducing alphabet Y, we denote by $e(f, \varphi, P, \Delta)$ the probability that the k-length message X^k of a DMS with generic distribution P is not reproduced within distortion Δ:

$$e(f, \varphi, P, \Delta) = e(f, \varphi, P, d, \Delta) \triangleq P^k(\{\mathbf{x} : d(\mathbf{x}, \varphi(f(\mathbf{x}))) > \Delta\}). \qquad \bigcirc$$

We shall show that for appropriate k-length block codes of rate converging to R the probability $e(f_k, \varphi_k, P, \Delta)$ converges to zero exponentially whenever $R(P, \Delta) < R$, with exponent → 9.5

$$F(P, R, \Delta) \triangleq \inf_{Q:R(Q,\Delta)>R} D(Q||P).$$

Further, Theorem 9.5 also asserts that this result is the best possible.

THEOREM 9.5 To every $R < \log |\mathsf{X}|$ and distortion measure d on $\mathsf{X} \times \mathsf{Y}$ there exists a sequence of k-length block codes for sources with alphabet X and reproduction alphabet Y such that

(i) $(1/k) \log ||f_k|| \to R$;
(ii) for every distribution P on X, $\Delta \geqq 0$ and $\delta > 0$

$$\frac{1}{k} \log e(f_k, \varphi_k, P, \Delta) \leqq -F(P, R, \Delta) + \delta$$

whenever $k \geqq k_0(|\mathsf{X}|, \delta)$.

Further, for every sequence of codes satisfying (i) and every distribution P on X

$$\varliminf_{k \to \infty} \frac{1}{k} \log e(f_k, \varphi_k, P, \Delta) \geqq -F(P, R, \Delta). \qquad \bigcirc$$

Proof In order to prove the existence part, consider the sets

$$\mathsf{U}_k \triangleq \bigcup_{Q:R(Q,\Delta)>R} \mathsf{T}_Q^k.$$

By Lemma 2.6 and the type counting lemma we have

$$P^k(\mathsf{U}_k) \leqq (k+1)^{|\mathsf{X}|} \exp\{-kF(P, R, \Delta)\}. \tag{9.12}$$

Now, by Lemma 4.1 we can find a sequence $\varepsilon_k \to 0$ such that to every type Q of sequences in X^k there is a set $\mathsf{B}_Q \subset \mathsf{Y}^k$ satisfying

$$\frac{1}{k} \log |\mathsf{B}_Q| \leqq R + \varepsilon_k \tag{9.13}$$

and

$$d(\mathbf{x}, \mathsf{B}_Q) \leqq \Delta(Q, R) \quad \text{for every} \quad \mathbf{x} \in \mathsf{T}_Q^k;$$

here $\Delta(Q, R)$ is the *distortion-rate function* of the DMS with generic distribution Q, i.e.,

$$\Delta(Q, R) \triangleq \min_{W: I(Q,W) \leqq R} d(Q, W).$$

Setting

$$\mathsf{B} \triangleq \bigcup_{Q} \mathsf{B}_Q$$

we see from (9.13) and the type counting lemma that

$$\frac{1}{k} \log |\mathsf{B}| \leqq R + \varepsilon'_k \quad \text{with} \quad \varepsilon'_k \to 0.$$

Further, since $R(Q, \Delta) \leqq R$ implies $\Delta(Q, R) \leqq \Delta$, we have

$$d(\mathbf{x}, \mathsf{B}) \leqq \Delta \quad \text{for every} \quad \mathbf{x} \in \mathsf{X}^k - \mathsf{U}_k.$$

The last two inequalities and (9.12) establish the existence part of the theorem.

Turning to the converse, consider any distribution Q on X such that $R(Q, \Delta) > R$. (If no such Q exists, the statement of the converse is void.) Fix some $\delta > 0$ with $R(Q, \Delta) > R + \delta$. Recall that when proving the rate-distortion theorem we have shown (see the lines preceding formula (7.7)) that for every k-length block code (f_k, φ_k) for sources with alphabet X and reproduction alphabet Y the condition

$$\frac{1}{k} \log ||f_k|| \leqq R(Q, \Delta) - \delta \tag{9.14}$$

implies

$$e(f_k, \varphi_k, Q, \Delta) \geqq \frac{1}{2}, \tag{9.15}$$

whenever $k \geqq k_0(d, \delta)$.

Since $R(Q, \Delta) > R + \delta$, assumption (i) implies (9.14) for sufficiently large integers k. Hence (9.15) holds, and thus by Corollary 1.2 for k large enough

$$e(f_k, \varphi_k, P, \Delta) \geqq \exp\{-k[D(Q||P) + \delta]\}.$$

As Q was arbitrary with $R(Q, \Delta) > R$ and $\delta > 0$ can be made arbitrarily small, the converse part of the theorem follows. □

→ 9.6 It is not hard to establish a result similar to Theorem 9.5 for rates below $R(P, \Delta)$. Analogous results for channel codes will be proved in Chapter 10.

Problems

9.1. Show that Lemma 9.1 implies the existence part of the rate-distortion theorem, Theorem 7.3.

9.2. (*Universal coding*) By the rate-distortion theorem, to any $R > 0$ and DMS with alphabet X there exist k-length block codes (f_k, φ_k) depending on the generic distribution P such that $(1/k) \log ||f_k|| \to R$ and

$$\Pr\{d(X^k, \varphi_k(f_k(X^k))) \leqq \Delta(P, R)\} \to 1, \tag{*}$$

where $\Delta(P, R)$ is the inverse of the rate-distortion function $R(P, \Delta)$ (with P fixed). Show the existence of codes depending only on the distortion measure, but not on P, which have rates converging to R and satisfy (*) for every P, with uniform convergence. (This result is partially contained in a general theorem of Neuhoff, Gray and Davisson (1975); for a sharper result, see Theorem 9.5.)

Hint Apply Lemma 9.1 to all possible types of sequences in X^k.

9.3. (*Compound DMS*) If the generic distribution of a DMS is an unknown element of a family $\mathcal{P} = \{P_s, s \in \mathsf{S}\}$ of PDs on X, one speaks of a *compound DMS* defined by $\mathcal{P}$. Define $R_\varepsilon(\Delta)$ and $R(\Delta)$ for a compound DMS and show that

$$R_\varepsilon(\Delta) = R(\Delta) = \sup_{P \in \mathcal{P}} R(P, \Delta) \quad \text{for every} \quad 0 < \varepsilon < 1.$$

Observe that the existence part of this result is weaker than Problem 9.2.

9.4. (*Average fidelity criterion*) Show that for both an AVS and a compound DMS the Δ-distortion rate remains the same if instead of the ε-fidelity criterion (d, Δ) the average fidelity criterion (d, Δ) is imposed, provided that $\Delta > 0$.

9.5. (*Discussion of* $F(P, R, \Delta)$)

(a) Note that $F(P, R, \Delta) > 0$ if $R > R(P, \Delta)$ and $F(P, R, \Delta) = 0$ if $R < R(P, \Delta)$. Further, $F(P, R, \Delta)$ is finite iff R is less than the zero-error rate $R_0(\Delta)$ of the DMS with generic distribution P.

(b) Show that for fixed P and Δ, $F(P, R, \Delta)$ is continuous at every R which is not a local maximum of $R(Q, \Delta)$. ($\tilde{R}$ is a local maximum of $R(Q, \Delta)$ if there exists a $\tilde{Q}$ with $R(\tilde{Q}, \Delta) = \tilde{R}$ such that $R(Q, \Delta) \leqq \tilde{R}$ for every Q in some neighborhood of $\tilde{Q}$.)

(c) Show that for $\mathsf{X} = \mathsf{Y}$ and the error frequency fidelity criterion, $F(P, R, \Delta)$ is a continuous function of R (see Problem 7.9). (Marton (1974).)

(d)* Show by an example that, in general, $R(Q, \Delta)$ may have local maxima different from the global maximum, and then $F(P, R, \Delta)$ as a function of R need not be everywhere continuous where it is finite. (Ahlswede (1990).)

9.6. (*Rates below* $R(P, \Delta)$) Given a DMS with generic distribution P, denote by $e_k(R, \Delta)$ the minimum of $e(f_k, \varphi_k, P, \Delta)$ for codes of rate $(1/k) \log ||f_k|| \leqq R$. Show that for $0 \leqq R < R(P, \Delta)$ we have

$$\lim_{k \to \infty} \left[-\frac{1}{k} \log(1 - e_k(R, \Delta)) \right] = G(R, \Delta),$$

where

$$G(R, \Delta) \triangleq \min_Q [D(Q||P) + |R(Q, \Delta) - R|^+].$$

Hint Show that $|G(R'', \Delta) - G(R', \Delta)| \leqq |R'' - R'|$, establishing the continuity of $G(R, \Delta)$. Then proceed similarly to Theorem 9.5. (A weaker existence result was proved by Omura (1975).)

9.7. (*Zero-error rate*) Show that for a DMS with generic distribution of support X_0

$$R_0(\Delta) = -\min_{Q}\min_{\delta \geqq 0}\max_{y\in\mathsf{Y}}\log\sum_{x\in\mathsf{X}} Q(x)\exp[\delta(\Delta - d(x,y))]$$

if $\Delta > 0$ and

$$R_0(0) = -\min_{Q}\max_{y\in\mathsf{Y}}\log\sum_{x:d(x,y)=0} Q(x),$$

where Q ranges over the distributions vanishing outside X_0.
Hint Use Theorem 9.2 and Corollary 8.7.

9.8. (a) Verify that the Δ-distortion rate of an AVS does not decrease if the code is allowed to depend on $\mathbf{s} \in \mathsf{S}^k$.
(b) Prove the analogous statement for a compound DMS.

9.9. An AVS defined by a family $\mathcal{P}$ of distributions on X is the class of all sequences of independent RVs $\{X_i\}_{i=1}^{\infty}$ such that $P_{X_i} \in \mathcal{P}, i = 1, 2, \ldots$. Consider the larger class of all sequences of not necessarily independent RVs such that $P_{X_i|X_1,\ldots,X_{i-1}} \in \mathcal{P}$, $i = 1, 2, \ldots$, and show that Theorem 9.3 remains valid even for this class. (Berger (1971).)

9.10.* (*Speed of convergence of average distortion*) Given a DMS $\{X_i\}_{i=1}^{\infty}$ and a distortion measure d on $\mathsf{X} \times \mathsf{Y}$, let

$$D_k(R) \triangleq \min_{\frac{1}{k}\log\|f_k\| \leqq R} Ed(X^k, \varphi_k(f_k(X^k)))$$

be the minimum average distortion achievable by k-length block codes of rate at most R. Show that for some constant $c > 0$

$$\Delta(R) \leqq D_k(R) \leqq \Delta(R) + c\frac{\log k}{k},$$

where $\Delta(R)$ is the distortion-rate function, i.e., the inverse of the rate-distortion function. (Pilc (1968). For a sharper result see Linder, Lugosi and Zeger (1995) and Zhang, Yang and Wei (1997).)

Graph entropy and convex corners

9.11. Interpret a graph G with vertex set X as specifying distinguishability: two elements of X are distinguishable iff they are adjacent in G. Accordingly, two sequences $\mathbf{x} = x_1 \ldots x_k$ and $\mathbf{x}' = x_1' \ldots x_k'$ are distinguishable iff x_i and x_i' are adjacent for some $1 \leqq i \leqq k$.
(a) Check that the minimum rate of source encoders of block length k that assign different codewords to distinguishable sequences in X^k is equal to $(1/k)\log\chi(G^k)$; see Problem 6.24 for notation. Similarly, the minimum rate of encoders having this property "with large probability," with respect to a given DMS with generic distribution P, is

$$\frac{1}{k} \log \min_{\substack{\mathsf{F} \subset \mathsf{X}^k \\ P^k(\mathsf{F}) \geqq 1-\varepsilon}} \chi(\tilde{\mathsf{F}}) \quad (\varepsilon \in (0,1)),$$

where $\tilde{\mathsf{F}}$ is the subgraph of G^k with vertex set F. The limit as $k \to \infty$ of the latter quantity (which exists and does not depend on ε; see part (b)) is called the *graph entropy* $H(G, P)$.

(b) Show that $H(G, P)$ is well defined (the limit exists and does not depend on ε) and equals the 0-distortion rate $R(P, 0)$ for a suitable reproduction alphabet and distortion measure. Namely, take as the reproduction alphabet the family $\mathcal{I}$ of those subsets I of X that do not contain adjacent elements and cannot be enlarged retaining this property, and let the distortion $d(x, \mathsf{I})$ be zero if $x \in \mathsf{I}$ and > 0 otherwise. Show also that in the definition of graph entropy the minimization could be dispensed with, just taking $\mathsf{F} \triangleq \mathsf{T}^k_{[P]}$, and that the limit as $k \to \infty$ of $(1/k) \log \chi(G^k)$ is equal to $\max_P H(G, P)$.

Hint For the last assertions, use Lemma 9.1 and Theorem 9.2.

(c) Show that, more explicitly,

$$H(G, P) = -\min_Q \{D(P||Q) + \max_{\mathsf{I} \in \mathcal{I}} \log \sum_{x \in \mathsf{I}} Q(x)\}$$

and

$$\lim_{k \to \infty} \frac{1}{k} \log \chi(G^k) = -\min_Q \max_{\mathsf{I} \in \mathcal{I}} \log \sum_{x \in \mathsf{I}} Q(x).$$

Hint Use (b) and Corollary 8.7.

(The concept of graph entropy has been introduced by Körner (1973a); he gave its representation in (b). The limit of $(1/k) \log \chi(G^k)$ was determined by McEliece and Posner (1971).)

9.12. A set $\mathcal{A}$ of $|\mathsf{X}|$-tuples $A = \{A(x),\ x \in \mathsf{X}\}$ of non-negative numbers is a *convex corner* if $\mathcal{A}$ represents a convex, compact set in the $|\mathsf{X}|$-dimensional Euclidean space , with non-empty interior, such that if $A \in \mathcal{A}$ and $0 \leqq A' \leqq A$ (componentwise) then also $A' \in \mathcal{A}$. The *entropy relative to the convex corner* $\mathcal{A}$ of a PD P is defined by

$$H_{\mathcal{A}}(P) \triangleq \min_{A \in \mathcal{A}} \sum_{x \in \mathsf{X}} P(x) \log \frac{1}{A(x)}.$$

Clearly, $H_{\mathcal{A}}(P) \geqq 0$ for each P iff $\mathcal{A}$ is a subset of the unit cube.

(a) For $G(\delta)$ in Theorems 8.6 and 8.7, verify that

$$G(\delta) = H_{\mathcal{A}}(P) = H(P) - H_{\mathcal{B}}(P), \qquad (*)$$

where the convex corners $\mathcal{A}$ and $\mathcal{B}$ consist of those $|\mathsf{X}|$-tuples that satisfy

$$A(x) \leqq \sum_y Q(y) \exp[-\delta d(x, y)], \quad x \in \mathsf{X},$$

for some PD Q on Y, respectively the inequalities in (8.19).

Hint See Corollary 8.5 and Theorem 8.7.

(b) Show that the 0-distortion rate $R(P, 0)$ can be represented as in (*), this time with $\mathcal{A}$ and $\mathcal{B}$ consisting of those $|\mathsf{X}|$-tuples that satisfy

$$A(x) \leqq \sum_{y:\ d(x,y)=0} Q(y), \quad x \in \mathsf{X},$$

for some PD Q on Y, respectively

$$\sum_{x:\ d(x,y)=0} B(x) \leqq 1, \quad y \in \mathsf{Y}.$$

In particular, the graph entropy $H(G, P)$ can be represented in this way, with $\mathcal{A}$ and $\mathcal{B}$ defined by the inequalities

$$A(x) \leqq \sum_{\mathsf{I}\in\mathcal{I}\ :\ x\in\mathsf{I}} Q(\mathsf{I}), \quad x \in \mathsf{X},$$

for some PD Q on $\mathcal{I}$, respectively

$$\sum_{x\in\mathsf{I}} B(x) \leqq 1, \quad \mathsf{I} \in \mathcal{I}.$$

Hint Use (a) and (8.6), or Problem 8.7(c).

(c)* Prove the following general result, and relate it to (a) and (b) above: for any convex corner $\mathcal{A}$ and its *antiblocker* $\mathcal{B}$, consisting of those non-negative $|\mathsf{X}|$-tuples $\{B(x), x \in \mathsf{X}\}$ that have inner product at most 1 with each $|\mathsf{X}|$-tuple in $\mathcal{A}$, the equality

$$H_{\mathcal{A}}(P) + H_{\mathcal{B}}(P) = H(P)$$

holds. Moreover, A^* and B^* attaining the minimum in the definition of $H_{\mathcal{A}}(P)$ and $H_{\mathcal{B}}(P)$ satisfy

$$A^*(x)B^*(x) = P(x), \quad x \in \mathsf{X}.$$

(The concept of entropy relative to a convex corner and the results in (b) and (c) are due to Csiszár *et al.* (1990).)

9.13. (*Subadditivity and additivity of graph entropy*)

(a) Let F and G be two graphs with the same vertex set X, and let $F \cup G$ denote the graph with vertex set X whose edge set is the union of those of F and G. Show that for each PD P of X

$$H(F \cup G,\ P) \leqq H(F,\ P) + H(G,\ P).$$

(Körner (1986).)

(b) For any graph G and its complement $\bar{G}$, (a) gives for each PD P

$$H(P) \leqq H(G,\ P) + H(\bar{G},\ P).$$

Show that the equality holds for every P iff G is perfect, see Problem 6.25. (Conjectured by Körner and Marton (1988b) and proved by Csiszár *et al.* (1990).)

Hint By (b) of Problem 9.12,

$$H(G, P) = H_{\mathcal{A}}(P) = H(P) - H_{\mathcal{B}}(P),$$

where $\mathcal{A}$ and $\mathcal{B}$ are the convex corners defined there. In graph theory, the former is known as the *vertex packing polytope* of the graph G, and the latter is the *fractional vertex packing polytope* of the graph $\bar{G}$ (note that the sets $\mathsf{I} \in \mathcal{I}$ are the cliques of $\bar{G}$.) The assertion follows from the theorem (Fulkerson (1973)) that a graph is perfect iff its vertex packing and fractional vertex packing polytopes coincide.

(c)* Let the graphs G_i, $i = 1, \dots, l$, partition the edge set of the complete graph with vertex set X. Show that

$$H(P) = \sum_{i=1}^{l} H(G_i, P)$$

holds for every PD P on X iff each G_i is perfect and for no triple of vertices do all three edges between them belong to different graphs G_i. (Körner, Simonyi and Tuza (1992).)

Story of the results

The key Lemma 9.1 of this chapter is due to Berger (1971), except for the case $\Delta = 0$ which was settled independently by Körner (1973b, unpublished) and Marton (1974). The method of random coding has been introduced by Shannon (1948); see Problem 6.9. An early appearance of the technique of proving existence results by random selection is Szekeres and Turán (1937). Theorem 9.2 is contained in Berger (1971) (for $\Delta > 0$). Theorem 9.3 is the authors' transcription of a similar result of Dobrušin (1970) and Berger (1971), who used the average fidelity criterion. The present simple proof is that of Berger (1971). Theorem 9.5 was proved by Marton (1974). The problem was investigated independently by Blahut (1974), who derived an exponential upper bound. A previous bound of this kind is implicit in Jelinek (1968b), Th. 11.1.

10 A packing lemma and the error exponent in channel coding

In this chapter we revisit the coding theorem for a DMC. By definition, for any $R > 0$ below capacity, there exists a sequence of n-length block codes (f_n, φ_n) with rates converging to R and maximum probability of error converging to zero as $n \to \infty$. On the other hand, by Theorem 6.5, for codes of rate converging to a number above capacity, the maximum probability of error converges to unity. Now we look at the speed of these convergences. This problem is far more complex than its source coding analog and it has not been fully settled yet.

We saw in Chapter 6 that the capacity of a DMC can be achieved by codes, all codewords of which have approximately the same type. In this chapter we shall con-
→ 6.19 centrate attention on *constant composition codes*, i.e., codes all codewords of which have the very same type. We shall investigate the asymptotics of the error probability for codes from this special class. The general problem reduces to this one in a simple manner.

Our present approach will differ from that in Chapter 6. In that chapter channel codes were constructed by defining the encoder and the decoder simultaneously, in
→ 6.2 a successive manner. Here, attention will be focused on finding suitable encoders; the decoder will be determined by the encoder in a way to be specified later. As in
→ 6.20 Chapter 9, we shall use the method of random selection. The error probability bounds will be derived by simple counting arguments, using the lemmas of the first part of Chapter 2. For convenience let us recapitulate some basic estimates on types proved there.

Denote by $\mathcal{V}(P) = \mathcal{V}_n(P)$ the family of stochastic matrices $V : \mathsf{X} \to \mathsf{Y}$ for which the V-shell of a sequence of type P in X^n is not empty. By the type counting lemma,

$$|\mathcal{V}_n(P)| \leqq (n+1)^{|\mathsf{X}|\cdot|\mathsf{Y}|}. \tag{10.1}$$

Further, by Lemma 2.5, for $V \in \mathcal{V}_n(P)$, $\mathbf{x} \in \mathsf{T}_P^n = \mathsf{T}_P$, we have

$$(n+l)^{-|\mathsf{X}||\mathsf{Y}|} \exp[nH(V|P)] \leqq |\mathsf{T}_V(\mathbf{x})| \leqq \exp[nH(V|P)]. \tag{10.2}$$

If $W : \mathsf{X} \to \mathsf{Y}$ is an arbitrary stochastic matrix, $\mathbf{x} \in \mathsf{T}_P$ and $\mathbf{y} \in \mathsf{T}_V(\mathbf{x})$, then by Lemma 2.6

$$W^n(\mathbf{y}|\mathbf{x}) = \exp\{-n[D(V\|W|P) + H(V|P)]\}, \tag{10.3}$$

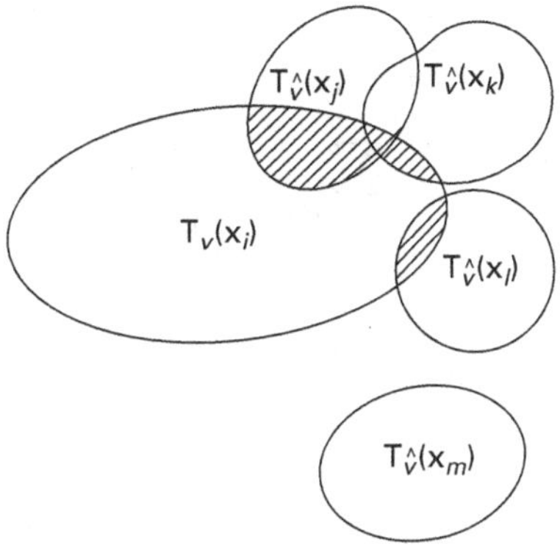

Figure 10.1 Intersection of V-shells.

where

$$D(V\|W|P) \triangleq \sum_{x\in\mathsf{X}} \sum_{y\in\mathsf{Y}} P(x)V(y|x)\log\frac{V(y|x)}{W(y|x)}. \tag{10.4}$$

By the same lemma, we also have

$$W^n(\mathsf{T}_V(\mathbf{x})|\mathbf{x}) \leqq \exp\left[-nD(V\|W|P_{\mathbf{x}})\right]. \tag{10.5}$$

One feels that the codewords of a "good" code must be far from each other, though it is not at all clear what mathematical meaning should be given to this statement. We shall select a prescribed number of sequences in X^n so that the shells around them have possibly small intersections. A good selection is provided by the following lemma, which plays a role similar to that of Lemma 9.1.

LEMMA 10.1 (*Packing*) For every $R > \delta > 0$ and every type P of sequences in X^n satisfying $H(P) > R$, there exist at least $\exp[n(R-\delta)]$ distinct sequences $\mathbf{x}_i \in \mathsf{X}^n$ of type P such that for every pair of stochastic matrices $V : \mathsf{X} \to \mathsf{Y}$, $\hat{V} : \mathsf{X} \to \mathsf{Y}$ and every i → 10.1

$$|\mathsf{T}_V(\mathbf{x}_i) \cap \bigcup_{j\neq i} \mathsf{T}_{\hat{V}}(\mathbf{x}_j)| \leqq |\mathsf{T}_V(\mathbf{x}_i)| \cdot \exp\left[-n|I(P,\hat{V}) - R|^+\right] \tag{10.6}$$

provided that $n \geqq n_0(|\mathsf{X}|, |\mathsf{Y}|, \delta)$. ○ → 10.2

REMARK Of course, if $P\hat{V} \neq PV$ then every $\mathsf{T}_V(\mathbf{x}_i) \cap \mathsf{T}_{\hat{V}}(\mathbf{x}_j)$ is void. By (10.6) and (10.2), $R < I(P,\hat{V}) - H(V|P)$ also implies $\mathsf{T}_V(\mathbf{x}_i) \cap \mathsf{T}_{\hat{V}}(\mathbf{x}_j) = \emptyset$ for every $i \neq j$. ○

Proof We shall use the method of random selection. For fixed positive integers n, m and fixed type P of sequences in X^n, let $\mathcal{C}_m$ be the family of all ordered collections $\mathsf{C} = (\mathbf{x}_1, \mathbf{x}_2, \ldots, \mathbf{x}_m)$ of m not necessarily distinct sequences of type P in X^n. Note that if some $\mathsf{C} = (\mathbf{x}_1, \mathbf{x}_2, \ldots, \mathbf{x}_m) \in \mathcal{C}_m$ satisfies (10.6) for every i, V, $\hat{V}$, then the $\mathbf{x}_i$ are necessarily distinct. This can be seen by choosing some $V = \hat{V} \in \mathcal{V}(P)$ such that $I(P, V) > R$.

For any collection $\mathsf{C} \in \mathcal{C}_m$, denote the left-hand side of (10.6) by $u_i(\mathsf{C}, V, \hat{V})$. On account of (10.2), a $\mathsf{C} \in \mathcal{C}_m$ certainly satisfies (10.6) for every i, V, $\hat{V}$ if

$$u_i(\mathsf{C}) \triangleq (n+1)^{|\mathsf{X}||\mathsf{Y}|} \sum_{V \in \mathcal{V}(P)} \sum_{\hat{V} \in \mathcal{V}(P)} u_i(\mathsf{C}, V, \hat{V}) \exp\{n[I(P, \hat{V}) - R - H(V|P)]\} \tag{10.7}$$

is at most 1, for every i.

Note that if for some $\mathsf{C} \in \mathcal{C}_m$

$$\frac{1}{m} \sum_{i=1}^{m} u_i(\mathsf{C}) \leqq \frac{1}{2} \tag{10.8}$$

then $u_i(\mathsf{C}) \leqq 1$ for at least $m/2$ indices i. Further, if C' is the subcollection of C with the above indices then $u_i(\mathsf{C}') \leqq u_i(\mathsf{C}) \leqq 1$ for every such index i. Hence, the lemma will be proved if for an m with

$$2\exp[n(R-\delta)] \leqq m \leqq \exp\left[n\left(R - \frac{\delta}{2}\right)\right] \tag{10.9}$$

we find a $\mathsf{C} \in \mathcal{C}_m$ satisfying (10.8).

Choose $\mathsf{C} \in \mathcal{C}_m$ at random, according to uniform distribution. In other words, let $Z^m = (Z_1, Z_2, \ldots, Z_m)$ be a sequence of independent RVs, each uniformly distributed over $\mathsf{T}_P = \mathsf{T}_P^n$. In order to prove that (10.8) holds for some $\mathsf{C} \in \mathcal{C}_m$, it suffices to show that

$$Eu_i(Z^m) \leqq \frac{1}{2}, \quad i = 1, 2, \ldots, m. \tag{10.10}$$

To this end, we bound $Eu_i(Z^m, V, \hat{V})$. Recalling that $u_i(\mathsf{C}, V, \hat{V})$ denotes the left-hand side of (10.6), we have

$$Eu_i(Z^m, V, \hat{V}) = \sum_{y \in \mathsf{Y}^n} \Pr\{\mathbf{y} \in \mathsf{T}_V(Z_i) \cap \bigcup_{j \neq i} \mathsf{T}_{\hat{V}}(Z_j)\}. \tag{10.11}$$

As the Z_j are independent and identically distributed, the probability under summation is less than or equal to

$$\sum_{j:j \neq i} \Pr\{\mathbf{y} \in \mathsf{T}_V(Z_i) \cap \mathsf{T}_{\hat{V}}(Z_j)\} = (m-1) \cdot \Pr\{\mathbf{y} \in \mathsf{T}_V(Z_1)\} \cdot \Pr\{\mathbf{y} \in \mathsf{T}_{\hat{V}}(Z_1)\}. \tag{10.12}$$

As the Z_j are uniformly distributed over T_P, we have for every fixed $\mathbf{y} \in \mathsf{Y}^n$

$$\Pr\{\mathbf{y} \in \mathsf{T}_V(Z_1)\} = \frac{|\{\mathbf{x} : \mathbf{x} \in \mathsf{T}_P, \mathbf{y} \in \mathsf{T}_V(\mathbf{x})\}|}{|\mathsf{T}_P|}.$$

The set in the numerator is non-void only if $\mathbf{y} \in \mathsf{T}_{PV}$. In this case it can be written as $T_{\bar{V}}(\mathbf{y})$, where $\bar{V} : \mathsf{Y} \to \mathsf{X}$ is such that

$$P(a)V(b|a) = PV(b)\bar{V}(a|b).$$

Thus by (10.2) and Lemma 2.3

$$\Pr\{\mathbf{y} \in \mathsf{T}_V(Z_1)\} \leqq \frac{\exp[nH(\bar{V}|PV)]}{(n+1)^{-|\mathsf{X}|}\exp[nH(P)]}$$
$$= (n+1)^{|\mathsf{X}|}\exp[-nI(PV, \bar{V})] = (n+1)^{|\mathsf{X}|}\exp[-nI(P, V)]$$

if $\mathbf{y} \in \mathsf{T}_{PV}$, and $\Pr\{\mathbf{y} \in \mathsf{T}_V(Z_1)\} = 0$ otherwise. Hence, upper bounding $|\mathsf{T}_{PV}|$ by Lemma 2.3, from (10.11), (10.12) and (10.9) we obtain

$$Eu_i(Z^m, V, \hat{V}) \leqq |\mathsf{T}_{PV}|(m-1)(n+1)^{2|\mathsf{X}|}\exp[-n(I(P, V) + I(P, \hat{V}))]$$
$$\leqq (n+1)^{2|\mathsf{X}|}\exp\left[n\left(R - \frac{\delta}{2} + H(V|P) - I(P, \hat{V})\right)\right].$$

On account of (10.7) and (10.1) this results in

$$Eu_i(Z^m) \leqq (n+1)^{2|\mathsf{X}|+3|\mathsf{X}||\mathsf{Y}|}\exp\left(-n\frac{\delta}{2}\right).$$

This establishes (10.10) for $n \geqq n_0(|\mathsf{X}|, |\mathsf{Y}|, \delta)$. □

As a first application of the packing lemma, we derive an upper bound on the maximum probability of error achievable by good codes on a DMC.

To this end, with every ordered collection $\mathsf{C} = (\mathbf{x}_1, \ldots, \mathbf{x}_m)$ of sequences in X^n we associate an n-length block code (f, φ) with $\mathsf{M}_f = \{1, \ldots, m\}$. Let f be defined by $f(i) \triangleq \mathbf{x}_i$; further, let φ be a *maximum mutual information (MMI) decoder* defined as follows. Recall that $I(\mathbf{x} \wedge \mathbf{y})$ means for $\mathbf{x} \in \mathsf{X}^n$, $\mathbf{y} \in \mathsf{Y}^n$ the mutual information corresponding to the joint type $P_{\mathbf{x},\mathbf{y}}$, i.e.,

$$I(\mathbf{x} \wedge \mathbf{y}) \triangleq I(P_{\mathbf{x}}, V) \quad \text{if } \mathbf{y} \in \mathsf{T}_V(\mathbf{x}).$$

Now let the decoder φ be any function $\varphi : \mathsf{Y}^n \to \{1, \ldots, m\}$ such that $\varphi(\mathbf{y}) = i$ satisfies

$$I(\mathbf{x}_i \wedge \mathbf{y}) = \max_{1 \leqq j \leqq m} I(\mathbf{x}_j \wedge \mathbf{y}). \tag{10.13}$$

A remarkable feature of the following theorem is that the same codes achieve the bound for every DMC.

THEOREM 10.2 (*Random coding bound for constant composition codes*) For every $R > \delta > 0$ and every type P of sequences in X^n there exists an n-length block code (f, φ) of rate

$$\frac{1}{n}\log|\mathsf{M}_f| \geqq R - \delta$$

such that all codewords $f(m)$, $m \in \mathsf{M}_f$ are of type P and

$$e(W^n, f, \varphi) \leqq \exp[-n(E_r(R, P, W) - \delta)] \tag{10.14}$$

for every DMC $\{W : \mathsf{X} \to \mathsf{Y}\}$, whenever $n \geqq n_0(|\mathsf{X}|, |\mathsf{Y}|, \delta)$. Here → 10.3

$$E_r(R, P, W) \triangleq \min_V (D(V\|W|P) + |I(P, V) - R|^+), \tag{10.15}$$

V ranging over all channels $V : \mathsf{X} \to \mathsf{Y}$. ○

REMARK $E_r(R, P, W)$ is called the *random coding exponent function* of channel W with input distribution P. ○

Proof Let $\mathsf{C} = \{\mathbf{x}_1, \ldots, \mathbf{x}_m\} \subset \mathsf{T}_P^n$ with $m \geqq \exp[n(R - \delta)]$ be any collection that satisfies (10.6) for every i, V and $\hat{V}$. We claim that if $f(i) \triangleq \mathbf{x}_i$, and φ is a corresponding MMI-decoder, then the code (f, φ) satisfies (10.14). This means that Theorem 10.2 follows from Lemma 10.1 (one may assume that $R < I(P, W) \leqq H(P)$, otherwise the bound (10.14) is trivial).

By (10.13), if $\mathbf{y} \in \mathsf{Y}^n$ leads to an erroneous decoding of the ith message, then

$$\mathbf{y} \in \mathsf{T}_V(\mathbf{x}_i) \cap \mathsf{T}_{\hat{V}}(\mathbf{x}_j) \quad \text{with} \quad I(P, \hat{V}) \geqq I(P, V)$$

for some $j \neq i$ and stochastic matrices $V, \hat{V} \in \mathcal{V}(P)$. Hence the probability of erroneous transmission of message i is bounded as

$$e_i = W^n(\{\mathbf{y} : \varphi(\mathbf{y}) \neq i\}|\mathbf{x}_i) \leqq \sum_{\substack{V, \hat{V} \in \mathcal{V}(P) \\ I(P, \hat{V}) \geqq I(P, V)}} W^n\Big(\mathsf{T}_V(\mathbf{x}_i) \cap \bigcup_{j \neq i} \mathsf{T}_{\hat{V}}(\mathbf{x}_j) \Big| \mathbf{x}_i\Big). \quad (10.16)$$

On account of (10.6), (10.2) and (10.3),

$$\begin{aligned} &W^n\Big(\mathsf{T}_V(\mathbf{x}_i) \cap \bigcup_{j \neq i} \mathsf{T}_{\hat{V}}(\mathbf{x}_j) \Big| \mathbf{x}_i\Big) \\ &\leqq \exp[-n(D(V||W|P) + |I(P, \hat{V}) - R|^+)]. \end{aligned} \quad (10.17)$$

Thus (10.16) may be continued as

$$\begin{aligned} e_i &\leqq \sum_{\substack{V, \hat{V} \in \mathcal{V}(P) \\ I(P, V) \geqq I(P, V)}} \exp[-n(D(V||W|P) + |I(P, V) - R|^+)] \\ &\leqq (n+1)^{2|\mathsf{X}||\mathsf{Y}|} \cdot \exp[-nE_r(R, P, W)], \end{aligned}$$

where the last step follows from (10.1). □

As the codes in Theorem 10.2 did not depend on W, one would expect that to every particular DMC codes having significantly smaller error probability can be found. (Note that for every encoder f, the decoder φ minimizing $e(W^n, f, \varphi)$ essentially depends on W.) Rather surprisingly, it turns out that for every DMC $\{W\}$ the above construction yields the best asymptotic performance in a certain rate interval.

THEOREM 10.3 (*Sphere packing bound for constant composition codes*) For every $R > 0, \delta > 0$ and every DMC $\{W : \mathsf{X} \to \mathsf{Y}\}$, every constant composition code (f, φ) of block length n and rate

$$\frac{1}{n} \log |\mathsf{M}_f| \geqq R + \delta$$

has maximum probability of error

$$e(W^n, f, \varphi) \geqq \frac{1}{2} \exp[-nE_{sp}(R, P, W)(1 + \delta)] \quad (10.18)$$

whenever $n \geqq n_0(|\mathsf{X}|, |\mathsf{Y}|, \delta)$. Here P is the common type of the codewords and

$$E_{sp}(R, P, W) \triangleq \min_{V: I(P,V) \leqq R} D(V||W|P). \tag{10.19}$$

○

REMARKS $E_{sp}(R, P, W)$ is called the *sphere packing exponent function* of channel W with input distribution P. This name, similarly to "random coding exponent function," is sanctioned by tradition and refers to those techniques by which similar bounds were first obtained. Note that in some cases $E_{sp}(R, P, W)$ may be infinite. Let $R_\infty(P, W)$ be the infimum of those R for which $E_{sp}(R, P, W) < +\infty$. For $R < R_\infty(P, W)$ the assertion of the theorem is void. → 10.4

○

Proof Consider an arbitrary DMC $\{V : \mathsf{X} \to \mathsf{Y}\}$ such that $I(P, V) \leqq R$. By Corollary 6.4, for $n \geqq n_1(|\mathsf{X}|, |\mathsf{Y}|, \delta)$ any n-length block code (f, φ) as above has maximum probability of error at least $1 - \delta/2$ (say) for the DMC $\{V\}$ (suppose that $\delta < 1$). This means that for some $m \in \mathsf{M}_f$ the set $\mathsf{S}_m \triangleq \{\mathbf{y} : \varphi(\mathbf{y}) \neq m\}$ satisfies

$$V^n(\mathsf{S}_m | f(m)) \geqq 1 - \frac{\delta}{2}. \tag{10.20}$$

Knowing that the $V^n(\cdot|f(m))$-probability of S_m is large, we conclude that its $W^n(\cdot|f(m))$-probability cannot be too small. In fact, note that for any probability distributions Q_1 and Q_2 on a finite set Z and any $\mathsf{S} \subset \mathsf{Z}$ the log-sum inequality implies

$$Q_1(\mathsf{S}) \log \frac{Q_1(\mathsf{S})}{Q_2(\mathsf{S})} + Q_1(\bar{\mathsf{S}}) \log \frac{Q_1(\bar{\mathsf{S}})}{Q_2(\bar{\mathsf{S}})} \leqq D(Q_1||Q_2).$$

Hence

$$Q_1(\mathsf{S}) \log \frac{1}{Q_2(\mathsf{S})} \leqq D(Q_1||Q_2) + h(Q_1(\mathsf{S}))$$

and thus

$$Q_2(\mathsf{S}) \geqq \exp\left[-\frac{D(Q_1||Q_2) + h(Q_1(\mathsf{S}))}{Q_1(\mathsf{S})}\right]. \tag{10.21}$$

Applying this to $V^n(\cdot|f(m))$, $W^n(\cdot|f(m))$ and S_m in the role of Q_1, Q_2 and S, respectively, we get by (10.20)

$$W^n(\mathsf{S}_m|f(m)) \geqq \exp\left[-\frac{nD(V||W|P) + h\left(1 - \frac{\delta}{2}\right)}{1 - \frac{\delta}{2}}\right]$$
$$\geqq \frac{1}{2}\exp[-nD(V||W|P)(1 + \delta)],$$

if δ satisfies $h(1 - \delta/2) < 1 - \delta/2$.

Choosing the channel V to achieve the minimum in (10.19), this gives

$$e(W^n, f, \varphi) \geqq W^n(\mathsf{S}_m|f(m)) \geqq \frac{1}{2}\exp[-nE_{sp}(R, P, W)(1 + \delta)].$$

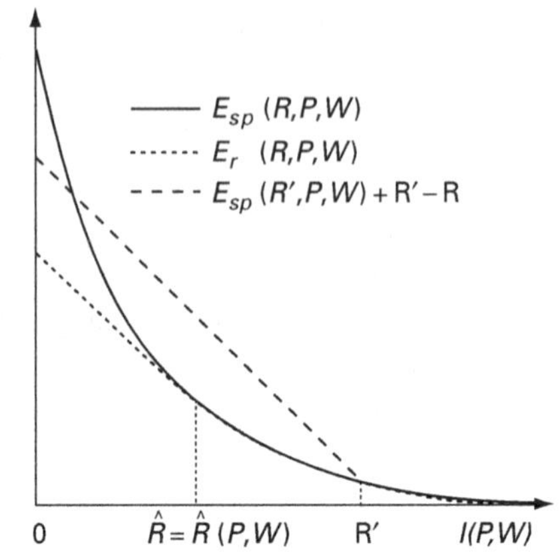

Figure 10.2 Sphere packing and random coding exponent functions ($R_\infty(P, W) = 0$).

→ 10.5

Clearly, the condition $h\,(1 - \delta/2) < 1 - \delta/2$ is no real restriction, as the validity of the theorem for some δ_0 implies the same for every $\delta > \delta_0$. □

→ 10.6

The following lemma clarifies the relation between the sphere-packing and random coding exponent functions, allowing us to compare the bounds of Theorems 10.2 and 10.3.

LEMMA 10.4 For fixed P and W, $E_{\text{sp}}(R, P, W)$ is a convex function of $R \geqq 0$, positive for $R < I(P, W)$ and vanishing otherwise. Further,

$$E_{\text{r}}(R, P, W) = \min_{R' \geqq R} (E_{\text{sp}}(R', P, W) + R' - R). \tag{10.22}$$

○

COROLLARY 10.4

$$E_{\text{r}}(R, P, W) = \begin{cases} E_{\text{sp}}(R, P, W) & \text{if} \quad R \geqq \hat{R} \\ E_{\text{sp}}(\hat{R}, P, W) + \hat{R} - R & \text{if} \quad 0 \leqq R \leqq \hat{R}, \end{cases}$$

where $\hat{R} = \hat{R}(P, W)$ is the smallest R at which the convex curve $E_{\text{sp}}(R, P, W)$ meets its supporting line of slope -1. ○

REMARK It follows from Corollary 10.4 and Theorem 10.3 that the codes of Theorem 10.2 are asymptotically optimal (among constant composition codes of type P) for every DMC $\{W\}$ such that $\hat{R}(P, W) \leqq R < I(P, W)$. ○

Proof Let $R_1 \neq R_2$ be arbitrary non-negative numbers and $0 < \alpha < 1$. Let $V_i, i = 1, 2$ achieve the minimum in the definition of $E_{\text{sp}}(R_i, P)$ and put

$$V_\alpha(y|x) \triangleq \alpha V_1(y|x) + (1 - \alpha)V_2(y|x).$$

Then, by the convexity lemma, Lemma 3.5,

$$\begin{aligned} D(V_\alpha||W|P) &\leqq \alpha D(V_1||W|P) + (1 - \alpha)D(V_2||W|P) \\ &= \alpha E_{\text{sp}}(R_1, P) + (1 - \alpha)E_{\text{sp}}(R_2, P) \end{aligned}$$

and

$$I(P, V_\alpha) \leqq \alpha I(P, V_1) + (1-\alpha) I(P, V_2) \leqq \alpha R_1 + (1-\alpha) R_2,$$

proving that

$$E_{\text{sp}}(\alpha R_1 + (1-\alpha) R_2, P) \leqq \alpha E_{\text{sp}}(R_1, P) + (1-\alpha) E_{\text{sp}}(R_2, P).$$

Obviously, $E_{\text{sp}}(R, P) = 0$ if $R \geqq I(P, W)$. On the other hand, if $I(P, V) < I(P, W)$ for a $V : \mathsf{X} \to \mathsf{Y}$, then for some $x \in \mathsf{X}$ we have $P(x) > 0$ and $V(\cdot|x) \neq W(\cdot|x)$, implying $D(V||W|P) > 0$. Hence $E_{\text{sp}}(R, P) > 0$ if $R < I(P, W)$.

By its convexity and positivity properties, $E_{\text{sp}}(R, P)$ is a strictly decreasing function of R in the interval where it is finite and positive. Hence

$$E_{\text{sp}}(R, P) = \min_{V : I(P,V) = R} D(V||W|P) \tag{10.23}$$

for $R_\infty(P, W) \leqq R \leqq I(P, W)$. As a V achieving the minimum in the definition of $E_{\text{r}}(R, P)$ certainly satisfies $R_\infty(P, W) \leqq I(P, V) \leqq I(P, W)$, (10.23) implies

$$\begin{aligned} E_{\text{r}}(R, P) &= \min_{R' : R_\infty(P,W) \leqq R' \leqq I(P,W)} \; \min_{V : I(P,V) = R'} (D(V||W|P) + |R' - R|^+) \\ &= \min_{R_\infty(P,W) \leqq R' \leqq I(P,W)} (E_{\text{sp}}(R', P) + |R' - R|^+) \\ &= \min_{R'} (E_{\text{sp}}(R', P) + |R' - R|^+), \end{aligned}$$

where the last step follows as $E_{\text{sp}}(R', P) = 0$ for $R' \geqq I(P, W)$. This proves (10.22) by the monotonicity of E_{sp}.

The corollary is immediate. □

In order to extend the previous results to arbitrary block codes, we have to look at the continuity properties of the exponent functions.

LEMMA 10.5 For every fixed $W : \mathsf{X} \to \mathsf{Y}$, consider $E_{\text{r}}(R, P, W)$ as a function of the pair R, P. This family of functions is uniformly equicontinuous. ○

Proof Fix an $\eta > 0$. For given R, P, W, let V achieve minimum in (10.15), and set

$$\tilde{V}(\cdot|x) \triangleq \begin{cases} V(\cdot|x) & \text{if} \quad P(x) \geqq \eta \\ W(\cdot|x) & \text{if} \quad P(x) < \eta. \end{cases}$$

As $D(V||W|P) \leqq E_{\text{r}}(R, P, W) \leqq I(P, W) \leqq \log|\mathsf{X}|$, for every $x \in \mathsf{X}$ with $P(x) \geqq \eta$ we have

$$D(\tilde{V}(\cdot|x)||W(\cdot|x)) = D(V(\cdot|x)||W(\cdot|x)) \leqq \frac{\log|\mathsf{X}|}{\eta}, \tag{10.24}$$

while otherwise the left-hand side of (10.24) is zero. It follows that for every R' and every P' satisfying $\sum_{x\in\mathsf{X}} |P'(x) - P(x)| \leqq \eta^2$

$$\begin{aligned} E_r(R', P', W) &\leqq D(\tilde{V}||W|P') + |I(P', \tilde{V}) - R'|^+ \\ &= D(\tilde{V}||W|P) + \sum_{x\in\mathsf{X}} (P'(x) - P(x)) D(\tilde{V}(\cdot|x)||W(\cdot|x)) + |I(P', \tilde{V}) - R'|^+ \\ &\leqq D(V||W|P) + \eta \log|\mathsf{X}| + |I(P', \tilde{V}) - R'|^+ \leqq E_r(R, P, W) + \eta \log|\mathsf{X}| \\ &+ |I(P', \tilde{V}) - I(P, \tilde{V})| + |I(P, \tilde{V}) - I(P, V)| + |R' - R|. \end{aligned} \tag{10.25}$$

By the uniform continuity of $I(P, V)$, to every $\varepsilon > 0$ there is an $\eta > 0$ such that $\sum_x |P'(x) - P(x)| < \eta^2$ implies $|I(P', \tilde{V}) - I(P, \tilde{V})| < \varepsilon$, whereas by the definition of $\tilde{V}$, $|I(P, \tilde{V}) - I(P, V)| < \varepsilon$ also holds for sufficiently small η. Thus (10.25) gives by symmetry

$$|E_r(R', P', W) - E_r(R, P, W)| \leqq \eta \log|\mathsf{X}| + 2\varepsilon + |R' - R|. \qquad \square$$

Now we are ready to analyze the best asymptotic performance of block codes for a given DMC.

A number $E \geqq 0$ will be called an *attainable error exponent at rate R* for a DMC $\{W\}$ if to every $\delta > 0$, for every sufficiently large n there exist n-length block codes of rate at least $R - \delta$ and having maximum probability of error not exceeding $\exp[-n(E - \delta)]$. The largest attainable error exponent at rate R, as a function of R, is called the *reliability function* of the DMC, denoted by $E(R) = E(R, W)$.

The previous results enable us to bound the reliability function, by optimizing the bounds of Theorems 10.2 and 10.4 with respect to P. To this end, define

$$E_{sp}(R) = E_{sp}(R, W) \triangleq \max_P E_{sp}(R, P, W)$$

and

$$E_r(R) = E_r(R, W) \triangleq \max_P E_r(R, P, W).$$

These functions are called the *sphere-packing* resp. *random coding exponent function* of channel W. Further, let $R_\infty = R_\infty(W) \triangleq \max_P R_\infty(P, W)$ be the smallest R to the right of which $E_{sp}(R)$ is finite.

→ 10.7

THEOREM 10.6 (*Random coding and sphere packing bounds*) For every DMC $\{W\}$ and $R > 0$

$$E_r(R) \leqq E(R) \leqq E_{sp}(R),$$

except that the second inequality is not claimed for $R = R_\infty$. ○

→ 10.8

COROLLARY 10.6 Let $R_{cr} = R_{cr}(W)$ be the smallest R at which the convex curve $E_{sp}(R)$ meets its supporting line of slope -1. Then

$$E(R) = E_{sp}(R) = E_r(R) \quad \text{if} \quad R \geqq R_{cr}. \qquad ○$$

REMARK R_{cr} is called the *critical rate* of the DMC $\{W\}$. ○

Proof Let P be a distribution maximizing $E_r(R, P, W)$. For every n, pick a type P_n of sequences in X^n such that $P_n \to P$. Then by Lemma 10.5 we have

$$E_r(R, P_n, W) \to E_r(R). \tag{10.26}$$

Given any $\delta > 0$, by Theorem 10.2 there exist constant composition codes (f_n, φ_n) with codeword-type P_n, rate

$$\frac{1}{n} \log |\mathsf{M}_{f_n}| \geqq R - \delta$$

and maximum probability of error

$$e(W^n, f_n, \varphi_n) \leqq \exp[-n(E_r(R, P_n, W) - \delta)],$$

provided that n is large enough. This and (10.26) prove that $E_r(R)$ is an attainable exponent at rate R.

To verify the upper bound on $E(R)$, observe that by the type counting lemma any code (f_n, φ_n) satisfying $|\mathsf{M}_{f_n}| \geqq \exp[n(R - \delta)]$ has a constant composition subcode $(\hat{f}_n, \hat{\varphi}_n)$ of rate

$$\frac{1}{n} \log |\mathsf{M}_{\hat{f}_n}| \geqq R - \delta - \frac{\log(n+1)}{n} |\mathsf{X}| \geqq R - 2\delta$$

if n is sufficiently large. Hence, by Theorem 10.3,

$$e(W^n, f_n, \varphi_n) \geqq e(W^n, \hat{f}_n, \hat{\varphi}_n) \geqq \frac{1}{2} \exp[-n(1+\delta) E_{sp}(R - 3\delta, W)].$$

Using the continuity of E_{sp} implied by its convexity, the last inequality completes the proof of the theorem.

The corollary follows by Lemma 10.4 as (10.22) implies → 10.9

$$E_r(R) = \min_{R' \geqq R} (E_{sp}(R') + R' - R). \tag{10.27}$$

□

Let us reconsider the hitherto obtained results from the point of view of universal coding; see the Discussion of Chapter 2. In the present context, this amounts to evaluating the performance of a code by the spectrum of its maximum probability of error for every DMC with the given input and output alphabets.

DEFINITION 10.7 A function $E^*(W)$ of W is a *universally attainable error exponent* at rate $R > 0$ for the family of DMCs $\{W : \mathsf{X} \to \mathsf{Y}\}$ if for every $\delta > 0$ and $n \geqq n_0(|\mathsf{X}|, |\mathsf{Y}|, R, \delta)$ there exist n-length block codes (f, φ) of rate at least $R - \delta$ and having maximum probability of error

$$e(W^n, f, \varphi) \leqq \exp[-n(E^*(W) - \delta)]$$

for every DMC $\{W : \mathsf{X} \to \mathsf{Y}\}$. Such a function $E^*(W)$ is called *maximal* if, to every other $E^{**}(W)$ which is universally attainable at rate R, there is some W_0 with → 10.10

$$E^{**}(W_0) < E^*(W_0).$$

○

The natural generalization of the error exponent problem for a single channel, i.e., of determining the quantity $E(R)$, is to determine all universally attainable error exponent functions (or, which is the same thing, all maximal ones) for every fixed R. This problem appears to be very difficult. Obvious examples of universally attainable error exponent functions are obtained by setting $E^*(W) = E(R, W)$ for some fixed W and $E^*(W) = 0$ otherwise. Theorem 10.2 leads to a more interesting family of universally attainable error exponent functions.

→ 10.11

THEOREM 10.8 For every fixed distribution P on X, the random coding exponent $E_r(R, P, W)$ is universally attainable at rate R. ○

Proof The assertion follows from Theorem 10.2, as by Lemma 10.5 to every distribution P on X and every $R > 0$ there is a sequence $P_n \to P$ such that P_n is a type of sequence in X^n and $E_r(R, P_n, W) \to E_r(R, P, W)$ uniformly in W. □

Another way of evaluating code performance for a family of channels is to consider the largest error probability the code achieves for the channels in the given family. Clearly, this is a less ambitious approach than universal coding. It is justified only if the involved family of channels is not too large. For example, for the family of all DMCs $\{W : \mathsf{X} \to \mathsf{Y}\}$ this largest error probability approaches unity for every $R > 0$ as $n \to \infty$.

DEFINITION 10.9 A *compound channel* with input set X and output set Y is a (not necessarily finite) family $\mathcal{W}$ of channels $W : \mathsf{X} \to \mathsf{Y}$. The *maximum probability of error* of a code (f, φ) over the compound channel $\mathcal{W}$ is defined as

$$e = e(\mathcal{W}, f, \varphi) \triangleq \sup_{W \in \mathcal{W}} e(W, f, \varphi).$$

Similarly, the *average probability of error* is

$$\bar{e} = \bar{e}(\mathcal{W}, f, \varphi) \triangleq \sup_{W \in \mathcal{W}} \bar{e}(W, f, \varphi).$$ ○

A *compound DMC* with input alphabet X and output alphabet Y is a sequence of compound channels $\mathcal{W}_n \triangleq \{W^n : W \in \mathcal{W}\}$, where $\mathcal{W}$ is some given set of stochastic matrices $W : \mathsf{X} \to \mathsf{Y}$. The *ε-capacity, capacity, attainable error exponents* and the *reliability function* of a compound DMC are defined analogously to those of a single DMC.

For any family $\mathcal{W}$ of stochastic matrices $W : \mathsf{X} \to \mathsf{Y}$, consider the following functions:

$$I(P, \mathcal{W}) \triangleq \inf_{W \in \mathcal{W}} I(P, W);$$

$$E_r(R, P, \mathcal{W}) \triangleq \inf_{W \in \mathcal{W}} E_r(R, P, W); \quad E_{sp}(R, P, \mathcal{W}) \triangleq \inf_{W \in \mathcal{W}} E_{sp}(R, P, W).$$

Define

$$C(\mathcal{W}) \triangleq \max_P I(P, \mathcal{W});$$

$$E_r(R, \mathcal{W}) \triangleq \max_P E_r(R, P, \mathcal{W}); \quad E_{sp}(R, \mathcal{W}) \triangleq \max_P E_{sp}(R, P, \mathcal{W}).$$

THEOREM 10.10 $E_r(R, \mathcal{W})$ is a lower bound and $E_{sp}(R, \mathcal{W})$ is an upper bound for the reliability function of the compound DMC determined by $\mathcal{W}$. ○

COROLLARY 10.10 (*Compound channel coding theorem*) For every $0 < \varepsilon < 1$, the ε-capacity of the compound DMC determined by $\mathcal{W}$ equals $C(\mathcal{W})$. ○ → 10.12

Proof By Theorem 10.8, for every PD P on X, $E_r(R, P, W)$ is universally attainable at rate R. This implies, by definition, that $E_r(R, P, \mathcal{W})$ is an attainable error exponent at rate R for the compound DMC. Hence $E_r(R, \mathcal{W})$ is a lower bound for the reliability function. The fact that $E_{sp}(P, \mathcal{W})$ is an upper bound follows from Theorem 10.3 in the same way as the analogous result of Theorem 10.6.

To prove the corollary, observe that $C_\varepsilon \leqq \max_P I(P, \mathcal{W})$ is a consequence of Corollary 6.4, while the present theorem immediately gives the opposite inequality. □ → 10.13

→ 10.14

The packing lemma is useful also for a more subtle analysis of the decoding error. Recall that given a channel $W : \mathsf{X} \to \mathsf{Y}$ and a code (f, φ), i.e., a pair of mappings $f : \mathsf{M}_f \to \mathsf{X}$, $\varphi : \mathsf{Y} \to \mathsf{M}' \supset \mathsf{M}_f$, the probability of erroneous transmission of message $m \in \mathsf{M}_f$ is

$$e_m = e_m(W, f, \varphi) \triangleq 1 - W(\varphi^{-1}(m)|f(m)).$$

So far we have paid no attention to a conceptual difference between two kinds of errors. If $\varphi(y) \in \mathsf{M}' - \mathsf{M}_f$ then it is obvious that an error has occurred, while if $\varphi(y)$ is some element of M_f different from the actual message m then the error remains *undetected*. From a practical point of view, such a confusion of messages usually creates more harm than the previous type of error, which means just the *erasure* of the message.

If message m has been transmitted, the *probability of undetected error* is

$$\hat{e}_m = \hat{e}_m(W, f, \varphi) \triangleq W\left(\bigcup_{m' \in \mathsf{M}_f - \{m\}} \varphi^{-1}(m') \middle| f(m)\right),$$

while the *probability of erasure* is

$$\tilde{e}_m = \tilde{e}_m(W, f, \varphi) \triangleq W\left(\bigcup_{m' \in \mathsf{M}' - \mathsf{M}_f} \varphi^{-1}(m') \middle| f(m)\right) = e_m - \hat{e}_m.$$

The maximum probability of undetected error is

$$\hat{e}(W, f, \varphi) \triangleq \max_{m \in \mathsf{M}_f} \hat{e}_m(W, f, \varphi).$$

Our aim is to give simultaneously attainable exponential upper bounds for the (maximum) probability of error resp. undetected error for n-length block codes over a DMC. We shall see that $\hat{e}(W^n, f, \varphi)$ can be made significantly smaller than the least possible value of $e(W^n, f, \varphi)$, at the expense of admitting a larger probability of erasure, i.e., by not insisting on $e(W^n, f, \varphi)$ to be least possible. Theorem 10.11 below is a corresponding generalization of Theorem 10.2. Its Corollary 10.11A emphasizes the universal character of the result, while Corollary 10.11B gives an upper bound of the attainable $\hat{e}(W^n, f, \varphi)$ for a single DMC, if the only condition on $e(W^n, f, \varphi)$ is that it should tend to zero.

Define the *modified random coding exponent function* as

$$E_{r,\lambda}(R, P, W) \triangleq \min_V (D(V||W|P) + \lambda|I(P, V) - R|^+), \tag{10.28}$$

where $\lambda > 0$ is an arbitrary parameter. Then, similarly to Lemma 10.4 and its corollary,

$$\begin{aligned} E_{r,\lambda}(R, P, W) &= \min_{R' \geqq R} (E_{sp}(R', P, W) + \lambda(R' - R)) \\ &= \begin{cases} E_{sp}(R, P, W) & \text{if} \quad R \geqq \hat{R}_\lambda \\ E_{sp}(\hat{R}_\lambda, P, W) + \lambda(\hat{R}_\lambda - R) & \text{if} \quad 0 \leqq R \leqq \hat{R}_\lambda, \end{cases} \end{aligned} \tag{10.29}$$

where $\hat{R}_\lambda = \hat{R}_\lambda(P, W)$ is the smallest R at which the convex curve $E_{sp}(R, P, W)$ meets its supporting line of slope $-\lambda$.

To the analogy of Definition 10.7 we shall say that a pair of functions $(\hat{E}^*(W), E^*(W))$ of W is a *universally attainable pair of error exponents* at rate $R > 0$ for the family of DMCs $\{W : \mathsf{X} \to \mathsf{Y}\}$, if for every $\delta > 0$ and $n \geqq n_0(|\mathsf{X}|, |\mathsf{Y}|, R, \delta)$ there exist n-length block codes (f, φ) of rate

$$\frac{1}{n} \log |\mathsf{M}_f| \geqq R - \delta$$

which for every DMC $\{W : \mathsf{X} \to \mathsf{Y}\}$ yield

$$\hat{e}(W^n, f, \varphi) \leqq \exp[-n(\hat{E}^*(W) - \delta)],$$

$$e(W^n, f, \varphi) \leqq \exp[-n(E^*(W) - \delta)].$$

THEOREM 10.11 For every $\tilde{R} \geqq R > 0$, $\lambda > 1, \delta > 0$ and every type P of sequences in X^n there exists an n-length block code (f, φ) of rate

$$\frac{1}{n} \log |\mathsf{M}_f| \geqq R - \delta$$

such that all codewords are of type P, and for every DMC $\{W : \mathsf{X} \to \mathsf{Y}\}$

$$\hat{e}(W^n, f, \varphi) \leqq \exp\{-n[E_{r,\lambda}(R, P, W) + \tilde{R} - R - \delta]\}, \tag{10.30}$$

$$e(W^n, f, \varphi) \leqq \exp\{-n[E_{r,1/\lambda}(\tilde{R}, P, W) - \delta]\}, \tag{10.31}$$

provided that $n \geqq n_0(|\mathsf{X}|, |\mathsf{Y}|, \delta)$. ○

COROLLARY 10.11A For every distribution P on X and $\tilde{R} \geqq R > 0, \lambda > 1$

$$\left(E_{r,\lambda}(R, P, W) + \tilde{R} - R, \quad E_{r,1/\lambda}(\tilde{R}, P, W)\right)$$

is a universally attainable pair of error exponents at rate R. ○

COROLLARY 10.11B For every DMC $\{W : \mathsf{X} \to \mathsf{Y}\}$ and $0 < R < C(W) = \max_P I(P, W)$ there exists a sequence of n-length block codes $\{(f_n, \varphi_n)\}_{n=1}^\infty$ of rates converging to R such that $e(W^n, f_n, \varphi_n) \to 0$ and

$$\hat{e}(W^n, f_n, \varphi_n) \leqq \exp[-n\hat{E}(R, W)], \tag{10.32}$$

→ 10.15

where

$$\hat{E}(R, W) \triangleq \max_P \left[E_{\rm sp}(R, P, W) + I(P, W) - R\right]. \tag{10.33}$$

○

REMARK $\hat{E}(R, W)$ is finite iff $E_{\rm sp}(R, W)$ is finite. If $\hat{E}(R, W) = \infty$, the inequality (10.32) means $\hat{e}(W^n, f_n, \varphi_n) = 0$. ○

Proof We shall consider the same encoder f as in the proof of Theorem 10.2, with a different decoder. Recalling that $\mathsf{M}_f = \{1, 2, \ldots, m\}$, define the decoder mapping $\varphi : \mathsf{Y}^n \to \mathsf{M}' \triangleq \{0, 1, \ldots, m\}$ by

$$\varphi(\mathbf{y}) \triangleq \begin{cases} i & \text{if} \quad I(\mathbf{x}_i \wedge \mathbf{y}) > \tilde{R} + \lambda|I(\mathbf{x}_j \wedge \mathbf{y}) - R|^+ \quad \text{for} \quad j \neq i \\ 0 & \text{else.} \end{cases}$$

Since $\tilde{R} \geqq R$, this definition is unambiguous.

Using this code (f, φ), if message i was transmitted, an undetected error can occur only if the received sequence $\mathbf{y} \in \mathsf{Y}^n$ is such that

$$I(\mathbf{x}_j \wedge \mathbf{y}) > \tilde{R} + \lambda|I(\mathbf{x}_i \wedge \mathbf{y}) - R|^+ \quad \text{for some} \quad j \neq i,$$

i.e., if $\mathbf{y}$ is contained in the set

$$\bigcup_{\substack{V, \hat{V} \in \mathcal{V}(P) \\ I(P,\hat{V}) > \tilde{R} + \lambda|I(P,V) - R|^+}} \left(\mathsf{T}_V(\mathbf{x}_i) \cap \bigcup_{j \neq i} \mathsf{T}_{\hat{V}}(\mathbf{x}_j) \right).$$

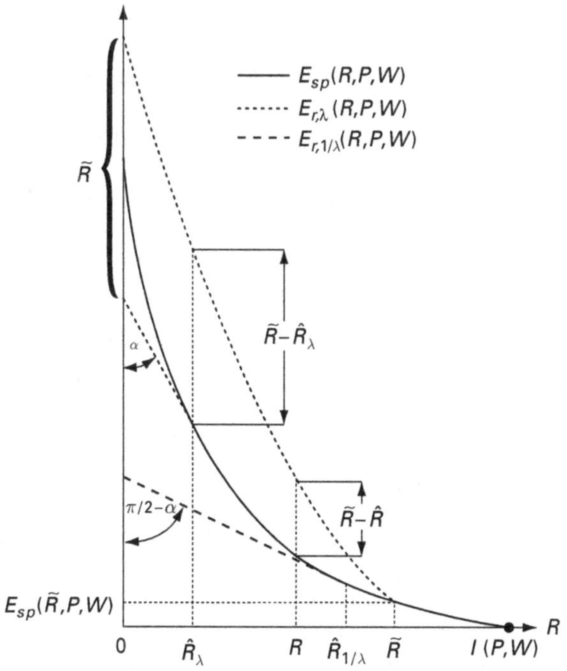

Figure 10.3 Attainable pairs of error exponents of Corollary 10.11A at various rates R (dotted lines).

As the codeword set has been chosen according to Lemma 10.1, the bound (10.17) applies and we obtain

$$\hat{e}_i \leqq \sum_{\substack{V,\hat{V}\in\mathcal{V}(P)\\ I(P,\tilde{V})>\tilde{R}+\lambda|I(P,V)-R|^+}} \exp\{-n[D(V||W|P)+|I(P,\tilde{V})-R|^+]\}$$

$$\leqq \sum_{V,\hat{V}\in\mathcal{V}(P)} \exp\{-n[D(V||W|P)+\lambda|I(P,V)-R|^+ + \tilde{R}-R]\}.$$

By the definition (10.28) of $E_{\mathrm{r},\lambda}$, this and (10.1) prove (10.30).

Further, if message i was transmitted, an error (undetected or erasure) occurs iff the received sequence $\mathbf{y}\in\mathsf{Y}^n$ satisfies

$$I(\mathbf{x}_i\wedge\mathbf{y})\leqq\tilde{R}+\lambda|I(\mathbf{x}_j\wedge\mathbf{y})-R|^+ \quad \text{for some} \quad j\neq i.$$

This happens iff either $I(\mathbf{x}_i\wedge\mathbf{y})\leqq\tilde{R}$ or

$$\tilde{R}<I(\mathbf{x}_i\wedge\mathbf{y})\leqq\tilde{R}+\lambda(I(\mathbf{x}_j\wedge\mathbf{y})-R) \quad \text{for some} \quad j\neq i,$$

i.e., iff $\mathbf{y}\in\mathsf{A}_i\cup\mathsf{B}_i$, where

$$\mathsf{A}_i \triangleq \bigcup_{\substack{V,\hat{V}\in\mathcal{V}(P)\\ I(P,V)\leqq\tilde{R}}} \mathsf{T}_V(\mathbf{x}_i), \qquad \mathsf{B}_i \triangleq \bigcup_{\substack{V,\hat{V}\in\mathcal{V}(P)\\ \tilde{R}<I(P,V)\leqq\tilde{R}+\lambda(I(P,\hat{V})-R)}} \left(\mathsf{T}_V(\mathbf{x}_i)\cap\bigcup_{j\neq i}\mathsf{T}_{\hat{V}}(\mathbf{x}_j)\right).$$

The condition $\tilde{R}<I(P,V)\leqq\tilde{R}+\lambda(I(P,\hat{V})-R)$ implies

$$I(P,\hat{V})\geqq R+\frac{1}{\lambda}(I(P,V)-\tilde{R}) = R+\frac{1}{\lambda}\,|\,I(P,V)-\tilde{R}\,|^+.$$

Thus, applying (10.5) and the consequence (10.17) of the packing lemma we see that

$$e_i \leqq W^n(\mathsf{A}_i|\mathbf{x}_i)+W^n(\mathsf{B}_i|\mathbf{x}_i)\leqq|\mathcal{V}(P)|\exp[-nE_{\mathrm{sp}}(\tilde{R},P,W)]$$
$$+\sum_{V,\hat{V}\in\mathcal{V}(P)}\exp\{-n[D(V||W|P)+\frac{1}{\lambda}|I(P,V)-\tilde{R}|^+]\}.$$

On account of (10.1) and (10.29), this proves (10.31).

Corollary 10.11A follows from the theorem in the same way as did Theorem 10.8 from Theorem 10.2.

To prove Corollary 10.11B, suppose first that $\hat{E}(R,W)<\infty$, and let P be a distribution achieving

$$E_{\mathrm{sp}}(R,P,W)+I(P,W)-R=\hat{E}(R,W).$$

By the assumption $R<C(W)$ we have $\hat{E}(R,W)>0$, whence necessarily $R<I(P,W)$. As a particular case of Corollary 10.11A, to every $(R',\tilde{R}')$ with $R'<R<\tilde{R}'<I(P,W)$, $\lambda>1$ and $\delta>0$, for sufficiently large n there exist n-length block codes with $(1/n)\log|\mathsf{M}_f|\geqq R'-\delta$ and

$$\hat{e}(W^n, f, \varphi) \leqq \exp\{-n[E_{\mathrm{r},\lambda}(R', P, W) + \tilde{R}' - R' - \delta]\},$$
$$e(W^n, f, \varphi) \leqq \exp\{-n[E_{\mathrm{r},\frac{1}{\lambda}}(\tilde{R}', P, W) - \delta]\}.$$

Setting $R' \triangleq R - \eta$, $\tilde{R}' \triangleq I(P, W) - \eta$ with arbitrarily small $\eta > 0$, by (10.29) and the strict monotonicity of $E_{\mathrm{sp}}(R, P, W)$ one can choose λ so large and δ so small that

$$E_{\mathrm{r},\lambda}(R', P, W) + \tilde{R}' - R' - \delta > \hat{E}(R, W), \quad E_{\mathrm{r},\frac{1}{\lambda}}(\tilde{R}', R, W) - \delta > 0.$$

This proves Corollary 10.11B for $\hat{E}(R, W) < \infty$. If $\hat{E}(R, W) = \infty$, a similar argument shows that for every $K > 0$ there exist n-length block codes of the desired rate such that $e(W^n, f, \varphi)$ is arbitrarily small and

$$\hat{e}(W^n, f, \varphi) \leqq \exp(-nK).$$

For a sufficiently large K, the preceding relation implies $\hat{e}(W^n, f, \varphi) = 0$. □

Discussion

In Chapter 6 we considered ε-capacities of a DMC for fixed ε. We have seen that $C_\varepsilon = C$ for every $\varepsilon > 0$ but, in general, $C_0 < C$. One aim of this chapter was to see what happens in the intermediate case when the tolerated probability of error is positive and has some specified order of magnitude depending on the block length n. To every such order of magnitude one might define a capacity. In particular, to the order of magnitude exp $(-nE)$ the corresponding capacity is the largest R with $E(R) \geqq E$. If the tolerated error probability tends to zero slower than exponentially, the corresponding capacity equals the ordinary one, as seen from Theorem 10.6. Of course, for a DMC, a sequence of positive error probabilities cannot converge to zero faster than exponentially. → 10.16

Theorem 10.6 provides only a partial solution to this generalized capacity problem. Note that a complete solution would involve the determination of the zero-error capacity, which in itself is one of the longest standing open problems in information theory. A problem of equal mathematical, if not practical, interest is to refine the converse part of the noisy channel coding theorem, i.e., to determine how fast the probability of correct decoding converges to zero for codes of rate above capacity. This problem is less difficult than the previous one and is almost completely solved. → 10.17

The source and channel coding results of the present chapter and Chapter 9 are similar in many respects. One essential difference is, however, that while for sources to every rate R there is a sequence of codes, the performance of which approaches the optimum uniformly for every DMS, for channels this is no longer so. As, for channels, the problem of universal coding is genuinely different from coding for a single DMC, we have given special emphasis to the universal coding implications of the packing lemma, including the more refined analysis of the decoding error (Corollary 10.11A).

Final results are not yet available. Both for a single channel and for universal coding, low-rate improvements on the bounds in the text are treated in Problems 10.11, 10.18–10.24 and 10.30. These improvements are shown in Fig. P10.30.

Some modifications of the model, such as the availability of feedback, which at the first level did not change the results, do influence the error exponent; see Problems 10.36–10.38.

Combining the results of the present chapter and Chapter 9, a more subtle analysis of the LMTR problem is possible. We have seen that if for the transmission of k-length messages of a DMS over a DMC an ε-fidelity criterion (d, Δ) is imposed, where $\varepsilon \in (0, 1)$ does not depend on k, then the limiting minimum transmission ratio can be achieved by composing "good" source and channel codes. It turns out that if ε depends on the block length k, then, in general, this is no longer true. → 10.32

From a practical point of view, merely asymptotic error estimates are of little value. Rather, one would be interested in error bounds for finite block lengths. By analyzing our proofs, it is easy to derive such bounds, though they may be poor for small block lengths. It was not our aim, however, to get sharp bounds for every block length n. → 10.33

An issue deserving interest is the proper choice of the decoder. For a fixed (known) DMC, maximum likelihood decoding provides the smallest average probability of error, and the joint typicality decoder is often used for simplifying calculations; see Problems 6.2 and 6.9. For universality, however, the decoder has to be channel-independent, and in this chapter the MMI decoder is used. When all codewords are of the same type, it is also a *mimimum entropy decoder*, i.e., the codeword $\mathbf{x}_i$ attaining the maximum in (10.13) is also a minimizer of $H(\mathbf{x}_i, \mathbf{y})$. When the codewords are of different types, the choice between a MMI and a minimum entropy decoder is a genuine alternative. For special purposes, a different decoder might be needed, such as to achieve the pair of error exponents in Corollary 10.11A. Proofs of some general results for channel models treated in Chapter 12 also require sophisticated decoders; see Problems 12.5 and 12.10. Needless to say, the above comments refer to the choice of the decoder for the theoretical purpose of proving achievability results. For practical purposes, a primary consideration is computational complexity that should be low enough to make the decoding actually feasible. This issue is highly important but is not addressed here. ○

Problems

10.1. (*General Gilbert bound*)

(a) Supposing $\mathsf{Y} \supset \mathsf{X}$, show that the sequences $\mathbf{x}_i$ of the packing lemma satisfy $I(\mathbf{x}_i \wedge \mathbf{x}_j) \leqq R$ if $i \neq j$.

(b) Prove directly the existence of $\exp\{n(R - \delta)\}$ sequences of type P that have the preceding property. (Blahut (1977).)

(c) Conclude that for even n there exist $\exp\{n(R - \delta)\}$ binary sequences of length n and type $P = (1/2, 1/2)$ with $R \geqq 1 - h\left(\frac{1}{n}\min_{i \neq j} d_{\mathrm{H}}(\mathbf{x}_i, \mathbf{x}_j)\right)$. Compare this result with Gilbert's bound; see Problem 5.5.

Hint Use Problem 6.20(b).

10.2. (*Simpler form of the packing lemma*) Show that the following assertion, not involving Y, is equivalent to Lemma 10.1: for every $R > \delta > 0$ and type P of sequences in X^n with $H(P) > R$, there exists a set $\mathsf{C} \subset \mathsf{T}_P$ of size at least

$\exp[n(R-\delta)]$ such that, for every $\mathbf{x} \in \mathsf{C}$ and pair of dummy RVs X, $\tilde{X}$ whose joint distribution equals the joint type $P_{\mathbf{x},\tilde{\mathbf{x}}}$ for some $\tilde{\mathbf{x}} \neq \mathbf{x}$ in C, we have

$$|\mathsf{T}_{\tilde{V}}(\mathbf{x}) \cap \mathsf{C}| \leqq \exp[n(R - I(X \wedge \tilde{X}))], \quad \tilde{V} = P_{\tilde{X}|X}.$$

Hint Lemma 10.1 implies the above assertion, taking $\mathsf{Y} = \mathsf{X}$, $V = \tilde{V}$, $\hat{V} =$ the identity matrix. Conversely, suppose $\mathsf{C} \subset \mathsf{T}_P$ satisfies that assertion. Then, choosing the sequences $\mathbf{x}_i$ as those in C, the left-hand side of (10.6) is bounded above for any V and $\tilde{V}$ as in Lemma 10.1 by

$$\sum_{\tilde{V}} \sum_{\mathbf{x}_j \in \mathsf{T}_{\tilde{V}}(\mathbf{x}_i) \cap \mathsf{C}} |\mathsf{T}_V(\mathbf{x}_i) \cap \mathsf{T}_{\hat{V}}(\mathbf{x}_j)|,$$

where the summation for $\tilde{V}$ is over all $\tilde{V} = P_{\tilde{X}|X}$ as in the hypothesis. Here the inner sum has no more than $\exp\{n(R - I(X \wedge \tilde{X})\}$ terms, by hypothesis, and each term is bounded above by $\exp\{n(\max H(Y|X\tilde{X}) + \varepsilon_n)\}$, see Problem 2.10, where the maximization refers to dummy RVs Y with $P_{Y|X} = V$, $P_{Y|\tilde{X}} = \hat{V}$. Hence the claim follows since

$$-I(X \wedge \tilde{X}) + H(Y|X\tilde{X}) = H(Y|X) - I(XY \wedge \tilde{X}) \leqq H(V|P) - I(P, \hat{V}).$$

10.3. Show that $E_{\mathrm{r}}(R, P, W) > 0$ iff $R < I(P, W)$, so that Theorem 10.2 implies Corollary 6.3 and thus the existence part of the noisy channel coding theorem. Observe that inequality (10.14) automatically holds if $R \geqq I(P, W)$.

10.4. (*Finiteness of the sphere-packing exponent*)

(a) Prove that $R_\infty(P, W)$ equals the minimum of $I(P, V)$ taken over those V for which $V(y|x) = 0$ whenever $W(y|x) = 0$. In particular, $R_\infty(P, W) > 0$ iff to every $y \in \mathsf{Y}$ there exists an $x \in \mathsf{X}$ with $P(x) > 0$, $W(y|x) = 0$.

(b) Show that $E_{\mathrm{sp}}(R, P, W)$ is finite and continuous from the right at $R = R_\infty(P, W)$, and that so is $E_{\mathrm{sp}}(R, W) \triangleq \max_P E_{\mathrm{sp}}(R, P, W)$ at $R_\infty \triangleq \max_P R_\infty(P, W)$.

(c) Prove that

$$R_\infty = -\min_P \max_y \log \sum_{x:W(y|x)>0} P(x).$$

Conclude that for DMCs with positive zero-error capacity, $R_\infty = C_{0,\mathrm{f}}$; see Problem 6.30. (Shannon, Gallager and Berlekamp (1967). See also Problem 10.28(d).)

Hint Use Problem 9.7.

10.5. (a) Show that except for the uniformity in W, for codes as in Theorem 10.3 the lower bound $e(W^n, f, \varphi) \geqq \exp\{-n[E_{\mathrm{sp}}(R, P, W) + \delta]\}$ could be proved for sufficiently large n also if instead of (10.20) only $V^n(\mathsf{S}_m | f(m)) > \varepsilon$ were known (with some fixed ε).

Hint Use Theorem 1.2 as in the proof of Theorem 9.5.

(b) Prove the lower bound of Theorem 9.5 by the method of Theorem 10.3.

10.6. (*Alternative derivation of the sphere-packing bound*)

(a) Given two finite sets X, Y and positive numbers R and δ, show that if P is any type of sequences in X^n, and $\{\mathbf{x}_i : i \in \mathsf{M}\}$ is any subset of T^n_P such that $(1/n)\log|\mathsf{M}| \geqq R + \delta$, then for every stochastic matrix $V \in \mathcal{V}(P)$ and every mapping $\varphi : \mathsf{Y}^n \to \mathsf{M}$ we have

$$\frac{1}{|\mathsf{M}|}\sum_{i \in \mathsf{M}} \frac{|\mathsf{T}_V(\mathbf{x}_i) \cap \varphi^{-1}(i)|}{|\mathsf{T}_V(\mathbf{x}_i)|} \leqq \exp[-n|R - I(P, V)|^+]$$

whenever $n \geqq n_0(|\mathsf{X}|, |\mathsf{Y}|, \delta)$. (Dueck and Körner (1979).)

(b) Conclude that for each DMC $\{W : \mathsf{X} \to \mathsf{Y}\}$, and n-length block code (f, φ) of rate $(1/n)\log|\mathsf{M}_f| \geqq R + \delta$ with codewords of type P, one has

$$\frac{1}{n}\log \bar{e}(W^n, f, \varphi) \geqq - \min_{\substack{V \in \mathcal{V}(P) \\ I(P,V) \leqq R - \frac{1}{n}}} D(V||W|P) - \frac{1}{n}$$

whenever $n \geqq n_0(|\mathsf{X}|, |\mathsf{Y}|, \delta)$.

10.7. Let

$$e(n, R) \triangleq \min_{\frac{1}{n}\log|\mathsf{M}_f| \geqq R} e(W^n, f, \varphi)$$

designate the maximum probability of error of the "best" n-length block code of rate R for the DMC $\{W\}$.

(a) Verify that Theorem 10.6 is equivalent to the pair of statements

(i) $$\overline{\lim_{n\to\infty}} \frac{1}{n}\log e(n, R) \leqq - E_{\mathrm{r}}(R),$$

(ii) $$\overline{\lim_{n\to\infty}} \frac{1}{n}\log e(n, R) \geqq - E_{\mathrm{sp}}(R) \quad \text{if} \quad R \neq R_\infty.$$

(b) Check that (ii) holds also if $\overline{\lim}$ is replaced by $\underline{\lim}$. Conclude that

$$\lim_{n\to\infty} \frac{1}{n}\log e(n, R) = E_{\mathrm{r}}(R) = E_{\mathrm{sp}}(R) \quad \text{if} \quad R \geqq R_{\mathrm{cr}}.$$

(For $R < R_{\mathrm{cr}}$ it is not even known whether the limit exists.)

10.8. (*Critical rate*) Show that $R_{\mathrm{cr}} < C$ whenever for some $x \in \mathsf{X}$ (i) $P(x) > 0$ for a distribution P maximizing $I(P, W)$ and (ii) the row $W(\cdot|x)$ of W has at least two different positive entries. Conclude that $R_{\mathrm{cr}} = C$ iff $R_\infty = C$; see Problem 10.4 (Gallager (1965).)

Hint If $\hat{R}(P, W) = I(P, W)$ for some P then $E_{\mathrm{r}}(R, P, W)$ is a straight line of slope -1 in the interval $0 \leqq R \leqq I(P, W)$. Setting $R = 0$, this means that $I(P, V) + D(V||W|P)$ is minimized for $V = W$. Show that in this case for $P(x) > 0$ the positive entries of $W(\cdot|x)$ must be the same.

10.9. (*Input constraint*) Prove the analog of Theorem 10.6 for a DMC with input constraint (c, Γ). Show that the lower bound holds also if instead of $c(f(m)) \leqq \Gamma$ for every m only $|\mathsf{M}_f|^{-1}\sum_{m\in\mathsf{M}_f} c(f(m)) \leqq \Gamma$ is required.

10.10. (*Non-existence of universally optimal codes*) For a fixed R, consider the pointwise supremum of the universally attainable error exponents $E^*(R)$. Show that this equals $E(R, W)$, but that $E(R, W)$ is not universally attainable. In particular, there is no "best" universally attainable error exponent. Interpret this result to the effect that for channels, unlike for sources, the problem of universal coding is essentially different from coding for a single channel.

10.11. (*Universal improvement of the random coding bound*)

(a) Show that to every maximal universally attainable error exponent $E^*(W)$ at rate R there exists a PD P on X such that $E^*(W) = E_r(R, P, W)$ for every W with $\hat{R}(P, W) \leqq R$ (see Corollary 10.4).

(b) Show that for each fixed distribution P on X

$$\tilde{E}_r(R, P, W) \triangleq \min_{V: I(P,V) - H(V|P) \geqq R} (D(V||W|P) + |I(P, V) - R|^+)$$

is a universally attainable error exponent at rate R. Give an example where $\tilde{E}_r(R, P, W) > E_r(R, P, W)$.

Hint Use the Remark to Lemma 10.1. (For a sharper result and further discussion, see Csiszár and Körner (1981a).)

Compound DMCs (Problems 10.12–10.14)

10.12. Prove that if for a compound DMC the ε-capacity and capacity are defined in terms of average rather than maximum probability of error, then the capacity remains the same, but the ε-capacity may become larger.

(The surprising fact that for compound DMCs with average probability of error the strong converse to the coding theorem is false was noticed by Ahlswede (1968).)

10.13. (*Maximal codes for compound channels*)

(a) Prove Corollary 10.10 directly, if $|\mathcal{W}| < \infty$.

Hint Show that Corollary 6.3 generalizes to any finite set $\mathcal{W}$ of channels as follows. For $n \geqq n_0(|\mathsf{X}|, |\mathsf{Y}|, \tau, \varepsilon, |\mathcal{W}|)$ if (f, φ) is an (n, ε)-code for each DMC $\{W\}$, $W \in \mathcal{W}$ with $f(m) \in \mathsf{T}_{[P]}, \varphi^{-1}(m) \subset \bigcup_{W \in \mathcal{W}} \mathsf{T}_{[W]}(f(m))$ for every $m \in \mathsf{M}_f$ and the code (f, φ) has no extension with the same properties, then

$$\frac{1}{n} \log |\mathsf{M}_f| \geqq I(P, \mathcal{W}) - 2\tau.$$

(b) Get rid of the assumption $|\mathcal{W}| < \infty$ in (a).

Hint Approximate the channels in $\mathcal{W}$ by a finite number of channels (depending on n) so that to every $W \in \mathcal{W}$ there is an approximating channel W^* satisfying $|W(y|x) - W^*(y|x)| < |\mathsf{X}| \cdot n^{-4}$ for every $x \in \mathsf{X}$, $y \in \mathsf{Y}$. Show that (a) applies also to sets of DMCs whose size depends on n provided it grows slower than exponentially. (Wolfowitz (1960), (1964).)

10.14. (*Compound channel with encoder or decoder informed*)

Given a compound DMC, suppose that the encoder may be chosen depending on $W \in \mathcal{W}$ (the actual channel in operation) while the decoder cannot. Defining capacity as usual, show that it then equals the infimum of the capacities of the channels $W \in \mathcal{W}$. Note that this may be larger than $C(\mathcal{W})$. Show, however, that if the decoder but not the encoder depends on W, the capacity always remains $C(\mathcal{W})$. (Wolfowitz (1960).)

10.15. (*Undetected error*)
(a) Check that $\hat{E}(R, W)$ is a decreasing convex function of R, strictly greater than $E_{\text{sp}}(R, W)$, in the interval $R_\infty \leqq R < C$.
(b) Show by an example that the bound of Corollary 10.11B involving $\hat{E}(R, W)$ is not tight, in general.
Hint For the binary erasure channel (Problem 6.10) $\hat{E}(R, W)$ is finite for every $R \geqq 0$, and still one can attain for every rate below capacity

$$e(W^n, f_n, \varphi_n) \to 0, \qquad \hat{e}(W^n, f_n, \varphi_n) = 0.$$

10.16. (*Generalized capacity*) For a DMC $\{W\}$, set $E_{\text{cr}} \triangleq E(R_{\text{cr}})$. Show that if $E \leqq E_{\text{cr}}$, the "capacity" for tolerated probability of error $\exp(-nE)$ (i.e., the largest R with $E(R) \geqq E$) equals

$$\max_P \min_{V: D(V\|W|P) \leqq E} I(P, V).$$

10.17. (*Probability of error for $R > C$*)
(a) Show that for every DMC $\{W: \mathsf{X} \to \mathsf{Y}\}$, for $n \geqq n_1(|\mathsf{X}|, |\mathsf{Y}|, \delta)$ every n-length block code (f, φ) of rate $(1/n) \log |\mathsf{M}_f| \geqq R + \delta$ has average probability of error

$$\bar{e}(W^n, f, \varphi) \geqq 1 - \exp\{-n[K(R, W) - \delta]\},$$

where

$$K(R, W) \triangleq \min_P \min_V (D(V\|W|P) + |R - I(P, V)|^+).$$

Hint Use the inequality of Problem 10.6(a).
(b) Prove that the result of (a) is exponentially tight for every $R > C$; i.e., for sufficiently large n there exist n-length block codes (f, φ) of rate $(1/n) \log |\mathsf{M}_f| > R - \delta$ satisfying

$$\bar{e}(W^n, f, \varphi) \leqq 1 - \exp\{-n[K(R, W) + \delta]\}.$$

Hint As in the proof of Theorem 10.2, show that for $n \geqq n_0(|\mathsf{X}|, |\mathsf{Y}|, \delta)$ to every type P of sequences in X^n there exists a code (f, φ) with codewords of type P and of rate $(1/n) \log |\mathsf{M}_f| \geqq R - \delta$ having maximum probability of error

$$e(W^n, f, \varphi) \leqq 1 - \exp\{-n[\min_{V: I(P,V) \geqq R} D(V\|W|P) + \delta]\}.$$

Further, note that any code of rate R with $e(W^n, f, \varphi) \leqq 1 - \exp(-nA)$ can be extended to a code $(\tilde{f}, \tilde{\varphi})$ of rate $R' > R$ satisfying $\bar{e}(W^n, \tilde{f}, \tilde{\varphi}) \leqq 1 - \exp\{-n[A - R' + R]\}$ (let $\tilde{\varphi} = \varphi$).

(Dueck and Körner (1979); the proof of (b) uses an idea of Omura (1975). The result of (a) was first obtained – in another form – by Arimoto (1973).)

(c) Show that feedback does not increase the exponent of correct decoding at rates above capacity, i.e., the result of (a) holds even for block codes with complete feedback; see Problem 6.29.

Hint It suffices to prove the following analog of the inequality in Problem 10.6(a):

$$\frac{1}{|\mathsf{M}|}\sum_{i\in\mathsf{M}} |\mathsf{A}(i, P, V) \cap \varphi^{-1}(i)| \leqq \exp\{n(H(V|P) - |R - I(P, V)|^{+})\},$$

where $\mathsf{A}(i, P, V)$ denotes the set of those $\mathbf{y} = y_1 \dots y_n \in \mathsf{Y}^n$ for which the sequence $\mathbf{x} = x_1 \dots x_n$ with $x_j = f_j(i, y_1 \dots y_{j-1})$ has type P and $\mathbf{y} \in \mathsf{T}_V(\mathbf{x})$. Since

$$\sum_{i\in\mathsf{M}} |\mathsf{A}(i, P, V) \cap \varphi^{-1}(i)| \leqq |\mathsf{T}_{PV}| \leqq \exp\{nH(PV)\},$$

it remains to check that

$$|\mathsf{A}(i, P, V)| \leqq \exp\{nH(V|P)\}.$$

This, however, follows from the identity

$$\sum_{\mathbf{y}\in\mathsf{Y}^n} V_f(\mathbf{y}|m) = 1,$$

where V_f is constructed from V as in Problem 6.29. (Csiszár and Körner (1982); an equivalent result was obtained earlier by Augustin (1978, unpublished) and A. Yu. Ševerdyaev, personal communication. The weaker result that for DMCs with complete feedback the strong converse holds appears in Wolfowitz (1964), attributed to independent unpublished works of J. H. B. Kemperman and H. Kesten.)

10.18. (*Improving the random coding bound*) Given a DMC $\{W : \mathsf{X} \to \mathsf{Y}\}$, define a (not necessarily finite-valued) distortion measure on $\mathsf{X} \times \mathsf{X}$ by

$$d_W(x, \tilde{x}) \triangleq -\log \sum_{y\in\mathsf{Y}} \sqrt{W(y|x)W(y|\tilde{x})}.$$

Set

$$E_{\mathrm{x}}(R, P, W) \triangleq \min_{\substack{P_X = P_{\tilde{X}} = P \\ I(X\wedge\tilde{X}) \leqq R}} [Ed_W(X, \tilde{X}) + I(X \wedge \tilde{X}) - R],$$

$$E_{\mathrm{x}}(R) = E_{\mathrm{x}}(R, W) \triangleq \max_P E_{\mathrm{x}}(R, P, W).$$

(a) Show that for every $R > \delta > 0$, for sufficiently large n, to any type P of sequences in X^n there exists a code with codewords of type P that has rate at least $R - \delta$ and maximum probability of error

$$e(W^n, f, \varphi) \leqq \exp\{-n[E_{\mathrm{x}}(R, P, W) - \delta]\}$$

(Csiszár, Körner and Marton (1977, unpublished).)

(b) Conclude that $E(R) \geqq E_{\mathrm{x}}(R)$. (A result equivalent to this was first established by Gallager (1965) who called it the *expurgated bound*. See Problem 10.24.)

(c) Show that for every DMC with positive capacity, $E_{\mathrm{x}}(R) > E_{\mathrm{r}}(R)$ if R is sufficiently small. More precisely, prove that $E_{\mathrm{x}}(0, P, W) > E_{\mathrm{r}}(0, P, W)$ whenever $I(P, W) > 0$.

Hint Take a codeword set $\mathsf{C} \subset \mathsf{T}_P$ as in Problem 10.2 and use a maximum likelihood decoder; see Problem 6.2. Then for each message i, with codeword $\mathbf{x}_i$,

$$\begin{aligned} e_i &\leqq \sum_{j \neq i} \sum_{\mathbf{y}:\, W^n(\mathbf{y}|\mathbf{x}_j) > W^n(\mathbf{y}|\mathbf{x}_i)} W^n(\mathbf{y}|\mathbf{x}_i) \\ &\leqq \sum_{j \neq i} \sum_{\mathbf{y} \in \mathsf{Y}^n} \sqrt{W^n(\mathbf{y}|\mathbf{x}_i) W^n(\mathbf{y}|\mathbf{x}_j)}. \end{aligned}$$

Here the inner sum equals $\exp[-nEd_W(X, \tilde{X})]$ for RVs X and $\tilde{X}$ of joint distribution $P_{\mathbf{x}_i, \mathbf{x}_j}$; thus, using the assumed property of the codeword set, assertion (a) follows.

For (c), note that

$$\begin{aligned} E_{\mathrm{r}}(0, P, W) &= \min_V [D(V||W|P) + I(P, V)] \\ &= \min_{V, Q} 2 \sum_{x, y} P(x) V(y|x) \log \frac{V(y|x)}{\sqrt{W(y|x) Q(y)}} \\ &= 2 \min_Q \left[-\sum_x P(x) \log \sum_y \sqrt{W(y|x) Q(y)} \right]. \end{aligned}$$

One sees by differentiation that if Q^* achieves this minimum then $Q^*(y) > 0$ and

$$\sum_x P(x) \frac{\sqrt{W(y|x)}}{\sqrt{Q^*(y)} \sum_{y'} \sqrt{W(y'|x) Q^*(y')}}$$

has a constant value for every $y \in \mathsf{Y}$ with $PW(y) > 0$. As Q^* is a PD on Y, this constant must be 1. By the concavity of the log function it follows that

$$\begin{aligned} &E_{\mathrm{x}}(0, P, W) - E_{\mathrm{r}}(0, P, W) \\ &= -\sum_x \sum_{\tilde{x}} P(x) P(\tilde{x}) \log \frac{\sum_y \sqrt{W(y|x) W(y|\tilde{x})}}{\sum_y \sqrt{W(y|x) Q^*(y)} \sum_y \sqrt{W(y|\tilde{x}) Q^*(y)}} \end{aligned}$$

$$\geqq -\log \sum_y \sum_x \sum_{\tilde{x}} \frac{P(x)P(\tilde{x})\sqrt{W(y|x)W(y|\tilde{x})}}{\sum_{y'} \sqrt{W(y'|x)Q^*(y')} \sum_{y'} \sqrt{W(y'|\tilde{x})Q^*(y')}}$$

$$= -\log \sum_y Q^*(y) = 0.$$

The inequality is strict whenever $W(y|x)$ depends on x for some y with $PW(y) > 0$, i.e., whenever $I(P, W) > 0$.

10.19. (*Expurgated bound and zero-error capacity*)

(a) Observe that the distortion measure d_W in Problem 10.18 is finite-valued iff the DMC $\{W\}$ has zero-error capacity $C_0 = 0$. Check that for P fixed $E_x(R, P)$ is a decreasing continuous convex function of R in the left-closed interval where it is finite, and that so is $E_x(R)$.

Hint For the first assertion, see Problem 6.4.

(b) Let

$$R^*_\infty(P) \triangleq \min_{\substack{P_X = P_{\tilde{X}} = P \\ Ed_W(X,\tilde{X}) < \infty}} I(X \wedge \tilde{X})$$

denote the smallest $R \geqq 0$ with $E_x(R, P) < \infty$, and let $R^*_\infty \triangleq \max_P R^*_\infty(P)$ denote the smallest $R \geqq 0$ with $E_x(R) < \infty$. Prove that $R^*_\infty = \log \omega(G(W))$; see Problem 6.23 for notation. (Gallager (1965), (1968); Korn (1968).)

Hint First show that the minimum in the formula for $R^*_\infty(P)$ is achieved iff

$$P_{X\tilde{X}}(x, \tilde{x}) = \begin{cases} cQ(x)Q(\tilde{x}) & \text{if } d_W(x, \tilde{x}) < \infty \\ 0 & \text{else,} \end{cases}$$

where the distribution Q and the constant c are uniquely determined by the condition $P_X = P_{\tilde{X}} = P$. Conclude that

$$R^*_\infty(P) = \log c - 2D(P||Q) \leqq \log c$$
$$\leqq -\log \min_Q \sum_{x,\tilde{x}: d_W(x,\tilde{x}) < \infty} Q(x)Q(\tilde{x}).$$

If $d_W(x_1, x_2) < \infty$, thus x_1 and x_2 are not adjacent in the graph $G(W)$, the last sum is a linear function of $(Q(x_1), Q(x_2))$ if the remaining $Q(x)$ are fixed. Thus the minimum is achieved for a Q concentrated on a subset of X consisting of mutually adjacent vertices of $G(W)$. Hence $R^*_\infty \leqq \log \omega(G(W))$ readily follows, while the opposite inequality is obvious.

(c) Let $E_x(R, W^n)$ denote the analog of $E_x(R, W)$ for $W^n: \mathsf{X}^n \to \mathsf{Y}^n$. Show that $(1/n)E_x(nR, W^n)$ is also a lower bound for the reliability function of the DMC $\{W\}$ and that $(1/n)E_x(nR, W^n) \geqq E_x(R, W)$. Using the result of (b), give examples for the strict inequality.

(d) Let $R^*_{\infty,n}$ be the analog of R^*_∞ for the channel $W^n\colon \mathsf{X}^n \to \mathsf{Y}^n$. Conclude from (b) and Problem 6.23 that

$$\lim_{n\to\infty} \frac{1}{n} R^*_{\infty,n} = C_0.$$

Reliability at $R = 0$ (Problems 10.20–10.23)

10.20. (*Error probability for two messages*) For a given DMC $\{W\colon \mathsf{X} \to \mathsf{Y}\}$ and two sequences $\mathbf{x} \in \mathsf{X}^n$, $\tilde{\mathbf{x}} \in \mathsf{X}^n$ let

$$e(\mathbf{x}, \tilde{\mathbf{x}}) \triangleq \min_{\mathsf{B}\subset\mathsf{Y}^n} \max\,(W^n(\mathsf{B}|\mathbf{x}), W^n(\bar{\mathsf{B}}|\tilde{\mathbf{x}}))$$

be the smallest maximum probability of error of codes with two messages having codewords $\mathbf{x}, \tilde{\mathbf{x}}$. For every $s \in [0,1]$ set

$$d_s(x, \tilde{x}) \triangleq -\log \sum_{y\in\mathsf{Y}} W^s(y|x)W^{1-s}(y|\tilde{x})$$

where (only in this problem) $0^0 \triangleq 0$. Further, write

$$d_s(\mathbf{x}, \tilde{\mathbf{x}}) \triangleq \frac{1}{n} \sum_{i=1} d_s(x_i, \tilde{x}_i).$$

(a) Show that for every $s \in [0, 1]$

$$e(\mathbf{x}, \tilde{\mathbf{x}}) \leqq \exp[-nd_s(\mathbf{x}, \tilde{\mathbf{x}})].$$

Hint For $\mathsf{B} \triangleq \{\mathbf{y} : W^n(\mathbf{y}|\mathbf{x}) < W^n(\mathbf{y}|\tilde{\mathbf{x}})\}$ clearly

$$W^n(\mathsf{B}|\mathbf{x}) \leqq \sum_{\mathbf{y}\in\mathsf{Y}^n} [W^n(\mathbf{y}|\mathbf{x})]^s [W^n(\mathbf{y}|\tilde{\mathbf{x}})]^{1-s}$$

and the same holds for $W^n(\bar{\mathsf{B}}|\tilde{\mathbf{x}})$.

(b) Prove that for every $\delta > 0$ and $n \geqq n_0(\delta, W)$

$$e(\mathbf{x}, \tilde{\mathbf{x}}) \geqq \exp\{-n[\max_{0\leqq s\leqq 1} d_s(\mathbf{x}, \tilde{\mathbf{x}}) + \delta]\}.$$

(Shannon, Gallager and Berlekamp (1967), generalizing a result of Chernoff (1952).)

Hint Consider the distributions $P_1, \ldots, P_n$ on Y defined by

$$P_i(y) \triangleq W^s(y|x_i)W^{1-s}(y|\tilde{x}_i) \left[\sum_{y'} W^s(y'|x_i)W^{1-s}(y'|\tilde{x}_i)\right]^{-1},$$

where $s \in [0, 1]$ maximizes $d_s(\mathbf{x}, \tilde{\mathbf{x}})$. Apply Theorem 1.2 with $\varepsilon = 1/2$ and $W(\cdot|x_1)$ resp. $W(\cdot|\tilde{x}_1)$ in the role of $M_i(\cdot)$. (This proof combines those of Csiszár and Longo (1971) and Blahut (1977).)

(c) In the case $|\mathsf{X}| = 2$, check that if $\mathbf{x}$ and $\tilde{\mathbf{x}}$ have the same type then $d_s(\mathbf{x}, \tilde{\mathbf{x}})$ is maximized for $s = 1/2$, and thus the lower bound of (b) holds with the distortion measure of Problem 10.18. Show further that for $|\mathsf{X}| \geqq 3$ this is no longer so.

Hint For a counterexample, let $W : \{0, 1, 2\} \to \{0, 1, 2\}$ be a channel with additive noise, with $P(0) > P(1) > P(2) = 0$; see Problem 6.13(a). Consider two sequences $\mathbf{x}$ and $\tilde{\mathbf{x}}$ of length $n = 3k$ such that

$$N(0, 1|\mathbf{x}, \tilde{\mathbf{x}}) = N(1, 2|\mathbf{x}, \tilde{\mathbf{x}}) = N(2, 0|\mathbf{x}, \tilde{\mathbf{x}}) = k.$$

(This example appears in Shannon, Gallager and Berlekamp (1967).)

10.21. (*Reliability at $R = 0$*) Prove that for every sequence of codes (f_n, φ_n) for a DMC $\{W : \mathsf{X} \to \mathsf{Y}\}$ the condition $|\mathsf{M}_{f_n}| \to \infty$ implies

$$\varliminf_{n\to\infty} \frac{1}{n} \log e(W^n, f_n, \varphi_n) \geqq -E_{\mathrm{x}}(0).$$

Note that, although $E(R)$ is unknown for $0 < R < R_{\mathrm{cr}}$, this result and Problem 10.18 imply

$$\lim_{R\to 0} E(R) = E_{\mathrm{x}}(0).$$

(Berlekamp (1964, unpublished); published, with a slight error not present in the original, in Shannon, Gallager and Berlekamp (1967).)

Hint

(i) By Problem 10.20(b), it suffices to show for every $\delta > 0$ that each subset of X^n of size $m = m(\delta)$ contains sequences $\mathbf{x} \neq \tilde{\mathbf{x}}$ with

$$\max_{0\leqq s\leqq 1} d_s(\mathbf{x}, \tilde{\mathbf{x}}) \leqq E_{\mathrm{x}}(0) + \delta.$$

One may suppose that d_s is finite-valued, for otherwise $E_{\mathrm{x}}(0) = \infty$. Note further that $d_s(\mathbf{x}, \tilde{\mathbf{x}})$ is a convex function of $s \in [0, 1]$.

(ii) Set

$$d'(a, b) \triangleq \frac{\partial}{\partial s} d_s(a, b)\Big|_{s=\frac{1}{2}}; \quad d'(\mathbf{x}, \tilde{\mathbf{x}}) \triangleq \frac{1}{n} \sum_{a,b} N(a, b|\mathbf{x}, \tilde{\mathbf{x}}) d'(a, b).$$

Since $d'(\mathbf{x}, \tilde{\mathbf{x}}) = -d'(\tilde{\mathbf{x}}, \mathbf{x})$ for every $\mathbf{x}, \tilde{\mathbf{x}}$, one sees that each set of $m = \exp k$ sequences in X^n contains an ordered subset $(\mathbf{x}_1, \ldots, \mathbf{x}_k)$ such that $d'(\mathbf{x}_i, \mathbf{x}_j) \geqq 0$ for $i < j$. Check that in this case

$$\max_{0\leqq s\leqq 1} d_s(\mathbf{x}_i, \mathbf{x}_j) \leqq \bar{d}(\mathbf{x}_i, \mathbf{x}_j) \quad \text{if } i < j,$$

where $\bar{d}$ denotes the distortion measure defined by

$$\bar{d}(a, b) \triangleq d_{\frac{1}{2}}(a, b) + \frac{1}{2} d'(a, b).$$

Conclude that it suffices to prove

$$\min_{i<j} \bar{d}(\mathbf{x}_i, \mathbf{x}_j) \leqq E_{\mathrm{x}}(0) + \delta \qquad (*)$$

for every $\mathsf{C} = \{\mathbf{x}_1, \ldots, \mathbf{x}_k\} \subset \mathsf{X}^n$ whenever $k \geqq k(\delta)$.

(iii) For every $\mathsf{C} = \{\mathbf{x}_1, \ldots, \mathbf{x}_k\} \subset \mathsf{X}^n$ define

$$V(\mathsf{C}) \triangleq \frac{1}{n} \sum_{l=1}^{n} \sum_{a \in \mathsf{X}} P_l^2(a),$$

where P_l denotes the type of the column vector of length k consisting of the lth components of the sequences $\mathbf{x}_i \in \mathsf{C}$. Supposing that $k = 2k_1$, let $\mathsf{C}_1 \subset \mathsf{X}^{2n}$ denote the ordered set of the juxtapositions $\mathbf{x}_i' \triangleq \mathbf{x}_i \mathbf{x}_{i+k_1}$, $i = 1, \ldots, k_1$. Show that

$$m(\mathsf{C}) \triangleq \min_{i<j} \bar{d}(\mathbf{x}_i, \mathbf{x}_j) \leqq E_{\mathsf{x}}(0) + 2d_{\max} |\mathsf{X}|^{\frac{1}{2}} (V(\mathsf{C}_1) - V(\mathsf{C}))^{\frac{1}{2}},$$

where $d_{\max} \triangleq \max_{a,b} \bar{d}(a, b)$.

To this end, using an idea of Plotkin (see Problem 5.5), bound the minimum $m(\mathsf{C})$ by the average

$$\frac{1}{k_1^2} \sum_{i=1}^{k_1} \sum_{j=k_1+1}^{2k_1} \bar{d}(\mathbf{x}_i, \mathbf{x}_j) = \frac{1}{n} \sum_{l=1}^{n} \sum_{a,b} P_l'(a) P_{n+l}'(b) \bar{d}(a, b),$$

where P_l' is the analog of P_l for C_1. To get the desired bound, check that

$$\frac{1}{n} \sum_{l=1}^{n} \sum_{a,b} P_l(a) P_l(b) \bar{d}(a, b)$$
$$= \frac{1}{n} \sum_{l=1}^{n} \sum_{a,b} P_l(a) P_l(b) d_{\frac{1}{2}}(a, b) \leqq E_{\mathsf{x}}(0)$$

and

$$\sum_{a,b} |P_l'(a) P_{n+1}'(b) - P_l(a) P_l(b)| \leqq \sum_{a} |P_l'(a) - P_l(a)|$$
$$+ \sum_{b} |P_{n+1}'(b) - P_l(b)| = 2 \sum_{a} |P_l'(a) - P_l(a)|,$$

while

$$V(\mathsf{C}_1) - V(\mathsf{C}) = \frac{1}{n} \sum_{l=1}^{n} \sum_{a} (P_l'(a) - P_l(a))^2;$$

the last two relations follow since $P_l = (1/2)(P_l' + P_{n+l}')$.

(iv) Set $k = \exp r$. Construct from $\mathsf{C} = \{\mathbf{x}_1, \ldots, \mathbf{x}_k\} \subset \mathsf{X}^n$ the set $\mathsf{C}_1 \subset \mathsf{X}^{2n}$ as in (iii), from C_1 construct $\mathsf{C}_2 \subset \mathsf{X}^{4n}$ by the same operation, etc., until finally $\mathsf{C}_r \subset \mathbf{X}^{kn}$ consisting of a single sequence is obtained. Show that

$$m(\mathsf{C}) \leqq m(\mathsf{C}_1) \leqq m(\mathsf{C}_2) \leqq \ldots \leqq m(\mathsf{C}_{r-1}).$$

Further $V(\mathsf{C}_i) \leqq 1$ for every i, thus $V(\mathsf{C}_{i+1}) - V(\mathsf{C}_i) \leqq 1/r$ for at least one i. Applying the bound in (iii) to this i, (*) follows if r is sufficiently large.

10.22. (*Reliability at $R = 0$, constant composition codes*)

(a) For any sequence of codes (f_n, φ_n) such that every codeword of the nth code has the same type P_n, show that $|\mathsf{M}_{f_n}| \to \infty$, $P_n \to P$ imply

$$\underline{\lim}_{n\to\infty} \frac{1}{n} \log e(W^n, f_n, \varphi_n) \geqq -E_{\mathrm{x}}^*(0, P, W),$$

where $E_{\mathrm{x}}^*(0, P, W)$ is the concave upper envelope of $E_{\mathrm{x}}(0, P, W)$ considered as a function of P.

Hint For sequences $\mathbf{x}_i$ of the same type, the average of the types P_l in part (iii) of the hint in Problem 10.21 equals the common type of the $\mathbf{x}_i$.

(b) Prove that the bound in (a) is tight; more precisely, given any sequence $\{m_n\}_{n=1}^{\infty}$ with $(1/n)\log m_n \to 0$, there exist codes (f_n, φ_n) with $|\mathsf{M}_{f_n}| = m_n$ and with codewords of types $P_n \to P$ such that

$$\overline{\lim}_{n\to\infty} \frac{1}{n} \log e(W^n, f_n, \varphi_n) \leqq -E_{\mathrm{x}}^*(0, P, W).$$

Hint If $P = (1/2)(P^{(1)} + P^{(2)})$, consider $(f_n^{(1)}, \varphi_n^{(1)})$ and $(f_n^{(2)}, \varphi_n^{(2)})$ with $\mathsf{M}_{f_n^{(1)}} = \mathsf{M}_{f_n^{(2)}}$, codeword compositions $P_n^{(i)} \to P^{(i)}$ and

$$\overline{\lim}_{n\to\infty} \frac{1}{n} \log e(W^n, f_n^{(i)}, \varphi_n^{(i)}) \leqq -E_{\mathrm{x}}(0, P^{(i)}, W), \quad i = 1, 2.$$

Such codes exist by Problem 10.18. Let $f_{2n}(m)$ be the juxtaposition of $f_n^{(1)}(m)$ and $f_n^{(2)}(m)$. Then for a maximum likelihood decoder φ_{2n}

$$e(W^{2n}, f_{2n}, \varphi_{2n}) \leqq e(W^n, f_n^{(1)}, \varphi_n^{(1)}) e(W^n, f_n^{(2)}, \varphi_n^{(2)}),$$

hence

$$\overline{\lim}_{n\to\infty} \frac{1}{2n} \log e(W^{2n}, f_{2n}, \varphi_{2n}) \leqq -\frac{1}{2}[E_{\mathrm{x}}(0, P^{(1)}, W) + E_{\mathrm{x}}(0, P^{(2)}, W)].$$

Generalize this argument to any convex combination of distributions.

10.23. (*Expurgated bound and reliability at $R = 0$ under input constraint*) Formulate and prove the analogs of Problems 10.18(b) and 10.21 for DMCs with input constraints.

Hint Use Problems 10.18(a) and 10.22.

10.24. (*Alternative forms of the sphere-packing and expurgated exponent functions*)

(a) Prove that

$$E_{\mathrm{sp}}(R, P) \geqq \max_{\delta \geqq 0} \left\{ -\delta R - \log \sum_y \left(\sum_x P(x) W^{\frac{1}{1+\delta}}(y|x) \right)^{1+\delta} \right\}$$

and for a P achieving $E_{\mathrm{sp}}(R) \triangleq \max_P E_{\mathrm{sp}}(R, P)$ the equality holds, so that

$$E_{\mathrm{sp}}(R) = \max_P \max_{\delta \geqq 0} \left\{ -\delta R - \log \sum_y \left(\sum_x P(x) W^{\frac{1}{1+\delta}}(y|x) \right)^{1+\delta} \right\}.$$

($E_{\text{sp}}(R)$ was defined in this form by Shannon, Gallager and Berlekamp (1967). An algorithm for computing $E_{\text{sp}}(R)$ based on this formula was given by Arimoto (1976).)

Hint

(i) Show as in Lemma 8.1 that

$$E_{\text{sp}}(R, P) = \max_{\delta \geqq 0} \left\{ -\delta R + \min_V [D(V||W|P) + \delta I(P, V)] \right\}.$$

(ii)
$$\begin{aligned} &\min_V [D(V||W|P) + \delta I(P, V)] \\ &= \min_{V,Q} \sum_{x,y} P(x)V(y|x) \left[\log \frac{V(y|x)}{W(y|x)} + \delta \log \frac{V(y|x)}{Q(y)} \right] \\ &= \min_Q \left\{ -(1+\delta) \sum_x P(x) \log \sum_y W^{\frac{1}{1+\delta}}(y|x) Q^{\frac{\delta}{1+\delta}}(y) \right\}. \end{aligned}$$

(iii) By convexity, the bracketed term is lower-bounded by

$$-(1+\delta) \log \sum_x P(x) \sum_y W^{\frac{1}{1+\delta}}(y|x) Q^{\frac{\delta}{1+\delta}}(y).$$

This expression is minimized for

$$Q(y) \triangleq c \left[\sum_x P(x) W^{\frac{1}{1+\delta}}(y|x) \right]^{1+\delta},$$

where c is a norming constant.

(iv) Show that

$$-\log \sum_y \left(\sum_x P(x) W^{\frac{1}{1+\delta}}(y|x) \right)^{1+\delta}$$

is maximized iff P satisfies

$$\begin{aligned} &\sum_y W^{\frac{1}{1+\delta}}(y|x) \left[\sum_{x'} P(x') W^{\frac{1}{1+\delta}}(y|x') \right]^{\delta} \\ &\geqq \sum_y \left(\sum_{x'} P(x') W^{\frac{1}{1+\delta}}(y|x') \right)^{1+\delta}, \end{aligned}$$

with equality for every $x \in \mathsf{X}$ such that $P(x) > 0$. To upperbound the minimum in (ii) choose Q as in (iii) but with the maximizing P.

(b) Prove that $E_x(R)$ can be expressed as

$$\max_P \max_{\delta \geqq 1} \left\{ -\delta R - \delta \log \sum_{x,\tilde{x}} P(x)P(\tilde{x}) \left(\sum_y \sqrt{W(y|x)W(y|\tilde{x})} \right)^{1/\delta} \right\}.$$

($E_x(R)$ was defined in this form by Gallager (1965).)

Hint

(i) Show that

$$E_x(R, P) = \max_{\delta \geqq 1} \{-\delta R + \min_{P_X = P_{\tilde{X}} = P} [E d_W(X, \tilde{X}) + \delta I(X \wedge \tilde{X})]\}.$$

(ii) For any fixed δ, the minimum is achieved for a joint distribution of form

$$P(x, \tilde{x}) = c Q(x) Q(\tilde{x}) \exp \left\{ -\frac{1}{\delta} d_W(x, \tilde{x}) \right\},$$

where Q is a distribution on X and c is a norming constant; this gives

$$\delta^{-1} E d_W(X, \tilde{X}) + I(X \wedge \tilde{X}) = \log c - 2D(P||Q) \leqq \log c$$
$$\leqq \max_Q \left\{ -\log \sum_x \sum_{\tilde{x}} Q(x) Q(\tilde{x}) \exp \left[-\frac{1}{\delta} d_W(x, \tilde{x}) \right] \right\}.$$

(iii) A Q^* achieving the maximum given in (ii) satisfies

$$\sum_{\tilde{x}} Q^*(\tilde{x}) \exp \left[-\frac{1}{\delta} d_W(x, \tilde{x}) \right] = \text{const}$$

for $Q^*(x) > 0$, so that the marginals of the joint distribution defined by this Q^* as in (ii) equal Q^*. Thus for $P = Q^*$ the inequalities in (ii) hold with equality.

(c) Conclude that $E_x(R_{cr}) = E_{sp}(R_{cr}) = E_r(R_{cr})$, where R_{cr} is the critical rate of the DMC $\{W\}$. (Gallager (1965).)

10.25. (*Expurgated exponent and distortion-rate functions*)

(a) Let $\Delta(R, P)$ be the distortion-rate function (i.e., the inverse of the rate-distortion function) of a DMS with generic distribution P with respect to the distortion measure $d_W(x, \tilde{x})$ of Problem 10.18. Observe that in the interval where the slope of the curve $E_x(R, P)$ is less than -1, we have $E_x(R, P) \geqq \Delta(R, P)$. In particular, denoting by R^*_{cr} the right endpoint of the interval where the slope of $E_x(R)$ is less than -1, we have

$$E_x(R) \geqq \Delta_0(R) \triangleq \max_P \Delta(R, P) \quad \text{if} \quad R \leqq R^*_{cr}.$$

(The first to relate the expurgated bound to distortion-rate functions was Omura (1974).)

(b) Show that if the distribution P achieving

$$E_x(R) \triangleq \max_P E_x(R, P)$$

consists of positive probabilities, then the same P achieves $\Delta_0(R)$, and we have $E_x(R) = \Delta_0(R)$. Conclude that if for some n a strictly positive distribution on X^n achieves $E_x(nR, W^n)$, then, see Problem 10.19(c),

$$\frac{1}{n} E_x(nR, W^n) = E_x(R).$$

(Blahut (1977).)

Hint Use Problem 10.24(b) and the fact that $\Delta_0(R)$ does not increase in product space, on account of Theorem 9.2.

(c) A DMC $\{W\}$ is *equidistant* if $d_W(x, \tilde{x}) = \text{const}$ for $x \neq \tilde{x}$. Show that for such channels $E_x(R, P)$ is maximized by the uniform distribution on X and $E_x(R) = (1/n)E_x(nR, W^n)$ for every n. Note that each binary input DMC is equidistant. (Jelinek (1968a).)

10.26. (*Error probability bounds for a BSC*) Consider a BSC with cross-over probability $p < 1/2$ (see Problem 6.10).

(a) Show that for $R < C = 1 - h(p)$ we have

$$E_{sp}(R) = q \log \frac{q}{p} + (1 - q) \log \frac{1 - q}{1 - p},$$

where $q \leqq 1/2,\ h(q) = 1 - R$.

(b) Conclude that the critical rate is given by

$$R_{cr} = 1 - h\left(\frac{\sqrt{p}}{\sqrt{p} + \sqrt{1 - p}}\right).$$

(c) Show that

$$E_x(R) = \begin{cases} E_r(R) = 1 - R - 2\log(\sqrt{p} + \sqrt{1 - p}) & \text{if } R^*_{cr} \leqq R \leqq R_{cr} \\ -\frac{1}{2} q \log(4p(1 - p)) & \text{if } R \leqq R^*_{cr}, \end{cases}$$

where q is defined as in (a), and

$$R^*_{cr} = 1 - h\left(\frac{1}{(\sqrt{p} + \sqrt{1 - p})^2}\right).$$

(Gallager (1965).)

(d)* Improving the result that $E(R) = E_r(R) > 0$ if $R_{cr} \leqq R < C$, show that the same holds also if $R' \leqq R < C$, for some $R' < R_{cr}$. (Barg and McGregor (2005); it is open whether $R' = R^*_{cr}$ can be taken.)

10.27. (*Expurgated bound and Gilbert bound*)

(a) Show that for a DMC $\{W : \{0, 1\} \to \mathsf{Y}\}$, every constant composition code (f, φ) of block length $n > n_0(\delta, W)$ has maximum probability of error

$$e(W^n, f, \varphi) \geqq \exp[-d_{min}(d_W(0, 1) + \delta)],$$

where d_{min} is the minimum Hamming distance between codewords and d_W is as in Problem 10.18.

Hint Use Problem 10.20(c), noticing that

$$d_{\frac{1}{2}}(\mathbf{x}, \tilde{\mathbf{x}}) = \frac{1}{n} d_{\mathrm{H}}(\mathbf{x}, \tilde{\mathbf{x}}) d_W(0, 1) \quad \text{if } P_{\mathbf{x}} = P_{\tilde{\mathbf{x}}}.$$

(b) Conclude that the conjecture that Gilbert's bound is asymptotically tight, i.e.,

$$R \leqq 1 - h\left(\frac{d_{\min}}{n}\right) + \delta$$

for every set of $\exp(nR)$ binary sequences of length $n \geqq n_0(\delta)$, would imply that the asymptotic tightness of the expurgated bound for DMCs with binary input alphabet for $R < R_{\mathrm{cr}}$. (Blahut (1977); McEliece and Omura (1977).)

Hint The conclusion is straightforward if $R \leqq R^*_{\mathrm{cr}}$, see Problem 10.25(a). If $R^*_{\mathrm{cr}} < R < R_{\mathrm{cr}}$ then use Problem 10.30.

10.28. (*List codes*) A list code for a channel W is a code (f, φ) such that the range of φ consists of subsets of M_f. Intuitively, the decoder produces a list of messages and an error occurs if the true message is not on the list. If each set in the range of φ has cardinality $\leqq l$, one speaks of a *list code with list size l*.

(a) Let $e(n, R, L)$ denote the analog of $e(n, R)$, see Problem 10.7, for list codes of list size $\exp(nL)$. Show that for a DMC

$$\lim_{n\to\infty} e(n, R, L) = \begin{cases} 0 & \text{if} \quad R < C + L \\ 1 & \text{if} \quad R > C + L. \end{cases}$$

(b) Show that for $R < C + L$

$$\overline{\lim_{n\to\infty}} \frac{1}{n} \log e(n, R, L) \leqq -E_{\mathrm{r}}(R - L),$$

$$\underline{\lim_{n\to\infty}} \frac{1}{n} \log e(n, R, L) \geqq -E_{\mathrm{sp}}(R - L).$$

(Shannon, Gallager and Berlekamp (1967).)

(c) Find the analogs of Theorems 10.2–10.6 for list codes of constant list size l, replacing $E_{\mathrm{r}}(R, P)$ by $E_{\mathrm{r},l}(R, P)$, see (10.28). Conclude that the reliability function of a DMC for list size l equals $E_{\mathrm{sp}}(R)$ for every $R \geqq R_{\mathrm{cr},l}$, where $R_{\mathrm{cr},l}$ is the smallest R at which the curve $E_{\mathrm{sp}}(R)$ meets its supporting line of slope $-l$.

Hint Generalize the packing lemma, Lemma 10.1, replacing inequality (10.6) by

$$\left| \mathsf{T}_V(\mathbf{x}_i) \cap \left(\bigcap_{k=1}^{l} \bigcup_{j \neq i} \mathsf{T}_{\hat{V}_k}(\mathbf{x}_j) \right) \right| \leqq |\mathsf{T}_V(\mathbf{x}_i)| \exp\left[-n \sum_{k=1}^{l} |I(P, \hat{V}_k) - R|^+ \right]$$

for arbitrary $V, \hat{V}_1, \ldots, \hat{V}_l$.

(d) Define the *list code zero-error capacity* of a DMC as the largest R for which there exist zero-error list codes of some constant list size with rates converging to R. Show that it equals R_∞ determined in Problem 10.4.

(e) Define the zero-error capacity of a DMC for list size l as the largest R for which there exist zero-error list codes with list size at most l and rates converging to R. Show that there exist channels with zero-error capacity 0 for list size $l = 1$ while having positive zero-error capacity for list size $l = 2$.

Hint Consider the 3×3 matrix W with diagonal entries equal to zero and off-diagonal elements equal to 1/2.

(List codes were first considered by Elias (1957). The result of (d) is due to Elias (1958, unpublished). Concerning simultaneous upper bounds on the average list size and the probability of error, see Forney (1968).)

10.29.* (*Zero-error capacity and perfect hashing*)

(a) (*Trifference*) Let us call the ternary sequences $\mathbf{x}$, $\mathbf{y}$ and $\mathbf{z} \in \{0, 1, 2\}^n$ trifferent if there exists at least one coordinate $i \in \{1, 2, \ldots, n\}$ such that $\{x_i, y_i, z_i\} = \{0, 1, 2\}$. Let $T(n)$ denote the largest cardinality of a subset of $\{0, 1, 2\}^n$ in which any three sequences are trifferent. Show that

$$C_{2,0} \triangleq \overline{\lim_{n\to\infty}} \frac{1}{n} \log T(n)$$

equals the zero-error capacity for list size $l = 2$ of the DMC in the hint to Problem 10.28(e). Show also that $T(1) = 3$, $T(2) = 4$, $T(3) = 6$, $T(4) = 9$,

$$T(n) \leqq \frac{3}{2} T(n-1),$$

and deduce that $C_{2,0} \leqq \log 3 - 1$. (The same bound also follows from Problem 10.28(d).) Show finally that $C_{2,0} \geqq (1/4) \log 9/5$. (Körner and Marton (1988a).)

(b) (*Perfect hashing*) Generalizing the problem of trifference, we say that a set of sequences from a b-ary alphabet is k-separated if to any k of its elements there is a coordinate in which the k sequences have k different values. Denote by $N(n, b, k)$ the largest cardinality of a set of k-separated b-ary sequences of length n. Show that

$$\frac{1}{k-1} \log \frac{1}{1 - \frac{b^k}{b(b-1)\cdots(b-k+1)}} \lessapprox \frac{1}{n} \log N(n, b, k)$$
$$\lessapprox \min_{0 \leqq j \leqq k-2} \frac{b(b-1)\cdots(b-j)}{b^{j+1}} \log \frac{b-j}{k-j-1}.$$

(Körner and Marton (1988a), using earlier results of Fredman and Komlós (1984) and Körner (1986).)

10.30.* (*Straight-line improvement of the sphere-packing bound*) Show that if the reliability function of a DMC has an upper bound E_0 at some $R_0 \geqq 0$ then for

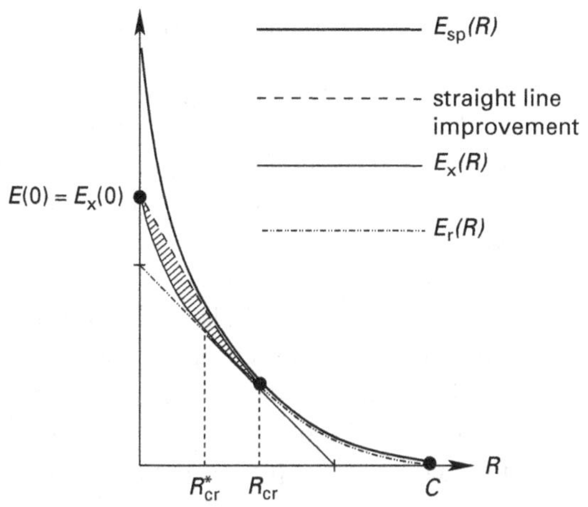

Figure P10.30 Various bounds on the reliability function of a DMC. For $R \in (0, R_{cr})$ the curve $E(R)$ lies in the dashed area.

every $R_1 > R_0$ the straight line connecting (R_0, E_0) with $(R_1, E_{sp}(R_1))$ is an upper bound of $E(R)$ in the interval (R_0, R_1). (Shannon, Gallager and Berlekamp (1967).)

Hint Let $\bar{e}(n, m, l)$ denote the minimum average probability of error for list codes of block length n, message set size m and list size l. The main step of the proof is to establish the inequality

$$\bar{e}(n_1 + n_2, m, 1) \geqq \bar{e}(n_1, m, l)\, \bar{e}(n_2, l + 1, 1).$$

10.31. (*Messages with different reliabilities*) Given a DMC $\{W : \mathsf{X} \to \mathsf{Y}\}$, show that for sufficiently large n and message set M equal to the disjoint union of sets M_i, $i = 1, \ldots, m_n$, a suitable code attains for the messages $m \in \mathsf{M}_i$ the random coding exponent corresponding to $R_i = (1/n) \log |\mathsf{M}_i|$, simultaneously for $i = 1, \ldots, m_n$, providing $(1/n) \log m_n \to 0$. More specifically, under this condition, to any $\delta > 0$ and given (not necessarily distinct) types P_i with $H(P_i) > R_i + \delta$, $i = 1, \ldots, m_n$, show that there exist codes (f, φ) such that f maps M_i into T_{P_i} and that the probability of decoding error is bounded above for each $m \in \mathsf{M}_i$ by $\exp[-n(E_r(R_i, P_i, W) - \delta)]$, for $i = 1, \ldots, m_n$, and any DMC $\{W : \mathsf{X} \to \mathsf{Y}\}$. If universality is not required, one can choose the types P_i depending on the given DMC and obtain the bound $\exp[-n(E_r(R_i, W) - \delta)]$. (Csiszár (1980).)

Hint Establish first, via random selection, the following extension of the packing lemma (see its equivalent version in Problem 10.2): given positive integers n, m, $N_1, \ldots, N_m$, and types $P_1, \ldots, P_m$ with $H(P_i) > (1/n) \log N_i + \eta$, where $\eta \triangleq (2/n)(|\mathsf{X}|^2 \log(n + 1) + \log m + 1)$, there exist m disjoint sets $\mathsf{C}_i \subset \mathsf{X}^n$ of size N_i consisting of sequences of type P_i, $i = 1, \ldots, m$, such that for every i, j and channel $\tilde{V} : \mathsf{X} \to \mathsf{X}$

$$|\mathsf{T}_{\tilde{V}}(\mathbf{x}) \cap \mathsf{C}_j| \leqq N_j \exp[-n(I(P_i, \tilde{V}) - \eta)] \text{ if } \mathbf{x} \in \mathsf{C}_i$$

unless $i = j$ and $\tilde{V}$ is the identity matrix. Next, given disjoint message sets M_i of size N_i, $i = 1, \dots, m_n$, take an encoder f that maps each M_i onto C_i. Use a generalized MMI decoder, assigning to the received sequence $\mathbf{y}$ a maximizer $\mathbf{x}$ of $I(\mathbf{x} \wedge \mathbf{y}) - r(\mathbf{x})$ in the union of the codeword sets C_i, where $r(\mathbf{x}) = R_i$ if $\mathbf{x} \in \mathsf{C}_i$.

10.32. (*Source–channel error exponent*)

(a) Assume a DMS with alphabet S and generic distribution P and a DMC $\{W : \mathsf{X} \to \mathsf{Y}\}$ with reliability function $E(R)$. For a code with encoder $f_k : \mathsf{S}^k \to \mathsf{X}^{n_k}$ and decoder $\varphi_k : \mathsf{Y}^{n_k} \to \mathsf{S}^k$, denote the overall probability of error (see Problem 6.1) by $e^*(f_k, \varphi_k) = e^*(f_k, \varphi_k, P, W)$. Show that for every sequence $\{(f_k, \varphi_k)\}_{k=1}^{\infty}$ of k-to-n_k block codes with $n_k/k \to L$,

$$\overline{\lim_{k\to\infty}} \frac{1}{k} \log e^*(f_k, \varphi_k) \geqq -\min_Q \left[L \cdot E\left(\frac{H(Q)}{L}\right) + D(Q||P) \right].$$

Show further that if the channel has positive zero-error capacity then the equality can be attained.

Hint $L \cdot E(H(Q)/L) + D(Q||P)$ is the exponent of the smallest possible contribution to the overall probability of error of the k-length messages of type Q. If $C_0 > 0$, a code (f_k, φ_k) can be obtained from optimal channel codes with respective message sets T_Q^k. To this end add a short prefix to the codewords so as to inform unambiguously the receiver on the type Q of the k-length source message.

(b) Show that one can always attain

$$\overline{\lim_{k\to\infty}} \frac{1}{k} \log e^*(f_k, \varphi_k) \leqq -\min_Q \left[L \cdot E_{\mathrm{r}}\left(\frac{H(Q)}{L}\right) + D(Q\|P) \right],$$

where E_{r} is the random coding exponent function of the given DMC. Conclude that this result is tight if there exists a minimizing Q with $H(Q)/L$ not less than the critical rate of the given DMC. (Csiszár (1980); for further results along these lines, see Csiszár (1982).)

Hint Use the result of Problem 10.31, giving the role of the sets M_i to the type classes $\mathsf{T}_Q \subset \mathsf{S}^k$.

10.33. (*Random coding bound without remainder term*)

(a) Show that for every DMC $\{W\}$, every block length n and every R there exist codes (f, φ) with $|\mathsf{M}_f| = \lceil \exp(nR) \rceil$ such that

$$\bar{e}(W^n, f, \varphi) \leqq \exp[-nE_{\mathrm{r}}(R)].$$

(Gallager (1965).)

Hint Choose $\lceil \exp(nR) \rceil$ codewords at random, independently of each other, with distribution $P^n(\mathbf{x}) \triangleq \prod_{i=1}^n P(x_i)$. Show that if maximum likelihood decoding is used then the expectation of $\bar{e}(W^n, f, \varphi)$ for this randomly selected code is at most

$$\exp(n\delta R) \cdot \left[\sum_y \left(\sum_x P(x) W^{\frac{1}{1+\delta}}(y|x)\right)^{1+\delta}\right]^n$$

for every $\delta \in (0, 1)$. Then use Problem 10.24.

(b) Conclude from (a) that there exist codes with $|\mathsf{M}_f| = \lceil \exp nR \rceil$ such that

$$e(W^n, f, \varphi) \leqq 4 \exp[-nE_r(R)].$$

(Gallager (1965).)

(c) Show that for a BSC the result of (a) holds even for linear codes; see Problem 6.13. (Elias (1955).)

10.34.* (*Expectation of error probability for randomly selected codes*)

(a) Show that for the random selection in the hint to Problem 10.33(a), the given bound on the expectation of $\bar{e}(W^n, f, \varphi)$ is exponentially tight. (Gallager (1973).)

(b) Let P be a type of sequences in X^n. Select $\lceil \exp nR \rceil$ codewords at random, independently of each other and with uniform distribution from T^n_P, where $0 < R < I(P, W)$. Show that for the resulting randomly selected encoder f and a corresponding maximum likelihood decoder φ, the expectation of $\bar{e}(W^n, f, \varphi)$ satisfies the bound (10.14); further, show that this bound of the expectation of $\bar{e}(W^n, f, \varphi)$ is exponentially tight. (Dyačkov (1980).)

10.35. (*Zero undetected error capacity*) For a DMC $\{W : \mathsf{X} \to \mathsf{Y}\}$, let $C_{0,\mathrm{u}}$ denote the largest R for which there exist n-length block codes of rates converging to R with $e(W^n, f_n, \varphi_n) \to 0$ and $\hat{e}(W^n, f_n, \varphi_n) = 0$.

(a) Conclude from Problem 10.15 that $R_\infty \leqq C_{0,\mathrm{u}} \leqq C$; show that both equalities are possible.

(b) Show that if for $l \geqq 2$ there do not exist distinct elements $x_1, \ldots, x_l$ of X and distinct elements $y_1, \ldots, y_l$ of Y such that $W(y_i|x_i) > 0$, $W(y_i|x_{i+1}) > 0$, $i = 1, \ldots, l$ (where $x_{l+1} \triangleq x_1$), then $C_{0,\mathrm{u}} = C$. (Pinsker and Ševerdyaev (1970).)

(c) Show that $C_{0,\mathrm{u}} = C$ holds for all DMCs such that $W(y|x)$ equals $a(x)b(y)$ if $(x, y) \in \mathsf{A}$, and is zero otherwise, for some positive numbers $a(x), b(y)$, $x \in \mathsf{X}$, $y \in \mathsf{Y}$, and a subset A of $\mathsf{X} \times \mathsf{Y}$. Verify that (b) is a special case. (Csiszár and Narayan (1995).)

Hint Use Problem 6.9(b). Check that the codes there have zero probability of undetected error for a channel as given above.

10.36. (*DMC with complete feedback, block codes*) Define attainable error exponents for a DMC as in the text but admitting block codes with feedback as in Problem 6.29.

(a) Show that while feedback does not increase capacity (Problem 6.29), it can yield larger attainable error exponents, at least for small rates. More specifically, show that for a DMC with positive zero-error capacity, $E_{\mathrm{sp}}(R)$

is an attainable error exponent for every $R \in (C_{0,\mathrm{f}}, C)$, where $C_{0,\mathrm{f}}$ is the zero-error capacity with feedback.

Hint By Problems 10.4(c) and 10.28(c), $E_{\mathrm{sp}}(R)$ is an attainable error exponent for list codes with some constant list size. As $C_0 > 0$, the true message can be specified within the list without asymptotically increasing the rate.

(b) Show that an attainable error exponent for a DMC $\{W\}$ with feedback cannot exceed

$$E^+(R) \triangleq \min_{V:C(V) \leqq R} \max_P D(V||W|P) = \min_{V:C(V) \leqq R} \max_{x \in \mathsf{X}} D(V(\cdot|x)||W(\cdot|x)),$$

where $C(V) = \max_P I(P, V)$ is the capacity of the DMC $\{V\}$. Verify that if W is symmetric (see Problem 6.12) then $E^+(R) = E_{\mathrm{sp}}(R)$. (Haroutunian (1977); for symmetric channels, see Dobrušin (1962b).)

Hint Proceed as in the proof of Theorem 10.3, using that the strong converse holds also in the presence of feedback; see Problem 10.17(c).

(c) Prove that for a BSC with crossover probability $p \in (0, 1/2)$ (see Problem 6.10) there is a number $R_{\mathrm{cr,f}} < R_{\mathrm{cr}}$ such that $E_{\mathrm{sp}}(R)$ is attainable at rate $R \geqq R_{\mathrm{cr,f}}$, while for $R < R_{\mathrm{cr,f}}$

$$E_1(R) \triangleq \max_{0<q<\frac{1}{2}, \lambda \leqq -1} \Bigg\{ \lambda R - \log \max \Bigg[p(2q)^{\lambda} + (1-p)(2(1-q))^{\lambda},\ p\left(\frac{q}{1-q}\right)^{\lambda} + (1-p)\left(\frac{1-q}{q}\right)^{\lambda} \Bigg] \Bigg\}$$

is attainable. Verify that

$$E_1(0) = -\log(p^{1/3}(1-p)^{2/3} + p^{2/3}(1-p)^{1/3}) > E(0).$$

(Zigangirov (1970); for arbitrary DMCs see Dyačkov (1975).)

Hint Define recursively the mappings $f_1, \dots, f_n$ which constitute the encoder by the aid of an auxiliary BSC with crossover probability $q < 1/2$ as follows. If $f_1, \dots, f_{j-1}$ are already defined, calculate the posterior probabilities $T_j(m)$ of the messages $m \in \mathsf{M}$ as functions of the received sequence $y_1 \dots y_{j-1}$, supposing that the auxiliary channel is used and the prior distribution is uniform (for $j = 1$, set $T_1(m) \triangleq 1/|\mathsf{M}|$). Arrange the messages successively, in the order of decreasing $T_j(m)$, into two groups, assigning each m to the group in which the sum of T_j is smaller (breaking ties by an arbitrary but fixed rule). Then let $f_j(m, y_1 \dots y_{j-1})$ be 0 or 1 according as to whether m belongs to the first or second group. Having defined $f_1, \dots, f_n$ in this way, let the decoding result in m iff $T_n(m) \cdot [1 - T_n(m)]^{-1} > 1$.

If $m \in \mathsf{M}$ was actually transmitted, write

$$U \triangleq \frac{T_n(m)}{1 - T_n(m)} = \frac{1}{|\mathsf{M}| - 1} \prod_{j=1}^{n-1} V_j,$$

$$V_j \triangleq \frac{T_{j+1}(m)}{1 - T_{j+1}(m)} \cdot \left[\frac{T_j(m)}{1 - T_j(m)} \right]^{-1}, \quad j = 1, \ldots, n-1.$$

Verify that V_j, conditioned on the first j received symbols, takes the values $2q(1 + (1 - 2q)t)^{-1}$ and $2(1 - q)(1 - (1 - 2q)t)^{-1}$ with probabilities p and $1 - p$, where $0 \leqq t \leqq 1$ depends on $\{T_j(m)\}_{m \in \mathsf{M}}$, i.e., on the received symbols in question. Conclude that for every $\lambda < 0$

$$e_m = \Pr\{U \leqq 1\} \leqq EU^{\lambda} \leqq (|\mathsf{M}| - 1)^{-\lambda} \left[\max_{0 \leqq t \leqq 1} f_{q,\lambda}(t) \right]^n,$$

where

$$f_{q,\lambda}(t) \triangleq p \left(\frac{2q}{1 + (1 - 2q)t} \right)^{\lambda} + (1 - p) \left(\frac{2(1 - q)}{1 - (1 - 2q)t} \right)^{\lambda}.$$

If $\lambda \leqq -1$, $f_{q,\lambda}(t)$ is maximized for $t = 0$ or 1, proving the second assertion. Further, if q is chosen as in Problem 10.26 and

$$\lambda \triangleq \{1 - \log[(1 - p)p^{-1}]\}\{\log[(1 - q)q^{-1}]\}^{-1},$$

then $\lambda R - \log f_{q,\lambda}(0) = E_{\text{sp}}(R)$; here $\lambda \gtreqless -1$ according as $R \gtreqless R_{\text{cr}}$. If $R_{\text{cr,f}} < R_{\text{cr}}$ is the unique R for which the above choice of q and λ yields $f_{q,\lambda}(0) = f_{q,\lambda}(1)$, then for $R \geqq R_{\text{cr,f}}$ we have $\max_{0 \leqq t \leqq 1} f_{q,\lambda}(t) = f_{q,\lambda}(0)$. Hence the first assertion follows.

(d) Conclude that for a BSC with feedback the best attainable error exponent at rate $R \geqq R_{\text{cr,f}}$ is $E_{\text{sp}}(R)$. (For small rates, the best attainable error exponent for a BSC with feedback (just as without feedback) is unknown. However, its limit as $R \to 0$ equals the $E_1(0)$ of part (c); see Berlekamp (1964, unpublished), Zigangirov (1978).)

10.37. *(Decision feedback)* For a DMC $\{W : \mathsf{X} \to \mathsf{Y}\}$, an n-length block code (f, φ) gives rise to a (variable-length) code with *decision feedback* as follows: if sending a codeword $f(m) \in \mathsf{X}^n$ ($m \in \mathsf{M}_f$) results in an erasure, i.e., $\varphi(\mathbf{y}) \notin \mathsf{M}_f$, then a repeat request is issued over a noiseless feedback link and the codeword $f(m)$ is sent once more; if another erasure occurs, $f(m)$ is repeated again, etc.

(a) Observe that the transmission length N is a RV with expectation $\bar{N} \triangleq EN$ satisfying

$$n \leqq \bar{N} \leqq n\,(1 - e(W^n, f, \varphi))^{-1}$$

so that the *rate* $\bar{N}^{-1} \log |\mathsf{M}_f|$ is arbitrarily close to the rate of the block code (f, φ) if $e(W^n, f, \varphi)$ is sufficiently small.

(b) Conclude from Corollary 10.11B that for a DMC to every $\delta > 0$ there exist codes with decision feedback having rate greater than $R - \delta$ and probability of error less than $\exp[-\bar{N}(\hat{E}(R, W) - \delta)]$. In other words, that $\hat{E}(R, W)$ is an attainable error exponent at rate R if decision feedback is permitted. (A similar result was obtained by Forney (1968). The phenomenon that for DMCs with feedback, variable-length codes can give a larger error exponent than the best block codes, was previously demonstrated by Horstein (1963).)

(c) Observe that for $R < C_{0,\mathrm{u}}$ (see Problem 10.35) one can achieve zero probability of error by using decision feedback.

10.38. (*Variable-length codes with complete feedback*) Consider a DMC with capacity C and with

$$C_1 \triangleq \max_{x_1, x_2} D(W(\,.\,|x_1)\|W(\,.\,|x_2)) < \infty.$$

(a) Show that if $0 < \delta < R < C$ and $|\mathsf{M}|$ is sufficiently large, there exist variable-length codes with complete feedback such that the average transmission length $\bar{N}$ and average probability of error $\bar{e}$ satisfy

$$\bar{N}^{-1} \log |\mathsf{M}| > R - \delta, \quad \bar{e} < \exp[-\bar{N}(E^*(R) - \delta)],$$

where

$$E^*(R) \triangleq \left(1 - \frac{R}{C}\right) C_1.$$

(Burnašev (1976); Yamamoto and Itoh (1979); for $C_1 = \infty$ see Problem 6.30(c).)

Hint Modify the scheme in Problem 10.37: if $f(m)$ is sent and the output $\mathbf{y}$ satisfies $\varphi(\mathbf{y}) = m$, then send k times x_1 to signal acceptance; otherwise send k times x_2, where x_1, x_2 attain the divergence maximum C_1. Based on the outputs $y_{n+1}, \ldots, y_{n+k}$, the receiver performs a hypothesis test to decide whether the sender has signaled acceptance or not. If the test confirms acceptance then stop, otherwise repeat the process. Using Stein's lemma (Corollary 1.2), verify that this scheme achieves the claim if the block code (f, φ) (with rate close to C) and k are suitably chosen.

(b)* Show that the exponent $E^*(R)$ in (a) is best possible, and could not be improved even if the sender could learn outcomes of random experiments performed by the receiver (*active feedback*) rather than only the previous DMC outputs (*passive feedback*). (Burnašev (1976); see also Berlin *et al.* (2009).)

Story of the results

Historically, the central problem in studying the exponential behavior of the error probability of codes for DMCs was to determine the reliability function $E(R, W)$ of a given DMC $\{W\}$. The standard results in this direction are summarized in Theorem 10.6 and

Problems 10.18(b), 10.21 and 10.30; in the literature these bounds appear in various algebraic forms. For a BSC, Theorem 10.6 was proved (in a different formulation) by Elias (1955). His results were extended to all symmetric channels by Dobrušin (1962a). For a general DMC, the lower bound of $E(R)$ in Theorem 10.6 is due to Fano (1961), in yet a different formulation. A simpler proof of his result and a new form of the bound was given by Gallager (1965). For a general DMC the upper bound of Theorem 10.6 was stated (with an incomplete proof) by Fano (1961), and proved by Shannon, Gallager and Berlekamp (1967). The present form of this upper bound was obtained by Haroutunian (1968), and independently by Blahut (1974).

In this chapter we have followed the approach of Csiszár, Körner and Marton (1977, unpublished), leading to universally attainable error exponents. The packing lemma, Lemma 10.1, the key tool of this approach, is theirs and so are Theorems 10.2, 10.8 and 10.11. An error bound for constant composition codes equivalent to that in Theorem 10.2 except for universality was first derived by Fano (1961). Maximum mutual information decoding was suggested by Goppa (1975). Theorem 10.3 (with an incomplete proof) appears in Fano (1961), in a different algebraic form. It was proved by Haroutunian (1968). Compound channels have been introduced independently by Blackwell, Breiman and Thomasian (1959), Dobrušin (1959a) and Wolfowitz (1960). These authors proved Corollary 10.10. Simultaneous exponential bounds on the probability of undetected error and the probability of erasure were first derived by Forney (1968). For other relevant references see the Problems section.

11 The compound channel revisited: zero-error information theory and extremal combinatorics

A basic common characteristic of almost all channel coding problems treated in this book is that an asymptotically vanishing probability of error in transmission is tolerated. This permits us to exploit the global knowledge of the statistics of sources and channels in order to enhance transmission speed. We see again and again that in the case of a correct tuning of the parameters most codes perform in the same manner and thus, in particular, optimal codes, instead of being rare, abound. This ceases to be true if we are dealing with codes that are error-free.

The zero-error capacity of a DMC or compound DMC has been defined in Chapters 6 and 10 as the special case $\varepsilon = 0$ of ε-capacity. To keep this chapter self-contained, we give an independent (of course, equivalent) definition below.

A *zero-error code* of block length n for a DMC will be defined by a (codeword) set $\mathsf{C} \subset \mathsf{X}^n$, rather than by an encoder–decoder pair (f, φ), understanding that the message set coincides with the codeword set and the encoder is the identity mapping. This definition makes sense because if to a codeword set C there exists a decoder $\varphi : \mathsf{Y}^n \to \mathsf{C}$ that yields probability of error equal to zero, this decoder is essentially unique. In fact, a decoder φ yields probability of error 0 for a DMC $\{W : \mathsf{X} \to \mathsf{Y}\}$ iff for each codeword $\mathbf{x} \in \mathsf{C}$ its full inverse image $\varphi^{-1}(\mathbf{x})$ contains all sequences $\mathbf{y} \in \mathsf{Y}^n$ for which $W^n(\mathbf{y}|\mathbf{x}) > 0$. Hence, for a given channel, the search for a zero-error code of block length n reduces to the search for a set $\mathsf{C} \subset \mathsf{X}^n$ to which a mapping $\varphi : \mathsf{Y}^n \to \mathsf{C}$ with the above property can be found.

Determining the asymptotics of the largest cardinality of a zero-error code $\mathsf{C} \subset \mathsf{X}^n$ for a DMC $\{W : \mathsf{X} \to \mathsf{Y}\}$ is a genuinely combinatorial problem. In order to see this better, consider for every $\mathbf{x} \in \mathsf{X}^n$ the set

$$\mathsf{S}_W(\mathbf{x}) \triangleq \{\mathbf{y} : \ W^n(\mathbf{y}|\mathbf{x}) > 0\}.$$

This is the set of output sequences a decoding function must map into $\mathbf{x}$ in order to guarantee zero probability of error. We note that $\mathbf{x} = x_1 x_2 \dots x_n$ implies the relation

$$\mathsf{S}_W(\mathbf{x}) = \bigtimes_{i=1}^{n} \mathsf{S}_W(x_i);$$

in other words, $\mathsf{S}_W(\mathbf{x})$ is a Cartesian product. It is obvious that C is a zero-error code iff for its different elements $\mathbf{x}$ and $\mathbf{z}$ the sets $\mathsf{S}_W(\mathbf{x})$ and $\mathsf{S}_W(\mathbf{z})$ are disjoint. The latter is achieved iff for some i with $1 \leqq i \leqq n$ the support sets $\mathsf{S}_W(x_i)$ and $\mathsf{S}_W(z_i)$ are disjoint. This disjointness relation between the elements of the input alphabet X can be expressed through a graph structure. Since this is the first time graphs appear in the text part of our book, it is appropriate to give some of the standard definitions.

DEFINITION 11.1 A (simple) graph G is a pair of finite sets $(V(G), E(G))$, where $E(G) \subset \binom{V(G)}{2}$. In other words, the elements of $E(G)$, called edges, are unordered pairs of elements of $V(G)$. The elements of $V(G)$ are called vertices. The graph F is a *subgraph* of G if $V(F) \subset V(G)$ and $E(F) \subset E(G)$. The subgraph F is *induced* if it contains every edge of G, both of whose vertices belong to $V(F)$. ○

We associate with the DMC $W : \mathsf{X} \to \mathsf{Y}$ a graph $G = G(W)$ with vertex set X. Two vertices of the graph $G(W)$ are adjacent if the corresponding support sets are disjoint. The largest cardinality of an n-length zero-error code for the DMC is therefore the largest cardinality of a set of vertices in $G(W^n)$, any two of which are adjacent. In the language of combinatorics, such a set is called a *clique*, and the largest possible cardinality of a clique is called the *clique number* of the graph G, denoted by $\omega(G)$. For any graph G, let G^n denote the graph whose vertex set is the nth Cartesian power of that of G, and in which two vertices (n-length sequences of vertices of G) are adjacent iff their ith components are adjacent for some $1 \leqq i \leqq n$. In particular, if $G = G(W)$ then $G^n = G(W^n)$. Thus $\omega(G^n)$ is the largest cardinality of a zero-error code of block length n for our DMC. (Traditionally, the zero-error problem is formulated in terms of the complementary graph of $G(W)$ defined here. For the generalizations in this chapter, the present terminology is more adequate.) Obviously, if $\mathsf{B} \subset \mathsf{X}^m$ and $\mathsf{C} \subset \mathsf{X}^n$ are cliques in the graphs G^m resp. G^n, then their Cartesian product $\mathsf{B} \times \mathsf{C} \subset \mathsf{X}^{m+n}$ is a clique in G^{m+n} , so that we have $\omega(G^{m+n}) \geqq \omega(G^m)\omega(G^n)$.

It is a well-known elementary fact that the preceding inequality implies the existence of the limit

$$C(G) = \lim_{n\to\infty} \frac{1}{n} \log \omega(G^n). \tag{11.1}$$

Indeed, this is an immediate corollary of the following lemma.

LEMMA 11.2 If the non-negative numbers a_n satisfy the inequality $a_{m+n} \geqq a_m + a_n$ for every m and n and a_n/n is bounded from above, then a_n/n converges to $\sup_n(a_n/n)$, as $n \to \infty$. ○

Proof Define $A \triangleq \sup_n(a_n/n)$. Then $A < \infty$. We claim that $A = \lim_{n\to\infty} (a_n/n)$. To see this, fix an arbitrary $\varepsilon > 0$ and an m such that $a_m/m > A - \varepsilon$. For every n there exists an integer $q_n \geqq 0$ such that $n = q_n m + r_n$ for some r_n with $0 \leqq r_n < m$. Then, by a repeated application of the super-additivity relation $a_{m+n} \geqq a_m + a_n$, we have

$$\frac{a_n}{n} \geqq \frac{q_n a_m + a_{r_n}}{q_n m + r_n} \geqq \frac{q_n}{q_n + 1} \cdot \frac{a_m}{m} > \frac{q_n}{q_n + 1}(A - \varepsilon).$$

Now, $n \to \infty$ implies that $q_n/(q_n + 1) \to 1$ whence $\liminf_{n\to\infty}(a_n/n) > A - \varepsilon$. On the other hand, by the definition of A we also have $\limsup_{n\to\infty}(a_n/n) \leqq A$, thus completing the proof. □

DEFINITION 11.3 The always existing limit

$$C(G) = \lim_{n\to\infty} \frac{1}{n} \log \omega(G^n)$$

is called the (logarithmic) *capacity of the graph* G. The *zero-error capacity* $C_0(W)$ of the DMC $\{W : \mathsf{X} \to \mathsf{Y}\}$ is defined as $C_0(W) \triangleq C(G(W))$, the capacity of the graph $C(G(W))$. ○

When constructing block codes to be used for transmission over a single known DMC, transmission speed can reach its theoretical maximum, channel capacity, even if one uses so-called constant composition codes. In fact, capacity can be attained by randomly selecting the codewords from the set of all the input sequences having the "right" composition, as close as possible to the input distribution that maximizes the mutual information between input and output in the capacity formula. Constant composition codes are asymptotically optimal in the sense of achieving capacity also for zero probability of error. In this chapter we will study to what extent this remains true if the DMC in use is not known to every participant in the communication. We have already encountered these models of information transmission in Chapter 10, where we have dealt with the asymptotic decay of the probability of error as the duration of the communication increases. We have seen that nothing particularly new and interesting arises when dealing with codes for a class of discrete memoryless channels rather than for a single channel known to the participants in the communication. In this chapter we are taking a second look at these models in the case for which no error in transmission is tolerated. From a mathematical point of view, this is a substantially different problem and we will have to start afresh in order to deal with it. We recall the definition of typical sequences from Chapter 2.

DEFINITION 11.4 Let us denote by $\mathsf{T}^n_{[P]_\delta}$ the set of all the sequences $\mathbf{x} \in \mathsf{X}^n$ whose type satisfies

$$|P_{\mathbf{x}}(x) - P(x)| < \delta$$

for every $x \in \mathsf{X}$, while $P_{\mathbf{x}}(x) = 0$ if $P(x) = 0$. ○

A key definition in this chapter will be a channel capacity-like notion for codes with codewords of type close to a fixed probability distribution on the input alphabet.

DEFINITION 11.5 Let $G^n[P, \delta]$ be the graph induced by G^n on the set $\mathsf{T}^n_{[P]_\delta}$. We write

$$C(G, P) \triangleq \lim_{\delta\to 0} \limsup_{n\to\infty} \frac{1}{n} \log \omega(G^n[P, \delta])$$

and call it the *zero-error capacity of the graph G within the fixed distribution P*. We write also that

→ 11.1

$$C_0(W, P) \triangleq C(G(W), P).$$ ○

It is obvious that

$$C_0(W) = \max_P C_0(W, P), \tag{11.2}$$

where W is the transmission probability matrix of the DMC $\{W : \mathsf{X} \to \mathsf{Y}\}$ and the distribution P is running over all input distributions on X.

A compound channel with an uninformed encoder is a DMC with a transmission probability matrix unknown to the encoder (also called the sender). All the encoder knows is that the matrix belongs to a finite set $\mathcal{W}$ of transmission matrices with the same input alphabet X and output alphabet Y. As we have seen, if a small probability of error is tolerated, then it is irrelevant whether or not the decoder knows the matrix governing the transmission, for even if it does not, the encoder can add a short (of length sublinear in transmission time) prefix to each codeword, thereby allowing the decoder to identify the transmission matrix within an arbitrarily low probability of error.

By Corollary 10.10, the capacity of the compound DMC is

$$C(\mathcal{W}) = \max_P \min_{W \in \mathcal{W}} I(P, W) \tag{11.3}$$

in which the mutual information $I(P, W)$ is the maximum achievable rate of transmission for the DMC $\{W : \mathsf{X} \to \mathsf{Y}\}$ with codes of codeword type close to the fixed distribution P. Intuitively, this formula is valid precisely because, as we have said, almost all randomly equiprobably selected codes of a given rate and with the same fixed type P of codewords achieve a small maximum probability of error for every DMC for which the mutual information $I(P, W)$ exceeds the given rate. Now, since for zero probability of error it is no longer true that "most" codes are good, it is very reasonable to expect no analogous formula to hold for the zero-error capacity of the compound channel. It is rather surprising that it still does. In order to see this, we will first give a mathematical definition of the zero-error capacity of the compound channel with an informed decoder. (We will see later on that for zero-error transmission it *does* matter whether or not the decoder is informed about the channel matrix in use.)

DEFINITION 11.6 A set $\mathsf{C} \subset \mathsf{X}^n$ is an *n-length zero-error block code for the compound channel* $\mathcal{W}$ if it is an n-length zero-error block code for every DMC $W : \mathsf{X} \to \mathsf{Y}$ with $W \in \mathcal{W}$. We denote the maximum cardinality $|\mathsf{C}|$ of such a code by $N(\mathcal{W}, n)$ and call

$$C_0(\mathcal{W}) \triangleq \sup_n \frac{1}{n} \log N(\mathcal{W}, n) = \lim_{n\to\infty} \frac{1}{n} \log N(\mathcal{W}, n)$$

the *zero-error capacity of the compound channel* $\mathcal{W}$ *with informed decoder.* ○

From a combinatorial point of view such an n-length zero-error block code C is simply a common clique in every graph $G(W^n)$ with $W \in \mathcal{W}$. The fact that the supremum is actually a limit follows from Lemma 11.2 as in case of the zero-error capacity of a DMC.

In analogy with the simple formula for the (maximum probability of error) capacity of the compound channel we immediately see the following.

PROPOSITION 11.7

$$C_0(\mathcal{W}) \leqq \max_P \min_{W \in \mathcal{W}} C_0(W, P). \qquad \bigcirc$$

Proof Consider for every n a set $\mathsf{C}_n \subset \mathsf{X}^n$ achieving the maximum cardinality of any n-length zero-error block code for the compound channel $\mathcal{W}$. Then, by definition,

$$C_0(\mathcal{W}) = \lim_{n\to\infty} \frac{1}{n} \log |\mathsf{C}_n|.$$

Let us partition X^n into type classes T^n_P, with P running over all those probability distributions P on X for which $\mathsf{T}^n_P \neq \emptyset$. Let $\mathcal{P}_n$ be the family of these possible types of sequences from X^n. We have

$$\mathsf{X}^n = \bigcup_{P \in \mathcal{P}_n} \mathsf{T}^n_P .$$

Consequently, $\mathsf{C}_n = \bigcup_{P \in \mathcal{P}_n} (\mathsf{C}_n \cap \mathsf{T}^n_P)$. By definition,

$$|\mathsf{C}_n \cap \mathsf{T}^n_P| \leqq \omega(G(W^n)[P, 0])$$

for every $W \in \mathcal{W}$. Note that since P is a type of sequences in X^n, we also have the upper bound $\omega(G(W^n)[P, 0]) \leqq \exp(nC_0(W, P))$. Thus

$$|\mathsf{C}_n \cap \mathsf{T}^n_P| \leqq \exp\left[n \min_{W \in \mathcal{W}} C_0(W, P)\right].$$

Since by the type counting lemma $|\mathcal{P}_n| < (n+1)^{|\mathsf{X}|}$, we obtain

$$|\mathsf{C}_n| \leqq \sum_{P \in \mathcal{P}_n} \exp\left[n \min_{W \in \mathcal{W}} C_0(W, P)\right] < (n+1)^{|\mathsf{X}|} \exp\left[n \max_P \min_{W \in \mathcal{W}} C_0(W, P)\right],$$

yielding

$$\frac{1}{n} \log |\mathsf{C}_n| < \frac{|\mathsf{X}| \log(n+1)}{n} + \max_P \min_{W \in \mathcal{W}} C_0(W, P).$$

We conclude that

$$C_0(\mathcal{W}) = \lim_{n\to\infty} \frac{1}{n} \log |\mathsf{C}_n| \leqq \max_P \min_{W \in \mathcal{W}} C_0(W, P). \qquad \square$$

We will show that the upper bound in Proposition 11.7 is tight. To this end we need some notation and preliminary lemmas. First of all we define the constant composition version of the zero-error capacity of the compound channel. For fixed $\delta > 0$ the intersection graph $\bigcap_{W \in \mathcal{W}} G(W^n)[P, \delta]$ is the graph induced by $\bigcap_{W \in \mathcal{W}} G(W^n)$ on the set $\mathsf{T}^n_{[P]_\delta}$. In other words, this is a graph whose vertex set $\mathsf{T}^n_{[P]_\delta}$ is the common vertex set of the $G^n[P, \delta]$ as G varies over the graphs $G(W)$ with $W \in \mathcal{W}$, while two vertices are adjacent in $\bigcap_{W \in \mathcal{W}} G(W^n)[P, \delta]$ iff they are adjacent in each of the graphs $G(W^n)$, for every $W \in \mathcal{W}$. We introduce the following shorthand notation for the clique number of this intersection graph:

$$N(\mathcal{W}, n, P, \delta) \triangleq \omega\left(\bigcap_{W \in \mathcal{W}} G(W^n)[P, \delta]\right),$$

and define

$$C_0(\mathcal{W}, P) \triangleq \lim_{\delta \to 0} \limsup_{n \to \infty} \frac{1}{n} \log N(\mathcal{W}, n, P, \delta). \tag{11.4}$$

We need an interesting and important combinatorial lemma which we shall treat in the context of *hypergraphs*.

DEFINITION 11.8 A (simple) *hypergraph* G is a pair of sets $(\mathsf{V}, \mathcal{E})$, where V is an arbitrary finite set, while $\mathcal{E}$ is a subset of 2^{V}, the power set of V. In other words, $\mathcal{E}$ is an arbitrary but fixed family of subsets of V. The elements of V are called vertices, while the elements of $\mathcal{E}$ are called hyperedges.

The *degree of a vertex* $x \in \mathsf{V}(G)$ is the number $d_G(x)$ of hyperedges containing it. The maximum of these degrees is denoted by $d(G)$ and is called the *degree of the hypergraph.* A hypergraph is regular if all its vertices have the same degree. A hypergraph is called *r-uniform* if all of its hyperedges contain r vertices. We denote by $r(G)$ the maximum cardinality of a hyperedge in G and call it the *rank* of the hypergraph. ○

Hypergraphs are a straightforward generalization of graphs. In fact, a graph is a 2-uniform hypergraph.

DEFINITION 11.9 A subset $\mathcal{F} \subset \mathcal{E}(G)$ is an *edge cover* if every vertex is contained in at least one hyperedge in $\mathcal{F}$. The minimum number of hyperedges in an edge cover is called the covering number of G, and is denoted by $\tau(G)$. ○

DEFINITION 11.10 A non-negative-valued function t defined on $\mathcal{E}(G)$ is a *fractional edge cover* of G if for every vertex $v \in \mathsf{V}(G)$ the sum of its values satisfies

$$\sum_{v \in \mathsf{E} \in \mathcal{E}(G)} t(\mathsf{E}) \geqq 1.$$

The *value* of the fractional edge cover t is the sum over all the hyperedges $\sum_{\mathsf{E} \in \mathcal{E}(G)} t(\mathsf{E})$. The *fractional edge covering number* $\tau^*(G)$ is the (always existing) minimum value of a fractional cover of G. ○

DEFINITION 11.11 A set of vertices is *i-independent* in G if none of the hyperedges of G contains more than i of its vertices; $\nu_i(G)$ is the maximum number of vertices in an i-independent set of G. A 1-independent set is called an *independent set.*

A non-negative real-valued function m defined on the vertex set of G is a *fractional independent set* of G if for every $\mathsf{E} \in \mathcal{E}(G)$ the sum of the values of m on the vertices contained in E satisfies

$$\sum_{v \in \mathsf{E}} m(v) \leqq 1.$$

The sum of the values of m over all the vertices of G is called the *size* of the fractional independent set m. We denote by $\nu^*(G)$ the (always existing) maximum size of a fractional independent set of G. ○

LEMMA 11.12 (*Hypergraph covering*) For every hypergraph G the edge covering number can be bounded from above in terms of the fractional independence number as

$$\tau(G) \leqq \nu^*(G) \sum_{i=1}^{r(G)} \frac{1}{i} < \nu^*(G)(1 + \ln r(G)). \qquad ○$$

Proof Let us define a hyperedge cover $\mathcal{C}$ of G by the greedy algorithm in which at each step we select a new edge containing the maximum number of vertices not contained in (or, in another terminology, covered by) any of the previously selected edges. We stop when such hyperedges no longer exist; in other words, when the edges already chosen form a cover of the hypergraph. Let us denote by t_i the number of those edges in $\mathcal{C}$ which, at the point of being selected, cover i new vertices. Let G_i be the hypergraph having as vertices those of G which are not contained in any of the previously selected $t_r + t_{r-1} + \cdots + t_{i+1}$ hyperedges and having as hyperedges the non-empty intersections of the original hyperedges with the new vertex set. Then $r(G_i) \leqq i$. We see that, by definition, the vertices of G_i represent an i-independent set of G, whence

$$|\mathsf{V}(G_i)| \leqq \nu_i(G).$$

The vertices of G_i are all the vertices of G yet to be covered by some hyperedge, whereby we have

$$|\mathsf{V}(G_i)| = it_i + (i-1)t_{i-1} + \cdots + 2t_2 + t_1.$$

This implies, for $i = 1, 2, \ldots, r(G)$, the inequality

$$it_i + (i-1)t_{i-1} + \cdots + 2t_2 + t_1 \leqq \nu_i(G).$$

Dividing the first of these inequalities (the one with $i = 1$) with $1 \cdot 2$, the second one with $2 \cdot 3$, the ith one with $i(i+1)$ for $i < r = r(G)$ and dividing the last one by r, we can sum the resulting new inequalities. Then on the left-hand side we obtain an expression in which the coefficient of t_i will be

$$i\left(\frac{1}{i(i+1)} + \frac{1}{(i+1)(i+2)} + \cdots + \frac{1}{(r-1)r} + \frac{1}{r}\right)$$

$$= i\left(\frac{1}{i} - \frac{1}{i+1} + \frac{1}{i+1} - \frac{1}{i+2} + \cdots + \frac{1}{r-1} - \frac{1}{r} + \frac{1}{r}\right) = 1.$$

Thus

$$t = \sum_{i=1}^{r} t_i \leqq \sum_{i=1}^{r-1} \frac{\nu_i(G)}{i(i+1)} + \frac{\nu_r(G)}{r}.$$

Next consider for every i the characteristic function of an i-independent set of cardinality $\nu_i(G)$. In other words, consider the function m_i defined on $\mathsf{V}(G)$ for which $m_i(v) = 1$ if v is in the independent set and zero elsewhere. It is obvious that the function obtained from m_i by dividing each of its values by i is a fractional independent set of G, whence we have, for every i, the inequality

$$\frac{\nu_i(G)}{i} \leqq \nu^*(G).$$

Substituting each of these inequalities into the preceding upper bound on t we see that

$$t \leqq \nu^*(G)\left(\sum_{i=1}^{r(G)} \frac{1}{i}\right).$$

The missing statement is obvious since

$$\sum_{i=2}^{r(G)} \frac{1}{i} < \ln r(G). \qquad \square$$

By the duality theorem of linear programming,

$$\nu^*(G) = \tau^*(G) \tag{11.5}$$

for every hypergraph G. Although we will not make use of this relation in this book, we mention the following corollary it immediately implies.

COROLLARY 11.12

$$\tau(G) \leqq \tau^*(G)(1 + \ln r(G)). \qquad \bigcirc$$

We can say more about uniform regular hypergraphs.

LEMMA 11.13 (*Edge cover*) If the hypergraph G is regular and r-uniform then the minimum number $\tau(G)$ of hyperedges needed to cover all its vertices is bounded by → 11.2

$$\frac{|\mathsf{V}(G)|}{r} \leqq \tau(G) \leqq (1 + \ln r)\frac{|\mathsf{V}(G)|}{r}. \qquad \bigcirc$$

Proof Let m be an optimal fractional independent set in G. If G is d-regular, we have

$$\sum_{\mathsf{E}\in\mathcal{E}(G)} \sum_{v\in\mathsf{E}} m(v) = d(G) \sum_{v\in V(G)} m(v) = d(G)\nu^*(G),$$

while at the same time

$$\sum_{\mathsf{E}\in\mathcal{E}(G)} \sum_{v\in\mathsf{E}} m(v) \leqq |\mathcal{E}(G)|,$$

whence we get

$$\nu^*(G) \leqq \frac{|\mathcal{E}(G)|}{d(G)}. \tag{11.6}$$

Note, however, that by double-counting the pairs (v, E) such that the vertex v is contained in the hyperedge E we obtain

$$|\mathcal{E}(G)|r = |\mathsf{V}(G)|d(G).$$

Comparing this with (11.6) gives the upper bound. The lower bound is obvious. □

We still need a technical lemma. For any partition $\mathcal{R}$ of a set Q we denote by $|\mathcal{R}|$ the number of its non-empty classes.

LEMMA 11.14 (*Twin partitions*) Let $\mathcal{S}'$ and $\mathcal{T}'$ be two arbitrary partitions of the same finite set Q. Then there exist two other partitions $\mathcal{S}$ and $\mathcal{T}$ such that $\mathcal{S}$ refines $\mathcal{S}'$ and $\mathcal{T}$ refines $\mathcal{T}'$,

$$|\mathcal{S}| = |\mathcal{T}| \leqq |\mathcal{S}'| + |\mathcal{T}'|,$$

and the two new partitions have the same "shape" in the sense of the existence of a bijection between the classes of the new partitions in which the corresponding classes have the same cardinality. ○

Proof The proof proceeds by induction over $|\mathcal{S}'| + |\mathcal{T}'|$. For this purpose we shall prove a formal generalization of the statement. We will drop the condition that $\mathcal{S}'$ and $\mathcal{T}'$ partition the same finite set. Rather, we suppose that $\mathcal{S}'$ partitions Q and $\mathcal{T}'$ partitions R where the ground sets Q and R have the same cardinality.

If $|\mathcal{S}'| + |\mathcal{T}'| = 2$, there is nothing to prove. Suppose that the statement is true for any pair of partitions as above with $|\mathcal{S}'| + |\mathcal{T}'| < k$ and consider two partitions with $|\mathcal{S}'| + |\mathcal{T}'| = k$. Without loss of generality, suppose that

$$|\mathcal{S}'| \leqq |\mathcal{T}'|.$$

We create an arbitrary matching between the classes of $\mathcal{S}'$ and a corresponding number of classes of $\mathcal{T}'$. For each pair of classes we match the elements of the smaller one with a corresponding number of elements of the larger one, and then divide both Q and R into two; on the one side we put the elements involved in the previous matchings, and on the other we put the unmatched ones. The original partitions generate a partition of each of the four sets so obtained. The matched parts of Q and R are partitioned into $|\mathcal{S}'|$ classes each. Let i be the number of those of the matched class pairs for which the class of partition $\mathcal{S}'$ is strictly larger than the corresponding one of $\mathcal{T}'$ and eliminate all the matched pairs of elements from Q and R, respectively. The resulting ground sets Q^* and R^* have equal cardinality. Partition $\mathcal{S}'$ generates a partition of Q^* into i classes, while $\mathcal{T}'$ generates a partition of R into $|\mathcal{T}'| - i$ classes, since of the original $|\mathcal{T}'|$ exactly i have been entirely eliminated. Thus the total number of classes of these partitions is $|\mathcal{T}'|$ and this is strictly smaller than $k = |\mathcal{S}'| + |\mathcal{T}'|$, whereby the induction hypothesis applies and we can obtain a partition into at most $|\mathcal{T}'|$ classes of both Q^* and R^*, where the two resulting partitions have the same shape. Both of these can be completed into a partition of the original sets Q and R so that the resulting overall partitions still have the same shape and both have a number of classes at most $|\mathcal{T}'| + i$. Since by definition $i \leqq |\mathcal{S}'|$, the statement of the lemma follows. □

Now we are ready to explain the promised lower bound for $C_0(\mathcal{W})$, i.e., to prove that in Proposition 11.7 the inequality holds with equality.

THEOREM 11.15 Let $\mathcal{V}$ and $\mathcal{W}$ be two finite sets of transmission matrices all with the same input alphabet X. Then for every input distribution P on X

$$C_0(\mathcal{V} \cup \mathcal{W}, P) \geqq \min\{C_0(\mathcal{V}, P), C_0(\mathcal{W}, P)\}. \qquad \bigcirc$$

It should be clear that the validity of this statement is sufficient to solve our problem. However, we will postpone the proof, preferring, for later purposes, to carry it out in more generality. The subsequent more general approach will allow us to prove an analogous result for zero-error capacity also in the case of the compound channel with both the encoder and the decoder uninformed. At the same time, we will be able to treat several well-known problems in extremal combinatorics as zero-error problems in information theory. All this is based on generalizing the problem of zero-error capacity of the DMC which, as we have seen, is a question about the asymptotic growth of the clique number in powers of graphs, to the case of directed graphs.

To motivate our generalization, we invoke the following classical problem of extremal set theory. (This theory is often called the theory of hypergraphs.) Consider an arbitrary finite set of n elements, e.g., the set

$$[n] \triangleq \{1, 2, \ldots, n\}$$

of the first n positive integers. This notation is very common in combinatorics and will be used throughout this chapter. A family $\mathcal{S} \subset 2^{[n]}$ of subsets of $[n]$ is called a *Sperner family* if neither of two different sets from $\mathcal{S}$ contains the other. In extremal set theory one is interested in the largest cardinality $I(n)$ of a Sperner family of subsets of $[n]$. Sperner's theorem states that the largest Sperner family is essentially unique and its cardinality is the middle binomial coefficient $\binom{n}{\lfloor \frac{n}{2} \rfloor}$. From an information-theoretic, → 11.3
asymptotic viewpoint this problem is fairly trivial. No family of subsets of $[n]$ can have more than 2^n elements, while, on the other hand, the family of all subsets of cardinality $\lfloor \frac{n}{2} \rfloor$ has the Sperner property and its cardinality is the middle binomial, so that we have

$$\lim_{n \to \infty} \frac{1}{n} \log I(n) = 1.$$

We will show that this limit is a special case of the natural generalization of graph capacity to directed graphs.

DEFINITION 11.16 A finite *directed graph (or digraph)* D is a couple of sets (V, E), where V is a finite set whose elements are called vertices and E is a subset of the set $\mathsf{V} \times \mathsf{V}$ of ordered couples of vertices of D. If not otherwise stated, E does not contain pairs (x, x) of identical vertices (loop arcs). The elements of E are called arcs (or directed edges). A set $\mathsf{S} \subset \mathsf{V}$ is a *symmetric clique* in D if every ordered couple of distinct vertices in S is an arc in D. (The vertices of a symmetric clique are called symmetrically adjacent.) The largest cardinality of such a set is the *symmetric clique number* $\omega_s(D)$ of the digraph D. Given any subset $\mathsf{R} \subset \mathsf{V}$, the subgraph of D *induced* by R is the digraph whose vertex set is R and whose set of edges consists of all the ordered couples of vertices from R which form an arc in D. ○

We are interested in the asymptotic growth of symmetric cliques in the powers of a digraph in the sense to be defined next.

DEFINITION 11.17 The *(co-normal) product of the digraphs* $F = (\mathsf{V}(F), E(F))$ and $G = (\mathsf{V}(G), E(G))$ is the digraph $F \cdot G$ with vertex set $\mathsf{V}(F \cdot G) \triangleq \mathsf{V}(\mathsf{F}) \times \mathsf{V}(G)$, where there is an arc from vertex (v, w) to (v', w') if at least one of the following two things happens: (i) there is an arc from v to v' in F; (ii) there is an arc from w to w' in G. The *power digraph* D^n is then the co-normal product of D with D^{n-1} and $D^1 \triangleq D$. ○

DEFINITION 11.18 The always existing limit

$$C(D) \triangleq \lim_{n\to\infty} \frac{1}{n} \log \omega_s(D^n)$$

is called the (logarithmic) *Sperner capacity* of the digraph D. ○

The existence of the limit is immediate by Lemma 11.2 in the same way as it is for undirected graphs. An undirected graph G can be considered equivalently as a digraph if we replace the edges between its adjacent vertices by two directed edges pointing in the two opposite directions. Clearly, the Sperner capacity of this directed graph is the same as the capacity of the simple graph defining it. However, the problem of determining the Sperner capacity of a digraph is a genuine generalization of → 11.4 the analogous question about graph capacity. To understand and verify this statement, it suffices to note that for the two non-isomorphic orientations of its three edges the → 11.5 complete graph on three vertices gives rise to two digraphs whose Sperner capacities are different.

→ 11.6 Let us now consider the directed graph S with the two-element vertex set $\{0, 1\}$ and the sole arc $(0, 1)$. The vertex set of S^n consists of all the binary strings of length n. → 11.8 There is an arc from $\mathbf{x} \in \{0, 1\}^n$ to $\mathbf{y} \in \{0, 1\}^n$ precisely when the set $\mathsf{B} \subset [n]$ of the positions of the 1's in the string $\mathbf{y}$ is not contained in the set $\mathsf{A} \subset [n]$ of the positions of the 1's in $\mathbf{x}$. In other words, considering $\mathbf{x}$ as the characteristic vector of a set A and $\mathbf{y}$ as that of B, we see that $\mathbf{x}$ and $\mathbf{y}$ are symmetrically adjacent precisely when neither of the sets A and B contains the other. Thus a set $\mathsf{C} \subset \{0, 1\}^n$ is a symmetric clique in S^n iff its elements are the characteristic vectors of the member sets of a Sperner family. In other words,

$$I(n) = \omega_s(S^n).$$

We are going to extend all the previous concepts to families of digraphs.

DEFINITION 11.19 Let $\mathcal{G}$ be a finite family of digraphs with common vertex set X. A set $\mathsf{C} \subset \mathsf{X}^n$ is a *symmetric clique for the digraph family* $\mathcal{G}$ if it is a symmetric clique in the digraph G^n for every $G \in \mathcal{G}$. Let $N(\mathcal{G}, n)$ denote the largest cardinality of such a symmetric clique $\mathsf{C} \subset \mathsf{X}^n$. The (always existing) limit

$$C(\mathcal{G}) \triangleq \lim_{n\to\infty} \frac{1}{n} \log N(\mathcal{G}, n)$$

is called the *Sperner capacity of the digraph family* $\mathcal{G}$. ○

The next few definitions are literal repetitions of those for families of undirected graphs. Let P be an arbitrary probability distribution on the common vertex set X of the digraphs of $\mathcal{G}$. For fixed $\delta > 0$ the intersection digraph $\bigcap_{G\in\mathcal{G}} G^n[P, \delta]$ is the digraph induced by $\bigcap_{G\in\mathcal{G}} G^n$ on the set $\mathsf{T}^n_{[P]_\delta}$. In other words, this is a digraph whose vertex set $\mathsf{T}^n_{[P]_\delta}$ is the common vertex set of the $G^n[P, \delta]$ as G varies over the digraphs $G \in \mathcal{G}$, while an ordered pair of vertices forms an arc in $\bigcap_{G\in\mathcal{G}} G^n[P, \delta]$ iff it does in each of the digraphs G^n with $G \in \mathcal{G}$. We introduce the following shorthand notation for the symmetric clique number of this intersection digraph:

$$N(\mathcal{G}, n, P, \delta) \triangleq \omega_s\left(\bigcap_{G\in\mathcal{G}} G^n[P, \delta]\right).$$

We further define

$$C(\mathcal{G}, P) \triangleq \lim_{\delta\to 0} \limsup_{n\to\infty} \frac{1}{n} \log N(\mathcal{G}, n, P, \delta).$$

PROPOSITION 11.20

$$C(\mathcal{G}) \leqq \max_P \min_{G\in\mathcal{G}} C(G, P).$$ ○

Proof The proof of Proposition 11.7 literally applies. □

The core result in this chapter is the following theorem.

THEOREM 11.21 (*Core theorem*) Let $\mathcal{F}$ and $\mathcal{G}$ be two families of digraphs with common vertex set X. For every distribution P on X we have

$$C(\mathcal{F} \cup \mathcal{G}, P) \geqq \min\{C(\mathcal{F}, P), C(\mathcal{G}, P)\}.$$ ○

Proof Let P be an arbitrary probability distribution on X. Pick a sequence $\{P_n\}_{n=1}^{\infty}$ of types of sequences in X^n for which → 11.1

$$\max_{x\in\mathsf{X}} |P(x) - P_n(x)| \to 0$$

and corresponding sets $\mathsf{A}_n \subset \mathsf{T}_{P_n}$, $\mathsf{B}_n \subset \mathsf{T}_{P_n}$ such that

$$\lim_{n\to\infty} \frac{1}{n} \log |\mathsf{A}_n| = C(\mathcal{F}, P), \qquad \lim_{n\to\infty} \frac{1}{n} \log |\mathsf{B}_n| = C(\mathcal{G}, P),$$

where the sets A_n are symmetric cliques for the digraph family $\mathcal{F}$ and the sets B_n are symmetric cliques for the digraph family $\mathcal{G}$.

Let S_n be the set of all permutations of $[n]$. Then for a $\pi \in \mathsf{S}_n$ and any $\mathbf{x} \in \mathsf{X}^n$ we write

$$\pi(\mathbf{x}) \triangleq x_{\pi(1)}x_{\pi(2)} \ldots x_{\pi(n)}$$

and

$$\pi(\mathsf{A}_n) \triangleq \{\pi(\mathbf{x}) : \mathbf{x} \in \mathsf{A}_n\}.$$

Consider the hypergraph whose vertices are the elements of T_{P_n} and whose hyperedges are the images $\pi(\mathsf{A}_n)$ of the set A_n for the various $\pi \in \mathsf{S}_n$. By virtue of the edge cover lemma, Lemma 11.13, there is a set $\mathsf{S}_n^* \subset \mathsf{S}_n$ of permutations such that

$$\bigcup_{\pi\in\mathsf{S}_n^*} \pi(\mathsf{A}_n) = \mathsf{T}_{P_n}, \text{ while } |\mathsf{S}_n^*| \leqq \frac{|\mathsf{T}_{P_n}|}{|\mathsf{A}_n|}(1 + \ln |\mathsf{A}_n|).$$

But then we can, by making our sets disjoint, construct a partition of T_{P_n} into the disjoint sets

$$\mathsf{Q}_\pi \subset \pi(\mathsf{A}_n) \text{ for } \pi \in \mathsf{S}_n^*.$$

Likewise, there exists a set $\mathsf{S}_n^{**} \subset \mathsf{S}_n$ of permutations of $[n]$ with

$$\bigcup_{\pi\in\mathsf{S}_n^{**}} \pi(\mathsf{B}_n) = \mathsf{T}_{P_n}, \text{ while } |\mathsf{S}_n^{**}| \leqq \frac{|\mathsf{T}_{P_n}|}{|\mathsf{B}_n|}(1 + \ln |\mathsf{B}_n|),$$

and corresponding disjoint sets $\mathsf{R}_\pi \subset \pi(\mathsf{B}_n)$ for $\pi \in \mathsf{S}_n^{**}$ partitioning T_{P_n}.

Let us apply the twin partitions lemma, Lemma 11.14, to these two partitions of T_{P_n}. We obtain two new partitions, $\{\mathsf{C}_i\}_{i=1}^t$ and $\{\mathsf{D}_i\}_{i=1}^t$ of T_{P_n}, with equal number of classes t, where

$$t \leqq 2\max\left\{\frac{|\mathsf{T}_{P_n}|}{|\mathsf{A}_n|}(1 + \ln |\mathsf{A}_n|), \frac{|\mathsf{T}_{P_n}|}{|\mathsf{B}_n|}(1 + \ln |\mathsf{B}_n|)\right\}. \tag{11.7}$$

Furthermore, each C_i is a symmetric clique for the digraph family $\mathcal{F}$, while the classes D_i are symmetric cliques for the digraph family $\mathcal{G}$. Also, the classes C_i and D_i of the same index have the same cardinality.

We will use our twin partitions to construct a symmetric clique for the graph family $\mathcal{F} \cup \mathcal{G}$. The vertices of this clique will be sequences from $[\mathsf{T}_{P_n}]^m$ for some integer m to be specified later. We will consider these sequences of length nm of elements of X as sequences of length m of elements of T_{P_n}. Let $\mathbf{x}$ be such a sequence. Then we can write $\mathbf{x} = \mathbf{x}_1\mathbf{x}_2 \ldots \mathbf{x}_m$ with $\mathbf{x}_i \in \mathsf{T}_{P_n}$ for every $i \in [m]$. Let M be the set of those sequences of this kind for which $\mathbf{x}_i \in \mathsf{C}_j$ implies $\mathbf{x}_{i+1} \in \mathsf{D}_j$, for every $i < m$ and every $j \in [t]$.

Further, for every $\mathbf{y} \in \mathsf{T}_{P_n}$ and $\mathbf{z} \in \mathsf{T}_{P_n}$, let the set $\mathsf{M}(\mathbf{y}, \mathbf{z})$ consist of those elements of M whose initial segment of length n is $\mathbf{y}$ and the last segment of length n is $\mathbf{z}$. Then for at least one pair of not necessarily distinct $\mathbf{y}$ and $\mathbf{z}$ we have

$$|\mathsf{M}(\mathbf{y}, \mathbf{z})| \geqq \frac{|\mathsf{M}|}{|\mathsf{T}_{P_n}|^2}. \tag{11.8}$$

We claim that $\mathsf{M}(\mathbf{y}, \mathbf{z})$ is a symmetric clique in $\mathcal{F} \cup \mathcal{G}$, no matter how we choose $\mathbf{y}$ and $\mathbf{z}$. To verify this, let $\mathbf{w}$ and $\mathbf{x}$ be two arbitrary but different sequences from $\mathsf{M}(\mathbf{y}, \mathbf{z})$. Write $\mathbf{w} = \mathbf{w}_1\mathbf{w}_2 \ldots \mathbf{w}_m$ and $\mathbf{x} = \mathbf{x}_1\mathbf{x}_2 \ldots \mathbf{x}_m$. Since the first segments of these two sequences coincide, there is an $i > 1$ for which $\mathbf{w}_h = \mathbf{x}_h$ whenever $h < i$ while $\mathbf{w}_i \neq \mathbf{x}_i$. This means that for some $j \in [t]$ the $(i-1)$th segment of both sequences satisfies $\mathbf{w}_{i-1} = \mathbf{x}_{i-1} \in \mathsf{C}_j$ and thus, by the rule governing the construction of these sequences, $\mathbf{w}_i$ and $\mathbf{x}_i$ both belong to the same class D_j of the other partition. This has the implication that the two sequences $\mathbf{x}$ and $\mathbf{w}$ satisfy in their ith segments the conditions for being symmetrically adjacent for the graph family $\mathcal{G}$. Reading the same sequences from right to left, we will see that they also satisfy the analogous conditions for the graph family $\mathcal{F}$. As a matter of fact, the last segment of the two sequences coincide. Thus we know that the last time they differ is in the ith segment for some $i < m$. Since they necessarily coincide in their $(i+1)$th segment, their respective $(i+1)$th segments are in the same class D_j of the second partition of T_{P_n}. But this can happen only if their (differing) ith segments both belong to the same class C_j of the first partition. Hence the sequences satisfy the condition for being symmetrically adjacent for the graph family $\mathcal{F}$ because they do so already in their ith segments. It remains to show that for a proper choice of $\mathbf{y}$ and $\mathbf{z}$ the symmetric clique $\mathsf{M}(\mathbf{y}, \mathbf{z})$ is large enough to yield the assertion of the theorem. To this purpose, recalling (11.8), a lower bound on the cardinality of M is needed.

It is not easy to count the sequences in M with precision. We will use information theory to do it approximately. To this end, we will define a Markov chain with a transition probability matrix W whose rows and columns are indexed by the elements of T_{P_n}. We define a stochastic matrix (transition probability matrix) $W : \mathsf{T}_{P_n} \to \mathsf{T}_{P_n}$ by setting $W(\mathbf{z}|\mathbf{y}) = 1/|\mathsf{D}_j|$ if $\mathbf{y} \in \mathsf{C}_j$, $\mathbf{z} \in \mathsf{D}_j$ for some $j \in [t]$ while all the other entries of the matrix W are equal to zero. It is easy to see that the uniform distribution on T_{P_n} is an invariant distribution for W. We consider the stationary Markov chain formed by a sequence of random variables X_i, $i \in [m]$ in which X_1 is uniformly distributed over T_{P_n} and the transition probability matrix from any variable X_i to the successive one is W. Note that the sequence $X_1X_2 \ldots X_m$ has support M. Hence, by Corollary 1.1, we have

$$H(X_1X_2 \ldots X_m) \leqq \log |\mathsf{M}|. \tag{11.9}$$

By the chain rule,

$$H(X_1X_2 \ldots X_m) = H(X_1) + \sum_{i=2}^{m} H(X_i|X_{i-1}X_{i-2} \ldots X_1),$$

where $H(X_1) = H(X_2) \geqq H(X_2|X_1)$ while, because of the Markov chain structure of our random variables, we also have $H(X_i|X_{i-1}X_{i-2}\dots X_1) = H(X_i|X_{i-1})$ for every $i > 1$. Noting that $H(X_i|X_{i-1})$ does not depend on i for $i > 1$ we conclude that

$$mH(X_2|X_1) \leqq \log|\mathsf{M}|.$$

Next we lower bound $H(X_2|X_1)$ as follows:

$$H(X_2|X_1) = \sum_{j=1}^{t} \frac{|\mathsf{C}_j|}{|\mathsf{T}_{P_n}|}\log|\mathsf{C}_j| \geqq \frac{1}{|\mathsf{T}_{P_n}|}\left(\sum_{j=1}^{t}|\mathsf{C}_j|\right)\log\frac{\sum_{j=1}^{t}|\mathsf{C}_j|}{t} = \log\frac{|\mathsf{T}_{P_n}|}{t}, \tag{11.10}$$

where the inequality is a direct consequence of the log-sum inequality Lemma 3.1. Replacing t by its upper bound (11.7), we finally see from (11.10) that

$$H(X_2|X_1) \geqq \log\min\left\{\frac{|\mathsf{A}_n|}{2(1+\log|\mathsf{A}_n|)}, \frac{|\mathsf{B}_n|}{2(1+\log|\mathsf{B}_n|)}\right\},$$

whence by (11.9) we have

$$\log|\mathsf{M}| \geqq mH(X_2|X_1) \geqq m\log\min\left\{\frac{|\mathsf{A}_n|}{2(1+\ln|\mathsf{A}_n|)}, \frac{|\mathsf{B}_n|}{2(1+\ln|\mathsf{B}_n|)}\right\}.$$

We choose $m = n$ and evaluate asymptotically. This yields

$$C(\mathcal{F}\cup\mathcal{G}, P) \geqq \limsup_{n\to\infty}\frac{1}{n^2}\log|M(\mathbf{y},\mathbf{z})| \geqq \limsup_{n\to\infty}\frac{1}{n^2}\log\frac{|M|}{|\mathsf{T}_{P_n}|^2}$$

$$\geqq \limsup_{n\to\infty}\left[\frac{1}{n}\min\left\{\log\frac{|\mathsf{A}_n|}{2(1+\ln|\mathsf{A}_n|)}, \log\frac{|\mathsf{B}_n|}{2(1+\ln|\mathsf{B}_n|)}\right\} - \frac{2}{n}\log|\mathsf{X}|\right]$$

$$\geqq \min\left\{\limsup_{n\to\infty}\frac{1}{n}\log|\mathsf{A}_n|, \limsup_{n\to\infty}\frac{1}{n}\log|\mathsf{B}_n|\right\} = \min\{C(\mathcal{F}, P), C(\mathcal{G}, P)\}.$$

□

Proposition 11.20 and Theorem 11.21 put together yield the following conclusion.

THEOREM 11.22 For an arbitrary family $\mathcal{G}$ of directed graphs with common vertex set X and an arbitrary probability distribution P on X

$$C(\mathcal{G}, P) = \min_{G\in\mathcal{G}} C(G, P)$$

and in particular, for the Sperner capacity of the graph family, we have the formula

$$C(\mathcal{G}) = \max_P \min_{G\in\mathcal{G}} C(G, P).$$ ○

Proof The missing element for the proof is to show that for every probability distribution P on X we have

$$C(\mathcal{G}, P) \geqq \min_{G\in\mathcal{G}} C(G, P).$$

This is, however, immediate by an iterative application of Theorem 11.21 to graph families $\mathcal{F}_i \triangleq \{G_1, G_2, \dots, G_{i-1}\}$ and $\mathcal{G}_i \triangleq \{G_i\}$, where $i \geqq 2$ and the G_i are the members of the graph family $\mathcal{G}$, considered in an arbitrary order. □

COROLLARY 11.22

$$C_0(\mathcal{W}) = \max_P \min_{W \in \mathcal{W}} C_0(W, P). \qquad \bigcirc$$

One of the consequences of this result is that if we had an explicit formula for the zero-error capacity of any DMC, or better for any DMC and codes of any prescribed composition, then we would automatically have one for the zero-error capacity of the compound channel. Theorem 11.22 on the Sperner capacity of digraph families permits us to obtain a formula for the compound channel also for the uninformed encoder and decoder. To see this, we will associate a family of directed graphs to any compound channel $\mathcal{W}$.

DEFINITION 11.23 Let $\mathcal{W}$ be a family of stochastic matrices with common input alphabet X and output alphabet Y. To any ordered pair of matrices (V, W) we associate the digraph $G_{(V,W)}$ with vertex set X. In this digraph there is an edge (a, b) pointing from vertex a to vertex b if for the support sets we have

$$\mathsf{S}_V(a) \cap \mathsf{S}_W(b) = \emptyset.$$

We call the family of these digraphs the digraph family associated with $\mathcal{W}$. ○

Let us give a formal definition of the concept of capacity we are discussing here. A pair (C, φ) is an n-length zero-error block code for the compound channel $\mathcal{W}$ with uninformed encoder and decoder if $\mathsf{C} \subset \mathsf{X}^n$ and the function $\varphi : \mathsf{Y}^n \to \mathsf{C}$ satisfies

$$W^n(\varphi^{-1}(\mathbf{x})|\mathbf{x}) = 1 \text{ for every } \mathbf{x} \in \mathsf{C},\ W \in \mathcal{W}. \tag{11.11}$$

It is worth noticing how it continues to be true that for any given codeword set the zero-error decoder is essentially unique. In fact, every output sequence in the union of the support sets of a codeword, with the union taken for all $W \in \mathcal{W}$, must be mapped by the decoder to the codeword itself. However, this union of the support sets does not generally have a product structure.

DEFINITION 11.24 A set $\mathsf{C} \subset \mathsf{X}^n$ is an n-length zero-error block code for the compound channel $\mathcal{W}$ with uninformed encoder and decoder if for a function $\varphi : \mathsf{Y}^n \to \mathsf{C}$ it satisfies (11.11). Let $N^*(\mathcal{W}, n)$ be the largest cardinality of such a codeword set. We call

$$C_0^*(\mathcal{W}) \triangleq \sup_n \frac{1}{n} \log N^*(\mathcal{W}, n) = \lim_{n\to\infty} \frac{1}{n} \log N^*(\mathcal{W}, n)$$

the *zero-error capacity of the compound channel with uninformed encoder and decoder.* ○

It is easy to establish the relationship between the present problem for the compound channel and the Sperner capacity of a corresponding family of digraphs.

THEOREM 11.25 The zero-error capacity of the compound channel $\mathcal{W}$ with an uninformed encoder–decoder pair is the Sperner capacity of the associated family of digraphs. ○

Proof The set $\mathsf{C} \subset \mathsf{X}^n$ is an n-length zero-error block code for the compound channel $\mathcal{W}$ if for every ordered pair of its distinct elements $(\mathbf{x}, \mathbf{z})$ and the ordered pair of not necessarily distinct stochastic matrices (V, W) contained in $\mathcal{W}$ we have

$$\mathsf{S}_V(\mathbf{x}) \cap \mathsf{S}_W(\mathbf{z}) = \emptyset.$$

Let us write $\mathbf{x} = x_1 x_2 \dots x_n$ and $\mathbf{z} = z_1 z_2 \dots z_n$. Since

$$\mathsf{S}_V(\mathbf{x}) = \bigtimes_{i=1}^{n} \mathsf{S}_V(x_i) \quad \text{and} \quad \mathsf{S}_W(\mathbf{z}) = \bigtimes_{i=1}^{n} \mathsf{S}_W(z_i),$$

for the disjointness of these two sets it is necessary and sufficient, for at least one coordinate $i \in [n]$, to have

$$\mathsf{S}_V(x_i) \cap \mathsf{S}_W(z_i) = \emptyset.$$

This means that the sequences $\mathbf{x}$ and $\mathbf{z}$ must be symmetrically adjacent in the digraph $G^n_{(V,W)}$. We conclude that $\mathsf{C} \subset \mathsf{X}^n$ is an n-length zero-error block code for the compound channel $\mathcal{W}$ with uninformed encoder and decoder iff it is a symmetric clique for the digraph family associated with $\mathcal{W}$. □

COROLLARY 11.25

$$C_0^*(\mathcal{W}) = \max_P \min_{(V,W) \in \mathcal{W}^2} C(G_{(V,W)}, P),$$

where P runs over all the probability distributions on the common input alphabet X of the channels in $\mathcal{W}$. ○

An important application of Theorem 11.22 on the Sperner capacity of a family of digraphs is concerned with *qualitatively independent partitions.*

DEFINITION 11.26 Two partitions of a finite set are called *qualitatively independent* if any two classes belonging to different partitions have a non-empty intersection. A family of partitions of a finite set is called qualitatively 2-independent if any two classes belonging to two different partitions have a non-empty intersection. ○

In other words, a family of partitions of a set is called qualitatively 2-independent if any two of the partitions are qualitatively independent. Qualitative independence is closely related to stochastic independence. Two partitions of a set can be generated by two independent random variables iff they are qualitatively independent. We will determine the exponential asymptotics of the largest families of qualitatively 2-independent k-partitions of a ground set of n elements, for any fixed value of k. The result is an immediate consequence of the next special case of Theorem 11.22. We begin with a technical lemma.

LEMMA 11.27 Let G be a directed graph with finite vertex set X and a single directed edge $(a, b) \in \mathsf{X}^2$ with $a \neq b$. Let P be an arbitrary probability distribution on X. Then

$$C(G, P) = [\,P(a) + P(b)\,]\, h\left(\frac{P(a)}{P(a) + P(b)}\right),$$

where h is the binary entropy function. ○

Proof Consider a sequence of probability distributions P_n converging to P and such that $\mathsf{T}_{P_n} \neq \emptyset$. Let C_n be the subset of T_{P_n} consisting of all those sequences which have only a and b terms in their first $n[P_n(a) + P_n(b)]$ coordinates. We have, for every $\delta > 0$ and sufficiently large n,

$$\omega_s(G^n[P, \delta]) \geqq |\mathsf{C}_n| = \binom{n[P_n(a) + P_n(b)]}{nP_n(a)},$$

whence, by Lemma 2.3 and the continuity of the binary entropy function, we obtain

$$C(G, P) \geqq [\,P(a) + P(b)\,]h\left(\frac{P(a)}{P(a) + P(b)}\right).$$

For the upper bound we use the following simple argument.

For any type Q of sequences in X^n, let $\mathsf{D} \subset \mathsf{T}_Q$ be a symmetric clique in G^n. For any sequence $\mathbf{x} \in \mathsf{D}$, let $\mathsf{D}(\mathbf{x})$ be the set of those sequences from T_Q that have their a and b terms in the same position as $\mathbf{x}$. Then, for different elements $\mathbf{x}$ and $\mathbf{y}$ of D, the sets $\mathsf{D}(\mathbf{x})$ and $\mathsf{D}(\mathbf{y})$ are disjoint. The cardinality of any of these sets does not depend on the particular $\mathbf{x}$ we consider and hence, for every $\mathbf{x} \in \mathsf{D}$,

$$|\mathsf{D}(\mathbf{x})| = \frac{|\mathsf{T}_Q|}{\binom{n[Q(a)+Q(b)]}{nQ(a)}}.$$

Since all of these sets are contained in T_Q, we obtain

$$\sum_{\mathbf{x} \in \mathsf{D}} |\mathsf{D}(\mathbf{x})| \leqq |\mathsf{T}_Q|$$

and conclude by Lemma 2.3 that

$$|\mathsf{D}| \leqq \binom{n[Q(a) + Q(b)]}{nQ(a)} \leqq \exp\left\{n[\,Q(a) + Q(b)\,]h\left(\frac{Q(a)}{Q(a) + Q(b)}\right)\right\}.$$

Applying the final inequality to any sequence $\{P_n\}$ of distributions on X with $P_n \to P$ the upper bound follows. □

The key to the solution of our problem on qualitatively independent partitions is the following special case of Theorem 11.22 for families of single-edge digraphs. Let G be a simple graph on vertex set X and consider a family $\mathcal{G}(G)$ of single-edge digraphs where every member of the family has as its only edge a different edge of G with an arbitrary orientation. We will say that the digraph family $\mathcal{G}(G)$ is *generated* by the graph G. While this family is not uniquely determined by the graph G, according to Theorem 11.28 its particular choice does not affect its Sperner capacity.

THEOREM 11.28 Let $\mathcal{G}(G)$ be a family of single-edge digraphs generated by G. Then

$$C(\mathcal{G}(G)) = \max_{P} \min_{\{a,b\}\in E(G)} [\,P(a) + P(b)\,]h\left(\frac{P(a)}{P(a)+P(b)}\right). \qquad \bigcirc$$

Proof The assertion follows from Theorem 11.22 by Lemma 11.27. □

COROLLARY 11.28 Let K be the complete graph on k vertices. Then

$$C(\mathcal{G}(K)) = \frac{2}{k}. \qquad \bigcirc$$

Proof Let P be the uniform distribution on $[k]$. We see by Lemma 11.27 that

$$C(\mathcal{G}(K)) \geqq \frac{2}{k}.$$

It remains to prove that this lower bound is tight. For this, note that for any probability distribution P on $[k]$ the sum of its smallest two probabilities is at most $2/k$. Considering that the binary entropy is at most 1, we see that

$$\min_{\{a,b\}\in E(K)} [\,P(a) + P(b)\,]h\left(\frac{P(a)}{P(a)+P(b)}\right) \leqq \min_{\{a,b\}\in E(K)} [\,P(a) + P(b)\,] \leqq \frac{2}{k}.$$

□

THEOREM 11.29 (*Qualitative independence*) Let $M(n,k)$ be the largest cardinality of a set of pairwise qualitatively independent k-partitions of an n-set. Then, for every fixed integer $k \geqq 2$,

$$\lim_{n\to\infty} \frac{1}{n} \log M(n,k) = \frac{2}{k}. \qquad \bigcirc$$

Proof Without loss of generality we can suppose that our n-set is $[n]$. We represent a k-partition of $[n]$ by an n-length sequence of elements of $[k]$. A sequence $\mathbf{x} \in [k]^n$ defines the partition in which the element $i \in [n]$ is in the class of index $a \in [k]$ provided that the ith coordinate x_i of $\mathbf{x}$ is a. We then say that $\mathbf{x}$ is a characteristic sequence of the partition it defines. A partition has several characteristic sequences. It is, however, obvious that if two partitions are qualitatively independent, then their characteristic sequences are symmetrically adjacent for the graph family $\mathcal{G}(K)$. This implies that

$$M(n,k) \leqq \omega_s\left(\bigcap_{G\in\mathcal{G}(K)} G^n\right),$$

whence, using Corollary 11.28, the left-hand side of the claimed equality is not larger
→ 11.9 than the right-hand side.

For inequality in the other direction, consider a sequence $\mathsf{C}_n \subset [k]^n$ of symmetric cliques of maximum cardinality for the graph family $\mathcal{G}(K)$ and let us consider each of the k-ary sequences in C_n as characteristic sequences of k-partitions of $[n]$. These partitions are not necessarily qualitatively 2-independent, for classes having the same index in two different partitions are not guaranteed to intersect. To fix this problem, let us add to each of the sequences in C_n the prefix $1, 2, \dots, k$. The resulting new sequences of length $(n+k)$ represent 2-independent k-partitions of a ground set of cardinality $(n+k)$. This implies

$$\omega_s\left(\bigcap_{G\in\mathcal{G}(K)} G^n\right) \leqq M(n+k, k),$$

which yields the missing inequality. □

It is easy to see that several intriguing problems in extremal set theory can be solved with the same proof technique.

Discussion

Our primary concern in this book is with the asymptotic analysis of the performance of block codes for sources and channels, as block length goes to infinity, within the stochastic framework introduced by Shannon. A main feature of the latter is that a small (asymptotically vanishing) probability of erroneous performance is tolerated, hence usually it is not considered to be a central problem what happens if this probability is set to zero (which in most cases leads to a discontinuity in the solution). Still, the analysis of the zero-error problem does provide an important insight into the exponential decrease of error probability in the stochastic theory. In Chapter 12 we will see that the study of channel coding within the stochastic framework, for time-varying channels, may lead to a zero-error problem.

More importantly, since the 1990s information theory has been used to solve celebrated problems in extremal combinatorics, while at the same time has offered a unified framework in which the collection of diverse combinatorial problems can be treated together, so as to clarify their relationships and provide mathematical methods to deal with them. The analysis of the asymptotic behavior of invariants (clique number, chromatic number, etc.) of appropriately defined powers of combinatorial structures such as graphs, digraphs and hypergraphs or families thereof leads to concepts of digraph or hypergraph capacity, graph entropy and the like (for graph entropy, see Problems 9.11–9.13). The limits of such appropriately normalized invariants seem to furnish interesting intrinsically information-theoretic tools to obtain structural characterizations in combinatorics and beyond. ○

Problems

11.1. Show that if P is an arbitrary distribution on $\mathsf{V}(G)$ and $P_n \to P$, where P_n is a type of sequences from $[\mathsf{V}(G)]^n$ having the same support as P, then $(1/n)\log\omega(G^n[P_n, 0]) \to C(G, P)$. Generalize this statement to the Sperner capacity of any finite family of digraphs.

Hint It is enough to show that the limit of $(1/n)\log\omega(G^n[P_n, 0])$ does not depend on the sequence P_n considered. To this end, note that if P_n' and P_n''

are types of sequences in $[\mathsf{V}(G)]^n$ having the same support and their variation distance converges to zero, then sequences of type P'_n and length n can be transformed to sequences of type P''_n of length $n + n\epsilon_n$ by an appropriate suffix, whereby

$$\omega(G^n[P'_n, 0]) \leqq \omega(G^{n+n\epsilon_n}[P''_n, 0]).$$

11.2. Show by the method of random selection that for an r-uniform regular hypergraph

$$\tau(G) \leqq \frac{|V(G)|}{r} \ln |V(G)|.$$

(The bound in the edge cover lemma, given for its mathematical interest, is sharper than this one if $r < (1/e)|V(G)|$. This weaker bound admitting a very simple proof is, however, equally suitable for the purposes of this chapter, namely for the proof of the core theorem, Theorem 11.21.)

Hint Randomly select M hyperedges uniformly from all the M-tuples of not necessarily distinct hyperedges. Show that the expected number of the vertices not covered by the selected hyperedges is less than 1 if $M \geqq (|V(G)|/r) \ln |V(G)|$. Proceed as in the proof of Lemma 9.1, which is, in effect, an instance of this result.

11.3. (*LYM inequality*)

(a) Prove that if $\mathcal{F} \subset [n]$ is a Sperner family then the following Kraft-like inequality holds for the cardinality of its member sets:

$$\sum_{\mathsf{F} \in \mathcal{F}} \frac{1}{\binom{n}{|\mathsf{F}|}} \leqq 1.$$

(b) Deduce Sperner's theorem, and for n even conclude that the optimal solution is unique.

Hint To any set $\mathsf{F} \in \mathcal{F}$ associate those permutations of $[n]$ which put its elements in the first $|\mathsf{F}|$ positions. Show that the sets of permutations associated with the different members of a Sperner family are disjoint. (The name LYM refers to the authors Lubell (1966), Yamamoto (1954) and Meshalkin (1963). A more general inequality (see Problem 11.9) is due to Bollobás (1965).)

11.4. A *transitive tournament* is a digraph whose vertices can be ordered in such a way that (a, b) is an arc precisely when vertex a precedes b. For an arbitrary digraph D let $\omega_t(D)$ be the largest cardinality of the vertex set of a (not necessarily induced) subgraph of D which is a transitive tournament. Show that $\omega_t(D^n)$ has the same exponential asymptotics as $\omega_s(D^n)$.

Hint Show that the Sperner capacity of a transitive tournament G equals the Shannon capacity $\log |\mathsf{V}(G)|$ of the underlying complete graph. In order to prove this, number the vertices of G in an arbitrary manner, and associate with any sequence of vertices the sum of the values in its coordinates. Note that if two sequences yield the same sum, then they are symmetrically adjacent in the corresponding power graph. Observing that the sums for n-length sequences can take less than n^2 different values, conclude that there is a set of cardinality $|\mathsf{V}(G)|^n/n^2$ with all its elements yielding the same value. This gives a sufficiently large symmetric clique for every n. (Calderbank *et al.* (1993).)

11.5. (*Cyclic triangle*) The complete graph on three vertices is often called a triangle. It gives rise to two non-isomorphic oriented graphs: the transitively oriented triangle and the cyclic triangle. (A digraph is called *oriented* if there is at most one arc between every pair of vertices.)

(a) Show that the transitively oriented triangle has Sperner capacity log 3.

Hint See Problem 11.4.

(b) Show that the cyclically oriented triangle C has Sperner capacity 1.

Hint Let $\{0, 1, 2\}$ be the vertex set of the triangle and let the arcs point from i to $i+1$ modulo 3. Associate to sequences $\mathbf{x} = x_1x_2 \ldots x_n \in [\mathrm{GF}(3)]^n$ the polynomials $P_{\mathbf{x}}(z_1, z_2, \ldots z_n) \triangleq \prod_{i=1}^n [z_i - (x_i + 1)]$ over GF(3). If S is a symmetric clique in C^n, for $\mathbf{x}$ and $\mathbf{z}$ in S we have $P_{\mathbf{x}}(\mathbf{z}) = 0$ if they are different, while $P_{\mathbf{x}}(\mathbf{x}) = (-1)^n$. Conclude that the polynomials assigned to the sequences in S are linearly independent. Since these polynomials are linear in each of their variables, the dimension of the vector space they generate over GF(3) is at most 2^n, and by linear independence this is also an upper bound for the cardinality of S. (Calderbank *et al.* (1993). The present proof is due to Blokhuis (1993).)

11.6.* (*Local chromatic number*) A coloring of a digraph is a coloring of the underlying undirected graph; see Problem 6.24. In a digraph, the outneighborhood of a vertex consists of the vertex itself and of all the endpoints of arcs pointing away from it. The largest number of colors assigned by a coloring to the outneighborhood of any vertex is a characteristic of the coloring. Its minimum over all colorings is called the local chromatic number of the digraph G and is denoted by $\Psi(G)$.

(a) Show that $C(G) \leqq \log \Psi(G)$. (Körner, Pilotto, Simonyi (2005).)

(b) The local chromatic number of an undirected graph is the local chromatic number of the corresponding symmetrically directed digraph. Show that it can be significantly smaller than the chromatic number of the same graph. (Erdős *et al.* (1986).)

11.7. (a) A cycle is a graph whose vertices are ordered, all the vertices are adjacent to their neighbors in the ordering, and the last one is also adjacent to the first. A cycle is odd if it has an odd number of vertices. Show that for all its orientations, with the possible exception of the quasi-alternating orientation, an oriented odd cycle has Sperner capacity 1. An orientation is quasi-alternating if all but one of its vertices has out-degree 0 or 2. (The out-degree of a vertex is the number of arcs pointing away from it.)

Hint Use the upper bound from Problem 11.5. (Körner, Pilotto and Simonyi (2005).)

11.8. (a)* Show that for every self-complementary graph there is an orientation of its edges such that an isomorphism of the graph and its complement can be extended to an isomorphism of the resulting digraphs for which the union of the two digraphs is the transitively oriented complete graph.

(b) Show that the quasi-alternating orientation gives the pentagon graph a Sperner capacity equal to the Shannon capacity of the underlying undirected graph. (Galluccio *et al.* (1994).)

Hint Note that for the pentagon the quasi-alternating orientation satisfies the property required in (a). (Sali and Simonyi (1999).)

11.9. (*Bollobás pairs*)

(a) Let A_i and B_i, $i \in [k]$, be finite subsets of a ground set X with the property that $\mathsf{A}_i \cap \mathsf{B}_j = \emptyset$ iff $i = j$. Such a family of set pairs is sometimes referred to as Bollobás pairs. Show that

$$\sum_{i=1}^{k} \binom{|\mathsf{A}_i| + |\mathsf{B}_i|}{|\mathsf{A}_i|}^{-1} \leqq 1.$$

Hint The inequality generalizes the one in Problem 11.3. and the proofs are in the same relation. Let us associate with every couple of sets $(\mathsf{A}_i, \mathsf{B}_i)$ those permutations of the elements of the ground set X which place all the elements of A_i before any element of B_i and note that every permutation of X will be associated with at most one couple of sets from the above family. (Bollobás (1965).)

(b) Use this inequality to deduce the upper bound on the Sperner capacity of a family of single-edge graphs.

11.10. Let the set pairs $(\mathsf{A}_i, \mathsf{B}_i)$, $i \in [t]$, form a family of Bollobás pairs, while at the same time let the union sets $\mathsf{A}_i \cup \mathsf{B}_i$ form a Sperner family. Let $t(n)$ be the maximum number of pairs in such a set family, in the case $\mathsf{A}_i \cup \mathsf{B}_i \subset [n]$ for every i. Show that

$$\lim_{n\to\infty} \frac{1}{n} \log t(n) = q,$$

where q is the unique root of the equation $h(q) = q$. (Gargano, Körner and Vaccaro (1994).)

11.11. (*Different capacities of a compound channel*)

(a) Show that there exist compound channels for which the zero-error capacity with an uninformed encoder–decoder pair is strictly less than the zero-error capacity with informed decoder (and uninformed encoder).

Hint Consider the compound channel $\mathcal{W} = \{A, B\}$ with input alphabet $\{1, 2\}$ and output alphabet $\{1, 2, 3, 4\}$ for which $A(j|1) > 0$ if $j \in \{1, 2\}$ and zero otherwise, while $A(j|2) > 0$ if $j \in \{3, 4\}$ and zero otherwise. In B we have $B(j|1) > 0$ if j is odd and zero otherwise, while $B(j|2) > 0$ if j is even and zero otherwise. Show that the zero-error capacity of this compound channel is zero when the decoder is uninformed, and unity when the decoder is informed.

(b) Define the zero-error capacities of a compound DMC with informed encoder and informed or uninformed decoder. Clearly, the first one equals the minimum of the zero-error capacities of the individual channels. Show that the second one is the same, if positive, and zero iff the zero-error capacity with uninformed encoder and decoder is zero. (Nayak and Rose (2005).)

11.12. (*Robust capacity*) Given a simple graph G we call a subset $\mathsf{C} \subset [\mathsf{V}(G)]^n$ a robust clique in G^n if for any pair of sequences $\mathbf{x}$ and $\mathbf{y}$ from C there is an edge $\{a, b\} \in E(G)$ and coordinates $i \in [n]$, $j \in [n]$ such that

$$(x_i, y_i) = (y_j, x_j) = (a, b).$$

Let $NN(G, n)$ denote the largest cardinality of a robust clique in G^n. We call the always existing limit

$$\lim_{n\to\infty} \frac{1}{n} \log NN(G, n)$$

the robust capacity of G. Show that it equals the Sperner capacity of the family of digraphs consisting of all the digraphs obtained from G by orienting its edges in an arbitrary manner. (Robust capacity was defined by Körner and Simonyi (1992). The formula hinted at is from Calderbank *et al.* (1993).)

11.13. Show that $C(\mathcal{G}(G)) = 2/k$ iff the graph G contains a spanning 2-matching. A graph is a 2-matching if all its connected components are either odd cycles or single edges. It is spanning in G if it contains all its vertices. (Greco (1998).)

11.14. (a) We call two permutations π and σ of the elements of $[n]$ colliding if they map some element $i \in [n]$ into consecutive integers. In other words, we have $|\pi(i) - \sigma(i)| = 1$. Let $T(n)$ be the maximum cardinality of a set of pairwise colliding permutations of $[n]$. Show that $T(n) \leqq \binom{n}{\lceil \frac{n}{2} \rceil}$. This upper bound is conjectured to be tight. Verify the conjecture for $n \leqq 7$. (Körner and Malvenuto (2006).)

(b) Show that

$$\lim_{n\to\infty} \sqrt[n]{T(n)} \geqq 1.815.$$

(Brightwell *et al.* (2010).)

11.15.* The permutations π and σ of the elements of $[n]$ are called very different if they map some element $i \in [n]$ into different and non-consecutive integers. In other words, we have $|\pi(i) - \sigma(i)| > 1$. Let $U(n)$ be the maximum cardinality of a set of pairwise very different permutations of $[n]$. Show that $U(n) = \frac{n!}{2^{\lfloor \frac{n}{2} \rfloor}}$ for every n. (Körner, Simonyi and Sinaimeri (2009).)

11.16. (*Chromatic number of product graphs*) A fractional coloring of a (simple) graph G is an assignment of non-negative numbers to the stable sets (edgeless subsets of the vertex set) of G in such a way that the sum of the values assigned to the stable sets containing any specific vertex is at least 1. The *fractional chromatic number* $\chi^*(G)$ is the minimum over all fractional colorings of the sum of values assigned to all the stable sets.

Show that $\lim_{k\to\infty} \sqrt[k]{\chi(G^k)} = \chi^*(G)$ for every simple graph G. (Berge and Simonovits (1974). Compare this result with the formula in

Problem 9.11(c) for the same limit. See also Problem 9.7. Further connections with the error exponent in channel coding are hinted at in Problems 10.4 and 10.28(d).)

Story of the results

The study of the zero-error capacity of the compound channel was initiated by Cohen, Körner and Simonyi (1990). They proved Proposition 11.7. The concept of capacity within a fixed distribution goes back to Csiszár and Körner (1981b). The auxiliary Lemma 11.2 is due to Fekete (1923). The hypergraph covering lemma was proved by Lovász (1975). Sperner's theorem is from Sperner (1928). Sperner capacity was introduced by Gargano, Körner and Vaccaro (1992) in continuation of earlier work by Körner and Simonyi (1992). The latter authors introduced the concept of robust capacity (Problem 11.10) to deal with Rényi's problem on qualitatively independent partitions. The twin partition lemma and the core theorem giving the formula for the Sperner capacity of a family of directed graphs were proved by Gargano, Körner and Vaccaro (1994).

The capacity formula, Theorem 11.4, for the compound channel with uninformed encoder and decoder was found by Nayak and Rose (2005). The capacity of graph families consisting of single-edge graphs, Theorem 11.28, is due to Gargano, Körner and Vaccaro (1993), who used the result to solve Rényi's problem on qualitatively 2-independent k-partitions of an n-set, the present qualitative independence theorem, Theorem 11.29.

The latter problem has a long history. It is implicit in Alfréd Rényi's posthumous book (Rényi (1970)). There, while qualitatively independent k-partitions are mentioned, their maximum number is addressed only for $k = 2$, in which case the exact number for a given n is of interest, its asymptotics as $n \to \infty$ being trivial. For the early history of this problem and non-asymptotic results for $k = 2$ we refer to Katona (2004). For $k \geqq 3$ some lower and upper bounds were proved by members of the Prague school of combinatorics; see, e.g., Poljak and Rödl (1980) and Poljak, Pultr and Rödl (1983). The upper bound in Theorem 11.29 was obtained by Poljak and Tuza (1989). For $k \geqq 3$ no asymptotically tight lower bound was known before Gargano, Körner and Vaccaro (1993).

Many more problems in extremal combinatorics are rooted in information theory; see Körner and Orlitsky (1998).

12 Arbitrarily varying channels

So far we have analyzed the performance of codes for a DMC resp. for a family of DMCs (compound DMC). In this chapter we are interested in codes performing well for families of channels which do not exclusively consist of DMCs.

The compound DMC is a model for communication devices of which the unknown parameters are constant during transmission of a codeword. Another model of no less practical and certainly of greater mathematical interest describes the case when the parameters of the channel may vary from letter to letter, in an arbitrary unknown manner. While in source coding the arbitrarily varying source was well within the scope of the methods of Chapter 9, arbitrarily varying channels cannot be treated by the analogous approach of Chapter 10.

Let $\mathcal{W}$ be a not necessarily finite family of channels $W : \mathsf{X} \to \mathsf{Y}$. This family is considered as a model of a single (engineering device) channel which has several possible states. The entries of the matrices $W \in \mathcal{W}$ will often be written as $W(y|x, s)$, where the *state* $s \in \mathsf{S}$ is an index identifying the particular $W \in \mathcal{W}$. For n-length sequences, the transition probabilities corresponding to a sequence of states $\mathbf{s} = s_1 \dots s_n$ are

$$W^n(\mathbf{y}|\mathbf{x}, \mathbf{s}) \triangleq \prod_{i=1}^{n} W(y_i|x_i, s_i). \tag{12.1}$$

The family of channels $W^n(\cdot|\cdot, \mathbf{s}) : \mathsf{X}^n \to \mathsf{Y}^n,\ \mathbf{s} \in \mathsf{S}^n$ will be denoted by $\mathcal{W}^n$.

DEFINITION 12.1 An *arbitrarily varying channel* (AVC) with input alphabet X and output alphabet Y is a sequence $\{\mathcal{W}^n\}_{n=1}^{\infty}$ as above. It will be denoted by $\{\mathcal{W} : \mathsf{X} \to \mathsf{Y}\}$ or simply $\{\mathcal{W}\}$. ○

When defining the capacity of a DMC (or of a compound DMC), it made no difference whether the code performance was evaluated in terms of maximum or average probability of error. It will turn out that for an AVC the two approaches may lead to different results. → 10.12

DEFINITION 12.2 The ε-*capacity* of an AVC $\{\mathcal{W} : \mathsf{X} \to \mathsf{Y}\}$ for maximum resp. average probability of error is the largest R such that for every $\delta > 0$ and $n \geqq n_0(\mathcal{W}, \delta)$ there exists an n-length block code (f, φ) having rate

$$\frac{1}{n} \log |\mathsf{M}_f| > R - \delta$$

and satisfying

$$e(\mathcal{W}^n, f, \varphi) \leqq \varepsilon \quad \text{resp.} \quad \bar{e}(\mathcal{W}^n, f, \varphi) \leqq \varepsilon;$$

that is, the code (f, φ) has maximum resp. average probability of error at most ε for every channel in the family $\mathcal{W}^n$ (cf. Definition 10.9).

The *m-capacity* $C_m = C_m(\mathcal{W})$ resp. *a-capacity* $C_a = C_a(\mathcal{W})$ is the infimum of the corresponding ε-capacities for $\varepsilon > 0$. ○

We designate by $\overline{\mathcal{W}}$ resp. $\overline{\overline{\mathcal{W}}}$ the closure of the set of matrices $V : \mathsf{X} \to \mathsf{Y}$ with entries of form

$$V(y|x) = \sum_{s \in \tilde{\mathsf{S}}} P(s) W(y|x, s) \tag{12.2}$$

resp.

$$V(y|x) = \sum_{s \in \tilde{\mathsf{S}}} P(s|x) W(y|x, s), \tag{12.3}$$

where $\tilde{\mathsf{S}}$ is any finite subset of S and $P(\cdot)$ resp. $P(\cdot|\cdot)$ is any distribution resp. stochastic matrix; $\overline{\mathcal{W}}$ is called the *convex closure* and $\overline{\overline{\mathcal{W}}}$ the *row-convex closure* of $\mathcal{W}$. Instead of $(\overline{\mathcal{W}})^n$ resp. $(\overline{\overline{\mathcal{W}}})^n$ we shall write $\overline{\mathcal{W}}^n$ resp. $\overline{\overline{\mathcal{W}}}^n$.

LEMMA 12.3 For every n-length block code (f, φ)

$$e(\mathcal{W}^n, f, \varphi) = e(\overline{\overline{\mathcal{W}}}^n, f, \varphi), \quad \bar{e}(\mathcal{W}^n, f, \varphi) = \bar{e}(\overline{\mathcal{W}}^n, f, \varphi). \qquad ○$$

COROLLARY 12.3 The ε-capacity of the AVC $\{\mathcal{W} : \mathsf{X} \to \mathsf{Y}\}$ for maximum resp. average probability of error equals, for every $0 < \varepsilon < 1$, the corresponding ε-capacity of the AVC $\{\overline{\overline{\mathcal{W}}}\}$ resp. $\{\overline{\mathcal{W}}\}$. It does not exceed the minimum of $C(W)$ as W ranges over $\overline{\overline{\mathcal{W}}}$ resp. $\overline{\mathcal{W}}$, where $C(W)$ is the capacity of the DMC $\{W\}$. ○

Proof For any matrices $V_1, \dots, V_n$ of form (12.3) consider the channel $\tilde{V} : \mathsf{X}^n \to \mathsf{Y}^n$ with transition probabilities

$$\tilde{V}(\mathbf{y}|\mathbf{x}) \triangleq \prod_{i=1}^{n} V_i(y_i|x_i) = \prod_{i=1}^{n} \sum_{s_i \in \tilde{\mathsf{S}}} P_i(s_i|x_i) W(y_i|x_i, s_i) = \sum_{\mathbf{s} \in \tilde{\mathsf{S}}^n} \tilde{P}(\mathbf{s}|\mathbf{x}) W^n(\mathbf{y}|\mathbf{x}, \mathbf{s}),$$

where $\tilde{P}(\mathbf{s}|\mathbf{x}) \triangleq \prod_{i=1}^{n} P_i(s_i|x_i)$. Then for every code (f, φ) and every $m \in \mathsf{M}_f$

$$\tilde{V}(\varphi^{-1}(m)|f(m)) = \sum_{\mathbf{s} \in \tilde{\mathsf{S}}^n} \tilde{P}(\mathbf{s}|f(m)) W^n(\varphi^{-1}(m)|f(m), \mathbf{s}), \tag{12.4}$$

implying

$$e(\tilde{V}, f, \varphi) \leqq \max_{\mathbf{s} \in \mathsf{S}^n} e(W^n(\cdot|\cdot, \mathbf{s}), f, \varphi) = e(\mathcal{W}^n, f, \varphi).$$

As $\overline{\overline{\mathcal{W}}}^n$ is the closure of the set of channels $\tilde{V}$ as above, the first assertion follows. If $V_1, \ldots, V_n$ are of form (12.2), then (12.4) holds with $\tilde{P}(\mathbf{s}|f(m))$ replaced by $\tilde{P}(\mathbf{s}) \triangleq \prod_{i=1}^n P_i(s_i)$. Averaging over $m \in \mathsf{M}_f$ we obtain

$$\bar{e}(\tilde{V}, f, \varphi) = \sum_{\mathbf{s} \in \tilde{\mathsf{S}}^n} \tilde{P}(\mathbf{s}) \bar{e}(W^n(\cdot|\cdot, \mathbf{s}), f, \varphi) \leqq \bar{e}(\mathcal{W}^n, f, \varphi).$$

This proves the second assertion.

The corollary is immediate, using the fact that the ε-capacity of a DMC (no matter whether for maximum or average probability of error) equals its capacity. As $C(W)$ is continuous due to the uniform joint continuity of $I(P, W)$, it attains its minimum on the compact sets $\overline{\mathcal{W}}$ resp. $\overline{\overline{\mathcal{W}}}$. □

→ 12.1
→ 12.2

REMARKS Recall the notations

$$I(P, \mathcal{W}) \triangleq \inf_{W \in \mathcal{W}} I(P, W), \qquad C(\mathcal{W}) \triangleq \max_P I(P, \mathcal{W}),$$

introduced in Chapter 10 for any family $\mathcal{W}$ of stochastic matrices. If the set $\mathcal{W}$ is convex and compact, we have

$$\min_{W \in \mathcal{W}} \max_P I(P, W) = \max_P \min_{W \in \mathcal{W}} I(P, W),$$

by the minimax theorem (see, e.g., Karlin (1959), Th. 1.1.5), because the mutual information $I(P, W)$ is concave in P and convex in W. In particular, this identity always holds for $\overline{\mathcal{W}}$ or $\overline{\overline{\mathcal{W}}}$ in the role of $\mathcal{W}$. The resulting identities

$$\min_{W \in \overline{\mathcal{W}}} C(W) = C(\overline{\mathcal{W}}), \qquad \min_{W \in \overline{\overline{\mathcal{W}}}} C(W) = C(\overline{\overline{\mathcal{W}}}),$$

will be used freely in the following. ○

First we consider the problem of determining the m-capacity of an AVC. We shall deal with the case $|\mathsf{Y}| = 2$; if the size of the output alphabet is larger than 2, the problem is unsolved in general.

→ 12.3

Temporarily, we restrict our attention to AVCs with $\mathsf{X} = \mathsf{Y} = \{0, 1\}$. Channels $W : \{0, 1\} \to \{0, 1\}$ are called *binary channels*. The key observation will be that for any family $\mathcal{W}$ of binary channels, $\overline{\overline{\mathcal{W}}}$ contains a "worst" channel W in the sense that "good" codes for W^n have even smaller error probabilities for any other channel in $\overline{\overline{\mathcal{W}}}^n$.

For binary channels, given an encoder $f : \{1, \ldots, M\} \to \{0, 1\}^n$, the corresponding *standard minimum distance* (*SMD*) *decoder* is the mapping $\varphi : \{0, 1\}^n \to \{1, \ldots, M\}$ defined by the property that $\varphi(\mathbf{y})$ is the smallest m for which $d_{\mathrm{H}}(\mathbf{y}, f(m))$ achieves its minimum over $m \in \{1, \ldots, M\}$.

LEMMA 12.4 Let $\mathcal{W}$ be any set of binary channels and let $\widetilde{W}$ be any binary channel such that

$$W(0|0) \geqq \widetilde{W}(0|0), \quad W(1|1) \geqq \widetilde{W}(1|1), \quad \text{for every} \quad W \in \mathcal{W}. \tag{12.5}$$

Then for every encoder $f : \{1, \ldots, M\} \to \{0, 1\}^n$ and the corresponding SMD decoder, the error probabilities e_m, $m \in \{1, \ldots, M\}$ over any channel in $\overline{\overline{\mathcal{W}}}^n$ are upper bounded by those over $\widetilde{W}^n$. ○

Proof For notational convenience, index $\widetilde{W}$ and the matrices in $\mathcal{W}$ by the elements of a set S and denote the entries of the matrix indexed by $s \in \mathsf{S}$ as $W(\cdot|\cdot, s)$. We claim that given any code (f, φ), where φ is an SMD decoder, the error probabilities e_m are maximized over the channels $W^n(\cdot|\cdot, \mathbf{s})$, $\mathbf{s} \in \mathsf{S}^n$ for $\mathbf{s} = \tilde{s} \ldots \tilde{s}$, where $\tilde{s}$ is the index of $\widetilde{W}$.

It suffices to show that for every $\mathbf{s} = s_1 \ldots s_n \in \mathsf{S}^n$, changing an $s_i \neq \tilde{s}$ (if any) to $s_i' = \tilde{s}$, the resulting sequence $\mathbf{s}' = s_1 \ldots s_{i-1}\tilde{s}s_{i+1} \ldots s_n$ satisfies

$$W^n(\varphi^{-1}(m)|f(m), \mathbf{s}) \geqq W^n(\varphi^{-1}(m)|f(m), \mathbf{s}'). \tag{12.6}$$

Fixing $1 \leqq i \leqq n$, we shall denote by $\hat{\mathbf{x}}, \hat{\mathbf{y}}, \hat{\mathbf{s}}$ the sequences of length $n-1$ obtained from the n-length sequences $\mathbf{x}, \mathbf{y}, \mathbf{s}$ by deleting their ith component. Fix some m and write

$$\mathsf{B}_j \triangleq \{\hat{\mathbf{y}} : \mathbf{y} \in \varphi^{-1}(m), y_i = j\}, \quad j = 0, 1.$$

As φ is an SMD decoder, we have $\mathsf{B}_1 \subset \mathsf{B}_0$ or $\mathsf{B}_0 \subset \mathsf{B}_1$ according as to whether the ith component of $\mathbf{x} \triangleq f(m)$ is $x_i = 0$ or $x_i = 1$. Thus if $x_i = 0$ then

$$\begin{aligned} W^n(\varphi^{-1}(m)|f(m), \mathbf{s}) &= W^{n-1}(\mathsf{B}_0|\hat{\mathbf{x}}, \hat{\mathbf{s}})W(0|0, s_i) + W^{n-1}(\mathsf{B}_1|\hat{\mathbf{x}}, \hat{\mathbf{s}})W(1|0, s_i) \\ &= W^{n-1}(\mathsf{B}_1|\hat{\mathbf{x}}, \hat{\mathbf{s}}) + W^{n-1}(\mathsf{B}_0 - \mathsf{B}_1|\hat{\mathbf{x}}, \hat{\mathbf{s}})W(0|0, s_i), \end{aligned}$$

and similarly, if $x_i = 1$, then

$$W^n(\varphi^{-1}(m)|f(m), \mathbf{s}) = W^{n-1}(\mathsf{B}_0|\hat{\mathbf{x}}, \hat{\mathbf{s}}) + W^{n-1}(\mathsf{B}_1 - \mathsf{B}_0|\hat{\mathbf{x}}, \hat{\mathbf{s}})W(1|1, s_i).$$

Replacing $\mathbf{s}$ by $\mathbf{s}'$, the factor $W(0|0, s_i)$ resp. $W(1|1, s_i)$ will be replaced by $W(0|0, \tilde{s})$ resp. $W(1|1, \tilde{s})$. On account of assumption (12.5), this proves (12.6). □

LEMMA 12.5 Let W be any binary channel with $W(0|0) + W(1|1) > 1$. Then the capacity of the DMC $\{W\}$ can be attained by codes with standard minimum distance decoders, i.e., for every $\varepsilon \in (0, 1)$, $\delta > 0$ and sufficiently large n there exist (n, ε)-codes with SMD decoder, having rate greater than $C(W) - \delta$. ○

Proof The assumption on W implies that $W(0|0) > W(0|1)$ and $W(1|1) > W(1|0)$. It follows that if all codewords $f(m)$, $1 \leqq m \leqq M$, have the same type and $\mathbf{y} \in \{0, 1\}^n$ is fixed then $W^n(\mathbf{y}|f(m))$ is a decreasing function of $d_{\mathrm{H}}(\mathbf{y}, f(m))$, so that the SMD decoder φ is a maximum likelihood decoder, i.e.,

$$W^n(\mathbf{y}|f(\varphi(\mathbf{y}))) = \max_{1 \leqq m \leqq M} W^n(\mathbf{y}|f(m)) \quad \text{for every} \quad \mathbf{y} \in \{0, 1\}^n. \tag{12.7}$$

Now let (f, φ') be any constant composition code of block length n with message set $\mathsf{M}_f \triangleq \{1, \ldots, 2M\}$,

$$M \geqq \exp\{n(C(W) - \delta)\},$$

and with average probability of error less than $\varepsilon/2$. Replacing φ' by the SMD decoder φ corresponding to f does not increase the average probability of error, because of (12.7). Restricting the domain of f to the M messages in M_f with smallest probability of error, the new f and the corresponding SMD decoder provide a code as required. □

THEOREM 12.6 The m-capacity of an AVC $\{\mathcal{W} : \{0, 1\} \to \{0, 1\}\}$ is

$$C_{\mathrm{m}}(\mathcal{W}) = C(\overline{\overline{\mathcal{W}}}). \tag{12.8}$$

Moreover, the ε-capacity for maximum probability of error equals $C_{\mathrm{m}}(\mathcal{W})$ for every $0 < \varepsilon < 1$. ○

Proof On account of Corollary 12.3, it suffices to show that

$$C_{\mathrm{m}}(\mathcal{W}) \geqq C(\widetilde{W}) \tag{12.9}$$

for some $\widetilde{W} \in \overline{\overline{\mathcal{W}}}$.

Exchanging the rows of the matrices in $\mathcal{W}$ if necessary, we may assume that $\max_{W \in \overline{\overline{\mathcal{W}}}}(W(0|0) + W(1|1)) \geqq 1$. Then if $\min_{W \in \overline{\overline{\mathcal{W}}}}(W(0|0) + W(1|1)) \leqq 1$, it follows by convexity that $\widetilde{W}(0|0) + \widetilde{W}(1|1) = 1$ for some $\widetilde{W} \in \overline{\overline{\mathcal{W}}}$, so that $C(\widetilde{W}) = 0$ and we have nothing to prove. In the opposite case the matrix $\widetilde{W} \in \overline{\overline{\mathcal{W}}}$ with

$$\widetilde{W}(0|0) = \inf_{W \in \mathcal{W}} W(0|0), \quad \widetilde{W}(1|1) = \inf_{W \in \mathcal{W}} W(1|1)$$

satisfies the condition of Lemma 12.5. Thus Lemmas 12.4 and 12.5 imply (12.9). □

THEOREM 12.7 The assertion of Theorem 12.6 holds for any AVC with $|\mathsf{Y}| = 2$. ○

Proof Write $\mathsf{Y} = \{0, 1\}$ and

$$a(x) \triangleq \inf_{W \in \mathcal{W}} W(0|x), \quad b(x) \triangleq \sup_{W \in \mathcal{W}} W(0|x) \quad (x \in \mathsf{X}).$$

As W ranges over $\overline{\overline{\mathcal{W}}}$, the column vector $\{W(0\,|x) : x \in \mathsf{X}\}$ ranges over the Cartesian product of the closed intervals $[a(x), b(x)]$, $x \in \mathsf{X}$. If no pair of these intervals is disjoint then they all have a common point w, say. In this case the matrix with identical rows $(w, 1 - w)$ belongs to $\overline{\overline{\mathcal{W}}}$, and

$$C_{\mathrm{m}}(\mathcal{W}) = C(\overline{\overline{\mathcal{W}}}) = 0.$$

In the opposite case, if x_0 and x_1 denote elements of X with largest $a(x)$ resp. smallest $b(x)$, we have $a(x_0) > b(x_1)$. By the proof of Theorem 12.6, the AVC with input alphabet $\{x_0, x_1\}$ determined by the corresponding rows of the matrices $W \in \mathcal{W}$ has m-capacity equal to $C(\widetilde{W})$, where

$$\widetilde{W} \triangleq \begin{pmatrix} a(x_0) & 1 - a(x_0) \\ b(x_1) & 1 - b(x_1) \end{pmatrix}.$$

This means that $C_{\mathrm{m}}(\mathcal{W}) \geqq C(\widetilde{W})$. Thus, by Corollary 12.3 we shall be done if we find a matrix $W \in \overline{\overline{\mathcal{W}}}$ with $C(W) \leqq C(\widetilde{W})$. By the choice of x_0 and x_1, every interval

$[a(x), b(x)]$, $x \in \mathsf{X}$, intersects $[b(x_1), a(x_0)]$. Hence there exists a $W \in \overline{\overline{\mathcal{W}}}$ such that $W(0|x) \in [b(x_1), a(x_0)]$ for every $x \in \mathsf{X}$. Each row of such a W is a convex combination of the rows of $\widetilde{W}$, with weights $\alpha(x)$, $1 - \alpha(x)$, say. It follows by the convexity lemma, Lemma 3.5 that if P is any distribution on X and $\tilde{P} = (\tilde{P}(x_0), \tilde{P}(x_1))$ is defined by

$$\tilde{P}(x_0) \triangleq \sum_{x \in \mathsf{X}} \alpha(x) P(x), \qquad \tilde{P}(x_1) \triangleq 1 - \tilde{P}(x_0)$$

→ 12.4 then we have $I(P, W) \leqq I(\tilde{P}, \widetilde{W})$. This proves $C(W) \leqq C(\widetilde{W})$. □

Next we turn to the problem of determining the a-capacity of an AVC with arbitrary (finite) alphabets. We shall use the method of random code selection in a more sophisticated manner than earlier.

Given finite sets X, Y and $\mathsf{M} \subset \mathsf{M}'$, let $\mathcal{C}(\mathsf{M} \to \mathsf{X}, \mathsf{Y} \to \mathsf{M}')$ denote the family of pairs of mappings (f, φ), $f : \mathsf{M} \to \mathsf{X}$, $\varphi : \mathsf{Y} \to \mathsf{M}'$. A *random code* is a RV taking values in such a family. If this RV has distribution Q, with some abuse of terminology we shall speak of the random code Q. For any channel $W : \mathsf{X} \to \mathsf{Y}$ a random code defines a new channel $T : \mathsf{M} \to \mathsf{M}'$ by

$$T(m'|m) \triangleq \sum_{(f,\varphi)} Q(f, \varphi) W(\varphi^{-1}(m')|f(m)). \tag{12.10}$$

Note that the channel (12.10) describes a realizable transmission scheme only in those cases when the result of the random experiment selecting (f, φ) can be observed at both ends of the (engineering device) channel described by the matrix W. Although this seldom occurs, random codes often provide a useful proof technique.

→ 12.6 The error probabilities of random codes are defined analogously to those of an ordinary code, replacing (6.1) by (12.10). In particular, we shall write

$$\begin{aligned} e_m(W, Q) &\triangleq 1 - T(m|m), \qquad e(W, Q) \triangleq \max_{m \in \mathsf{M}} e_m(W, Q), \\ \bar{e}(W, Q) &\triangleq \frac{1}{|\mathsf{M}|} \sum_{m \in \mathsf{M}} e_m(W, Q). \end{aligned} \tag{12.11}$$

Similarly, for a family of channels $\mathcal{W}$ we write

$$\begin{aligned} e(\mathcal{W}, Q) &\triangleq \max_{W \in \mathcal{W}} e(W, Q) = \max_{m \in \mathsf{M}, W \in \mathcal{W}} e_m(W, Q), \\ \bar{e}(\mathcal{W}, Q) &\triangleq \max_{W \in \mathcal{W}} \bar{e}(W, Q). \end{aligned} \tag{12.12}$$

Clearly, for a single channel any random code has a "value" (an ordinary code) which has average probability of error not larger than that of the random code itself. For families of channels the situation is more complex. The hard core of our determining the a-capacity of an AVC will be just the construction of "good" ordinary codes from "good" random ones. For this purpose we shall use the following random code reduction lemma.

LEMMA 12.8 (*Random code reduction*) Let $\mathcal{W}$ be a finite set of channels $W : \mathsf{X} \to \mathsf{Y}$ and let Q be a probability distribution on $\mathcal{C}(\mathsf{M} \to \mathsf{X}, \mathsf{Y} \to \mathsf{M}')$. Then for any ε and K satisfying

$$\varepsilon > 2\log(1 + e(\mathcal{W}, Q)), \quad K > \frac{2}{\varepsilon}(\log|\mathsf{M}| + \log|\mathcal{W}|), \tag{12.13}$$

there exist K codes $(f_i, \varphi_i) \in \mathcal{C}(\mathsf{M} \to \mathsf{X}, \mathsf{Y} \to \mathsf{M}')$ such that

$$\frac{1}{K}\sum_{i=1}^{K} e_m(W, f_i, \varphi_i) < \varepsilon \quad \text{for every} \quad m \in \mathsf{M}, \quad W \in \mathcal{W}. \tag{12.14}$$

○

REMARK The assertion means that to every random code there exists another one, with uniform distribution over K codes and with maximum probability of error less than ε, provided that (12.13) holds. ○

Proof Consider K repetitions of the random experiment of code selection with distribution Q, i.e., let $\{(F_i, \Phi_i)\}_{i=1}^{K}$ be independent RVs with values in $\mathcal{C}(\mathsf{M} \to \mathsf{X}, \mathsf{Y} \to \mathsf{M}')$ and common distribution Q. Then for every $m \in \mathsf{M}$ and channel $W \in \mathcal{W}$,

$$\Pr\left\{\frac{1}{K}\sum_{i=1}^{K} e_m(W, F_i, \Phi_i) \geqq \varepsilon\right\} = \Pr\left\{\exp\sum_{i=1}^{K} e_m(W, F_i, \Phi_i) \geqq \exp(K\varepsilon)\right\}$$
$$\leqq \exp(-K\varepsilon) E \exp\sum_{i=1}^{K} e_m(W, F_i, \Phi_i),$$

by Markov's inequality.

As the RVs $e_m(W, F_i, \Phi_i), i = 1, \dots, K$, are independent and identically distributed, the obvious inequality $\exp t \leqq 1 + t$ $(0 \leqq t \leqq 1)$ gives for the last expectation the upper bound

$$[E \exp e_m(W, F_1, \Phi_1)]^K \leqq [1 + E e_m(W, F_1, \Phi_1)]^K$$
$$= [1 + e_m(W, Q)]^K \leqq [1 + e(\mathcal{W}, Q)]^K.$$

Hence

$$\Pr\left\{\frac{1}{K}\sum_{i=1}^{K} e_m(W, F_i, \Phi_i) \geqq \varepsilon\right\} \leqq \exp\{-K[\varepsilon - \log(1 + e(\mathcal{W}, Q))]\}. \tag{12.15}$$

Thus, on account of condition (12.13), we have

$$\Pr\left\{\frac{1}{K}\sum_{i=1}^{K} e_m(W, F_i, \Phi_i) < \varepsilon \quad \text{for every} \quad m \in \mathsf{M}, W \in \mathcal{W}\right\}$$
$$\geqq 1 - |\mathsf{M}||\mathcal{W}| \exp\{-K[\varepsilon - \log(1 + e(\mathcal{W}, Q))]\} > 0.$$

This means that some realization $\{(f_i, \varphi_i)\}_{i=1}^{K}$ of the sequence of RVs (F_i, Φ_i) satisfies (12.14). □

First we determine the "random code capacity" of an AVC. We shall consider codes $(f, \varphi) \in \mathcal{C}(\mathsf{M} \to \mathsf{X}, \mathsf{Y} \to \mathsf{M}')$ such that the decoder φ is defined in terms of a non-negative-valued function $d(x, y)$ on $\mathsf{X} \times \mathsf{Y}$ by

$$\varphi(y) \triangleq \begin{cases} m & \text{if } \max_{m' \neq m} d(f(m'), y) < d(f(m), y) \\ m_0 \in \mathsf{M}' - \mathsf{M} & \text{if there is no such } m \in \mathsf{M}. \end{cases} \tag{12.16}$$

The family of such codes will be denoted by $\mathcal{C}_d = \mathcal{C}_d(\mathsf{M} \to \mathsf{X}, \mathsf{Y} \to \mathsf{M}')$.

LEMMA 12.9 Let X be a RV with values in X and let $d(x, y)$ be a non-negative-valued function on $\mathsf{X} \times \mathsf{Y}$ such that

$$Ed(X, y) = 1 \quad \text{for every} \quad y \in \mathsf{Y}. \tag{12.17}$$

Then there exists a distribution Q on $\mathcal{C}_d(\mathsf{M} \to \mathsf{X}, \mathsf{Y} \to \mathsf{M}')$ such that for every channel $W : \mathsf{X} \to \mathsf{Y}$, every $\varepsilon > 0$ and every $m \in \mathsf{M}$

$$e_m(W, Q) \leqq \Pr\left\{d(X, Y) < \frac{|\mathsf{M}|}{\varepsilon}\right\} + \varepsilon,$$

where Y is a RV connected with X by the channel W. ○

Proof Denote the distribution of X by P and let X_m, $m \in \mathsf{M}$, be independent RVs with distribution P. Set $F(m) \triangleq X_m$ for every $m \in \mathsf{M}$ and define Φ by (12.16), with F playing the role of f. Then (F, Φ) is a RV with values in $\mathcal{C}_d$. Denoting its distribution by Q, we have

$$e_m(W, Q) = Ee_m(W, F, \Phi) = E \sum_{y \in \mathsf{Y}} W(y|X_m) Z(y, m), \tag{12.18}$$

where $Z(y, m) = 0$ if $\Phi(y) = m$ and $Z(y, m) = 1$ otherwise. Evaluate the expectation in (12.18) by first taking conditional expectation given $X_m = x$. On account of (12.16) and the independence of the X_m we obtain

$$\begin{aligned} e_m(W, Q) &= \sum_x P(x) \sum_y W(y|x) E(Z(y, m)|X_m = x) \\ &= \sum_{x,y} P(x) W(y|x) \Pr\{\Phi(y) \neq m | X_m = x\} \\ &= \sum_{x,y} P(x) W(y|x) \Pr\{\max_{m' \neq m} d(X_{m'}, y) \geqq d(x, y)\}. \end{aligned}$$

Bound the last probability by 1 if $d(x, y) < (1/\varepsilon)|\mathsf{M}|$. In the opposite case note that (12.17) and Markov's inequality yield

$$\sum_{m' \neq m} \Pr\left\{d(X_{m'}, y) \geqq \frac{1}{\varepsilon} |\mathsf{M}|\right\} \leqq \varepsilon.$$

Thus we obtain

$$e_m(W, Q) \leqq \sum_{x, y : d(x,y) < \frac{1}{\varepsilon}|\mathsf{M}|} P(x) W(y|x) + \varepsilon. \qquad \square$$

LEMMA 12.10 Given any AVC $\{\mathcal{W} : \mathsf{X} \to \mathsf{Y}\}$ and distribution P on X, there exist random n-length block codes Q_n (where Q_n is a distribution on the set of pairs of mappings $f : \mathsf{M}_n \to \mathsf{X}^n, \varphi : \mathsf{Y}^n \to \mathsf{M}'_n$), satisfying

$$e(\mathcal{W}^n, Q_n) \to 0$$

and → 12.7

$$\frac{1}{n} \log |\mathsf{M}_n| \to I(P, \overline{\mathcal{W}}).$$ ○

Proof We shall assume without any loss of generality that all entries of the matrices $W \in \mathcal{W}$ are bounded below by a positive constant η. Indeed, one could always change each output to either $y \in \mathsf{Y}$ with probability η, resulting in a family $\mathcal{W}_\eta$ of modified channels formally defined by $W_\eta(y|x) \triangleq (1 - |\mathsf{Y}|\eta)W(y|x) + \eta$. Replacing $\mathcal{W}$ by $\mathcal{W}_\eta$ causes a negligible change of $I(P, \mathcal{W})$ if η is small, and any random code can be trivially modified at the decoder to give the same error probabilities for the AVC $\{\mathcal{W}\}$ as the original one did for the AVC $\{\mathcal{W}_\eta\}$.

Fixing P, let $\widetilde{W} \in \overline{\mathcal{W}}$ be a channel minimizing $I(P, W)$ for $W \in \overline{\mathcal{W}}$. Then for every $W \in \mathcal{W}$ and $0 \leqq \alpha \leqq 1$ we have

$$I(P, \alpha W + (1 - \alpha)\widetilde{W}) \geqq I(P, \widetilde{W})$$

whence

$$\lim_{\alpha \to 0} \frac{\partial}{\partial \alpha} I(P, \alpha W + (1 - \alpha)\widetilde{W}) \geqq 0.$$

As

$$\frac{\partial}{\partial \alpha} I(P, \alpha W + (1 - \alpha)\widetilde{W}) = \sum_{x,y} P(x)(W(y|x) - \widetilde{W}(y|x)) \log \frac{\alpha W(y|x) + (1 - \alpha)\widetilde{W}(y|x)}{\alpha PW(y) + (1 - \alpha)P\widetilde{W}(y)},$$

it follows that

$$\sum_{x,y} P(x)W(y|x) \log \frac{\widetilde{W}(y|x)}{P\widetilde{W}(y)} \geqq \sum_{x,y} P(x)\widetilde{W}(y|x) \log \frac{\widetilde{W}(y|x)}{P\widetilde{W}(y)} = I(P, \widetilde{W}) = I(P, \overline{\mathcal{W}}). \tag{12.19}$$

Let $X^n = X_1 \dots X_n$ be a sequence of independent RVs with common distribution P. Apply Lemma 12.9 with this X^n in the role of X and with

$$d(\mathbf{x}, \mathbf{y}) \triangleq \frac{\widetilde{W}^n(\mathbf{y}|\mathbf{x})}{(P\widetilde{W})^n(\mathbf{y})}.$$

Then condition (12.17) is satisfied, and thus we obtain that for any message set M_n there exists an n-length random block code Q_n such that for every channel $V \in \mathcal{W}^n$, every $\varepsilon > 0$ and every $m \in \mathsf{M}_n$

$$e_m(V, Q_n) \leqq \Pr\left\{\frac{\widetilde{W}^n(Y^n|X^n)}{(P\widetilde{W})^n(Y^n)} < \frac{1}{\varepsilon}|\mathsf{M}_n|\right\} + \varepsilon, \qquad (12.20)$$

where $Y^n = Y_1 \dots Y_n$ is connected with X^n by the channel V.

Introduce the RVs

$$Z_i \triangleq \log \frac{\widetilde{W}(Y_i|X_i)}{P\widetilde{W}(Y_i)}, \quad i = 1, \dots, n.$$

As the pairs (X_i, Y_i) are mutually independent, $P_{X_i} = P$, and the RVs X_i and Y_i are connected by some channel $W \in \mathcal{W}$ (depending on i), the RVs Z_i are also independent and have expectation

$$EZ_i = \sum_{x,y} P(x)W(y|x) \log \frac{\widetilde{W}(y|x)}{(P\widetilde{W})(y)} \quad \text{for some} \quad W \in \mathcal{W}. \qquad (12.21)$$

Further, as $|Z_i| \leqq -\log m_{\widetilde{W}}$ with probability 1, where $m_{\widetilde{W}}$ is the smallest positive entry of $\widetilde{W}$,

$$\operatorname{var}(Z_i) \leqq [\log m_{\widetilde{W}}]^2. \qquad (12.22)$$

Supposing $|\mathsf{M}_n| \leqq \exp[n(I(P, \overline{\mathcal{W}}) - \varepsilon)]$, for $n \geqq (2/\varepsilon)\log(1/\varepsilon)$ we obtain from (12.20), (12.21), (12.19) and (12.22) by Chebyshev's inequality

$$\begin{aligned} e_m(V, Q_n) &\leqq \Pr\left\{\sum_{i=1}^n Z_i < n\left(I(P, \overline{\mathcal{W}}) - \frac{\varepsilon}{2}\right)\right\} + \varepsilon \\ &\leqq \Pr\left\{\left|\sum_{i=1}^n (Z_i - EZ_i)\right| > \frac{n\varepsilon}{2}\right\} + \varepsilon \leqq \frac{4}{n\varepsilon^2}[\log m_{\widetilde{W}}]^2 + \varepsilon. \end{aligned} \qquad (12.23)$$

As $\varepsilon > 0$ was arbitrary, this proves Lemma 12.10. □

Now we are ready to prove the following.

→ 12.8 THEOREM 12.11 The a-capacity of an AVC $\{\mathcal{W} : \mathsf{X} \to \mathsf{Y}\}$ equals either zero or $C(\overline{\mathcal{W}})$. ○

→ 12.9 *Proof* Since by Corollary 12.3

$$C_a(\mathcal{W}) \leqq C(\overline{\mathcal{W}}), \qquad (12.24)$$

it suffices to prove that if $C_a(\mathcal{W}) > 0$ then $C_a(\mathcal{W}) \geqq C(\overline{\mathcal{W}})$.

The idea of the proof is the following: by Lemma 12.10, there exist random n-length block codes such that

$$\frac{1}{n} \log |\mathsf{M}_n| \to C(\overline{\mathcal{W}}), \quad e(\mathcal{W}^n, Q_n) \to 0 \quad (n \to \infty). \qquad (12.25)$$

Using Lemma 12.8, we shall conclude that there exist random codes attaching positive weights only to "exponentially few" codes and still satisfying (12.25). If $C_a(\mathcal{W}) > 0$, it will be possible to reduce the random code to an ordinary one by adding short prefixes to the original codewords so as to specify which code has actually been selected.

Turning to the formal proof, suppose first that $|\mathcal{W}| < \infty$. Apply Lemma 12.8 to the family of channels $\mathcal{W}^n$ and the random codes satisfying (12.25), with $K = n^2$, say. It follows that for any $\varepsilon > 0$ and sufficiently large n there exist codes

$$(f_{ni}, \varphi_{ni}) \in \mathcal{C}(\mathsf{M}_n \to \mathsf{X}^n, \mathsf{Y}^n \to \mathsf{M}'_n) \quad i = 1, \dots, n^2$$

such that (with the notation (12.1))

$$\frac{1}{n^2} \sum_{i=1}^{n^2} e_m(W^n(\cdot|\cdot, \mathbf{s}), f_{ni}, \varphi_{ni}) \leqq \varepsilon \quad \text{for every } m \in \mathsf{M}_n, \mathbf{s} \in \mathsf{S}^n. \tag{12.26}$$

Further, supposing that $C_{\mathrm{a}}(\mathcal{W}) > 0$, there exists a sequence of codes $(\hat{f}_n, \hat{\varphi}_n)$ such that $\hat{f}_n : \{1, \dots, n^2\} \to \mathsf{X}^{k_n}$, $\hat{\varphi}_n : \mathsf{Y}^{k_n} \to \{1, \dots, n^2\}$, where $k_n/n \to 0$ and

$$\bar{e}(W^{k_n}(\cdot|\cdot, \hat{\mathbf{s}}), \hat{f}_n, \hat{\varphi}_n) \leqq \varepsilon \quad \text{for every } \hat{\mathbf{s}} \in \mathsf{S}^{k_n}. \tag{12.27}$$

Now define new codes $(\tilde{f}_n, \tilde{\varphi}_n)$ of block length $k_n + n$ with message sets $\{1, \dots, n^2\} \times \mathsf{M}_n$: let $\tilde{f}_n(i, m)$ be the juxtaposition of the codewords $\hat{f}_n(i)$ and $f_{ni}(m)$. Decomposing the sequences $\tilde{\mathbf{y}} \in \mathsf{Y}^{k_n+n}$ as $\tilde{\mathbf{y}} = \hat{\mathbf{y}}\mathbf{y}$, $\hat{\mathbf{y}} \in \mathsf{Y}^{k_n}$, $\mathbf{y} \in \mathsf{Y}^n$, set $\tilde{\varphi}(\tilde{\mathbf{y}}) \triangleq (i, \varphi_{ni}(\mathbf{y}))$ with $i \triangleq \hat{\varphi}_n(\hat{\mathbf{y}})$. Then for any fixed $\tilde{\mathbf{s}} = \hat{\mathbf{s}}\mathbf{s}$, $\hat{\mathbf{s}} \in \mathsf{S}^{k_n}$, $\mathbf{s} \in \mathsf{S}^n$, writing

$$e_i \triangleq e_i(W^{k_n}(\cdot|\cdot, \hat{\mathbf{s}}), \hat{f}_n, \hat{\varphi}_n), \quad e_m(i) \triangleq e_m(W^n(\cdot|\cdot, \mathbf{s}), f_{ni}, \varphi_{ni}), \tag{12.28}$$

we have by (12.26) and (12.27)

$$\begin{aligned}
\bar{e}(W^{k_n+n}(\cdot|\cdot, \tilde{\mathbf{s}}), \tilde{f}_n, \tilde{\varphi}_n) &\leqq \frac{1}{n^2|\mathsf{M}_n|} \sum_{i=1}^{n^2} \sum_{m \in \mathsf{M}_n} (e_i + e_m(i)) \\
&= \frac{1}{n^2} \sum_{i=1}^{n^2} e_i + \frac{1}{|\mathsf{M}_n|} \sum_{m \in M_n} \left(\frac{1}{n^2} \sum_{i=1}^{n^2} e_m(i) \right) \leqq 2\varepsilon,
\end{aligned}$$

i.e., $\bar{e}(\mathcal{W}^{k_n+n}, \tilde{f}_n, \tilde{\varphi}_n) \leqq 2\varepsilon$. As the limit of the rates of the codes $(\tilde{f}_n, \tilde{\varphi}_n)$ equals $\lim_{n\to\infty}(1/n) \log |\mathsf{M}_n|$, this completes the proof in the case $|\mathcal{W}| < \infty$.

If $|\mathcal{W}| = \infty$, an approximation argument is needed. Subdividing the $|\mathsf{X}||\mathsf{Y}|$-dimensional unit cube into cubes of edge length n^{-4}, pick a $W \in \mathcal{W}$ in each cube containing at least one such matrix. Let $\mathsf{S}' \subset \mathsf{S}$ be the set of indices of the matrices picked in this way. Then to each $s \in \mathsf{S}$ there exists an $s' \in \mathsf{S}'$ with the property that $|W(y|x, s) - W(y|x, s')| \leqq n^{-4}$ and hence

$$W(y|x, s) \leqq \left(1 + \frac{1}{n^2}\right) W(y|x, s') \quad \text{for every } x, y \quad \text{with } W(y|x, s) \geqq \frac{1}{n^2}.$$

As $\left(1 + 1/n^2\right)^n \leqq 2$, it follows that to every $\mathbf{s} \in \mathsf{S}^n$ there exists an $\mathbf{s}' \in \mathsf{S}'^n$ such that $W^n(\mathbf{y}|\mathbf{x}, \mathbf{s}) \leqq 2W^n(\mathbf{y}|\mathbf{x}, \mathbf{s}')$ unless $W(y_i|x_i, s_i) < 1/n^2$ for some i. Hence for every code

$$e_m(W^n(\cdot|\cdot, \mathbf{s}), f, \varphi) \leqq 2e_m(W^n(\cdot|\cdot, \mathbf{s}'), f, \varphi) + \frac{|\mathsf{Y}|}{n}. \tag{12.29}$$

Now we can apply Lemma 12.8 to the family of channels $\{W^n(\cdot|\cdot, \mathbf{s}) : \mathbf{s} \in \mathsf{S}'^n\}$ of size $|\mathsf{S}'|^n < n^{4|\mathsf{X}||\mathsf{Y}|n}$ and the random codes satisfying (12.25), again with $K = n^2$. It follows that for suitable codes (f_{ni}, φ_{ni}) (12.26) is valid for every $m \in \mathsf{M}_n$ and $\mathbf{s} \in \mathsf{S}'^n$. On account of (12.29) a similar inequality holds also for every $\mathbf{s} \in \mathsf{S}^n$, with ε replaced by 3ε, say. The rest of the proof is the same as in the case $|\mathcal{W}| < \infty$. □

→ 12.11

REMARK The essence of Theorem 12.11 is that the a-capacity of an AVC is either zero or equals the random code capacity. Note that the proof of this depends neither on the actual value of the random code capacity nor on the specific product structure of an AVC. ○

Recall that a general random code describes a realizable transmission scheme only if the outcome of the random experiment involved is observable at both ends of the channel. There is no such problem, however, with random codes in which the decoder is non-random. In the latter case, the probabilities $T(m'|m)$ of (7.10) are uniquely determined by $|\mathsf{M}|$ distributions on X. For each $m \in \mathsf{M}$ the selection of the codeword of message m is governed by the corresponding distribution. The joint distribution of codewords of different messages m is irrelevant. This motivates the following definition.

DEFINITION 12.12 A code with *stochastic encoder* is a pair (F, φ), where F is a stochastic matrix $F : \mathsf{M} \to \mathsf{X}$ and φ is a mapping $\varphi : \mathsf{Y} \to \mathsf{M}'$. For any given channel $W : \mathsf{X} \to \mathsf{Y}$, the pair (F, φ) defines a new channel $T : \mathsf{M} \to \mathsf{M}'$ by

$$T(m'|m) \triangleq \sum_{x \in \mathsf{X}} F(x|m) W(\varphi^{-1}(m')|x). \tag{12.30}$$

The different error probabilities of (F, φ) are defined, using this T, in the same way as those of an ordinary code (f, φ). ○

One defines n-length block codes with stochastic encoder as pairs (F, φ) where the roles of X and Y of Definition 12.12 are played by X^n and Y^n. Admitting such codes in Definition 12.2, we arrive at the concepts of m-capacity and a-capacity of an AVC for codes with stochastic encoder.

THEOREM 12.13 Both the m- and a-capacities of an AVC $\{\mathcal{W} : \mathsf{X} \to \mathsf{Y}\}$ for codes with stochastic encoder equal $C_{\mathrm{a}}(\mathcal{W})$. ○

Proof Denote the m-capacity resp. a-capacity for codes with stochastic encoder by C'_{m} resp. C'_{a}.

First we prove $C'_{\mathrm{a}} = C_{\mathrm{a}}$. Since $C'_{\mathrm{a}} \geqq C_{\mathrm{a}}$ is obvious, the assertion will follow if we show that for every $R > 0$, $\varepsilon > 0$ and sufficiently large n, to any n-length block code with stochastic encoder (F, φ) satisfying

$$\frac{1}{n} \log |\mathsf{M}| \geqq R, \quad \bar{e}(\mathcal{W}^n, F, \varphi) \leqq \varepsilon$$

there exists an encoder $f : \mathsf{M} \to \mathsf{X}^n$ such that

$$\bar{e}(W^n(\cdot\,|\,\cdot\,,\mathbf{s}), f, \varphi) \leqq a\varepsilon \quad \text{for every } \mathbf{s} \in \mathsf{S}^n, \tag{12.31}$$

where a is an absolute constant.

Consider independent RVs X_m, $m \in \mathsf{M}$, with values in X^n and with distributions $P_{X_m} = F(\cdot\,|m)$. Fixing an $\mathbf{s} \in \mathsf{S}^n$, the RVs

$$Z_m = Z_m(\mathbf{s}) \triangleq 1 - W^n(\varphi^{-1}(m)|X_m, \mathbf{s})$$

are mutually independent and, by assumption,

$$\frac{1}{|\mathsf{M}|} \sum_{m \in \mathsf{M}} EZ_m = \bar{e}(W^n(\cdot\,|\,\cdot\,,\mathbf{s}), F, \varphi) \leqq \varepsilon.$$

Note that the inequalities $2^t \leqq 1 + t \leqq 3^t$ $(0 \leqq t \leqq 1)$ imply $E \exp Z_m \leqq 3^{EZ_m}$.

Thus we have

$$\begin{aligned}
\Pr\left\{\frac{1}{|\mathsf{M}|} \sum_{m \in \mathsf{M}} Z_m \geqq a\varepsilon\right\} &= \Pr\left\{\exp \sum_{m \in \mathsf{M}} Z_m \geqq \exp(|\mathsf{M}|a\varepsilon)\right\} \\
&\leqq \exp(-|\mathsf{M}|a\varepsilon) \prod_{m \in \mathsf{M}} E \exp Z_m \leqq \exp\left[-|\mathsf{M}|a\varepsilon + \log 3 \cdot \sum_{m \in \mathsf{M}} EZ_m\right] \\
&\leqq \exp\left[-|\mathsf{M}|\varepsilon(a - \log 3)\right] \leqq \exp\left[-\exp(nR) \cdot \varepsilon(a - \log 3)\right].
\end{aligned}$$

Choosing $a = 2$, say, it follows that if S is finite then

$$\Pr\left\{\frac{1}{|\mathsf{M}|} \sum_{m \in \mathsf{M}} Z_m(\mathbf{s}) \geqq 2\varepsilon \quad \text{for some } \mathbf{s} \in \mathsf{S}^n\right\} < 1$$

for sufficiently large n. This means that for some realization $X_m = \mathbf{x}_m$, $m \in \mathsf{M}$, of the RVs X_m we have

$$\frac{1}{|\mathsf{M}|} \sum_{m \in \mathsf{M}} (1 - W^n(\varphi^{-1}(m)|\mathbf{x}_m, \mathbf{s})) \leqq 2\varepsilon \quad \text{for every } \mathbf{s} \in \mathsf{S}^n.$$

Setting $f(m) \triangleq \mathbf{x}_m$, this establishes (12.31) with $a = 2$. If S is infinite, the same approximation argument can be used as in the proof of Theorem 12.11, cf. (12.29).

Since clearly $C'_{\mathrm{m}} \leqq C'_{\mathrm{a}}$, and we already know that $C'_{\mathrm{a}} = C_{\mathrm{a}}$, we are left to prove that $C'_{\mathrm{m}} \geqq C_{\mathrm{a}}$. To this end, suppose that $C_{\mathrm{a}} = C > 0$. By the proof of Theorem 12.11, for every $\varepsilon > 0$ and sufficiently large n there exist codes

$$(f_{ni}, \varphi_{ni}) \in \mathcal{C}(\mathsf{M}_n \to \mathsf{X}^n, \mathsf{Y}^n \to \mathsf{M}'_n), \quad i = 1, \ldots, n^2$$

with $(1/n)\log|\mathsf{M}_n| \to C$ and satisfying (12.26). Also, there exist codes $(\hat{f}_n, \hat{\varphi}_n)$ with $\hat{f}_n : \{1, \ldots, n^2\} \to \mathsf{X}^{k_n}$, $\hat{\varphi}_n : \mathsf{Y}^{k_n} \to \{1, \ldots, n^2\}$, $k_n/n \to 0$ and satisfying (12.27). Consider the codes $(\tilde{f}_n, \tilde{\varphi}_n)$ built from the above ones as in the proof of Theorem 12.11. Define the stochastic encoder $F_n : \mathsf{M}_n \to \mathsf{X}^{k_n+n}$ by letting $F_n(\cdot\,|m)$ be the distribution on X^{k_n+n} attaching weights $1/n^2$ to the codewords $\tilde{f}_n(i, m)$, $i = 1, \ldots, n^2$. Further, for every $\tilde{\mathbf{y}} \in \mathsf{Y}^{k_n+n}$, set $\varphi_n(\tilde{\mathbf{y}}) \triangleq m$ if $\tilde{\varphi}_n(\tilde{\mathbf{y}}) = (i, m)$ for some $1 \leqq i \leqq n^2$. Then the codes with stochastic encoder (F_n, φ_n) have rate tending to C, while for every $\tilde{\mathbf{s}} \in \mathsf{S}^{k_n+n}$ and $m \in \mathsf{M}_n$, with the notation (12.28),

$$e_m(W^{k_n+n}(\cdot \mid \cdot, \tilde{\mathbf{s}}), F_n, \varphi_n) \leqq \frac{1}{n^2} \sum_{i=1}^{n^2} (e_i + e_m(i)) \leqq 2\varepsilon,$$

by (12.27) and (12.26). □

Discussion

In the study of arbitrarily varying channels, we encountered phenomena not present in simpler models: in case of AVCs (i) the capacities for maximum and average probability of error may be different and (ii) stochastic encoding may increase m-capacity. It should be clear that both $C_{\mathrm{a}}(\mathcal{W})$ and $C_{\mathrm{m}}(\mathcal{W})$ are relevant quantities for (different) source–channel transmission problems. No general source–channel transmission theorem for AVSs and AVCs can be formulated in terms of the ratio of two numbers characterizing the AVS and the AVC, respectively, if one sticks to ordinary codes. Admitting codes with stochastic encoder, however, this inconvenience disappears. One may argue that for AVCs the "natural" code concept involves stochastic encoding and the "true" capacity of the AVC is the common value of both capacities for codes with stochastic encoder. The latter equals $C_{\mathrm{a}}(\mathcal{W})$ by Theorem 12.13. Note that stochastic decoders offer no advantage.We have not considered exponential error bounds for AVCs because at present very little is known in this direction.

→ 12.12

→ 12.13

It is instructive to look at the engineering problem of transmitting messages over a channel with varying states from a game-theoretic point of view (see e.g. Karlin (1959)). This is helpful for a certain systematization of the large spectrum of information-theoretic problems which correspond to different communication situations involving a channel subject to unpredictable state changes during transmission. Although the problems treated in the text represent but a small fraction of this spectrum, the scope of the presented methods is quite large.

Given a family of channels $\mathcal{W} = \{W(\cdot \mid \cdot, s) : s \in \mathsf{S}\}$, for any message set M and block length n consider a two-person zero-sum game between a "code selector" and a "state selector." The game is determined by the sets of permissible (pure) strategies of the players and the corresponding pay-off. We shall always assume that the pay-off to the state selector is either the maximum or the average probability of erroneous transmission, i.e.,

$$\max_{m \in \mathsf{M}} (1 - T(m|m)) \quad \text{or} \quad \frac{1}{|\mathsf{M}|} \sum_{m \in \mathsf{M}} (1 - T(m|m)),$$

where $T(m'|m)$ is the probability (corresponding to the selected strategies) that the message m is decoded as m'. A *communication model* is a family of such games (for every n and M). For this model, a number $R > 0$ is an *achievable rate* if for every $\varepsilon > 0, \delta > 0$ and sufficiently large n, there exist message sets M with $(1/n) \log |\mathsf{M}| \geqq R - \delta$ such that some permissible strategy of the code selector guarantees a pay-off at most ε, no matter what strategy the state selector adopts. The largest achievable rate is the *capacity* for the considered model.

If for given M and n the permissible strategies of the code selector are codes (f, φ) in $\mathcal{C}(\mathsf{M} \to \mathsf{X}^n, \mathsf{Y}^n \to \mathsf{M}')$ and those of the state selector are sequences $\mathbf{s} \in \mathsf{S}^n$, we arrive at the concepts of m- and a-capacities of the AVC $\{\mathcal{W}\}$ as defined in the text. → 12.14

Enlarging the set of the code selector's permissible strategies one can get models with *channel state information* (CSI), where the actual states of the channel are (completely or partially) known at the input or at the output or both. For example, if the states are known only at the output, one defines the code selector's permissible strategies as pairs (f, φ), where $f : \mathsf{M} \to \mathsf{X}^n$ is an encoder in the usual sense, while φ is a mapping $\varphi : \mathsf{Y}^n \times \mathsf{S}^n \to \mathsf{M}'$. Given such an (f, φ) and an $\mathbf{s} \in \mathsf{S}^n$, the probability that message m will be decoded as m' is → 12.15

$$T(m'|m) \triangleq \sum_{\mathbf{y}:\varphi(\mathbf{y},\mathbf{s})=m'} W^n(\mathbf{y}|f(m), \mathbf{s}). \tag{12.32}$$

Defining the maximum and average probabilities of error via (12.32), one arrives at the concepts of m- and a-capacities of an AVC with states known at the output. → 12.16

One can proceed analogously if the states are known at the input, letting the code selector's permissible strategies be pairs (f, φ), where $\varphi : \mathsf{Y}^n \to \mathsf{M}'$ is a decoder in the usual sense, while the encoder f maps $\mathsf{M} \times \mathsf{S}^n$ into X^n. Using such an (f, φ) when the state sequence is $\mathbf{s} \in \mathsf{S}^n$, the probability that message m will be decoded as m' is

$$T(m'|m) \triangleq \sum_{\mathbf{y}:\varphi(\mathbf{y})=m'} \prod_{i=1}^{n} W(y_i|f_i(m, \mathbf{s}), s_i), \tag{12.33}$$

where $f_i(m, \mathbf{s})$ denotes the ith component of $f(m, \mathbf{s}) \in \mathsf{X}^n$. It may be that the sender has causal CSI (as in Problem 6.31 where the states are assumed random rather than arbitrarily varying), thus at each instant i only the previous states $s_1 \dots s_{i-1}$ or the previous and present states $s_1 \dots s_i$ are known. This means, formally, that only such encoders $f : \mathsf{M} \times \mathsf{S}^n \to \mathsf{S}^n$ are permitted for which $f_i(m, \mathbf{s}) = f_i(m, s_1, \dots, s_{i-1})$ or $f_i(m, \mathbf{s}) = f_i(m, s_1, \dots, s_i)$, $i = 1, \dots, n$. In the case of non-causal CSI, when the sender knows all states in advance, there are no such restrictions on the permissible encoders. In this way one arrives at the concepts of m- and a-capacities of an AVC with previous resp. previous and present states, or all states, known at the input. → 12.17 → 12.18 → 12.25

The assumption of the knowledge of previous (and present) states may be replaced by or combined with the assumption that at each time instant the previous output symbols are known at the input (through a noiseless feedback link). The formal definitions of the code selector's permissible strategies for these cases are obvious and so are the corresponding analogs of (12.33). Thus one arrives at different concepts of capacity of an AVC with complete feedback. Needless to say, all the mentioned models can be modified by allowing stochastic encoders; this does not lead, however, to new capacity values, since the analog of Theorem 12.13 is valid for all these models. → 12.20

It is also possible to enlarge the set of permissible strategies of the state selector. In this way one gets new channel models which may be more suitable for describing certain engineering situations than the AVC. For example, it is physically reasonable to assume that at each time instant i the state s_i of the channel may depend on the previous input symbols $x_1, \dots, x_{i-1}$ in an unknown way. This can be formalized by letting the state selector's permissible strategies be mappings $\sigma : \mathsf{X}^n \to \mathsf{S}^n$ defined by sequences of mappings $\sigma_i : \mathsf{X}^{i-1} \to \mathsf{S}$ so that $\sigma(\mathbf{x}) = \sigma_1\sigma_2(x_1)\dots\sigma_n(x_1 \dots x_{n-1})$. Every such strategy defines a channel $W^n(\cdot\,|\,\cdot, \sigma)$ by

$$W^n(\mathbf{y}|\mathbf{x}, \sigma) \triangleq W^n(\mathbf{y}|\mathbf{x}, \sigma(\mathbf{x})). \tag{12.34}$$

We designate by $\mathcal{W}^{n*}$ the family of channels of this form as σ runs over the strategies as above. The sequence $\{\mathcal{W}^{n*}\}_{n=1}^{\infty}$ is a channel model which allows for an even greater freedom in the variability of states than the AVC. This model is no longer a memoryless one. The sequence $\{\mathcal{W}^{n*}\}_{n=1}^{\infty}$ will be called an *arbitrarily "star" varying channel* or A*VC. The communication model with state selector's strategies as above and code selector's strategies $(f, \varphi) \in \mathcal{C}(\mathsf{M} \to \mathsf{X}^n, \mathsf{Y}^n \to \mathsf{M}')$ leads to the concepts of m- and
→ 12.21 a-capacities of an A*VC.

One might permit for the state selector also such strategies where each state s_i depends on the present input symbol x_i or on the previous and present input symbols
→ 12.22 $x_1, \dots, x_i$. In this way, however, no really new models arise.

Enlarging the set of the state selector's strategies as above, the m-capacity of the AVC $\{\mathcal{W}\}$ does not decrease but the a-capacity does, in general. In the extreme case when every mapping of X^n into S^n is permissible, the m-capacity is still equal to $C_m(\mathcal{W})$;
→ 12.23 further, in this case the a-capacity becomes equal to the m-capacity.

Some pairs of strategies of the code selector and the state selector described above are compatible while others are not. For example, the code selector's strategies with previous states known at the input are compatible with the state selector's strategies depending on previous inputs. The probabilities $T(m'|m)$ corresponding to a pair of such strategies are still given by (12.33), with $f_i(m, \mathbf{s})$ replaced by $f_i(m, s_1, \dots, s_{i-1})$, the states s_i now being defined recursively by

$$s_i = \sigma_i(x_1, \dots, x_{i-1}), \quad x_i = f_i(m, s_1, \dots, s_{i-1}).$$

An example of incompatible pairs is obtained if the code selector's strategy lets the ith input symbol depend on the ith state while the state selector's strategy lets the ith state depend on the ith input symbol.

The communication models hinted at above have various degrees of practical and mathematical interest. While much effort has been spent on some of them, others have not been tackled at all.

Progress since the first edition of this book is illustrated by some new items added to the Problems part. In particular, Problem 12.5 extends the main m-capacity result Theorem 12.7 to a large class of AVCs, and Problem 12.10 points at an almost complete solution of the a-capacity problem also for AVCs with a state constraint, to which the proof technique of Theorem 12.11 does not apply. These new results were proved employing the method of types as in Chapter 9. Attention is called also to

Problems 12.24 and 12.25. The former addresses channels with random rather than arbitrarily varying states when the sender has non-causal CSI, completing results in Problem 6.31; the latter is about determining the capacity of an AVC with non-causal CSI, using the previous result and more. ○

Problems

12.1. (*Positivity of* m-*capacity*)

(a) Show that if to every pair x_1, x_2 of elements of X there exists a $W \in \overline{\overline{\mathcal{W}}}$ such that $W(y|x_1) = W(y|x_2)$ for every $y \in \mathsf{Y}$ then every n-length block code with at least two messages has $e(\mathcal{W}^n, f, \varphi) \geqq 1/2$.

(b) Prove that $C_m(\mathcal{W}) > 0$ iff $\mathcal{W}$ does not have the property stated in (a). (Kiefer and Wolfowitz (1962).)

Hint To prove the sufficiency part of (b), note that the rows of matrices in $\overline{\overline{\mathcal{W}}}$ corresponding to any fixed $x \in \mathsf{X}$ form a compact convex set. Thus if for some pair x_1, x_2 these sets are disjoint, they are separated by a hyperplane, i.e., for some constants $\alpha(y)$, $y \in \mathsf{Y}$ and γ we have

$$\max_{W \in \overline{\overline{\mathcal{W}}}} \sum_{y \in \mathsf{Y}} \alpha(y) W(y|x_1) < \gamma < \min_{W \in \overline{\overline{\mathcal{W}}}} \sum_{y \in \mathsf{Y}} \alpha(y) W(y|x_2).$$

Use $\mathbf{x}_1 \triangleq x_1 \dots x_1 \in \mathsf{X}^k$ and $\mathbf{x}_2 \triangleq x_2 \dots x_2 \in \mathsf{X}^k$ as building blocks of codewords of n-length block codes (k fixed, $n \to \infty$), having in mind that the set $\mathsf{A} \triangleq \{\mathbf{y} : \mathbf{y} \in \mathsf{Y}^k, \sum_{b \in \mathsf{Y}} N(b|\mathbf{y})\alpha(b) < k\gamma\}$ satisfies $W^k(\mathsf{A}|\mathbf{x}_1, \mathbf{s}) > 3/4$, $W^k(\mathsf{A}|\mathbf{x}_2, \mathbf{s}) < 1/4$ (say), for every $\mathbf{s} \in \mathsf{S}^k$ if k is sufficiently large.

(c) Show that if $\overline{\mathcal{W}} = \overline{\overline{\mathcal{W}}}$ then $C_{\mathrm{m}}(\mathcal{W})$ is positive whenever $C_{\mathrm{a}}(\mathcal{W})$ is. (Ahlswede (1978).)

Hint Use Problem 12.9(b).

12.2. Give examples of AVCs such that

(a) $C(\overline{\overline{\mathcal{W}}}) > 0$ but $C_{\mathrm{m}}(\mathcal{W}) = 0$,

(b) $C(\overline{\mathcal{W}}) > 0$ but $C_{\mathrm{a}}(\mathcal{W}) = 0$.

Hint For (b), set $\mathsf{X} = \mathsf{S} = \{0, 1\}$, $\mathsf{Y} = \{0, 1, 2\}$, $W(y|x, s) = 1$ if $y = x + s$ and zero otherwise. Check that for every n-length block code (f, φ)

$$\bar{e}(\mathcal{W}^n, f, \varphi) = \max_{\mathbf{s} \in \{0,1\}^n} \frac{1}{|\mathsf{M}|} |\{m : \varphi(f(m) + \mathbf{s}) \neq m\}|.$$

Considering state vectors $\mathbf{s}$ equal to codewords, conclude that

$$\bar{e}(\mathcal{W}^n, f, \varphi) \geqq \frac{1}{|\mathsf{M}|^2} |\{(m, m') : \varphi(f(m) + f(m')) \neq m\}| \geqq \frac{1}{2} \frac{|\mathsf{M}| - 1}{|\mathsf{M}|}.$$

(Blackwell, Breiman and Thomasian (1960).)

12.3. (m-*capacity and zero-error capacity*) Let $W : \mathsf{X} \to \mathsf{Y}$ be any channel and let $\mathcal{W}$ be the family of all stochastic matrices $V : \mathsf{X} \to \mathsf{Y}$ with entries 0 and 1 such that $W(y|x) = 0$ implies $V(y|x) = 0$. Check that $C_{\mathrm{m}}(\mathcal{W})$ equals the zero-error capacity of the DMC $\{W\}$. This example shows that it must be very hard to give a computable characterization of the m-capacity of general AVCs. (Ahlswede (1970).)

12.4. Let $\{\mathcal{W} : \mathsf{X} \to \mathsf{Y}\}$ be an AVC with $|\mathsf{Y}| = 2$. For every pair $x', x'' \in \mathsf{X}$ let $\{\mathcal{W}(x', x'') : \{x', x''\} \to \mathsf{Y}\}$ be the AVC defined by the corresponding rows of the matrices $W \in \mathcal{W}$. Conclude from the proof of Theorem 12.7 that

(a) $C_{\mathrm{m}}(\mathcal{W}) = \max_{x',x''} C_{\mathrm{m}}(\mathcal{W}(x', x''))$,

(b) the maximum is achieved if (x', x'') maximizes

$$\inf_{W \in \mathcal{W}} W(0|x') + \inf_{W \in \mathcal{W}} W(1|x''),$$

provided that this maximum is greater than 1; otherwise $C_{\mathrm{m}}(\mathcal{W}) = 0$. (Ahlswede and Wolfowitz (1970).)

12.5. (m-*capacity of AVCs with non-binary output*) For an AVC $\{\mathcal{W} : \mathsf{X} \to \mathsf{Y}\}$, write $x_1 \sim x_2$ if there exists $W \in \overline{\overline{\mathcal{W}}}$ such that $W(y|x_1) = W(y|x_2)$ for all $y \in \mathsf{Y}$. Suppose the class $\mathcal{W}$ consists of a finite number of matrices $W(\,.\,|\,.\,s)$, $s \in \mathsf{S}$.

(a) Show that if $x_1 \sim x_2$ never holds when $x_1 \neq x_2$ then $C_{\mathrm{m}}(\mathcal{W}) = C(\overline{\overline{\mathcal{W}}})$. (Ahlswede (1980).)

(b) Show that always $C_{\mathrm{m}}(\mathcal{W}) \geqq \max_P \min[I(P, \overline{\overline{\mathcal{W}}}), D(P)]$, where

$$D(P) \triangleq \min\{I(X \wedge \tilde{X}) : P_X = P_{\tilde{X}} = P,\ \Pr\{X \sim \tilde{X}\} = 1\}.$$

In particular, show that the result of (a) holds also under the weaker condition that some P achieving $I(P, \overline{\overline{\mathcal{W}}}) = C(\overline{\overline{\mathcal{W}}})$ satisfies $D(P) \leqq C(\overline{\overline{\mathcal{W}}})$. (Csiszár and Körner (1981b).)

Hint Employ the method of types. First prove, by random selection, the following extension of the packing lemma (see its version in Problem 10.2): for any $R > \delta > 0$ and $n \geqq n_0(|\mathsf{X}|, |\mathsf{S}|, \delta)$, any type class $\mathsf{T}_P \subset \mathsf{X}^n$ with $H(P) \geqq R - \delta$ has a subset C such that

(i) $(1/n) \log |\mathsf{C}| \geqq R - \delta$,

(ii) $|\mathsf{T}_V(\mathbf{x}, \mathbf{s}) \cap \mathsf{C}| \leqq 3(n+1)^{|\mathsf{X}|} \exp\{n\,|R - I(P_{\mathbf{x},\mathbf{s}}, V)|^+\}$ for all V-shells $\mathsf{T}_V(\mathbf{x}, \mathbf{s}) \subset \mathsf{X}^n$, $V : \mathsf{X} \times \mathsf{S} \to \mathsf{X}$,

(iii) $I(\mathbf{x} \wedge \tilde{\mathbf{x}}) < R$ for $\mathbf{x} \neq \tilde{\mathbf{x}}$ in C.

The next key step is to find a good decoder to the codeword set C. Show that if $R < \min[I(P, \overline{\overline{\mathcal{W}}}), D(P)] - \delta$, a decoder constructed as follows yields abitrarily small maximum probability of error. Assign to each received sequence $\mathbf{y}$ a set of "candidate codewords" consisting of those $\mathbf{x} \in \mathsf{C}$ that satisfy $D(P_{\mathbf{y}|\mathbf{x},\mathbf{s}} \| W | P_{\mathbf{x},\mathbf{s}}) < \eta$ for some $\mathbf{s} \in \mathbf{S}^n$. Then choose a candidate $\mathbf{x}$ such that $I(\tilde{\mathbf{x}} \wedge \mathbf{y}|\mathbf{x}, \mathbf{s}) < \eta$ for each $\mathbf{s}$ as above and all other candidates $\tilde{\mathbf{x}}$ (if such candidate $\mathbf{x}$ exists; otherwise an error is declared). The last crucial step is to show that this decoder is well defined if η is sufficiently small.

12.6. (*Random codes*) Given the sets X, Y and $\mathsf{M} \subset \mathsf{M}'$, observe that random codes are equivalent to triples (U, f, φ), where U is a RV with a finite range U

and $f : \mathsf{M} \times \mathsf{U} \to \mathsf{X}$, $\varphi : \mathsf{Y} \times \mathsf{U} \to \mathsf{M}'$ are some mappings, in the following sense. Every (U, f, φ) defines a random code, i.e., a distribution Q on $\mathcal{C}(\mathsf{M} \to \mathsf{X}, \mathsf{Y} \to \mathsf{M}')$ and every Q can be obtained in this way.

(a) Show that if for a RV M independent of U and a channel $W : \mathsf{X} \to \mathsf{Y}$

$$X \triangleq f(M, U), \; MU \multimap X \multimap Y, \; P_{Y|X} = W, \; M' \triangleq \varphi(Y, U),$$

then $\Pr\{M' = m' | M = m\}$ is given by (12.10). In particular, if M is uniformly distributed over M then $\Pr\{M' \neq M\} = \bar{e}(W, Q)$.

(b) Show that for random codes it does not matter whether maximum or average probability of error is used. In other words, to every random code Q there exists another random code $\tilde{Q}$ such that $e_m(W, \tilde{Q}) = \bar{e}(W, Q)$ for every channel $W : \mathsf{X} \to \mathsf{Y}$ and every $m \in \mathsf{M}$.

Hint Let Π be the set of all permutations Π of M; let Π be a RV independent of U and uniformly distributed on Π. Replace the triple (U, f, φ) by $(\tilde{U}, \tilde{f}, \tilde{\varphi})$, where

$$\tilde{U} \triangleq (U, \Pi), \quad \tilde{f}(m, u, \pi) \triangleq f(\pi(m), u), \quad \tilde{\varphi}(y, u, \pi) \triangleq \pi^{-1}(\varphi(y, u)).$$

12.7. (*Random code capacity of an AVC*) Show that the "random code capacity" of an AVC $\{\mathcal{W}\}$ equals $C(\overline{\mathcal{W}})$, i.e., $R \leqq C(\overline{\mathcal{W}})$ is necessary and sufficient for the existence of random n-length block codes with $\underline{\lim}_{n\to\infty}(1/n) \log |\mathsf{M}_n| \geqq R$ and with (maximum or average) probability of error tending to zero. (Blackwell, Breiman and Thomasian (1960).)

Hint Observe that the random code capacity of a DMC equals its ordinary capacity. Then use Lemmas 12.3 and 12.10.

12.8. Motivated by Theorem 12.11 one might conjecture that $C_{\mathrm{m}}(\mathcal{W})$ is either zero or it equals $C(\overline{\overline{\mathcal{W}}})$. Disprove this conjecture by a counterexample. Note that any counterexample as above is also an example for $C_{\mathrm{m}}(\mathcal{W}) < C_{\mathrm{a}}(\mathcal{W})$. (R. Ahlswede, personal communication in 1977.)

12.9. (*Positivity of* a*-capacity*)

(a) Show that if there exist PDs P_0 and P_1 on X such that the distributions $\{\sum_{x,s} P_i(x) Q_i(s) W(y|x, s) : y \in \mathsf{Y}\}$, $i = 0, 1$, are different for every pair of PDs Q_0 and Q_1 on S having finite support, then $C_{\mathrm{a}}(\mathcal{W}) > 0$.

Hint The solution depends on Theorem 12.13. Check that any code for the AVC $\{\widehat{\mathcal{W}} : \{0, 1\} \to \mathsf{Y}\}$ defined by

$$\widehat{W}(y|i, s) \triangleq \sum_{x \in \mathsf{X}} P_i(x) W(y|x, s), \quad i = 0, 1,$$

gives rise to a code with stochastic encoder for the AVC $\{\mathcal{W}\}$. Use Problem 12.1 for the AVC $\{\widehat{\mathcal{W}}\}$.

(b) Show that $C_{\mathrm{a}}(\mathcal{W})$ is positive iff for some $n \geqq 1$ the condition of (a) is valid for $\mathcal{W}^n$ in the role of $\mathcal{W}$. (Ahlswede (1978).)

(c) Show that if $\mathcal{W}$ is *symmetric*, i.e., $\mathsf{S} = \mathsf{X}$ and $W(y|x, x') = W(y|x', x)$ for all x,x' in X and y in Y, then $C_{\mathrm{a}}(\mathcal{W}) = 0$; moreover, then $\bar{e}(\mathcal{W}^n, f, \varphi)$ is bounded below as in the hint to Problem 12.2. Show the same is also true for *symmetrizable* AVCs, namely when a "symmetrizing channel" $U : \mathsf{X} \to \mathsf{S}$ exists such that

$$W'(y|x, x') \triangleq \sum_{s \in \mathsf{S}} W(y|x, s)U(s|x')$$

defines a symmetric AVC. (Ericson (1985).)

(d)* Show that when $\mathcal{W}$ is a finite class of channels, non-symmetrizability is necessary and sufficient for positivity of $C_{\mathrm{a}}(\mathcal{W})$. (Csiszár and Narayan (1988). Sufficiency follows from Problem 12.10; a simple direct proof is still elusive.)

12.10.* (*AVC with state constraint*) Given a non-negative function l on S, suppose only those state sequences $\mathbf{s} \in \mathsf{S}^n$ are admissible for which $l(\mathbf{s}) \triangleq (1/n)\sum_{i=1}^{n} l(s_i)$ does not exceed a given constant Λ. Suppose S is finite, and denote by $\mathcal{P}(\Lambda)$ the set of those PDs P on X for which no symmetrizing channel $U : \mathsf{X} \to \mathsf{S}$ (see Problem 12.9(c)) satisfies $\sum_{x \in \mathsf{X}} \sum_{s \in \mathsf{S}} P(x)U(s|x)l(s) < \Lambda$.

(a) Show that the a-capacity under the above state constraint, denoted by $C_{\mathrm{a}}(\mathcal{W}, \Lambda)$, is zero if $\mathcal{P}(\Lambda) = \emptyset$ and positive if $\mathcal{P}(\Lambda')$ is non-empty for some $\Lambda' < \Lambda$. Moreover, in the latter case show that

$$C_{\mathrm{a}}(\mathcal{W}, \Lambda) = \max_{P \in \mathcal{P}(\Lambda)} \quad \min_{Q :\ \sum Q(s)l(s) \leqq \Lambda} I(P, W_Q),$$

where $W_Q(y|x) \triangleq \sum_{s \in \mathsf{S}} Q(s)W(y|x, s)$. (Csiszár and Narayan (1988). The idea of proof is similar to that in Problem 12.5: a suitable packing lemma is needed, and a key part is again to find a good decoder. A special case of this result is that the a-capacity of a non-symmetrizable AVC without state constraint equals $C(\overline{\mathcal{W}})$; this is equivalent to the assertion of Problem 12.9(d), due to Theorem 12.11.)

(b) Consider the "mod 2 adder" AVC with $\mathsf{X} = \mathsf{Y} = \mathsf{S} = \{0, 1\}$ and $W(y|x, s) = 1$ if $y = x + s \bmod 2$, otherwise zero, and let $l(s) \triangleq s$. Show that in this case a-capacity equals random code capacity, which is $1 - h(\Lambda)$ if $\Lambda < 1/2$, and zero otherwise. Moreover, show that the m-capacity is zero if $\Lambda \geqq 1/4$, and is at least $1 - h(2\Lambda)$ if $\Lambda < 1/4$.

Hint The assertion about a-capacity follows from (a), and that about m-capacity follows from Problem 5.5. Check that the maximum probability of error of any code is either unity or zero, the latter (with a suitable decoder) iff the codeword set has minimum distance $d_{\min} > 2\lfloor \Lambda n \rfloor$.

(c) Let $\{\mathcal{W}\}$ be the "arithmetic adder" AVC in the hint to Problem 12.2, again with $l(0) \triangleq 0$, $l(1) \triangleq 1$. Deduce from (a) that if $1/2 < \Lambda < 1$ then $C_{\mathrm{a}}(\mathcal{W}, \Lambda)$ equals $I(X \wedge X + S)$ for independent RVs X, S with $P_X = P_S = (1 - \Lambda, \Lambda)$. Show also that the random code capacity under the given state constraint is $1/2$. (Csiszár and Narayan (1998). This example shows that the dichotomy in Theorem 12.11 need not hold for AVCs with state constraint.)

Hint The only symmetrizing channel $U : \mathsf{X} \to \mathsf{S}$ is the identity matrix.

12.11. Formulate a communication model involving block codes with active feedback (see Problem 10.38) for an AVC. Show that if active feedback lets the sender learn the outcome of a random experiment performed at the receiver, and codes depending on it are admitted, both m- and a-capacities will be equal to the random code capacity $C(\overline{\mathcal{W}})$, even if the capacity of the feedback link is arbitrarily small.

Hint Use Lemmas 12.8 and 12.10 to show that if a random experiment with at least n^2 equiprobable outcomes is accessible to both sender and receiver and codes depending on its outcome may be used, there exist such codes with rate approaching $C(\overline{\mathcal{W}})$ and maximum probability of error approaching zero. This idea works also if only passive feedback is available; see Problem 12.16.

12.12. (*Source–channel transmission theorems for AVCs*)

(a) Show that to any sequence of distributions P_n on sets M_n such that

$$\overline{\lim_{n\to\infty}} \frac{1}{n} \log |\mathsf{M}_n| < C_{\mathrm{a}}(\mathcal{W}), \qquad \lim_{n\to\infty} n^2 \cdot \max_{m\in\mathsf{M}_n} P_n(m) = 0,$$

there exists a sequence of n-length block codes with message sets M_n and overall probability of error

$$\max_{\mathbf{s}\in\mathsf{S}^n} \sum_{m\in\mathsf{M}_n} P_n(m) e_m(W^n(\cdot\,|\,\cdot\,, \mathbf{s}), f_n, \varphi_n) \to 0.$$

Hint Assume without any loss of generality that $|\mathsf{M}_n|$ is divisible by n^2. Represent M_n in the form $\{1, \dots, n^2\} \times \tilde{\mathsf{M}}_n$ so that the probabilities of any two sets $\{i\} \times \tilde{\mathsf{M}}_n$ differ by at most $\max_m P_n(m)$. Then the probabilities of these sets asymptotically equal $1/n^2$ and the codes constructed in the proof of Theorem 12.11 (with $\tilde{\mathsf{M}}_n$ playing the role of the M_n of (12.25)) will do.

(b) Given a source $\{Z_i\}_{i=1}^{\infty}$ with alphabet Z, the error probability of a k-to-n block code $(f, \varphi) \in \mathcal{C}(\mathsf{Z}^k \to \mathsf{X}^n, \mathsf{Y}^n \to \mathsf{Z}^k)$ for the AVC $\{\mathcal{W} : \mathsf{X} \to \mathsf{Y}\}$ is defined as

$$\max_{\mathbf{s}\in\mathsf{S}^n} \left[1 - \sum_{\mathbf{z}\in\mathsf{Z}^k} P_{Z^k}(\mathbf{z}) W^n(\varphi^{-1}(\mathbf{z}) | f(\mathbf{z}), \mathbf{s}) \right].$$

Show that for a DMS with generic distribution P and an AVC $\{\mathcal{W}\}$ there exist k-to-n block codes with $n/k \to L$ and error probability tending to 0 iff $L \geqq \frac{H(P)}{C_{\mathrm{a}}(\mathcal{W})}$.

Hint The sufficiency follows from (a), setting $\mathsf{M}_n \triangleq \mathsf{T}^k_{[P]}$. The necessity is easily seen for the a-capacity for codes with stochastic encoder (partition $\mathsf{T}^k_{[P]}$ into $\exp[k(H(P) - \delta)]$ subsets of asymptotically equal probability; any encoder $f : \mathsf{Z}^k \to \mathsf{X}^n$ gives rise to a stochastic encoder $F : \tilde{\mathsf{M}}_n \to \mathsf{X}^n$, where

$\tilde{\mathsf{M}}_n$ is the family of atoms of the above partition). But the latter capacity equals $C_{\mathrm{a}}(\mathcal{W})$ by Theorem 12.11.

(c) Observe that there exist k-to-n block codes with $n/k \to L$ and with error probability tending to zero uniformly for every source with alphabet Z iff $L \geqq \log|\mathsf{Z}|/C_{\mathrm{m}}(\mathcal{W})$. The same holds also for the smaller family of all sources with independent outputs.

(d) Conclude from (b) and (c) that no source–channel transmission theorem can be formulated for AVSs and AVCs in terms of the ratio of two numbers, characterizing the AVS and the AVC, respectively, if ordinary codes are used. If, however, codes with stochastic encoder are admitted, then for any AVS (defined by a set of PDs $\mathcal{P}$) and any AVC $\{\mathcal{W}\}$, the greatest lower bound of L equals $(1/C)\max_{P\in\overline{\mathcal{P}}} H(P)$, where $C = C_{\mathrm{a}}(\mathcal{W})$ is the common value of the m- and a-capacities for codes with stochastic encoder.

12.13. (*Stochastic decoding does not increase capacity*) A code with stochastic encoder and decoder is a pair of stochastic matrices $F : \mathsf{M} \to \mathsf{X}$, $\Phi : \mathsf{Y} \to \mathsf{M}'$. Given a channel $W : \mathsf{X} \to \mathsf{Y}$ and such a code, the analog of the matrix T of (12.30) is the product matrix $T \triangleq FW\Phi$.

(a) Observe that, from the point of view of error probabilities, codes with stochastic encoder and decoder are equivalent to random codes defined by those distributions Q on $\mathcal{C}(\mathsf{M} \to \mathsf{X}, \mathsf{Y} \to \mathsf{M}')$ which have a product form

$$Q(f, \varphi) = Q_1(f)Q_2(\varphi).$$

(b) Check that Theorem 12.13 is valid also for codes with stochastic encoder and decoder.

(c) Show that admitting codes (f, Φ) with ordinary encoder and stochastic decoder does not increase the m-capacity of an AVC.

Hint Given (f, Φ) with $e(\mathcal{W}^n, f, \Phi) < \varepsilon$, verify the existence of a decoder φ with $e(\mathcal{W}^n, f, \varphi) < 2\varepsilon$ by random selection of φ. (Let the RVs $\varphi(\mathbf{y})$, $\mathbf{y} \in \mathsf{Y}^n$ be independent with distributions $\Phi(\cdot\,|\mathbf{y})$.)

12.14. (*Game-theoretic approach*) If the permissible (pure) strategies of the code selector are codes $(f, \varphi) \in \mathcal{C} = \mathcal{C}(\mathsf{M} \to \mathsf{X}^n, \mathsf{Y}^n \to \mathsf{M}')$ then the mixed strategies are random codes, i.e., distributions Q on $\mathcal{C}$.

(a) Note that if the pay-off is the average probability of error then for a mixed strategy Q of the code selector and a state sequence $\mathbf{s} \in \mathsf{S}^n$ the expected pay-off equals $\bar{e}(W^n(\cdot\,|\cdot, \mathbf{s}), Q)$. For the maximum probability of error pay-off, however, the expected pay-off is, in general, larger than $e(W^n(\cdot\,|\cdot, \mathbf{s}), Q)$.

(b) If the permissible (pure) strategies of the state selector are sequences $\mathbf{s} \in \mathsf{S}^n$, then for a mixed strategy, i.e., a distribution P on S^n, write

$$W^n(\mathbf{y}|\mathbf{x}, P) \triangleq \sum_{\mathbf{s}} W^n(\mathbf{y}|\mathbf{x}, \mathbf{s})P(\mathbf{s}).$$

Conclude from the minimax theorem (see Karlin (1959), Th. 1.4.1) that

$$\min_{Q} \bar{e}(\mathcal{W}^n, Q) = \max_{P} \min_{(f,\varphi)} \bar{e}(W^n(\cdot\,|\,\cdot, P), f, \varphi).$$

Use this result to determine the random code capacity of $\{\mathcal{W}\}$. (Blackwell, Breiman and Thomasian (1960).)

12.15. (*States known both at input and output*) Show that if the permissible strategies of the code selector are codes depending on the whole state sequence $\mathbf{s} \in \mathsf{S}^n$ then both the m- and a-capacities are equal to $\inf_{W \in \mathcal{W}} C(W)$.
Hint Use the result of Problem 6.8.
(From a game-theoretic point of view, this model can also be formulated in terms of the same sets of (pure) strategies as in Problem 12.14, but assuming that the state selector uses the minimax strategy, while the model of Definition 12.2 involves the code selector's minimax strategy. Note the somewhat surprising fact that if $\mathcal{W}$ is convex then the code selector's knowledge of states does not increase the a-capacity, provided that $C_{\mathrm{a}}(\mathcal{W}) > 0$.)

12.16. (*States known at the output*)
(a) Conclude from Theorem 12.11 that the a-capacity of an AVC $\{\mathcal{W}\}$ with states known at the output is either zero or $C(\mathcal{W})$.
Hint Similarly to Problem 6.31(c), the a- resp. m-capacity with states known at the output of an AVC $\{\mathcal{W} : \mathsf{X} \to \mathsf{Y}\}$ with $\mathcal{W}$ consisting of stochastic matrices $W(.\,|\,., s)$, $s \in \mathsf{S}$ is the same as the ordinary a- resp. m-capacity of the AVC $\{\mathcal{W}' : \mathsf{X} \to \mathsf{Y} \times \mathsf{S}\}$ defined by stochastic matrices W' as in the hint of Problem 6.31(c). Show that for any PD P on X, the infimum of $I(P, W')$ subject to $W' \in \overline{\mathcal{W}'}$ equals that of $I(P, W)$ subject to $P \in \mathcal{W}$.
(b) Prove the stronger statement that the above a-capacity always equals $C(\mathcal{W})$, i.e., the capacity of the compound DMC defined by $\mathcal{W}$. (Stambler (1975).)
Hint To any encoder $f : \{1, \dots, M_n\} \to \mathsf{X}^n$ define the decoder φ by

$$\varphi(\mathbf{y}, \mathbf{s}) \triangleq \begin{cases} \text{smallest } i \text{ for which } \mathbf{y} \in \mathsf{T}^n_{[W]}(f(i), \mathbf{s}) \\ 0 \quad \text{if} \quad \mathbf{y} \notin \bigcup_{i=1}^{M_n} \mathsf{T}^n_{[W]}(f(i), \mathbf{s}). \end{cases}$$

Supposing $|\mathsf{S}| < \infty$, it suffices to prove the following: if the codewords are chosen at random, setting $f(i) \triangleq X_i^n$, where the RVs $X_i^n, i = 1, \dots, M_n$, are independent with common distribution P^n, then for $(1/n) \log M_n \to R$, $0 < R < I(P, \mathcal{W})$, the probability $\Pr\{\bar{e}(W^n(\cdot\,|\,\cdot, \mathbf{s}), f, \varphi(\cdot, \mathbf{s})) > \varepsilon\}$ tends to zero faster than exponentially (as $n \to \infty$), uniformly in $\mathbf{s}$. The last probability equals

$$\Pr\left\{\frac{1}{M_n} \sum_{i=1}^{M_n} Z_i(\mathbf{s}) > \varepsilon\right\},$$

where

$$Z_i(\mathbf{s}) \triangleq W^n \left(\left(\mathsf{Y}^n - \mathsf{T}^n_{[W]}(X_i^n, \mathbf{s})\right) \cup \left(\bigcup_{j<i} \mathsf{T}^n_{[W]}\left(X_j^n, \mathbf{s}\right) \right) \Big| X_i^n, \mathbf{s} \right).$$

One easily checks that, for any fixed values of $X_1^n, \dots, X_{i-1}^n$, the conditional expectation of $Z_i(\mathbf{s})$ is less than $\varepsilon/2$ (if n is sufficiently large). Hence the assertion can be deduced.

12.17. (*Previous states known at the input*)

(a) Show that the m-capacity of an AVC with previous states known at the input does not exceed $C(\overline{\overline{\mathcal{W}}})$. In particular, if $|\mathsf{Y}| = 2$ show that the knowledge of previous states at the input does not increase m-capacity.

Hint Let $V : \mathsf{X} \to \mathsf{Y}$ be any channel of form (12.3), i.e.

$$V(y|x) = \sum_{s \in \tilde{\mathsf{S}}} P(s|x) W(y|x, s),$$

and let (f, φ) be any permissible strategy of the code selector, where $f = f_1 \dots f_n$, $f_i : \mathsf{M} \times \mathsf{S}^{i-1} \to \mathsf{X}$. Let M be a RV uniformly distributed on M and let the RVs $X_i, S_i, Y_i, i = 1, \dots, n$, be defined recursively so that

$$X_i \triangleq f_i(M, S^{i-1}), \quad MX^{i-1}S^{i-1}Y^{i-1} \multimap X_i \multimap S_i Y_i, \quad P_{S_i|X_i} \triangleq P$$

and $P_{Y_i|X_i S_i} \triangleq W$. Show that

$$\Pr\{\varphi(Y^n) \neq M\} \leqq e(\mathcal{W}^n, f, \varphi)$$

and

$$I(M \wedge \varphi(Y^n)) \leqq \sum_{i=1}^{n} I(M \wedge Y_i | Y^{i-1}) \leqq \sum_{i=1}^{n} I(X_i \wedge Y_i | Y^{i-1}) \leqq nC(V).$$

Hence the first assertion follows by Fano's inequality. The second assertion follows from Theorem 12.7.

(b) Show that the a-capacity of an AVC with previous states known at the input is either zero or $C(\overline{\mathcal{W}})$. In particular, if $C_a(\mathcal{W}) > 0$ show that the knowledge of previous states at the input does not increase a-capacity.

Hint In order to check that the a-capacity with previous states known at the input never exceeds $C(\overline{\mathcal{W}})$, let $V : \mathsf{X} \to \mathsf{Y}$ be any channel of form (12.2) and let M, X_i, S_i, Y_i be RVs as above but now let S_i be independent of X_i, with distribution P. Show that in this case $\Pr\{\varphi(Y^n) \neq M\} \leqq \bar{e}(\mathcal{W}^n, f, \varphi)$.

(c) Show that the a-capacity with previous states known at the input can be positive (and then equal to) $C(\overline{\mathcal{W}})$) even if $C_a(\mathcal{W}) = 0$.

Hint Let $\tilde{\mathsf{X}}_i$ be the family of all mappings $\tilde{x}_i : \mathsf{S}^i \to \mathsf{X}$ and consider the AVC $\{\widetilde{\mathcal{W}}_k : \mathsf{X} \times \tilde{\mathsf{X}}_1 \times \dots \times \tilde{\mathsf{X}}_{k-1} \to \mathsf{Y}^k\}$ with set of states S^k, defined by

$$\widetilde{W}_k(\mathbf{y}|\tilde{\mathbf{x}}, \mathbf{s}) \triangleq W(y_1|x, s_1) \prod_{i=1}^{k} W(y_i|\tilde{x}_{i-1}(s_1, \ldots, s_{i-1}), s_i),$$

$$\tilde{\mathbf{x}} = x\tilde{x}_1 \ldots \tilde{x}_{k-1}, \quad \tilde{x}_i \in \tilde{\mathsf{X}}_i.$$

Then every n-length block code for the AVC $\{\widetilde{\mathcal{W}}_k\}$ defines a permissible strategy of the code selector for $\{\mathcal{W}\}$ with previous states known at the input (of block length kn) so that the average probabilities of error are the same. Applying Problem 12.9(a) to $\{\widetilde{\mathcal{W}}_k\}$ we obtain a sufficient condition for the positivity of the a-capacity in question. Verify that if $\mathcal{W}$ is the AVC of the hint to Problem 12.2 then this condition is fulfilled for $k = 2$. (It is a plausible conjecture that the a-capacity with previous states known at the input always equals $C(\overline{\mathcal{W}})$. Apparently this problem has not been tackled yet.)

(d) Check that Theorem 12.13 remains valid if all capacities are replaced by the corresponding ones with previous states known at the input.

12.18. (*Present state known at the input*) Let the code selector's permissible strategies be pairs (f, φ), where f is a sequence of mappings $f_i : \mathsf{M} \times \mathsf{S} \to \mathsf{X}$, $i = 1, \ldots, n$, and $\varphi : \mathsf{Y}^n \to \mathsf{M}'$ is an ordinary decoder.

(a) Show that the m- and a-capacities of $\{\mathcal{W}\}$ admitting such strategies are equal to $C_{\mathrm{m}}(\widetilde{\mathcal{W}})$ resp. $C_{\mathrm{a}}(\widetilde{\mathcal{W}})$, where $\{\widetilde{\mathcal{W}} : \tilde{\mathsf{X}} \to \mathsf{Y}\}$ is the AVC defined on letting $\tilde{\mathsf{X}}$ be the family of all mappings $\tilde{x} : \mathsf{S} \to \mathsf{X}$ and setting $\widetilde{W}(y|\tilde{x}, s) \triangleq W(y|\tilde{x}(s), s)$; see Problem 6.31. Conclude that knowledge of the present state at the input may increase the capacities of an AVC.

(b) Show that the capacities of the AVC $\{\mathcal{W}\}$ with previous and present states known at the input are equal to the corresponding capacities of the AVC $\{\widetilde{\mathcal{W}}\}$ with previous states known at the input; see Problem 12.17.

12.19. (*List codes*) Define the *list code* m-*capacity* of an AVC $\{\mathcal{W}\}$ as the largest R such that for every $\varepsilon \in (0, 1)$, $\delta > 0$ and every sufficiently large n there exist list codes of block length n, list size $\exp\{n\delta\}$, rate $(1/n) \log |\mathsf{M}_f| > R - \delta$ and having maximum probability of error less than ε for every channel in $\mathcal{W}^n$. The *list code* a-*capacity* is defined similarly. Define also m- and a-capacities for list codes with constant list size l; denote them by $C_{\mathrm{m},l}(\mathcal{W})$ and $C_{\mathrm{a},l}(\mathcal{W})$.

(a) Note that Lemma 12.3 is valid for list codes too.

(b)* Show that the list code m-capacity of $\{\mathcal{W}\}$ equals $C(\overline{\overline{\mathcal{W}}})$. Moreover, show that $C_{\mathrm{m},l}(\mathcal{W})$ approaches this capacity as $l \to \infty$. (Ahlswede (1973b), (1993).)

(c)* Show that the list code a-capacity of $\{\mathcal{W}\}$ equals $C(\overline{\mathcal{W}})$. Moreover, show that $C_{\mathrm{a},l}(\mathcal{W})$ approaches this capacity as $l \to \infty$. (Ahlswede and Cai (1991).)

(d)* Improving the result of (c), show that there exists $L = L(\mathcal{W})$ such that $C_{\mathrm{a},L}(\mathcal{W}) = C(\overline{\mathcal{W}})$ while $C_{\mathrm{a},l}(\mathcal{W}) = 0$ if $l < L$. Characterize this L. (Blinovsky, Narayan and Pinsker (1995), Hughes (1997).)

12.20. (*Feedback*) A code selector's strategy using complete feedback is a pair (f, φ), where f is a sequence of mappings $f_i : \mathsf{M} \times \mathsf{Y}^{i-1} \to \mathsf{X}$, for $i = 1, \ldots, n$, and $\varphi : \mathsf{Y}^n \to \mathsf{M}'$ is a decoder in the usual sense as in Problem 6.29. Active feedback as in Problem 12.11 is not allowed. If previous states are also known at the input, f will be a sequence of mappings $f_i : \mathsf{M} \times \mathsf{S}^{i-1} \times \mathsf{Y}^{i-1} \to \mathsf{X}$. Then, (12.33) changes, replacing $W(y_i|f_i(m, s_1, \ldots, s_{i-1}), s_i)$ by $W(y_i|f_i(m, y_1, \ldots, y_{i-1}), s_i)$ in the first case, and by $W(y_i|f_i(m, s_1, \ldots, s_{i-1}, y_1, \ldots, y_{i-1}), s_i)$ in the second case.

(a) Show that m-capacity with feedback, if positive, is not less than $C(\overline{\overline{\mathcal{W}}})$, and that it may be larger.

Hint The first assertion follows from Problem 12.19(b). For the second, set $\mathsf{X} = \mathsf{Y} = \mathsf{S} \triangleq \{0, 1\}$,

$$W(0|0, 0) = W(1|1, 1) \triangleq 1, \quad W(i|0, 1) = W(i|1, 0) \triangleq \frac{1}{2}, \quad i = 0, 1.$$

Define a new AVC $\{\widetilde{\mathcal{W}}\}$ with input alphabet X, output alphabet Y^2 and set of states S^2 by

$$\widetilde{W}(y_1, y_2|0, s_1, s_2) \triangleq W(y_1|0, s_1)W(y_2|y_1, s_2),$$
$$\widetilde{W}(y_1, y_2|1, s_1, s_2) \triangleq W(y_1|1, s_1)W(y_2|y_1, s_2).$$

Every code of block length n for the AVC $\{\widetilde{\mathcal{W}}\}$ defines a code selector's strategy with feedback for the AVC $\{\mathcal{W}\}$ (of block length $2n$) which has the same maximum probability of error. Note that $C(\overline{\overline{\mathcal{W}}}) = 0$, while $C_{\mathrm{m}}(\widetilde{\mathcal{W}})$ is positive by Problem 12.1. (This example is due to Ahlswede (1973c).)

(b) Show that the a-capacity with feedback of an AVC, even with also previous states known at the input, does not exceed $C(\overline{\mathcal{W}})$.

Hint See the hint to Problem 12.17(b).

(c) Show that a-capacity with feedback always equals $C(\overline{\mathcal{W}})$. (Ahlswede and Csiszár (1998).)

Hint Use the idea of Problem 12.11, dispensing with active feedback. To show that the a-capacity with feedback is not less than $C(\overline{\mathcal{W}})$, the latter may be assumed positive. Then there exists an input RV X such that the entropy of the output Y (which depends on $s \in \mathsf{S}$) is bounded away from zero. In m time instances, a negligible fraction of the block length n, send i.i.d. repetitions of X. A crucial fact is that there exists a mapping κ of Y^m into a set K whose size is exponential in m, such that $\kappa(Y^n)$ is nearly uniformly distributed simultaneously for all $\mathbf{s} \in \mathsf{S}^m$; see Problem 17.6(b). Then $\kappa(Y^n)$ may be regarded as the outcome of a random experiment known to both sender and receiver, and they can set up a good code as in Problem 12.11, to be used in the remaining $n - m$ instances, no longer needing feedback. This gives rise to a code of block length n with stochastic encoder (as the first m transmissions were random). To get a deterministic code, check that Theorem 12.13

extends to AVCs with feedback, or proceed directly as in the proof of Theorem 12.11.

(d) Show also that for the m-capacity with feedback to equal $C(\overline{\mathcal{W}})$, a sufficient condition is the existence of an $x_0 \in \mathsf{X}$ with $H(W(\,.\,|x_0, s))$ bounded away from zero. (For the general case, m-capacity with feedback was determined by Ahlswede and Cai (2000).)

Hint Proceed as in (c). This time the first m transmissions need not be random, just send m times x_0.

(e) Show that for a DMC $\{W : \mathsf{X} \to \mathsf{Y}\}$ the zero-error capacity with feedback equals either zero or the minimum of $C(V)$ for channels $V : \mathsf{X} \to \mathsf{Y}$ such that in case $W(y|x) = 0$ also $V(y|x) = 0$. (Conjectured by Shannon (1956), proved by Ahlswede (1973c); Shannon gave another formula for the zero-error capacity with feedback of a DMC – see Problem 6.30.)

Hint As in Problem 12.3, the zero-error capacity with feedback of the DMC $\{W\}$ is equal to the m-capacity with feedback of the AVC $\{\mathcal{W}\}$ defined there. Check that $\overline{\overline{\mathcal{W}}}$ and $\overline{\mathcal{W}}$ are both equal to the set of channels V in the assertion, and verify via (a) and (b) that the m-capacity with feedback of the AVC $\{\mathcal{W}\}$, if non-zero, is equal to $C(\overline{\mathcal{W}})$.

12.21. (*States depending on previous inputs*)

(a) Show that the random code capacity of the A*VC $\{\mathcal{W}^{n*}\}_{n=1}^{\infty}$ (see (12.34)) equals that of the AVC $\{\mathcal{W}\}$. (Blackwell, Breiman and Thomasian (1960).)

Hint If Y^n is connected with X^n by any channel in $\mathcal{W}^{n*}$ then for the RVs Z_i in the proof of Lemma 12.10 the conditional expectations $E(Z_i|X^{i-1} = \mathbf{x}^{i-1})$ have the form (12.21). Thus, the inequality (12.23) still follows if one replaces the RVs $Z_i - EZ_i$ by the uncorrelated ones $Z_i - E(Z_i|X^{i-1})$.

(b) Let $\mathcal{W}^{n,k}$ be the subset of $\mathcal{W}^{n*}$ consisting of those channels $W^n(\cdot\,|\,\cdot\,, \sigma)$ which correspond to state selector's strategies of "depth of memory k," i.e., where the states $s_i = \sigma_i(\mathbf{x}^{i-1})$ depend only on the last k symbols of $\mathbf{x}^{i-1}$. Show that for every fixed k, the a-capacity of the channel $\{\mathcal{W}^{n,k}\}_{n=1}^{\infty}$ is either zero or $C(\overline{\mathcal{W}})$, and also that the analog of Theorem 12.13 is true for these channels. (It is unknown whether the same holds also for the A*VC $\{\mathcal{W}^{n*}\}_{n=1}^{\infty}$. The cardinality of the family $\mathcal{W}^{n*}$ grows doubly exponentially with n, causing the proofs used for the AVC to break down for this channel.)

(c) Suppose that the states may depend both on previous input and output symbols, i.e., the state selector's strategies are given by mappings $\sigma_i : \mathsf{X}^{i-1} \times \mathsf{Y}^{i-1} \to \mathsf{S}$. For such a strategy $\sigma = (\sigma_1, \ldots, \sigma_n)$ define

$$W^n(\mathbf{y}|\mathbf{x}, \sigma) \triangleq \prod_{i=1}^{n} W(y_i|x_i, \sigma_i(\mathbf{x}^{i-1}, \mathbf{y}^{i-1})).$$

Show that replacing $\mathcal{W}^{n*}$ by the larger family of channels $W^n(\cdot\,|\,\cdot\,, \sigma)$ as above, the analogs of the assertions of (a) and (b) still hold.

12.22. (*States depending on present input*)

(a) Let the state selector's permissible strategies be sequences of mappings $\sigma_i : \mathsf{X} \to \mathsf{S}$ giving rise to $\sigma(\mathbf{x}) \triangleq \sigma_1(x_1)\ldots\sigma_n(x_n)$, and let the code selector's strategies be codes $(f, \varphi) \in \mathcal{C}(\mathsf{M} \to \mathsf{X}^n, \mathsf{Y}^n \to \mathsf{M}')$. Show that according to whether the pay-off is the maximum resp. average probability of error, the capacity for the obtained model equals the m- resp. a-capacity of the AVC $\{\tilde{\mathcal{W}} : \mathsf{X} \to \mathsf{Y}\}$, defined by the family $\tilde{\mathcal{W}}$ of channels $W(\cdot\,|\,\cdot, \sigma)$ with "states" equal to mappings $\sigma : \mathsf{X} \to \mathsf{S}$, where $W(y|x, \sigma) \triangleq W(y|x, \sigma(x))$.

(b) Suppose that the states may depend on the previous and present inputs, i.e., the state selector's permissible strategies are defined by sequences of mappings $\sigma_i : \mathsf{X}^i \to \mathsf{S}$ as $\sigma(\mathbf{x}) = \sigma_1(x_1)\ldots\sigma_n(x_1, \ldots, x_n)$. Show that the resulting model is equivalent to the A*VC $\{\tilde{\mathcal{W}}^{n*}\}_{n=1}^{\infty}$, where $\tilde{\mathcal{W}}$ is defined as in (a).

12.23. (*States depending on the whole input sequence*)
Suppose the state selector's permissible strategies are all the mappings $\sigma : \mathsf{X}^n \to \mathsf{S}^n$, and the code selector's ones are the codes as in Problem 12.22. Show that for this model the maximum and average probability of error pay-offs lead to the same capacity value, equal to the m-capacity of the AVC $\{\mathcal{W}\}$. (Ahlswede and Wolfowitz (1969).)

12.24. (*DMC with random states, non-causal CSI*)
As in Problem 6.31, given a DMC $\{W : \mathsf{X} \times \mathsf{S} \to \mathsf{Y}\}$ let $s \in \mathsf{S}$ represent a random state, the consecutive states being i.i.d. with known distribution Q. Unlike in Problem 6.31, suppose the sender fully knows the actual state sequence $\mathbf{s} \in \mathsf{S}^n$. Formally, let the admissible encoders be all the mappings $f : \mathsf{M} \times \mathsf{S}^n \to \mathsf{X}^n$. Show that the corresponding channel capacity equals

$$C_Q \triangleq \max\,[I(U \wedge Y) - I(U \wedge S)],$$

over triples of RVs (U, S, Y) such that (i) $P_S = Q$, (ii) U is an auxiliary RV whose range satisfies $|\mathsf{U}| \leqq |\mathsf{X}| + |\mathsf{S}|$, (iii) $P_{Y|US}(y|u, s) = W(y|s, h(u, s))$, for some mapping $h : \mathsf{U} \times \mathsf{S} \to \mathsf{X}$. (Gelfand and Pinsker (1980b), motivated by a model of defects and errors in computer memories, Kuznetsov and Tsybakov (1974).)

Hint Show first that for the achievability of C_Q, it suffices to prove that of $I(X \wedge Y) - I(X \wedge S)$ for triples of RVs X, S, Y satisfying $P_S = Q$, $P_{Y|XS} = W$ (then the role of W can be given to $W' : \mathsf{U} \times \mathsf{S} \to \mathsf{Y}$ defined by $W'(y|u, s) \triangleq W(y|h(u, s), s)$). To prove the latter, select randomly sequences $\mathbf{x}_{lm}$, $l \in \mathsf{L}$, $m \in \mathsf{M}$ from the uniform distribution on $\mathsf{T}_{[X]}$, where

$$|\mathsf{L}| = \exp\{n[I(X \wedge S) + \delta]\}, \quad |\mathsf{M}| = \exp\{n[I(X \wedge Y) - I(X \wedge S) - 2\delta]\}.$$

Check that with probability close to 1, for each $m \in \mathsf{M}$ and $\mathbf{s} \in \mathsf{T}_{[S]}$ some of the sequences $\mathbf{x}_{lm}$, $l \in \mathsf{L}$ are jointly typical with $\mathbf{s}$ (show as in the proof of Lemma 9.1 that the probability that some $\mathbf{s} \in \mathsf{T}_{[S]}$ is not jointly typical with any of $\exp\{n[I(X \wedge S) + \delta]\}$ randomly selected X-typical sequences, is doubly exponentially small). Take an encoder that assigns to each $m \in \mathsf{M}$ and $\mathbf{s} \in \mathsf{T}_{[S]}$ an $\mathbf{x}_{lm}$ jointly typical with $\mathbf{s}$, and a joint typicality decoder that assigns $m \in \mathsf{M}$ to output $\mathbf{y} \in \mathsf{Y}^n$ if the latter is jointly typical with $\mathbf{x}_{lm}$ for some $l \in \mathsf{L}$, declaring an error unless a unique such $m \in \mathsf{M}$ exists. Then show that the expectation of the average probability of error of this randomly selected code goes to zero as $n \to \infty$.

For the converse, given an encoder $f : \mathsf{M} \times \mathsf{S}^n \to \mathsf{X}^n$, let M be a RV uniformly distributed on M as usual, independent of the random state sequence S^n, and let Y^n be the channel output for input $X^n \triangleq f(M, S^n)$ and states S^n. If a decoder $\varphi : \mathsf{Y}^n \to \mathsf{M}$ has a small probability of error, the rate $(1/n) \log |\mathsf{M}|$ is close to $(1/n) I(M \wedge Y^n)$, by Fano's inequality, hence this mutual information has to be bounded above.

Next, techniques developed in Chapter 15 are needed. By (15.34), whose derivation does not depend on the hypotheses of Lemma 15.7, see also (15.40) and the preceding notation,

$$\frac{1}{n}[H(S^n|M) - H(Y^n|M)] = H(S_J|U) - H(Y_J|U),$$

where J is a RV uniformly distributed on $\{1, \dots, n\}$, independent of the triple (M, S^n, Y^n), and $U \triangleq (J, M, Y^{J-1} S_{J+1} \dots S_n)$. Since

$$\frac{1}{n} H(S^n|M) = \frac{1}{n} H(S^n) = H(S_J),$$

$$\frac{1}{n} H(Y^n) \leqq \frac{1}{n} \sum_{j=1}^{n} H(Y_j) = H(Y_J|J) \leqq H(Y_J),$$

a simple calculation gives that $(1/n) I(M \wedge Y^n) \leqq I(U \wedge Y_J) - I(U \wedge S_J)$.

Here $P_{S_J} = Q$, $P_{Y_J|X_J S_J} = W$, and the Markov relation $U \multimap X_J S_J \multimap Y_J$ holds. The RVs U, X_J, S_J, Y_J can be replaced by another U, and X, S, Y that have the same properties and also (ii) is met, retaining the value of the mutual information difference, due to the support lemma, Lemma 15.4. Finally, as $I(U \wedge Y)$ is a convex function of $P_{X|US}$ when P_{US} is fixed, this difference is maximized for X equal to a function of (U, S); then also (iii) is met.

12.25. Modifying the model in Problem 12.24, suppose the states are arbitrarily varying rather than random. Thus, consider the AVC $\{\mathcal{W}\}$, where $\mathcal{W}$ is the (finite) set of channels $W(.|., s)$, $s \in \mathsf{S}$. As in Problem 12.24, suppose the sender knows the state sequence $\mathbf{s} \in \mathsf{S}^n$ before transmission starts.

(a) Show that for this AVC with non-causal CSI, the m- and a-capacity are both equal to the minimum for Q of C_Q defined in Problem 12.24. (Ahlswede (1986).)

Hint Step (i). Verify that as the sender knows the states, stochastic encoders have no advantage over deterministic ones. Draw the conclusion, see the proof of Theorem 12.13, that m- and a-capacity coincide; denote this by C.

Show that $C > 0$ iff each channel $W \in \mathcal{W}$ has positive capacity (non-trivial!). Recalling the proof of Theorem 12.11, it remains to show that if $C > 0$, there exist good random codes (with states known to sender) of rates arbitrarily close to $\min C_Q$.

Step (ii). Using the result of Problem 12.24, for each type Q of sequences in S^n take a code (f_Q, φ_Q), where f_Q maps $\mathsf{M} \times \mathsf{S}^n$ into X^n with rate $(1/n)\log|\mathsf{M}| = \min C_Q - \delta$, such that, if the states were Q-i.i.d., the maximum probability of error would be small, i.e.,

$$\sum_{\mathbf{s}\in\mathsf{S}^n} W^n(\{\mathbf{y}\colon \varphi_Q(\mathbf{y}) \neq m\}|f_Q(m,\mathbf{s}),\mathbf{s})Q^n(\mathbf{s}) \leqq \varepsilon_n, \quad m \in \mathsf{M}.$$

One sees from Problem 12.24 that ε_n can be assumed to go to zero, uniformly in Q, faster than any negative power of n (actually, exponentially fast, as one could see with a little more effort). Conclude that

$$\frac{1}{|\mathsf{T}_Q|}\sum_{\mathbf{s}\in\mathsf{T}_Q} W^n(\{\mathbf{y}\colon \varphi_Q(\mathbf{y}) \neq m\}|f_Q(m,\mathbf{s}),\mathbf{s}) \leqq (n+1)^{|\mathsf{S}|}\varepsilon_n.$$

Step (iii). Show that the preceding inequality also holds with f_Q, φ_Q replaced by suitable f, φ not depending on Q (as $C > 0$, the sender may use a small fraction of the rate to tell the receiver the type of the state sequence $\mathbf{s}$). Verify that, with this replacement, the left-hand side of the preceding inequality is equal to

$$\frac{1}{n!}\sum_{\pi} W^n(\{\mathbf{y}\colon \varphi(\mathbf{y}) \neq m\}|f(m,\pi\mathbf{s}),\pi\mathbf{s}),$$

where $\mathbf{s} \in \mathsf{T}_Q$ is arbitrary and π runs over all permutations of $\{1,\ldots,n\}$.

Step (iv). Verify that the preceding expression equals

$$\frac{1}{n!}\sum_{\pi} W^n(\{\mathbf{y}\colon \varphi(\pi\mathbf{y}) \neq m\}|\pi^{-1}f(m,\pi\mathbf{s}),\mathbf{s}),$$

and interpret this as the probability of error for message m when the state sequence is $\mathbf{s}$, under the random code defined by the uniform distribution on the $n!$ deterministic codes with encoders $\pi^{-1}f(m,\pi\mathbf{s})$ and decoders $\varphi(\pi\mathbf{y})$.

(b)* Show that the capacity in (a) is equal to $\min_{W\in\mathcal{W}} C(W)$ if the matrices $W \in \mathcal{W}$ have only 0 and 1 entries, but can be smaller in general. (Ahlswede (1986).)

Story of the results

Arbitrarily varying channels were introduced by Blackwell, Breiman and Thomasian (1960). The main content of this chapter is a somewhat streamlined presentation of the results of Ahlswede and Wolfowitz (1970) and Ahlswede (1978). The results from Lemma 12.3 to Theorem 12.7 are those of Ahlswede and Wolfowitz (1970).

Theorems 12.11 and 12.13 belong to Ahlswede (1978), and the random code reduction lemma contains the key idea of his proof. Lemma 12.10 was established by Blackwell, Breiman and Thomasian (1960). Their proof was simplified by Stiglitz (1966); the proof here is that of Ahlswede and Wolfowitz (1969). For a certain class of AVCs, the a-capacity was determined previously by Dobrušin and Stambler (1975).

Other relevant references, and also results obtained since the first edition of this book, are given in the Problems section.

Part III

Multi-terminal systems

13 Separate coding of correlated sources

Let us be given a sequence $\{(X_i, Y_i)\}_{i=1}^{\infty}$ of independent copies of a pair of RVs (X, Y) taking values in finite sets X and Y, respectively. Considered separately, the sequences $\{X_i\}_{i=1}^{\infty}$ and $\{Y_i\}_{i=1}^{\infty}$ are two DMSs whose ith outputs X_i and Y_i are not necessarily independent. Therefore, the sequence $(X_i, Y_i)_{i=1}^{\alpha}$ will be called a *discrete memoryless multiple source* (DMMS) with two components and *generic variables* (X, Y) or simply a *2-source.* If the two sources are encoded and decoded separately as if the other source were not there, then, of course, the minimum rate needed for a nearly perfect reproduction is $H(X) + H(Y)$. If X and Y are not independent,

$$H(X, Y) < H(X) + H(Y).$$

This means that separate encoding–decoding of the two component sources yields significantly worse codes of the joint source than joint encoding–decoding. This is not surprising at all, and makes it interesting to look at in-between cases. These are the typical problems in multi-terminal source coding.

Let us see first how the presence of the source $\{Y_i\}_{i=1}^{\infty}$ affects the encoding–decoding of the source $\{X_i\}_{i=1}^{\infty}$. Consider the DMS $\{X_i\}_{i=1}^{\infty}$ as the "main source" and the DMS $\{Y_i\}_{i=1}^{\infty}$ as "side information" available to the encoder and the decoder of the main source. Clearly, whatever fidelity criterion we consider, knowledge of the actual outputs of the DMS $\{Y_i\}_{i=1}^{\infty}$ can be helpful in defining the best possible way of encoding and decoding the source $\{X_i\}_{i=1}^{\infty}$. In fact, for every value $\mathbf{y} \in \mathsf{Y}^k$ of Y^k the RVs $X_1 X_2 \dots X_k$ → 1.5
can have a different conditional joint distribution, upon which the most appropriate way of encoding and decoding depends.

The problem becomes more interesting if the side information is available only at the decoder. After having formalized the problem, we shall show that the minimum rate at which X^k must be encoded is asymptotically the same in the two aforementioned situations, if nearly perfect reproduction of X^k is required. This is a rather surprising result. We shall see in Chapter 16 that it does not generalize to arbitrary fidelity criteria.

DEFINITION 13.1 A *k-length block code with side information at the decoder* for sources with main source alphabet X and side information alphabet Y is a pair of mappings $f : \mathsf{X}^k \to \mathsf{M}$, $\varphi : \mathsf{M} \times \mathsf{Y}^k \to \mathsf{X}^k$. Here M is an arbitrary finite set. The error probability of this code for a 2-source $\{(X_i, Y_i)\}_{i=1}^{\infty}$ is

$$e(f, \varphi) \triangleq \Pr\{\varphi(f(X^k), Y^k) \neq X^k\}.$$

○

→ 13.1

THEOREM 13.2 For every ε, $\delta \in (0, 1)$ and 2-source with arbitrary generic variables (X, Y) there exists a k-length block code with side information at the decoder, (f, φ), such that

(i) $\frac{1}{k} \log \|f\| \leqq H(X|Y) + \delta$,

(ii) $e(f, \varphi) \leqq \varepsilon$,

whenever $k \geqq k_0(|\mathsf{X}|, |\mathsf{Y}|, \varepsilon, \delta)$. ○

Proof Consider a stochastic matrix $W : \mathsf{X} \to \mathsf{Y}$ such that whenever $\Pr\{Y = y\,|X = x\}$ is defined it equals $W(y|x)$. Then Y^k can be regarded as the output of channel W^k corresponding to input X^k. Temporarily, let $f : \mathsf{X}^k \to \mathsf{M}$ be any mapping of X^k onto $\mathsf{M} \triangleq \{1, 2, \ldots, m\}$ such that, for some $\hat{\varepsilon}$ to be specified later, $f^{-1}(i)$ is the codeword set of a $(k, \hat{\varepsilon})$-code for the DMC $\{W\}$ for every $i < m$, and $\Pr\{f(X^k) = m\} \leqq 2\hat{\varepsilon}$.

We suppose that for every channel code in question the message set coincides with the set of codewords. Thus the decoder of the $(k, \hat{\varepsilon})$-code with codeword set $f^{-1}(i)$ is some mapping $\varphi^{(i)} : \mathsf{Y}^k \to f^{-1}(i)$. It is quite clear that mappings f with the above properties always exist.

If $\hat{\varepsilon} < \varepsilon/3$ then the decoders $\varphi^{(i)}$ of our $(k, \hat{\varepsilon})$-codes can be combined into a mapping $\varphi : \mathsf{M} \times \mathsf{Y}^k \to \mathsf{X}^k$ such that (f, φ) meets condition (ii) of the theorem. In fact, putting

$$\varphi(i, \mathbf{y}) \triangleq \begin{cases} \varphi^{(i)}(\mathbf{y}) & \text{if } \ i < m \\ \text{arbitrary} & \text{if } \ i = m \end{cases}$$

we have

$$\begin{aligned} \Pr\{\varphi(f(\mathbf{x}), Y^k) \neq \mathbf{x}|X^k = \mathbf{x}\} &= \Pr\{\varphi^{(i)}(Y^k) \neq \mathbf{x}|X^k = \mathbf{x}\} \\ &= 1 - W^k({\varphi^{(i)}}^{-1}(\mathbf{x})|\mathbf{x}) < \hat{\varepsilon} \end{aligned}$$

for every $\mathbf{x}$ with $i = f(\mathbf{x}) < m$. Thus

$$\begin{aligned} e(f, \varphi) \leqq &\sum_{\mathbf{x}:\, f(\mathbf{x})<m} \Pr\{X^k = \mathbf{x}\}\Pr\{\varphi(f(\mathbf{x}), Y^k) \neq \mathbf{x}|X^k = \mathbf{x}\} \\ &+ \Pr\{f(X^k) = m\} \leqq 3\hat{\varepsilon}. \end{aligned}$$

Hence the proof will be complete if we can find a mapping f as above meeting condition (i). In other words, it suffices to show that there exist $m - 1$ $(k, \hat{\varepsilon})$-codes for the DMC $\{W\}$ with disjoint codeword sets C_i, $i = 1, \ldots, m - 1$, such that

$$\frac{1}{k} \log m \leqq H(X|Y) + \delta \tag{13.1}$$

and

$$\Pr\left\{X^k \notin \bigcup_{i=1}^{m-1} \mathsf{C}_i\right\} < 2\hat{\varepsilon}. \tag{13.2}$$

We shall construct such codes by means of Theorem 6.10.

Fixing $0 < \hat{\varepsilon} < 1/2$ and $\delta > 0$, let k be so large that

$$\Pr\{X^k \in \mathsf{T}^k_{[X]}\} \geqq 1 - \hat{\varepsilon} \tag{13.3}$$

and that for every set $\mathsf{A} \subset \mathsf{T}^k_{[X]}$ with $\Pr\{X^k \in \mathsf{A}\} \geqq \hat{\varepsilon}$ there exists a $(k, \hat{\varepsilon})$-code for the DMC $\{W\}$ with codeword set $\mathsf{C} \subset \mathsf{A}$ of size

$$|\mathsf{C}| \geqq \exp\{k[I(X \wedge Y) - 3\hat{\delta}]\}. \tag{13.4}$$

Put $\mathsf{A}_1 \triangleq \mathsf{T}^k_{[X]} = \mathsf{T}_{[X]}$ and let $\mathsf{C}_1 \subset \mathsf{A}_1$ be the codeword set of a $(k, \hat{\varepsilon})$-code satisfying (13.4). After having constructed the disjoint codeword sets $\mathsf{C}_1, \ldots, \mathsf{C}_{i-1}$ of $(k, \hat{\varepsilon})$-codes satisfying (13.4) we can still construct the ith codeword set

$$\mathsf{C}_i \subset \mathsf{A}_i \triangleq \mathsf{T}_{[X]} - \bigcup_{j=1}^{i-1} \mathsf{C}_j$$

having the same property, unless $\Pr\{X^k \in \mathsf{A}_i\} < \hat{\varepsilon}$. Suppose that the construction stops after C_{m-1}. Then, since

$$\Pr\{X^k \notin \bigcup_{i=1}^{m-1} \mathsf{C}_i\} \leqq \Pr\{X^k \notin \mathsf{T}_{[X]}\} + \Pr\{X^k \in \mathsf{A}_m\},$$

condition (13.2) is fulfilled; see (13.3). Further, from (13.4),

$$|\mathsf{T}_{[X]}| \geqq \sum_{i=1}^{m-1} |\mathsf{C}_i| \geqq (m-1)\exp\{k[I(X \wedge Y) - 3\hat{\delta}]\}.$$

Setting $\hat{\delta} \triangleq \delta/4$, this implies by Lemma 2.13 that also condition (13.1) is met for k large enough. → 13.2 □

COROLLARY 13.2 (*Rate slicing*) For every $\delta > 0$, $\varepsilon \in (0, 1)$, to any 2-source with generic variables (X, Y) there exist encoders $f : \mathsf{X}^k \to \mathsf{M}_\mathsf{X}$, $g : \mathsf{Y}^k \to \mathsf{M}_\mathsf{Y}$, and a decoder $\varphi : \mathsf{M}_\mathsf{X} \times \mathsf{M}_\mathsf{Y} \to \mathsf{X}^k \times \mathsf{Y}^k$ such that

(i) $\frac{1}{k} \log \|f\| \leqq H(X|Y) + \delta$,

(ii) $\frac{1}{k} \log \|g\| \leqq H(Y) + \delta$

and

(iii) $\Pr\{\varphi(f(X^k), g(Y^k)) \neq (X^k, Y^k)\} \leqq \varepsilon$,

whenever $k \geqq k_0(|\mathsf{X}|, |\mathsf{Y}|, \delta, \varepsilon)$. Moreover, if g is the encoder of any k-length block code of the Y-source with error probability less than $\varepsilon/2$ then there exist f and φ such that (i) and (iii) hold, whenever $k \geqq k_0(|\mathsf{X}|, |\mathsf{Y}|, \delta, \varepsilon)$. ○

Proof On account of Theorem 1.1 it suffices to prove the last assertion. By Theorem 13.2 there exist functions $f : \mathsf{X}^k \to \mathsf{M}_\mathsf{X}$, $\tilde{\varphi} : \mathsf{M}_\mathsf{X} \times \mathsf{Y}^k \to \mathsf{X}^k$ such that (i) holds and

$$\Pr\{\tilde{\varphi}(f(X^k), Y^k) \neq X^k\} < \frac{\varepsilon}{2}.$$

If (g, ψ) is a code for the Y-source, where $g : \mathsf{Y}^k \to \mathsf{M}_\mathsf{Y}$, $\psi : \mathsf{M}_\mathsf{Y} \to \mathsf{Y}^k$, with $\Pr\{\psi(g(Y^k)) \neq Y^k\} < \varepsilon/2$, then f, g and

$$\varphi(i, j) \triangleq (\tilde{\varphi}(i, \psi(j)), \psi(j)), \quad i \in \mathsf{M}_\mathsf{X}, \quad j \in \mathsf{M}_\mathsf{Y}$$

satisfy (i) and (iii). □

The above theorem plays a fundamental role in the theory of multi-terminal memoryless systems, especially in proving coding theorems for networks involving several information sources.

Let $X_\mathsf{A} \triangleq \{X_a\}_{a \in A}$ be an r-tuple of RVs ranging over finite sets X_a, where A is an index set of size $|\mathsf{A}| = r$. A sequence $\{X_{\mathsf{A},i}\}_{i=1}^{\infty}$ of independent copies $X_{\mathsf{A},i} \triangleq \{\mathrm{X}_{a,i}\}_{a \in \mathsf{A}}$ of X_A is called a *discrete memoryless multiple source* (DMMS) with r components, or simply an *r-source.* The r DMSs $\{X_{a,i}\}_{i=1}^{\infty}$ are called the *component sources* of the DMMS. The RVs X_a, $a \in \mathsf{A}$ are called the *generic variables* of the DMMS.

Suppose that separately block-encoded versions of the r components of a DMMS are available to a decoder. The task of the latter is to reproduce the r source output blocks within small probability of error. Because of the visual pattern describing the connection of the encoders to the decoder, this setup will be called a *fork network.* Defining the rates of the r encoders as usual, we ask how small these rates can be. We are interested in an asymptotic characterization. In order to avoid the inconvenience of involved notation, we shall temporarily restrict our attention to 3-sources.

DEFINITION 13.3 Let $\{(X_i, Y_i, Z_i)\}_{i=1}^{\infty}$ be a DMMS with three components and generic variables X, Y, Z ranging over X, Y and Z, respectively. A *k-length block code for the corresponding fork network* is a quadruple of mappings (f, g, h, φ) with

$$f : \mathsf{X}^k \to \mathsf{M}_\mathsf{X}, \quad g : \mathsf{Y}^k \to \mathsf{M}_\mathsf{Y}, \quad h : \mathsf{Z}^k \to \mathsf{M}_\mathsf{Z},$$
$$\varphi : \mathsf{M}_\mathsf{X} \times \mathsf{M}_\mathsf{Y} \times \mathsf{M}_\mathsf{Z} \to \mathsf{X}^k \times \mathsf{Y}^k \times \mathsf{Z}^k,$$

where M_X, M_Y and M_Z are arbitrary finite sets. The error probability of this code is

$$e(f, g, h, \varphi) \triangleq \Pr\{\varphi(f(X^k), g(Y^k), h(Z^k)) \neq (X^k, Y^k, Z^k)\}. \tag{13.5}$$

The *rate triple* of the code is

$$\mathbf{R}(f, g, h) \triangleq \left(\frac{1}{k}\log\|f\|, \frac{1}{k}\log\|g\|, \frac{1}{k}\log\|h\|\right).$$

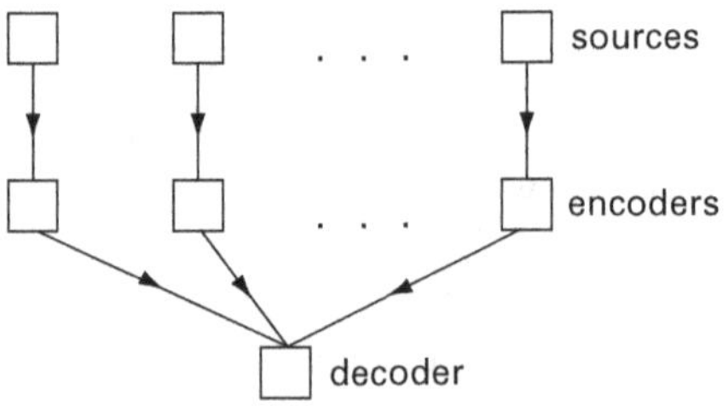

Figure 13.1 Fork network.

A triple of non-negative real numbers $(R_{\mathsf{X}}, R_{\mathsf{Y}}, R_{\mathsf{Z}})$ is an *ε-achievable rate triple* for the fork network if for every $\delta > 0$ and sufficiently large k there exist k-length block codes of error probability at most ε and rate triple

$$\frac{1}{k}\log\|f\| \leqq R_{\mathsf{X}} + \delta, \quad \frac{1}{k}\log\|g\| \leqq R_{\mathsf{Y}} + \delta, \quad \frac{1}{k}\log\|h\| \leqq R_{\mathsf{Z}} + \delta.$$

Further, $(R_{\mathsf{X}}, R_{\mathsf{Y}}, R_{\mathsf{Z}})$ is an *achievable rate triple* if it is ε-achievable for every $\varepsilon > 0$. The set of ε-achievable (achievable) rate triples is the *ε-achievable (achievable) rate region*, denoted by $\mathcal{R}_\varepsilon(X, Y, Z)$ and $\mathcal{R}(X, Y, Z)$, respectively. ○

Clearly,

$$\mathcal{R}(X, Y, Z) = \bigcap_{\varepsilon>0} \mathcal{R}_\varepsilon(X, Y, Z).$$

REMARK Considering the 3-source with generic variables X, Y, Z as a single DMS with generic variable (X, Y, Z), the fork network code (f, g, h, φ) can also be regarded as a code in the usual sense with encoder $(f, g, h) : \mathsf{X}^k \times \mathsf{Y}^k \times \mathsf{Z}^k \to \mathsf{M}_{\mathsf{X}} \times \mathsf{M}_{\mathsf{Y}} \times \mathsf{M}_{\mathsf{Z}}$ and decoder φ. In particular, the error probability $e(f, g, h, \varphi)$ equals the error probability of the latter code. The rate of the code (f, g, h, φ) is the sum of the components of the triple $\mathbf{R}(f, g, h)$. ○

Obviously, if $(R_{\mathsf{X}}, R_{\mathsf{Y}}, R_{\mathsf{Z}})$ belongs to $\mathcal{R}(X, Y, Z)$ then so also does each $(\hat{R}_{\mathsf{X}}, \hat{R}_{\mathsf{Y}}, \hat{R}_{\mathsf{Z}}) \geqq (R_{\mathsf{X}}, R_{\mathsf{Y}}, R_{\mathsf{Z}})$. Hence the problem of real interest is to determine the *optimal points* of $\mathcal{R}(X, Y, Z)$; these are the triples $(R_{\mathsf{X}}, R_{\mathsf{Y}}, R_{\mathsf{Z}})$ in $\mathcal{R}(X, Y, Z)$ for which there is no other point $(\hat{R}_{\mathsf{X}}, \hat{R}_{\mathsf{Y}}, \hat{R}_{\mathsf{Z}})$ in $\mathcal{R}(X, Y, Z)$ such that $(\hat{R}_{\mathsf{X}}, \hat{R}_{\mathsf{Y}}, \hat{R}_{\mathsf{Z}}) \leqq (R_{\mathsf{X}}, R_{\mathsf{Y}}, R_{\mathsf{Z}})$. In the case of a single source, the achievable rates were the points of an infinite interval, the left endpoint of which was the entropy of the source (or, more generally, the Δ-distortion rate). The role of this endpoint is now taken by the optimal points of the achievable rate region. Once these points have been determined, the whole rate region can be obtained.

Clearly, $\mathcal{R}(X, Y, Z)$ is closed and contains all the triples of real numbers which are coordinate-wise lower-bounded by some optimal rate triple. A further important property of the achievable rate region is its convexity.

LEMMA 13.4 (*Time sharing principle*) $\mathcal{R}(X, Y, Z)$ is convex. ○

Proof Let k_1 and k_2 be arbitrary positive integers and let $(f^{(i)}, g^{(i)}, h^{(i)}, \varphi^{(i)})$ $(i = 1, 2)$ be a k_i-length block code for the fork network involving our 3-source. The *juxtaposition* (f, g, h, φ) of the above two codes is a $(k_1 + k_2)$-length block code for the same network defined as follows: set $k \triangleq k_1 + k_2$ and consider $f : \mathsf{X}^k \to \mathsf{M}_{\mathsf{X}}^{(1)} \times \mathsf{M}_{\mathsf{X}}^{(2)}$ obtained as

$$f(X_1 X_2 \ldots X_{k_1} X_{k_1+1} \ldots X_h) \triangleq (f^{(1)}(X_1 X_2 \ldots X_{k_1}), f^{(2)}(X_{k_1+1} \ldots X_k)).$$

Let the mappings $g : \mathsf{Y}^k \to \mathsf{M}_{\mathsf{Y}}^{(1)} \times \mathsf{M}_{\mathsf{Y}}^{(2)}$ and $h : \mathsf{Z}^k \to \mathsf{M}_{\mathsf{Z}}^{(1)} \times \mathsf{M}_{\mathsf{Z}}^{(2)}$ be defined similarly by means of $g^{(i)} : \mathsf{Y}^{k_i} \to \mathsf{M}_{\mathsf{Y}}^{(i)}$ and $h^{(i)} : \mathsf{Z}^{k_i} \to \mathsf{M}_{\mathsf{Z}}^{(i)}$, respectively. Finally, we define the decoder

$$\varphi : \mathsf{M}_{\mathsf{X}}^{(1)} \times \mathsf{M}_{\mathsf{X}}^{(2)} \times \mathsf{M}_{\mathsf{Y}}^{(1)} \times \mathsf{M}_{\mathsf{Y}}^{(2)} \times \mathsf{M}_{\mathsf{Z}}^{(1)} \times \mathsf{M}_{\mathsf{Z}}^{(2)} \to \mathsf{X}^k \times \mathsf{Y}^k \times \mathsf{Z}^k$$

by the juxtaposition of the corresponding components of $\varphi^{(1)}$ and $\varphi^{(2)}$. We have the relations

$$e(f, g, h, \varphi) \leqq e(f^{(1)}, g^{(1)}, h^{(1)}, \varphi^{(1)}) + e(f^{(2)}, g^{(2)}, h^{(2)}, \varphi^{(2)})) \tag{13.6}$$

and

$$\mathbf{R}(f, g, h) = \frac{k_1}{k}\mathbf{R}(f^{(1)}, g^{(1)}, h^{(1)}) + \frac{k_2}{k}\mathbf{R}(f^{(2)}, g^{(2)}, h^{(2)}). \tag{13.7}$$

Let $\mathbf{R}^{(i)} \triangleq (R_{\mathsf{X}}^{(i)},\ R_{\mathsf{Y}}^{(i)},\ R_{\mathsf{Z}}^{(i)})$ be achievable rate triples for $i = 1, 2$, and let $0 < \lambda < 1$ be arbitrary. We shall show that $\mathbf{R} \triangleq \lambda\mathbf{R}^{(1)} + (1 - \lambda)\mathbf{R}^{(2)}$ is also an achievable rate triple. In fact, for every integer k put $k^{(1)} \triangleq [\lambda k]$ and $k^{(2)} \triangleq k - k^{(1)}$. By assumption, there is a sequence of codes $\{(f^{(i)},\ g^{(i)},\ h^{(i)}, \varphi^{(i)})\}$ achieving $\mathbf{R}^{(i)}$ for $i = 1, 2$. Juxtaposing the code of block length $k^{(1)}$ of the sequence achieving $\mathbf{R}^{(1)}$ and the code of block length $k^{(2)}$ of the other sequence we get a code of block length k exhibiting all the desired properties, as can be seen from (13.6) and (13.7). □

THEOREM 13.5 (*Fork network coding theorem*) $\mathcal{R}(X, Y, Z)$ consists of those triples $(R_{\mathsf{X}}, R_{\mathsf{Y}}, R_{\mathsf{Z}})$ which satisfy the inequalities

$$R_{\mathsf{X}} \geqq H(X|Y, Z), \quad R_{\mathsf{Y}} \geqq H(Y|X, Z), \quad R_{\mathsf{Z}} \geqq H(Z|X, Y),$$
$$R_{\mathsf{X}} + R_{\mathsf{Y}} \geqq H(X, Y|Z), \quad R_{\mathsf{Y}} + R_{\mathsf{Z}} \geqq H(Y, Z|X), \quad R_{\mathsf{X}} + R_{\mathsf{Z}} \geqq H(X, Z|Y),$$
$$R_{\mathsf{X}} + R_{\mathsf{Y}} + R_{\mathsf{Z}} \geqq H(X, Y, Z).$$

→ 13.3 Moreover, for every $0 < \varepsilon < 1$, $\quad \mathcal{R}_\varepsilon(X, Y, Z) = \mathcal{R}(X, Y, Z)$. ○

Proof Let us denote by $\mathcal{R}^*$ the set of all triples $(R_{\mathsf{X}}, R_{\mathsf{Y}}, R_{\mathsf{Z}})$ satisfying all the inequalities of the theorem. Suppose that $(R_{\mathsf{X}}, R_{\mathsf{Y}}, R_{\mathsf{Z}})$ is an ε-achievable rate triple for some fixed $\varepsilon > 0$. We shall prove that it belongs to $\mathcal{R}^*$. To this end, consider for every k a k-length block code $(f_k, g_k, h_k, \varphi_k)$ such that

$$\Pr\{\varphi_k(f_k(X^k), g_k(Y^k), h_k(Z^k)) \neq (X^k, Y^k, Z^k)\} \leqq \varepsilon \tag{13.8}$$

and

$$R_{\mathsf{X}} \geqq \overline{\lim_{k\to\infty}} \frac{1}{k} \log \|f_k\|, \quad R_{\mathsf{Y}} \geqq \overline{\lim_{k\to\infty}} \frac{1}{k} \log \|g_k\|, \quad R_{\mathsf{Z}} \geqq \overline{\lim_{k\to\infty}} \frac{1}{k} \log \|h_k\|.$$

It is clear from Theorem 1.1 that

$$R_{\mathsf{X}} + R_{\mathsf{Y}} + R_{\mathsf{Z}} \geqq H(X, Y, Z).$$

To prove the inequality

$$R_{\mathsf{Y}} + R_{\mathsf{Z}} \geqq H(Y, Z|X),$$

replace f_k by $\hat{f}_k$ defined as

$$\hat{f}_k(\mathbf{x}) \triangleq \begin{cases} f_k(\mathbf{x}) & \text{if} \quad \mathbf{x} \in \mathsf{T}^k_{[X]} \\ \text{an arbitrary constant otherwise.} \end{cases}$$

Then (13.8) implies for any $\hat{\varepsilon} > \varepsilon$ and sufficiently large k

$$\Pr\{\varphi_k(\hat{f}_k(X^k), g_k(Y^k), h_k(Z^k)) \neq (X^k, Y^k, Z^k)\} < \hat{\varepsilon}.$$

Thus, again by Theorem 1.1,

$$\overline{\lim_{k\to\infty}} \left(\frac{1}{k} \log \|\hat{f}_k\| + \frac{1}{k} \log \|g_k\| + \frac{1}{k} \log \|h_k\| \right) \geqq H(X, Y, Z).$$

Since by definition

$$\overline{\lim_{k\to\infty}} \frac{1}{k} \log \|\hat{f}_k\| \leqq H(X),$$

we obtain

$$R_\mathsf{Y} + R_\mathsf{Z} \geqq H(X, Y, Z) - H(X) = H(Y, Z|X).$$

The inequality

$$R_\mathsf{X} \geqq H(X|Y, Z)$$

can be proved similarly, replacing the pair of functions (g_k, h_k) by a new function l_k with domain $\mathsf{Y}^k \times \mathsf{Z}^k$ defined as

$$l_k(\mathbf{y}, \mathbf{z}) \triangleq \begin{cases} (g_k(\mathbf{y}), h_k(\mathbf{z})) & \text{if} \quad (\mathbf{y}, \mathbf{z}) \in \mathsf{T}^k_{[YZ]} \\ \text{an arbitrary constant otherwise.} \end{cases}$$

Then the remaining inequalities follow by symmetry. Thus we have proved

$$\mathcal{R}_\varepsilon(X, Y, Z) \subset \mathcal{R}^* \quad \text{for} \quad \text{every} \quad \varepsilon \in (0, 1).$$

It remains to establish $\mathcal{R}(X, Y, Z) \supset \mathcal{R}^*$. It is enough to show that the points of $\mathcal{R}^*$ satisfying $R_\mathsf{X} + R_\mathsf{Y} + R_\mathsf{Z} = H(X, Y, Z)$ are achievable, for any other point of $\mathcal{R}^*$ is above some point with this property. Note that $\mathcal{R}^*$ intersects the plane of points with coordinate sum $H(X, Y, Z)$ in a convex set, the boundary of which consists of at most six straight lines: These lines are the intersections of the plane $R_\mathsf{X} + R_\mathsf{Y} + R_\mathsf{Z} = H(X, Y, Z)$ with each of the planes determined by the equality in the six remaining inequalities. Consequently, the extremal points of this intersection are the following (not necessarily distinct) points:

$$\begin{aligned} \mathbf{R}_1 &= (H(X), H(Y|X), H(Z|X, Y)), & \mathbf{R}_4 &= (H(X|Y, Z), H(Y), H(Z|Y)), \\ \mathbf{R}_2 &= (H(X), H(Y|X, Z), H(Z|X)), & \mathbf{R}_5 &= (H(X|Y, Z), H(Y|Z), H(Z)), \\ \mathbf{R}_3 &= (H(X|Y), H(Y), H(Z|X, Y)), & \mathbf{R}_6 &= (H(X|Z), H(Y|X, Z), H(Z)). \end{aligned}$$

As $\mathcal{R}(X, Y, Z)$ is convex by the time sharing principle, it suffices to show that these extremal points are achievable or, because of symmetry, that $\mathbf{R}_3$, say, is achievable. This can be done by iterated application of the rate slicing corollary, Corollary 13.2. Given any $\varepsilon > 0$ and $\delta > 0$, applying this corollary first to X and Y, we see that for sufficiently large k there exist functions f, g, ψ satisfying (i), (ii) and (iii) of Corollary 13.2 with ψ in the role of φ. Now apply the lemma for (X, Y) in the role of Y and Z in the role of X. It follows that on keeping the encoders f, g there exist $h : \mathsf{Z}^k \to \mathsf{M}_\mathsf{Z}$ and $\varphi : \mathsf{M}_\mathsf{X} \times \mathsf{M}_\mathsf{Y} \times \mathsf{M}_\mathsf{Z} \to \mathsf{X}^k \times \mathsf{Y}^k \times \mathsf{Z}^k$ satisfying

$$\frac{1}{k} \log \|h\| < H(Z|X, Y) + \delta,$$

$$\Pr\{\varphi(f(X^k), g(Y^k), h(Z^k)) \neq (X^k, Y^k, Z^k)\} \leqq 2\varepsilon.$$

→ 13.4

This proves the achievability of $\mathbf{R}_3$. □

COMMENT As noted earlier, the DMMS $\{(X_i, Y_i, Z_i)\}_{i=1}^{\infty}$ can be considered as a single DMS with generic distribution P_{XYZ}. From a formal point of view, a block code (f, g, h, φ) for the fork network is just a block code for the above DMS satisfying an additional restriction, namely that the encoder has to be of the form (f, g, h), where the three mappings depend on $(\mathbf{x}, \mathbf{y}, \mathbf{z}) \in (\mathsf{X} \times \mathsf{Y} \times \mathsf{Z})^k$ only through $\mathbf{x}$, $\mathbf{y}$ and $\mathbf{z}$, respectively. By Shannon's source coding theorem (Theorem 1.1), one needs rate $H(X, Y, Z)$ in order to reproduce (X^k, Y^k, Z^k) with small probability of error (if $k \to \infty$). Theorem 13.5 means that the above restriction on the encoder, i.e., separate encoding of the three component sources, does not increase the minimal rate. Also, the theorem gives an account of all the possible ways of "slicing" the minimal coding rate, i.e., joint entropy, into rate triples achievable for the fork network. As one would expect, if a triple $(R_\mathsf{X}, R_\mathsf{Y}, R_\mathsf{Z})$ is in the interior of $\mathcal{R}(X, Y, Z)$ then the error probability of optimal k-length block codes with these rates tends exponentially to zero. It is also true that there exists a sequence of codes performing well universally for every DMMS, the rate region of which contains the given rate triple. ○

→ 13.5
→ 13.6
→ 13.7

The problem solved by Theorem 13.5 is just the tip of an iceberg of unknown dimensions. We now formulate a general graph-theoretic model of noiseless communication networks. This will enable us to present solved and unsolved problems in a simple logical way so that the reader becomes aware of their natural connections.

A *directed graph* is a finite set of points, called *vertices,* together with a set of ordered pairs of vertices called *directed edges.* A sequence of consecutive directed edges $(b_0, b_1), (b_1, b_2), \dots, (b_{l-1}, b_l)$ is called *a path* (of length l); a path with $b_l = b_0$ is a *circuit.* If a directed graph has no circuits, the *depth of a vertex* is defined as the length of the longest path to that vertex. Clearly, all edges connect vertices of different depth, pointing to the direction of increasing depth. The *depth of the graph* is the maximum vertex depth, i.e., the length of the longest path in the graph. For every vertex b we denote by S_b the set of starting points of the directed edges leading to b.

DEFINITION 13.6 A *network* is a directed graph without circuits and isolated vertices. The *inputs* resp. *outputs* of the network are the vertices without incoming resp. outgoing edges; the remaining ones are the *intermediate vertices.* A *code associated with a network* is a family of mappings $f_b : \times_{a\in S_b} M_a \to M_b$, where M_b is a finite set associated with vertex b and f_b is defined for every vertex b which is not an input. ○

The set M_a corresponding to an input a of the network will be interpreted as the set of possible messages at this input; as a rule, it will consist of the k-length messages of a component source of a DMMS. The functions f are describing either encoding or decoding operations and we shall call them simply *coders.* For networks of depth 2, however, we shall retain the terms *encoder* and *decoder* for the mappings assigned to vertices of depth 1 and 2, respectively. The coders assigned to the vertices of depth 1 operate on the messages of one or several inputs. For $i \geqq 2$, the coders assigned to vertices of depth i are operating on messages and (or) images of messages provided by coders of depth $\leqq i-1$. By this recursion, to every vertex b of depth at least 1 there corresponds a composite function f_b^*, the domain of which is the Cartesian product of sets M_a corresponding to some inputs a. In the following this f_b^* will be considered as being defined on the Cartesian product of the sets M_a corresponding to all inputs, though it actually depends only on those inputs from which there is a path to b. In particular, the functions f_c^* corresponding to the output vertices c describe the response of the network to an arbitrary vector of input messages. In the following, unless otherwise stated, the set of input, intermediate resp. output vertices of a network will be denoted by A, B, C, respectively.

DEFINITION 13.7 A *source network* is given by a network, a DMMS $\{X_{A,i}\}_{i=1}^{\infty}$ with component sources $\{X_{a,i}\}_{i=1}^{\infty}$ assigned to the inputs $a \in A$, reproduction alphabets V_c assigned to the outputs $c \in C$ and distortion measures $d_c(.,.)$, $c \in C$ on $X_A \times V_c$, where

$$X_A \triangleq \mathop{\times}_{a\in A} X_a$$

is the Cartesian product of the alphabets of the given DMMS. A *k-length block code* for this source network is a code as in Definition 13.6 such that $M_a \triangleq X_a^k$ for each input a and $M_c \triangleq V_c^k$ for each output c. ○

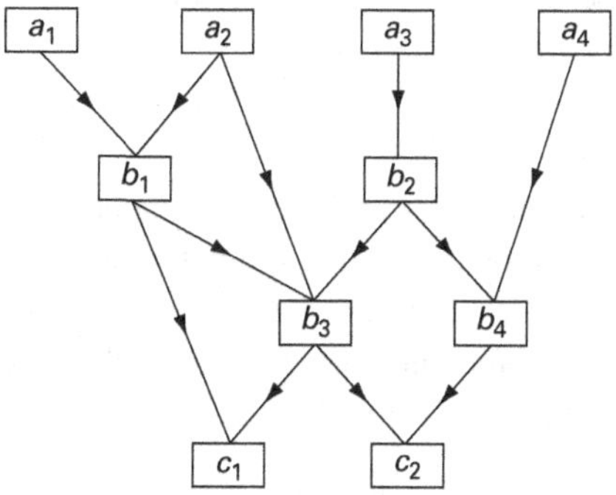

Figure 13.2 A typical network. In a code associated with this network, the coders at b_3 and c_2, say, have domain $M_{b_1} \times M_{a_2} \times M_{b_2}$ resp. $M_{b_3} \times M_{b_4}$. The composite mapping $f_{b_3}^*$ resp. $f_{c_2}^*$ depends on the messages at the inputs a_1, a_2, a_3 resp. at all inputs.

Given a vector $\mathbf{\Delta} = \{\Delta_c\}_{c \in \mathsf{C}}$ and an $\varepsilon > 0$, we say that a block code as in Definition 13.7 meets the ε-fidelity criterion $(\mathbf{d}, \mathbf{\Delta})$ if

$$\Pr\{d_c(X_{\mathsf{A}}^k, f_c^*(X_{\mathsf{A}}^k)) \leqq \Delta_c \text{ for each } c \in \mathsf{C}\} \geqq 1 - \varepsilon, \tag{13.9}$$

where f_c^* denotes the composite function defined by the code at output c, and the distortions between k-length sequences are understood in the usual averaging sense; see (6.1).

DEFINITION 13.8 For a given source network, a vector $\mathbf{R} = \{R_b\}_{b \in \mathsf{B}}$ with non-negative components is an *ε-achievable rate vector at distortion level* $\mathbf{\Delta}$ if for every $\delta > 0$ and sufficiently large k there exists a k-length block code meeting the ε-fidelity criterion $(\mathbf{d}, \mathbf{\Delta})$ such that the coders f_b belonging to the intermediate vertices have rates

$$\frac{1}{k} \log \|f_b\| \leqq R_b + \delta, \quad b \in \mathsf{B}.$$

→ 13.8 An $\mathbf{R} = \{R_b\}_{b \in \mathsf{B}}$ is an *achievable rate vector* if it is ε-achievable for every $0 < \varepsilon < 1$. The set of all ε-achievable resp. achievable rate vectors is called the *ε-achievable* resp. *achievable rate region* (at distortion level $\mathbf{\Delta}$). ○

A network will be given by drawing its graph, the vertices being drawn as boxes. The assignment of the component sources to the inputs will be visualized by putting their generic variables into the corresponding boxes.

As a rule, we shall consider fidelity criteria demanding the exact reproduction of some components with small probability of error, with no requirement on the others. Formally, for each output $c \in \mathsf{C}$ let D_c be a subset of the set of those inputs from which there is a path to c, and let

$$\mathsf{V}_c \triangleq \mathsf{X}_{\mathsf{D}_c} = \mathop{\mathsf{X}}_{a \in \mathsf{D}_c} \mathsf{X}_a;$$

further, for $u \triangleq \{x_a\}_{a \in \mathsf{A}} \in \mathsf{X}_{\mathsf{A}}$ and $v \in \mathsf{V}_c$ set

$$d_c(u, v) = 0 \quad \text{iff} \quad \{x_a\}_{a \in \mathsf{D}_c} = v,$$

and let $\mathbf{\Delta}$ be the zero vector. Then the ε-fidelity criterion (13.9) requires that the *probability of error* be at most ε, i.e.,

$$e \triangleq \Pr\{f_c^*(X_{\mathsf{A}}^k) \neq X_{\mathsf{D}_c}^k \quad \text{for some} \quad c \in \mathsf{C}\} \leqq \varepsilon. \tag{13.10}$$

A fidelity criterion of this type will be called a *probability of error fidelity criterion* and will be visualized by putting the symbol $\hat{X}$ into an output box whenever the component source with generic variable X is to be reproduced at that output.

In order to become familiar with this symbolism, look at Fig. 13.3, which shows the
→ 13.9 models of Definitions 13.1 and 13.3.

REMARK For a source network with a probability of error fidelity criterion, the *probability of error at output* $c \in \mathsf{C}$ is

$$e_c \triangleq \Pr\{f_c^*(X_{\mathsf{A}}^h) \neq X_{\mathsf{D}_c}^k\}.$$

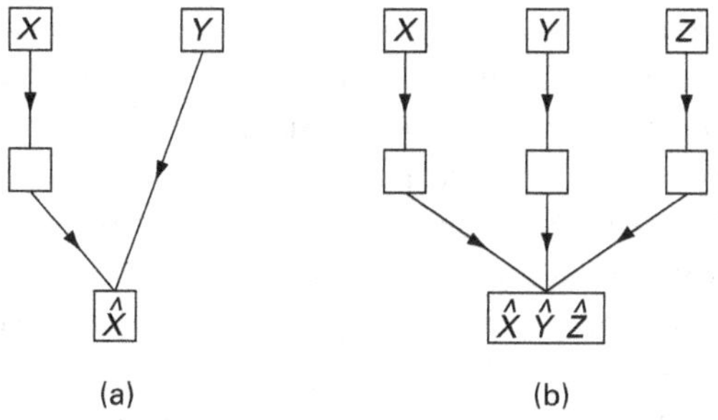

Figure 13.3 The source networks of Definitions 13.1 and 13.3.

Whereas in Definition 13.8 the role of the ε-fidelity criterion $(\mathbf{d}, \mathbf{\Delta})$ is now played by (13.10), when determining the achievable rate region we shall often use the conditions $e_c \leqq \varepsilon$ for every $c \in \mathsf{C}$, rather than (13.10). This is justified by the inequalities

$$e_c \leqq e \leqq \sum_{c \in \mathsf{C}} e_c. \qquad \bigcirc$$

Most of the source networks for which a computable characterization of the achievable rate region is known involve a network of depth 2 and a probability of error fidelity criterion. The following simple observations will reduce the number of those source networks within this class which have to be treated individually.

LEMMA 13.9 To every network there exists another one such that

(i) the two networks have the same sets of input and output vertices;
(ii) in the new network, edges connect vertices of depth differing by 1;
(iii) for every DMMS and fidelity criterion, the achievable rate region of the original source network is a projection of that of the corresponding new one. ○

Proof Replace every edge connecting vertices of depth difference $l > 1$ by a path of length l. □

LEMMA 13.10 To every source network of depth 2 without edges from inputs to outputs and with a probability of error fidelity criterion there exists another source network of the same kind such that

(i) in the new network the edges between inputs and intermediate vertices establish a one-to-one correspondence;
(ii) the set B of intermediate vertices of the original network may be identified with a subset of the set $\tilde{\mathsf{B}}$ of intermediate vertices of the new network so that a vector $\mathbf{R} = \{R_b\}_{b \in \mathsf{B}}$ belongs to the original achievable rate region iff setting

$$\tilde{R}_{\tilde{b}} \triangleq \begin{cases} R_{\tilde{b}} & \text{if } \tilde{b} \in \mathsf{B} \\ 0 & \text{otherwise,} \end{cases}$$

the vector $\tilde{\mathbf{R}} \triangleq \{\tilde{R}_{\tilde{b}}\}_{\tilde{b}} \in \tilde{\mathsf{B}}$ is an element of the new achievable rate region. ○

Proof From the point of view of an encoder resp. an encoder–decoder pair of a given source network it is natural to consider as a single source the collection of those component sources which are encoded resp. both encoded and reproduced together. The idea of the proof is to introduce a new DMMS with these sources as components.

Formally, let E be the set of those edges (b, c) of the given network for which $\mathsf{S}_b \cap \mathsf{D}_c$ is non-void and differs from S_b. Let $\mathcal{A}$ be the class of subsets of A representable as $\mathsf{S}_b \cap \mathsf{D}_c$, $(b, c) \in \mathsf{E}$. Let the set of intermediate vertices of the new network be $\tilde{\mathsf{B}} \triangleq \mathsf{B} \cup \mathcal{A}$. Let the input vertices of the new network be in a one-to-one correspondence with its intermediate vertices; connect each new input vertex with the corresponding intermediate one. Further, let the set of new output vertices be identical with the original C. Whenever $b \in \mathsf{B}$ and $c \in \mathsf{C}$ are connected with an edge in the original network, let in the new one both b and $\mathsf{S}_b \cap \mathsf{D}_c$ (provided that $(b, c) \in \mathsf{E}$) be connected with c.

Finally, consider a DMMS with generic variables

$$Y_{\tilde{b}} \triangleq \begin{cases} X_{\mathsf{S}_b} & \text{if } \tilde{b} = b \in \mathsf{B} \\ X_{\mathsf{S}_b \cap \mathsf{D}_c} & \text{if } \tilde{b} = \mathsf{S}_b \cap \mathsf{D}_c \in \mathcal{A}. \end{cases}$$

Assign the component source with generic variable $Y_{\tilde{b}}$ to the input vertex connected with $\tilde{b}$. Let the new fidelity criterion require at each output c the reproduction of the component source with generic variable $Y_{\tilde{b}}$ iff the vertex $\tilde{b}$ is connected with c and either $\tilde{b} = b \in \mathbf{B}$ with $\mathsf{S}_b \subset \mathsf{D}_c$ or $\tilde{b} \in \mathcal{A}$.

This new source network satisfies the assertions of the lemma. In order to see this, note that if $\tilde{\mathbf{R}} = \{\tilde{R}_{\tilde{b}}\}_{\tilde{b} \in \tilde{\mathsf{B}}}$ is any achievable rate vector for the new source network, then we get an achievable rate vector $\mathbf{R}$ for the original one by setting

$$R_b \triangleq \tilde{R}_b + \sum_{\tilde{b} \in \mathcal{A},\, \tilde{b} \subset \mathsf{S}_b} \tilde{R}_{\tilde{b}}. \qquad \square$$

LEMMA 13.11 Given any source network with a probability of error fidelity criterion, one can modify the fidelity criterion without changing the achievable rate region so that after the modification, for every pair of outputs, $\mathsf{S}_{c'} \subset \mathsf{S}_{c''}$ implies $\mathsf{D}_{c'} \subset \mathsf{D}_{c''}$. After this modification, if there are several outputs with the same S_c, one can delete all but one of them, without changing the achievable rate region. ○

Proof For a fixed pair of outputs with $\mathsf{S}_{c'} \subset \mathsf{S}_{c''}$ the addition of the component sources to be reproduced at c' to those to be reproduced at c'' does not change the achievable rate region. Then proceed iteratively. The last assertion is obvious. □

Summarizing Lemmas 13.9 to 13.11 we see that in order to determine the achievable rate region of source networks of depth 2 with a probability of error fidelity criterion, one may restrict attention to source networks as in the following definition.

DEFINITION 13.12 A *normal source network* (NSN) is a source network of depth 2 with a probability of error fidelity criterion such that

(i) there are no edges from inputs to outputs;
(ii) $|\mathsf{A}| = |\mathsf{B}|$, and the edges from A to B define a one-to-one correspondence between input and intermediate vertices;
(iii) every output vertex is connected with a different set of intermediate vertices;
(iv) $\mathsf{S}_{c'} \subset \mathsf{S}_{c''}$ implies $\mathsf{D}_{c'} \subset \mathsf{D}_{c''}$. ○

On account of the one-to-one correspondence between input and intermediate vertices, in a normal source network the component sources of the underlying DMMS may be considered as being assigned to the intermediate vertices $b \in \mathsf{B}$. In the following, we shall speak of component sources with generic variables X_b, $b \in \mathsf{B}$, and write $X_{\mathsf{L}} \triangleq \{X_b\}_{b\in\mathsf{L}}$ for subsets L of B.

For a NSN it is significant whether there exist intermediate vertices b such that for some output c connected with b the component source assigned to b need not be reproduced at the output c. Such a vertex b will be called a *helper*. Intuitively, the information provided by the encoder at b about the corresponding component source is not needed at c for reproducing this source; rather, its only role is to help the decoder at c in meeting the fidelity criterion. For example, the NSN of Fig. 13.4 contains one helper, whereas the NSN of Fig. 13.3(b) contains no helper.

Normal source networks without helpers are characterized by the property that at each output $c \in \mathsf{C}$ all the component sources assigned to vertices $b \in \mathsf{S}_c$ have to be reproduced. In particular, these source networks are uniquely determined by the network and the assignment of the component sources to the elements of B. The achievable rate region of normal source networks without helpers can be easily determined, using a sharper version of the existence part of the fork network coding theorem. This sharper result (Lemma 13.13 below) says, intuitively, that almost all randomly selected codes are good.

Consider the fork network of Fig. 13.1 with a DMMS having generic variables X_b, $b \in \mathsf{B}$. Associate with each collection $f = \{f_b\}_{b\in\mathsf{B}}$ of mappings $f_b : \mathsf{X}_b^k \to \mathsf{M}_b$ considered as k-length block encoders for this fork network, a typicality decoder $\varphi = \varphi_f$ corresponding to the distribution P_{X_B}, defined as follows. Identify the collection f with the mapping of $\mathsf{X}_\mathsf{B}^k \triangleq \mathsf{X}_{b\in\mathsf{B}} \mathsf{X}_b^k$ into $\mathsf{M} \triangleq \mathsf{X}_{b\in\mathsf{B}} \mathsf{M}_b$ given by

$$f(\mathbf{u}) \triangleq \{f_b(\mathbf{x}_b)\}_{b\in\mathsf{B}} \quad \text{if} \quad \mathbf{u} = \{\mathbf{x}_b\}_{b\in\mathsf{B}};\ \mathbf{x}_b \in \mathsf{X}_b^k,$$

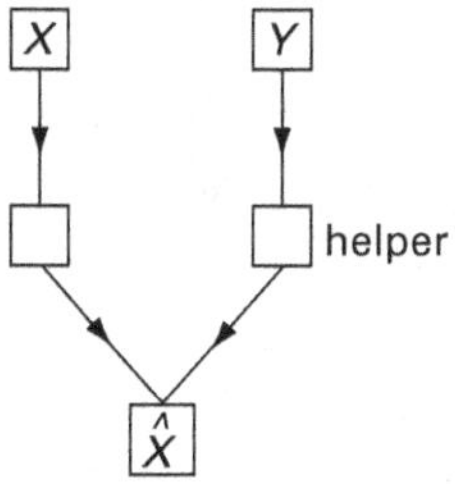

Figure 13.4 Normal source network with one helper.

and set

$$\varphi(\mathbf{m}) \triangleq \begin{cases} \mathbf{u} & \text{if} \quad f^{-1}(\mathbf{m}) \cap \mathsf{T}^k_{[X_\mathsf{B}]} = \{\mathbf{u}\} \\ \text{arbitrary} & \text{if} \quad |f^{-1}(\mathbf{m}) \cap \mathsf{T}^k_{[X_\mathsf{B}]}| \neq 1. \end{cases} \tag{13.11}$$

Denote by $e(f)$ the probability of error of the code (f, φ), i.e.,

$$e(f) \triangleq \Pr\{\varphi(f(X^k_\mathsf{B})) \neq X^k_\mathsf{B}\}.$$

LEMMA 13.13 For a DMMS with generic variables X_B, the inequalities

$$\sum_{b\in\mathsf{L}} \frac{1}{k} \log|\mathsf{M}_b| \geqq H(X_\mathsf{L}|X_{\mathsf{B}-\mathsf{L}}) + \delta \quad \text{for every} \quad \mathsf{L} \subset \mathsf{B} \tag{13.12}$$

imply

$$\frac{1}{|\mathcal{F}|} \sum_{f\in\mathcal{F}} e(f) \leqq \varepsilon \tag{13.13}$$

for every $\varepsilon \in (0, 1)$ and $k \geqq k_0(\{|\mathsf{X}_b|\}_{b\in\mathsf{B}}, \varepsilon, \delta)$. Here $\mathcal{F}$ is the family of all collections of mappings $f = \{f_b\}_{b\in\mathsf{B}}$, $f_b : \mathsf{X}^k_b \to \mathsf{M}_b$. ○

Proof For any pair of $|\mathsf{B}|$-tuples $\mathbf{u} = \{\mathbf{x}_b\}_{b\in\mathsf{B}}$, $\mathbf{u}' = \{\mathbf{x}'_b\}_{b\in\mathsf{B}}$ denote by $\mathsf{L}(\mathbf{u}, \mathbf{u}')$ the set of those indices $b \in \mathsf{B}$ for which $\mathbf{x}_b \neq \mathbf{x}'_b$. Note that if $\mathsf{L}(\mathbf{u}, \mathbf{u}') = \mathsf{L}$ then

$$\frac{1}{|\mathcal{F}|}|\{f : f(\mathbf{u}) = f(\mathbf{u}')\}| = \prod_{b\in\mathsf{L}} |\mathsf{M}_b|^{-1}.$$

Thus, writing $\mathsf{T} \triangleq \mathsf{T}^k_{[X_\mathsf{B}]}$ and

$$\chi_f(\mathbf{u}, \mathbf{u}') \triangleq \begin{cases} 1 & \text{if } f(\mathbf{u}) = f(\mathbf{u}') \\ 0 & \text{otherwise,} \end{cases}$$

we have

$$\begin{aligned} \frac{1}{|\mathcal{F}|} \sum_{f\in\mathcal{F}} e(f) &\leqq \Pr\{X^k_\mathsf{B} \notin \mathsf{T}\} + \frac{1}{|\mathcal{F}|} \sum_{f\in\mathcal{F}} \sum_{\mathbf{u}\in\mathsf{T}} P^k_{X_\mathsf{B}}(\mathbf{u}) \sum_{\substack{\mathbf{u}'\in\mathsf{T} \\ \mathbf{u}'\neq\mathbf{u}}} \chi_f(\mathbf{u}, \mathbf{u}') \\ &= \Pr\{X^k_\mathsf{B} \notin \mathsf{T}\} + \sum_{\mathbf{u}\in\mathsf{T}} P^k_{X_\mathsf{B}}(\mathbf{u}) \sum_{\substack{\mathbf{u}'\in\mathsf{T} \\ \mathbf{u}'\neq\mathbf{u}}} \frac{1}{|\mathcal{F}|}|\{f : f(\mathbf{u}) = f(\mathbf{u}')\}| \\ &= \Pr\{X^k_\mathsf{B} \notin \mathsf{T}\} + \sum_{\mathbf{u}\in\mathsf{T}} P^k_{X_\mathsf{B}}(\mathbf{u}) \sum_{\substack{\mathsf{L}\subset\mathsf{B} \\ \mathsf{L}\neq\emptyset}} \left|\{\mathbf{u}' : \mathbf{u}' \in \mathsf{T}, \mathsf{L}(\mathbf{u}, \mathbf{u}') = \mathsf{L}\}\right| \prod_{b\in\mathsf{L}} |\mathsf{M}_b|^{-1}. \end{aligned} \tag{13.14}$$

Here the first term goes to zero as $k \to \infty$, by Lemma 2.12. Further, if $\mathsf{L}(\mathbf{u}, \mathbf{u}') = \mathsf{L}$ then the X_B-typical $\mathbf{u}'$ (identified with an element of X^k_B) is also $X_\mathsf{B}|X_{\mathsf{B}-\mathsf{L}}$-generated by the $X_{\mathsf{B}-\mathsf{L}}$-typical tuple $\{\mathbf{x}_b\}_{b\in\mathsf{B}-\mathsf{L}}$ (regarded as an element of $\mathsf{X}^k_{\mathsf{B}-\mathsf{L}}$). It follows by Lemma 2.13 that

$$\left|\{\mathbf{u}' : \mathbf{u}' \in \mathsf{T}, \mathsf{L}(\mathbf{u}, \mathbf{u}') = \mathsf{L}\}\right| \leqq \exp\{k[H(X_{\mathsf{B}}|X_{\mathsf{B}-\mathsf{L}}) + \varepsilon_k]\}$$
$$= \exp\{k[H(X_{\mathsf{L}}|X_{\mathsf{B}-\mathsf{L}}) + \varepsilon_k]\},$$

where $\varepsilon_k \to$ as $k \to \infty$. This proves that the second term of (13.14) goes to zero as well. □

COMMENT Lemma 13.13 says that on randomly selecting encoders $f_b : \mathsf{X}_b^k \to \mathsf{M}_b$ and assigning to them a suitable decoder, the expected value of the probability of error of the resulting code is arbitrarily small, providing the constraints (13.12) hold. In particular, a randomly selected code has small probability of error with probability close to 1. The selection of a mapping f_b as above may be visualized as putting each sequence $\mathbf{x}_b \in \mathsf{X}_b^k$ into one of $|\mathsf{M}_b|$ bins. Hence this kind of random selection is commonly referred to as *random binning*. ○

THEOREM 13.14 The achievable rate region of a normal source network without helpers equals the set of those vectors $\mathbf{R} = \{R_b\}_{b\in\mathsf{B}}$ which satisfy the inequalities

$$\sum_{b\in\mathsf{L}} R_b \geqq H(X_{\mathsf{L}}|X_{\mathsf{S}_c-\mathsf{L}}) \tag{13.15}$$

for every output $c \in \mathsf{C}$ and set $\mathsf{L} \subset \mathsf{S}_c$. Moreover, the ε-achievable rate region is the same for every $0 < \varepsilon < 1$. ○

Proof The converse part immediately follows from the fork network coding theorem, Theorem 13.5. In fact, if $\mathbf{R}$ is an ε-achievable rate vector then

$$\left(\sum_{b\in\mathsf{L}} R_b, \sum_{b\in\mathsf{S}_c-\mathsf{L}} R_b\right)$$

is an ε-achievable rate pair for the fork network with two inputs and the 2-source with generic variables X_{L} and $X_{\mathsf{S}_c-\mathsf{L}}$.

In order to prove the direct part, consider a vector $\mathbf{R}$ satisfying the inequalities (13.15) and fix some $\delta > 0$. Let M_b, $b \in \mathsf{B}$ be sets of size

$$|\mathsf{M}_b| \geqq \exp[k(R_b + \delta)], \quad b \in \mathsf{B}. \tag{13.16}$$

For any collection $g = \{f_b\}_{b\in\mathsf{B}}$ of encoders $f_b : \mathsf{X}_b^k \to \mathsf{M}_b$, $b \in \mathsf{B}$, its subcollection $\{f_b\}_{b\in\mathsf{S}_c}$ corresponding to an output vertex c may be considered as a collection f of encoders for a fork network with intermediate vertices $b \in \mathsf{S}_c$ and output vertex c. Let φ_c denote the corresponding decoder in the sense of (13.11), and let $e_c(g)$ be the probability of error of the resulting code for the fork network. The mappings $\{f_b\}_{b\in\mathsf{B}}$, $\{\varphi_c\}_{c\in\mathsf{C}}$ define a k-length block code for our source network. Using this code, the probability of error at the output c equals $e_c(g)$. Since $e_c(g)$ depends on g only through the subcollection $\{f_b\}_{b\in\mathsf{S}_c}$, from Lemma 13.13 it follows that for any fixed $\hat{\varepsilon} > 0$

$$\frac{1}{|\mathcal{G}|}\sum_{g\in\mathcal{G}} e_c(g) < \hat{\varepsilon} \quad \text{for every} \quad c \in \mathsf{C},$$

if k is sufficiently large; here $\mathcal{G}$ denotes the family of all collections $g = \{f_b\}_{b\in\mathsf{B}}$ with $f_b : \mathsf{X}_b^k \to \mathsf{M}_b$. Applying this for $\hat{\varepsilon} \triangleq \varepsilon/|\mathsf{C}|$ and summing over $c \in \mathsf{C}$ it follows that there exists $g \in \mathcal{G}$ with $\sum_{c\in\mathsf{C}} e_c(g) < \varepsilon$. □

Since the system of inequalities (13.15) is, in general, redundant, for the actual description of the achievable rate region, the following corollary is of interest.

COROLLARY 13.14 The achievable rate region of a normal source network without helpers equals the set of those vectors $\mathbf{R} = \{R_b\}_{b\in\mathsf{B}}$ which satisfy the inequalities (13.15) for every output $c \in \mathsf{C}$ and set $\mathsf{L} \subset \mathsf{S}_c$ such that L is disjoint from every $\mathsf{S}_{c'} \subset \mathsf{S}_c$ $(c' \in \mathsf{C})$. ○

Proof It is enough to prove that these inequalities imply the remaining ones. This can be shown recursively, using the fact that for every $\mathsf{L} = \mathsf{L}_1 \cup \mathsf{L}_2$, $\mathsf{L}_1 \subset \mathsf{S}_{c'} \subset \mathsf{S}_c$ and $\mathsf{L}_2 \subset \mathsf{S}_c - \mathsf{S}_{c'}$ the inequality

$$H(X_{\mathsf{L}}|X_{\mathsf{S}_c-\mathsf{L}}) \leqq H(X_{\mathsf{L}_1}|X_{\mathsf{S}_c-\mathsf{L}_1}) + H(X_{\mathsf{L}_2}|X_{\mathsf{S}_c-\mathsf{L}_2})$$

holds. □

In the case of arbitrary NSNs no computable characterization of the achievable rate region is known. However, the characterization given below (Theorem 13.15) is useful for many purposes. In particular, it reduces the problem of characterizing achievable rate regions to a mathematical problem of rather simple structure. We shall return to the latter in Chapter 15.

With some abuse of notation, in the following theorem the set of those intermediate vertices which correspond to input vertices in D_c will also be denoted by D_c. Then the set of helpers $\mathsf{K} \subset \mathsf{B}$ is given by $\mathsf{K} = \bigcup_{c\in\mathsf{C}} (\mathsf{S}_c - \mathsf{D}_c)$.

THEOREM 13.15 (*Helpers theorem*) The achievable rate region of a normal source network with set of helpers $\mathsf{K} \subset \mathsf{B}$ is equal to the closure of the set of those vectors $\mathbf{R} = \{R_b\}_{b\in\mathsf{B}}$ to which for some $n \geqq 1$ there exist functions f_b, $b \in \mathsf{K}$ with domain X_b^n (and arbitrary range) such that the following inequalities hold:

$$R_b \geqq \frac{1}{n} H(f_b(X_b^n)), \quad b \in \mathsf{K} - \bigcup_{c\in\mathsf{C}} \mathsf{D}_c \tag{13.17}$$

and, for every $c \in \mathsf{C}$,

$$\sum_{b\in\mathsf{L}} R_b \geqq \frac{1}{n} H(X_{\mathsf{L}}^n|X_{\mathsf{D}_c-\mathsf{L}}^n, \{f_b(X_b^n)\}_{b\in\mathsf{S}_c\cap\mathsf{K}}) + \frac{1}{n} \sum_{b\in\mathsf{L}\cap\mathsf{K}} H(f_b(X_b^n)) \tag{13.18}$$

for those $\mathsf{L} \subset \mathsf{D}_c$ which are disjoint from each $\mathsf{D}_{c'}$ with $c' \neq c$, $\mathsf{S}_{c'} \subset \mathsf{S}_c$. ○

Proof To prove the converse part, fix some $\delta > 0$ and achievable rate vector $\mathbf{R}' = \{R'_b\}_{b\in\mathsf{B}}$. Consider an n-length block code $\{f_b, \varphi_c\}_{b\in\mathsf{B},c\in\mathsf{C}}$ with rates

$$\frac{1}{n} \log \|f_b\| \leqq R'_b + \delta, \quad b \in \mathsf{B}, \tag{13.19}$$

and error probability

$$e \triangleq \Pr\{\varphi_c(\{f_b(X_b^n)\}_{b\in \mathsf{S}_c}) \neq X^n_{\mathsf{D}_c}, \quad \text{for some} \quad c \in \mathsf{C}\} \leqq \varepsilon. \tag{13.20}$$

It suffices to show that on setting $R_b \triangleq R'_b + 2\delta$ and properly choosing ε, the encoders $f_b, b \in \mathsf{K}$ satisfy the inequalities (13.17) and (13.18).

Clearly, it is enough to see that

$$\sum_{b\in \mathsf{L}-\mathsf{K}} R_b \geqq \frac{1}{n} H(X^n_{\mathsf{L}} | X^n_{\mathsf{D}_c - \mathsf{L}}, \{f_b(X^n_b)\}_{b\in \mathsf{S}_c \cap \mathsf{K}}) \tag{13.21}$$

for every $c \in \mathsf{C}$ and $\mathsf{L} \subset \mathsf{D}_c$. On account of (13.19) and the relation $\mathsf{S}_c - \mathsf{D}_c \subset \mathsf{S}_c \cap \mathsf{K}$, we have

$$\begin{aligned}\sum_{b\in \mathsf{L}-\mathsf{K}} (R'_b + \delta) &\geqq \frac{1}{n} H(\{f_b(X^n_b)\}_{b\in \mathsf{L}-\mathsf{K}}) \\ &\geqq \frac{1}{n} H(\{f_b(X^n_b)\}_{b\in \mathsf{L}-\mathsf{K}} | X^n_{\mathsf{D}_c-\mathsf{L}}, \{f_b(X^n_b)\}_{b\in \mathsf{S}_c\cap \mathsf{K}}) \\ &= \frac{1}{n} H(\{f_b(X^n_b)\}_{b\in \mathsf{S}_c} | X^n_{\mathsf{D}_c-\mathsf{L}}, \{f_b(X^n_b)\}_{b\in \mathsf{S}_c\cap \mathsf{K}}).\end{aligned}$$

Using (13.20) and Fano's inequality, the last term is lower-bounded by

$$\frac{1}{n} H(X^n_{\mathsf{D}_c} | X^n_{\mathsf{D}_c-\mathsf{L}}, \{f_b(X^n_b)\}_{b\in \mathsf{S}_c\cap \mathsf{K}}) - \varepsilon \log |\mathsf{X}_{\mathsf{D}_c}| - \frac{1}{n} h(\varepsilon),$$

whence (13.21) follows.

Turning to the direct part of the theorem, it is enough to prove that if $\mathbf{R}$ satisfies the inequalities (13.17) and (13.18) for some n and mappings f_b, $b \in \mathsf{K}$, with domains X^n_b, then $\mathbf{R}$ is an achievable rate vector. Fixing some $n \geqq 1$ and a family of mappings $f_b, b \in \mathsf{K}$, we introduce an auxiliary NSN without helpers. For brevity, we denote the given source network by $\mathcal{S}$ and the auxiliary one by $\hat{\mathcal{S}}$. The latter is defined as the following modification of $\mathcal{S}$.

Write $\mathsf{D} \triangleq \bigcup_{c\in \mathsf{C}} \mathsf{D}_c$, where the sets D_c are understood as subsets of B. To every b in $\mathsf{K} \cap \mathsf{D}$ introduce a new intermediate vertex b^+ and a new input vertex connected with b^+, and connect b^+ with every c such that $b \in \mathsf{D}_c$. Further, to every $b \in \mathsf{K}$ introduce a new output vertex and connect it with b. Define a new DMMS with generic variables assigned to the intermediate vertices of the modified network, setting

$$Y_b \triangleq \begin{cases} X^n_b & \text{if} \quad b \notin \mathsf{K} \\ f_b(X^n_b) & \text{if} \quad b \in \mathsf{K}, \end{cases}$$

$$Y_{b^+} \triangleq X^n_b, \quad b \in \mathsf{K} \cap \mathsf{D}.$$

Let $\hat{\mathcal{S}}$ be the NSN without helpers uniquely determined by this assignment.

We claim that if a vector $\hat{\mathbf{R}}$ with components $\hat{R}_b, b \in \mathsf{B}$ and $\hat{R}_{b+}, b \in \mathsf{K} \cap \mathsf{D}$ is an achievable rate vector for $\hat{\mathcal{S}}$, then

$$R_b \triangleq \begin{cases} \frac{1}{n}\hat{R}_b & \text{if} \quad b \notin \mathsf{K} \cap \mathsf{D} \\ \frac{1}{n}(\hat{R}_b + \hat{R}_{b+}) & \text{if} \quad b \in \mathsf{K} \cap \mathsf{D} \end{cases} \tag{13.22}$$

defines an achievable rate vector $\mathbf{R} \triangleq \{R_b\}_{b \in \mathsf{B}}$ for $\mathcal{S}$. To see this, observe first that every l-length block code for $\hat{\mathcal{S}}$ gives rise to an ln-length block code with at most the same error probability for $\mathcal{S}$; further, the rates of this code for $\mathcal{S}$ are related to the rates of the code for $\hat{\mathcal{S}}$ as in (13.22). In fact, the encoders $\hat{f}_b, b \in \mathsf{B}$ and $\hat{f}_{b+}, b \in \mathsf{K} \cap \mathsf{D}$ of any length block code for $\hat{\mathcal{S}}$ define mappings F_b resp. F_{b+} with domain X_b^{ln} such that

$$F_b(X^{ln}) = \hat{f}_b(Y_b^l), \quad F_{b+}(X^{ln}) = \hat{f}_{b+}(Y_{b+}^l).$$

Consider the mappings $F_b, b \notin \mathsf{K} \cap \mathsf{D}$ resp. the pairs $(F_b, F_{b+}), b \in \mathsf{K} \cap \mathsf{D}$ as encoders of an ln-length block code for $\mathcal{S}$. Further, for every $c \in \mathsf{C}$, the decoder $\hat{\varphi}_c$ of the code for $\hat{\mathcal{S}}$ maps into a Cartesian product containing all the sets X_b^{ln}, $b \in \mathsf{D}_c$ as factors. Thus one can define the corresponding decoder of the code for $\mathcal{S}$ by projecting the values taken by $\hat{\varphi}_c$ to the Cartesian product of the sets X_b^{ln}, $b \in \mathsf{D}_c$. Clearly, the so obtained code for $\mathcal{S}$ does have the announced properties. This proves that if $\hat{\mathbf{R}}$ is an achievable rate vector for $\hat{\mathcal{S}}$ and $\mathbf{R}$ is defined by (13.22) then to every $\varepsilon > 0$, $\delta > 0$ and sufficiently large integral multiple $\tilde{k} = ln$ of n there exists a $\tilde{k}$-length block code for $\mathcal{S}$ with rates

$$\frac{1}{\tilde{k}} \log \|\tilde{f}_b\| \leqq R_b + \delta, \quad b \in \mathsf{B} \tag{13.23}$$

and error probability (se (13.20))

$$e \leqq \varepsilon. \tag{13.24}$$

It is easy to see that this implies the same for every sufficiently large k, whereby $\mathbf{R}$ is an achievable rate vector for $\mathcal{S}$ as claimed. In fact, supposing $(l-1)n < k < ln$, consider for $\mathcal{S}$ a code of block length $\tilde{k} \triangleq ln$ with encoders $\hat{f}_b$, $b \in \mathsf{B}$, and decoders $\tilde{\varphi}_c$, $c \in \mathsf{C}$, with the properties (13.23), (13.24). One can apply this code to k-length sequences by adding to the latter arbitrary suffixes of length $\tilde{k} - k$. Formally, for a fixed $\mathbf{z} = \{\mathbf{z}_b\}_{b \in \mathsf{B}}$, $\mathbf{z}_b \in \mathsf{X}_b^{\tilde{k}-k}$, define k-length block encoders $f_{b,\mathbf{z}}$ by

$$f_{b,\mathbf{z}}(\mathbf{x}_b) \triangleq \tilde{f}_b(\mathbf{x}_b \mathbf{z}_b) \quad (\mathbf{x}_b \in \mathsf{X}_b^k);$$

further, for every $c \in \mathsf{C}$, let φ_c be defined by deleting the last $\tilde{k} - k$ symbols from every component sequence of the values of $\tilde{\varphi}_c$. Denoting by $e_{\mathbf{z}}$ the error probability of the code with encoders $f_{b,\mathbf{z}}$ and decoders φ_c, (13.24) implies

$$\sum_{\mathbf{z} \in \mathsf{X}_{\mathsf{B}}^{\tilde{k}-k}} \Pr\{X_{\mathsf{B}}^{\tilde{k}-k} = \mathbf{z}\} e_{\mathbf{z}} \leqq \varepsilon.$$

It follows that for some $\mathbf{z} \in \mathsf{X}_{\mathsf{B}}^{\tilde{k}-k}$ we have $e_{\mathbf{z}} \leqq \varepsilon$, completing the proof of our claim concerning (13.22).

Now the direct part of the theorem will follow if we show that whenever $\mathbf{R}$ satisfies the inequalities (13.17) and (13.18) then

$$\hat{R}_b \triangleq \begin{cases} H(f_b(X_b^n)) & \text{if} \quad b \in \mathsf{K} \\ nR_b & \text{if} \quad b \notin \mathsf{K}, \end{cases} \tag{13.25}$$

$$\hat{R}_{b^+} \triangleq nR_b - H(f_b(X_b^n)) \quad b \in \mathsf{K} \cap \mathsf{D}$$

defines an achievable rate vector $\hat{\mathbf{R}}$ for $\hat{\mathcal{S}}$. Note that $\hat{\mathcal{S}}$ is an NSN without helpers in which the set of intermediate vertices connected with an output vertex $c \in \mathsf{C}$ equals $\mathsf{S}_c \cap \mathsf{D}_c^+$, where $\mathsf{D}_c^+ \triangleq \{b^+ : b \in \mathsf{D}_c \cap \mathsf{K}\}$; further, the outputs of $\hat{\mathcal{S}}$ not belonging to C are in a one-to-one correspondence with the elements of K, each being connected only with the corresponding $b \in \mathsf{K}$. Thus by Corollary 13.14, $\hat{\mathbf{R}}$ is an achievable rate vector for $\hat{\mathcal{S}}$ if

$$\hat{R}_b \geqq H(Y_b) \quad \text{for every} \quad b \in \mathsf{K} \tag{13.26}$$

and

$$\sum_{b \in \mathsf{L}} \hat{R}_b + \sum_{b^+ \in \mathsf{L}^+} \hat{R}_{b^+} \geqq H(Y_{\mathsf{L}}, Y_{\mathsf{L}^+} | Y_{\mathsf{S}_c - \mathsf{L}}, Y_{\mathsf{D}_c^+ - \mathsf{L}^+}) \tag{13.27}$$

for every $\mathsf{L} \subset \mathsf{S}_c - \mathsf{K}$ and $\mathsf{L}^+ \subset \mathsf{D}_c^+$, such that $\mathsf{L} \cup \mathsf{L}^+$ is disjoint from every $\mathsf{S}_{c'} \cup \mathsf{D}_{c'}^+$ contained in $\mathsf{S}_c \cup \mathsf{D}_c^+$. Since $\mathsf{S}_{c'} \cup \mathsf{D}_c^+ \subset \mathsf{S}_c \cup \mathsf{D}_c^+$ iff $\mathsf{S}_{c'} \subset \mathsf{S}_c$, further $\mathsf{S}_c - \mathsf{K} = \mathsf{D}_c - \mathsf{K}$, the inequalities (13.26), (13.27) follow from (13.17), (13.18) and (13.25) by the definition of the RVs Y_b and Y_{b^+}. □

At this point we note only one consequence of Theorem 13.15, namely that for source networks of depth 2 and a probability of error fidelity criterion, the "optimist's point of view" leads to the same achievable rate region as the pessimist's one adopted in Definition 13.8. This is the content of Theorem 13.16.

THEOREM 13.16 If for a source network of depth 2 with a probability of error fidelity criterion there exists a sequence of k_n-length block codes of rates converging to the components of a vector $\mathbf{R}$ and error probabilities converging to zero, then $\mathbf{R}$ is an achievable rate vector for this source network. ○

Proof Similarly to the reduction of the problem of achievable rate regions to the special case of normal source networks, one sees that it is enough to prove the theorem for NSNs. Then, however, the statement is contained in the proof of Theorem 13.15, as the converse part of that proof literally applies to the present case. □

COMMENTS *(on the computability of characterizations)*
The typical results in this book are characterizations of the achievable rates (or of the attainable error exponents at fixed rates) for various coding problems for sources, channels and their networks.

As a rule, it is easy to characterize an achievable rate region as a countable union of sets, where the nth set is defined in terms of information quantities involving n-tuples of RVs. These *product space characterizations* are not very valuable. Most of them (in fact, each one appearing in this book) fail to meet the basic criterion of computability and therefore cannot be accepted as solutions to the problem of determining the region in question. Still, some product space characterizations do have mathematical interest and give considerable insight into the problem, such as the helpers theorem, Theorem 13.15.

We say that a characterization of a closed subset of an Euclidean space is *computable* if it leads to an algorithm for deciding whether an arbitrary point of the space belongs to the ε-neighborhood of this set, and the number of arithmetic operations needed to this end is bounded by a function of ε. The computable characterizations in this book are given in terms of Shannon's information quantities involving RVs with values in finitely many fixed finite sets (rather than in terms of n-tuples of RVs with $n \to \infty$). In the coding theorems treated so far, the ranges of these RVs were the very alphabets pertinent to the problem. This will not be typical in what follows; rather, RVs taking values in auxiliary sets will also enter the characterizations. In the literature of our subject, a problem is considered settled if a computable characterization of this kind, called a *single-letter characterization*, has been found.

In spite of its fundamental role, the concept of a single-letter characterization has not been formalized as yet. The above description covering the coding theorems in this book would be too narrow for the purpose of defining single-letter characterizations. In fact, it is clearly too restrictive to admit Shannon's information measures exclusively, if only because similar characterizations involving some other functionals may be found in the future. On the other hand, the class of permissible functionals must be delimited in some way for else we would have to accept, e.g., the very definition of the achievable rate region as a single-letter characterization. While at present an attempt at a general definition appears premature, there is a property of single-letter characterizations which, in our opinion, should be implied by any reasonable definition. Namely, a single-letter characterization should be computable and it should even lead to an algorithm that decides within less than $(1/\varepsilon)^K$ arithmetic operations whether an arbitrary point of the space belongs to the ε-neighborhood of the region to be characterized. It is unknown whether such a "polynomial" algorithm always exists, e.g., for the achievable rate region of an arbitrary source network. A negative answer to this question could be interpreted as proof of the non-existence of a single-letter characterization.

While a single-letter characterization always suggests some algorithm for numerical evaluation, this usually is too complex for actual computation. Thus it happens quite often that a single-letter characterization solves the problem only in a theoretical sense. As an example, compare the single-letter characterization of the capacity of a DMC resp. of the Δ-distortion rate of a DMS (given in Chapter 6 resp. 7) with the efficient algorithm for their actual computation treated in Chapter 8.

Note, finally, that product space characterizations may also be computable in principle. In fact, could one tell how fast the union of the first n sets in the given representation approximates the countable union, and an algorithm for deciding (within a bounded number of operations) the membership of points in the ε-neighborhood of the latter could also be given. Normally this bound would grow faster than any power of $1/\varepsilon$ as $\varepsilon \to 0$.

Discussion

The key phenomenon underlying the results of this chapter is contained in Theorem 13.2 resp. its rate slicing corollary. The proof we have given relies on the same mathematical basis as did the noisy channel coding theorem of Chapter 6. Recall that in Chapter 6 we have been looking for a single large set of typical sequences which is the codeword set of a code for some DMC, whereas now we have partitioned almost all typical sequences into such sets. The two problems can therefore be considered as "duals" in some vague sense. This vague concept of "duality" will continue to connect many other problems of Part III in a manner that will be exploited in our treatment. The proof of the fork network coding theorem, Theorem 13.5 gives an example of how more complex problems can be solved via "rate slicing." An alternative approach is random selection of codes, more specifically random binning, see the comment to Lemma 13.13. The two techniques have different advantages; the first one provides more insight into the structure of the codes, while the second one is usually more convenient for generalizations. → 13.2

We have introduced a general set-up for multi-terminal source coding problems. In the following, we concentrate on determining achievable rate regions, and we will not consider more subtle questions such as the speed of convergence of error probabilities to zero (but in the Problems part of this chapter, there are items about error exponents). Further, we shall be content with characterizations which are computable in principle, without considering the problem of actual computations.

We have attempted to outline a general theory of source networks in which the coding operations are performed simultaneously, rather than in several successive steps (source networks of depth 2). In doing so, we have restricted attention to probability of error fidelity criteria. It was shown that within this class it is sufficient to consider the subclass of "normal source networks." Theorem 13.14 gives a computable characterization of the achievable rate region of those source networks in which the fidelity criterion requires that at every output all the component sources be reproduced which are located at inputs from which there is a path to this output.

Theorem 13.15 characterizes the achievable rate region of arbitrary normal source networks. Although this characterization is not a computable one, it exhibits the common mathematical background of all these problems. The theorem shows that an essential parameter of NSNs is the number of "helpers," i.e., of those encoders which provide information to some decoders about a component source they are not bound to reproduce. In particular, it follows from Theorem 13.15 that the achievable rate region of every NSN with one helper can be determined if we can solve the following *entropy characterization problem.* Given an arbitrary DMMS with generic variables $X_a, a \in \mathsf{A}$, characterize in a computable manner the closure of the set of all vectors $\{(1/n)H(X_a^n|f(X_{a_0}^n))\}_{a\in\mathsf{A}}$, where a_0 is some fixed element of A, n ranges over the positive integers and f ranges over all mappings with domain $\mathsf{X}_{a_0}^n$. Although this entropy characterization problem for arbitrary $|\mathsf{A}|$ is unsolved, the special case $|\mathsf{A}| = 3$ will → 13.10

be solved in Chapter 15. Of course, for determining the achievable rate region of an individual source network, partial results for the corresponding entropy characterization problem may be sufficient; see the Problems to Chapter 16. One can generalize the entropy characterization problem to cover the case of normal source networks with several helpers, and these authors believe that such problems represent the mathematical essence of the theory of source networks. These more general problems, however, are beyond our reach.

The computable characterization of the achievable rate region in Theorem 13.14 resp. the non-computable one in the helpers theorem, Theorem 13.15 are examples of what is usually called a single-letter characterization resp. a product space characterization; see the comments preceding this discussion. These intuitive terms, to which it is hard to give a rigorous mathematical meaning, are widely used in the information theory of discrete memoryless systems.

→ 13.8 While computable characterizations of achievable rate regions are known only in a few cases, some simple properties of such rate regions have been established: every achievable rate region is convex and, at least for source networks of depth 2 with a probability of error fidelity criterion, the optimist's and pessimist's definitions of the achievable rate region are equivalent (Theorem 13.16). Note, however, that even these properties have not been proved for the ε-achievable rate regions, although we do not know of any example where the ε-achievable rate region of a source network is different from the achievable one. For normal source networks without helpers the equality of the ε-achievable and achievable rate region has been shown in Theorem 13.14. We shall return to this question in Chapter 16.

The mathematical framework developed in this chapter is suitable for source networks of any depth (see also Problem 13.14), but networks of depth larger than 2 had hardly been studied when the first edition of this book was published. Since then, progress concerning general networks has been substantial, at least subject to the probability of error fidelity criterion, or zero error. Coding theorems for networks of depth larger than 2 are considered to belong to a new direction of multi-user information theory, called *network coding*, which is not discussed here. ○

Problems

13.1. (a) Consider k-length block codes with side information both at the encoder and decoder, i.e., $f_k : \mathsf{X}^k \times \mathsf{Y}^k \to \mathsf{M}$, $\varphi_k : \mathsf{M} \times \mathsf{Y}^k \to \mathsf{X}^k$; let the error probability $e(f_k, \varphi_k)$ be defined similarly to Definition 13.1. Prove that if

$$\overline{\lim_{k\to\infty}}\, e(f_k, \varphi_k) < 1 \text{ then } \underline{\lim_{k\to\infty}}\, \frac{1}{k} \log \|f_k\| \geqq H(X|Y).$$

(b) Conclude that the result of Theorem 13.2 is best possible and the lack of knowledge of side information at the encoder asymptotically does not matter.

13.2. Prove that if (f, φ) is a k-length block code with side information at the decoder satisfying (i) and (ii) of Theorem 13.2, then there exists a function $\tilde{f} : \mathsf{X}^k \to \mathsf{M}$

such that $(\tilde{f}, \varphi)$ also satisfies those conditions, whereas $\tilde{f}$ has the following additional properties:

(i) $\Pr\{\tilde{f}(X^k) = f(X^k)\} \geqq 1 - \tilde{\varepsilon}$,

(ii) for every $i < m$, $\tilde{f}^{-1}(i)$ is the codeword set of a $(k, \sqrt{\varepsilon})$-code for channel W^k such that $\tilde{f}^{-1}(i) \subset \mathsf{T}^k_{[X]}$.

13.3. (*Zero-error rate region*)

(a) Show that for a 2-source with generic variables (X, Y) the zero-error rate region $\mathcal{R}_0(X, Y)$ of the corresponding fork network is, in general, smaller than the achievable rate region $\mathcal{R}(X, Y)$.

(b) Give a graph-theoretic formulation of the problem of finding the projection of $\mathcal{R}_0(X, Y)$ corresponding to $R_\mathsf{Y} = \log|\mathsf{Y}|$. More precisely, denoting by $\chi(G)$ the chromatic number of a graph G and by G^n the nth power of G (see Problem 6.23), show that the problem of finding this projection for every DMMS includes the graph-theoretic problem of determining

$$\chi_c(G) \triangleq \lim_{n\to\infty} \frac{1}{n} \log \chi(\overline{G^n})$$

for an arbitrary graph G, where the overbar denotes complementation.

Hint If $R_\mathsf{Y} = \log|\mathsf{Y}|$, then finding the smallest R_X for which $(R_\mathsf{X}, R_\mathsf{Y}) \in \mathcal{R}_0(X, Y)$ is equivalent to finding $\chi_c(G)$ for the graph G assigned to the channel $W = P_{Y|X}$ as described in Problem 6.23. (Witsenhausen (1976).)

(c) If G is the pentagon, show that $\chi_c(G) = (1/2)\log 5$.

Hint Use Lovász's result hinted at in Problem 6.23.

(d) Let us consider a DMMS with $r + 1$ component sources having generic variables $(X, Y^{(1)}, \ldots, Y^{(r)})$, and the smallest R for which there exist encoders $f_k : \mathsf{X}^k \to \mathsf{M}_k$ with rates $(1/k)\log|\mathsf{M}_k| \to R$ such that X^k can be determined in an error-free manner from $f_k(X^k)$ and the first k outputs of either of the r sources with generic variable $Y^{(i)}$, $i = 1, \ldots, r$. Show that this R is the maximum for $i = 1, \ldots, r$ of the smallest rates R_X for the 2-sources with generic variables $(X, Y^{(i)})$, as in the hint to (b).

Hint The analogous result for non-zero-error achievable rates follows by random selection; see Problem 13.4. To prove the present assertion, apply Theorem 11.22, noting that the required f_k can be identified with a partition of X^k into zero-error channel codes for the compound channel (with uninformed encoder and informed decoder) whose channel matrices are the conditional probability matrices $P_{Y^{(i)}|X}$, $i = 1, \ldots, r$. (Simonyi (2003).)

13.4. (*Random selection of rate-slicing codes*)

(a) Prove by random selection that for k large enough "most" of the k-length block codes (f, g, φ) with rate pair $(H(X|Y), H(Y))$ fulfil condition (iii) of the rate slicing corollary. More precisely, for fixed k and $\delta > 0$ let M_X and M_Y be arbitrary sets satisfying

$$\frac{1}{k}\log|\mathsf{M}_\mathsf{X}| \geqq H(X|Y) + \delta, \quad \frac{1}{k}\log|\mathsf{M}_\mathsf{Y}| \geqq H(Y) + \delta.$$

Let $\mathcal{F}$ denote the family of all mappings $f : \mathsf{X}^k \to \mathsf{M}_\mathsf{X}$ and let $\mathcal{G}$ denote the family of all mappings $g : \mathsf{Y}^k \to \mathsf{M}_\mathsf{Y}$. For every pair $(f, g) \in \mathcal{F} \times \mathcal{G}$ and values $i \in \mathsf{M}_\mathsf{X}$, $j \in \mathsf{M}_\mathsf{Y}$ denote by $\mathsf{T}(i, j)$ the set of those sequences $(\mathbf{x}, \mathbf{y}) \in \mathsf{T}^k_{[XY]}$ for which $f(\mathbf{x}) = i$, $g(\mathbf{y}) = j$. Associate with every pair of encoders $f \in \mathcal{F}$, $g \in \mathcal{G}$ a typicality decoder $\varphi_{fg} : \mathsf{M}_\mathsf{X} \times \mathsf{M}_\mathsf{Y} \to \mathsf{X}^k \times \mathsf{Y}^k$, by the rule

$$\varphi_{fg}(i, j) \triangleq \begin{cases} (\mathbf{x}, \mathbf{y}) & \text{if} \quad (\mathbf{x}, \mathbf{y}) \in \mathsf{T}(i, j) \quad \text{and} \quad |\mathsf{T}(i, j)| = 1 \\ \text{arbitrary otherwise.} \end{cases}$$

Prove that if (F, G) is a RV uniformly distributed over $\mathcal{F} \times \mathcal{G}$, then with probability converging to 1 as $k \to \infty$, the random triple (F, G, φ_{FG}) will satisfy the rate slicing corollary.

(b) Generalizing (a), prove Theorem 13.5 directly, without time-sharing between extremal points.

Hint See Lemma 13.13. (Cover (1975b).)

13.5. (*Exponential error bounds*) Consider the fork network for a 2-source with generic variables X, Y. Show that if $(R_\mathsf{X}, R_\mathsf{Y}) \in \mathcal{R}(X, Y)$ then to every positive δ and sufficiently large k there exist k-length block codes (f, g, φ) for this network such that $\mathbf{R}(f, g) \leqq (R_\mathsf{X}, R_\mathsf{Y}) + (\delta, \delta)$ and

$$e(f, g, \varphi) \leqq \exp\{-k(E_1(R_\mathsf{X}, R_\mathsf{Y}, X, Y) - \delta)\}.$$

Here

$$E_1(R_\mathsf{X}, R_\mathsf{Y}, X, Y) \triangleq \min_{\hat{X}, \hat{Y}}\{D(P_{\hat{X},\hat{Y}} \| P_{XY}) + |\min[R_\mathsf{X} + R_\mathsf{Y} - H(\hat{X}, \hat{Y}),$$
$$R_\mathsf{X} - H(\hat{X}|\hat{Y}), R_\mathsf{Y} - H(\hat{Y}|\hat{X})]|^+\},$$

where $(\hat{X}, \hat{Y})$ runs over all the RVs with range $\mathsf{X} \times \mathsf{Y}$.

Hint See Problem 13.6. (Exponential error bounds for this network were first obtained by Gallager (1976, unpublished) and Koševel (1977); the above bound, proved by the authors and K. Marton in 1977, was published in Csiszár and Körner (1980). For a partial sharpening see Csiszár and Körner (1981a).)

13.6. (*Universal coding for the fork network*)

(a) Show that given the finite sets X, Y and any $\delta > 0$, there exist k-length block codes (f, g, φ) for the fork network with the properties

$$\frac{1}{k} \log \|f\| \leqq R_\mathsf{X} + \delta,$$
$$\frac{1}{k} \log \|g\| \leqq R_\mathsf{Y} + \delta,$$

such that for every DMMS with arbitrary generic variables (X, Y) ranging over X and Y, respectively, we have

$$e(f, g, \varphi) \leqq \exp[-k(E_1(R_\mathsf{X}, R_\mathsf{Y}, X, Y) - \delta)],$$

provided that $k \geqq k_0(|\mathsf{X}|, |\mathsf{Y}|, \delta)$. Here E_1 is the same as in Problem 13.5.

(b) Note that if $R_X + R_Y > H(X, Y)$, $R_X > H(X|Y)$, $R_Y > H(Y|X)$, then $E_1(R_X, R_Y, X, Y) > 0$. Thus the above sequence of codes has an error probability converging to zero exponentially for every DMMS such that (R_X, R_Y) is an inner point of $\mathcal{R}(X, Y)$.

Hint Let M_X and M_Y be sets satisfying

$$\frac{1}{k} \log |\mathsf{M}_X| \geqq R_X + \delta, \quad \frac{1}{k} \log |\mathsf{M}_Y| \geqq R_Y + \delta.$$

Denote the family of all mappings $f : \mathsf{X}^k \to \mathsf{M}_X$ resp. $g : \mathsf{Y}^k \to \mathsf{M}_Y$ by $\mathcal{F}$ and $\mathcal{G}$, and associate with every pair of encoders $f \in \mathcal{F}$, $g \in \mathcal{G}$ a *minimum entropy decoder* $\varphi = \varphi_{f,g}$ defined by

$$\varphi(i, j) \triangleq \begin{cases} (\mathbf{x}, \mathbf{y}) \text{ if } f(\mathbf{x}) = i, \ g(\mathbf{y}) = j, \text{ and } H(\mathbf{x}, \mathbf{y}) < H(\mathbf{x}', \mathbf{y}') \\ \qquad\qquad \text{whenever } f(\mathbf{x}') = i, \ g(\mathbf{y}') = j, \ (\mathbf{x}', \mathbf{y}') \neq (\mathbf{x}, \mathbf{y}) \\ \text{arbitrary if there is no such } (\mathbf{x}, \mathbf{y}). \end{cases}$$

Denote by $\mathsf{E}(f, g)$ the set of those pairs $(\mathbf{x}, \mathbf{y})$ for which the code $(f, g, \varphi_{f,g})$ makes an error, i.e., $\varphi(f(\mathbf{x}), g(\mathbf{y})) \neq (\mathbf{x}, \mathbf{y})$. Show first that for each type P of pairs in $\mathsf{X}^k \times \mathsf{Y}^k$

$$\frac{1}{|\mathcal{F}|\,|\mathcal{G}|} \sum_{f \in \mathcal{F},\, g \in \mathcal{G}} |\mathsf{E}(f, g) \cap \mathsf{T}_P)|$$
$$\leqq \sum_{(\mathbf{x}, \mathbf{y}) \in \mathsf{T}_P} \frac{1}{|\mathcal{F}|\,|\mathcal{G}|} \Big| \{(f, g) \colon f(\mathbf{x}') = f(\mathbf{x}), \ g(\mathbf{y}') = g(\mathbf{y})$$
$$\text{for some } (\mathbf{x}', \mathbf{y}') \neq (\mathbf{x}, \mathbf{y}) \quad \text{with} \quad H(\mathbf{x}', \mathbf{y}') \leqq H(\mathbf{x}, \mathbf{y})\} \Big| .$$

Show via Lemmas 2.2–2.5 that this implies

$$\frac{1}{|\mathcal{F}|\,|\mathcal{G}|} \sum_{(f,g) \in \mathcal{F}} \frac{|\mathsf{E}(f, g) \cap \mathsf{T}_P|}{|\mathsf{T}_P|} \leqq \exp\Big\{-k \,\Big|\min[R_X + R_Y - H(\hat{X}, \hat{Y}),$$
$$R_X - H(\hat{X}|\hat{Y}),\ R_Y - H(\hat{Y}|\hat{X})]\Big|^+\Big\},$$

where $\hat{X}, \hat{Y}$ denote RVs with joint distribution P. Conclude that there exists an $(f, g) \in \mathcal{F} \times \mathcal{G}$ such that for every joint type P

$$\frac{|\mathsf{E} \cap \mathsf{T}_P|}{|\mathsf{T}_P|} \leqq \exp\Big\{-k \,\Big|\min[R_X + R_Y - H(\hat{X}, \hat{Y}),$$
$$R_X - H(\hat{X}|\hat{Y}),\ R_Y - H(\hat{Y}|\hat{X})]\Big|^+ + k\delta\Big\},$$

and verify that $(f, g, \varphi_{f,g})$ is a code as required. (This joint result of the authors and Marton from 1977 is published in Csiszár and Körner (1980); for a generalization see Problem 13.12. An improvement for large rates of the exponent $E_1(R_X, R_Y, X, Y)$ appears in Oohama and Han (1994), who also studied the speed of convergence to zero of the probability of correct decoding for $(R_{\mathsf{X}}, R_{\mathsf{Y}}) \notin \mathcal{R}(X, Y)$.)

(c) Show that universal codes as in (a) exist also with *linear encoders* f and g (if X and Y are Galois fields; see Problem 1.7). (See Csiszár (1982), where it was also shown that linear encoders and a non-universal decoder may yield a better error exponent than E_1, for large rates, related to the expurgated exponent for DMCs; see Problem 10.18.)

Hint Proceed as in (b), $\mathcal{F}$ and $\mathcal{G}$ now consisting of linear mappings.

13.7. (*Tightness of exponential error bounds*)

(a) Prove that for every $R_{\mathsf{X}} > 0,\ R_{\mathsf{Y}} > 0,\ \delta > 0$, every k-length block code (f, g, φ) of rates

$$\frac{1}{k}\log\|f\| \leqq R_{\mathsf{X}} - \delta, \quad \frac{1}{k}\log\|g\| \leqq R_{\mathsf{Y}} - \delta$$

for a fork network with a 2-source having generic variables X, Y has error probability

$$e(f, g, \varphi) \geqq \frac{1}{2}\exp[-kE_2(R_{\mathsf{X}}, R_{\mathsf{Y}}, X, Y)(1+\delta)],$$

where

$$E_2(R_{\mathsf{X}}, R_{\mathsf{Y}}, X, Y) \triangleq \min_{\hat{X},\hat{Y}:(R_{\mathsf{X}},R_{\mathsf{Y}})\notin\mathcal{R}(\hat{X},\hat{Y})} D(P_{\hat{X}\hat{Y}}\|P_{XY}),$$

whenever $k \geqq k_0(|\mathsf{X}|, |\mathsf{Y}|, \delta)$.

(b) Comparing (a) and Problem 13.5, prove that

$$E_2(R_{\mathsf{X}}, R_{\mathsf{Y}}, X, Y) = E_1(R_{\mathsf{X}}, R_{\mathsf{Y}}, X, Y)$$

in some neighborhood of the points $(R_{\mathsf{X}}, R_{\mathsf{Y}}) \in \mathcal{R}(X, Y)$ satisfying

$$R_{\mathsf{X}} + R_{\mathsf{Y}} = H(X, Y).$$

Hint See the proof of Theorem 10.3.

(Unpublished result of the authors and K. Marton from 1977. A partial result was obtained earlier by Gallager (1976, unpublished).)

13.8. (*Time sharing principle*) Show that the achievable rate region of an arbitrary source network is convex.

Hint Generalize the time sharing principle, Lemma 13.4.

13.9. (*Reduction of networks*) Prove that for every fidelity criterion, the problem of Fig. P13.9 is reducible to a fork network problem.

Hint Consider a fork network with three inputs and a 3-source with generic variables $\tilde{X} \triangleq (Z, Y), \tilde{Y} \triangleq (X, Z), \tilde{Z} \triangleq (X, Y)$.

13.10. (*Entropy characterization*) Express the achievable rate region of an arbitrary normal source network with one helper using the solution of an entropy characterization problem.

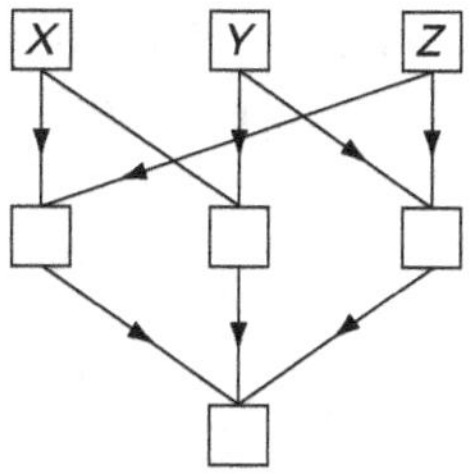

Figure P13.9

Hint Let b_0 be the helper. Write in (13.17) and (13.18) f for f_{b_0} and note that $H(f(X^n_{b_0})) = nH(X_{b_0}) - H(X^n_{b_0}|f(X^n_{b_0}))$. Further, rewrite the conditional entropies in (13.18) as $H(X^n_{\mathsf{D}_c}|f(X^n_{b_0})) - H(X^n_{\mathsf{D}_c-\mathsf{L}}|f(X^n_{b_0}))$. Introduce a new DMMS with generic variables X_{b_0} and X_{J}, with J ranging over the subsets of the D_c.

13.11. (*Helpers in arbitrary source networks*) Given any source network of depth 2 with a probability of error fidelity criterion, a vertex $b \in \mathsf{B}$ is called a *helper* if there exists a path (a, b), (b, c) such that $a \notin \mathsf{D}_c$. We have proved the existence of an NSN, the achievable rate region of which gives that of the original one by first intersecting it with a hyperplane spanned by some coordinate axes and then performing a projection. Show that this NSN can be chosen such that it contains no more helpers than the original source network.

13.12. (*Universally attainable error exponents*)

(a) Show that for a normal source network without helpers, for every vector $\mathbf{R} = \{R_b\}_{b\in\mathsf{B}}$ with non-negative components, every $\delta > 0$ and sufficiently large k there exists a k-length block code $\{f_b, \varphi_c\}_{b\in\mathsf{B},c\in\mathsf{C}}$ with

$$\frac{1}{k}\log\|f_b\| \leqq R_b + \delta \quad \text{for every} \quad b \in \mathsf{B},$$

such that for every output $c \in \mathsf{C}$ the error probability e_c is bounded above by

$$\exp\left\{-k\min_P\left[D(P\|Q) + \min_{\emptyset\neq\mathsf{L}\subset\mathsf{S}_c}\left|\sum_{b\in\mathsf{L}} R_b - H(\hat{X}_{\mathsf{L}}|\hat{X}_{\mathsf{S}_c-\mathsf{L}})\right|^+\right] + k\delta\right\}.$$

Here Q denotes the joint distribution of the generic variables of the underlying DMMS; the minimization is for distributions P on X_{B}, and $\hat{X}_b$, $b \in \mathsf{S}_c$ denote dummy RVs whose joint distribution equals the projection of P to $\mathsf{X}_{\mathsf{S}_c}$. Check that all the above exponents are positive iff $\mathbf{R}$ is an inner point of the achievable rate region. (For the tightness of these bounds, see Problem 13.13.)

(b) Show that the error exponents of (a) are universally attainable; i.e., that the same set of encoders $\{f_b\}_{b\in\mathsf{B}}$ achieves these bounds for every NSN without helpers having given source alphabets X_b, $b \in \mathsf{B}$, if φ_c is the minimum entropy decoder corresponding to the encoders f_b, $b \in \mathsf{S}_c$; see Problem 13.6.

Hint Let us denote by $\mathcal{F}$ the family of collections $f = \{f_b\}_{b \in \mathsf{B}}$ of mappings $f_b : \mathsf{X}_b^k \to \mathsf{M}_b$, where M_b, $b \in \mathsf{B}$ are given sets satisfying

$$\frac{1}{k} \log |\mathsf{M}_b| \geqq R_b + \delta \quad (b \in \mathsf{B}).$$

For sets $\mathsf{L} \subset \mathsf{S} \subset \mathsf{B}$ (where S stands for either S_c, $c \in \mathsf{C}$) let $\mathsf{E}(\mathsf{S}, \mathsf{L}, f)$ denote the set of those $|\mathsf{S}|$-tuples $\mathbf{u} = \{\mathbf{x}_b\}_{b \in \mathsf{S}}$ ($\mathbf{x}_b \in \mathsf{X}_b^k$) for which there exists an $|\mathsf{S}|$-tuple $\mathbf{v} = \{\mathbf{y}_b\}_{b \in \mathsf{S}}$ ($\mathbf{y}_b \in \mathsf{X}_b^k$) such that

(i) $\mathbf{y}_b \neq \mathbf{x}_b$ for $b \in \mathsf{L}$; $\quad \mathbf{y}_b = \mathbf{x}_b$ for $b \in \mathsf{S} - \mathsf{L}$;
(ii) $f_b(\mathbf{y}_b) = f_b(\mathbf{x}_b)$ for every $b \in \mathsf{S}$;
(iii) $H(\mathbf{v}) \leqq H(\mathbf{u})$.

Note that the minimum in the assertion does not change if the divergence of P and Q is replaced by that of their projections to $\mathsf{X}_{\mathsf{S}_c}$. Hence it suffices to show that for every $\delta > 0$ and $k \geqq k_0(\{|\mathsf{X}_b|\}_{b \in \mathsf{B}}, \delta)$ there exists $f \in \mathcal{F}$ such that, for every $\mathsf{S} \subset \mathsf{B}$, joint type P of $|\mathsf{S}|$-tuples $\{\mathbf{x}_b\}_{b \in \mathsf{S}}$ and every $\emptyset \neq \mathsf{L} \subset \mathsf{S} \subset \mathsf{B}$,

$$\frac{|\mathsf{E}(\mathsf{S}, \mathsf{L}, f) \cap \mathsf{T}_P|}{|\mathsf{T}_P|} \leqq \exp\left\{-k \left| \sum_{b \in \mathsf{L}} R_b - H(\hat{X}_\mathsf{L} | \hat{X}_{\mathsf{S}_c - \mathsf{L}}) \right|^+ \right\}.$$

This, however, is a straightforward generalization of the result in the hint of Problem 13.6. Indeed, denoting by $\mathsf{K}(\mathsf{S}, \mathsf{L}, \mathbf{u})$ the set of $|\mathsf{S}|$-tuples $\mathbf{v}$ satisfying (i) and (iii) for given S, L and $\mathbf{u} = \{\mathbf{x}_b\}_{b \in \mathsf{S}}$, we have for each joint type P of $|\mathsf{S}|$-tuples $\mathbf{u}$

$$\begin{aligned} &\frac{1}{|\mathcal{F}|} \sum_{f \in \mathcal{F}} \frac{|\mathsf{E}(\mathsf{S}, \mathsf{L}, f) \cap \mathsf{T}_P|}{|\mathsf{T}_P|} \\ &= \frac{1}{|\mathsf{T}_P|} \sum_{\mathbf{u} \in \mathsf{T}_P} \frac{1}{|\mathcal{F}|} |\{f : \text{some } \mathbf{v} \in \mathsf{K}(\mathsf{S}, \mathsf{L}, \mathbf{u}) \text{ satisfies (ii)}\}| \\ &\leqq \frac{1}{|\mathsf{T}_P|} \sum_{\mathbf{u} \in \mathsf{T}_P} \frac{|\mathsf{K}(\mathsf{S}, \mathsf{L}, \mathbf{u})|}{\prod_{b \in \mathsf{L}} |\mathsf{M}_b|}. \end{aligned}$$

Hence the assertion follows as in the proof of Lemma 13.13. (Csiszár and Körner (1980).)

13.13. (*Converse exponential error bounds*)

(a) Consider an arbitrary NSN with one output vertex. Denote by $\mathcal{R}(Q)$ the achievable rate region if the generic variables of the underlying DMMS have joint distribution Q. Show that for every sequence of k-length block codes with rate vector converging to $\mathbf{R}$, the error probability $e_k(Q)$ satisfies

$$\lim_{k \to \infty} \frac{1}{k} \log e_k(Q) \geqq - \inf_{P : \mathbf{R} \notin \mathcal{R}(P)} D(P \| Q).$$

Hint Consider any joint distribution P with $\mathbf{R} \notin \mathcal{R}(P)$. Then, by Theorem 13.16, there exists an $\varepsilon > 0$ such that for every sufficiently large k the error probability $e_k(P)$ is at least ε. Now apply Corollary 1.2. (This idea dates back to Haroutunian (1968) and Blahut (1974).)

(b) Conclude that the exponential error bounds given in Problem 13.12 are tight in a neighborhood of the boundary of the achievable rate region of an NSN without helpers, in the sense that they cannot be simultaneously improved.

13.14. (*Alternative definition of network code*) Definition 13.6 involves the assumption that an intermediate vertex b sends the same message over all outgoing edges (a function of the messages received by b from the vertices in S_b). Give a formal definition of an alternative model that permits the sending of different messages over different outgoing edges. Show that this alternative model is not genuinely different, indeed, it gives rise to the same achievable rate region as the model in the text with a suitably modified underlying graph.

Hint In the alternative model, a code associated with the network has as many coders at an intermediate vertex b as there are outgoing edges from b. Given a model of this kind, construct one similar to that in the text that has the same achievable rates, replacing each intermediate vertex b by as many vertices as there are outgoing edges from b and appropriately defining the new edges.

Story of the results

The study of multi-terminal systems is one of the most dynamic research areas in information theory. In order to get a good picture of its development, the reader is advised to consult the Story sections of the subsequent chapters.

The first results concerning separate coding of correlated sources were obtained by Gács and Körner (1973), see Problem 16.27, and Slepian and Wolf (1973a). The foundations of the theory of source networks were laid by the latter authors. In particular, they proved Theorem 13.2 and the fork network coding theorem, Theorem 13.5, for 2-sources. The present Theorem 13.5 appears in Wolf (1974). The proof of Theorem 13.2 given in the text is due to R. Ahlswede; see Ahlswede and Körner (1975). The extension of Theorem 13.5 to arbitrary fork networks as well as its simple proof reproduced in Lemma 13.13 are due to Cover (1975b).

An attempt towards a unified treatment of multi-terminal source coding problems was made by Körner (1975). A class of source networks of depth 2 in the sense of this chapter was treated in Han and Kobayashi (1980). Their approach led to direct results but not to single-letter converses.

The general set-up and results starting with Definition 13.6 are due to Csiszár and Körner (1980). The idea of connecting the concept of a single-letter characterization with computational complexity was suggested to us by P. Gács.

Addition. The first major results for source networks of arbitrary depth were the single-letter solutions obtained, for one-output networks, by Han (1980) and for one-input networks by Ahlswede *et al.* (2000). Readers interested in developments in the research direction called network coding are advised to visit the *Network Coding Homepage* (http://www.networkcoding.info).

14 Multiple-access channels

In Chapter 13 we formulated a fairly general model of noiseless communication networks. The absence of noise means that the coders located at the vertices of the network have direct access to the results of coding operations performed at immediately preceding vertices. By dropping this assumption, we now extend the model to cover communication in a noisy environment. We shall suppose that codewords produced at certain vertices are components of a vector input of a noisy channel, and it is the corresponding channel output that can be observed at some other vertex of the network.

The mathematical problems solved in this chapter will relate to the noisy version of the simplest multi-terminal network, the fork. In order to avoid clumsy notation, we give the formal definitions only for the case of two inputs.

Given finite sets $\mathsf{X}, \mathsf{Y}, \mathsf{Z}$, consider channels with input set $\mathsf{X} \times \mathsf{Y}$ and output set Z. A *multiple-access code* (MA code) for such channels is a triple of mappings $f : \mathsf{M}_1 \to \mathsf{X}$, $g : \mathsf{M}_2 \to \mathsf{Y}$, $\varphi : \mathsf{Z} \to \mathsf{M}_1 \times \mathsf{M}_2$, where M_1 and M_2 are arbitrary finite sets. The mappings f and g are called *encoders,* with *message sets* M_1 resp. M_2, while φ is the *decoder.* A MA code is also a code in the usual sense, with encoder $(f, g) : \mathsf{M}_1 \times \mathsf{M}_2 \to \mathsf{X} \times \mathsf{Y}$ and decoder φ. The error probabilities for this code and a channel $W : \mathsf{X} \times \mathsf{Y} \to \mathsf{Z}$ will be called the error probabilities of the MA code for channel W.

The performance of MA codes on DMCs $\{W : \mathsf{X} \times \mathsf{Y} \to \mathsf{Z}\}$ will be studied below. In this context the DMC $\{W : \mathsf{X} \times \mathsf{Y} \to \mathsf{Z}\}$ is called a (discrete memoryless) *multiple-access channel*, abbreviated as MAC; X and Y are called the *input alphabets* and Z is the *output alphabet* of the MAC $\{W\}$.

For a given MAC $\{W : \mathsf{X} \times \mathsf{Y} \to \mathsf{Z}\}$ and a DMMS with two component sources one can consider MA codes (f, g, φ), where f and g map the kth Cartesian powers of the alphabets of the component sources into X^{n_k} and Y^{n_k}, respectively, while φ maps Z^{n_k} into the product of the domains of f and g. Using such a code, the probability of erroneous reproduction of k-length messages of the DMMS is well-defined. Now one can ask what is the *limiting minimum transmission ratio*,

$$L \triangleq \min \overline{\lim_{k\to\infty}} \frac{n_k}{k}, \tag{14.1}$$

where the minimum refers to sequences of codes as above such that the probability of erroneous reproduction of k-length messages tends to zero (or, more generally, some other fidelity criterion is satisfied). The analogy of the two-terminal theory suggests that we look separately at a "source coding problem" (see Chapter 13) and a "channel coding problem" (formulated below), and then combine the results to obtain a source–channel transmission theorem such as Theorem 7.4. Unfortunately, this approach works only in special cases and the general problem is still unsolved. Nevertheless, the channel coding problem – which we are turning to now – is of independent interest, both from the mathematical and practical points of view. → 14.1

An *n-length block MA code* for MACs with aphabets X, Y and Z is a MA code (f, g, φ) with X^n, Y^n, Z^n playing the roles of X, Y, Z, respectively. For convenience, we shall denote the corresponding message sets by $\mathsf{M_X}$ and $\mathsf{M_Y}$ rather than M_1 and M_2. As a performance characteristic of an n-length MA block code, we shall use the *rate pair*

$$\left(\frac{1}{n}\log|\mathsf{M_X}|, \quad \frac{1}{n}\log|\mathsf{M_Y}|\right).$$

Note that the rate of (f, g, φ) considered as an ordinary block code is the sum of these two numbers. → 14.2

DEFINITION 14.1 A pair of non-negative numbers $(R_\mathsf{X}, R_\mathsf{Y})$ is an *ε-achievable rate pair for* the MAC $\{W : \mathsf{X} \times \mathsf{Y} \to \mathsf{Z}\}$ if to every $\delta > 0$, for all sufficiently large n, there exist n-length block MA codes (f, g, φ) such that

$$\frac{1}{n}\log|\mathsf{M_X}| \geqq R_\mathsf{X} - \delta, \quad \frac{1}{n}\log|\mathsf{M_Y}| \geqq R_\mathsf{Y} - \delta,$$

and the average probability of error satisfies

$$\bar{e}(W^n, f, g, \varphi) \leqq \varepsilon.$$

Further, $(R_\mathsf{X}, R_\mathsf{Y})$ is an *achievable rate pair* if it is ε-achievable for every $\varepsilon > 0$. The set of all achievable rate pairs is the *capacity region* of the MAC $\{W : \mathsf{X} \times \mathsf{Y} \to \mathsf{Z}\}$. ○

REMARK Achievable rate pairs and a capacity region can also be defined for maximum (rather than average) probability of error. Unlike in Chapter 6, the two approaches lead, in general, to different results. For multi-terminal models, the average probability of error is a more usual performance criterion than the maximum probability of error. Also, the former is easier to handle with the now available methods. As a further point, it should be mentioned that if block MA codes with stochastic encoders are also allowed then both performance criteria lead to the capacity region defined above. ○ → 14.3 → 14.4 → 14.5

LEMMA 14.2 (*Time sharing principle*) The capacity region of a MAC is a closed convex set. ○

Proof Closedness is obvious from the definition. Convexity follows analogously to the proof of Lemma 13.4. The *juxtaposition* of MA codes (f_i, g_i, φ_i) of block length n_i $(i = 1, 2)$ is the MA code (f, g, φ) of block length $n_1 + n_2$ constructed as follows: $\mathsf{M_X} \triangleq \mathsf{M_X^{(1)}} \times \mathsf{M_X^{(2)}}$ is the Cartesian product of the message sets of f_1 and f_2; for

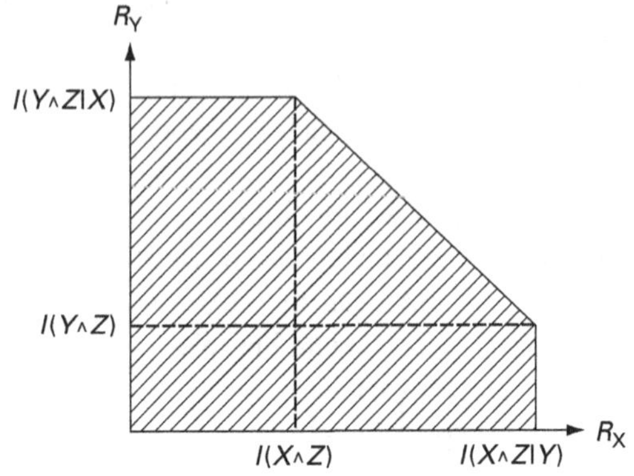

Figure 14.1 Set of achievable rate pairs for a MAC.

$(m_1, m_2) \in \mathsf{M}_\mathsf{X}$ let $f(m_1, m_2)$ be the juxtaposition of $f_1(m_1)$ and $f_2(m_2)$. The encoder g is defined similarly, and φ is defined by

$$\varphi(\mathbf{z}) \triangleq (\varphi_1(\mathbf{z}_1), \varphi_2(\mathbf{z}_2)) \quad \text{if} \quad \mathbf{z} = \mathbf{z}_1\mathbf{z}_2, \quad \mathbf{z}_i \in \mathsf{Z}^{n_i}, \quad i = 1, 2,$$

identifying $(\mathsf{M}_\mathsf{X}^{(1)} \times \mathsf{M}_\mathsf{Y}^{(1)}) \times (\mathsf{M}_\mathsf{X}^{(2)} \times \mathsf{M}_\mathsf{Y}^{(2)})$ with $\mathsf{M}_\mathsf{X} \times \mathsf{M}_\mathsf{Y}$ in the obvious way. □

Now we prove an existence result which, together with the time sharing principle, will actually yield the capacity region of multiple-access channels.

→ 14.7

THEOREM 14.3 For any pair of independent RVs X, Y with ranges X and Y, all pairs $(R_\mathsf{X}, R_\mathsf{Y})$ satisfying

$$0 \leqq R_\mathsf{X} \leqq I(X \wedge Z|Y), \quad 0 \leqq R_\mathsf{Y} \leqq I(Y \wedge Z|X), \quad R_\mathsf{X} + R_\mathsf{Y} \leqq I(X, Y \wedge Z) \tag{14.2}$$

belong to the capacity region of the MAC $\{W : \mathsf{X} \times \mathsf{Y} \to \mathsf{Z}\}$, where Z denotes a RV connected with (X, Y) by the channel W. ○

Proof The assertion is a rather simple consequence of the fork network coding theorem, Theorem 13.5. To see this, consider a DMMS with generic RVs X, Y, Z as above. Let $(\tilde{f}, \tilde{g}, \tilde{h}, \tilde{\varphi})$ be an n-length block code for the fork network associated with this DMMS such that

$$\tilde{f} : \mathsf{X}^n \to \{1, \ldots, K\}, \quad \tilde{g} : \mathsf{Y}^n \to \{1, \ldots, L\},$$

and $\tilde{h}$ is the identity mapping on Z^n. Accordingly, let the decoder $\tilde{\varphi}$ be of form

$$\tilde{\varphi}(i, j, \mathbf{z}) = (\hat{\mathbf{x}}, \hat{\mathbf{y}}, \mathbf{z}) \text{ for every } 1 \leqq i \leqq K, \ 1 \leqq j \leqq L, \ \mathbf{z} \in \mathsf{Z}^n. \tag{14.3}$$

Suppose further that for every i and j the sets

$$\mathsf{A}_i \triangleq \{\mathbf{x} : \tilde{f}(\mathbf{x}) = i\}, \quad \mathsf{B}_j \triangleq \{\mathbf{y} : \tilde{g}(\mathbf{y}) = j\}$$

consist of sequences of a single type. With any such $(\tilde{f}, \tilde{g}, \tilde{h}, \tilde{\varphi})$ we associate a family of n-length block MA codes $(f_i, g_j, \varphi_{ij}), i = 1, \ldots, K, j = l, \ldots, L$. Let f_i resp. g_j be the identity mapping of A_i resp. B_j onto itself and let φ_{ij} map $\mathbf{z}$ into the first two components of $\tilde{\varphi}(i, j, \mathbf{z})$, i.e.,

$$\varphi_{ij}(\mathbf{z}) \triangleq (\hat{\mathbf{x}}, \hat{\mathbf{y}}) \quad \text{if} \quad \tilde{\varphi}(i, j, \mathbf{z}) = (\hat{\mathbf{x}}, \hat{\mathbf{y}}, \mathbf{z}).$$

We claim that, for some weights, the weighted average of the average probabilities of error of the MA codes (f_i, g_j, φ_{ij}) equals the error probability of the fork network code $(\tilde{f}, \tilde{g}, \tilde{h}, \tilde{\varphi})$.

To see this, note first that since both A_i and B_j consist of sequences of a single type and X, Y are independent, we have

$$P^n_{XY}(\mathbf{x}, \mathbf{y}) = \frac{P^n_X(\mathsf{A}_i) P^n_Y(\mathsf{B}_j)}{|\mathsf{A}_i||\mathsf{B}_j|} \quad \text{for} \quad \mathbf{x} \in \mathsf{A}_i, \mathbf{y} \in \mathsf{B}_j.$$

This implies that the average probability of error of the MA code (f_i, g_j, φ_{ij}) over the MAC $\{W\}$ satisfies

$$\begin{aligned}\bar{e}(W^n, f_i, g_j, \varphi_{ij}) &= \frac{1}{|\mathsf{A}_i||\mathsf{B}_j|} \sum_{\mathbf{x} \in \mathsf{A}_i} \sum_{\mathbf{y} \in \mathsf{B}_j} W^n(\{\mathbf{z} : \varphi_{ij}(\mathbf{z}) \neq (\mathbf{x}, \mathbf{y})\} | \mathbf{x}, \mathbf{y}) \\ &= \frac{\Pr\{\tilde{f}(X^n) = i, \tilde{g}(Y^n) = j, \tilde{\varphi}(i, j, Z^n) \neq (X^n, Y^n, Z^n)\}}{P^n_X(\mathsf{A}_i) P^n_Y(\mathsf{B}_j)}.\end{aligned}$$

Hence, recalling the definition (13.5) of the error probability $e(\tilde{f}, \tilde{g}, \tilde{h}, \tilde{\varphi})$,

$$\sum_{i=1}^{K} \sum_{j=1}^{L} P^n_X(\mathsf{A}_i) P^n_Y(\mathsf{B}_j) \bar{e}(W^n, f_i, g_j, \varphi_{ij}) = e(\tilde{f}, \tilde{g}, \tilde{h}, \tilde{\varphi}). \tag{14.4}$$

This establishes our first claim.

Next we show that among the MA codes (f_i, g_j, φ_{ij}) there is one having "small probability of error and large rates." To this end, let $\mathsf{I} \subset \{1, \ldots, K\}$ denote the set of those indices i for which

$$\sum_{j=1}^{L} P^n_Y(\mathsf{B}_j) \bar{e}(W^n, f_i, g_j, \varphi_{ij}) \leqq 2e(\tilde{f}, \tilde{g}, \tilde{h}, \tilde{\varphi}). \tag{14.5}$$

It follows from (14.4) that

$$\sum_{i \in \mathsf{I}} P^n_X(\mathsf{A}_i) \geqq \frac{1}{2}.$$

Hence by Lemma 2.14, for any $\delta > 0$ and $n \geqq n_1(|\mathsf{X}|, \delta)$,

$$\frac{1}{n} \log |\bigcup_{i \in \mathsf{I}} \mathsf{A}_i| \geqq H(X) - \delta.$$

Thus there exists at least one $i \in \mathsf{I}$ for which

$$|\mathsf{A}_i| \geqq \frac{1}{|\mathsf{I}|}\left|\bigcup_{i\in\mathsf{I}} \mathsf{A}_i\right| \geqq \frac{1}{K}\exp\{n(H(X)-\delta)\}. \tag{14.6}$$

Fixing an $i \in \mathsf{I}$ satisfying (14.6), let $\mathsf{J} \subset \{1, \ldots, L\}$ be the set of those indices j for which

$$\bar{e}(W^n, f_i, g_j, \varphi_{ij}) \leqq 4e(\tilde{f}, g, \tilde{h}, \tilde{\varphi}). \tag{14.7}$$

Then from (14.5)

$$\sum_{j\in\mathsf{J}} P_Y^n(\mathsf{B}_j) \geqq \frac{1}{2},$$

and this implies as above (for $n \geqq n_2(|\mathsf{Y}|, \delta)$) the existence of a $j \in \mathsf{J}$ with

$$|\mathsf{B}_j| \geqq \frac{1}{L}\exp\{n(H(Y)-\delta)\}.$$

This and (14.6) prove that to any fork network code $(\tilde{f}, \tilde{g}, \tilde{h}, \tilde{\varphi})$ with the postulated properties there exists a multiple-access channel code (f_i, g_j, φ_{ij}) having average probability of error as in (14.7) and rate pair

$$\left(\frac{1}{n}\log|\mathsf{A}_i|, \frac{1}{n}\log|\mathsf{B}_j|\right) \geqq \left(H(X) - \frac{1}{n}\log\|\tilde{f}\| - \delta, \quad H(Y) - \frac{1}{n}\log\|\tilde{g}\| - \delta\right). \tag{14.8}$$

Now fix any $(R_\mathsf{X}, R_\mathsf{Y})$ satisfying (14.2) and set

$$\tilde{R}_\mathsf{X} \triangleq H(X) - R_\mathsf{X}, \quad \tilde{R}_\mathsf{Y} \triangleq H(Y) - R_\mathsf{Y}, \quad \tilde{R}_\mathsf{Z} \triangleq H(Z). \tag{14.9}$$

Then by (14.2) and the independence of X and Y we have

$$\tilde{R}_\mathsf{X} \geqq H(X|Y,Z), \quad \tilde{R}_\mathsf{Y} \geqq H(Y|X,Z), \quad \tilde{R}_\mathsf{X} + \tilde{R}_\mathsf{Y} \geqq H(X,Y|Z).$$

On account of Theorem 13.5, $(\tilde{R}_\mathsf{X}, \tilde{R}_\mathsf{Y}, \tilde{R}_\mathsf{Z})$ is an achievable rate triple for the fork network corresponding to the DMMS with generic RVs X, Y, Z. It follows that for every $\varepsilon > 0$, $\delta > 0$ and sufficiently large n there exists an n-length block code $(\hat{f}, \hat{g}, \hat{h}, \hat{\varphi})$ for this fork network satisfying

$$\frac{1}{n}\log\|\hat{f}\| \leqq \tilde{R}_\mathsf{X} + \frac{\delta}{2}, \quad \frac{1}{n}\log\|\hat{g}\| \leqq \tilde{R}_\mathsf{Y} + \frac{\delta}{2}, \quad e(\hat{f}, \hat{g}, \hat{h}, \hat{\varphi}) \leqq \varepsilon.$$

Clearly, this code can be modified to a code $(\tilde{f}, \tilde{g}, \tilde{h}, \tilde{\varphi})$ having the properties postulated at the beginning of the proof and satisfying

$$\frac{1}{n}\log\|\tilde{f}\| \leqq \tilde{R}_\mathsf{X} + \delta, \quad \frac{1}{n}\log\|\tilde{g}\| \leqq \tilde{R}_\mathsf{Y} + \delta, \quad e(\tilde{f}, \tilde{g}, \tilde{h}, \tilde{\varphi}) \leqq \varepsilon. \tag{14.10}$$

(To modify $\hat{f}$ and $\hat{g}$, juxtapose to each codeword the type of the encoded sequence and disregard this additional information at the decoder.) Hence, by the foregoing, there exists an n-length block MA code (f_i, g_j, φ_{ij}) with average probability of error at most 4ε, see (14.7), and having rate pair

$$\left(\frac{1}{n}\log|\mathsf{A}_i|, \frac{1}{n}\log|\mathsf{B}_j|\right) \geqq (R_\mathsf{X} - 2\delta, R_\mathsf{Y} - 2\delta),$$

see (14.8)–(14.10). As $\varepsilon > 0$ and $\delta > 0$ were arbitrary, this proves that $(R_\mathsf{X}, R_\mathsf{Y})$ is an achievable rate pair for the MAC. □ → 14.8

THEOREM 14.4 (*Multiple-access channel coding theorem*) The capacity region of a MAC $\{W : \mathsf{X} \times \mathsf{Y} \to \mathsf{Z}\}$ equals the convex closure of the set of those points $(R_\mathsf{X}, R_\mathsf{Y})$ which satisfy (14.2) for some triple of RVs X, Y, Z as in Theorem 14.3. ○

Proof The direct part of the theorem, i.e., that the capacity region contains the convex closure of the mentioned set, is an immediate consequence of Theorem 14.3 and Lemma 14.2. To prove the converse, consider any n-length block MA code (f, g, φ) with average probability of error

$$\bar{e}(W^n, f, g, \varphi) \leqq \varepsilon. \tag{14.11}$$

Let S and T be independent RVs uniformly distributed over the message sets M_X resp. M_Y of the code. Write

$$X^n \triangleq f(S), \quad Y^n \triangleq g(T)$$

and let Z^n be a RV connected with (X^n, Y^n) by the channel W^n such that $ST \multimap X^nY^n \multimap Z^n$. Then (14.11) can be rewritten as

$$\Pr\{\varphi(Z^n) \neq (\dot{S}, T)\} \leqq \varepsilon,$$

whence by Fano's inequality (Lemma 3.8)

$$H(S|Z^n) \leqq \varepsilon \log|\mathsf{M}_\mathsf{X}| + 1, \quad H(ST|Z^n) \leqq \varepsilon \log|\mathsf{M}_\mathsf{X} \times \mathsf{M}_\mathsf{Y}| + 1. \tag{14.12}$$

As, by definition,

$$H(S) = \log|\mathsf{M}_\mathsf{X}|, \quad H(ST) = \log|\mathsf{M}_\mathsf{X} \times \mathsf{M}_\mathsf{Y}| = \log|\mathsf{M}_\mathsf{X}| + \log|\mathsf{M}_\mathsf{Y}|,$$

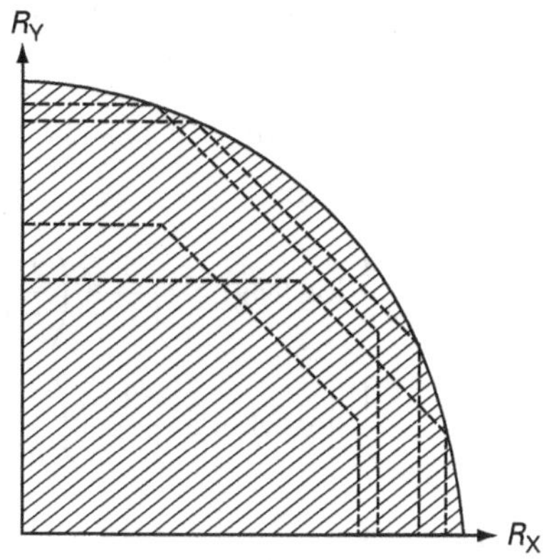

Figure 14.2 Capacity region of a MAC.

we obtain from (14.12)

$$(1-\varepsilon)\log|\mathsf{M}_\mathsf{X}| - 1 \leqq H(S) - H(S|Z^n) = I(S \wedge Z^n) \leqq I(X^n \wedge Y^n Z^n) \quad (14.13)$$

(where the last step follows from the data processing lemma).

Similarly,

$$(1-\varepsilon)(\log|\mathsf{M}_\mathsf{X}| + \log|\mathsf{M}_\mathsf{Y}|) - 1 \leqq H(ST) - H(ST|Z^n) \leqq I(X^n Y^n \wedge Z^n). \quad (14.14)$$

By the independence of X^n and Y^n, (14.13) yields

$$(1-\varepsilon)\log|\mathsf{M}_\mathsf{X}| - 1 \leqq I(X^n \wedge Z^n|Y^n). \quad (14.15)$$

Further, by symmetry, we also have

$$(1-\varepsilon)\log|\mathsf{M}_\mathsf{Y}| - 1 \leqq I(Y^n \wedge Z^n|X^n). \quad (14.16)$$

Note that for any RVs $X^n = X_1 \dots X_n$, $Y^n = Y_1 \dots Y_n$ and $Z^n = Z_1 \dots Z_n$ connected with (X^n, Y^n) by the channel W^n we have

$$H(Z^n|X^n Y^n) = \sum_{i=1}^n H(Z_i|X^n Y^n Z^{i-1}) = \sum_{i=1}^n H(Z_i|X_i Y_i).$$

Thus

$$I(X^n \wedge Z^n|Y^n) \leqq \sum_{i=1}^n H(Z_i|Y^n) - \sum_{i=1}^n H(Z_i|X_i Y_i) \leqq \sum_{i=1}^n I(X_i \wedge Z_i|Y_i) \quad (14.17)$$

and

$$I(X^n Y^n \wedge Z^n) \leqq \sum_{i=1}^n H(Z_i) - \sum_{i=1}^n H(Z_i|X_i Y_i) = \sum_{i=1}^n I(X_i Y_i \wedge Z_i). \quad (14.18)$$

Also, by symmetry,

$$I(Y^n \wedge Z^n|X^n) \leqq \sum_{i=1}^n I(Y_i \wedge Z_i|X_i). \quad (14.19)$$

Comparing (14.14)–(14.16) with (14.17)–(14.19) yields for the rate pair of the code (f, g, φ)

$$\frac{1}{n}\log|\mathsf{M}_\mathsf{X}| \leqq \frac{1}{1-\varepsilon}\left(\frac{1}{n}\sum_{i=1}^n I(X_i \wedge Z_i|Y_i) + \frac{1}{n}\right),$$

$$\frac{1}{n}\log|\mathsf{M}_\mathsf{Y}| \leqq \frac{1}{1-\varepsilon}\left(\frac{1}{n}\sum_{i=1}^n I(Y_i \wedge Z_i|X_i) + \frac{1}{n}\right),$$

$$\frac{1}{n}\log|\mathsf{M}_\mathsf{X}| + \frac{1}{n}\log|\mathsf{M}_\mathsf{Y}| \leqq \frac{1}{1-\varepsilon}\left(\frac{1}{n}\sum_{i=1}^n I(X_i Y_i \wedge Z_i) + \frac{1}{n}\right).$$

On account of the following lemma this proves that if (R_X, R_Y) is an achievable rate pair for the MAC $\{W\}$, then every neighborhood of (R_X, R_Y) contains convex combinations of points satisfying (14.2) for certain triples of RVs X_i, Y_i, Z_i as in Theorem 14.3. □

LEMMA 14.4+ Given n sets $\mathcal{C}_i$ of form (14.2), each determined by a pair of independent RVs (X_i, Y_i), a vector (R_X, R_Y) equals a convex combination with weights α_i of n vectors belonging to these sets iff

$$0 \leqq R_X \leqq \sum_{i=1}^{n} \alpha_i I(X_i \wedge Z_i | Y_i),$$

$$0 \leqq R_Y \leqq \sum_{i=1}^{n} \alpha_i I(Y_i \wedge Z_i | X_i),$$

$$R_X + R_Y \leqq \sum_{i=1}^{n} \alpha_i I(X_i Y_i \wedge Z_i).$$

○

Proof The sets $\mathcal{C}_i$, as well as the set $\mathcal{R}$ of all (R_X, R_Y) satisfying the above conditions, are pentagons. The vertices of $\mathcal{C}_i$ are

$$\mathbf{v}_i^{(0)} = (0, 0),\ \mathbf{v}_i^{(1)} = (I(X_i \wedge Z_i | Y_i), 0),\ \mathbf{v}_i^{(2)} = (I(X_i \wedge Z_i | Y_i), I(Y_i \wedge Z_i)),$$

$$\mathbf{v}_i^{(3)} = (I(X_i \wedge Z_i), I(Y_i \wedge Z_i | X_i)),\ \mathbf{v}_i^{(4)} = (0, I(Y_i \wedge Z_i | X_i));$$

see Fig. 14.1 (the pentagon $\mathcal{C}_i$ may be degenerate: these five vertices need not be all distinct). The vertices of $\mathcal{R}$ are the points $\sum_{i=1}^{n} \alpha_i \mathbf{v}_i^{(j)}$, $0 \leqq j \leqq 4$. As these vertices are contained in the (convex) set of convex combinations with weights α_i of vectors in the sets $\mathcal{C}_i$, the whole $\mathcal{R}$ is. The reverse inclusion is obvious. □

REMARK We have not proved that for a fixed $0 < \varepsilon < 1$ all the ε-achievable rate pairs belong to the capacity region characterized above. That statement (the strong converse to the multiple-access channel coding theorem) is true but will not be proved here. The proof of the last theorem does show, however, that the "optimistic" point of view (see the Discussion) leads to the same capacity region as the "pessimistic" one adopted in Definition 2.1. ○

The capacity region of a MAC can be described also in a slightly different way, introducing an auxiliary RV for the operation of taking convex closure. This will make the result more similar to forthcoming ones.

COROLLARY 14.4 The capacity region of a MAC $\{W : \mathsf{X} \times \mathsf{Y} \to \mathsf{Z}\}$ consists of those pairs (R_X, R_Y) to which there exist RVs X, Y, Z, U with values in $\mathsf{X}, \mathsf{Y}, \mathsf{Z}$ and some finite set U, respectively, such that $X \multimap U \multimap Y$, $U \multimap XY \multimap Z$, Z is connected with XY by the channel W, and

$$0 \leqq R_X \leqq I(X \wedge Z | UY), \quad 0 \leqq R_Y \leqq I(Y \wedge Z | UX), \quad R_X + R_Y \leqq I(XY \wedge Z | U). \tag{14.20}$$

Further, one may assume that $|\mathsf{U}| = 2$. ○

Proof Let $\mathcal{C}(P, Q)$ denote the set of points (R_X, R_Y) satisfying (14.2) with $P_X = P$, $P_Y = Q$. By Lemma 14.4+, (14.20) means that (R_X, R_Y) is a convex combination with weights $\Pr\{U = u\}$ of points belonging to the sets $\mathcal{C}(P_u, Q_u)$, where P_u and Q_u are the conditional distributions of X and Y given $U = u$. To verify that every such (R_X, R_Y) can already be obtained from two points belonging to some $\mathcal{C}(P_i, Q_i)$, $i = 1, 2$, consider for each $(P_1, Q_1), (P_2, Q_2)$ the convex closure of $\mathcal{C}(P_1, Q_1) \cup \mathcal{C}(P_2, Q_2)$. Clearly, the union of all such sets is closed and convex, hence it equals the convex closure of the union of all $\mathcal{C}(P, Q)$. □

We have solved a channel coding problem for a simple network. Although a general solution of the corresponding problem for more complex networks appears to be a long way off, we give the formal problem statement for networks of depth 2 (in the sense of Chapter 13). This will enable us to describe the available results and the currently most challenging open questions within the same framework and thereby clarify their interrelations.

Consider a network of depth 2 (see Definition 13.6). Let A be the set of inputs, B the set of intermediate vertices and C the set of outputs of this network. Recall that for every vertex c, S_c denotes the set of starting points of the directed edges leading to c.

We shall suppose that the decoder located at an output $c \in \mathsf{C}$ can observe the vector of codewords produced at the vertices $b \in \mathsf{S}_c \cap \mathsf{B}$ over a DMC, while the messages from the inputs $a \in \mathsf{S}_c \cap \mathsf{A}$ are directly observable at c. Further, the model will cover communication situations in which not all input messages are meant for all outputs. The addressing of the messages will be given by sets $\mathsf{D}_c \subset \mathsf{A}$, $c \in \mathsf{C}$, and we shall say that an error occurs at output c if the messages from some inputs $a \in \mathsf{D}_c$ are not correctly reproduced by the decoder at c. Since the messages from inputs $a \in \mathsf{S}_c$ are directly observable at c, one may assume that $\mathsf{D}_c \cap \mathsf{S}_c = \emptyset$.

The following definition is adapted to the typical kind of problems we shall consider which involve output distributions at the individual output vertices but no joint distributions for several output vertices. When such joint distributions are involved, as in Problem 14.21, a suitably modified definition is required.

DEFINITION 14.5 A (discrete memoryless) *channel network* is defined by a network of depth 2 (with sets A, B, C of input, intermediate and output vertices, respectively), a family of DMCs

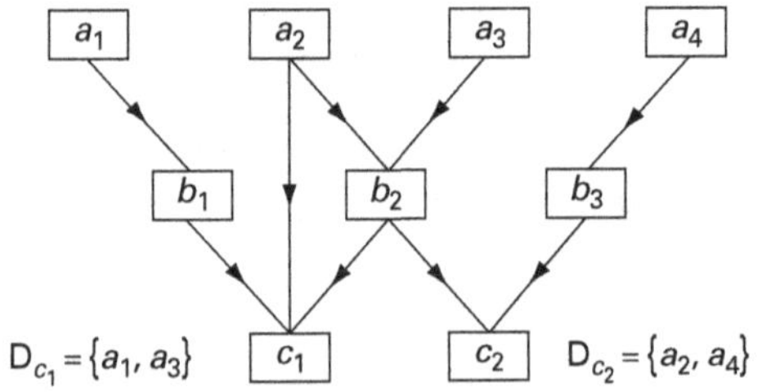

Figure 14.3 A typical channel network. In an n-length block code for this network, e.g., the encoder f_{b_2} maps $\mathsf{M}_{a_2} \times \mathsf{M}_{a_3}$ into $\mathsf{X}^n_{b_2}$, while the decoder φ_{c_1} maps $\mathsf{X}^n_{c_1} \times \mathsf{M}_{a_2}$ into a set $\mathsf{M}'_{c_1} \supset \mathsf{M}_{a_1} \times \mathsf{M}_{a_3}$.

$$\{W_c : \mathsf{X}_{b \in \mathsf{S}_c \cap \mathsf{B}} \mathsf{X}_b \to \mathsf{X}_c\}, \quad c \in \mathsf{C},$$

and a family of sets $\mathsf{D}_c \subset \mathsf{A} - \mathsf{S}_c$, $c \in \mathsf{C}$, called an *addressing*. An *n-length block code* for this channel network is a family of mappings

$$f_b : \mathsf{X}_{a \in \mathsf{S}_b} \mathsf{M}_a \to \mathsf{X}_b^n, \quad b \in \mathsf{B}; \quad \varphi_c : \mathsf{X}_c^n \times \left(\mathsf{X}_{a \in \mathsf{S}_c \cap \mathsf{A}} \mathsf{M}_a \right) \to \mathsf{M}'_c, \quad c \in \mathsf{C},$$

where $\mathsf{M}'_c \supset \mathsf{X}_{a \in \mathsf{D}_c} \mathsf{M}_a$. ○

REMARK In a more general context, the channel networks defined above should be called channel networks of depth 2. Since more general cases will not be considered in this book, our slight abuse of terminology will cause no ambiguity. ○

The DMCs $\{W_c\}$ are the *component channels* of the channel network. The sets M_a, $a \in \mathsf{A}$, are the *message sets*, and the mappings f_b resp. φ_c are the *encoders* resp. *decoders*. The vector with components $(1/n) \log |\mathsf{M}_a|$, $a \in \mathsf{A}$, will be called the *rate vector* of the code.

Channel networks will be visualized by drawing the underlying graph, as for source networks. As a rule, the inputs will be numbered, and the addressing will be given by writing numbers relating to inputs also into the output boxes. At each output, the decoder has to reproduce the messages from those inputs whose numbers are written into the respective output box, as shown in Fig. 14.4. Note that an n-length block MA code is an n-length block code for a channel network as shown in Fig. 14.5; with some abuse of terminology, such a channel network will be referred to as a MAC.

For any vector $\mathbf{z}$ (say) with components indexed by elements of some set E, and any subset F of E, denote the vector of components of $\mathbf{z}$ with indices in F by $\mathbf{z}_{\mathsf{F}}$.

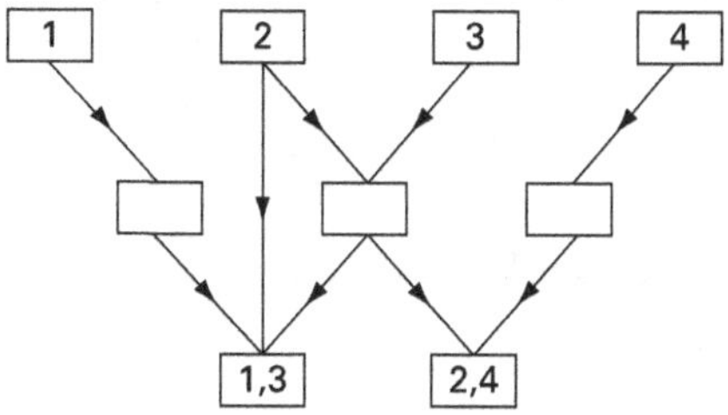

Figure 14.4 Simplified representation of the channel network of Fig. 14.3.

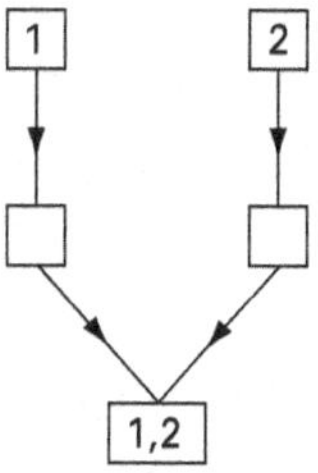

Figure 14.5 Multiple-access channel.

Given an n-length block code for a channel network, we shall consider the encoders f_b, $b \in \mathsf{B}$, as defined on the Cartesian product M of all message sets M_a, $a \in \mathsf{A}$, by setting

$$f_b(\mathbf{m}) \triangleq f_b(\mathbf{m}_{\mathsf{S}_b}), \quad \mathbf{m} \in \mathsf{M} \triangleq \bigtimes_{a \in \mathsf{A}} \mathsf{M}_a.$$

For any set $\mathsf{B}' \subset \mathsf{B}$ we shall write

$$f_{\mathsf{B}'}(\mathbf{m}) \triangleq \{f_b(\mathbf{m})\}_{b \in \mathsf{B}'}.$$

Further, for every message vector $\mathbf{m} \in \mathsf{M}$ and output vertex $c \in \mathsf{C}$ let $\mathsf{K}_c(\mathbf{m}) \subset \mathsf{X}_c^n$ be the set of those sequences $\mathbf{x} \in \mathsf{X}_c^n$ for which the decoder at c correctly reproduces the required messages, i.e.,

$$\mathsf{K}_c(\mathbf{m}) \triangleq \{\mathbf{x} : \mathbf{x} \in \mathsf{X}_c^n, \quad \varphi_c(\mathbf{x}, \mathbf{m}_{\mathsf{S}_c \cap \mathsf{A}}) = \mathbf{m}_{\mathsf{D}_c}\}.$$

Then for a given message vector $\mathbf{m}$ the *probability of error at output* c is

$$e_c(\mathbf{m}) \triangleq 1 - W_c^n(\mathsf{K}_c(\mathbf{m}) | f_{\mathsf{S}_c \cap \mathsf{B}}(\mathbf{m})). \tag{14.21}$$

The *average* resp. *maximum probability of error at output* c are given by

$$\bar{e}_c = \bar{e}(W_c^n, f, \varphi_c) \triangleq \frac{1}{|\mathsf{M}|} \sum_{\mathbf{m} \in \mathsf{M}} e_c(\mathbf{m}) \tag{14.22}$$

resp.

$$e_c = e(W_c^n, f, \varphi_c) \triangleq \max_{\mathbf{m} \in \mathsf{M}} e_c(\mathbf{m}). \tag{14.23}$$

DEFINITION 14.6 For a given channel network, a vector $\mathbf{R}$ with non-negative components R_a, $a \in \mathsf{A}$, is an *ε-achievable rate vector* if to any $\delta > 0$, for every sufficiently large n, there exists an n-length block code for the channel network with rates satisfying

$$\frac{1}{n} \log |\mathsf{M}_a| \geqq R_a - \delta, \quad a \in \mathsf{A},$$

and with average probabilities of error

$$\bar{e}_c \leqq \varepsilon, \quad c \in \mathsf{C}.$$

The *achievable rate vectors* are those which are ε-achievable for every $\varepsilon > 0$. The set of achievable rate vectors is the *capacity region* of the channel network. ○

→ 14.9 Note that Definition 14.6 is a straightforward generalization of the corresponding concepts for a MAC, see Definition 14.1, if we look at a MAC as the channel network
→ 14.11 shown in Fig. 14.5.

REMARK One can also define achievable rate vectors and the capacity region of a channel network by using maximum probabilities of error rather than average ones. The so obtained capacity region will be called the *m-capacity region*, whereas the one defined above is the *a-capacity region*. We have adopted a terminology distinguishing the average probability of error partly for convenience, as most of the available results refer to this performance criterion. Moreover, another natural approach – admitting → 14.12 stochastic encoders – would lead to the a-capacity region anyhow. ○

→ 14.13

Finally, let us sketch how the source–channel transmission problem can be formulated for multi-terminal networks (of depth 2).

DEFINITION 14.7 A (discrete memoryless) *source–channel network* consists of a DMMS and a channel network, the component sources of the DMMS being assigned in a one-to-one way to the inputs of the network. A k-to-n *block code* for the source–channel network is an n-length block code for the channel network such that at each input the message set is the kth Cartesian power of the corresponding source alphabet. ○

The error probability at output c for a given source–channel network and a k-to-n block code is defined in the obvious way as the average of the error probabilities $e_c(\mathbf{m})$ (see (14.21)), where $\mathbf{m}$ ranges over the k-length realization sequences of the given DMMS and the average is taken with respect to the joint distribution of the source RVs.

DEFINITION 14.8 The *limiting minimum transmission ratio* (LMTR) for a source–channel network is the minimum of

$$\overline{\lim_{k\to\infty}} \frac{n_k}{k}$$

for all sequences of k-to-n_k block codes such that at each output the error probability tends to zero. ○

REMARK In formulating the preceding definitions, we have assumed that at each output of the network a probability of error fidelity criterion is imposed, i.e., the messages of the component sources addressed to that output are to be reproduced within small probability of error. It is obvious how to extend the definitions to more general fidelity criteria. At present, however, almost no results in this direction exist. Unlike → 14.14 in the classical (two-terminal) theory, the LMTR for source–channel networks cannot be determined, in general, by solving a source coding problem and a channel coding problem. ○

Discussion

The main mathematical content of this chapter is the solution of a channel coding problem for a simple network, see Theorem 14.4. The direct part (Theorem 14.3) has a straightforward proof by random selection; we have preferred, however, to deduce → 14.8 it from the coding theorem for the fork (source) network, thereby emphasizing the intrinsic connection between the two problems.

We have also introduced a fairly general class of channel coding problems for multi-terminal networks, and developed a simple graphical notation for presenting such problems. The main limitation of our general setup is the assumption that all encoding operations can take place at the same time, rather than sequentially. It would be easy to extend the model to include, e.g., the possibility of feedback. Since very little is known about problems whose treatment requires such extensions (see Problems 14.20 and 14.21), we see no reason to formulate more general definitions.

The fundamental problem for a channel network – as, in particular, for a MAC – is to give a computable characterization of the capacity region. Several points deserve attention in this respect. First, the definitions of the capacity region (Definitions 14.1 and 14.7) reflect a "pessimistic point of view." Namely, achievable rate vectors were defined by postulating the existence of codes with small average probability of error and approximately the desired rates for *every* sufficiently large block length n. An optimist might be satisfied with the existence of such codes for an infinite sequence of block lengths. One can show, however (see Problem 14.24), that the "optimistic" approach always leads to the same capacity region as the one adopted in the text. Of course, for the MAC the equivalence of the two definitions is obvious from the proof of Theorem 14.4. While the choice between the "pessimistic" and "optimistic" approaches does not affect the capacity region of channel networks, using the maximum rather than the average probability of error leads, in general, to a smaller capacity
→ 14.3 region. This can already happen for the MAC. Admitting stochastic encoders, however, both perfomance criteria lead to the same capacity region. This equals the (a-) capacity
→ 14.12 region of Definition 14.6. As a third point, we note that, at least for channel networks with more than one output, the ε-capacity region (for average probability of error) can exceed the capacity region if ε is "large" (see Problems 14.9 and 10.12). It is unknown whether this can happen also for arbitrary ε or for channel networks with one output vertex.

→ 14.1 A further point in which multi-terminal problems differ from the classical (two-terminal) ones is that the source–channel transmission problem cannot be reduced to the combination of a source coding problem and a channel coding problem. In par-
→ 14.6 ticular, the LMTR problem for a general MAC and arbitrary DMMS appears to be very difficult to solve, although the achievable rate region resp. capacity region of the
→ 14.14 corresponding source resp. channel network is well-known.

Several channel networks are treated in the Problems section. One method of constructing achievable rate vectors is based on the intrinsic relationship of channel and source networks (see Theorem 14.3 and Problems 14.16–14.19). Another familiar technique is random selection. Unfortunately, a computable description of the capacity region is known for only a few cases. These are treated in the Problems section to this chapter and Chapter 16.

Since the first edition of this book was published, a variety of new models have been considered, although computable characterizations of capacity regions were obtained only in a few cases. Some of these developments are illustrated in the following and other Problems sections, while others, though also relevant, are not. For example, Problems 14.10, 12.24 and 12.25 address channels with variable states; the latter problems involve two-terminal channels but require typical multi-user techniques which also apply in multi-terminal contexts. Examples of results relevant but not included are those mentioned after Problem 14.18, as well as exponential error bounds for multi-terminal channels. Models involving secrecy are discussed in Chapter 17. ○

Problems

14.1. (*Independent channels*) Let $W_1 : \mathsf{X} \to \mathsf{Z}_1$ and $W_2 : \mathsf{Y} \to \mathsf{Z}_2$ be two given channels, and consider the MAC $\{W : \mathsf{X} \times \mathsf{Y} \to \mathsf{Z}_1 \times \mathsf{Z}_2\}$ with

$$W(z_1, z_2|x, y) \triangleq W_1(z_1|x) \cdot W_2(z_2|y).$$

Show that for this MAC and a DMMS with arbitrary generic RVs S,T, the LMTR defined by (14.1) equals

$$L = \min_{(R_1,R_2)\in\mathcal{R}(S,T)} \max\left(\frac{R_1}{C_1}, \frac{R_2}{C_2}\right),$$

where $\mathcal{R}(S, T)$ is the achievable rate region of the fork network with the given DMMS and C_i is the capacity of the DMC $\{W_i\}$, $i = 1, 2$.

Remark This problem is simple because the underlying MAC is an independent composition of DMCs. It is the "correlated noise" that makes the general problem difficult.

14.2. (*"MA capacity" of a DMC*) The MA capacity of a DMC $\{W : \mathsf{X} \times \mathsf{Y} \to \mathsf{Z}\}$ is the largest R such that, to every $\varepsilon > 0$ and $\delta > 0$, for sufficiently large n there exist n-length block MA codes which – considered as ordinary block codes – have rate greater than $R - \delta$ and average probability of error less than ε.

(a) Show that this MA capacity equals max $I(X, Y \wedge Z)$ for independent RVs X, Y where Z is connected with (X, Y) by the channel W.

Hint See Theorem 14.4.

(b) Show by an example that the MA capacity of $\{W : \mathsf{X} \times \mathsf{Y} \to \mathsf{Z}\}$ can be less than the capacity of this DMC.

14.3.* (*Maximum probability of error*) Define the m-capacity region of a MAC $\{W : \mathsf{X} \times \mathsf{Y} \to \mathsf{Z}\}$ using in Definition 14.1 the maximum rather than the average probability of error. Show that the m-capacity region may be smaller than the a-capacity region defined in the text. (Dueck (1978).)

14.4. (*Adder MAC*) Consider the noiseless MAC with $\mathsf{X} = \mathsf{Y} = \{0, 1\}$, $\mathsf{Z} = \{0, 1, 2\}$, $W(z|x, y) \triangleq 1$ if $z = x + y$ and zero otherwise.

(a) Show that the capacity region is given by $0 \leqq R_\mathsf{X} \leqq 1$, $0 \leqq R_\mathsf{Y} \leqq 1$, $R_\mathsf{X} + R_\mathsf{Y} \leqq 1,5$.

Hint Verify by calculus that the entropy of the sum of two independent binary RVs is maximum if both have distribution $(1/2,\ 1/2)$.

(b) Show that the m-capacity region of this MAC is equal to its zero-error capacity region.

(This MAC could be expected to give a simple answer to Problem 14.3. However, the available proof that its m-capacity or zero-error capacity region is a proper subset of the a-capacity region, Urbanke and Li (1998, unpublished), is quite involved. Apparently, it has not been proved that the sum-rate of zero-error codes for this MAC is bounded away from 1,5.)

14.5. (*Stochastic encoders*) An n-length block MA code with stochastic encoders is a triple (F, G, φ), where $F : \mathsf{M}_\mathsf{X} \to \mathsf{X}^n$ and $G : \mathsf{M}_\mathsf{Y} \to \mathsf{Y}^n$ are stochastic matrices and $\varphi : \mathsf{Z}^n \to \mathsf{M}' \supset \mathsf{M}_\mathsf{X} \times \mathsf{M}_\mathsf{Y}$ is an ordinary decoder. The probability of erroneous transmission of $(m_1, m_2) \in \mathsf{M}_\mathsf{X} \times \mathsf{M}_\mathsf{Y}$ over a MAC $\{W : \mathsf{X} \times \mathsf{Y} \to \mathsf{Z}\}$ using this code is

$$e_{m_1,m_2} \triangleq 1 - \sum_{\mathbf{x}\in\mathsf{X}^n} \sum_{\mathbf{y}\in\mathsf{Y}^n} F(\mathbf{x}|m_1)G(\mathbf{y}|m_2)W^n(\varphi^{-1}(m_1, m_2)|\mathbf{x}, \mathbf{y}).$$

Define the capacity region of a MAC admitting stochastic encoders, both with the average and maximum probability of error performance criteria. Prove that in both cases one gets the same capacity region as that of Definition 14.1. (R. Ahlswede, personal communication in 1977.)

Hint The proof is similar to that of Theorem 12.13.

(i) Show first that if $(R_\mathsf{X}, R_\mathsf{Y})$ is an achievable rate pair in the sense of Definition 14.1 then to every $\varepsilon > 0$ and $\delta > 0$ for sufficiently large n there exist n-length block MA codes $(\tilde{f}, \tilde{g}, \tilde{\varphi})$ such that both message sets are disjoint unions of subsets of equal sizes n^2:

$$\mathsf{M}_\mathsf{X} = \bigcup_{l=1}^{L} \mathsf{A}_l, \quad \mathsf{M}_\mathsf{Y} = \bigcup_{m=1}^{M} \mathsf{B}_m, \quad |\mathsf{A}_l| = |\mathsf{B}_m| = n^2,$$

where $(1/n)\log L > R_\mathsf{X} - \delta$, $(1/n)\log M > R_\mathsf{Y} - \delta$ and within each pair of sets A_i, B_m of messages the average probability of error is less than ε, i.e.,

$$\frac{1}{n^4} \sum_{m_1\in\mathsf{A}_l} \sum_{m_2\in\mathsf{B}_m} e_{m_1,m_2}(W^n, \tilde{f}, \tilde{g}, \tilde{\varphi}) < \varepsilon, \quad l = 1, \ldots, L, \quad m = 1, \ldots, M. \qquad (*)$$

To see this, let (f, g, φ) be a MA code of block length $n' < n$ with message sets $\{1, \ldots, L\}$, $\{1, \ldots, M\}$, where

$$\frac{1}{n'} \log L > R_\mathsf{X} - \frac{\delta}{2}, \quad \frac{1}{n'} \log M > R_\mathsf{Y} - \frac{\delta}{2},$$

and write

$$e(l, m) \triangleq 1 - W^n(\varphi^{-1}(l, m)|f(l), g(m)).$$

Let $S_1, \ldots, S_K, T_1, \ldots, T_K$ be mutually independent RVs uniformly distributed over $\{1, \ldots, L)$ resp. $\{1, \ldots, M)$. Then, for every fixed l and m,

$$Ee(l + S_i, m + T_j) = \frac{1}{LM} \sum_{l=1}^{L} \sum_{m=1}^{M} e(l, m) = \bar{e}(W^{n'}, f, g, \varphi)$$

(where the addition among messages is understood mod L resp. mod M). For fixed k, defining $T_{i+k} \triangleq T_{i+k-K}$ if $i + k > K$, the RVs

$e(l+S_i, m+T_{i+k})$ are mutually independent. It follows, as in the proof of Lemma 12.8, that

$$\Pr\left\{\frac{1}{K}\sum_{i=1}^{K} e(l+S_i, m+T_{i+k}) \geqq \varepsilon\right\}$$
$$\leqq \exp\{-K[\varepsilon - \log(1+\bar{e}(W^{n'}, f, g, \varphi))]\},$$

whence

$$\Pr\left\{\frac{1}{K^2}\sum_{i=1}^{K}\sum_{j=1}^{K} e(l+S_i, m+T_j) < \varepsilon \quad \text{for every} \quad 1 \leqq l \leqq L, 1 \leqq m \leqq M\right\}$$
$$\geqq 1 - KLM\exp\{-K[\varepsilon - \log(1+\bar{e}(W^{n}, f, g, \varphi))]\}.$$

Setting $K \triangleq n^2$, for large n the preceding probability is positive if $\bar{e}(W^{n'}, f, g, \varphi) < \exp(\varepsilon/2) - 1$, say. This can be achieved for sufficiently large n'. One concludes that there exist $s_1, \dots, s_{n^2}$, $t_1, \dots, t_{n^2}$ such that

$$\frac{1}{n^4}\sum_{i=1}^{n^2}\sum_{j=1}^{n^2} e(l+s_i, m+t_j) < \varepsilon \quad \text{for} \quad 1 \leqq l \leqq L,\ 1 \leqq m \leqq M.$$

Now let $n' < n$ with $n'/n \to 1$ be chosen so that there exist MA codes $(\hat{f}, \hat{g}, \varphi)$ of block length $n - n'$, message sets $\{1, \dots, n^2\}$ and average probability of error less than ε. Set

$$\mathsf{M_X} \triangleq \{1, \dots, n^2\} \times \{1, \dots, L\}, \quad \mathsf{M_Y} \triangleq \{1, \dots, n^2\} \times \{1, \dots, M\},$$
$$\tilde{f}(i, l) \triangleq \hat{f}(i) f(l+s_i), \quad \tilde{g}(j, m) \triangleq \hat{g}(j) g(m+t_j).$$

Then, with the obvious definition of $\tilde{\varphi}$, we obtain a code as desired, with $\mathsf{A}_l \triangleq \{1, \dots, n^2\} \times \{l\}$, $\mathsf{B}_m \triangleq \{1, \dots, n^2\} \times \{m\}$; (*) will hold with 2ε instead of ε.

(ii) One sees from (i) that every $(R_\mathsf{X}, R_\mathsf{Y})$ achievable in the sense of Definition 14.1 is also achievable in the sense of maximum probability of error if stochastic encoders are permitted (let F and G be stochastic encoders with message sets $\{1, \dots, L\}$ resp. $\{1, \dots, M\}$, where $F(\cdot|l)$ resp. $G(\cdot|m)$ attaches weights $1/n^2$ to the codewords $\tilde{f}(m_1)$, $m_1 \in \mathsf{A}_l$, resp. $\tilde{g}(m_2)$, $m_2 \in \mathsf{B}_m$). To show that every $(R_\mathsf{X}, R_\mathsf{Y})$ achievable in the sense of average probability of error with stochastic encoders is also achievable with deterministic encoders, proceed exactly as in the proof of Theorem 12.13.

14.6. (*Transmission of independent sources over a MAC*)

(a) Show that the capacity region of a MAC determines the LMTR (see (14.1)) for 2-sources with independent components. More exactly, for a two-source with independent generic RVs S, T, the LMTR equals

$$L = \min_{(R_\mathsf{X}, R_\mathsf{Y})} \max \left(\frac{H(S)}{R_\mathsf{X}}, \frac{H(T)}{R_\mathsf{Y}} \right),$$

where $(R_\mathsf{X}, R_\mathsf{Y})$ ranges over the capacity region of the given MAC.

Hint If $(\alpha H(S), \alpha H(T))$ is an achievable rate pair for some $\alpha > 0$, the existence of k-to-n block MA codes with $k/n \to \alpha$ and arbitrarily small probability of erroneous reproduction of k-length messages follows from part (i) of the hint in Problem 14.5, just as in Problem 12.12 (partitioning $\mathsf{T}^k_{[S]}$ and $\mathsf{T}^k_{[T]}$ into n^2 subsets of nearly equal probability). Conversely, if (f, g, φ) is any k-to-n block MA code, restricting the domain of f resp. g to sequences of type P resp. Q yields an MA code (f_P, g_Q, φ) with message sets T_P and T_Q; the probability of erroneous reproduction of k-length messages is a weighted average of $\bar{e}(W^n, f_P, g_Q, \varphi)$ over the possible pairs (P, Q). Hence, if this probability tends to zero and $k/n \to \alpha$, it follows that $(aH(S), \alpha H(T))$ is an achievable rate pair for the given MAC.

(b) In analogy to (a) and Problem 14.1 one might infer that for every 2-source and MAC the LMTR equals

$$\min_{R_1, R_2, R_\mathsf{X}, R_\mathsf{Y}} \max \left(\frac{R_1}{R_\mathsf{X}}, \frac{R_2}{R_\mathsf{Y}} \right),$$

where (R_1, R_2) resp. $(R_\mathsf{X}, R_\mathsf{Y})$ range over the achievable rate region for the given DMMS resp. the capacity region of the given MAC. Give a counterexample.

Hint Consider a two-source, the component sources of which are equal with probability 1 and a DMC $\{W : \mathsf{X} \times \mathsf{Y} \to \mathsf{Z}\}$ having MA capacity less than capacity; see Problem 14.2.

14.7. Give an example of a MAC for which some achievable rate pairs do not satisfy (14.2) for any pair of independent RVs X, Y.

(Such an example was given independently by the authors (see the following hint) and by Bierbaum and Wallmeier (1979).)

Hint Set $\mathsf{X} = \mathsf{Y} = \mathsf{Z} \triangleq \{0, 1\}$ and $W(0|0, 0) = W(1|0, 1) = W(1|1, 0) \triangleq 1$, $W(0|1, 1) \triangleq 1/2$. Then (1,0) and (0,1) are achievable rate pairs, whence by the time sharing principle every $(R_\mathsf{X}, R_\mathsf{Y})$ with $R_\mathsf{X} + R_\mathsf{Y} = 1$ is achievable. Out of these pairs, those with strictly positive components cannot satisfy (14.2) because if X and Y are independent and non-degenerate then $I(XY \wedge Z) < 1$.

14.8. (*Random selection of MA codes*)

(a) The proof of Theorem 14.3 given in the text considers an average of $\bar{e}(W^n, f, g, \varphi)$ over a family of codes, see (14.4), which is the same idea as that of random code selection. Prove Theorem 14.3 directly using the latter method.

Hint Given (R_X, R_Y) as in (14.2), consider message sets M_X, M_Y of sizes $\lceil \exp\{n(R_X - \delta)\}\rceil$ and $\lceil \exp\{n(R_Y - \delta)\}\rceil$, respectively. Choose encoders f and g at random, selecting the codewords corresponding to different messages $m_1 \in \mathsf{M}_X$ resp. $m_2 \in \mathsf{M}_Y$ independently and with uniform distribution from $\mathsf{T}^n_{[X]}$ resp. $\mathsf{T}^n_{[Y]}$. To given f and g, let φ be defined by $\varphi(\mathbf{z}) \triangleq (m_1, m_2)$ if $\mathbf{z} \in \mathsf{T}^n_{[W]}(f(m_1), g(m_2))$ and (m_1, m_2) is the only pair with this property; otherwise set $\varphi(\mathbf{z}) \triangleq m' \notin \mathsf{M}_X \times \mathsf{M}_Y$. Show that the expectation of $\bar{e}(W^n, f, g, \varphi)$ over this random code selection tends to zero as $n \to \infty$.

(b) Prove directly, without using the time sharing principle, that if X, Y, Z, U are RVs satisfying $X \multimap U \multimap Y$, $U \multimap XY \multimap Z$, and Z is connected with $X\,Y$ by the channel W, then every (R_X, R_Y) satisfying the inequalities (14.20) is an achievable rate pair for the MAC $\{W\}$.

Hint Proceed similarly as in (a), with the difference that the codewords are now chosen (at random) from $\mathsf{T}^n_{[X|U]}(\mathbf{u})$ resp. $\mathsf{T}^n_{[Y|U]}(\mathbf{u})$, for some fixed sequence $\mathbf{u} \in \mathsf{T}^n_{[U]}$. Accordingly, the typicality requirement in the definition of the decoder is modified to $\mathbf{z} \in \mathsf{T}^n_{[Z|UXY]}(\mathbf{u}, f(m_1), g(m_2))$.

14.9. (*Compound MAC*)

(a) Observe that the channel network of Fig. P14.9(a) represents a compound DMC with decoder informed (see Problem 10.14), defined by the two component channels of the network. In particular, the capacity region is the interval $[0, C(\mathcal{W})]$, where $C(\mathcal{W})$ is the capacity of this compound DMC.

(b) The channel network in Fig. P14.9(b) is called a *two-input–two-output MAC;* it may also be interpreted as a compound MAC with decoder informed. Let X, Y be independent RVs and let Z_1 resp. Z_2 be RVs connected with (X, Y) by the channels $W_1 : \mathsf{X} \times \mathsf{Y} \to \mathsf{Z}_1$ resp. $W_2 : \mathsf{X} \times \mathsf{Y} \to \mathsf{Z}_2$. Prove that all pairs (R_X, R_Y) satisfying the inequalities

$$0 \leqq R_X \leqq \min(I(X \wedge Z_1|Y), I(X \wedge Z_2|Y)),$$
$$0 \leqq R_Y \leqq \min(I(Y \wedge Z_1|X), I(Y \wedge Z_2|X)),$$
$$R_X + R_Y \leqq \min(I(X, Y \wedge Z_1), I(X, Y \wedge Z_2))$$

are achievable rate pairs for the channel network of Fig. P14.9(b).

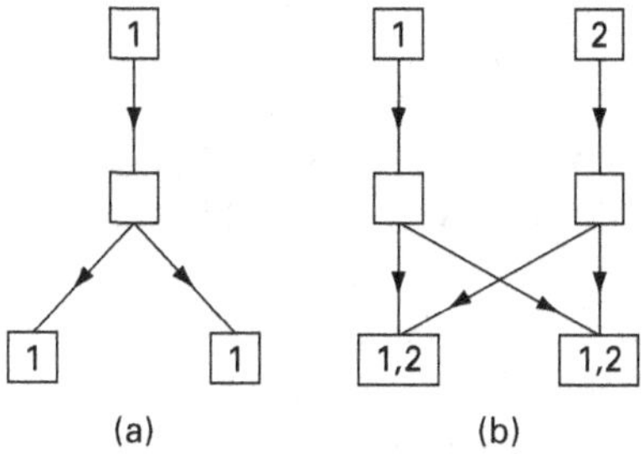

Figure P14.9 (a) Compound DMC. (b) Two-input–two-output MAC.

Hint One sees by the random selection of 14.8(a) that the pairs (R_X, R_Y) satisfying the given inequalities are achievable.

(c) Show that the capacity region of the channel network of Fig. P14.9(b) equals the set of those pairs (R_X, R_Y) which, for some RVs X, Y, U, Z_1, Z_2 with $X \multimap U \multimap Y$, $U \multimap XY \multimap Z_1$, $U \multimap XY \multimap Z_2$ and such that Z_1 resp. Z_2 is connected with (X, Y) by the channel W_1 resp. W_2, satisfy the following inequalities:

$$0 \leqq R_X \leqq \min(I(X \wedge Z_1|YU), I(X \wedge Z_2|YU)),$$
$$0 \leqq R_Y \leqq \min(I(Y \wedge Z_1|XU), I(Y \wedge Z_2|XU)),$$
$$R_X + R_Y \leqq \min(I(XY \wedge Z_1|U), I(XY \wedge Z_2|U)).$$

Further, one may assume that $|\mathsf{U}| \leqq 6$. (Ahlswede (1974).)

Hint The direct part follows by the random selection of Problem 14.8(b). The converse can be seen by the argument of the proof of Theorem 14.4, applied to both component channels. Denoting the output sequence of the jth component channel by $Z_j^n = Z_{j1} \dots Z_{jn}$ $(j = 1, 2)$, note that, e.g.,

$$\frac{1}{n}\sum_{i=1}^{n} I(X_i \wedge Z_{ji}|Y_i) = I(X \wedge Z_j|YU),$$

where U is a RV uniformly distributed on $\{1, \dots, n\}$ and independent of all the other RVs, while $X \triangleq X_U$, $Y \triangleq Y_U$, $Z_j \triangleq Z_{jU}$. To prove that $|U| \leqq 6$ may be assumed, see Lemma 15.4 and the subsequent remark.

(d) Show by an example that the above capacity region may be larger than the convex hull of the union of all sets defined by independent RVs X, Y as in (b).

Hint Set $\mathsf{X} = \mathsf{Y} \triangleq \{0, 1, 2, 3\}$, $\mathsf{Z} \triangleq \{0, 1\}$,

$$W_1(z|x, y) \triangleq \begin{cases} 1 & \text{if} \quad x \in \{0,1\},\ y \in \{0,1\},\ z = x \quad \text{or} \\ & \phantom{\text{if}} \quad x \in \{2,3\},\ y \in \{2,3\},\ z = y - 2 \\ \frac{1}{2} & \text{else,} \end{cases}$$

$$W_2(z|x, y) \triangleq \begin{cases} 1 & \text{if} \quad x \in \{0,1\},\ y \in \{0,1\},\ z = y \quad \text{or} \\ & \phantom{\text{if}} \quad x \in \{2,3\},\ y \in \{2,3\},\ z = x - 2 \\ \frac{1}{2} & \text{else.} \end{cases}$$

Check that $(1/2, 1/2)$ is an achievable rate pair for the two-input–two-output MAC defined by W_1, W_2, although it does not belong to the convex hull in question.

14.10. (*Arbitrarily varying MAC*) Extend the main results of Chapter 12 about the a-capacity of AVCs to arbitrarily varying MACs, denoted below as AVMACs.

(a) Define formally the AVMAC $\{\mathcal{W}\}$ determined by a (perhaps infinite) family $\mathcal{W}$ of channels $W : \mathsf{X} \times \mathsf{Y} \to \mathsf{Z}$ and its capacity region (meaning, as before for MACs, the a-capacity region). Extending Corollary 12.3, show that the AVMACs $\{\mathcal{W}\}$ and $\{\overline{\mathcal{W}}\}$ have the same capacity region, where the overbar denotes convex closure.

(b) Define the random code capacity region of the AVMAC $\{\mathcal{W}\}$, with the understanding that only random codes with *independently chosen* encoders $f : \mathsf{M}_\mathsf{X} \to \mathsf{X}^n$, $g : \mathsf{M}_\mathsf{Y} \to \mathsf{Y}^n$ are admitted. Show that this random code capacity region contains all rate pairs

$$(R_\mathsf{X}, R_\mathsf{Y}) \in \bigcap_{W \in \overline{\mathcal{W}}} \mathcal{C}(X, Y, W),$$

where (X, Y) is any pair of independent input RVs and $\mathcal{C}(X, Y, W)$ denotes the set in (14.2).

(c) Prove the analog of Lemma 12.8: if $|\mathcal{W}| < \infty$ and some random codes of block length $n \to \infty$ with rate pair approaching $(R_\mathsf{X}, R_\mathsf{Y})$ have asymptotically vanishing probability of error, then such random codes also exist with encoders chosen by uniform distribution from sets of deterministic encoders of size n^2.

Hint See Problem 14.5.

(d) Prove the analog of Theorem 12.11: the capacity region of an AVMAC $\{\mathcal{W}\}$ either has empty interior, or it equals the convex hull of the union for all independent pairs (X, Y) of the intersections in part (b).

Hint Each rate pair in this convex hull is achievable by random codes, by (b) and time sharing. It follows from (c), as in the proof of Theorem 12.11, that this rate pair is achievable also by deterministic codes if the capacity region has a non-empty interior. For the converse, use (a) and proceed as in the proof of the converse part of Theorem 14.4.

(The results of (a)–(d) are due to Jahn (1981).)

(e) Suppose for convenience that $\mathcal{W}$ is a finite set of matrices $W(\,.\,|\,.\,, .\,, s)$, $s \in \mathsf{S}$. Call it (X, Y)-symmetrizable if $\mathcal{W}$ is symmetrizable in the sense of Problem 12.9(c) (with $\mathsf{X} \times \mathsf{Y}$ in the role of X) and X-symmetrizable if there exists a channel $U : \mathsf{X} \to \mathsf{S}$ such that

$$\sum_{s \in \mathsf{S}} W(\,.\,|x, y, s)\, U(s|x') = \sum_{s \in \mathsf{S}} W(\,.\,|x', y, s)\, U(s|x)$$

for all $x \in \mathsf{X}$, $x' \in \mathsf{X}$, $y \in \mathsf{Y}$; define Y-symmetrizability similarly. Show that either kind of symmetrizability implies that the interior of the capacity region of the AVMAC $\{\mathcal{W}\}$ is empty. (Gubner (1990).)

(f)* Complementing part (e), show that if neither kind of symmetrizability holds then the interior of the capacity region is non-empty. (Conjectured by Gubner (1990); proved by Ahlswede and Cai (1999).)

14.11. (*Two-way channel*) The channel network of Fig. P14.11 is called a *two-way channel.* The boxes on the left- resp. right-hand side are visualized as being located at the "left" resp. "right" terminal of an (engineering device) channel through which simultaneous communication is possible in both directions.

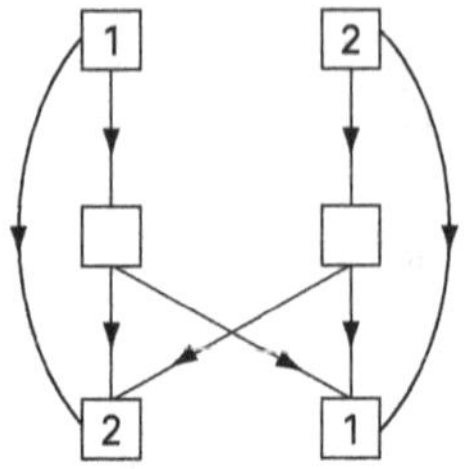

Figure P14.11 Two-way channel.

Show that from the point of view of capacity region, this channel network is equivalent to one of the type shown in Fig. P14.9(b). Conclude that the capacity region of the two-way channel equals the convex closure of the set of those pairs $(R_\mathsf{X}, R_\mathsf{Y})$ to which there exist RVs X, Y, Z_1, Z_2 such as in Problem 14.9(b) satisfying

$$0 \leqq R_\mathsf{X} \leqq I(X \wedge Z_2|Y), \quad 0 \leqq R_\mathsf{Y} \leqq I(Y \wedge Z_1|X).$$

(Shannon (1961) raised a more complex problem, see Problem 14.21. Shannon determined the above capacity region directly, without relating the problem to a MAC.)

Hint Define the channel $\widetilde{W}_1 : \mathsf{X} \times \mathsf{Y} \to \mathsf{Z}_1 \times \mathsf{X}$ by $\widetilde{W}_1\ (z,\ x'|x, y) \triangleq W_1(z|x, y)\delta(x, x')$ (where $\delta(x, x) = 1, \delta(x, x') = 0$ if $x \neq x'$), and define similarly $\widetilde{W}_2 : \mathsf{X} \times \mathsf{Y} \to \mathsf{Z}_2 \times \mathsf{Y}$. Then every code for the two-input–two-output MAC with component channels $\widetilde{W}_1$, $\widetilde{W}_2$ gives rise to a code (with the same probabilities of error) for the two-way channel with component channels W_1, W_2. For codes with distinct codewords the converse implication is also true, and to obtain the capacity region it clearly suffices to consider codes with distinct codewords. The claimed characterization of the capacity region follows from Problem 14.9(c) since the first two inequalities of Problem 14.9(c) with (Z_1, X) resp. (Z_2, Y) instead of Z_1 resp. Z_2 reduce to $0 \leqq R_\mathsf{X} \leqq I(X \wedge Z_2|YU)$, $0 \leqq R_\mathsf{Y} \leqq I(Y \wedge Z_1|XU)$, while the third one is now implied by the first two.

14.12. (*Stochastic encoders*) Define block codes with stochastic encoders for an arbitrary channel network, replacing the mappings f_b of Definition 14.5 by stochastic matrices F_b. Then (14.21) should be replaced by

$$e_c(\mathbf{m}) \triangleq 1 - \sum_{\mathbf{x}} W_c^n(\mathsf{K}_c(\mathbf{m})|\mathbf{x}) \prod_{b \in \mathsf{S}_c \cap \mathsf{B}} F_b(\mathbf{x}_b|\mathbf{m}_{\mathsf{S}_b}),$$

where $\mathbf{x} \triangleq \{\mathbf{x}_b\}$ ranges over the Cartesian product of the sets X_b^n, $b \in \mathsf{S}_c \cap \mathsf{B}$. Define achievable rate vectors and capacity region admitting codes with stochastic encoders, both with respect to average and maximum probability of error. Show that both capacity regions are the same as the (a-) capacity region of Definition 14.6.

Hint See Problem 14.5. Show first the result analogous to (*) there that to every achievable rate vector there exist n-length block codes of the required rates such that the message sets M_a, $a \in \mathsf{A}$, are disjoint unions of sets $\mathsf{E}_i^{(a)}$, $i = 1, \dots, L_a$, of size n^2 satisfying

$$n^{-2|\mathsf{A}|} \sum_{\mathbf{m} \in \mathsf{E}_\mathbf{i}} e_c(\mathbf{m}) < \varepsilon \quad \text{for every} \quad \mathbf{i} \in \bigtimes_{a \in \mathsf{A}} \{1, \dots, L_a\} \quad \text{and} \quad c \in \mathsf{C},$$

where

$$\mathsf{E}_\mathbf{i} \triangleq \bigtimes_{a \in \mathsf{A}} \mathsf{E}_{i_a}^{(a)} \quad \text{if} \quad \mathbf{i} = \{i_a\}_{a \in \mathsf{A}}.$$

14.13. (*Broadcast channels: m- and a-capacity regions are equal*)

(a) Show that if a channel network has but one intermediate vertex then its m-capacity region (not admitting stochastic encoders) equals the a-capacity region. Channel networks with one intermediate vertex and at least two input and two output vertices are called *broadcast channels* (see Chapter 16).

Hint Use Problem 14.12. Note that for $|\mathsf{B}| = 1$ the expression for $e_c(\mathbf{m})$ in Problem 14.12 reduces to

$$e_c(\mathbf{m}) = 1 - \sum_{\mathbf{x} \in \mathsf{X}^n} W_c^n(\mathsf{K}_c(\mathbf{m})|\mathbf{x}) F(\mathbf{x}|\mathbf{m});$$

thus, if a code with stochastic encoder has maximum probability of error less than ε, the $F(\cdot|\ \mathbf{m})$-probability of the set $\{\mathbf{x} : 1 - W_c^n(\mathsf{K}_c(\mathbf{m})|\mathbf{x}) \geqq |\mathsf{C}|\varepsilon\}$ must be less than $|\mathsf{C}|^{-1}$. Let $f(\mathbf{m}) \in \mathsf{X}^n$ be any sequence not belonging to either of these $|\mathsf{C}|$ sets; thus we obtain a code with (deterministic) encoder f which has maximum probability of error less than $|\mathsf{C}|\varepsilon$.

(b) Show that the result of (a) does not extend to the ε-capacity regions. In particular, in the sense of maximum probability of error the ε-capacity region may be equal to the capacity region, without the same being true in the sense of average probability of error.

Hint This is already the case for the network of Fig. P14.9(a), see Problem 10.12.

14.14. (*Transmission of independent sources over a channel network*) Show that for a source–channel network with component sources having independent generic RVs S_a, $a \in \mathsf{A}$, the LMTR equals

$$L = \min_{\mathbf{R}} \max_{a \in \mathsf{A}} \frac{H(S_a)}{R_a},$$

where $\mathbf{R} = \{R_a\}_{a \in \mathsf{A}}$ ranges over the capacity region of the underlying channel network.

Hint See Problem 14.6.

14.15. (*General multiple-access channels*)

(a) The channel network shown in Fig. P14.15 with component channels $W_i : \bigtimes_{j=1}^{s} \mathsf{X}_j \to \mathsf{Y}_i$, $i = 1, \dots, r$, is called a *MAC with s inputs and r outputs* or s senders and r receivers. Generalizing Problem 14.9(c), determine the capacity region of this channel network. For any RV U with values

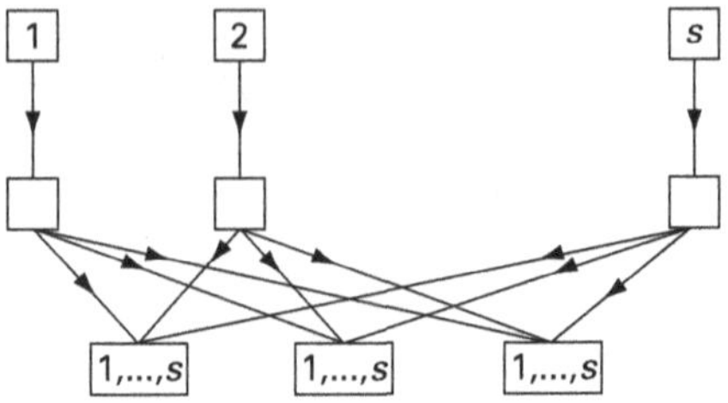

Figure P14.15 Multiple-access channel with s senders and r receivers.

in a finite set U and any s-tuple of RVs $X_1, \dots, X_s$ such that the X_i are conditionally independent given U, denote by $\mathcal{C}(U, X_1, \dots, X_s)$ the set of all vectors $\mathbf{R} = (R_1, \dots, R_s)$ satisfying, for every $1 \leqq i \leqq r$ and every set $\mathsf{L} \subset \{1, \dots, s\}$, the inequalities

$$0 \leqq \sum_{j \in \mathsf{L}} R_j \leqq I(\{X_j\}_{j \in \mathsf{L}} \wedge Y_i | \{X_j\}_{j \notin \mathsf{L}}, U);$$

here Y_i stands for a RV connected with $X_1, \dots, X_s$ by the channel W_i and such that $U \multimap X_1 \dots X_s \multimap Y_i$. Prove that the capacity region equals the union of all the sets $\mathcal{C}(U, X_1, \dots, X_s)$. Further, show that it suffices to consider $|\mathsf{U}| \leqq r(2^s - 1)$. (Ulrey (1975).)

Hint See Problem 14.9(c).

(b) If $\mathsf{Y}_i = \mathsf{Y}, i = 1, \dots, r$, the MAC with s senders and r receivers of part (a) can be interpreted as a *compound multiple-access channel* with s senders and one receiver, the receiver being informed on the actual channel in operation; see Problem 10.13. Show that just as for two-terminal channels, the capacity region remains the same if the receiver is uninformed, i.e., if the same decoder φ must be used for $i = 1, \dots, r$.

14.16. (*Channel networks with one output vertex*) Determine the capacity region of an arbitrary channel network with one output and two non-terminal vertices, see Fig. P14.16(a). This channel network is called a *MAC with common messages*, as the component m_0 of the message triple (m_0, m_1, m_2) is known at both intermediate vertices, i.e., at both inputs of the MAC.

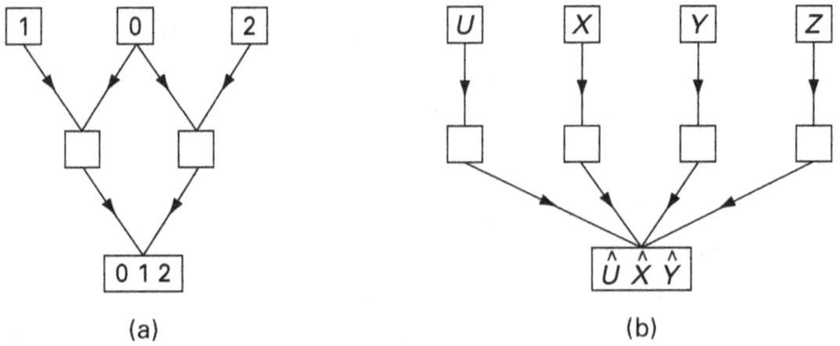

Figure P14.16 (a) General channel network with one output and two intermediate vertices. (b) Auxiliary source network.

(a) Show that $(R_0, R_1, R_2) \geqq \mathbf{0}$ is an achievable rate triple iff there exist RVs U, X, Y, Z, where $X \multimap U \multimap Y$, $U \multimap XY \multimap Z$ and Z is connected with XY by the channel W, such that

$$R_1 \leqq I(X \wedge Z|YU), \quad R_2 \leqq I(Y \wedge Z|XU).$$
$$R_1 + R_2 \leqq I(XY \wedge Z|U), \quad R_0 + R_1 + R_2 \leqq I(XY \wedge Z).$$

(Slepian and Wolf (1973b).)

(b) The characterization (a) of the achievable rate region would not be computable if one could not bound the range of the auxiliary RV U. Show, however, that $|\mathsf{U}| \leqq |\mathsf{Z}| + 3$ can be assumed.

(c) Conclude from (a) that it may be possible to send common messages at a positive rate even if separate messages are sent at an optimal rate pair, i.e., $R_0 > 0$ is possible even if (R_1, R_2) is on the boundary of the capacity region of the MAC $\{W : \mathsf{X} \times \mathsf{Y} \to \mathsf{Z}\}$.

(d) Show that to determine the capacity region of any channel network with one output vertex and k intermediate vertices, one may assume that the number of input vertices is less than 2^k. (A computable characterization of the capacity region of a general channel network with one output was given by Han (1979).)

Hint For the direct part of (a), note that if (R_0, R_1, R_2) is an achievable rate triple, and $R_1' \leqq R_1$, $R_2' \leqq R_2$, $R_0' \leqq R_0 + (R_1 - R_1') + (R_2 - R_2')$, then (R_0', R_1', R_2') is also achievable. Hence it suffices to verify that those triples satisfying the inequalities in (a) are achievable for which, in the last two, the equality holds, i.e., $R_1 + R_2 = I(XY \wedge Z|U)$, $R_0 = I(U \wedge Z)$.

This can be achieved either by random selection as in Problem 14.8(b), where, instead of a single $\mathbf{u} \in \mathsf{T}^n_{[U]}$, now $|\mathsf{M}_0| = \lceil \exp\{n(R_0 - \delta)\} \rceil$ sequences $\mathbf{u}_i$ randomly selected from $\mathsf{T}^n_{[U]}$ representing the possible values of the common message m_0 are required. Or one can proceed analogously to the proof of Theorem 14.3, as follows.

Consider n-length block codes with error probability less than ε for the source network of Fig. P14.16(b) such that the encoders of U^n, X^n, Y^n have rates close to R_{U}, R_{X}, R_{Y} with

$$R_{\mathsf{U}} = H(U|Z), R_{\mathsf{X}} \geqq H(X|UYZ), R_{\mathsf{Y}} \geqq H(Y|UXZ),$$
$$R_{\mathsf{X}} + R_{\mathsf{Y}} = H(XY|UZ),$$

while the Z-encoder is the identity mapping. Refine the partitions of U^n, X^n, Y^n defined by the source encoders so that in the new partition $\mathsf{U}^n = \bigcup_{m=1}^{M} \mathsf{C}_m$ of U^n each C_m consists of sequences of the same type. Further, for every $\mathbf{u} \in \mathsf{U}^n$ let the refined partitions $\mathsf{X}^n = \bigcup_{i=1}^{K} \mathsf{A}_i(\mathbf{u})$ and $\mathsf{Y}^n = \bigcup_{j=1}^{L} \mathsf{B}_j(\mathbf{u})$ have the property that each $\mathsf{A}_i(\mathbf{u})$ resp. $\mathsf{B}_j(\mathbf{u})$ consists of sequences having the same joint type with $\mathbf{u}$. This can be done so that we still have

$$\left|\frac{1}{n}\log M - R_{\mathsf{U}}\right| < \delta, \quad \left|\frac{1}{n}\log K - R_{\mathsf{X}}\right| < \delta, \quad \left|\frac{1}{n}\log L - R_{\mathsf{Y}}\right| < \delta.$$

Denote by $\varphi_{m,i,j}(\mathbf{z})$ the triple $(\hat{\mathbf{u}}, \hat{\mathbf{x}}, \hat{\mathbf{y}})$, which is the result of the source network decoding for $\mathbf{u} \in \mathsf{C}_m$, $\mathbf{x} \in \mathsf{A}_i(\mathbf{u})$, $\mathbf{y} \in \mathsf{B}_j(\mathbf{u})$ and the given $\mathbf{z}$; it follows that

$$\sum_{m=1}^{M} P_U^n(\mathsf{C}_m)\left[\frac{1}{|\mathsf{C}_m|}\sum_{\mathbf{u}\in\mathsf{C}_m}\sum_{i=1}^{K}\sum_{j=1}^{L} P_{X|U}^n(\mathsf{A}_i(\mathbf{u})|\mathbf{u})P_{Y|U}^n(\mathsf{B}_j(\mathbf{u})|\mathbf{u})e(\mathbf{u},i,j)\right] < \varepsilon,$$

where

$$e(\mathbf{u},i,j) \triangleq \frac{1}{|\mathsf{A}_i|(\mathbf{u})||\mathsf{B}_j(\mathbf{u})|}\sum_{\mathbf{x}\in\mathsf{A}_j(\mathbf{u})}\sum_{\mathbf{y}\in\mathsf{B}_j(\mathbf{u})} W^n(\{\mathbf{z} : \varphi_{m,i,j}(\mathbf{z}) \neq (\mathbf{u},\mathbf{x},\mathbf{y})\}|\mathbf{x},\mathbf{y}).$$

Using the bound on M, it follows that there exists an m with $\mathsf{C}_m \subset \mathsf{T}^n_{[U]}$, $|\mathsf{C}_m| \geqq \exp[n(H(U) - R_{\mathsf{U}} - 2\delta)]$, for which the term in square brackets is less than 2ε. Then, for some $\mathsf{C}'_m \subset \mathsf{C}_m$ of size $|\mathsf{C}'_m| \geqq (1/2)|\mathsf{C}_m|$, the inner double sum is less than 4ε for every $\mathbf{u} \in \mathsf{C}'_m$. Now, using the bounds on K and L, the argument in the proof of Theorem 14.3 shows that to every $\mathbf{u} \in \mathsf{C}'_m$ there exist indices $i = i(\mathbf{u})$, $j = j(\mathbf{u})$, such that $e(\mathbf{u}, i, j) < 16\varepsilon$ and

$$|\mathsf{A}_i(\mathbf{u})| > \exp[n(H(X|U) - R_{\mathsf{X}} - 2\delta)],$$
$$|\mathsf{B}_j(\mathbf{u})| > \exp[n(H(Y|U) - R_{\mathsf{Y}} - 2\delta)].$$

Choosing subsets $\mathsf{A}(\mathbf{u}) \subset \mathsf{A}_i(\mathbf{u})$, $\mathsf{B}(\mathbf{u}) \subset \mathsf{B}_j(\mathbf{u})$ of size not depending on $\mathbf{u}$, still satisfying the above inequalities and

$$\frac{1}{|\mathsf{A}(\mathbf{u})||\mathsf{B}(\mathbf{u})|}\sum_{\mathbf{x}\in\mathsf{A}(\mathbf{u})}\sum_{\mathbf{y}\in\mathsf{B}(\mathbf{u})} W^n(\{\mathbf{z} : \varphi_{m,i,j}(\mathbf{z}) \neq (\mathbf{u},\mathbf{x},\mathbf{y})\}|\mathbf{x},\mathbf{y}) \leqq e(\mathbf{u}, i.j) < 16\varepsilon$$

for every $\mathbf{u} \in \mathbf{C}_m$, the proof of the direct part is complete.

Note now that the set of vectors (R_0, R_1, R_2) defined in (a) is convex and closed (the latter follows from (b)). Thus the converse can be proved by the argument of the proof of Theorem 14.4, letting U be a RV uniformly distributed on the message set M_0.

To prove (b), see Lemma 15.4 and the remark to it. For (c), see Corollary 14.4 and Problem 14.7.

14.17. (*Interference channel*) The channel network of Fig. P14.17(b) with component channels $W_i : \mathsf{X}_1 \times \mathsf{X}_2 \to \mathsf{Y}_i$, $i = 1, 2$, is called an *interference channel.* Note that it differs from the two-input–two-output MAC only in the addressing.

(a) Show that if $R_1 \leqq I(X_1 \wedge Y_1)$, $R_2 \leqq I(X_2 \wedge Y_2)$ for some pair of independent RVs X_1, X_2 and RVs Y_i connected with (X_1, X_2) by the channel W_i, $i = 1, 2$, then (R_1, R_2) is an achievable rate pair.

(b) By the time sharing principle, the convex closure of the set of all pairs (R_1, R_2) as in (a) is a subset of the capacity region of the interference channel. Show by an example that it may be a proper subset.

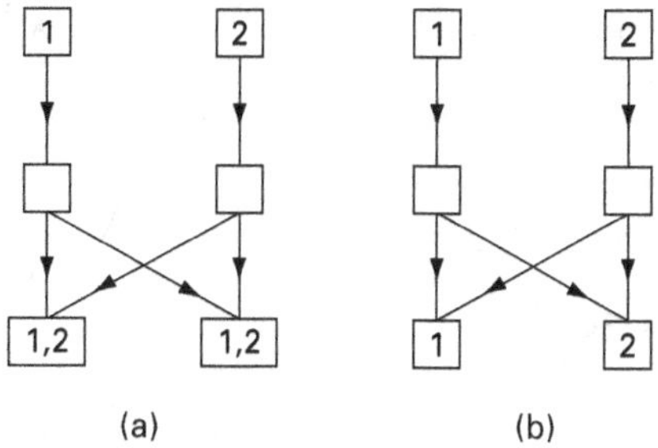

Figure P14.17 (a) Two-input–two-output MAC. (b) Interference channel.

(c) Prove that the capacity region is the closure of the set of those pairs (R_1, R_2), which for some n, independent RVs X_1^n, X_2^n and RVs Y_i^n connected with (X_1^n, X_2^n) by the channel W_i^n, $i = 1, 2$, satisfy

$$R_1 \leqq \frac{1}{n} I(X_1^n \wedge Y_1^n), \qquad R_2 \leqq \frac{1}{n} I(X_2^n \wedge Y_2^n).$$

(A computable characterization of the capacity region is not known; the largest known achievable rate region with a single-letter description is due to Han and Kobayashi (1981). The product space characterization (c) appears in Ahlswede (1973a) as a joint result with U. Augustin and J. Wolfowitz; a counterexample needed in (b) was given by Ahlswede (1974).)

Hint Codes achieving $R_1 = I(X_1 \wedge Y_1)$, $R_2 = I(X_2 \wedge Y_2)$ can be obtained either by the method of Theorem 14.3, from codes with side information at the decoder for the 2-sources with generic variables (X_1, Y_1) resp. (X_2, Y_2), or directly by random selection. The direct part of (c) follows from (a), while the converse is obvious. For (b), note that if $W_1 = W_2 = W$ then the capacity region equals that of the MAC $\{W : \mathsf{X}_1 \times \mathsf{X}_2 \to \mathsf{Y}\}$.

14.18. (a) Another channel network differing from a two-input–two-output MAC only in the addressing is shown in Fig. P14.18A. Show that for this network a pair of non-negative numbers (R_1, R_2) is an achievable rate pair if for some independent RVs X_1, X_2 and for Y_i connected with (X_1, X_2) by channel W_i, $i = 1, 2$,

$$R_1 \leqq I(X_1 \wedge Y_1 | Y_2), \quad R_2 \leqq \min[I(X_2 \wedge Y_1 | X_1),\ I(X_2 \wedge Y_2)],$$

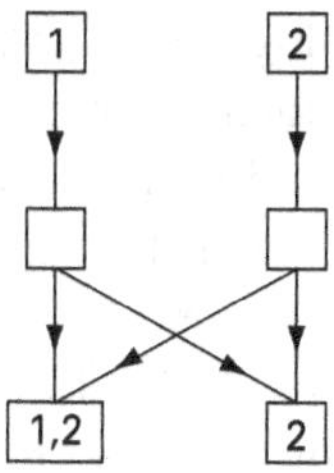

Figure P14.18A

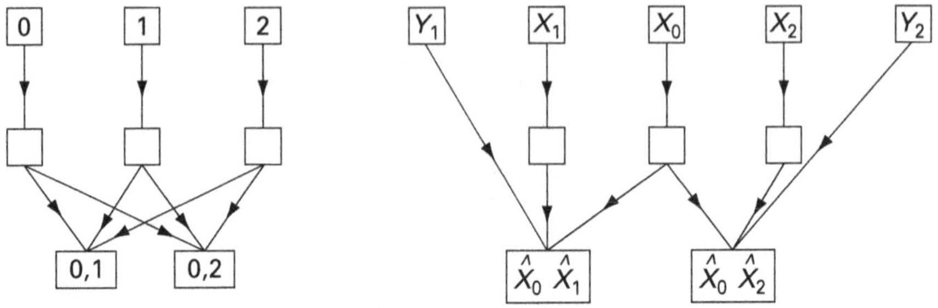

Figure P14.18B

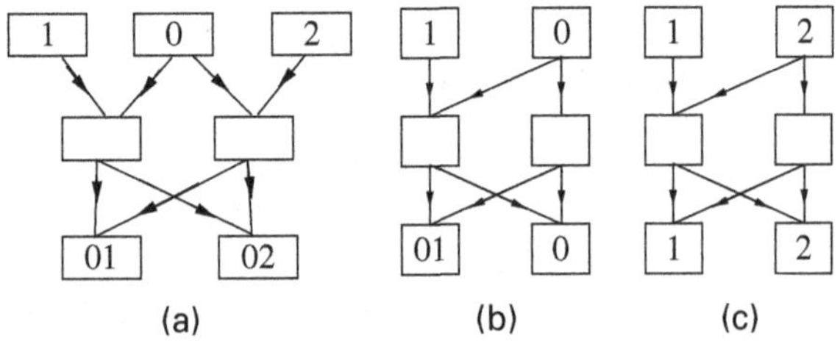

Figure P14.18C (a) Interference channel with common messages. (b) Cognitive interference channel. (c) Interference channel with encoder cooperation.

$$R_1 + R_2 \leqq I(X_1X_2 \wedge Y_1).$$

(b) Show that for the channel network of Fig. P14.18B (a) with component channels $W_i : \mathsf{X}_0 \times \mathsf{X}_1 \times \mathsf{X}_2 \to \mathsf{Y}_i, i = 1, 2$, every $(R_0, R_1, R_2) \geqq 0$ is an achievable rate triple which, for some independent RVs X_0, X_1, X_2 and Y_i connected with (X_0, X_1, X_2) by channel $W_i, i = 1, 2$, satisfies

$$R_0 \leqq I(X_0 \wedge Y_1|X_1), \quad R_1 \leqq I(X_1 \wedge Y_1|X_0), \quad R_0 + R_1 \leqq I(X_0X_1 \wedge Y_1),$$
$$R_0 \leqq I(X_0 \wedge Y_2|X_2), \quad R_2 \leqq I(X_2 \wedge Y_2|X_0), \quad R_0 + R_2 \leqq I(X_0X_2 \wedge Y_2).$$

Note that (a) is a particular case of (b). A computable characterization of the capacity region is not known even for (a).

Hint Suitable codes can be obtained from codes for the source network of Fig. P14.18B(b) or directly by random selection.

Remark Figure P14.18C shows related channel networks which attracted considerable attention in the years preceding 2010. For the *cognitive interference channel* in Fig. 14.18C(b), a single-letter characterization of the capacity region has been found: Jiang, Xin and Garg (2008) improved previous achievability results for Fig. P14.18C(a), and as a special case they obtained an achievable rate region for Fig. P14.18C(b). Liang *et al.* (2009) proved the converse for that region.

14.19. (*Two-observer DMC*) Consider a DMC $\{W : \mathsf{X} \to \mathsf{Y}\}$ such that Y is a product set $\mathsf{Y}_1 \times \mathsf{Y}_2$. *A two-observer n-length block code* for this DMC is a quadruple of mappings $(f, \varphi_1, \varphi_2, \psi)$ such that

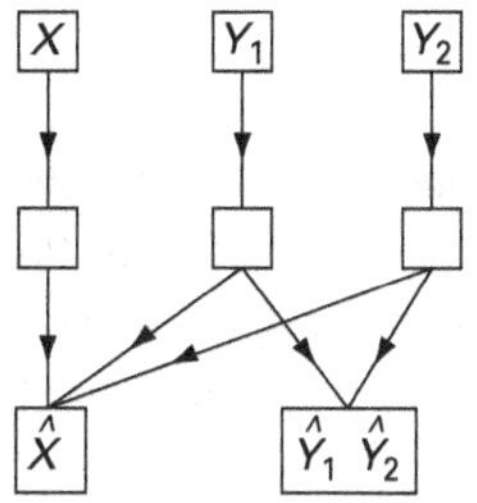

Figure P14.19 Auxiliary source network for a two-observer DMC.

$$f : \mathsf{M}_f \to \mathsf{X}^n, \ \varphi_i : \mathsf{Y}_i^n \to \mathsf{M}_i, \ i = 1, 2, \ \psi : \mathsf{M}_1 \times \mathsf{M}_2 \to \mathsf{M}', \ \mathsf{M}' \supset \mathsf{M}_f.$$

Then the composite function $\tilde{\psi} : \mathsf{Y}^n \to \mathsf{M}'$ defined by

$$\tilde{\psi}(\mathbf{y}_1, \mathbf{y}_2) \triangleq \psi(\varphi_1(\mathbf{y}_1), \varphi_2(\mathbf{y}_2))$$

gives rise to an n-length block code $(f, \tilde{\psi})$ for the DMC $\{W\}$ in the usual sense. The average probability of error of the two-observer code is defined as the average probability of error $\bar{e}(W^n, f, \tilde{\psi})$. A triple (R_X, R_1, R_2) is an *ε-achievable rate triple* for the two-observer DMC if for every $\delta > 0$ and n large enough there exists a two-observer n-length block code $(f, \varphi_1, \varphi_2, \psi)$ such that $\bar{e}(W^\mathbf{n}, f, \tilde{\psi}) < \varepsilon$ and the inequalities

$$\frac{1}{n} \log |\mathsf{M}_f| \geqq R_\mathsf{X} - \delta, \quad \frac{1}{n} \log \|\varphi_i\| \leqq R_i + \delta, \quad i = 1, 2,$$

hold. Achievable rate triples, etc., are defined as usual.

Prove that for every triple of RVs (X, Y_1, Y_2) such that (Y_1, Y_2) is connected with X by the channel W, every (R_X, R_1, R_2) with

$$0 \leqq R_\mathsf{X} \leqq I(X \wedge Y_1 Y_2),$$
$$R_1 \geqq H(Y_1|Y_2), \quad R_2 \geqq H(Y_2|Y_1), \quad R_1 + R_2 \geqq H(Y_1 Y_2)$$

is an achievable rate triple.

Hint Consider codes for the source network of Fig. P14.19.

14.20. (*Multiple-access channel with feedback*) For a MAC $\{W : \mathsf{X} \times \mathsf{Y} \to \mathsf{Z}\}$, *complete feedback to* X means, informally, that at each time i all the previous output symbols $z_1, \dots, z_{i-1}$ are known at the intermediate vertex of the network to which the alphabet X is assigned. In the case of feedback, encoders should be defined as in Problem 6.29, letting the ith symbol of the codeword also depend, in addition to the message, on the previous output symbols.

Give a formal definition of an n-length block code for a MAC with complete feedback to X resp. to both X and Y. Define the corresponding capacity regions. Show that feedback can increase the capacity region of a MAC by proving the following statement: for every quadruple of RVs (U, X, Y, Z) as in Problem 14.16, each rate pair (R_1, R_2) satisying the inequalities

$$0 \leqq R_1 \leqq I(X \wedge Z|YU), \quad 0 \leqq R_2 \leqq I(Y \wedge Z|XU),$$
$$R_1 + R_2 \leqq I(XY \wedge Z)$$

belongs to the capacity region of the MAC $\{W : \mathsf{X} \times \mathsf{Y} \to \mathsf{Z}\}$ with complete feedback to X (or to Y).

(Gaarder and Wolf (1975) showed that feedback increases the capacity region of the adder MAC; see Problem 14.4. The result here was proved by Cover and Leung (1981) for the case of complete feedback to both senders. That assumption was relaxed independently by the authors, Carleial (1982) and Willems and Van der Meulen (1983). The converse holds for a special class of MACs that includes the adder MAC, Willems (1982), but not in general, Kramer (2003).)

Hint If $R_1 + R_2 \leqq I(XY \wedge \mathsf{Z}|U)$ then (R_1, R_2) is achievable without feedback; otherwise write $\Delta \triangleq R_1 + R_2 - I(XY \wedge Z|U) > 0$. It suffices to prove the existence of list codes for the channel network of Fig. P14.16(a) with message sets M_0, M_1 and $\mathsf{M}_2 = \mathsf{M}'_2 \times \mathsf{M}''_2$ such that

(i) $\frac{1}{n} \log |\mathsf{M}_0| = \frac{1}{n} \log |\mathsf{M}''{}_2| > \Delta - \delta$,

$\frac{1}{n} \log |\mathsf{M}_1| > R_1 - \delta, \quad \frac{1}{n} \log |\mathsf{M}'{}_2| > R_2 - \Delta - \delta$;

(ii) the decoder reconstructs m_0 with small probability of error, and

(iii) if the decoder knew, in addition, either m_1 or m''_2, the whole triple (m_0, m_1, m_2) could be reconstructed with small probability of error.

In fact, from such list codes of block length n one can build codes for the MAC with feedback of block length ln, say, with message sets $\mathsf{M}_1^{l-1}, \mathsf{M}_2^{l-1}$: send in each sub-block both new information (i.e., some $m_1 \in \mathsf{M}_1, m_2 \in \mathsf{M}_2$) and complementary information needed for the decoding of the previous sub-block (taking as the present m_0 the previous m''_2 , which is also known at X by assumption (iii) and the feedback).

Codes for the channel network of Fig. P14.16(a) with properties (i)–(iii) can be obtained either by random selection or, as in the hint of Problem 14.16, taking $R_{\mathsf{X}} \triangleq H(X|U) - R_1$, $R_{\mathsf{Y}} \triangleq H(Y|U) - R_2 + \Delta$. A modification needed in the construction of that hint is that the Y-encoder of the source network should yield a refinement of a partition $\mathcal{B}$ of Y^n. This $\mathcal{B}$ is defined by an encoder of rate $H(Y|UXZ) + \delta$, which allows us to reproduce Y^n if $U^n X^n Z^n$ is available at the decoder as side information. After having selected a proper m, for each $\mathbf{u} \in \mathsf{C}_m$ one now has to find $\exp[n(\Delta - \delta)]$ sets $\mathsf{B}_j(\mathbf{u})$ (rather than just one), each being a subset of the same atom (depending on $\mathbf{u}$) of the partition $\mathcal{B}$.

14.21. (*Two-way channel*) The intuitive interpretation of the channel network of Problem 14.11 suggests that both encoders should be allowed to "use the previous output symbols received at the same terminal." This means considering the channel network of Fig. P14.11 with feedback (from each output to the corresponding intermediate vertex); then one speaks of an *unrestricted two-way channel.*

(a) Give a formal definition of n-length block codes for this case. Note that the error probabilities cannot be defined in terms of the component channels W_1 and W_2. Rather, to this end a channel $W : \mathsf{X} \times \mathsf{Y} \to \mathsf{Z}_1 \times \mathsf{Z}_2$

must be given (so that the distributions $W(\cdot, \cdot|x, y)$ have marginals $W_i(\cdot|x, y)$, $i = 1, 2$). Given such a W, define the error probabilities of a code, as well as achievable rate pairs and the capacity region of the unrestricted two-way channel.

(b)* Give an example of an unrestricted two-way channel having a larger capacity region than the corresponding channel network of Fig. P14.11. (Dueck (1979).)

(c) Show that every achievable rate pair belongs to the convex closure of the set of pairs satisfying the inequalities of Problem 14.11 for some not necessarily independent X and Y and the corresponding Z_1, Z_2. (Shannon (1961).)

Reduction of channel network problems (Problems 14.22–14.24)

14.22. Define a *normal channel network* as one having an underlying network with the following properties: (i) no edges connect input and output vertices; (ii) every intermediate vertex is connected with every output; (iii) the edges connecting input and intermediate vertices establish a one-to-one correspondence. For such networks, the sets A and B of input resp. intermediate vertices will be identified according to (iii).

(a) Observe that every channel network can be modified to meet conditions (i) and (ii), without changing the capacity region.

(b) Given any channel network with the properties (i) and (ii), consider arbitrary "pre-channels" $V_b : \mathsf{X}_{\mathsf{S}_b} \to \mathsf{X}_b$, $b \in \mathsf{B}$, where X_a, $a \in \mathsf{A}$, are arbitrary alphabets and $\mathsf{X}_\mathsf{L} \triangleq \times_{a \in \mathsf{L}} \mathsf{X}_a$ for every set L. Define a normal channel network with component channels $\widetilde{W}_c : \mathsf{X}_\mathsf{A} \to \mathsf{X}_c$, $c \in \mathsf{C}$, where

$$\widetilde{W}_c(x_c|x_\mathsf{A}) \triangleq \sum_{x_\mathsf{B} \in \mathsf{X}_\mathsf{B}} W_c(x_c|x_\mathsf{B}) \prod_{b \in \mathsf{B}} V_b(x_b|x_{\mathsf{S}_b})$$

($x_\mathsf{A} \triangleq \{x_a\}_{a \in \mathsf{A}}$, $x_\mathsf{B} \triangleq \{x_b\}_{b \in \mathsf{B}}$), and with the addressing of the original channel network. Show that every achievable rate vector for this new channel network belongs to the capacity region of the original one.

Hint Every code for the new (normal) channel network gives rise to a code with stochastic encoders for the original channel network. Use Problem 14.12.

14.23. (*Achievable rate vectors for general channel networks*)

(a) Let us be given a normal channel network (see Problem 14.22), a family of independent RVs $X_\mathsf{A} \triangleq \{X_a\}_{a \in \mathsf{A}}$ and corresponding RVs X_c connected with X_A by the channels W_c, $c \in \mathsf{C}$. Show that if a vector $\mathbf{R} \triangleq \{R_a\}_{a \in \mathsf{A}}$ with non-negative components satisfies the inequalities

$$\sum_{a \in \mathsf{L}} R_a \leqq I(X_\mathsf{L} \wedge X_c | X_{\mathsf{D}_c - \mathsf{L}}) \text{ for every } c \in \mathsf{C} \text{ and } \mathsf{L} \subset \mathsf{D}_c,$$

then $\mathbf{R}$ belongs to the capacity region of the given channel network.

Hint Proceed as in the proof of Theorem 14.3, looking at codes for a suitable NSN without helpers, or directly by random code selection as in Problem 14.8.

(b) Consider any channel network satisfying (i) and (ii) of Problem 14.22. Let $U_{\mathsf{A}} \triangleq \{U_a\}_{a \in \mathsf{A}}$ be any family of independent RVs (with arbitrary finite ranges), let X_b, $b \in \mathsf{B}$, be RVs with ranges X_b, conditionally independent given U_{A} and such that $U_{\mathsf{A}} \multimap U_{\mathsf{S}_b} \multimap X_b$ for every $b \in \mathsf{B}$. Further, let the RVs X_c, $c \in \mathsf{C}$, be connected with $X_{\mathsf{B}} \triangleq \{X_b\}_{b \in \mathsf{B}}$ by the channel W_c and satisfy $U_{\mathsf{A}} \multimap X_{\mathsf{B}} \multimap X_c$. Show that a vector $\mathbf{R} \triangleq \{R_a\}_{a \in \mathsf{A}}$ with non-negative components is an achievable rate vector provided that the following inequalities hold:

$$\sum_{a \in \mathsf{L}} R_a \leqq I(U_{\mathsf{L}} \wedge X_c | U_{\mathsf{D}_c - \mathsf{L}}) \quad \text{for every} \quad c \in \mathsf{C} \quad \text{and} \quad \mathsf{L} \subset \mathsf{D}_c.$$

Hint See (a) and Problem 14.22 (b).

(c) Let U_{A} be as in (b), but now let the RVs X_b resp. X_c have range X_b^n resp. X_c^n, X_c being connected with X_{B} by the channel W_c^n. Show that if $\sum_{a \in \mathsf{L}} R_a \leqq (1/n) I(U_{\mathsf{L}} \wedge X_c | U_{\mathsf{D}_c - \mathsf{L}})$ for every $c \in \mathsf{C}$ and $\mathsf{L} \subset \mathsf{D}_c$ then $\mathbf{R} \triangleq \{R_a\}_{a \in \mathsf{A}}$ is an achievable rate vector.

14.24. (*Optimists and pessimists*) Given any channel network show that to every $\delta > 0$ there exists an $\varepsilon > 0$ such that if some block code of rate vector $\mathbf{R}$ has average error probabilities $\bar{e}_c \leqq \varepsilon$ for every $c \in \mathsf{C}$ then $\mathbf{R} - (\delta, \ldots, \delta)$ is an achievable rate vector. Conclude that the "optimistic" definition of the capacity region leads to the same region as the "pessimistic" one adopted in the text.
Hint Similarly to Problem 14.22(a), one can assume that (i) and (ii) hold. Then apply Problem 14.23(c), letting U_a be uniformly distributed on the message set M_a of the given code, and use Fano's inequality. (This shows that the capacity region actually equals the closure of the set of all vectors $\mathbf{R}$ of form as in Problem 14.23(c). This is a non-computable "product space characterization" of the capacity region.)

Story of the results

The information theory of multi-terminal networks was initiated by Shannon (1961). He formulated several problems and proved the first coding theorem for a multi-terminal network, see Problem 14.11. In the cited paper he mentioned the multiple-access channel as "a case for which a complete and simple solution of the capacity region has been found." Unfortunately, Shannon wrote no further papers on multi-terminal networks, and a computable characterization of the MAC capacity region was not put forward until 1971, Ahlswede (1973a); such a result is contained also in Liao (1972, unpublished), quoted in Slepian and Wolf (1973b).

Ahlswede (1973a) proved the MAC coding theorem in a slightly different form. The present formulation appears as a special case in Slepian and Wolf (1973b) and Ahlswede (1974). The idea of constructing channel network codes from source network codes came up in discussions between the authors and R. Ahlswede. Our proof of the direct part of the MAC coding theorem is an elaboration of this idea.

Channel networks of depth 2 constitute a natural common generalization of the models treated in the above-mentioned papers and of the "broadcast channel" of Cover (1972), to be discussed in Chapter 16.

Addition. A formal framework for channel networks more general than Definition 14.6 appears in Kramer (2003), who attributes the first such framework to Carleial (1975, unpublished).

15 Entropy and image size characterization

This chapter is of a technical nature. The probabilistic problems treated here are the common core of various source and channel coding results for multi-terminal systems.

We saw in Chapter 13 that for a large class of source networks the problem of determining the achievable rate region can be reduced to an "entropy characterization problem." In this chapter we shall solve the latter for the case of 3-sources. Throughout this chapter, X, Y and Z will denote RVs with finite ranges X, Y and Z, respectively, and $V : \mathsf{X} \to \mathsf{Y}$, $W : \mathsf{X} \to \mathsf{Z}$ will denote stochastic matrices for which

$$\Pr\{Y = y|X = x\} = V(y|x), \qquad \Pr\{Z = z|X = x\} = W(z|x)$$

whenever the left-hand probabilities are defined. Along with RVs X, Y, Z we shall always consider a DMMS $\{(X_i, Y_i, Z_i)\}_{i=1}^{\infty}$ with these generic variables.

The *entropy characterization problem* for a 3-source with generic variables (X, Y, Z), as originally formulated, consists in determining the closure of the set of all vectors of the following form:

$$\left(\frac{1}{k}H(X^k|f(X^k)),\ \frac{1}{k}H(Y^k|f(X^k)),\ \frac{1}{k}H(Z^k|f(X^k))\right), \tag{15.1}$$

with k running over the positive integers and f running over all functions with domain X^k. It will be one of our basic aims to give a computable characterization of the closure of the set of entropy vectors (15.1). It is natural to compare this set with an apparently larger one which we now describe. For every positive integer k write

$$\begin{aligned}\mathcal{H}_k &= \mathcal{H}_k(X; Y; Z|X)\\ &\triangleq \left\{\left(\frac{1}{k}H(X^k|U),\ \frac{1}{k}H(Y^k|U),\ \frac{1}{k}H(Z^k|U)\right) : U \multimap X^k \multimap Y^k Z^k\right\},\end{aligned} \tag{15.2}$$

where U is running over all RVs with finite range satisfying the above Markov condition. Define

$$\mathcal{H} = \mathcal{H}(X; Y; Z|X) \triangleq \text{closure of } \bigcup_{k=1}^{\infty} \mathcal{H}_k(X; Y; Z|X). \tag{15.3}$$

Obviously, the closure of the set of entropy vectors (15.1) is contained in $\mathcal{H}$. It will turn out, however, that this closure actually equals $\mathcal{H}$. Moreover, we shall see that

$\mathcal{H}$ can be obtained in a simple manner from $\mathcal{H}_1$. This fact will lead to a computable characterization of $\mathcal{H}$, and thus to the solution of the entropy characterization problem.

The proofs of strong converse theorems for source and channel networks will rely on a computable characterization of the size of sets and their images through two DMCs. Actually, this question is closely related to the preceding ones, so they will be discussed in an interwoven manner. However, those readers who are not interested in strong converse theorems may skip a large part of the text subsequent to Theorem 15.9.

Let us turn to the formal problem statements. Recall from Chapter 6 that a set $\mathsf{B} \subset \mathsf{Y}^n$ has been called an η-image of $\mathsf{A} \subset \mathsf{X}^n$ over the channel $V^n : \mathsf{X}^n \to \mathsf{Y}^n$ if $V^n(\mathsf{B}|\mathbf{x}) \geqq \eta$ for every $\mathbf{x} \in \mathsf{A}$. The minimum size of such a set B, denoted by $g_{V^n}(\mathsf{A}, \eta)$, was a relevant quantity in constructing codes for the DMC $\{V\}$.

DEFINITION 15.1 Given the RVs X, Y, Z, a triple of non-negative numbers (a, b, c) is called an *achievable entropy triple* for the corresponding entropy characterization problem if for every $\delta > 0$ and every n sufficiently large there exists a function f with domain X^n such that

$$\text{(i)} \quad \left|\frac{1}{n}H(X^n|f(X^n)) - a\right| \leqq \delta, \quad \left|\frac{1}{n}H(Y^n|f(X^n)) - b\right| \leqq \delta, \quad \left|\frac{1}{n}H(Z^n|f(X^n)) - c\right| \leqq \delta.$$

The set of such triples will be denoted by $\mathcal{F}(X; Y; Z|X)$.

Further, a triple (a, b, c) is called an *achievable exponent triple* for the corresponding image size problem if for every $0 < \varepsilon < 1$, $\delta > 0$ and sufficiently large n there exist sets $\mathsf{A} \subset \mathsf{T}^n_{[X]}$ such that

$$\text{(ii)} \quad \left|\frac{1}{n}\log|\mathsf{A}| - a\right| \leqq \delta, \quad \left|\frac{1}{n}\log g_{V^n}(\mathsf{A}, 1-\varepsilon) - b\right| \leqq \delta, \quad \left|\frac{1}{n}\log g_{W^n}(\mathsf{A}, 1-\varepsilon) - c\right| \leqq \delta.$$

The set of such triples will be denoted by $\mathcal{G}(X; Y; Z|X)$. ○

REMARKS (1) It is easily seen that $\mathcal{F}(X; Y; Z|X)$ equals the closure of the set of entropy vectors of form (15.1). Obviously, $\mathcal{F}(X; Y; Z|X)$ is contained in the said closure. On the other hand, every vector (a, b, c) of form (15.1) satisfies (i) for every n which is an integral multiple of k, whence one easily concludes that (a, b, c) satisfies (i) for every sufficiently large n. Thus the entropy characterization problem is equivalent to the determination of $\mathcal{F}(X; Y; Z|X)$.

(2) Clearly, the sets $\mathcal{F}(X; Y; Z|X)$, $\mathcal{H}(X; Y; Z|X)$ and $\mathcal{G}(X; Y; Z|X)$ depend on the joint distribution of X, Y, Z only through the distributions of the pairs (X, Y) and (X, Z). ○

We shall derive computable characterizations of the sets $\mathcal{F}$, $\mathcal{G}$, $\mathcal{H}$ and discuss their relation. In this context it is important to see how image sizes are related to information quantities.

LEMMA 15.2 For any set $\mathsf{A} \subset \mathsf{X}^n$, consider a RV $\hat{X}^n = \hat{X}_1 \dots \hat{X}_n$ distributed over A and let the RV $\hat{Y}^n = \hat{Y}_1 \dots \hat{Y}_n$ be connected with $\hat{X}^n$ by the channel $V^n : \mathsf{X}^n \to \mathsf{Y}^n$. Then, for every $\delta > 0,\ 0 < \eta < 1$,

$$\frac{1}{n} H(\hat{Y}^n) - \delta \leqq \frac{1}{n} \log g_{V^n}(\mathsf{A}, \eta), \tag{15.4}$$

whenever $n \geqq n_0(|\mathsf{X}|, |\mathsf{Y}|, \delta, \eta)$.

Moreover, if $\mathsf{A} \subset \mathsf{T}^n_{[X]}$ is the codeword set of an (n, ε)-code for the DMC $\{V\}$, and $\hat{X}^n$ has a uniform distribution over A, then one also has

$$\frac{1}{n} H(\hat{Y}^n) + \delta + \varepsilon \log |\mathsf{Y}| \geqq \frac{1}{n} \log g_{V^n}(\mathsf{A}, \eta), \tag{15.5}$$

provided that $n \geqq n_0(|\mathsf{X}|, |\mathsf{Y}|, \delta, \eta)$. ○

Proof To prove (15.4), on account of Lemma 6.6 it is sufficient to show that for some $\eta_0 = \eta_0(|\mathsf{Y}|, \delta)$ we have

$$\frac{1}{n} H(\hat{Y}^n) - \frac{\delta}{2} \leqq \frac{1}{n} \log g_{V^n}(\mathsf{A}, \eta_0) \quad \text{if} \quad n \geqq n_1(|\mathsf{Y}|, \delta). \tag{15.6}$$

Let $\mathsf{B} \subset \mathsf{Y}^n$ be an η_0-image of A that achieves $g_{V^n}(\mathsf{A}, \eta_0)$. Then for $p_n \triangleq \Pr\{\hat{Y}^n \in \mathsf{B}\}$ we have

$$p_n = \sum_{\mathbf{x} \in \mathsf{A}} \Pr\{\hat{X}^n = \mathbf{x}\} \Pr\{\hat{Y}^n \in \mathsf{B} | \hat{X}^n = \mathbf{x}\} \geqq \eta_0.$$

Hence, if J is a RV taking the value 1 if $\hat{Y}^n \in \mathsf{B}$ and 0 otherwise, we obtain

$$\begin{aligned} H(\hat{Y}^n) = H(J) + H(\hat{Y}^n | J) &\leqq 1 + (1 - p_n) n \log |\mathsf{Y}| + p_n \log |\mathsf{B}| \\ &\leqq 1 + (1 - \eta_0) n \log |\mathsf{Y}| + \log g_{V^n}(\mathsf{A}, \eta_0). \end{aligned}$$

This proves (15.6) for $\eta_0 = 1 - \delta/4 \log |\mathsf{Y}|$, say.
In order to establish (15.5), let us recall that for $n \geqq n_2(|\mathsf{X}|, |\mathsf{Y}|, \eta, \delta)$ the set

$$\mathsf{T}_{[V]}(\mathsf{A}) \triangleq \bigcup_{\mathbf{x} \in \mathsf{A}} \mathsf{T}_{[V]}(\mathbf{x}) \tag{15.7}$$

is an η-image of A by Lemma 2.12. Hence by Lemma 2.13, for $n \geqq n_3$,

$$\log g_{V^n}(\mathsf{A}, \eta) \leqq \log |\mathsf{T}_{[V]}(\mathsf{A})| \leqq \log |\mathsf{A}| + nH(Y|X) + \frac{n\delta}{3}. \tag{15.8}$$

Now

$$\begin{aligned} H(\hat{Y}^n | \hat{X}^n) - nH(Y|X) &= \sum_{i=1}^{n} (H(\hat{Y}_i | \hat{X}_i) - H(Y|X)) \\ &= \sum_{i=1}^{n} \sum_{x \in \mathsf{X}} (P_{\hat{X}_i}(x) - P_X(x)) H(V(\cdot|x)) \\ &= n \sum_{x \in \mathsf{X}} (\hat{P}(x) - P_X(x)) H(V(\cdot|x)), \end{aligned}$$

where

$$\hat{P}(x) \triangleq \frac{1}{n}\sum_{i=1}^{n} P_{\hat{X}_i}(x) = \frac{1}{n|\mathsf{A}|}\sum_{\mathbf{x}\in\mathsf{A}} N(x|\mathbf{x}).$$

As $\mathsf{A} \subset \mathsf{T}_{[X]} = \mathsf{T}_{[X]\delta_n}$ implies $|\hat{P}(\mathbf{x}) - P_X(x)| \leqq \delta_n$, it follows that

$$|H(\hat{Y}^n|\hat{X}^n) - nH(Y|X)| \leqq \frac{n\delta}{3} \quad \text{for} \quad n \geqq n_4(|X|, |Y|, \delta).$$

On account of this and $\log|\mathsf{A}| = H(\hat{X}^n)$, inequality (15.8) yields

$$\frac{1}{n}\log g_{V^n}(\mathsf{A}, \eta) \leqq \frac{1}{n}H(\hat{X}^n, \hat{Y}^n) + \frac{2}{3}\delta.$$

Since A is the codeword set of an (n, ε)-code for the DMC $\{V\}$, by Fano's inequality

$$\frac{1}{n}H(\hat{X}^n|\hat{Y}^n) \leqq \frac{1}{n}h(\varepsilon) + \varepsilon\log|\mathsf{Y}|.$$

The preceding two inequalities prove (15.5) if $n \geqq n_5$. □

A primary form of the basic constructions of this chapter is contained in the following lemma and its corollary.

LEMMA 15.3 $\mathcal{H}_1(X; Y; Z|X) \subset \mathcal{G}(X; Y; Z|X)$. More precisely, for every $\eta > 0$, any pair of RVs U, X and set $\mathsf{E} \subset \mathsf{X}^n$ with $P_X^n(\mathsf{E}) \geqq \eta$, there exists a subset $\mathsf{A} \subset \mathsf{E}$ and a sequence $\mathbf{u} \in \mathsf{T}^n_{[U]}$ such that

$$\mathsf{A} \subset \mathsf{T}_{[X|U]}(\mathbf{u}), \quad P^n_{[X|U]}(\mathsf{A}|\mathbf{u}) \geqq \frac{\eta}{2}, \tag{15.9}$$

whenever $n \geqq n_0$ $(|\mathsf{U}|, |\mathsf{X}|, \eta)$. Further, if a set A and a sequence $\mathbf{u} \in \mathsf{T}^n{}_{[U]}$ satisfy (15.9), then

$$\left|\frac{1}{n}\log|\mathsf{A}| - H(X|U)\right| \leqq \delta \tag{15.10}$$

and for every $V: \mathsf{X} \to \mathsf{Y}$ and RV Y with $U \multimap X \multimap Y$, $P_{Y|X} = V$,

$$\left|\frac{1}{n}\log g_{V^n}(\mathsf{A}, 1-\varepsilon) - H(Y|U)\right| \leqq \delta, \tag{15.11}$$

whenever $n \geqq n_1(|\mathsf{U}|, |\mathsf{X}|, |\mathsf{Y}|, \varepsilon, \delta, \eta)$. ○

Proof We first show that for sufficiently large n there exists a set $\mathsf{F} \subset \mathsf{T}^n_{[U]}$ having probability bounded away from 0, such that

$$\mathsf{A}(\mathbf{u}) \triangleq \mathsf{E} \cap \mathsf{T}_{[X|U]}(\mathbf{u})$$

satisfies (15.9) for every $\mathbf{u} \in \mathsf{F}$. By Lemma 2.12, $P^n_{X|U}(\mathsf{T}_{[X|U]}(\mathbf{u})|\mathbf{u}) \geqq 1 - \eta/4$ if $n \geqq n_2$, whence

$$\sum_{\mathbf{u}\in\mathsf{U}^n} P^n_U(\mathbf{u})P^n_{X|U}(\mathbf{E} \cap \mathbf{T}_{[X|U]}(\mathbf{u})|\mathbf{u}) \geqq P^n_X(\mathsf{E}) - \frac{\eta}{4} \geqq \frac{3\eta}{4}.$$

It follows that if $n \geqq n_3 \geqq n_2$ is such that $P^n_{[U]}(\mathsf{T}_{[U]}) \geqq 1 - \eta/8$ then

$$\mathsf{F} \triangleq \mathsf{T}_{[U]} \cap \left\{ \mathbf{u} : P^n_{X|U}(\mathsf{E} \cap \mathsf{T}_{[X|U]}(\mathbf{u})|\mathbf{u}) \geqq \frac{\eta}{2} \right\} \quad \text{has} \quad P^n_U(\mathsf{F}) \geqq \frac{\eta}{8}. \tag{15.12}$$

By definition, for $\mathbf{u} \in \mathsf{F}$ the set $\mathsf{A} \triangleq \mathsf{A}(\mathbf{u})$ satisfies (15.9).

Further, any set satisfying (15.9) has size as in (15.10) by Lemmas 2.13 and 2.14. To verify (15.11), note that for every $n \geqq n_4$ and $\mathsf{A} \subset \mathsf{T}_{[X|U]}(\mathbf{u})$ the set

$$\mathsf{B} \triangleq \bigcup_{\mathbf{x} \in \mathsf{A}} \mathsf{T}_{[Y|XU]}(\mathbf{x}, \mathbf{u})$$

is an $(1 - \varepsilon)$-image of A over the channel V^n since the Markov property $U \multimap X \multimap Y$ and Lemma 2.12 imply

$$V^n(\mathsf{B}|\mathbf{x}) = P^n_{Y|XU}(\mathsf{B}|\mathbf{x}, \mathbf{u}) \geqq 1 - \varepsilon \quad \text{for} \quad \mathbf{x} \in \mathsf{A}.$$

To the analogy of Lemma 2.10 we have

$$\mathsf{T}_{[Y|XU]}(\mathbf{x}, \mathbf{u}) \subset \mathsf{T}_{[Y|U]}(\mathbf{u}) \quad \text{if} \quad \mathbf{x} \in \mathsf{T}_{[X|U]}(\mathbf{u})$$

provided that the sets $\mathsf{T}_{[Y|U]}(\mathbf{u})$ are suitably defined; see Convention 2.11. Thus it follows that

$$\begin{aligned} \frac{1}{n} \log g_{V^n}(\mathsf{A}, 1 - \varepsilon) &\leqq \frac{1}{n} \log |\mathsf{B}| \leqq \frac{1}{n} \log |\mathsf{T}_{[Y|U]}(\mathbf{u})| \\ &\leqq H(Y|U) + \delta \end{aligned} \tag{15.13}$$

whenever $\mathsf{A} \subset \mathsf{T}_{[X|U]}(\mathbf{u})$. On the other hand, if $\mathsf{B} \subset \mathsf{Y}^n$ is any $(1 - \varepsilon)$-image of a set A satisfying (15.9), then

$$P^n_{[Y|U]}(\mathsf{B}|\mathbf{u}) \geqq \sum_{\mathbf{x} \in \mathsf{A}} P^n_{[X|U]}(\mathbf{x}|\mathbf{u}) V^n(\mathsf{B}|\mathbf{x}) \geqq \frac{\eta}{2}(1 - \varepsilon).$$

Hence by Lemma 2.14 we obtain for $n \geqq n_5$

$$\frac{1}{n} \log g_{V^n}(\mathsf{A}, 1 - \varepsilon) \geqq H(Y|U) - \delta. \qquad \square$$

COROLLARY 15.3 For every $\delta > 0, \varepsilon > 0$ and $n \geqq n_0(|\mathsf{U}|, |\mathsf{X}|, |\mathsf{Y}|, \delta, \varepsilon)$, to every pair of RVs U, X there exists a function $f : \mathsf{X}^n \to \mathsf{U}^n$ such that

(i) with the exception of a value $\mathbf{u}_0$ taken with probability

$$P^n_X(f^{-1}(\mathbf{u}_0)) < \varepsilon, \tag{15.14}$$

every $\mathbf{u} \in \mathsf{U}^n$ in the range of f satisfies

$$f^{-1}(\mathbf{u}) \subset \mathsf{T}^n_{[X|U]}(\mathbf{u}), \quad \mathbf{u} \in \mathsf{T}^n_{[U]}; \tag{15.15}$$

(ii) for every RV Y with $U \multimap X \multimap Y$

$$\frac{1}{n} H(Y^n|f(X^n)) \leqq H(Y|U) + \delta \tag{15.16}$$

while

$$\left|\frac{1}{n}H(X^n|f(X^n)) - H(X|U)\right| \leqq \delta. \tag{15.17}$$

○

Proof It is sufficient to prove the assertion for $\varepsilon < \varepsilon_0(|\mathsf{Y}|, \delta)$. By an iterative application of Lemma 15.3 (with $\eta = \varepsilon$) one can find disjoint subsets $\mathsf{A}_1, \ldots, \mathsf{A}_M$ of X^n such that

$$P_X^n\left(\mathsf{X}^n - \bigcup_{i=1}^{M} \mathsf{A}_i\right) < \varepsilon \tag{15.18}$$

and, for some sequences $\mathbf{u}_i \in \mathsf{T}^n_{[U]}$, $i = 1, \ldots, M$,

$$\mathsf{A}_i \subset \mathsf{T}_{[X|U]}(\mathbf{u}_i), \quad P^n_{X|U}(\mathsf{A}_i|\mathbf{u}_i) \geqq \frac{\varepsilon}{2}, \tag{15.19}$$

provided that $n \geqq n_1$. By Lemma 15.3, the preceding relations imply

$$\left|\frac{1}{n}\log|\mathsf{A}_i| - H(X|U)\right| \leqq \frac{\delta}{2} \tag{15.20}$$

and for every Y with $U \multimap X \multimap Y$, $P_{Y|X} = V$

$$\left|\frac{1}{n}\log g_{V^n}\left(\mathsf{A}_i, \frac{1}{2}\right) - H(Y|U)\right| \leqq \frac{\delta}{2}, \tag{15.21}$$

whenever $n \geqq n_2$.

Define

$$f(\mathbf{x}) \triangleq \begin{cases} \mathbf{u}_i & \text{if } \mathbf{x} \in \mathsf{A}_i, \quad 1 \leqq i \leqq M \\ \mathbf{u}_0, & \text{an arbitary element of } \mathsf{U}^n\text{, different} \\ & \text{from the } \mathbf{u}_i \text{ otherwise.} \end{cases}$$

Then (15.14) holds by (15.18). Relations (15.15) hold by definition. Further,

$$\frac{1}{n}H(X^n|f(X^n)) = H(X) - \frac{1}{n}H(f(X^n)) \geqq H(X) - \frac{1}{n}\log(M+1). \tag{15.22}$$

Since by (15.19) f is the decoder of an $(n, 1 - \varepsilon/2)$-code for the DMC connecting U with X, Corollary 6.4 gives

$$\frac{1}{n}\log(M+1) \leqq I(U \wedge X) + \delta$$

for $n \geqq n_3$. Hence (15.22) yields

$$\frac{1}{n}H(X^n|f(X^n)) \geqq H(X|U) - \delta. \tag{15.23}$$

Further, we have

$$H(Y^n|f(X^n)) = P_X^n(f^{-1}(\mathbf{u}_0)H(Y^n|f(X^n) = \mathbf{u}_0) + \sum_{i=1}^{M} P_X^n(f^{-1}(\mathbf{u}_i))H(Y^n|f(X^n) = \mathbf{u}_i)$$

$$\leqq n\varepsilon \log|\mathsf{Y}| + \sum_{i=1}^{M} P_X^n(f^{-1}(\mathbf{u}_i))H(Y^n|f(X^n) = \mathbf{u}_i). \quad (15.24)$$

As (15.19) implies $\mathsf{A}_i \subset \mathsf{T}^n_{[X]}$, Lemma 15.2 gives

$$H(Y^n|f(X^n) = \mathbf{u}_i) \leqq \log g_{V^n}\left(\mathsf{A}_i, \frac{1}{2}\right) + \frac{n\delta}{4}$$

for $i = 1, \dots, M$ if $n \geqq n_4$. This, (15.21) and (15.24) prove

$$\frac{1}{n}H(Y^n|f(X^n)) \leqq H(Y|U) + \frac{3}{4}\delta + \varepsilon \log|\mathsf{Y}|. \quad (15.25)$$

If $\varepsilon < \delta/4 \log|\mathsf{Y}|$ then (15.25) implies (15.16). Observe that the missing part of (15.17) is a particular case of (15.16). □

We have already mentioned that our computable characterizations of $\mathcal{F}$, $\mathcal{G}$ and $\mathcal{H}$ will rely on the set $\mathcal{H}_1$ defined in (15.2). We now establish some simple properties of $\mathcal{H}_1 = \mathcal{H}_1\,(X; Y; Z|X)$. Note that the definition of $\mathcal{H}_1$ does not give automatically a computable characterization of $\mathcal{H}_1$, since the range of the auxiliary RV U is not specified and hence it may be arbitrarily large. By means of the next lemma we shall characterize $\mathcal{H}_1$ so that all the involved RVs have a fixed range.

LEMMA 15.4 (*Support*) Let $\mathcal{P}(\mathsf{X})$ be the family of all probability distributions on the set X, and let f_j, $j = 1, \dots, k$ be real-valued continuous functions on $\mathcal{P}(\mathsf{X})$. Then to any probability measure μ on the Borel σ-algebra of $\mathcal{P}(\mathsf{X})$ there exist k elements P_i of $\mathcal{P}(\mathsf{X})$ and non-negative numbers $\alpha_1, \dots, \alpha_k$ with $\sum_{i=1}^{k} \alpha_i = 1$ such that for every $j = 1, \dots, k$

$$\int_{\mathcal{P}(\mathsf{X})} f_j(P)\mu(\mathrm{d}P) = \sum_{i=1}^{k} \alpha_i f_j(P_i). \quad \bigcirc$$

REMARK The result also holds, by the same proof, if $\mathcal{P}(\mathsf{X})$ is replaced by any connected compact set $\mathcal{P} \subset \mathcal{P}(\mathsf{X})$. In applications when X is a product set, it is convenient to take the family of all product distributions for $\mathcal{P}$. ○

Proof The assertion follows from the Carathéodory–Fenchel theorem, stating that in the k-dimensional Euclidean space each point of the convex closure of a connected compact set $\mathcal{K}$ can be represented as the convex combination of at most k points of $\mathcal{K}$. (Carathéodory's theorem asserts this for any $\mathcal{K}$, but with $k+1$ rather than k points.) Now let $\mathcal{K}$ be the image of $\mathcal{P}(\mathsf{X})$ under the continuous mapping f defined by $f(P) \triangleq (f_1(P), \dots, f_k(P))$. As $\mathcal{P}(\mathsf{X})$ is a connected compact set, so is $\mathcal{K}$. Further, the point $\mathbf{r}$ with coordinates

$$r_j \triangleq \int_{\mathcal{P}(\mathsf{X})} f_j(P)\mu(\mathrm{d}P), \quad j = 1, \ldots, k,$$

clearly belongs to the convex closure of $\mathcal{K}$, and thus by the Carathéodory–Fenchel theorem there exist k points of $\mathcal{K}$, say $f(P_1), \ldots, f(P_k)$, such that $\mathbf{r}$ is a convex combination of those. □

LEMMA 15.5 $\mathcal{H}(X; Y; Z|X)$ equals the set of triples

$$(H(X|U),\ H(Y|U),\ H(Z|U))$$

for RVs U satisfying, in addition to the Markov condition

$$U \multimap X \multimap YZ, \tag{15.26}$$

also the range constraint

$$|\mathsf{U}| \leqq |\mathsf{X}| + 2. \tag{15.27}$$

○

Proof Let us be given a RV U satisfying the Markov condition (15.26). We show that there exists another RV $\bar{U}$ with range $\bar{\mathsf{U}} \triangleq \{1, 2, \ldots, |\mathsf{X}| + 2\}$ such that $\bar{U} \multimap X \multimap YZ$ and

$$(H(X|\bar{U}),\ H(Y|\bar{U}),\ H(Z|\bar{U})) = (H(X|U),\ H(Y|U),\ H(Z|U)).$$

Consider the following $|\mathsf{X}| + 2$ real-valued continuous functions on $\mathcal{P}(\mathsf{X})$:

$$\begin{aligned} &f_x(P) \triangleq P(x) \quad \text{for all elements } x \in \mathsf{X} \text{ but one,}\\ &f_{\mathsf{X}}(P) \triangleq H(P), \quad f_V(P) \triangleq H(PV), \quad f_W(P) \triangleq H(PW), \end{aligned}$$

where PV and PW are the output distributions of the channels $V : \mathsf{X} \to \mathsf{Y}$ resp. $W : \mathsf{X} \to \mathsf{Z}$ corresponding to the input distribution P. Apply Lemma 15.4 to this set-up, with the distribution of U in the role of μ. More precisely, consider the distribution $P_{X|U}(\cdot|u)$ as an element of $\mathcal{P}(\mathsf{X})$ having μ-measure $\Pr\{U = u\}$. Clearly, the corresponding averages are

$$\begin{aligned} &\textstyle\sum_u \Pr\{U = u\} f_x(P_{X|U}(\cdot|u)) = \Pr\{X = x\},\\ &\textstyle\sum_u \Pr\{U = u\} f_{\mathsf{X}}(P_{X|U}(\cdot|u)) = H(X|U),\\ &\textstyle\sum_u \Pr\{U = u\} f_V(P_{X|U}(\cdot|u)) = H(Y|U),\\ &\textstyle\sum_u \Pr\{U = u\} f_W(P_{X|U}(\cdot|u)) = H(Z|U). \end{aligned}$$

According to Lemma 15.4, there exist $|\mathsf{X}| + 2$ elements $P_i, i = 1, \ldots, |\mathsf{X}| + 2$, of $\mathcal{P}(\mathsf{X})$ and numbers $\alpha_i \geqq 0$ with $\sum_{i=1}^{|\mathsf{X}|+2} \alpha_i = 1$ such that

$$\sum_{i=1}^{|\mathsf{X}|+2} \alpha_i f_x(P_i) = \Pr\{X = x\}, \tag{15.28a}$$

$$\sum_{i=1}^{|\mathsf{X}|+2} \alpha_i f_{\mathsf{X}}(P_i) = H(X|U), \tag{15.28b}$$

$$\sum_{i=1}^{|\mathsf{X}|+2} \alpha_i f_V(P_i) = H(Y|U), \tag{15.28c}$$

$$\sum_{i=1}^{|\mathsf{X}|+2} \alpha_i f_W(P_i) = H(Z|U). \tag{15.28d}$$

By the condition (15.28a) there exists a RV $\bar{U}$ with values in $\bar{\mathsf{U}} \triangleq \{1, \ldots, |\mathsf{X}| + 2\}$ such that

$$P_{\bar{U}XYZ}(i, x, y, z) = \alpha_i P_i(x) P_{YZ|X}(y, z|x)$$

for every $i \in \bar{\mathsf{U}}$, $x \in \mathsf{X}$, $y \in \mathsf{Y}$, $z \in \mathsf{Z}$. Then (U, X, Y, Z) satisfies the Markov condition (15.26), and by (15.28) we have

$$H(X|\bar{U}) = H(X|U), \quad H(Y|\bar{U}) = H(Y|U), \quad H(Z|\bar{U}) = H(Z|U). \qquad \square$$

LEMMA 15.6 $\mathcal{H}_1 = \mathcal{H}_1(X; Y; Z|X)$ is a closed convex subset of the three-dimensional interval

$$\{(a, b, c) : 0 \leqq a \leqq H(X),\ H(Y|X) \leqq b \leqq H(Y),\ H(Z|X) \leqq c \leqq H(Z)\}. \qquad \bigcirc$$

Proof One sees from the definition of $\mathcal{H}_1$ that it is contained in the above interval. The closedness of $\mathcal{H}_1$ follows from Lemma 15.5 by the continuity of information quantities. To verify its convexity, consider two RVs U_1, U_2 satisfying $U_i \multimap X \multimap YZ$, $i = 1, 2$. We claim that for any $0 < \alpha < 1$

$$(\tilde{a}, \tilde{b}, \tilde{c}) \in \mathcal{H}_1, \text{ where } \begin{aligned} \tilde{a} &\triangleq \alpha H(X|U_1) + (1-\alpha) H(X|U_2), \\ \tilde{b} &\triangleq \alpha H(Y|U_1) + (1-\alpha) H(Y|U_2), \\ \tilde{c} &\triangleq \alpha H(Z|U_1) + (1-\alpha) H(Z|U_2). \end{aligned} \tag{15.29}$$

Let J be a RV taking values 1 and 2 with probabilities α, $1 - \alpha$ and independent of the other RVs. Then, putting $U \triangleq (U_j, J)$, the point in (15.29) equals $(H(X|U), H(Y|U), H(Z|U))$. Here U satisfies the Markov condition (15.26) because

$$H(YZ|XU) = \alpha H(YZ|XU_1) + (1-\alpha) H(YZ|XU_2) = H(YZ|X). \qquad \square$$

Next we describe the set that will provide the computable characterization of both $\mathcal{F}$ and $\mathcal{H}$. This set will be obtained from $\mathcal{H}_1$ by a simple geometric transformation, completing the latter by parallel straight-line segments consecutively in two different directions. More precisely, with each point $(a, b, c) \in \mathcal{H}_1(X; Y; Z|X)$ consider also the points $(\hat{a}, b, c)$ with

$$\max[b - H(Y|X),\ c - H(Z|X)] \leqq \hat{a} \leqq a.$$

Further, for each $(\hat{a}, b, c)$ obtained in this way, consider the points

$$(\hat{a}+t,\ b+t,\ c+t) : 0 \leqq t \leqq \min[H(Y) - b,\ H(Z) - c]. \tag{15.30}$$

The set of these points will be denoted by $\mathcal{H}^* = \mathcal{H}^*(X; Y; Z|X)$.

In other words, we define

$$\begin{aligned}\mathcal{H}^* &= \mathcal{H}^*(X; Y; Z|X)\\ &\triangleq \left\{(a, b, c) : \begin{array}{c}\max[b - H(Y|X), c - H(Z|X)] \leqq a \leqq H(X|U) + t,\\ b = H(Y|U) + t,\ c = H(Z|U) + t\end{array}\right\},\end{aligned} \tag{15.31}$$

where U runs over RVs satisfying (15.26) and (15.27) in Lemma 15.5 and

$$0 \leqq t \leqq \min[I(U \wedge Y),\ I(U \wedge Z)]. \tag{15.32}$$

COROLLARY 15.6 $\mathcal{H}^* = \mathcal{H}^*(X; Y; Z|X)$ is a closed convex subset of the three-dimensional interval

$$\{(a, b, c) : 0 \leqq a \leqq H(X),\ H(Y|X) \leqq b \leqq H(Y),\ H(Z|X) \leqq c \leqq H(Z)\}. \qquad \bigcirc$$

The following lemma is the basis of all converse results in this chapter.

LEMMA 15.7 (*Single-letterization*) Let $\hat{X}^n$ be any RV ranging over X^n, and let $\hat{U} -\!\!\circ\!\!- \hat{X}^n -\!\!\circ\!\!- \hat{Y}^n\hat{Z}^n$ be arbitrary RVs such that $\hat{Y}^n$ and $\hat{Z}^n$ are connected with $\hat{X}^n$ via the channels V^n and W^n, respectively. Then

$$\left(\frac{1}{n}H(\hat{X}^n|\hat{U}), \frac{1}{n}H(\hat{Y}^n|\hat{U}), \frac{1}{n}H(\hat{Z}^n|\hat{U})\right) \in \mathcal{H}^*(\bar{X}; \bar{Y}; \bar{Z}|\bar{X}),$$

where the RVs $\bar{X}, \bar{Y}, \bar{Z}$ are such that

$$P_{\bar{X}} = \frac{1}{n}\sum_{i=1}^{n} P_{\hat{X}_i} \quad \text{and} \quad P_{\bar{Y}|\bar{X}} = V, \quad P_{\bar{Z}|\bar{X}} = W. \qquad \bigcirc$$

COROLLARY 15.7

$$\mathcal{H}(X; Y; Z|X) \subset \mathcal{H}^*(X; Y; Z|X). \qquad \bigcirc$$

Proof Assume without restricting generality that $\hat{Y}^n -\!\!\circ\!\!- \hat{X}^n -\!\!\circ\!\!- \hat{Z}^n$. Put

$$a \triangleq \frac{1}{n}H(\hat{X}^n|\hat{U}), \quad b \triangleq \frac{1}{n}H(\hat{Y}^n|\hat{U}), \quad c \triangleq \frac{1}{n}H(\hat{Z}^n|\hat{U}). \tag{15.33}$$

We can write

$$
\begin{aligned}
n(c-b) &= H(\hat{Z}^n|\hat{U}) - H(\hat{Y}^n|\hat{U}) \\
&= H(\hat{Z}^n|\hat{U}) - H(\hat{Y}_1\hat{Z}_2\ldots\hat{Z}_n|\hat{U}) + H(\hat{Y}_1\hat{Z}_2\ldots\hat{Z}_n|\hat{U}) - H(\hat{Y}^n|\hat{U}) \\
&= [H(\hat{Z}_1|Z_2\ldots\hat{Z}_n\hat{U}) - H(\hat{Y}_1|\hat{Z}_2\ldots\hat{Z}_n\hat{U})] \\
&\quad + [H(\hat{Z}_2\ldots\hat{Z}_n|\hat{Y}_1\hat{U}) - H(\hat{Y}_2\ldots\hat{Y}_n|\hat{Y}_1\hat{U})].
\end{aligned}
$$

Rewriting the second difference in a similar manner and iterating this operation, we finally obtain

$$
c-b = \frac{1}{n}\sum_{i=1}^{n}[H(\hat{Z}_i|\hat{Y}^{i-1}\hat{Z}_{i-1}\ldots\hat{Z}_n\hat{U}) - H(\hat{Y}_i|\hat{Y}^{i-1}\hat{Z}_{i+1}\ldots\hat{Z}_n\hat{U})]. \tag{15.34}
$$

Similarly

$$
a-b = \frac{1}{n}\sum_{i=1}^{n}[H(\hat{X}_i|\hat{Y}^{i-1}\hat{X}_{i+1}\ldots\hat{X}_n\hat{U}) - H(\hat{Y}_i|\hat{Y}^{i-1}\hat{X}_{i+1}\ldots\hat{X}_n\hat{U})] \tag{15.35}
$$

and

$$
c-a = \frac{1}{n}\sum_{i=1}^{n}[H(\hat{Z}_i|\hat{X}^{i-1}\hat{Z}_{i+1}\ldots\hat{Z}_n\hat{U}) - H(\hat{X}_i|\hat{X}^{i-1}\hat{Z}_{i+1}\ldots\hat{Z}_n\hat{U})]. \tag{15.36}
$$

Note that

$$
H(\hat{X}_i|\hat{Y}^{i-1}\hat{X}_{i+1}\ldots\hat{X}_n\hat{U}) = H(\hat{X}_i|\hat{Y}^{i-1}\hat{X}_{i+1}\ldots\hat{X}_n\hat{Z}_{i+1}\ldots\hat{Z}_n\hat{U}).
$$

In fact, the assumptions $\hat{U} \multimap \hat{X}^n \multimap \hat{Y}^n\hat{Z}^n$, $\hat{Y}^n \multimap \hat{X}^n \multimap \hat{Z}^n$, $P_{\hat{Z}^n|\hat{X}^n} = W^n$ imply that the conditional distribution of $\hat{Z}^n$ given $\hat{U}, \hat{X}^n, \hat{Y}^n$ equals W^n, whence, given $\hat{X}_{i+1}, \ldots, \hat{X}_n$, the RVs $\hat{Z}_{i+1}, \ldots, \hat{Z}_n$ are conditionally independent of the remaining RVs. Similarly,

$$
H(\hat{Y}_i|\hat{Y}^{i-1}\hat{X}_{i+1}\ldots\hat{X}_n\hat{U}) = H(\hat{Y}_i|\hat{Y}^{i-1}\hat{X}_{i+1}\ldots\hat{X}_n\hat{Z}_{i+1}\ldots\hat{Z}_n\hat{U}).
$$

Hence (15.35) becomes

$$
\begin{aligned}
a-b = \frac{1}{n}\sum_{i=1}^{n}[&H(\hat{X}_i|\hat{Y}^{i-1}\hat{X}_{i+1}\ldots\hat{X}_n\hat{Z}_{i+1}\ldots\hat{Z}_n\hat{U}) \\
&- H(\hat{Y}_i|\hat{Y}^{i-1}\hat{X}_{i+1}\ldots\hat{X}_n\hat{Z}_{i+1}\ldots\hat{Z}_n\hat{U})].
\end{aligned} \tag{15.37}
$$

Analogously, we obtain from (15.36)

$$
c-a = \frac{1}{n}\sum_{i=1}^{n}[H(\hat{Z}_i|\hat{X}^{i-1}\hat{Y}^{i-1}\hat{Z}_{i+1}\ldots\hat{Z}_n\hat{U}) - H(\hat{X}_i|\hat{X}_i^{i-1}\hat{Y}^{i-1}\hat{Z}_{i+1}\ldots\hat{Z}_n\hat{U})]. \tag{15.38}
$$

Further, we can write

$$c = \frac{1}{n} H(\hat{Z}^n | \hat{U}) = \frac{1}{n} \sum_{i=1}^{n} H(\hat{Z}_i | \hat{Z}_{i+1} \dots \hat{Z}_n \hat{U}) \geqq \frac{1}{n} \sum_{i=1}^{n} H(\hat{Z}_i | \hat{Y}^{i-1} \hat{Z}_{i+1} \dots \hat{Z}_n \hat{U}). \tag{15.39}$$

Let J be a RV uniformly distributed over the integers 1, 2, ..., n and independent of $(\hat{U}, \hat{X}^n, \hat{Y}^n, \hat{Z}^n)$. Put

$$\bar{X} \triangleq \hat{X}_J, \quad \bar{Y} \triangleq \hat{Y}_J, \quad \bar{Z} \triangleq \hat{Z}_J, \quad U \triangleq (J, \hat{U}, \hat{Y}^{J-1}, \hat{Z}_{J+1} \dots \hat{Z}_n),$$
$$S \triangleq \hat{X}_{J+1} \dots \hat{X}_n, \quad T \triangleq \tilde{X}^{J-1}.$$

With this notation, (15.34), (15.37), (15.38) and (15.39) can be rewritten as follows:

$$c - b = H(\bar{Z}|U) - H(\bar{Y}|U), \tag{15.40}$$
$$a - b = H(\bar{X}|SU) - H(\bar{Y}|SU), \tag{15.41}$$
$$c - a = H(\bar{Z}|TU) - H(\bar{X}|TU), \tag{15.42}$$
$$c \geqq H(\bar{Z}|U). \tag{15.43}$$

Now, (15.40), (15.43) and the obvious inequalities $b \leqq H(\bar{Y})$, $c \leqq H(\bar{Z})$ imply the existence of a number t with $0 \leqq t \leqq \min[I(U \wedge \bar{Y}), I(U \wedge \bar{Z})]$ such that

$$b = H(\bar{Y}|U) + t, \; c = H(\bar{Z}|U) + t. \tag{15.44}$$

Comparing this with (15.42) we get

$$a = H(\bar{X}|TU) + I(\bar{Z} \wedge T|U) + t. \tag{15.45}$$

Clearly,

$$P_{\bar{X}} = \frac{1}{n} \sum_{i=1}^{n} P_{\hat{X}_i}, \quad P_{\bar{Y}|\bar{X}} = V, \quad P_{\bar{Z}|\bar{X}} = W.$$

Further,

$$UST \multimap \bar{X} \multimap \bar{Y}\bar{Z} \tag{15.46}$$

since

$$H(\bar{Y}\bar{Z}|\bar{X}UST) = \frac{1}{n} \sum_{i=1}^{n} H(\hat{Y}_i \hat{Z}_i | \hat{X}_i \hat{U} \hat{X}^{i-1} \hat{X}_{i+1} \dots \hat{X}_n Y^{i-1} \hat{Z}_{i+1} \dots \hat{Z}_n)$$
$$= \frac{1}{n} \sum_{i=1}^{n} H(\hat{Y}_i \hat{Z}_i | \hat{X}_i) = H(\bar{Y}\bar{Z}|\bar{X}).$$

Thus by (15.44) the numbers b and c have the required form. In order to prove that $(a, b, c) \in \mathcal{H}^* (\bar{X}; \bar{Y}; \bar{Z}|\bar{X})$, it remains to check (see (15.31)) that

$$\max[b - H(\bar{Y}|\bar{X}), \; c - H(\bar{Z}|\bar{X})] \leqq a \leqq H(\bar{X}|U) + t.$$

In order to prove the second inequality, note that

$$I(\bar{Z} \wedge T|U) \leqq I(\bar{Z}\bar{X} \wedge T|U) = I(\bar{X} \wedge T|U),$$

where the last step follows from (15.46). Thus (15.45) yields

$$a \leqq H(\bar{X}|TU) + I(\bar{X} \wedge T|U) + t = H(\bar{X}|U) + t.$$

The remaining inequality follows from the identities (15.41), (15.42) and the inequalities

$$H(\bar{Y}|\bar{X}) + H(\bar{X}|SU \geqq H(\bar{Y}|SU),$$
$$H(\bar{Z}|\bar{X}) + H(\bar{X}|TU) \geqq H(\bar{Z}|TU).$$

completing the proof of the lemma. The corollary is obvious. □

To prove the existence results of this chapter, we need an analog of the maximal code lemma, Lemma 6.3 for two DMCs.

LEMMA 15.8 For every ε, τ, η in (0, 1), every quadruple of RVs U, X, Y, Z satisfying the Markov condition $U \multimap X \multimap YZ$, every sequence $\mathbf{u} \in \mathsf{T}^n_{[U]}$ and set $\mathsf{A} \subset \mathsf{X}^n$ satisfying

$$P^n_{X|U}(\mathsf{A}|\mathbf{u}) \geqq \eta,$$

there exist (n, ε)-codes for the DMCs $\{V : \mathsf{X} \to \mathsf{Y}\}$ resp. $\{W : \mathsf{X} \to \mathsf{Z}\}$ having the same encoder $f : \mathsf{M}_f \to \mathsf{A}$ and decoders φ resp. ψ such that

$$\frac{1}{n} \log |\mathsf{M}_f| \geqq \min[I(X \wedge Y|U), I(X \wedge Z|U)] - 2\tau \qquad (15.47)$$

and

$$\varphi^{-1}(m) \subset \mathsf{T}_{[V]}(f(m)), \quad \psi^{-1}(m) \subset \mathsf{T}_{[W]}(f(m)) \quad \text{for every} \quad m \in \mathsf{M}_f, \qquad (15.48)$$

whenever $n \geqq n_0(|\mathsf{U}|, |\mathsf{X}|, |\mathsf{Y}|, |\mathsf{Z}|, \varepsilon, \tau, \eta)$. ○

COROLLARY 15.8 For ε, τ, η in (0, 1), to every set $\mathsf{A} \subset \mathsf{X}^n$ satisfying $P^n_X(\mathsf{A}) \geqq \eta$ there exist (n, ε)-codes for the DMCs $\{V : \mathsf{X} \to \mathsf{Y}\}$ and $\{W : \mathsf{X} \to \mathsf{Z}\}$ having the same encoder $f : \mathsf{M}_f \to \mathsf{A}$ and rate

$$\frac{1}{n} \log |\mathsf{M}_f| \geqq \min[I(X \wedge Y), I(X \wedge Z)] - 2\tau$$

whenever $n \geqq n_0(|\mathsf{X}|, |\mathsf{Y}|, |\mathsf{Z}|, \varepsilon, \tau, \eta)$. ○

Proof Consider (n, ε)-codes (f, φ) and (f, ψ) for the two DMCs with codewords $f(m) \in \mathsf{A}$ such that (15.48) holds. We claim that if these codes have no extensions with the same properties, then (15.47) holds. In fact, write

$$\mathsf{B} \triangleq \bigcup_{m \in \mathsf{M}_f} \varphi^{-1}(m), \qquad \mathsf{C} \triangleq \bigcup_{m \in \mathsf{M}_f} \psi^{-1}(m).$$

Then, as in the proof of the maximal code lemma, Lemma 6.3, for every $\mathbf{x} \in \mathsf{A}$

$$V^n(\mathsf{T}_{[V]}(\mathbf{x}) - \mathsf{B}|\mathbf{x}) \leqq 1 - \varepsilon \quad \text{or} \quad W^n(\mathsf{T}_{[W]}(\mathbf{x}) - \mathsf{C}|\mathbf{x}) \leqq 1 - \varepsilon,$$

implying for $n \geqq n_1$

$$V^n(\mathsf{B}|\mathbf{x}) > \varepsilon - \tau \quad \text{or} \quad W^n(\mathsf{C}|\mathbf{x}) > \varepsilon - \tau \quad \text{if} \quad \mathbf{x} \in \mathsf{A}.$$

Let A_1 resp. A_2 be the set of those $\mathbf{x} \in \mathsf{A}$ for which the first resp. second inequality holds. We have

$$P^n_{X|U}(\mathsf{A}_1|\mathbf{u}) \geqq \frac{1}{2}\eta \quad \text{or} \quad P^n_{X|U}(\mathsf{A}_2|\mathbf{u}) \geqq \frac{1}{2}\eta.$$

In the first case

$$P^n_{Y|U}(\mathsf{B}|\mathbf{u}) \geqq \frac{1}{2}\eta(\varepsilon - \tau),$$

thus by Corollary 2.14 for $n \geqq n_2$

$$|\mathsf{B}| \geqq \exp\{n(H(Y|U) - \tau)\}.$$

But if $n \geqq n_3$ then

$$|\mathsf{B}| \leqq \sum_{m \in \mathsf{M}_f} |\varphi^{-1}(m)| \leqq \sum_{m \in \mathsf{M}_f} |\mathsf{T}_{[V]}(f(m))| \leqq |\mathsf{M}_f| \exp\{n(H(Y|X) + \tau)\},$$

whence

$$|\mathsf{M}_f| \geqq \exp\{n(I(X \wedge Y|U) - 2\tau)\}.$$

Similarly, in the case $P^n_{X|U}(\mathsf{A}_2|\mathbf{u}) \geqq \eta/2$ we get

$$|\mathsf{M}_f| \geqq \exp\{n(I(X \wedge Z|U) - 2\tau)\}.$$

□

Now we turn to the characterization of $\mathcal{G}(X; Y; Z|X)$. Readers interested only in the entropy characterization problem may proceed after Theorem 15.9 directly to Lemma 15.15.

First we give a computable characterization of the set $\mathcal{G}(Y; Z|X)$ of those pairs (b, c) to which for every $\varepsilon \in (0, 1)$, $\delta > 0$ and sufficiently large n there exists a set $\mathsf{A} \subset \mathsf{T}^n_{[X]}$ satisfying the second and third conditions in (ii) of Definition 15.1. To this end, denote by $\mathcal{H}^*(Y; Z|X)$ the projection of $\mathcal{H}^*(X; Y; Z|X)$ onto the (b, c)-plane. In other words, let $\mathcal{H}^*(Y; Z|X)$ be the set of points (b, c) of form

$$b = H(Y|U) + t, \quad c = H(Z|U) + t, \quad 0 \leqq t \leqq \min[I(U \wedge Y), I(U \wedge Z)], \tag{15.49}$$

where the RV U satisfies the conditions in (15.26) and (15.27), i.e., → 15.1

$$U \multimap X \multimap YZ, \quad |\mathsf{U}| \leqq |\mathsf{X}| + 2. \tag{15.50}$$

We shall prove that

$$\mathcal{G}(Y; Z|X) = \mathcal{H}^*(Y; Z|X).$$

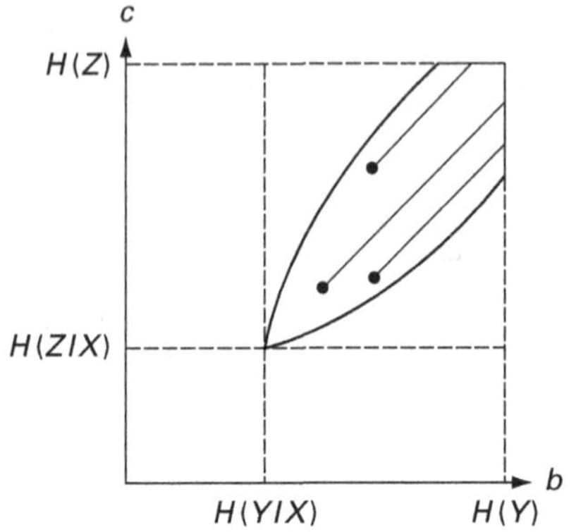

Figure 15.1 Typical shape of $\mathcal{H}^*(Y;Z|X)$. With each of its points, it contains a straight-line segment of slope 1 upwards from that point.

We begin with the following theorem.

THEOREM 15.9

$$\mathcal{H}^*(Y;Z|X) \subset \mathcal{G}(Y;Z|X).$$

More exactly, for every $\delta, \varepsilon, \eta$ in $(0,1)$, to every $(b,c) \in \mathcal{H}^*(Y;Z|X)$ and $\mathsf{E} \subset \mathsf{T}^n_{[X]}$ with $P^n_X(\mathsf{E}) \geqq \eta$ there exists a set $\mathsf{A} \subset \mathsf{E}$ satisfying

$$\left|\frac{1}{n}\log g_{V^n}(\mathsf{A}, 1-\varepsilon) - b\right| \leqq \delta, \qquad \left|\frac{1}{n}\log g_{W^n}(\mathsf{A}, 1-\varepsilon) - c\right| \leqq \delta,$$

whenever $n \geqq n_0(|\mathsf{X}|, |\mathsf{Y}|, |\mathsf{Z}|, \delta, \varepsilon, \eta)$. ○

Proof Fix some $(b,c) \in \mathcal{H}^*(Y;Z|X)$, given by (15.49), with U satisfying (15.50). Consider the DMCs $\{\hat{V} : \mathsf{U} \to \mathsf{Y}\}$, $\{\hat{W} : \mathsf{U} \to \mathsf{Z}\}$, where $\hat{V}$ resp. $\hat{W}$ are channels connecting U with Y resp. Z, and define F as in (15.12):

$$\mathsf{F} \triangleq \mathsf{T}^n_{[U]} \cap \left\{\mathbf{u} : P^n_{X|U}(\mathsf{E} \cap \mathsf{T}_{[X|U]}(\mathbf{u})|\mathbf{u}) \geqq \frac{\eta}{2}\right\}.$$

By (15.12) and Corollary 15.8, to every $\varepsilon' > 0$ and n sufficiently large there exist (n, ε')-codes for the DMCs $\{\hat{V}\}$ and $\{\hat{W}\}$ with common encoder $f : \{1, \ldots, K\} \to \mathsf{F} \subset \mathsf{T}^n_{[U]}$ and respective decoders $\varphi : \mathsf{Y}^n \to 1, \ldots, K$, $\psi : \mathsf{Z}^n \to \{1, \ldots, K\}$ such that for t in (15.49)

$$\left|\frac{1}{n}\log K - t\right| \leqq \frac{\delta}{2}. \tag{15.51}$$

We claim that for suitable ε', the set

$$\mathsf{A} \triangleq \bigcup_{k=1}^{K} \hat{\mathsf{A}}_k, \quad \text{where} \quad \hat{\mathsf{A}}_k \triangleq \mathsf{E} \cap \mathsf{T}_{[X|U]}(\mathbf{u}_k), \quad \mathbf{u}_k \triangleq f(k),$$

has the properties required in the theorem. By symmetry, it suffices to prove that

$$\left| \frac{1}{n} \log g_{V^n}(\mathsf{A}, 1-\varepsilon) - H(Y|U) - t \right| \leqq \delta.$$

Now, for n sufficiently large we have

$$\begin{aligned}\frac{1}{n} \log g_{V^n}(\mathsf{A}, 1-\varepsilon) &\leqq \frac{1}{n} \log \sum_{k=1}^{K} g_{V^n}(\hat{\mathsf{A}}_k, 1-\varepsilon) \\ &\leqq \frac{1}{n} \log K + H(Y|U) + \frac{\delta}{2} \leqq t + H(Y|U) + \delta, \end{aligned} \tag{15.52}$$

where the second inequality holds by (15.13) and the third by (15.51).

On the other hand, by the definition of F and $\hat{\mathsf{A}}_k$ we have

$$P^n_{X|U}(\hat{\mathsf{A}}_k|\mathbf{u}_k) \geqq \frac{\eta}{2}.$$

Thus if $\mathsf{B} \subset \mathsf{Y}^n$ is any $(1-\varepsilon)$-image of A then

$$\hat{V}^n(\mathsf{B}|\mathbf{u}_k) \geqq \sum_{\mathbf{x} \in \hat{\mathsf{A}}_k} P^n_{X|U}(\mathbf{x}|\mathbf{u}_k) V^n(\mathsf{B}|\mathbf{x}) \geqq \frac{\eta}{2}(1-\varepsilon) \quad \text{for every } k.$$

Hence

$$\hat{V}^n(\mathsf{B} \cap \varphi^{-1}(k)|\mathbf{u}_k) \geqq \frac{\eta}{2}(1-\varepsilon) - \varepsilon' \quad \text{for every } k.$$

Thus, setting $\varepsilon' \triangleq (\eta/4)(1-\varepsilon)$, by Lemma 2.14 and (15.51) we obtain (for sufficiently large n)

$$|\mathsf{B}| \geqq \sum_{k=1}^{K} |\mathsf{B} \cap \varphi^{-1}(k)| \geqq \exp\{n(t + H(Y|U) - \delta)\}.$$

Since B was any $(1-\varepsilon)$-image of A, comparing the preceding inequality with (15.52) completes the proof. □

THEOREM 15.10

$$\mathcal{G}(Y; Z|X) \subset \mathcal{H}^*(Y; Z|X).$$

Moreover, for every $0 < \varepsilon < 1$, $\delta > 0$, to every $\mathsf{A} \subset \mathsf{T}^n_{[X]}$ there exists a point $(b, c) \in \mathcal{H}^*(Y; Z|X)$ such that

$$\left| \frac{1}{n} \log g_{V^n}(\mathsf{A}, 1-\varepsilon) - b \right| \leqq \delta, \qquad \left| \frac{1}{n} \log g_{W^n}(\mathsf{A}, 1-\varepsilon) - c \right| \leqq \delta, \tag{15.53}$$

whenever $n \geqq n_0(|\mathsf{X}|, |\mathsf{Y}|, |\mathsf{Z}|, \delta, \varepsilon)$. ○

Proof For brevity, in this proof we write $\mathcal{H}^* = \mathcal{H}^*(Y; Z|X)$. Observe first the following property of $\mathcal{H}^*$. For any two points $(b_i, c_i) \in \mathcal{H}^*, i = 1, 2$, the point (b, c) defined by

$$b \triangleq \max[b_1, b_2], \quad c \triangleq \max[c_1, c_2]$$

also belongs to $\mathcal{H}^*$. In fact, the straight-line segment connecting the points (b_1, c_1) and (b_2, c_2) is in $\mathcal{H}^*$ by convexity; choosing the point (b', c') of this segment satisfying $c' - b' = c - b$, the relation $(b, c) \in \mathcal{H}^*$ follows.

We claim that it is sufficient to prove that in the case $n \geqq n_0(|\mathsf{X}|, |\mathsf{Y}|, |\mathsf{Z}|, \delta, \varepsilon)$ for every $\mathsf{A} \subset \mathsf{T}^n_{[X]}$ there exists a point $(b_1, c_1) \in \mathcal{H}^*$ such that

$$\left|\frac{1}{n} \log g_{V^n}(\mathsf{A}, 1-\varepsilon) - b_1\right| \leqq \delta, \tag{15.54}$$

$$\frac{1}{n} \log g_{W^n}(\mathsf{A}, 1-\varepsilon) \geqq c_1 - \delta. \tag{15.55}$$

In fact, this implies by symmetry that there also exists a point $(b_2, c_2) \in \mathcal{H}^*$ with

$$\left|\frac{1}{n} \log g_{W^n}(\mathsf{A}, 1-\varepsilon) - c_2\right| \leqq \delta \tag{15.56}$$

and

$$\frac{1}{n} \log g_{V^n}(\mathsf{A}, 1-\varepsilon) \geqq b_2 - \delta. \tag{15.57}$$

Now, putting

$$b \triangleq \max[b_1, b_2], \quad c \triangleq \max[c_1, c_2],$$

the relations (15.54)–(15.57) imply (15.53), and by our first observation we have $(b, c) \in \mathcal{H}^*$. Hence, given an arbitrary set $\mathsf{A} \subset \mathsf{T}^n_{[X]}$, our aim will be to find a point $(b_1, c_1) \in \mathcal{H}^*$ satisfying (15.54) and (15.55).

Choose some $\varepsilon' > 0$ and $\delta' > 0$ to be specified later. Let $\mathsf{A}' \subset \mathsf{A}$ be the codeword set of an (n, ε')-code of maximum rate $C_V(\mathsf{A}, \varepsilon')$ for the DMC $\{V\}$. Then $C_V(\mathsf{A}, \varepsilon') = C_V(\mathsf{A}', \varepsilon')$, which by Lemma 6.8 implies for $n \geqq n_1(|\mathsf{X}|, |\mathsf{Y}|, \delta', \varepsilon, \varepsilon')$ that

$$\left|\frac{1}{n} \log g_{V^n}(\mathsf{A}, 1-\varepsilon) - \frac{1}{n} \log g_{V^n}(\mathsf{A}', 1-\varepsilon)\right| \leqq \delta'. \tag{15.58}$$

Of course,

$$\frac{1}{n} \log g_{W^n}(\mathsf{A}', 1-\varepsilon) \leqq \frac{1}{n} \log g_{W^n}(\mathsf{A}, 1-\varepsilon). \tag{15.59}$$

Let $\hat{X}^n$ be a RV uniformly distributed over the set A' and let the RVs $\hat{Y}^n$ and $\hat{Z}^n$ be connected with $\hat{X}^n$ by the channels V^n and W^n, respectively, their joint distribution being arbitrary otherwise. For $n \geqq n_2(|\mathsf{X}|, |\mathsf{Y}|, |\mathsf{Z}|, \delta', \varepsilon, \varepsilon')$ we have, by Lemma 15.2,

$$\frac{1}{n} \log g_{W^n}(\mathsf{A}', 1-\varepsilon) \geqq \frac{1}{n} H(\hat{Z}^n) - \delta' \tag{15.60}$$

and

$$\frac{1}{n} H(\hat{Y}^n) - \delta' \leqq \frac{1}{n} \log g_{V^n}(\mathsf{A}', 1-\varepsilon) \leqq \frac{1}{n} H(\hat{Y}^n) + \frac{\delta'}{2} + \varepsilon' \log|\mathsf{Y}|. \tag{15.61}$$

Choosing $\varepsilon' \triangleq \delta'/2 \log \mathsf{Y}|$, the inequalities (15.61) result in

$$\left|\frac{1}{n}H(\hat{Y}^n) - \frac{1}{n}\log g_{V^n}(\mathsf{A}', 1-\varepsilon)\right| \leqq \delta'. \tag{15.62}$$

Comparing this with relations (15.58)–(15.60) we see that

$$\left|\frac{1}{n}\log g_{V^n}(\mathsf{A}, 1-\varepsilon) - \frac{1}{n}H(\hat{Y}^n)\right| \leqq 2\delta', \tag{15.63}$$

$$\frac{1}{n}\log g_{W^n}(\mathsf{A}, 1-\varepsilon) \geqq \frac{1}{n}H(\hat{Z}^n) - 2\delta'. \tag{15.64}$$

By Lemma 15.7

$$\left(\frac{1}{n}H(\hat{Y}^n), \quad \frac{1}{n}H(\hat{Z}^n)\right) \in \mathcal{H}^*(\bar{Y}, \bar{Z}|\bar{X}), \tag{15.65}$$

where

$$P_{\bar{X}}(x) = \frac{1}{n}\sum_{i=1}^{n} P_{\hat{X}_i}(x) \quad \text{for every} \quad x \in \mathsf{X}.$$

The proof can be completed by a continuity argument. Since

$$\frac{1}{n}\sum_{i=1}^{n} P_{\hat{X}_i}(x) = \frac{1}{n|\mathsf{A}'|}\sum_{\mathbf{x}\in\mathsf{A}'} N(x|\mathbf{x}) \quad \text{for every} \quad x \in \mathsf{X},$$

the relation $\mathsf{A}' \subset \mathsf{T}^n_{[X]}$ implies for $n \geqq n_3$ that

$$|P_{\bar{X}}(x) - P_X(x)| < \delta' \quad \text{for every} \quad x \in \mathsf{X}. \tag{15.66}$$

Further, by the definition of $\bar{X}, \bar{Y}, \bar{Z}$ we can suppose that $P_{\bar{Y}\bar{Z}|\bar{X}} = P_{YZ|X}$. Now, if $\bar{U}$ is any RV satisfying $\bar{U} -\!\!\circ\!\!- \bar{X} -\!\!\circ\!\!- \bar{Y}\bar{Z}$ and the range constraint $|\mathsf{U}| \leqq |\mathsf{X}| + 2$, consider a RV U such that $U -\!\!\circ\!\!- X -\!\!\circ\!\!- YZ$ and $P_{U|X} = P_{\bar{U}|\bar{X}}$. Then, because of (15.66), the joint distributions of U, X, Y, Z and of $\bar{U}, \bar{X}, \bar{Y}, \bar{Z}$ are arbitrarily close if δ' is sufficiently small. By the uniform continuity of the involved information quantities as implied by Lemma 2.7, it follows that to every $(\bar{b}, \bar{c}) \in \mathcal{H}^*(\bar{Y}; \bar{Z}|\bar{X})$ there exists $(b, c) \in \mathcal{H}^*(Y; Z|X)$ such that

$$|\bar{b} - b| \leqq \frac{\delta}{2}, \quad |\bar{c} - c| \leqq \frac{\delta}{2}. \tag{15.67}$$

Then (15.63), (15.64), (15.65) and (15.67) imply (15.54) and (15.55). □

Summarizing Theorems 15.9 and 15.10 we have Theorem 15.11.

THEOREM 15.11

$$\mathcal{G}(Y; Z|X) = \mathcal{H}^*(Y; Z|X). \quad \bigcirc$$

Theorem 15.11 can be used to obtain more explicit characterizations of $\mathcal{G}(Y; Z|X)$ in various special cases. A case of independent interest is $Y = X$. Then $V : \mathsf{X} \to \mathsf{Y}$ is the identity matrix so that $g_{V^n}(\mathsf{A}, 1-\varepsilon) = |\mathsf{A}|$. Thus $\mathcal{G}(X; Z|X)$ gives an asymptotic characterization of the pairs

$$(|\mathsf{A}|,\ g_{W^n}(\mathsf{A}, 1-\varepsilon))$$

as A ranges over the subsets of $T^n_{[X]}$ and n goes to infinity.

LEMMA 15.12 (a, c) belongs to $\mathcal{G}(X; Z|X) = \mathcal{H}^*(X; Z|X)$ iff

$$c = H(Z|U), \quad c - H(Z|X) \leqq a \leqq H(X|U)$$

→ 15.2

for some RV U satisfying

$$U \multimap X \multimap Z, \quad |\mathsf{U}| \leqq |\mathsf{X}| + 2. \qquad \bigcirc$$

Proof We have that $\mathcal{G}(X; Z|X) = \mathcal{H}^*(X; Z|X)$ by Theorem 15.11. By (15.49), (15.50), $\mathcal{H}^*(X; Z|X)$ consists of those points (a, c) that can be represented as

$$a = H(X|U) + t, \quad c = H(Z|U) + t, \quad 0 \leqq t \leqq I(U \wedge Z),$$

for some RV U as above.

We have to show that this is the same as the set of points (a, c) satisfying

$$H(Z|X) \leqq c \leqq H(Z), \qquad a^-(c) \leqq a \leqq a^+(c),$$

where

$$a^-(c) \triangleq c - H(Z|X), \quad a^+(c) \triangleq \max_{\substack{H(Z|U)=c \\ U \multimap X \multimap Z \\ |\mathsf{U}| \leqq |\mathsf{X}|+2}} H(X|U).$$

(See Fig. 15.2, where the a-axis is drawn vertically.) Clearly, the latter set is contained in the convex set $\mathcal{H}^*(X; Z|X)$, as both $(a^-(c), c)$ and $(a^+(c), c)$ belong to this set for every $c \in [H(Z|X), H(Z)]$.

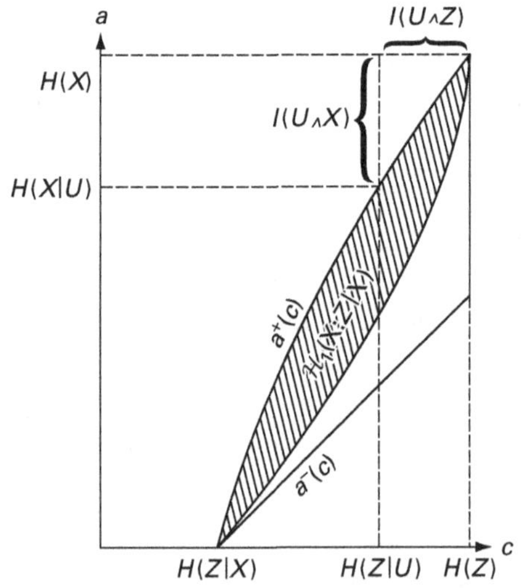

Figure 15.2 Shape of $\mathcal{H}^*(X; Z|X)$.

To check the opposite inclusion, note first that every $(a, c) \in \mathcal{H}^*(X; Z|X)$ satisfies $a \geqq a^-(c)$ since

$$H(Z|U) - H(Z|X) \leqq H(X|U).$$

It remains to prove that the "upper" boundary $a^+(c)$ of the set $\mathcal{H}_1(X; Z|X)$ is also the "upper" boundary of $\mathcal{H}^*(X; Z|X)$. This amounts to verifying the inequality → 15.3

$$a^+(c) + t \leqq a^+(c + t) \quad \text{for} \quad 0 \leqq t \leqq H(Z) - c. \tag{15.68}$$

As $\mathcal{H}_1(X; Z|X)$ is a convex set by Lemma 15.6, it is enough to check (15.68) for the largest possible value of t. For $t = H(Z) - c$, (15.68) becomes

$$a^+(H(Z)) \geqq a^+(c) + H(Z) - c,$$

where $a^+(H(Z)) = H(X)$. Thus we have to show that

$$H(X) \geqq H(X|U) + H(Z) - H(Z|U)$$

whenever $U \multimap X \multimap Z$. This, however, follows from the data processing lemma, Lemma 3.11. □

From Theorem 15.11 and Lemma 15.12 we can easily obtain the desired characterization of $\mathcal{G}(X; Y; Z|X)$. To this end, consider the lower and upper boundaries of $\mathcal{H}^* = \mathcal{H}^*(X; Y; Z|X)$ supposing that the a-axis is drawn vertically. These are given by the functions

$$\begin{aligned} a^-(b, c) &\triangleq \min_{(a,b,c)\in\mathcal{H}^*} a = \max\,[\,b - H(Y|X),\ c - H(Z|X)\,], \\ a^+(b, c) &\triangleq \max_{(a,b,c)\in\mathcal{H}^*} a = \max_{\substack{H(Y|U)+t=b \\ H(Z|U)+t=c}} (H(X|U) + t), \end{aligned} \tag{15.69}$$

where U is a RV such that $U \multimap X \multimap YZ$, $|\mathsf{U}| \leqq |\mathsf{X}| + 2$ and

$$0 \leqq t \leqq \min[I(U \wedge Y), I(U \wedge Z)].$$

Both $a^-(b, c)$ and $a^+(b, c)$ are defined for $(b, c) \in \mathcal{H}^*(Y; Z|X)$; see (15.49). Now define the function $a^{++}(b, c)$ with the same domain by

$$a^{++}(b, c) \triangleq \max_{\substack{H(Y|U)\leqq b \\ H(Z|U)\leqq c}} H(X|U), \tag{15.70}$$

where the RV U is as above. We designate by $\mathcal{G}^*(X; Y; Z|X)$ the following set:

$$\mathcal{G}^*(X; Y; Z|X) \triangleq \{(a, b, c) : (b, c) \in \mathcal{H}^*(Y; Z|X),\ a^-(b, c) \leqq a \leqq a^{++}(b, c)\}. \tag{15.71}$$

We shall prove that $\mathcal{G}(X; Y; Z|X) = \mathcal{G}^*(X; Y; Z|X)$. For this purpose we need the following lemma which, at the same time, shows the relation of the sets $\mathcal{G}^*(X; Y; Z|X)$ and $\mathcal{H}^*(X; Y; Z|X)$.

LEMMA 15.13

$$a^{++}(b,c) = \max_{\bar b \leqq b,\, \bar c \leqq c} a^{+}(\bar b, \bar c). \qquad \bigcirc$$

→ 15.4 COROLLARY 15.13 $\mathcal{G}^*(X;Y;Z|X)$ equals the set of points of form

$$(\max(a_1,a_2), \max(b_1,b_2), \max(c_1,c_2)),$$

where $(a_i,b_i,c_i) \in \mathcal{H}^*(X;Y;Z|X),\ i = 1,2.$ $\bigcirc$

Proof We have, by definition,

$$a^{++}(b,c) \leqq \max_{\bar b \leqq b,\, \bar c \leqq c} a^{+}(\bar b, \bar c).$$

To prove the opposite inequality, note that for every $\bar b \leqq b,\ \bar c \leqq c$

$$a^{+}(\bar b,\bar c) \leqq \max_{t \geqq 0} \left[\max_{\substack{H(Y|U) \leqq b-t \\ H(Z|U) \leqq c-t}} H(X|U) + t \right] \leqq \max_{t \geqq 0} [a^{++}(b-t, c-t) + t].$$

Hence it suffices to show that

$$a^{++}(b-t,c-t) + t \leqq a^{++}(b,c) \qquad \text{for } t \geqq 0.$$

The preceding inequality is equivalent to

$$a^{++}(b,c) + t \leqq a^{++}(b+t,c+t) \quad \text{for} \quad 0 \leqq t \leqq \min[H(Y)-b, H(Z)-c]. \tag{15.72}$$

By Lemma 15.6, the function $a^{++}(b+t,c+t)$ is concave in t. Hence it is enough to check (15.72) for the largest possible t; without loss of generality, suppose that this is $t = H(Y) - b$. Then

$$a^{++}(b+t,c+t) = a^{++}(H(Y), c+t) = \max_{H(Z|U) \leqq c+t} H(X|U).$$

Noting that the function $a^{+}(c)$ defined in the proof of Lemma 15.12 is strictly increasing by (15.68), the preceding identity can be rewritten as

$$a^{++}(b+t,c+t) = a^{+}(c+t).$$

Since we also have

$$a^{++}(b,c) \leqq \max_{H(Z|U) \leqq c} H(X|U) = a^{+}(c),$$

(15.72) now follows from (15.68), proving the lemma.

The corollary is an easy consequence, taking into account that $(b_i,c_i) \in \mathcal{H}^*(Y;Z|X),\ i = 1,2$, implies

$$(\max(b_1,b_2),\ \max(c_1,c_2)) \in \mathcal{H}^*(Y;Z|X)$$

as observed in the proof of Theorem 15.10. □

Next, as a first step to show that $\mathcal{G}(X; Y; Z|X) = \mathcal{G}^*(X; Y; Z|X)$, we prove the following theorem.

THEOREM 15.14

$$\mathcal{G}(X; Y; Z|X) \subset \mathcal{G}^*(X; Y; Z|X).$$

Moreover, for every $\varepsilon \in (0, 1)$, $\delta > 0$, to every set $\mathsf{A} \subset \mathsf{T}^n_{[X]}$ there exists a point $(a, b, c) \in \mathcal{G}^*(X; Y; Z|X)$ such that

→ 15.5

$$\left|\frac{1}{n}\log|\mathsf{A}| - a\right| \leqq \delta, \quad \left|\frac{1}{n}\log g_{V^n}(\mathsf{A}, 1-\varepsilon) - b\right| \leqq \delta,$$
$$\left|\frac{1}{n}\log g_{W^n}(\mathsf{A}, 1-\varepsilon) - c\right| \leqq \delta, \tag{15.73}$$

whenever $n \geqq n_0(|\mathsf{X}|, |\mathsf{Y}|, |\mathsf{Z}|, \delta, \varepsilon)$. ○

Proof To every $\mathsf{A} \subset \mathsf{T}^n_{[X]}$ we have to find a point $(b, c) \in \mathcal{H}^*(Y; Z|X)$ such that

$$\left|\frac{1}{n}\log g_{V^n}(\mathsf{A}, 1-\varepsilon) - b\right| \leqq \delta, \quad \left|\frac{1}{n}\log g_{W^n}(\mathsf{A}, 1-\varepsilon) - c\right| \leqq \delta, \tag{15.74}$$

$$a^-(b, c) - \delta \leqq \frac{1}{n}\log|\mathsf{A}| \leqq a^{++}(b, c) + \delta. \tag{15.75}$$

Let us fix some $\delta' > 0$ to be specified later. It follows from Lemma 6.8 that for $n \geqq n_1(|\mathsf{X}|, |\mathsf{Y}|, |\mathsf{Z}|, \delta', \varepsilon)$

$$\frac{1}{n}\log|\mathsf{A}| \geqq C_V(\mathsf{A}, 1-\varepsilon) \geqq \frac{1}{n}\log g_{V^n}(\mathsf{A}, 1-\varepsilon) - H(Y|X) - \delta', \tag{15.76}$$

$$\frac{1}{n}\log|\mathsf{A}| \geqq C_W(\mathsf{A}, 1-\varepsilon) \geqq \frac{1}{n}\log g_{W^n}(\mathsf{A}, 1-\varepsilon) - H(Z|X) - \delta'. \tag{15.77}$$

Further, by Theorem 15.10, for $n \geqq n_2(|\mathsf{X}|, |\mathsf{Y}|, |\mathsf{Z}|, \delta', \varepsilon)$ there exists a pair $(b', c') \in \mathcal{H}^*(Y; Z|X)$ such that

$$\left|\frac{1}{n}\log g_{V^n}(\mathsf{A}, 1-\varepsilon) - b'\right| \leqq \delta', \quad \left|\frac{1}{n}\log g_{W^n}(\mathsf{A}, 1-\varepsilon) - c'\right| \leqq \delta'. \tag{15.78}$$

Comparing this with (15.76) and (15.77), we get for $n \geqq \max(n_1, n_2)$ the inequalities

$$\frac{1}{n}\log|\mathsf{A}| \geqq b' - H(Y|X) - 2\delta', \quad \frac{1}{n}\log|\mathsf{A}| \geqq c' - H(Z|X) - 2\delta'. \tag{15.79}$$

On the other hand, it follows from Lemma 15.2 that if the RV $\hat{X}^n$ is uniformly distributed over A and the RVs $\hat{Y}^n$ and $\hat{Z}^n$ satisfy $P_{\hat{Y}^n\hat{Z}^n|\hat{X}^n} = P_{Y^nZ^n|X^n}$ then for $n \geqq n_3(|\mathsf{X}|, |\mathsf{Y}|, |\mathsf{Z}|, \delta', \varepsilon)$

$$\log g_{V^n}(\mathsf{A}, 1-\varepsilon) \geqq H(\hat{Y}^n) - n\delta', \tag{15.80}$$
$$\log g_{W^n}(\mathsf{A}, 1-\varepsilon) \geqq H(\hat{Z}^n) - n\delta', \tag{15.81}$$

whereas, of course, we also have

$$\log|\mathsf{A}| = H(\hat{X}^n). \tag{15.82}$$

By Lemma 15.7,

$$\left(\frac{1}{n}H(\hat{X}^n), \frac{1}{n}H(\hat{Y}^n), \frac{1}{n}H(\hat{Z}^n)\right) \in \mathcal{H}^*(\bar{X}; \bar{Y}; \bar{Z}|\bar{X}),$$

where $P_{\bar{X}} \triangleq (1/n)\sum_{i=1}^n P_{\hat{X}_i}$ and we can suppose that $P_{\bar{Y}\bar{Z}|\bar{X}} = P_{YZ|X}$. We have shown in the proof of Theorem 15.10 (see (15.66)) that for $n \geqq n_4(|\mathsf{X}|, |\mathsf{Y}|, |\mathsf{Z}|, \delta")$

$$|P_{\bar{X}}(x) - P_X(x)| \leqq \delta'' \quad \text{for every} \quad x \in \mathsf{X}.$$

Hence, by the same argument, we can conclude that there exists a point $(\tilde{a}, \tilde{b}, \tilde{c}) \in \mathcal{H}^*(X; Y; Z|X)$ such that

$$\left|\frac{1}{n}H(\hat{X}^n) - \tilde{a}\right| \leqq \delta', \quad \left|\frac{1}{n}H(\hat{Y}^n) - \tilde{b}\right| \leqq \delta', \quad \left|\frac{1}{n}H(\hat{Z}^n) - \tilde{c}\right| \leqq \delta'.$$

Compared with (15.80)–(15.82), these inequalities yield

$$\left|\frac{1}{n}\log|\mathsf{A}| - \tilde{a}\right| \leqq \delta', \quad \frac{1}{n}\log g_{V^n}(\mathsf{A}, 1-\varepsilon) \geqq \tilde{b} - 2\delta', \\ \frac{1}{n}\log g_{W^n}(\mathsf{A}, 1-\varepsilon) \geqq \tilde{c} - 2\delta'. \tag{15.83}$$

In particular, the preceding two inequalities and (15.78) give

$$\tilde{b} \leqq b' + 3\delta', \quad \tilde{c} \leqq c' + 3\delta'.$$

Hence by (15.83) and Lemma 15.13

$$\frac{1}{n}\log|\mathsf{A}| \leqq \max_{\substack{\bar{b} \leqq b'+3\delta' \\ \bar{c} \leqq c'+3\delta'}} a^+(\bar{b}, \bar{c}) + \delta' = a^{++}(b' + 3\delta', c' + 3\delta') + \delta'. \tag{15.84}$$

Setting $b \triangleq b' + 3\delta'$ and $c \triangleq c' + 3\delta'$, the preceding inequality results in

$$\frac{1}{n}\log|\mathsf{A}| \leqq a^{++}(b, c) + \delta'.$$

Comparing this with (15.78) and (15.79) we see that (15.74) and (15.75) hold if $\delta' \leqq \delta/5$. □

THEOREM 15.15

$$\mathcal{G}^*(X; Y; Z|X) \subset \mathcal{G}(X; Y; Z|X).$$

Moreover, every $(a, b, c) \in \mathcal{G}^*(X; Y; Z|X)$ has the property that for every $0 < \varepsilon < 1, \delta > 0, \eta > 0$ and to every $\mathsf{E} \subset \mathsf{T}^n_{[X]}$ with $P^n_X(\mathsf{E}) \geqq \eta$ there exists a set $\mathsf{A} \subset \mathsf{E}$ satisfying

$$\left|\frac{1}{n}\log|\mathsf{A}| - a\right| \leqq \delta, \quad \left|\frac{1}{n}\log g_{V^n}(\mathsf{A}, 1-\varepsilon) - b\right| \leqq \delta,$$
$$\left|\frac{1}{n}\log g_{W^n}(\mathsf{A}, 1-\varepsilon) - c\right| \leqq \delta, \tag{15.85}$$

whenever $n \geqq n_0(|\mathsf{X}|, |\mathsf{Y}|, |\mathsf{Z}|, \delta, \varepsilon, \eta)$. ○

Proof Fix some $(b, c) \in \mathcal{H}^*(Y; Z|X)$. First we prove that $(a^-(b, c), b, c)$ does have the asserted property. To this end, consider a set $\tilde{\mathsf{A}} \subset \mathsf{E}$ satisfying

$$\left|\frac{1}{n}\log g_{V^n}(\tilde{\mathsf{A}}, 1-\varepsilon) - b\right| \leqq \frac{\delta}{4}, \quad \left|\frac{1}{n}\log g_{W^n}(\tilde{\mathsf{A}}, 1-\varepsilon) - c\right| \leqq \frac{\delta}{4}. \tag{15.86}$$

Such sets exist by Theorem 15.9 if $n \geqq n_1$. Similarly to the proof of Theorem 15.10, let $\mathsf{A}' \subset \tilde{\mathsf{A}}$ be the codeword set of an (n, ε)-code of maximum rate $C_V(\tilde{\mathsf{A}}, \varepsilon)$ for the DMC $\{V\}$. Then, by (15.58) and (15.86), for $n \geqq n_2$ we have

$$\left|\frac{1}{n}\log g_{V^n}(\mathsf{A}', 1-\varepsilon) - b\right| \leqq \frac{\delta}{3}, \tag{15.87}$$

while, of course,

$$\frac{1}{n}\log g_{W^n}(\mathsf{A}', 1-\varepsilon) \leqq \frac{1}{n}\log g_{W^n}(\tilde{\mathsf{A}}, 1-\varepsilon) \leqq c + \frac{\delta}{4}. \tag{15.88}$$

Since $(1/n)\log|\mathsf{A}'| = C_V(\mathsf{A}',\varepsilon)$, by Lemma 6.8 we obtain from (15.87) for $n \geqq n_3$

$$\frac{1}{n}\log|\mathsf{A}'| - (b - H(Y|X)) \leqq \frac{\delta}{2}. \tag{15.89}$$

Similarly, if $\mathsf{A}'' \subset \tilde{\mathsf{A}}$ is the codeword set of an (n, ε)-code of maximum rate for the DMC $\{W\}$, then for $n \geqq n_4$

$$\left|\frac{1}{n}\log g_{W^n}(\mathsf{A}'', 1-\varepsilon) - c\right| \leqq \frac{\delta}{3}, \tag{15.90}$$

$$\frac{1}{n}\log g_{V^n}(\mathsf{A}'', 1-\varepsilon) \leqq b + \frac{\delta}{4}, \tag{15.91}$$

$$\left|\frac{1}{n}\log|\mathsf{A}''| - (c - H(Z|X))\right| \leqq \frac{\delta}{2}. \tag{15.92}$$

One immediately sees from (15.87)–(15.92) that the set $\mathsf{A}^- \triangleq \mathsf{A}' \cup \mathsf{A}''$ satisfies (15.85) if $n \geqq n_5$.

Next we prove the assertion for $(a^{++}(b, c), b, c)$. By Lemma 15.3, for $n \geqq n_6$ there exist sets $\mathsf{A}^* \subset \mathsf{E}$ satisfying

$$\left|\frac{1}{n}\log|\mathsf{A}^*| - a^{++}(b, c)\right| \leqq \frac{\delta}{2},$$
$$\frac{1}{n}\log g_{V^n}(\mathsf{A}^*, 1-\varepsilon) \leqq b + \frac{\delta}{2}, \quad \frac{1}{n}\log g_{W^n}(\mathsf{A}^*, 1-\varepsilon) \leqq c + \frac{\delta}{2}.$$

This means that $\mathsf{A}^+ \triangleq \mathsf{A}^- \cup \mathsf{A}^*$ satisfies (15.85) for $n \geqq n_7$. It follows that suitable sets A with $\mathsf{A}^- \subset \mathsf{A} \subset \mathsf{A}^+$ can be found to every (a, b, c) with $a^-(b, c) \leqq a \leqq a^{++}(b, c)$ so that (15.85) holds for sufficiently large n. □

THEOREM 15.16 (*Image size*)

$$\mathcal{G}(X; Y; Z|X) = \mathcal{G}^*(X; Y; Z|X). \qquad \bigcirc$$

Let us now return to the entropy characterization problem. We have promised to show that

$$\mathcal{F}(X; Y; Z|X) = \mathcal{H}(X; Y; Z|X) = \mathcal{H}^*(X; Y; Z|X).$$

We already know that $\mathcal{F} \subset \mathcal{H} \subset \mathcal{H}^*$. In the following, we shall establish the missing existence result

$$\mathcal{H}^*(X; Y; Z|X) \subset \mathcal{F}(X; Y; Z|X).$$

For this purpose we prove two technical lemmas generalizing Lemma 15.8, which concerns codes with common encoder for two DMCs. As these lemmas look rather complex, let us first sketch the idea of our proof technique, which is quite simple. We will have to construct functions f with domain X^n and prescribed conditional entropy $(1/n)H(Y^n|f(X^n))$, say. Let $\mathsf{X}^n = \bigcup_{i=1}^M \mathsf{A}_i$ be the partition of X^n generated by the function f. Obviously,

$$H(Y^n|f(X^n)) = \sum_{i=1}^{M} P_X^n(\mathsf{A}_i) H(Y^n|X^n \in \mathsf{A}_i),$$

where $H(Y^n|X^n \in \mathsf{A}_i)$ is a shorthand for the entropy of the conditional distribution of Y^n under the condition that $X^n \in \mathsf{A}_i$. Hence we can produce a function with prescribed entropy $H(Y^n|f(X^n))$ by constructing disjoint sets A_i which yield prescribed entropy

$$H(Y^n|X^n \in \mathsf{A}_i).$$

This entropy can be upper bounded in terms of the image size $g_{V^n}(\mathsf{A}_i, 1-\varepsilon)$ (where $V = P_{Y|X}$), by (15.4) of Lemma 15.2. We shall obtain our function f by constructing sets A_i with prescribed image sizes for which this upper bound is tight. This is the case, by the same lemma, if A_i consists of sequences of the same type and A_i is the codeword set of an (n, ε)-code for the DMC $\{V\}$. We shall see that the same is true if A is the disjoint union of such codeword sets, when the corresponding codes are nearly maximal, i.e., their rate is close to $C_V(\mathsf{A}, \varepsilon)$. Intuitively, we shall say that such a set
→ 15.6 A is *code-stuffed* for $\{V\}$. This notion will not be given a precise definition, as it will be used only in intuitive anticipatory descriptions of results to be stated and proved subsequently.

The content of the following lemma is that for any RV U with $U -\!\!\circ\!\!- X -\!\!\circ\!\!- YZ$ such that $I(X \wedge Y|U) \geqq I(X \wedge Z|U)$, for large n every set $\mathsf{A} \subset \mathsf{X}^n$ satisfying

$$P_{X|U}^n(\mathsf{A}|\mathbf{u}) \geqq \eta \quad \text{for some} \quad \mathbf{u} \in \mathsf{T}_{[U]}^n$$

has a subset $\tilde{\mathsf{A}}$ which is both the codeword set of a "large" (n, ε)-code for the DMC $\{V : \mathsf{X} \to \mathsf{Y}\}$ and a code-stuffed set for the DMC $\{W : \mathsf{X} \to \mathsf{Z}\}$.

LEMMA 15.17 For any $\tau > 0$, $\eta > 0$, $\varepsilon \in (0, 1)$ and $n \geqq n_0(|\mathsf{X}|, |\mathsf{Y}|, |\mathsf{Z}|, \tau, \eta, \varepsilon)$, for every quadruple of RVs (U, X, Y, Z) satisfying (15.50) and

$$I(X \wedge Y|U) \geqq I(X \wedge Z|U),$$

to every set $\mathsf{A} \subset \mathsf{X}^n$ and sequence $\mathbf{u} \in \mathsf{T}^n_{[U]}$ with $P^n_{X|U}(\mathsf{A}|\mathbf{u}) \geqq \eta$ there exists a subset $\tilde{\mathsf{A}}$ of A such that

(i) all sequences in $\tilde{\mathsf{A}}$ have the same type;
(ii) $\tilde{\mathsf{A}}$ is the codeword set of an (n, ε)-code for the DMC $\{V\}$;
(iii) $\tilde{\mathsf{A}}$ is the disjoint union of sets $\tilde{\mathsf{A}}^{(m)}$, $m = 1, \ldots, M$, where

$$\left|\frac{1}{n}\log|\tilde{\mathsf{A}}^{(m)}| - I(X \wedge Z|U)\right| \leqq \tau,$$

$$\left|\frac{1}{n}\log M - (H(X|UZ) - H(X|UY))\right| \leqq \tau,$$

and $\tilde{\mathsf{A}}^{(m)}$ is the codeword set of an (n, ε)-code for the DMC $\{W\}$. ○

Proof We shall apply the maximal code construction of Lemma 15.8 in two cycles. For all codes considered in this proof, the encoder will be the identity mapping on the codeword set. Accordingly, codes will be defined by the codeword set and the decoder mapping, rather than the encoder and the decoder. Write $\tau' \triangleq \tau/4$ and suppose without any loss of generality that $\varepsilon \leqq 1/2$. By Lemma 15.8, there exist (n, ε)-codes with common codeword set $\mathsf{A}^* \subset \mathsf{A}$ for the DMCs $\{V\}$ and $\{W\}$ such that

$$\left|\frac{1}{n}\log|\mathsf{A}^*| - I(X \wedge Z|U)\right| \leqq 2\tau' \tag{15.93}$$

and the corresponding decoders satisfy

$$\varphi^{-1}(\mathbf{x}) \subset \mathsf{T}^n_{[V]}(\mathbf{x}), \quad \psi^{-1}(\mathbf{x}) \subset \mathsf{T}^n_{[W]}(\mathbf{x}) \quad \text{for every} \quad \mathbf{x} \in \mathsf{A}^*, \tag{15.94}$$

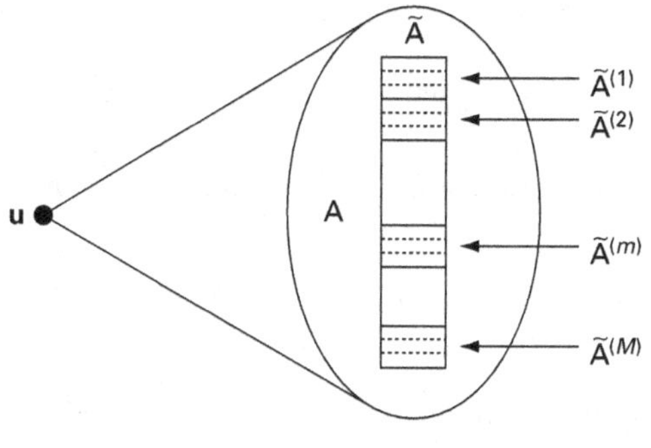

Figure 15.3

if n is large enough. For any such code write

$$\mathsf{B}^* \triangleq \bigcup_{\mathbf{x}\in\mathsf{A}^*} \varphi^{-1}(\mathbf{x}), \quad \mathsf{C}^* \triangleq \bigcup_{\mathbf{x}\in\mathsf{A}^*} \psi^{-1}(\mathbf{x}).$$

Let us consider a family of such pairs of codes, with disjoint codeword sets $\mathsf{A}^{(m)}$, $m = 1, \ldots, \hat{M}-1$, such that the corresponding sets B^*, denoted by $\mathsf{B}^{(m)}$, $m = 1, \ldots, \hat{M}-1$ are also disjoint. We claim that for large enough n such a family exists with

$$\frac{1}{n}\log \hat{M} \geqq H(X|UZ) - H(X|UY) - 4\tau'. \tag{15.95}$$

In fact, consider a maximal family of code pairs as above. This means that if a pair of (n, ε)-codes with common codeword set $\mathsf{A}^* \subset \mathsf{A}$ satisfies (15.94), with A^* disjoint from $\bigcup_{m<\hat{M}} \mathsf{A}^{(m)}$ and B^* disjoint from $\bigcup_{m<\hat{M}} \mathsf{B}^{(m)}$, then A^* cannot meet (15.93).

Let the (possibly void) set $\mathsf{A}^{(\hat{M})}$ be such an A^* which is not properly contained in any set with the same properties; let $\mathsf{B}^{(\hat{M})}$ and $\mathsf{C}^{(\hat{M})}$ be the corresponding sets B^* resp. C^*. Then

$$\frac{1}{n}\log|\mathsf{A}^{(\hat{M})}| < I(X \wedge Z|U) - 2\tau'. \tag{15.96}$$

Further, setting

$$\mathsf{B} \triangleq \bigcup_{m=1}^{\hat{M}} \mathsf{B}^{(m)},$$

it follows that

$$V^n(\mathsf{T}_{[V]}(\mathbf{x}) - \mathsf{B}|\mathbf{x}) < 1-\varepsilon \quad \text{or} \quad W^n(\mathsf{T}_{[W]}(\mathbf{x}) - \mathsf{C}^{\hat{M}}|\mathbf{x}) < 1-\varepsilon$$

for every $\mathbf{x} \in \mathsf{A} - \bigcup_{m=1}^{\hat{M}} \mathsf{A}^{(m)}$. Hence for $n \geqq n_1$

$$V^n(\mathsf{B}|\mathbf{x}) \geqq \varepsilon - \tau' \quad \text{or} \quad W^n(\mathsf{C}^{(\hat{M})}|\mathbf{x}) \geqq \varepsilon - \tau' \tag{15.97}$$

for every $\mathbf{x} \in \mathsf{A} - \bigcup_{m=1}^{\hat{M}} \mathsf{A}^{(m)}$. As for $\mathbf{x} \in \bigcup_{m=1}^{\hat{M}} \mathsf{A}^{(m)}$ we have $V^n(\mathsf{B}|\mathbf{x}) \geqq 1-\varepsilon$ by definition, $\varepsilon \leqq 1/2$ yields that (15.97) is valid for every $\mathbf{x} \in \mathsf{A}$. Denoting by A_1 resp. A_2 the set of those $\mathbf{x} \in \mathsf{A}$ for which the first resp. second inequality of (15.97) holds, the assumption $P^n_{X|U}(\mathsf{A}|\mathbf{u}) \geqq \eta$ gives

$$P^n_{X|U}(\mathsf{A}_1|\mathbf{u}) \geqq \frac{\eta}{2} \quad \text{or} \quad P^n_{X|U}(\mathsf{A}_2|\mathbf{u}) \geqq \frac{\eta}{2}.$$

As in the proof of Lemma 15.8, it follows for $n \geqq n_2$ that

$$\frac{1}{n}\log\left|\bigcup_{m=1}^{\hat{M}} \mathsf{A}^{(m)}\right| \geqq I(X \wedge Y|U) - 2\tau \quad \text{or} \quad \frac{1}{n}\log|\mathsf{A}^{(\hat{M})}| \geqq I(X \wedge Z|U) - 2\tau'.$$

But the second possibility is ruled out by (15.96). Thus, on account of (15.93) and (15.96), we obtain (15.95). The disjointness of the sets $\mathsf{B}^{(m)}$ means exactly that the union of the sets $\mathsf{A}^{(m)}$ is the codeword set of an (n, ε)-code for the DMC $\{V\}$. Thus the proof is complete except for the condition that all codewords have the same type.

This gap is filled easily. By the type counting lemma, for every $m < \hat{M}$ there is a type P_m such that

$$\left|\frac{1}{n}\log|\mathsf{A}^{(m)} \cap \mathsf{T}_{P_m}| - I(X \wedge Z|U)\right| \leqq 3\tau'$$

if $n \geqq n_3$. Denote $\tilde{\mathsf{A}}^{(m)} \triangleq \mathsf{A}^{(m)} \cap \mathsf{T}_{P_m}$. Similarly, denote for every type P by $\mathsf{M}(P)$ the set of those indices m for which $P_m = P$. Then, by the type counting lemma and the foregoing there exists a type $\tilde{P}$ such that

$$\frac{1}{n}\log|\mathsf{M}(\tilde{P})| \geqq H(X|UZ) - H(X|UY) - \tau'$$

if $n \geqq n_4$. Restricting the original codes to the codeword sets $\tilde{\mathsf{A}}^{(m)}$, $m \in \mathsf{M}(\tilde{P})$, the preceding inequality completes the proof. □

The previous construction is the basis of our final lemma. This says that if $(b, c) \in \mathcal{H}^*(Y; Z|X)$ is such that $a^-(b, c) = \max[b - H(Y|X), c - H(Z|X)] = b - H(Y|X)$, then for n large enough every set $\mathsf{E} \subset \mathsf{X}^n$ with

$$P_X^n(\mathsf{E}) \geqq \eta$$

has a subset $\tilde{\mathsf{A}}$ with the following properties: $\tilde{\mathsf{A}}$ is the disjoint union of sets $\mathsf{A}^{(l)}$ such that, whereas $|\tilde{\mathsf{A}}|$ is nearly $\exp[na^+(b, c)]$ and $|\mathsf{A}^{(l)}|$ is nearly $\exp[na^-(b, c)]$, the image sizes $g_{V^n}(\tilde{\mathsf{A}}, 1 - \varepsilon)$ and $g_{V^n}(A^{(l)}, 1 - \varepsilon)$ are nearly the same and about $\exp(nb)$, and similarly for the DMC $\{W\}$ these quantities are each around $\exp(nc)$. Further, for every l, $\mathsf{A}^{(l)}$ is the codeword set of an (n, ε)-code for the DMC $\{V\}$ and it is code-stuffed for the DMC $\{W\}$.

More specifically, we have the following.

LEMMA 15.18 (*Code-stuffing*) For every $\tau > 0$, $\eta > 0$, $0 < \varepsilon < 1$ and every $n \geqq n_0(|\mathsf{X}|, |\mathsf{Y}|, |\mathsf{Z}|, \tau, \eta, \varepsilon)$, for every quadruple of RVs (U, X, Y, Z) satisfying (15.50) and $I(X \wedge Y|U) \geqq I(X \wedge Z|U)$, and for every number t with

$$0 \leqq t \leqq \min[I(U \wedge Y), I(U \wedge Z)],$$

each set $\mathsf{E} \subset \mathsf{X}^n$ with $P_X^n(\mathsf{E}) \geqq \eta$ has a subset $\tilde{\mathsf{A}} \subset \mathsf{E} \cap \mathsf{T}_{[X]}^n$ such that

(i) $\tilde{\mathsf{A}}$ is the disjoint union of sets A^l, $l = 1, \ldots, L$,

$$\left|\frac{1}{n}\log L - H(X|UY)\right| \leqq \tau,$$

all sequences in A^l have the same type (for every fixed l) and A^l is the codeword set of an (n, ε)-code for the DMC $\{V\}$;

(ii) for every l, A^l is the disjoint union of sets $\mathsf{A}^{(l,m)}$, $m = 1, \ldots, M$,

$$\left|\frac{1}{n}\log M - (H(X|UZ) - H(X|UY))\right| \leqq \tau,$$

such that $\mathsf{A}^{(l,m)}$ is the codeword set of an (n, ε)-code for the DMC $\{W\}$, satisfying

$$\left|\frac{1}{n}\log g_{W^n}(\mathsf{A}^{(l,m)}, 1-\varepsilon) - (H(Z|U)+t)\right| \leqq \tau;$$

(iii) for every l,

$$\left|\frac{1}{n}\log |\mathsf{A}^{(l)}| - (I(X\wedge Y|U)+t)\right| \leqq \tau,$$
$$\left|\frac{1}{n}\log g_{V^n}(\mathsf{A}^{(l)}, 1-\varepsilon) - (H(Y|U)+t)\right| \leqq \tau,$$
$$\left|\frac{1}{n}\log g_{W^n}(\mathsf{A}^{(l)}, 1-\varepsilon) - (H(Z|U)+t)\right| \leqq \tau;$$

(iv)

$$\left|\frac{1}{n}\log |\tilde{\mathsf{A}}| - (H(X|U)+t)\right| \leqq \tau,$$
$$\left|\frac{1}{n}\log g_{V^n}(\tilde{\mathsf{A}}, 1-\varepsilon) - (H(Y|U)+t)\right| \leqq \tau,$$
$$\left|\frac{1}{n}\log g_{W^n}(\tilde{\mathsf{A}}, 1-\varepsilon) - (H(Z|U)+t)\right| \leqq \tau. \qquad \bigcirc$$

Proof Fix some ε', τ' to be specified later. By the proof of Theorem 15.9, for $n \geqq n_1(|\mathsf{X}|, |\mathsf{Y}|, |\mathsf{Z}|, \tau', \eta, \varepsilon, \varepsilon')$ there exist (n, ε')-codes for the DMCs $\{\hat{V}\}$ and $\{\hat{W}\}$ connecting the RVs U and Y resp. U and Z, with common codeword set $\{\mathbf{u}_k\}_{k=1}^{\hat{K}} \subset \mathsf{T}^n_{[U]}$, where

$$\left|\frac{1}{n}\log \hat{K} - t\right| \leqq \tau', \tag{15.98}$$

such that the sets $\hat{\mathsf{A}}_k \triangleq \mathsf{E} \cap \mathsf{T}^n_{[X|U]}(\mathbf{u}_k)$ satisfy

$$P^n_{X|U}(\hat{\mathsf{A}}_k|\mathbf{u}_k) \geqq \frac{\eta}{2} \tag{15.99}$$

and the set $\mathsf{A} \triangleq \bigcup_{k=1}^{\hat{K}} \hat{\mathsf{A}}_k \subset \mathsf{E} \cap \mathsf{T}^n_{[X]}$ satisfies

$$\begin{aligned} &\frac{1}{n}\log g_{V^n}(\mathsf{A}, 1-\varepsilon) \leqq H(Y|U)+t+\tau', \\ &\frac{1}{n}\log g_{W^n}(\mathsf{A}, 1-\varepsilon) \leqq H(Z|U)+t+\tau'. \end{aligned} \tag{15.100}$$

For the decoders $\varphi : \mathsf{Y}^n \to \{1, \ldots, \hat{K}\}$, $\psi : \mathsf{Z}^n \to \{1, \ldots, \hat{K}\}$ of the two (n, ε')-codes we have

$$\begin{aligned} \Pr\{\varphi(Y^n) = k | U^n = \mathbf{u}_k\} &= \sum_{\mathbf{x}\in\mathsf{X}^n} P^n_{X|U}(\mathbf{x}|\mathbf{u}_k) V^n(\varphi^{-1}(k)|\mathbf{x}) \geqq 1-\varepsilon', \\ \Pr\{\psi(Z^n) = k | U^n = \mathbf{u}_k\} &= \sum_{\mathbf{x}\in\mathsf{X}^n} P^n_{X|U}(\mathbf{x}|\mathbf{u}_k) W^n(\psi^{-1}(k)|\mathbf{x}) \geqq 1-\varepsilon.' \end{aligned} \tag{15.101}$$

Let A_k be the largest subset of $\hat{\mathsf{A}}_k$ such that

$$V^n(\varphi^{-1}(k)|\mathbf{x}) \geqq 1 - \frac{\varepsilon}{2}, \quad W^n(\psi^{-1}(k)|\mathbf{x}) \geqq 1 - \frac{\varepsilon}{2} \quad \text{for every} \quad \mathbf{x} \in \mathsf{A}_k. \tag{15.102}$$

It follows from (15.101) and (15.99) that

$$P^n_{X|U}(\mathsf{A}_k|\mathbf{u}_k) \geqq \frac{\eta}{4}, \tag{15.103}$$

provided that ε' (depending on ε and η) is sufficiently small. Further, by (15.102), the sets A_k are disjoint.

Now fix some k. On account of (15.103), Lemma 15.17 can be applied iteratively to the set A_k. It follows that for $n \geqq n_2(|\mathsf{X}|, |\mathsf{Y}|, |\mathsf{Z}|, \tau', \eta, \varepsilon)$ there exist disjoint subsets $\tilde{\mathsf{A}}^{(l,k)}$, $l = 1, \ldots, L$, of A_k with the following properties: $\tilde{\mathsf{A}}^{(l,k)}$ consists of sequences of the same type, it is the codeword set of an $(n, \varepsilon/2)$-code for the DMC $\{V\}$ and it is the disjoint union of sets $\tilde{\mathsf{A}}^{(l,k,m)}$, $m = 1, \ldots, M$, with

$$\begin{gathered} \left|\frac{1}{n}\log|\tilde{\mathsf{A}}^{(l,m,k)}| - I(X \wedge Z|U)\right| \leqq \tau', \\ \left|\frac{1}{n}\log M - (H(X|ZU) - H(X|UY))\right| \leqq \tau', \end{gathered} \tag{15.104}$$

every $\tilde{\mathsf{A}}^{(l,m,k)}$ being the codeword set of an $(n, \varepsilon/2)$-code for the DMC $\{W\}$. We may suppose that

$$\left|\frac{1}{n}\log L - H(X|UY)\right| \leqq 3\tau', \tag{15.105}$$

which is a consequence of (15.104) and Lemmas 6.13 and 6.14 if the $P^n_{X|U}(\cdot|\mathbf{u}_k)$-probability of the union of the sets $\tilde{\mathsf{A}}^{(l,k)}$, $l = 1, ..., L$, already exceeds $\eta/8$.

Note that since $\tilde{\mathsf{A}}^{(l,k)} \subset \mathsf{A}_k$ is the codeword set of an $(n, \varepsilon/2)$-code for the DMC $\{V\}$ for every k, (15.102) implies that

$$\hat{\mathsf{A}}^{(l)} \triangleq \bigcup_{k=1}^{\hat{K}} \tilde{\mathsf{A}}^{(l,k)}$$

is the codeword set of an (n, ε)-code for the DMC $\{V\}$. Similarly, the set

$$\hat{\mathsf{A}}^{(l,m)} \triangleq \bigcup_{k=1}^{\hat{K}} \tilde{\mathsf{A}}^{(l,m,k)}$$

is the codeword set of an (n, ε)-code for the DMC $\{W\}$. Since every fixed $\tilde{\mathsf{A}}^{(l,k)}$ consists of sequences of the same type, by the type counting lemma for $n \geqq n_3(|\mathsf{X}|, \tau')$ there exists a set of indices $\mathsf{K}_l \subset \{1, \ldots, \hat{K}\}$ such that

$$\left|\frac{1}{n}\log|\mathsf{K}_l| - \frac{1}{n}\log\hat{K}\right| < \tau' \tag{15.106}$$

and all sequences in $\bigcup_{k\in\mathsf{K}_l} \tilde{\mathsf{A}}^{(l,k)}$ have the same type.

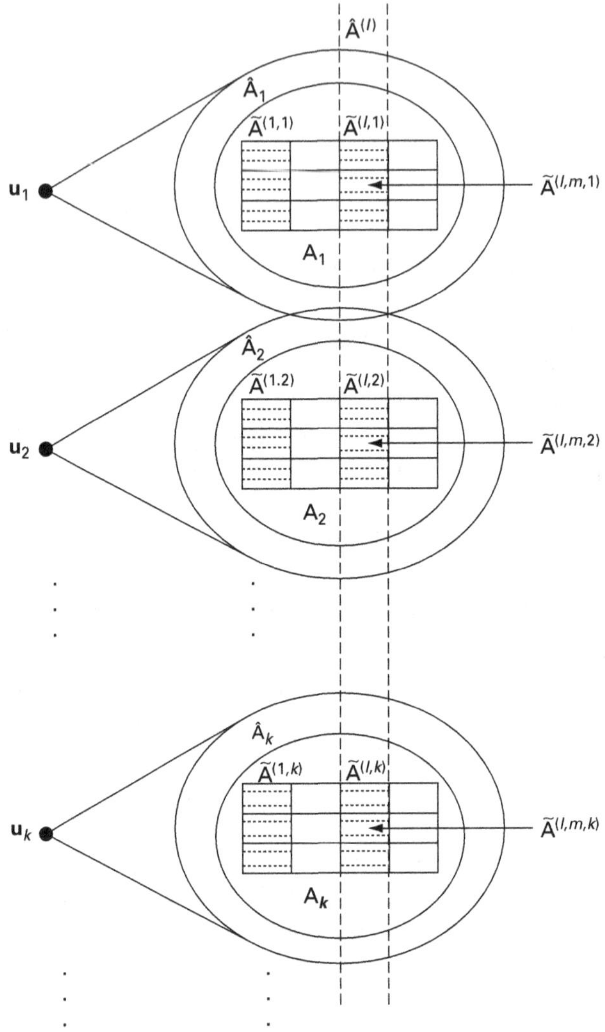

Figure 15.4 Construction underlying the code-stuffing lemma.

We claim that the sets

$$\mathsf{A}^{(l,m)} \triangleq \bigcup_{k \in \mathsf{K}_l} \tilde{\mathsf{A}}^{(l,m,k)}, \quad \mathsf{A}^{(l)} \triangleq \bigcup_{k \in \mathsf{K}_l} \tilde{\mathsf{A}}^{(l,k)}, \quad \mathsf{A} \triangleq \bigcup_{l=1}^{L} \mathsf{A}^{(l)}$$

satisfy the assertions of the lemma. To see this, it only remains to establish the inequalities of (ii), (iii) and (iv). It follows from (15.104), (15.98) and (15.106) that for every $1 \leqq l \leqq L$, $1 \leqq m \leqq M$ and $n \geqq n_4(|\mathsf{X}|, |\mathsf{Y}|, |\mathsf{Z}|, \tau', \eta, \varepsilon)$

$$\left| \frac{1}{n} \log |\mathsf{A}^{(l,m)}| - (I(X \wedge Z|U) + t) \right| \leqq 2\tau'. \tag{15.107}$$

This implies by (15.104) that

$$\left|\frac{1}{n}\log|\mathsf{A}^{(l)}| - (I(X \wedge Y|U) + t)\right| \leqq 3\tau'. \tag{15.108}$$

Further, by construction, $\mathsf{A}^{(l)} \subset \tilde{\mathsf{A}} \subset \mathsf{A}$ for every l $(1 \leqq l \leqq L)$ and hence, by (15.100),

$$\frac{1}{n}\log g_{V^n}(\mathsf{A}^{(l)}, 1-\varepsilon) \leqq \frac{1}{n}\log g_{V^n}(\tilde{\mathsf{A}}, 1-\varepsilon) \leqq H(Y|U) + t + \tau'. \tag{15.109}$$

Similarly, for every l and m

$$\begin{aligned}\frac{1}{n}\log g_{W^n}(\mathsf{A}^{(l,m)}, 1-\varepsilon) &\leqq \frac{1}{n}\log g_{W^n}(\mathsf{A}^{(l)}, 1-\varepsilon) \leqq \frac{1}{n}\log g_{W^n}(\tilde{\mathsf{A}}, 1-\varepsilon)\\ &\leqq H(Z|U) + t + \tau'.\end{aligned} \tag{15.110}$$

On the other hand, since the set $\mathsf{A}^{(l)}$ is the codeword set of an (n, ε)-code for the DMC $\{V\}$, by Lemma 6.8 we obtain for $n \geqq n_5(|\mathsf{X}|, |\mathsf{Y}|, |\mathsf{Z}|, \tau', \eta, \varepsilon)$ the relation

$$\frac{1}{n}\log g_{V^n}(\mathsf{A}^{(l)}, 1-\varepsilon) \geqq \frac{1}{n}\log|\mathsf{A}^{(l)}| + H(Y|X) - \tau'.$$

Thus by (15.108)

$$\frac{1}{n}\log g_{V^n}(\tilde{\mathsf{A}}, 1-\varepsilon) \geqq \frac{1}{n}\log g_{V^n}(\mathsf{A}^{(l)}, 1-\varepsilon) \geqq H(Y|U) + t - 4\tau'. \tag{15.111}$$

By the same argument, we obtain for $n \geqq n_6(|\mathsf{X}|, |\mathsf{Y}|, |\mathsf{Z}|, \tau', \eta, \varepsilon)$

$$\frac{1}{n}\log g_{W^n}(\mathsf{A}^{(l,m)}, 1-\varepsilon) \geqq \frac{1}{n}\log|\mathsf{A}^{(l,m)}| + H(Z|X) - \tau',$$

whence (15.107) gives

$$\begin{aligned}\frac{1}{n}\log g_{W^n}(\tilde{\mathsf{A}}, 1-\varepsilon) &\geqq \frac{1}{n}\log g_{W^n}(\mathsf{A}^{(l)}, 1-\varepsilon)\\ &\geqq \frac{1}{n}\log g_{W^n}(\mathsf{A}^{(l,m)}, 1-\varepsilon) \geqq H(Z|U) + t - 3\tau'.\end{aligned} \tag{15.112}$$

Choosing $\tau' \triangleq \tau/4$, the inequalities (15.104) and (15.108) – (15.112) complete the proof of the lemma. □

Now we are ready to prove Theorem 15.19.

THEOREM 15.19

$$\mathcal{H}^*(X; Y; Z|X) \subset \mathcal{F}(X; Y; Z|X).$$

More precisely, for every $\delta > 0$ and every point $(a, b, c) \in \mathcal{H}^*(X; Y; Z|X)$ there exists a function f with domain X^n such that

$$\left|\frac{1}{n}H(X^n|f(X^n)) - a\right| \leqq \delta, \tag{15.113}$$

$$\left|\frac{1}{n}H(Y^n|f(X^n)) - b\right| \leqq \delta, \quad \left|\frac{1}{n}H(Z^n|f(X^n)) - c\right| \leqq \delta, \tag{15.114}$$

whenever $n \geqq n_0(|\mathsf{X}|, |\mathsf{Y}|, |\mathsf{Z}|, \delta)$. ○

Proof By definition, see (15.31), to every $(a, b, c) \in \mathcal{H}^*(X; Y; Z|X)$ there exists a RV U with $|\mathsf{U}| \leqq |\mathsf{X}| + 2$ satisfying the Markov condition $U \multimap X \multimap YZ$ and a number t with

$$0 \leqq t \leqq \min[I(U \wedge Y), I(U \wedge Z)]$$

such that

$$b = H(Y|U) + t, \quad c = H(Z|U) + t, \tag{15.115}$$

$$\max[I(X \wedge Y|U), I(X \wedge Z|U)] + t \leqq a \leqq H(X|U) + t. \tag{15.116}$$

First we construct functions f^- and f that satisfy (15.113) and (15.114) for b, c as in (15.115), and a in (15.116) with the equality in the first resp. second inequality.

Suppose, without any loss of generality, that

$$I(X \wedge Y|U) \geqq I(X \wedge Z|U). \tag{15.117}$$

Fix some $\eta > 0$, $\tau > 0$, $0 < \varepsilon < 1$ to be specified later. By an iterative application of Lemma 15.18, for $n \geqq n_1(|\mathsf{X}|, |\mathsf{Y}|, |\mathsf{Z}|, \tau, \eta, \varepsilon)$ there exist disjoint sets $\tilde{\mathsf{A}}_j \subset \mathsf{T}^n_{[X]}$, $j = 1, \ldots, J$, such that

$$P^n_X\left(\bigcup_{j=1}^{J} \tilde{\mathsf{A}}_j\right) \geqq 1 - \eta \tag{15.118}$$

and each $\tilde{\mathsf{A}}_j$ has the properties of the set $\tilde{\mathsf{A}}$ of the said lemma. In particular, each $\tilde{\mathsf{A}}_j$ is the disjoint union of sets $\mathsf{A}_j^{(l)}$, which are the codeword sets of (n, ε)-codes for the DMC $\{V\}$ and each $\mathsf{A}_j^{(l)}$ is the disjoint union of sets $\mathsf{A}_j^{(l,m)}$, which are the codeword sets of (n, ε)-codes for the DMC $\{W\}$. Define

$$f^-(\mathbf{x}) \triangleq \begin{cases} (j, l) & \text{if } \mathbf{x} \in \mathsf{A}_j^{(l)} \text{ for some } j \text{ and } l \\ 0 & \text{otherwise;} \end{cases}$$

$$f(\mathbf{x}) \triangleq \begin{cases} j & \text{if } \mathbf{x} \in \tilde{\mathsf{A}}_j \text{ for some } j \\ 0 & \text{otherwise.} \end{cases}$$

Note that $f^-(\mathbf{x}) = (j, l)$ implies $f(\mathbf{x}) = j$. On account of (15.118) we have

$$\begin{aligned} \frac{1}{n} H(Y^n|f^-(X^n)) &\leqq \frac{1}{n} H(Y^n|f(X^n)) \\ &\leqq \frac{1}{n} \sum_{j=1}^{J} \Pr\left\{f(X^n) = j\right\} H(Y^n|f(X^n) = j) + \eta \log |\mathsf{Y}|. \end{aligned} \tag{15.119}$$

Here, by Lemma 15.2, for $n \geqq n_2(|\mathsf{X}|, |\mathsf{Y}|, |\mathsf{Z}|, \tau, \eta, \varepsilon)$,

$$H(Y^n|f(X^n) = j) = H(Y^n|X^n \in \tilde{\mathsf{A}}_j) \leqq \log g_{V^n}(\tilde{\mathsf{A}}_j, 1 - \varepsilon) + n\tau.$$

Now, by (iv) of Lemma 15.18, (15.119) and the preceding inequality result in

$$\frac{1}{n}H(Y^n|f^-(X^n)) \leqq \frac{1}{n}H(Y^n|f(X^n)) \leqq \eta \log|\mathsf{Y}| + H(Y|U) + t + 2\tau. \quad (15.120)$$

Similarly, we obtain

$$\begin{aligned}\frac{1}{n}H(Z^n|f^-(X^n)) \leqq \frac{1}{n}H(Z^n|f(X^n)) &\leqq \eta \log|\mathsf{Z}| + H(Z|U) + t + 2\tau.\\ \frac{1}{n}H(X^n|f(X^n)) &\leqq \eta \log|\mathsf{X}| + H(X|U) + t + 2\tau.\end{aligned} \quad (15.121)$$

Further, by the definition of f^-, (15.118) and (iii) of Lemma 15.18, we obtain

$$\frac{1}{n}H(X^n|f^-(X^n)) \leqq \eta \log|\mathsf{X}| + I(X \wedge Y|U) + t + \tau. \quad (15.122)$$

In order to prove the inequalities in the opposite direction, write

$$H(Y^n|f^-(X^n)) \geqq \sum_{j=1}^{J}\sum_{l=1}^{L} \Pr\left\{X^n \in \mathsf{A}_j^{(l)}\right\} H(Y^n|X^n \in \mathsf{A}_j^{(l)}). \quad (15.123)$$

Since $\mathsf{A}_j^{(l)}$ contains sequences of the same type, the conditional distribution of X^n, given that $X^n \in \mathsf{A}_j^{(l)}$, is uniform over $\mathsf{A}_j^{(l)}$ for every j, l. As $\mathsf{A}_j^{(l)}$ is the codeword set of an (n, ε)-code for the DMC $\{V\}$, it follows by Lemma 15.2 that for $n \geqq n_3(|\mathsf{X}|, |\mathsf{Y}|, |\mathsf{Z}|, \tau, \eta, \varepsilon)$ and every j, l

$$\frac{1}{n}H(Y^n|X^n \in \mathsf{A}_j^{(l)}) \geqq \frac{1}{n}\log g_{V^n}(\mathsf{A}_j^{(l)}, 1-\varepsilon) - \tau - \varepsilon \log|\mathsf{Y}|.$$

Hence by (iii) of Lemma 15.18 we obtain, using (15.123) and (15.118), that

$$\begin{aligned}\frac{1}{n}H(Y^n|f(X^n)) &\geqq \frac{1}{n}H(Y^n|f^-(X^n))\\ &\geqq (1-\eta)\left(\min_{j,l}\frac{1}{n}\log g_{V^n}(\mathsf{A}_j^{(l)}, 1-\varepsilon) - \tau - \varepsilon\log|\mathsf{Y}|\right)\\ &\geqq (1-\eta)(H(Y|U) + t - 2\tau - \varepsilon\log|\mathsf{Y}|). \end{aligned} \quad (15.124)$$

Similarly,

$$H(Z^n|f^-(X^n)) \geqq \sum_{j=1}^{J}\sum_{l=1}^{L}\sum_{m=1}^{M} \Pr\left\{X^n \in \mathsf{A}_j^{(l,m)}\right\} H\left(Z^n|X^n \in \mathsf{A}_j^{(l,m)}\right). \quad (15.125)$$

As $\mathsf{A}_j^{(l,m)}$ is the codeword set of an (n, ε)-code for the DMC $\{W\}$, by the same argument as before we have

$$\frac{1}{n}H\left(Z^n|X^n \in \mathsf{A}_j^{(l,m)}\right) \geqq \frac{1}{n}\log g_{W^n}\left(\mathsf{A}_j^{(l,m)}, 1-\varepsilon\right) - \tau - \varepsilon\log|\mathsf{Z}|,$$

provided that $n \geqq n_4(|\mathsf{X}|, |\mathsf{Y}|, |\mathsf{Z}|, \tau, \eta, \varepsilon)$. Thus (15.125), (15.118) and (ii) of Lemma 15.18 result in

$$\begin{aligned}\frac{1}{n}H(Z^n|f(X^n)) \geqq \frac{1}{n}H(Z^n|f^-(X^n))&\\ &\geqq (1-\eta)\,(\min_{j,l,m}\frac{1}{n}\log g_{W^n}(\mathsf{A}_j^{(l,m)}, 1-\varepsilon) - \tau - \varepsilon\log|\mathsf{Z}|)\\ &\geqq (1-\eta)(H(Z|U) + t - 2\tau - \varepsilon\log|\mathsf{Z}|).\end{aligned} \tag{15.126}$$

Finally,

$$\frac{1}{n}H(X^n|f^-(X^n)) = H(X) - \frac{1}{n}H(f^-(X^n)) \geqq H(X) - \frac{1}{n}\log\|f^-\|$$

and

$$\frac{1}{n}H(X^n|f(X^n)) = H(X) - \frac{1}{n}H(f(X^n)) \geqq H(X) - \frac{1}{n}\log\|f\|.$$

Here by (iii) and (iv) of Lemma 15.18 and by (15.118), for $n \geqq n_5$

$$\frac{1}{n}\log\|f^-\| \leqq H(X) - \min_{j,l}\frac{1}{n}\log|\mathsf{A}_j^{(l)}| + \tau \leqq H(X) - (I(X\wedge Y|U) + t) + 2\tau$$

and

$$\frac{1}{n}\log\|f\| \leqq H(X) - \min_{j}\frac{1}{n}\log|\tilde{\mathsf{A}}_j| + \tau \leqq H(X) - (H(X|U) + t) + 2\tau.$$

Thus

$$\begin{aligned}\frac{1}{n}H(X^n|f^-(X^n)) &\geqq I(X\wedge Y|U) + t - 2\tau,\\ \frac{1}{n}H(X^n|f(X^n)) &\geqq H(X|U) + t - 2\tau.\end{aligned} \tag{15.127}$$

As τ, η and ε may be arbitrarily small, comparing (15.120), (15.121), (15.122), (15.124), (15.126) and (15.127) proves our statements for those (a, b, c) that satisfy (15.115) and (15.116) with either equality in (15.116).

For arbitrary points (a, b, c) satisfying (15.115) and (15.116) one can proceed similarly, partitioning the index set L into subsets L_s of nearly equal size and considering a function $\hat{f}(\mathbf{x})$ which takes the value (j, s) if $\mathbf{x} \in \mathsf{A}_j^{(l)}$ for some $l \in \mathsf{L}_s$. □

Combining Corollary 15.7 and Theorem 15.19, we have proved Theorem 15.20.

THEOREM 15.20 (*Entropy characterization*)

$$\mathcal{F}(X; Y; Z|X) = \mathcal{H}(X; Y; Z|X) = \mathcal{H}^*(X; Y; Z|X).$$ □

Discussion

In this chapter we have solved the entropy characterization problem and the image size characterization problem in the special case of 3-sources resp. two DMCs. It was shown in Chapter 13 that a computable characterization of the achievable rate region of a source network with one helper can be obtained by solving an entropy characterization problem. We shall see in Chapter 16 that the corresponding image size characterizations can be used to prove strong converse results for source networks and also to solve channel network problems. In this respect, it is important that the sets of achievable entropy resp. exponent triples have the same two-dimensional projections see Theorems 15.11 and 15.20. The two sets, however, need not be equal; their relationship is described by Corollary 15.13. → 15.8

It seems to us that Theorems 15.16 and 15.20 represent the mathematical basis of the now available proof techniques for multi-terminal source and channel networks. Of course, the whole machinery is not required in the proof of particular coding theorems. Also, particular results can be obtained by different and sometimes simpler methods, as will be seen in the following chapter.

One might think that the main theorems easily generalize, and one can characterize in a similar way the achievable entropy vectors for arbitrary DMMSs resp. the exponential behavior of the η-images of subsets of $\mathsf{T}^n{}_{[X]}$ for several DMCs, as n tends to infinity. This, however, does not seem to be the case, and these general problems are still unsolved. For partial results, see Problems 15.16–15.21.

The proofs in this chapter rely on the single-letterization lemma, Lemma 15.7, as well as on the blowing up lemma (Chapter 5) and the maximal code lemma (Chapter 6). We have repeatedly exploited the facts that $(1/n)\log g_{V^n}(\mathsf{A}, \eta)$ does not essentially depend on η for large n (blowing up lemma) and that the smallest image of a set is only slightly greater than that of the largest codeword set of any (n, ε)-code it contains (a consequence of the maximal code lemma). We have essentially used a result from the theory of convexity, namely the support lemma, Lemma 15.4. It was by this lemma that we could bound the range of auxiliary RVs. Such bounds are necessary for the characterizations in this chapter to be computable. Note that achievability results for the entropy characterization problem could also be obtained without reliance on the maximal code lemma, via the method of random selection. ○

Problems

15.1. Show that in the definition of $\mathcal{H}^*(Y; Z|X)$, see (15.49), the range constraint can be improved to $|\mathsf{U}| \leqq |\mathsf{X}| + 1$.
Hint Apply the support lemma directly.

15.2. (*Degraded channels*) As in Problem 6.16, a channel $W : \mathsf{X} \to \mathsf{Z}$ is called a degraded version of a channel $V : \mathsf{X} \to \mathsf{Y}$ if there exists a channel $D : \mathsf{Y} \to \mathsf{Z}$ such that

$$W(z|x) = \sum_{y \in \mathsf{Y}} V(y|x) D(z|y) \quad \text{for every } x \in \mathsf{X},\ z \in \mathsf{Z}.$$

For such a pair of channels, and RVs X, Y, Z with $P_{Z|X} = V$, $P_{Z|X} = W$, show that $(b, c) \in \mathcal{H}^*(Y; Z|X)$ iff

$$H(Z|X) \leqq c \leqq H(Z), \quad b^-(c) \leqq b \leqq b^+(c),$$

where

$$b^-(c) \triangleq c + H(Y|X) - H(Z|X)$$

and $b^+(c)$ is the maximum of $H(Y|U)$ under the constraint $H(Z|U) \leqq c$ for RVs U with

$$U \multimap X \multimap YZ, \quad |\mathsf{U}| \leqq |\mathsf{X}| + 1.$$

Hint The proof of Lemma 15.12 literally applies; for the range constraint see Problem 15.1.

15.3. (*Slope of the boundary of* $\mathcal{H}^*(X; Z|X)$.) By Lemma 15.12, see (15.68), the slope of the "upper boundary" $a^+(c)$ of $\mathcal{H}^*(X; Z|X)$ is everywhere at least 1. Show that if the channel $W = P_{Z|X}$ has strictly positive entries then this slope is strictly greater than 1.

Hint It is enough to show that $\mathcal{H}^*(X; Z|X)$ has a supporting line of slope greater than 1 at its "upper-most" point

$$(a, c) \triangleq (H(X), H(Z))$$

(provided that the a-axis is drawn vertically, as in Fig. 15.2). Observe that this slope equals the infimum of $I(U \wedge X)/I(U \wedge Z)$ for RVs U such that $U \multimap X \multimap Z$, $|\mathsf{U}| \leqq |\mathsf{X}| + 2$, $I(U \wedge Z) > 0$. As

$$I(U \wedge X) = \sum_{u \in \mathsf{U}} P_U(u) D(P_{X|U=u} \| P_X),$$
$$I(U \wedge Z) = \sum_{u \in \mathsf{U}} P_U(u) D(P_{Z|U=u} \| P_Z),$$

this infimum is strictly greater than 1 by Problem 3.19. (For a stronger result see Problem 15.12.)

Image size of arbitrary sets (Problems 15.4–15.5)

15.4. (*Max-closure*) For an arbitrary set $\mathcal{E}$ of triples of real numbers, define its *max-closure*, denoted by $\max(\mathcal{E})$, as the set of those triples (a, b, c) which can be obtained from triples $(a_i, b_i, c_i) \in \mathcal{E}$, $i = 1, 2, 3$, setting

$$a \triangleq \max_{1 \leqq i \leqq 3} a_i, \quad b \triangleq \max_{1 \leqq i \leqq 3} b_i, \quad c \triangleq \max_{1 \leqq i \leqq 3} c_i.$$

(a) Prove that $\max(\mathcal{E})$ is a max-closed set, for every set $\mathcal{E}$ in the three-dimensional Euclidean space, i.e.,

$$\max(\max(\mathcal{E})) = \max(\mathcal{E}).$$

(b) Prove that $\mathcal{G}^*(X; Y; Z|X)$ is a max-closed set.

(c) Show that the union

$$\bigcup \mathcal{G}^*(X; Y; Z|X), \tag{*}$$

taken for all distributions P_{XYZ} with $P_{Y|X} = V$, $P_{Z|X} = W$, is not a max-closed set.

Hint Find channels V, W for which there exist distributions P and Q on X such that PV and QV are the uniform distribution on Y resp. Z, whereas P (say) is not uniform on X. Then note that

$$(\log|\mathsf{X}|, \log|\mathsf{Y}|, \log|\mathsf{Z}|)$$

is contained in the max-closure of the union (*), but not in the union itself.

15.5. (*Image size of arbitrary sets*) Given two DMCs with common input alphabet, $\{V : \mathsf{X} \to \mathsf{Y}\}$ and $\{W : \mathsf{X} \to \mathsf{Z}\}$, call the triple (a, b, c) an achievable exponent triple for the corresponding *unrestricted image size problem* if for every ε and δ in (0,1) and sufficiently large n there exists a set $\mathsf{A} \subset \mathsf{X}^n$ such that (ii) of Definition 15.1 holds, i.e.,

$$\left|\frac{1}{n}\log|\mathsf{A}| - a\right| \leqq \delta, \quad \left|\frac{1}{n}\log g_{V^n}(\mathsf{A}, 1-\varepsilon) - b\right| \leqq \delta,$$

$$\left|\frac{1}{n}\log g_{W^n}(\mathsf{A}, 1-\varepsilon) - c\right| \leqq \delta.$$

Denote the set of these triples by $\mathcal{G}(V, W)$. Show that this set is the max-closure (see Problem 15.4) of $\bigcup \mathcal{G}^*(X; Y; Z|X)$, where the union is taken as in Problem 15.4(c).

Hint The existence part is obvious. To prove the converse part, consider A as the disjoint union of sets of sequences of the same type and apply Theorem 15.14 to each of these.

15.6. (*Code-stuffed sets*) Generalize the bound (15.5) of Lemma 15.2 for sets A equal to the disjoint union of subsets A_i of $\mathsf{T}^n_{[X]}$ satisfying

$$\frac{1}{n}\log|\mathsf{A}_i| > C_V(\mathsf{A}, \varepsilon) - \delta,$$

each A_i being the codeword set of an (n, ε)-code for the DMC $\{V : \mathsf{X} \to \mathsf{Y}\}$, where $V = P_{Y|X}$. More exactly, show that if $\hat{X}^n$ is uniformly distributed on a set A as above and $\hat{Y}^n$ is connected with $\hat{X}^n$ by the channel V^n, then (15.5) holds, with δ replaced by 3δ, whenever $n \geqq n_0(|\mathsf{X}|, |\mathsf{Y}|, \varepsilon, \delta, \eta)$.

Hint By Lemma 6.8, the assumptions on A_i imply

$$\frac{1}{n}\log g_{V^n}(\mathsf{A}_i, \eta) > \frac{1}{n}\log g_{V^n}(\mathsf{A}, \eta) - 2\delta$$

for $n \geqq n_1(|\mathsf{X}|, |\mathsf{Y}|, \varepsilon, \delta, \eta)$. Thus, applying (15.5) to A_i, we obtain

$$\frac{1}{n}H(\hat{Y}^n|\hat{X}^n \in \mathsf{A}_i) + \delta + \varepsilon\log|\mathsf{Y}| \geqq \frac{1}{n}\log g_{V^n}(\mathsf{A}, \eta) - 2\delta.$$

15.7. Prove directly that for every $\delta > 0$, $n \geqq n_0(|\mathsf{X}|, |\mathsf{Y}|, |\mathsf{Z}|, \delta)$ and every point $(a, b, c) \in \mathcal{H}^*(X; Y; Z|X)$ there exists a function f with domain X^n such that

$$\frac{1}{n}H(X^n|f(X^n)) \leqq a + \delta, \quad \frac{1}{n}H(Y^n|f(X^n)) \leqq b + \delta,$$

$$\frac{1}{n}H(Z^n|f(X^n)) \leqq c + \delta$$

and, in addition, any prescribed inequality among the following ones holds:

$$\frac{1}{n}H(X^n|f(X^n)) \geqq a-\delta,\ \frac{1}{n}H(Y^n|f(X^n)) \geqq b-\delta,$$
$$\frac{1}{n}H(Z^n|f(X^n)) \geqq c-\delta.$$

Hint Use Theorem 15.15 and Lemma 15.2.

15.8. Give an example of a distribution P_{XYZ} for which

$$\mathcal{G}(X;Y;Z|X) \neq \mathcal{F}(X;Y;Z|X).$$

Hint We have to show that $\mathcal{G}^*(X;Y;Z|X) \neq \mathcal{H}^*(X;Y;Z|X)$.

(i) For every distribution P_{XYZ}, the upper boundary $a^{++}(b,c)$ of $\mathcal{G}^*$ satisfies

$$a^{++}(H(Y),c) = a^+(c) = \max_{H(Z|U)=c} H(X|U),$$

the maximum taken for RVs U satisfying $U \multimap X \multimap Z$ and $|\mathsf{U}| \leqq |\mathsf{X}|+2$.

(ii) If the channel $W = P_{Z|X}$ has strictly positive entries, then by Problem 15.3

$$a^+(c) > a^+(c-t)+t \quad \text{for} \quad t>0.$$

(iii) The upper boundary $a^+(b,c)$ of $\mathcal{H}^*$ satisfies

$$a^+(H(Y),c) = \max_{H(Z|U)+I(U\wedge Y)=c} [H(X|U)+I(U\wedge Y)],$$

where the maximum refers to RVs U such that $U \multimap X \multimap YZ$ and $|\mathsf{U}| \leqq |\mathsf{X}|+2$.

(iv) If $c < H(Z)$ then the minimum of $I(U\wedge Y)$ under the condition that the quadruple of RVs (U,X,Y,Z) achieves $a^{++}(H(Y),c)$ is strictly positive, provided that the matrix V is an invertible square matrix. Denote this minimum by t_0.

(v) Conclude from (iii) and (ii) that

$$a^+(H(Y),c) \leqq \max_{t\geqq t_0}(a^+(c-t)+t) < a^+(c).$$

By (i), this proves our claim, if P_{XYZ} satisfies the conditions that $V = P_{Y|X}$ has non-zero determinant and W has strictly positive entries.

15.9. (*Images over a BSC*)

(a) Show that if V is a binary symmetric channel with cross-over probability p then for RVs $\hat{X}^n$ and $\hat{Y}^n$ connected by the channel V^n (where the components $\hat{X}_i$ of $\hat{X}^n$ need not be independent)

$$\frac{1}{n}H(\hat{X}^n) \geqq h(\lambda) \quad \text{implies} \quad \frac{1}{n}H(\hat{Y}^n) \geqq h(\lambda(1-p)+(1-\lambda)p).$$

Hint It follows from Corollary 15.7 and Lemma 15.12 that

$$\min_{\frac{1}{n}H(\hat{X}^n)\geqq c} \frac{1}{n}H(\hat{Y}^n) \geqq \min_{H(X|U)\geqq c} H(Y|U),$$

where the right-hand minimum is taken for arbitrary RVs U, X, Y such that $U \multimap X \multimap Y$ and X and Y are connected by channel V. Observe that the function

$$g(c) \triangleq \min_{H(X|U)\geqq c} H(Y|U)$$

is the lower convex envelope of

$$f(c) \triangleq \min_{H(X)\geqq c} H(Y).$$

Hence it suffices to show that $f(c)$ is convex. Clearly,

$$f(c) = h(h^{-1}(c)(1-p) + (1-h^{-1}(c))p),$$

where h^{-1}: $[0, 1] \to \left[0, 1/2\right]$ is the inverse of the binary entropy function; the convexity is shown by taking the second derivative. (Wyner and Ziv (1973).)

(b) Prove that for sets A_n of n-length binary sequences with

$$\lim_{n\to\infty} \frac{1}{n} \log |\mathsf{A}_n| = c$$

it holds for every $0 < \varepsilon < 1$ that

$$\lim_{n\to\infty} \frac{1}{n} \log g_{V^n}(\mathsf{A}_n, 1-\varepsilon) \geqq f(c).$$

Hint Apply (a) and Lemma 15.2. (Ahlswede, Gács and Körner (1976).)

15.10. (*Images for binary channels*)

(a) Show that if V is a binary channel, i.e., $\mathsf{X} = \mathsf{Y} = \{0, 1\}$, then for any RVs $\hat{X}^n$ and $\hat{Y}^n$ connected by V^n

$$\frac{1}{n}H(\hat{X}^n) \geqq c \quad \text{implies} \quad \frac{1}{n}H(\hat{Y}^n) \geqq f(c),$$

where

$$f(c) \triangleq \min_{H(X)\geqq c} H(Y)$$

and the last minimum is taken for RVs X and Y connected by channel V. Prove the corresponding generalization of Problem 15.9(b).

Hint As in Problem 15.9(a) we have to show that $f(c)$ is a convex function. This follows if for

$$f_0(c) \triangleq \min_{H(X)=c} H(Y)$$

we prove that $f_0'(c) > 0$, $f_0''(c) > 0$. Show that if

$$V = \begin{pmatrix} \alpha & \bar{\alpha} \\ \beta & \bar{\beta} \end{pmatrix}$$

(where $\bar{\alpha} \triangleq 1 - \alpha$) then $f_0(c) = h(l(h^{-1}(c)))$, where the function $l(x)$ is defined by

$$l(x) \triangleq x(\alpha - \beta) + \bar{\alpha} \quad \left(0 \leqq x \leqq \frac{1}{2}\right).$$

Now, $f_0'(c) > 0$ follows by straightforward computation, whereas $f_0''(c) > 0$ amounts to proving for $0 \leqq x \leqq 1/2$ and $y \triangleq (\alpha - \beta)x + \bar{\alpha}$ that

$$\left(\frac{1-y}{y}\right)^{(1-y)y} \geqq \left(\frac{1-x}{x}\right)^{(\alpha-\beta)(1-x)x}. \qquad (*)$$

This clearly holds if $y \leqq x$. If $x \leqq y \leqq 1/2$, consider all the channels

$$W = \begin{pmatrix} \lambda & \bar{\lambda} \\ \mu & \bar{\mu} \end{pmatrix}$$

for which $y = (\lambda - \mu)x + \bar{\lambda}$. If (*) holds for the one with the largest $(\lambda - \mu)$ then it holds everywhere. However, $(\lambda - \mu)$ takes its maximum when W is a BSC, and Problem 15.9(a) applies.

(b) Show that for every X, Y with $\max(|\mathsf{X}|, |\mathsf{Y}|) > 2$ there exists a channel $V : \mathsf{X} \to \mathsf{Y}$ such that the corresponding function $f(c)$ is not convex.
Hint Consider the channels

$$\begin{pmatrix} \frac{3}{4} & \frac{1}{4} \\ \frac{3}{4} & \frac{1}{4} \\ 0 & 1 \end{pmatrix} \quad \text{and} \quad \begin{pmatrix} \frac{1}{2} & \frac{1}{2} & 0 \\ \frac{1}{8} & \frac{1}{8} & \frac{3}{4} \end{pmatrix}.$$

(Witsenhausen (1974); Ahlswede and Körner (1977).)

15.11. (*Comparison of channels*)

(a) Channel $V : \mathsf{X} \to \mathsf{Y}$ is called *less noisy* than channel $W : \mathsf{X} \to \mathsf{Z}$ if for every channel $F : \mathsf{U} \to \mathsf{X}$ (where U is arbitrary) $FV : \mathsf{U} \to \mathsf{Y}$ is more capable than FW (see Problem 6.18). Show that the description of $\mathcal{H}^*(Y; Z|X)$ given in Problem 15.2 is valid also in this case.

Hint See Problem 15.2.

(b) Show that V can be more capable than W without being less noisy.

Hint Let $\mathsf{X} \triangleq \{1, 2, 3\}$, $\mathsf{Y} = \mathsf{Z} \triangleq \{1, 2\}$ and consider the channels

$$V \triangleq \begin{pmatrix} 1 & 0 \\ 0 & 1 \\ \frac{1}{2} & \frac{1}{2} \end{pmatrix}, \qquad W \triangleq \begin{pmatrix} 1 & 0 \\ \frac{1}{2} & \frac{1}{2} \\ \frac{1}{2} & \frac{1}{2} \end{pmatrix}.$$

Put

$$P_X \triangleq \left(\frac{1}{4}, \frac{1}{4}, \frac{1}{2}\right), \qquad U = f(X) \triangleq |X - 2|^+.$$

Then

$$I(U \wedge Y) = 0 \quad \text{while} \quad I(U \wedge Z) > 0.$$

(c) Show that channel V is less noisy than channel W iff the latter is more divergence-contracting, i.e., for every pair of input distributions $P, \tilde{P}$

$$D(PV\|\tilde{P}V) \geqq D(PW\|\tilde{P}W).$$

(Körner and Marton (1977a); they attribute the counterexample of (b) to R. Ahlswede. For an operational characterization of the "less noisy" property, see Corollary 17.11.)

15.12.* (*Strong data-processing lemmas*) The joint distribution of the RVs X and Y is *indecomposable* if there are no functions f and g with respective domains X and Y so that

(i) $\Pr\{f(X) = g(Y)\} = 1$ and

(ii) $f(X)$ takes at least two values with non-zero probability.

(a) Show that the RVs (X, Y) have an indecomposable joint distribution iff the region $\mathcal{H}^*(X; Y|X)$ has a supporting line of slope strictly less than 1 at the point $(H(X), H(Y))$. (Unlike in Problem 15.3, the a-axis is now drawn horizontally. Thus the minimum slope of supporting lines at $(H(X), H(Y))$ is

$$\alpha_{XY} \triangleq \sup \frac{I(U \wedge Y)}{I(U \wedge X)},$$

where the supremum is taken for RVs U such that $U \multimap X \multimap Y$, $|\mathsf{U}| \leqq |\mathsf{X}| + 2$, $I(U \wedge X) > 0$.)

(b) For a channel $W : \mathsf{X} \to \mathsf{Y}$ define $\alpha_W \triangleq \max \alpha_{XY}$, where the maximum is taken over RVs X, Y connected by W. Prove that $\alpha_W < 1$ iff the zero-error capacity of the DMC $\{W\}$ is zero.

(c) Prove that, for every channel,

$$\alpha_W = \sup_{P_1 \neq P_2} \frac{D(P_1 W \| P_2 W)}{D(P_1 \| P_2)}.$$

(Joint results of Ahlswede, Gács and the authors, published in Ahlswede and Gács (1976).)

15.13. (*Quasi-images of a set*) Given a pair of RVs X, Y, a set $\mathsf{B} \subset \mathsf{Y}^n$ is said to be an *η-quasi-image* of a set $\mathsf{A} \subset \mathsf{X}^n$ over the channel V^n (where $V = P_{Y|X}$) if

$$\Pr\{Y^n \in \mathsf{B} | X^n \in \mathsf{A}\} > \eta.$$

Denote by $\bar{g}_{V^n}(\mathsf{A}, \eta)$ the minimum size of η-quasi-images of A over V^n. Show that the set of those triples (a, b, c) for which to every $\delta, \eta \in (0, 1)$ and sufficiently large n there exists a set $\mathsf{A} \subset \mathsf{T}^n_{[X]}$ such that

$$\left|\frac{1}{n}\log|\mathsf{A}| - a\right| \leqq \delta, \quad \left|\frac{1}{n}\log \bar{g}_{V^n}(\mathsf{A}, \eta) - b\right| \leqq \delta, \quad \left|\frac{1}{n}\log \bar{g}_{W^n}(\mathsf{A}, \eta) - c\right| \leqq \delta,$$

is equal to $\mathcal{H}(X; Y; Z|X)$.

15.14.* (*Mutual quasi-images*) Let X, Y be RVs with indecomposable joint distribution (see Problem 15.12).

(a) Show that if for some fixed $\eta \in (0, 1)$ the sets $\mathsf{A}_n \subset \mathsf{X}^n$ and $\mathsf{B}_n \subset \mathsf{Y}^n$ are mutually η-quasi-images of each other, i.e.,

$$\Pr\{X^n \in \mathsf{A}_n | Y^n \in \mathsf{B}_n\} \geqq \eta \quad \text{and} \quad \Pr\{Y^n \in \mathsf{B}_n | X^n \in \mathsf{A}_n\} \geqq \eta \qquad (*)$$

then

$$\frac{1}{n}\log \Pr\{X^n \in \mathsf{A}_n,\ Y^n \in \mathsf{B}_n\} \to 0.$$

(Gács and Körner (1973).)

(b) Prove that to every $0 < \sigma < 1/2$ there exists $\tau = \tau(P_{XY}, \sigma) < 1$ such that for every pair of sets $\mathsf{A}_n \subset \mathsf{X}^n$, $\mathsf{B}_n \subset \mathsf{Y}^n$ satisfying

$$\sigma < \Pr\{X^n \in \mathsf{A}_n\} < 1 - \sigma, \quad \sigma < \Pr\{Y^n \in \mathsf{B}_n\} < 1 - \sigma$$

we have

$$\Pr\{X^n \in \mathsf{A}_n, Y^n \in \mathsf{B}_n\} + \Pr\{X^n \notin \mathsf{A}_n, Y^n \notin \mathsf{B}_n\} < \tau.$$

(Witsenhausen (1975).)

(c) Show that there exist two numbers $\lambda, \mu \in (0, 1)$ (depending on P_{XY}) with $\lambda + \mu > 1$ such that for every n, $\mathsf{A}_n \subset \mathsf{X}^n$, $\mathsf{B}_n \subset Y^n$ we have

$$\Pr\{X^n \in \mathsf{A}_n, Y^n \in \mathsf{B}_n\} \leqq [\Pr\{X^n \in \mathsf{A}_n\}]^\lambda [\Pr\{Y^n \in B_n\}]^\mu.$$

Note that this implies (b). Prove that it also implies the following strengthening of (a): there exists a constant $\varepsilon = \varepsilon(P_{XY}, \eta) > 0$ such that for every n and sets $\mathsf{A}_n \subset \mathsf{X}^n$, $\mathsf{B}_n \subset \mathsf{Y}^n$ the conditions (*) imply

$$\Pr\{X^n \in \mathsf{A}_n, Y^n \in \mathsf{B}_n\} \geqq \varepsilon.$$

(Ahlswede and Gács (1976).)

15.15. (*Images and generated sequences*) Construct a DMC $\{W : \mathsf{X} \to \mathsf{Y}\}$ and sets $\mathsf{A}_n \subset \mathsf{X}^n$ such that, with the notation in (15.7),

$$\underline{\lim}_{n\to\infty} \frac{1}{n}(\log|\mathsf{T}^n_{[W]}(\mathsf{A}_n)| - \log g_{W^n}(\mathsf{A}_n, 1 - \varepsilon)) > 0.$$

Hint Consider a triple of RVs U, X, Y such that $U \multimap X \multimap Y$ and $P_{Y|X} = W$. Define

$$\mathsf{A}_n \triangleq \mathsf{T}^n_{[X|U]}(\mathbf{u}_n) \quad \text{for some} \quad \mathbf{u}_n \in \mathsf{T}^n_{[U]}.$$

Then $(1/n) \log g_{W^n}(\mathsf{A}_n, 1 - \varepsilon) \to H(Y|U)$ (see Lemma 15.3), whereas

$$\varliminf_{n\to\infty} \frac{1}{n} \log |\mathsf{T}^n_{[W]}(\mathsf{A}_n)| \geqq \max H(Y|\hat{U}),$$

where the maximum is taken for all RVs $\hat{U}$ such that $P_{\hat{U}X} = P_{UX}$ but the Markov property is not required.

More-than-three-component sources (Problems 15.16–15.21)

15.16. (*Sub-achievable entropy vectors*) Let us be given a DMMS with $r + 1$ component sources having generic variables $X, Y^{(1)}, \ldots, Y^{(r)}$. Call the vector $\mathbf{R}$ with non-negative components $R_i, 0 \leqq i \leqq r$, a sub-achievable entropy vector if for every $\delta > 0$ and sufficiently large n there exists a function f on X^n such that

$$\frac{1}{n} H(Y^{(i)n}|f(X^n)) \leqq R_i + \delta, \quad i = 1, \ldots, r,$$

and

$$\left| \frac{1}{n} H(X^n|f(X^n)) - R_0 \right| \leqq \delta.$$

Prove that $\mathbf{R}$ is a sub-achievable entropy vector iff there exists a RV U such that $U \multimap X \multimap Y^{(1)} \ldots Y^{(r)}$, the size of the range of U is at most $|\mathsf{X}| + r$ and

$$R_0 = H(X|U), \quad R_i \geqq H(Y^{(i)}|U), \quad i = l, \ldots, r.$$

Hint The sub-achievability of the vectors $\mathbf{R}$ as above follows from Corollary 15.3. To prove that no other vector is a sub-achievable entropy vector, note that

$$\begin{aligned} H(Y^n|f(X^n)) &= \sum_{j=1}^{n} H(Y_j|Y^{j-1}, f(X^n)) \\ &\geqq \sum_{j=1}^{n} H(Y_j|Y^{j-1}, X^{j-1}, f(X^n)) = \sum_{j=1}^{n} H(Y_j|X^{j-1}, f(X^n)), \end{aligned}$$

where Y stands for either $Y^{(i)}$ and

$$H(X^n|f(X^n)) = \sum_{j=1}^{n} H(X_j|X^{j-1}, f(X^n)).$$

Introduce a RV J independent of the rest and uniformly distributed over $\{1, \ldots, n\}$ and consider the RVs

$$\tilde{U} \triangleq X^{J-1} f(X^n), \quad \hat{X} \triangleq X_j, \quad \tilde{Y}^{(i)} \triangleq Y_J^{(i)}.$$

Bound the range of U via the support lemma. (Ahlswede and Körner (1974, unpublished); Gray and Wyner (1974).)

15.17. ($\mathcal{F} = \mathcal{H}$ *for every DMMS*)

(a) Under the hypotheses of Corollary 15.3, show the existence of a function $f : \mathsf{X}^n \to \mathsf{U}^n$ with the properties there but strengthening (15.15) to

$$\left| \frac{1}{n} H(Y^n | f(X^n)) - H(Y|U) \right| < \delta. \qquad (*)$$

Hint Show first that the conclusions (15.10) and (15.11) of Lemma 15.3 still hold if in (15.9) the condition $P^n_{X|U}(\mathsf{A}|\mathbf{u}) \geqq \eta/2$ is replaced by the weaker one

$$P^n_{X|U}(\mathsf{A}|\mathbf{u}) \geqq \eta_n, \quad \text{where} \quad \eta_n \to 0, \quad \frac{1}{n} \log \eta_n \to 0.$$

Conclude that each set A_i in the proof of Corollary 15.3 can be chosen to consist of sequences of the same conditional type given $\mathbf{u}_i$ so that (15.18)–(15.21) hold, except that the condition $P^n_{X|U}(\mathsf{A}_i|\mathbf{u}_i) \geqq \varepsilon/2$ of (15.19) is replaced by $P^n_{X|U}(\mathsf{A}_i|\mathbf{u}_i) \geqq \eta_n$. Defining $f : \mathsf{X}^n \to \mathsf{U}^n$ via these A_i, (*) will follow. In fact, given a RV Y with $U \multimap X \multimap Y$, for sufficiently large n each A_i can be represented as the union of codeword sets A_{ij}, $1 \leqq j \leqq N_i$, of (n, ε)-codes for the DMC $\{V\}$ and of a set A_{i0} so that $|\mathsf{A}_{i0}| < \delta |\mathsf{A}_i|$ and

$$\frac{1}{n} \log |\mathsf{A}_{ij}| \geqq C_V(\mathsf{A}_i, \varepsilon) - \delta, \quad 1 \leqq j \leqq N_i.$$

Setting

$$g(\mathbf{x}) \triangleq \begin{cases} 0 & \text{if} \quad \mathbf{x} \in \bigcup_{i=1}^{M} \mathsf{A}_{i0} \\ 1 & \text{otherwise,} \end{cases}$$

we get

$$H(Y^n | f(X^n)) \geqq (1 - \delta) H(Y^n | f(X^n), g(X^n) = 1).$$

Now the proof can be completed using Problem 15.6 to show that

$$\begin{aligned} \frac{1}{n} H(Y^n | f(X^n) = i, g(X^n) = 1) &= \frac{1}{n} H(Y^n | X^n \in \mathsf{A}_i - \mathsf{A}_{i0}) \\ &\geqq \frac{1}{n} \log g_{V^n}(\mathsf{A}_i, 1 - \varepsilon) - 3\delta - \varepsilon \log |\mathsf{Y}|. \end{aligned}$$

(b) Define for an $(r+1)$-source with generic variables $X, Y^{(1)}, \ldots, Y^{(r)}$ the sets $\mathcal{F}(X; Y^{(1)}; \ldots; Y^{(r)} | X)$ and $\mathcal{H}(X; Y^{(1)}; \ldots; Y^{(r)} | X)$ as the closure of the set of all vectors of form

$$\left(\frac{1}{n}H(X^n|f(X^n)), \frac{1}{n}H(Y^{(1)n}|f(X^n)), \ldots, \frac{1}{n}H(Y^{(r)n}|f(X^n))\right)$$

resp.

$$\left(\frac{1}{n}H(X^n|U), \frac{1}{n}H(Y^{(1)n}|U), \ldots, \frac{1}{n}H(Y^{(r)n}|U)\right),$$

where $U \multimap X^n \multimap (Y^{(1)} \ldots Y^{(r)})^n$. Show that $\mathcal{F} = \mathcal{H}$.

Hint Use (a).

15.18. (*Achievable entropy vectors*) Show that for an $(r+1)$-source with arbitrary generic variables $(X, Y^{(1)}, \ldots, Y^{(r)})$ the coordinate-wise minimum of the two vectors

$$(H(X|U) + t, H(Y^{(1)}|U) + t, \ldots, H(Y^{(r)}|U) + t),$$
$$(H(X), H(Y^{(1)}), \ldots, H(Y^{(r)}))$$

is contained in $\mathcal{F} = \mathcal{F}(X; Y^{(1)}; \ldots; Y^{(r)}|X)$, whenever

$$U \multimap X \multimap Y^{(1)} \ldots Y^{(r)} \quad \text{and} \quad t \geqq 0$$

where $\mathcal{F}$ is as in Problem 15.17.

Hint For $t = 0$, the statement is contained in Problem 15.17(a). For $t > 0$ apply the argument of Problem 15.17(a) to sets A_i constructed in the same way as the set A in the proof of Theorem 15.9.

15.19.* (*Achievable entropy vectors*)

(a) Show that for a 4-source with arbitrary generic variables (S, X, Y, Z) the coordinate-wise minimum of the four vectors

$$(H(X), H(Y), H(Z)),$$

$$(H(X|T) + t, H(Y|T) + t, H(Z|T) + t),$$

$$(H(X|TU) + t', H(Y|TU) + t', H(Z|TU) + t'),$$

$$H(X|TUV) + t'', H(Y|TUV) + t'', H(Z|TUV) + t'')$$

is an element of $\mathcal{H}(X, Y, Z|S)$ for any random variables T, U, V such that $TUV \multimap S \multimap XYZ$ and any non-negative numbers t, t' and t''.

(b) (*Symmetric description of* $\mathcal{H}(X, Y, Z|X)$) Show that $\mathcal{H}(X, Y, Z|X)$ consists of precisely those vectors which can be written as the coordinate-wise minimum of the three vectors

$$(H(X), H(Y), H(Z)),$$

$$(H(X|U) + t, H(Y|U) + t, H(Z|U) + t),$$

$$(t', H(Y|X) + t', H(Z|X) + t'),$$

where U is a random variable with range size at most $|\mathsf{X}|+2$ that satisfies $U \multimap X \multimap YZ$, and $t,\ t'$ are non-negative numbers. (Körner (1984).)

15.20. (*Super-achievable entropy vectors*) Let us be given a DMMS with $r+2$ component sources having generic variables $(X, Y^{(1)}, \ldots, Y^{(r)}, S)$. Call the vector $\mathbf{R}$ with non-negative components R_i, $0 \leqq i \leqq r+1$, a super-achievable entropy vector if for every $\delta > 0$ and every sufficiently large n there exists a function f with domain X^n such that

$$\frac{1}{n}H(X^n|f(X^n)) \geqq R_0 - \delta,$$

$$\frac{1}{n}H(Y^{(i)n}|f(X^n)) \geqq R_i - \delta, \quad 1 \leqq i \leqq r,$$

and

$$\left|\frac{1}{n}H(S^n|f(X^n)) - R_{r+1}\right| \leqq \delta.$$

Prove that if $X \multimap S \multimap Y^{(1)} \ldots Y^{(r)}$ holds then $\mathbf{R}$ is a super-achievable entropy vector iff there exists a RV U satisfying

$$U \multimap X \multimap S \multimap Y^{(1)} \ldots Y^{(r)}, \quad |\mathsf{U}| \leqq |\mathsf{X}| + r + 1,$$

and a number t with $0 \leqq t \leqq I(U \wedge S)$ such that

$$\begin{aligned} R_0 &\leqq H(X|U) + t, \\ R_i &\leqq \min\,[H(Y^{(i)}|U) + t,\ H(Y^{(i)})], \quad 1 \leqq i \leqq r, \\ R_{r+1} &= H(S|U) + t. \end{aligned}$$

Hint The super-achievability of these vectors $\mathbf{R}$ follows from Problem 15.18. To prove the converse result, we use the identity

$$H(T) - H(Z) = H(T|Z) - H(Z|T) \qquad (*)$$

and write Y for any of the RVs $Y^{(i)}$. We have

$$\begin{aligned} H(S^n|f(X^n)) - H(Y^n|f(X^n)) &= H(S^n|Y^n f(X^n)) - H(Y^n|S^n f(X^n)) \\ &= H(S^n|Y^n f(X^n)) - nH(Y|S) \\ &= \sum_{j=1}^{n} [H(S_j|S^{j-1}\bar{Y}_j Y_j f(X^n)) - H(Y_j|S_j)] \\ &\geqq \sum_{j=1}^{n} [H(S_j|\bar{S}_j \bar{Y}_j Y_j f(X^n)) - H(Y_j|S_j)], \end{aligned}$$

where $\bar{Y}_j \triangleq Y_1 \ldots Y_{j-1} Y_{j+1} \ldots Y_n$ and $\bar{S}_j$ is defined similarly. Here

$$H(S_j|\bar{S}_j \bar{Y}_j f(X^n) Y_j) = H(S_j|\bar{S}_j f(X^n) Y_j),$$

since $S_j \multimap Y_j \bar{S}_j f(X^n) \multimap \bar{Y}_j$ follows from our Markov condition. Similarly,

$$H(Y_j|S_j) = H(Y_j|S_j \bar{S}_j f(X^n)).$$

Using these identities and (*), the preceding sum equals

$$\sum_{j=1}^{n}[H(S_j|\bar{S}_j f(X^n)Y_j) - H(Y_j|S_j\bar{S}_j f(X^n))]$$
$$= \sum_{j=1}^{n}[H(S_j|\bar{S}_j f(X^n)) - H(Y_j|\bar{S}_j f(X^n))].$$

Thus we have proved

$$H(S^n|f(X^n)) - H(Y^{(i)n}|f(X^n))$$
$$\geqq \sum_{j=1}^{n}[H(S_j|\bar{S}_j f(X^n)) - H(Y_j^{(i)}|\bar{S}_j f(X^n))]. \qquad (**)$$

Further, by a repeated use of (*),

$$\begin{aligned} H(S^n|f(X^n)) - H(X^n|f(X^n)) &= H(S^n|X^n f(X^n)) - H(X^n|S^n f(X^n)) \\ &\geqq \sum_{j=1}^{n}[H(S_j|X_j) - H(X_j|S_j\bar{S}_j f(X^n)] \\ &= \sum_{j=1}^{n}[H(S_j|\bar{S}_j f(X^n)) - H(X_j|\bar{S}_j f(X^n))] \\ &= \sum_{j=1}^{n}[H(S_j|\bar{S}_j f(X^n)) - H(X_j|\bar{S}_j f(X_n))]. \end{aligned}$$

Introduce a RV J independent of the rest and uniformly distributed over $\{1, 2, \ldots, n\}$. Consider the RVs

$$\tilde{U} \triangleq \bar{S}_J f(X^n), \quad \tilde{X} \triangleq X_J, \quad \tilde{Y}^{(i)} \triangleq Y_J^i, \quad \tilde{S} \triangleq S_J.$$

Clearly, $\tilde{U} \multimap \tilde{X} \multimap \tilde{S} \multimap \tilde{Y}^{(1)} \ldots \tilde{Y}^{(i)}$. Introduce a RV U such that the joint distribution of the tilted variables equals that of the corresponding untilted ones. Then (**) and the preceding series of inequalities can be written as

$$H(S^n|f(X^n)) - H(Y^{(i)n}|f(X^n)) \geqq H(S|U) - H(Y^{(i)}|U)$$

and

$$H(S^n|f(X^n)) - H(X^n|f(X^n)) \geqq H(S|U) - H(X|U).$$

Further, the inequality

$$H(S^n|f(X^n)) \geqq \sum_{i=1}^{n} H(S_i|\bar{S}_i f(X^n))$$

gives

$$H(S^n|f(X^n)) \geqq H(S|U).$$

15.21. (*Three channels*) Let $F : \mathsf{X} \to \mathsf{S}$, $V : \mathsf{X} \to \mathsf{Y}$, $W : \mathsf{X} \to \mathsf{Z}$ be three channels such that both V and W are degraded versions of F; let $X \multimap S \multimap YZ$ be RVs such that $P_{S|X} = F$, $P_{Y|X} = V$, $P_{Z|X} = W$.

(a) Denote by $\mathcal{G}(S; Y; Z|X)$ the set of those triples (a, b, c) for which to every $\delta > 0$ and n large enough there exists a set $\mathsf{A} \subset \mathsf{T}^n_{[X]}$ satisfying

$$\left|\frac{1}{n}\log g_{F^n}(\mathsf{A}, 1-\varepsilon) - a\right| \leqq \delta, \quad \left|\frac{1}{n}\log g_{V^n}(\mathsf{A}, 1-\varepsilon) - b\right| \leqq \delta,$$
$$\left|\frac{1}{n}\log g_{W^n}(\mathsf{A}, 1-\varepsilon) - c\right| \leqq \delta.$$

Show that $\mathcal{G}(S; Y; Z|X)$ consists of those triples (a, b, c) for which $(b, c) \in \mathcal{H}^*(Y; Z|X)$ and $a^-(b, c) \leqq a \leqq a^{++}(b, c)$, where

$$a^-(b, c) \triangleq \max[b + H(S|X) - H(Y|X),\ c + H(S|X) - H(Z|X)]$$

and $a^{++}(b, c)$ is the maximum of $H(S|U)$ under the constraints $H(Y|U) \leqq b$, $H(Z|U) \leqq c$ taken for RVs U such that $U \multimap X \multimap S \multimap YZ$, $|\mathsf{U}| \leqq |\mathsf{X}| + 3$.

(b) Define $\mathcal{F}(S; Y; Z|X)$ analogously, as the set of triples (a, b, c) such that for every $\delta > 0$ and n large enough there exists a function f of X^n satisfying

$$\left|\frac{1}{n}H(S^n|f(X^n)) - a\right| \leqq \delta, \quad \left|\frac{1}{n}H(Y^n|f(X^n)) - b\right| \leqq \delta,$$
$$\left|\frac{1}{n}H(Z^n|f(X^n)) - c\right| \leqq \delta.$$

Show that $\mathcal{F}(S; Y; Z|X)$ consists of those triples (a, b, c) for which $(b, c) \in \mathcal{H}^*(Y; Z|X)$ and $a^-(b, c) \leqq a \leqq a^+(b, c)$. Here $a^-(b, c)$ is the same as in (a) and $a^+(b, c)$ is the maximum of $H(S|U) + t$ under the constraints $H(Y|U) + t = b$, $H(Z|U) + t = c$ for RVs U as in (a) and $0 \leqq t \leqq \min(I(U \wedge Y), I(U \wedge Z))$.

Hint The proofs of the corresponding theorems of the text apply almost literally.

Story of the results

The study of sizes of sets and their images in connection with coding problems was initiated by Gács and Körner (1973) and Ahlswede, Gács and Körner (1976). The relevance to coding problems of the joint description of entropies of sequences of RVs connected by a DMC was recognized (in a special case) in the very stimulating paper Wyner and Ziv (1973). The first major step towards a general technique of proving converse results was made by Gallager (1974a). An entropy characterization problem in the sense of (15.1) was first applied to prove a source network coding theorem by Ahlswede and Körner (1975).

An essential part of this chapter is based on Ahlswede, Gács and Körner (1976) and Körner and Marton (1977c). Our streamlined presentation uses cardinalities rather than probabilities of sets. This was helpful to obtain uniform bounds. The technique of bounding the range of auxiliary RVs based on the support lemma, Lemma 15.4 was suggested independently by Ahlswede and Körner (1975) and Wyner (1975b). The underlying Carathéodory–Fenchel theorem is, in this form, due to Fenchel (1929); see Eggleston (1958) for more on this. The single-letterization lemma, Lemma 15.7 generalizes a result of Körner and Marton (1977c). The proof of Lemma 15.8 uses an idea of Ahlswede (personal communication, 1974). Theorems 15.9–15.11 are due to Körner and Marton (1977c). Lemma 15.12 was proved earlier by Ahlswede, Gács and Körner (1976). Lemma 15.17 is taken from Csiszár and Körner (1978). The rest of the material in the text is original here; the results therein have been obtained jointly with K. Marton.

Addition. A random coding approach to the achievability part of the entropy characterization problem is developed in Körner (1984). It has led, in particular, to the results in Problem 15.19.

16 Source and channel networks

The results of Chapter 15 enable us to solve a number of coding problems for various source and channel networks. Most of the resulting coding theorems are presented as problems which can be solved more or less in the same way. As an illustration of the methods, we shall discuss in detail a channel network and a (normal) source network. In addition, we shall consider a source network with a more general fidelity criterion than probability of error.

Channel networks with a single intermediate vertex are called *broadcast channels.* → 16.1 The simplest case of a network with two outputs has been studied intensively. Without loss of generality, one can suppose that this network has three inputs. This two-output broadcast channel (BC) is illustrated in Fig. 16.1.

At present, a computable characterization of the capacity region of the two-output broadcast channel is available only in special cases. A model of independent interest is obtained if "either of inputs 1 and 2 of the network is idle." This corresponds to the new channel network in Fig. 16.2, the *asymmetric two-output broadcast channel* (ABC), which is treated as our next problem.

In the following, the DMCs corresponding to the outputs addressed by 10 and 0 will be denoted by $\{V : \mathsf{X} \to \mathsf{Y}\}$ resp. $\{W : \mathsf{X} \to \mathsf{Z}\}$.

THEOREM 16.1 (*Direct part of the ABC coding theorem*) If for a pair (R_1, R_0) of non-negative numbers there exists a quadruple of RVs (U, X, Y, Z) such that $U -\!\!\circ\!\!- X -\!\!\circ\!\!- YZ$, Y resp. Z are connected with X by the channels V resp. W, and the inequalities

$$R_1 + R_0 \leqq \min[I(X \wedge Y), I(X \wedge Y|U) + I(U \wedge Z)] \tag{16.1}$$

$$R_0 \leqq I(U \wedge Z) \tag{16.2}$$

hold, then (R_1, R_0) is an achievable rate pair for the asymmetric two-output broadcast channel defined by the DMCs $\{V : \mathsf{X} \to \mathsf{Y}\}$ and $\{W : \mathsf{X} \to \mathsf{Z}\}$. Moreover, every pair → 16.2 (R_1, R_0) as above is contained in the m-capacity region of the ABC. ○

Proof It is sufficient to prove the final statement. To this end, note the following consequences of the very definition of an ABC: with every achievable rate pair (R_1, R_0) also $(R_1 + \lambda R_0, (1 - \lambda)R_0)$ is achievable for every $0 \leqq \lambda \leqq 1$, and every $(R_1', R_0') \leqq (R_1, R_0)$ is achievable, too.

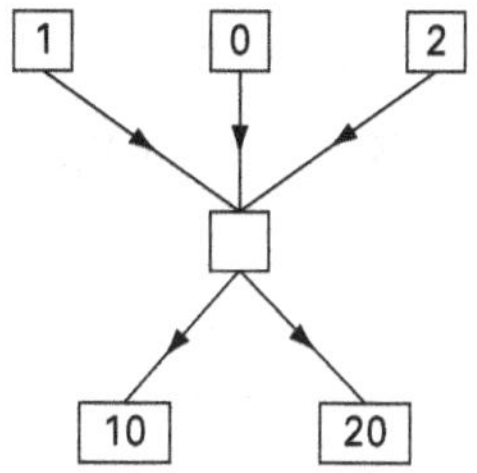

Figure 16.1 Two-output broadcast channel (BC).

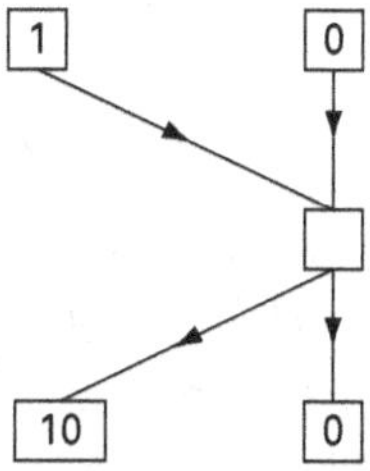

Figure 16.2 Asymmetric two-output broadcast channel (ABC).

Observe that if in (16.1)

$$I(X \wedge Y) \leqq I(X \wedge Z)$$

then

$$R_1 + R_0 \leqq I(X \wedge Y) \leqq I(X \wedge Z)$$

and thus (R_1, R_0) is achievable by Corollary 15.8 and the above remark. Thus it remains to consider the case

$$I(X \wedge Y) > I(X \wedge Z).$$

We claim that in this case the U appearing in (16.1) and (16.2) can be supposed to satisfy

$$I(X \wedge Y) \geqq I(X \wedge Y|U) + I(U \wedge Z).$$

In fact, if the original U does not have this property, then introduce a RV J_α independent of all the others with

$$\Pr\{J_\alpha = 0\} \triangleq \alpha, \quad \Pr\{J_\alpha = 1\} \triangleq 1 - \alpha.$$

Setting

$$U_\alpha \triangleq \begin{cases} (U, 0) & \text{if} \quad J_\alpha = 0 \\ (X, 1) & \text{if} \quad J_\alpha = 1, \end{cases}$$

we see that $U_\alpha \multimap X \multimap YZ$ and U_α satisfies (16.2). Further, $I(X \wedge Y|U_\alpha) + I(U_\alpha \wedge Z)$ is a linear function of α, whose value equals $I(X \wedge Z) < I(X \wedge Y)$ if $\alpha = 0$ and $I(X \wedge Y|U) + I(U \wedge Z) > I(X \wedge Y)$ if $\alpha = 1$. Hence for some $0 < \alpha < 1$

$$I(X \wedge Y) = I(X \wedge Y|U_\alpha) + I(U_\alpha \wedge Z)$$

holds, establishing our claim.

Thus, again using the starting remark, it suffices to prove that if

$$R_1 = I(X \wedge Y|U), \quad R_0 = I(U \wedge Z) \tag{16.3}$$

and

$$R_1 + R_0 \leqq I(X \wedge Y) \tag{16.4}$$

for some quadruple of RVs $U,\ X,\ Y,\ Z$ as in the theorem, then (R_1, R_0) is an achievable rate pair (in the sense of maximum probability of error). Note that (16.3) and (16.4) imply

$$I(U \wedge Z) \leqq I(U \wedge Y). \tag{16.5}$$

Fix some $\delta > 0$, $\varepsilon \in (0, 1/4)$. Let $\hat{V} : \mathsf{U} \to \mathsf{Y}$ and $\hat{W} : \mathsf{U} \to \mathsf{Z}$ be channels connecting Y resp. Z with U. By (16.5) and Corollary 15.8, to every $n \geqq n_1(|\mathsf{U}|, |\mathsf{Y}|, |\mathsf{Z}|, \varepsilon, \delta)$ there exist (n, ε)-codes $(\hat{f}, \hat{\varphi})$ resp. $(\hat{f}, \hat{\psi})$ for the respective DMCs $\{\hat{V}\}$ and $\{\hat{W}\}$ with a common encoder of rate

$$\frac{1}{n} \log |\mathsf{M}_{\hat{f}}| \geqq I(U \wedge Z) - \delta. \tag{16.6}$$

For every $m \in \mathsf{M}_{\hat{f}}$ we have, by definition, the inequalities

$$\begin{aligned} \hat{V}^n(\hat{\varphi}^{-1}(m)|\hat{f}(m)) &= \sum_{\mathbf{x} \in \mathsf{X}^n} V^n(\hat{\varphi}^{-1}(m)|\mathbf{x}) P^n_{X|U}(\mathbf{x}|\hat{f}(m)) \geqq 1 - \varepsilon, \\ \hat{W}^n(\hat{\psi}^{-1}(m)|\hat{f}(m)) &= \sum_{\mathbf{x} \in \mathsf{X}^n} W^n(\hat{\psi}^{-1}(m)|\mathbf{x}) P^n_{X|U}(\mathbf{x}|\hat{f}(m)) \geqq 1 - \varepsilon. \end{aligned} \tag{16.7}$$

Let $\mathsf{A}(m)$ be the largest subset of X^n such that

$$V^n(\hat{\varphi}^{-1}(m)|\mathbf{x}) \geqq 1 - 2\varepsilon, \quad W^n(\hat{\psi}^{-1}(m)|\mathbf{x}) \geqq 1 - 2\varepsilon \quad \text{for every} \quad \mathbf{x} \in \mathsf{A}(m). \tag{16.8}$$

By their definition, the sets $\mathsf{A}(m)$ are disjoint (recall that $\varepsilon < 1/4$), and (16.7) implies

$$P^n_{X|U}(\mathsf{A}(m)|\hat{f}(m)) \geqq \frac{1}{2} \quad \text{for} \quad m \in \mathsf{M}_{\hat{f}}.$$

Using Lemma 15.8, to every $m \in \mathsf{M}_{\hat{f}}$ we can construct an (n, ε)-code (f_m, φ_m) for the DMC $\{V\}$ with codewords in $\mathsf{A}(m)$, each code having the same message set M_1, where

$$\frac{1}{n} \log |\mathsf{M}_1| \geqq I(X \wedge Y|U) - \delta, \tag{16.9}$$

whenever $n \geqq n_2(|\mathsf{X}|, |\mathsf{Y}|, |\mathsf{U}|, \varepsilon, \delta)$. Define now the mappings $f : \mathsf{M}_1 \times \mathsf{M}_0 \to \mathsf{X}^n$, $\varphi : \mathsf{Y}^n \to \mathsf{M}_1 \times \mathsf{M}_0$ and $\psi : \mathsf{Z}^n \to \mathsf{M}_0$, where $\mathsf{M}_0 \triangleq \mathsf{M}_{\hat{f}}$, as follows:

$$\begin{aligned} &f(m', m'') \triangleq f_{m''}(m') \quad \text{for every} \quad m' \in \mathsf{M}_1, m'' \in \mathsf{M}_0, \\ &\varphi(\mathbf{y}) \triangleq (\varphi_m(\mathbf{y}), m) \quad \text{where} \quad m \triangleq \hat{\varphi}(\mathbf{y}), \\ &\psi(\mathbf{z}) \triangleq \hat{\psi}(\mathbf{z}). \end{aligned}$$

Recalling that (f_m, φ_m) is an (n, ε)-code for the DMC $\{V\}$, for every $m \in \mathsf{M}_0$, it follows by (16.8) that (f, φ, ψ) is an n-length block code with maximum probabilities of error less than 3ε for the given ABC. As ε and δ were arbitrary, (16.6) and (16.9) prove the achievability of (R_1, R_0). □

COMMENT A feature of the above proof deserves emphasis. First an auxiliary code has been constructed, then each auxiliary codeword $\hat{f}(m'')$, $m'' \in \mathsf{M}_0$ served, in a sense, as a "center" of a "cloud" of actual codewords, namely $\{f(m', m''),\ m' \in \mathsf{M}_1\}$. This technique, typical in multi-user information theory, is known as *superposition coding*, though the term usually refers to constructions which – unlike above – involve random selection; see Lemma 17.14. ○

In general, the m-capacity region of a channel network can differ from the a-capacity region. This, however, is not the case for the ABC.

LEMMA 16.2 To every ε-achievable rate pair (R_1, R_0) of an ABC with component channels $\{V : \mathsf{X} \to \mathsf{Y}\}$ and $\{W : \mathsf{X} \to \mathsf{Z}\}$, for every $\delta > 0$ and $n \geqq n_0(|\mathsf{X}|, \varepsilon, \delta)$ there exists an n-length block code (f, φ, ψ) with maximum probabilities of error less than $2\varepsilon + \delta$ and such that the rate vector $(\hat{R}_1, \hat{R}_0)$ of the code satisfies

$$\hat{R}_1 \geqq R_1 - 3\delta, \qquad \hat{R}_0 \geqq R_0 - 3\delta.$$

Moreover, it can be supposed that all the codewords of this code have the same type. ○

COROLLARY 16.2 The m-capacity region of an ABC equals the a-capacity region. ○ → 14.13

Proof Let $(\hat{f}, \varphi, \psi)$ be an n-length block code for the ABC with average probabilities of error less than ε, where φ resp. ψ are the decoders corresponding to the outputs addressed by 10 resp. 0:

$$\hat{f} : \hat{\mathsf{M}}_1 \times \hat{\mathsf{M}}_0 \to \mathsf{X}^n, \quad \varphi : \mathsf{Y}^n \to \mathsf{M}_\mathsf{Y} \supset \hat{\mathsf{M}}_1 \times \hat{\mathsf{M}}_0, \quad \psi : \mathsf{Z}^n \to \mathsf{M}_\mathsf{Z} \supset \hat{\mathsf{M}}_0,$$

$$\frac{1}{|\hat{\mathsf{M}}_1||\hat{\mathsf{M}}_0|} \sum_{m_1 \in \hat{\mathsf{M}}_1} \sum_{m_0 \in \hat{\mathsf{M}}_0} e_{10}(m_1, m_0) \leqq \varepsilon, \tag{16.10}$$

$$\frac{1}{|\hat{\mathsf{M}}_1||\hat{\mathsf{M}}_0|} \sum_{m_1 \in \hat{\mathsf{M}}_1} \sum_{m_0 \in \hat{\mathsf{M}}_0} e_0(m_1, m_0) \leqq \varepsilon,$$

where

$$e_{10}(m_1, m_0) \triangleq 1 - V^n(\varphi^{-1}(m_1, m_0)|\hat{f}(m_1, m_0))$$

and

$$e_0(m_1, m_0) \triangleq 1 - W^n(\psi^{-1}(m_0)|\hat{f}(m_1, m_0)).$$

Suppose further that

$$\frac{1}{n} \log |\hat{\mathsf{M}}_1| \geqq R_1 - \delta, \qquad \frac{1}{n} \log |\hat{\mathsf{M}}_0| \geqq R_0 - \delta. \tag{16.11}$$

Write

$$\bar{e}(m) \triangleq \frac{1}{|\hat{\mathsf{M}}_1|} \sum_{m_1 \in \hat{\mathsf{M}}_1} (e_{10}(m_1, m) + e_0(m_1, m)).$$

Since by (16.10)

$$\frac{1}{|\hat{\mathsf{M}}_0|} \sum_{m \in \hat{\mathsf{M}}_0} \bar{e}(m) \leqq 2\varepsilon,$$

it follows that there exists a set $\tilde{\mathsf{M}}_0 \subset \hat{\mathsf{M}}_0$ with

$$\frac{1}{n} \log |\tilde{\mathsf{M}}_0| \geqq \frac{1}{n} \log |\hat{\mathsf{M}}_0| - \delta \tag{16.12}$$

such that $\bar{e}(m) < 2\varepsilon + \delta/2$ for every $m \in \tilde{\mathsf{M}}_0$, if n is sufficiently large depending on δ. Similarly, for every $m \in \tilde{\mathsf{M}}_0$ there exists a set $\tilde{\mathsf{M}}_1 \triangleq \tilde{\mathsf{M}}_1(m)$ such that

$$\frac{1}{n} \log |\tilde{\mathsf{M}}_1(m)| \geqq \frac{1}{n} \log |\hat{\mathsf{M}}_1| - \delta \tag{16.13}$$

and

$$e_{10}(m_1, m) + e_0(m_1, m) < 2\varepsilon + \delta \quad \text{for every} \quad m \in \tilde{\mathsf{M}}_0, \quad m_1 \in \tilde{\mathsf{M}}_1(m), \tag{16.14}$$

if n is large enough.

By the type counting lemma, for $n \geqq n_1(|\mathsf{X}|, \delta)$ to every $m \in \tilde{\mathsf{M}}_0$ there exists a set $\mathsf{M}_1(m) \subset \tilde{\mathsf{M}}_1(m)$ such that

$$\frac{1}{n} \log |\mathsf{M}_1(m)| \geqq \frac{1}{n} \log |\tilde{\mathsf{M}}_1(m)| - \delta \tag{16.15}$$

and all sequences $\hat{f}(m_1, m)$, $m_1 \in \mathsf{M}_1(m)$ have the same type for every fixed m. Inequality (16.15) implies by (16.13) and (16.11) that for $n \geqq n_2(|\mathsf{X}|, \varepsilon, \delta)$ we have

$$\frac{1}{n} \log |\mathsf{M}_1(m)| \geqq R_1 - 3\delta. \tag{16.16}$$

We can suppose that all these message sets $\mathsf{M}_1(m)$ have the same cardinality, and that $\mathsf{M}_1 \triangleq \mathsf{M}_1(m)$ is the same set, whatever the value of m. By the type counting lemma, for $n \geqq n_3(|\mathsf{X}|, \delta)$ there exists a type P such that for a set $\mathsf{M}_0 \subset \tilde{\mathsf{M}}_0$ with

$$\frac{1}{n} \log |\mathsf{M}_0| \geqq \frac{1}{n} \log |\tilde{\mathsf{M}}_0| - \delta$$

the common type of the sequences $\{\hat{f}(m_1, m); m_1 \in \mathsf{M}_1\}$ is P for every $m \in \mathsf{M}_0$. The preceding inequality, (16.11) and (16.12) yield

$$\frac{1}{n} \log |\mathsf{M}_0| \geqq R_0 - 3\delta.$$

Recalling (16.14), and our definition of the set M_1, this and (16.16) establish the lemma. □

→ 16.3

THEOREM 16.3 (*Converse part of the ABC coding theorem*) If (R_1, R_0) is an ε-achievable rate vector with some $0 < \varepsilon < 1/2$ for the asymmetric two-output broadcast channel defined by the DMCs $\{V : \mathsf{X} \to \mathsf{Y}\}$ and $\{W : \mathsf{X} \to \mathsf{Z}\}$, then the inequalities (16.1) and (16.2) hold for some RVs U, X, Y, Z such that Y and Z are connected with X by the channels V and W, and

$$U \multimap X \multimap YZ, \qquad U \in \mathsf{U} \text{ with } |\mathsf{U}| \leqq |\mathsf{X}| + 2. \tag{16.17}$$

○

Proof Suppose that (R_1, R_0) is an ε-achievable rate vector and fix any $0 < \delta < 1/2 - \varepsilon$. By Lemma 16.2, for $n \geqq n_1$ there exists an n-length block code (f, φ, ψ) for the ABC, where

$$f : \mathsf{M}_1 \times \mathsf{M}_0 \to \mathsf{X}^n, \quad \varphi : \mathsf{Y}^n \to \mathsf{M}_\mathsf{Y} \supset \mathsf{M}_1 \times \mathsf{M}_0, \quad \psi : \mathsf{Z}^n \to \mathsf{M}_\mathsf{Z} \supset \mathsf{M}_0$$

such that all the codewords have the same type, P say, and

$$e_{10}(m_1, m_0) + e_0(m_1, m_0) < 2\varepsilon + \delta \text{ for every } m_1 \in \mathsf{M}_1, m_0 \in \mathsf{M}_0, \tag{16.18}$$

$$\frac{1}{n} \log |\mathsf{M}_1| \geqq R_1 - \delta, \quad \frac{1}{n} \log |\mathsf{M}_0| \geqq R_0 - \delta. \tag{16.19}$$

Let (X, Y, Z) be any triple of RVs such that $P_X = P$, $P_{Y|X} = V$, $P_{Z|X} = W$. By the noisy channel coding theorem, more precisely, by Corollary 6.4, (16.18) implies for $n \geqq n_2$ the inequality

$$\frac{1}{n} \log(|\mathsf{M}_1| \cdot |\mathsf{M}_0|) \leqq I(X \wedge Y) + \delta,$$

whence by (16.19)

$$R_0 + R_1 \leqq I(X \wedge Y) + 3\delta. \tag{16.20}$$

For every $m \in \mathsf{M}_0$ consider the sets

$$\mathsf{A}(m) \triangleq \{f(m_1, m) : m_1 \in \mathsf{M}_1\}, \quad \mathsf{C}(m) \triangleq \mathsf{T}^n_{[Z]} \cap \psi^{-1}(m).$$

Since $\psi^{-1}(m)$ is an $(1 - 2\varepsilon - \delta)$-image of $\mathsf{A}(m)$ by (16.18), and $\mathsf{T}^n_{[Z]}$ is an $(1 - \delta)$-image of $\mathsf{A}(m)$ by Lemmas 2.10 and 2.12 for $n \geqq n_3$, it follows that $\mathsf{C}(m)$ is an $1 - (2\varepsilon + 2\delta)$-image of $\mathsf{A}(m)$. Thus, putting $\eta \triangleq 1 - (2\varepsilon + 2\delta)$,

$$|\mathsf{C}(m)| \geqq g_{W^n}(\mathsf{A}(m), \eta) \quad \text{for} \quad m \in \mathsf{M}_0. \tag{16.21}$$

As the sets $\mathsf{C}(m)$ are disjoint subsets of $\mathsf{T}^n_{[Z]}$ by definition, we see by Lemma 2.13 that for $n \geqq n_4$

$$\sum_{m \in \mathsf{M}_0} |\mathsf{C}(m)| \leqq |\mathsf{T}^n_{[Z]}| \leqq \exp\{n(H(Z) + \delta)\}.$$

Hence, denoting by A resp. C some sets $\mathsf{A}(m)$ and $\mathsf{C}(m)$ for which $\mathsf{C}(m)$ has minimum cardinality, we have by (16.21)

$$\frac{1}{n} \log |\mathsf{M}_0| \leqq H(Z) + \delta - \frac{1}{n} \log |\mathsf{C}| \leqq H(Z) + \delta - \frac{1}{n} \log g_{W^n}(\mathsf{A}, \eta). \tag{16.22}$$

On the other hand, by (16.18) A is the codeword set of an $(n, 2\varepsilon + \delta)$-code for the DMC $\{V\}$ with message set M_1. As $\mathsf{A} \subset \mathsf{T}_P \subset \mathsf{T}^n_{[X]}$, by Lemma 6.8 we also have

$$\frac{1}{n}\log|\mathsf{M}_1| \leqq \frac{1}{n}\log g_{V^n}(\mathsf{A}, \eta) + \delta - H(Y|X), \tag{16.23}$$

whenever $n \geqq n_5$.

Now, by Theorem 15.10, for $n \geqq n_6$ there exists a pair $(b, c) \in \mathcal{H}^*(Y; Z|X)$ such that

$$\left|\frac{1}{n}\log g_{V^n}(\mathsf{A},\eta) - b\right| \leqq \delta, \qquad \left|\frac{1}{n}\log g_{W^n}(\mathsf{A},\eta) - c\right| \leqq \delta.$$

Comparing this with (16.23) and (16.22), we obtain

$$\frac{1}{n}\log|\mathsf{M}_0| \leqq H(Z) - c + 2\delta,$$
$$\frac{1}{n}\log|\mathsf{M}_1| \leqq b - H(Y|X) + 2\delta,$$

whence by (16.19)

$$R_1 \leqq b - H(Y|X) + 3\delta, \qquad R_0 \leqq H(Z) - c + 3\delta. \tag{16.24}$$

Recall (see (15.49) and (15.50)) that $\mathcal{H}^*(Y; Z|X)$ is the set of pairs (b, c) with

$$b = H(Y|U) + t, \quad c = H(Z|U) + t, \quad 0 \leqq t \leqq \min[I(U \wedge Z), I(U \wedge Z)],$$

where U satisfies (16.17). Thus (16.24) becomes

$$R_1 \leqq I(X \wedge Y|U) + t + 3\delta, \qquad R_0 \leqq I(U \wedge Z) - t + 3\delta.$$

Hence

$$R_1 + R_0 \leqq I(X \wedge Y|U) + I(U \wedge Z) + 6\delta,$$
$$R_0 \leqq I(U \wedge Z) + 3\delta.$$

As for every $\delta > 0$ there exists a quadruple of RVs (U, X, Y, Z) as above, satisfying the preceding inequalities and (16.20), the proof can be completed using the continuity of entropy. □

→ 16.4 COROLLARY 16.3 (*ABC coding theorem*) The region of ε-achievable rate pairs for the ABC defined by the DMCs $\{V : \mathsf{X} \to \mathsf{Y}\}$ and $\{W : \mathsf{X} \to \mathsf{Z}\}$ is the same for every
→ 16.5 $0 < \varepsilon < 1/2$ and it consists of those pairs (R_1, R_0) of non-negative numbers that satisfy the inequalities (16.1) and (16.2) for some quadruple of RVs (U, X, Y, Z) as in Theorem 16.3. ○

Our next target is to apply the results of Chapter 15 to source networks. It follows from the helpers theorem, Theorem 13.15, that the solution of the entropy characterization problem formulated in the Discussion of Chapter 13 would lead to a computable characterization of the achievable rate region of every normal source network with one helper. The entropy characterization problem has been solved for

3-sources (Theorem 15.20); this solution allows us to prove coding theorems for a number of source networks. Recall that Theorem 13.15 leaves open the question whether an ε-achievable rate region of a source network can differ from the achievable rate region. The solution of the image size problem given in Chapter 15 will enable us to prove for certain networks the strong converse, i.e., the equality of the ε-achievable and achievable rate regions.

As an illustration of the methods, we shall discuss one network in detail. Other networks will be treated in problems. In addition, we shall also consider a problem involving arbitrary fidelity criteria, the solution of which does not fully rely on Chapter 15.

We solved in Chapter 13 the coding problem for the fork network of Fig. 16.3, proving that the achievable rate region consists of those pairs $(R_{\mathsf{X}}, R_{\mathsf{Y}})$ which satisfy the inequalities

$$R_{\mathsf{X}} + R_{\mathsf{Y}} \geqq H(X, Y), \quad R_{\mathsf{X}} \geqq H(X|Y), \quad R_{\mathsf{Y}} \geqq H(Y|X).$$

At present, we shall consider source networks having the same underlying network, but different fidelity criteria. First, we shall discuss the source network of Fig. 16.4. Denote the corresponding ε-achievable rate region by $\mathcal{R}_\varepsilon(\langle X\rangle, Y)$, and the achievable rate region by $\mathcal{R}(\langle X\rangle, Y)$. Note that the problem of determining $\mathcal{R}(\langle X\rangle, Y)$ is a natural generalization of that of Theorem 13.2. Namely, considering the component source with generic variable Y as the main source and the other one as the side information source, in the present problem only *partial side information* on X^n is available at the decoder.

THEOREM 16.4 Given a 2-source with generic variables (X, Y), the pair $(R_{\mathsf{X}}, R_{\mathsf{Y}})$ is an element of $\mathcal{R}(\langle X\rangle, Y)$ iff there exists a RV U such that

$$U \multimap X \multimap Y, \quad |\mathsf{U}| \leqq |\mathsf{X}| + 2, \tag{16.25}$$

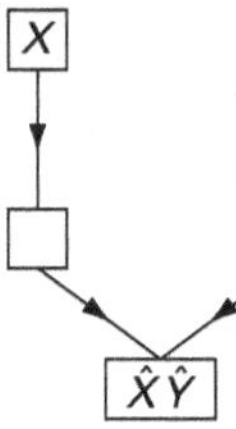

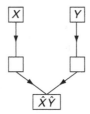

Figure 16.3 Fork network.

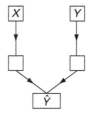

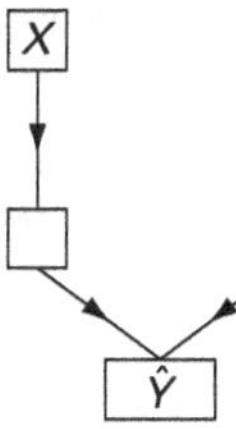

Figure 16.4 Partial side information at the decoder.

and

$$R_{\mathsf{X}} \geqq I(U \wedge X), \quad R_{\mathsf{Y}} \geqq H(Y|U).$$

→ 16.6 Moreover, for every $0 < \varepsilon < 1$

$$\mathcal{R}_\varepsilon(\langle X \rangle, Y) = \mathcal{R}(\langle X \rangle, Y).$$ ○

Proof The source network of Fig. 16.4 is a normal source network with one helper. Thus by Theorem 13.15, the achievable rate region is the closure of the set of those vectors $(R_{\mathsf{X}}, R_{\mathsf{Y}})$ to which for some $n \geqq 1$ there exists a function f with domain X^n such that

$$R_{\mathsf{X}} \geqq \frac{1}{n} H(f(X^n)), \quad R_{\mathsf{Y}} \geqq \frac{1}{n} H(Y^n | f(X^n)).$$

Hence by Theorem 15.20, $(R_{\mathsf{X}}, R_{\mathsf{Y}}) \in \mathcal{R}(\langle X \rangle, Y)$ iff

$$R_{\mathsf{X}} \geqq H(X) - a, \quad R_{\mathsf{Y}} \geqq b \quad \text{for some} \quad (a, b) \in \mathcal{H}^*(X; Y|X). \tag{16.26}$$

This is already a computable characterization of $\mathcal{R}(\langle X \rangle, Y)$. The simple characterization stated in the theorem follows using the form of $\mathcal{H}^*(X; Y|X)$ given in Lemma 15.12.

It remains to prove the strong converse. To this end, let (f_k, g_k, φ_k) be any k-length block code for this source network such that

$$\Pr\{\varphi_k(f_k(X^k), g_k(Y^k)) \neq Y^k\} < \varepsilon. \tag{16.27}$$

For every $\mathbf{x} \in \mathsf{X}^k$ denote by $\mathsf{B}(\mathbf{x})$ the set of those elements $\mathbf{y}$ of Y^k for which no decoding error occurs, i.e., for which

$$\varphi_k(f_k(\mathbf{x}), g_k(\mathbf{y})) = \mathbf{y}.$$

With this notation, (16.27) can be rewritten as

$$\sum_{\mathbf{x} \in \mathsf{X}^k} P_X^k(\mathbf{x}) V^k(\mathsf{B}(\mathbf{x})|\mathbf{x}) \geqq 1 - \varepsilon, \tag{16.28}$$

where $V = P_{Y|X}$. Denote by $\tilde{\mathsf{A}}$ the set of those sequences $\mathbf{x} \in \mathsf{T}^k_{[X]} \subset \mathsf{X}^k$ for which $V^k(\mathsf{B}(\mathbf{x})|\mathbf{x}) \geqq (1 - \varepsilon)/2$. By (16.28) and Lemma 2.12, for $k \geqq k_1(|\mathsf{X}|, \varepsilon)$ we have $P_X^k(\tilde{\mathsf{A}}) \geqq (1 - \varepsilon)/3$ and hence by Lemma 2.14 for $k \geqq k_2(|\mathsf{X}|, \varepsilon, \delta)$

$$\frac{1}{k} \log |\tilde{\mathsf{A}}| \geqq H(X) - \delta, \tag{16.29}$$

where $\delta > 0$ is arbitrary. Clearly, there exists a set $\mathsf{A} \subset \tilde{\mathsf{A}} \subset \mathsf{T}^k_{[X]}$ such that f_k is constant on A and

$$|\mathsf{A}| \geqq \frac{|\tilde{\mathsf{A}}|}{\|f_k\|}.$$

By (16.29) this set will satisfy

$$\frac{1}{k} \log |\mathsf{A}| \geqq H(X) - \delta - \frac{1}{k} \log \|f_k\| \tag{16.30}$$

for $k \geqq k_2$. Consider the set

$$\mathsf{B} \triangleq \bigcup_{\mathbf{x}\in\mathsf{A}} \mathsf{B}(\mathbf{x}).$$

By definition, B is an $((1-\varepsilon)/2)$-image of A over the channel V^k, and thus

$$|\mathsf{B}| \geqq g_{V^k}\left(\mathsf{A}, \frac{1-\varepsilon}{2}\right). \tag{16.31}$$

Furthermore, since f_k is constant on A, say $f_k = c$, and all the sequences $\mathbf{y} \in \mathsf{B}$ satisfy $\varphi_k(c, g_k(\mathbf{y})) = \mathbf{y}$, we must have $\|g_k\| \geqq |\mathsf{B}|$. Comparing this with (16.31) we obtain

$$\|g_k\| \geqq g_{V^k}\left(\mathsf{A}, \frac{1-\varepsilon}{2}\right). \tag{16.32}$$

Now, by Theorem 15.10, for $k \geqq k_3(|\mathsf{X}|, |\mathsf{Y}|, \varepsilon, \delta)$ there exists a pair (a, b) in $\mathcal{H}^*(X; Y|X)$ such that

$$\left|\frac{1}{k}\log|\mathsf{A}| - a\right| \leqq \delta, \quad \left|\frac{1}{k}\log g_{V^k}\left(\mathsf{A}, \frac{1-\varepsilon}{2}\right) - b\right| \leqq \delta.$$

Thus, (16.30) and (16.32) result in

$$\frac{1}{k}\log\|f_k\| \geqq H(X) - a - 2\delta, \quad \frac{1}{k}\log\|g_k\| \geqq b - \delta.$$

As δ was arbitrary, this proves the strong converse; see (16.26). □

More generally, consider now the source network of Fig. 16.5 with a 2-source with generic variables (X, Y) and an arbitrary distortion measure on $\mathsf{X} \times \mathsf{Y} \times \mathsf{Z}$, where Z is the reproduction alphabet. The problem of computable characterization of the achievable rate region of this source network is not within the reach of our method and seems very hard. The next results exhibit a boundary point of the achievable rate region.

THEOREM 16.5 Given a source network with a distortion measure as above, let (U, Z) be any pair of RVs such that

$$Z = h(U, X) \quad \text{for some mapping} \quad h : \mathsf{U} \times \mathsf{X} \to \mathsf{Z}, \tag{16.33}$$

$$U \multimap Y \multimap X, \tag{16.34}$$

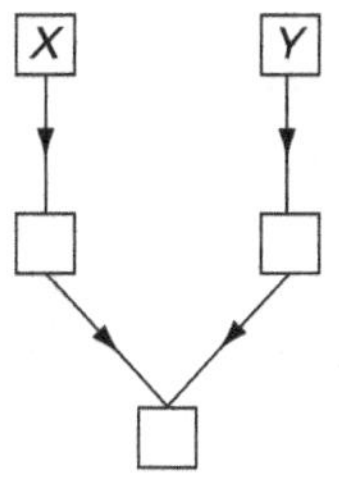

Figure 16.5 Fork network with arbitrary fidelity criterion.

$$Ed(X, Y, Z) < \Delta. \tag{16.35}$$

→ 16.7 Then $(H(X), I(U \wedge Y|X))$ is an achievable rate pair at distortion level Δ. ○

Proof Fix arbitrary numbers $\tau > 0$ and $0 < \varepsilon < 1$. Consider any pair of RVs (U, Z) satisfying (16.33), (16.34) and (16.35). For $k \geqq k_1(|\mathsf{U}|, |\mathsf{X}|, |\mathsf{Y}|, \tau)$, by Corollary 15.3 there exists a function $\psi : \mathsf{Y}^k \to \mathsf{U}^k$ such that

$$\frac{1}{k}H(\psi(Y^k)) \leqq I(U \wedge Y) + \tau, \tag{16.36}$$

$$\frac{1}{k}H(X^k|\psi(Y^k)) \leqq H(X|U) + \tau \tag{16.37}$$

and

$$\Pr\{(Y^k, \psi(Y^k)) \in \mathsf{T}^k_{[YU]}\} \geqq 1 - \frac{\tau}{2}.$$

This and the Markov property $U \multimap Y \multimap X$ imply for a suitable definition of $\mathsf{T}_{[XYU]} = \mathsf{T}^k_{[XYU]}$ (see Convention 2.11) that for $k \geqq k_2$

$$\Pr\{(X^k, Y^k, \psi(Y^k)) \in \mathsf{T}_{[XYU]}\} \geqq 1 - \tau. \tag{16.38}$$

Define for every $\mathbf{x} \in \mathsf{X}^k$, $\mathbf{u} \in \mathsf{U}^k$ the sequence $h(\mathbf{u}, \mathbf{x}) \in \mathsf{Z}^k$ so that its jth element is $h(u_j, x_j)$, where $\mathbf{x} = x_1x_2 \dots x_k$, $\mathbf{u} = u_1u_2 \dots u_k$ and $h : \mathsf{U} \times \mathsf{X} \to \mathsf{Z}$ is the function appearing in (16.33). Now, $(\mathbf{x}, \mathbf{y}, \mathbf{u}) \in \mathsf{T}_{[XYU]}$ implies $(\mathbf{x}, \mathbf{y}, h(\mathbf{u}, \mathbf{x})) \in \mathsf{T}_{[XYZ]}$, for a suitable definition of $\mathsf{T}_{[XYZ]} = \mathsf{T}^k_{[XYZ]\delta_k}$. Similarly to the proof of the rate-distortion theorem, Theorem 7.3, it follows that

$$d(\mathbf{x}, \mathbf{y}, h(\mathbf{u}, \mathbf{x})) \leqq \sum_{x \in \mathsf{X}} \sum_{y \in \mathsf{Y}} \sum_{z \in \mathsf{Z}} P_{XYZ}(x, y, z) d(x, y, z) + \delta_k |\mathsf{X}||\mathsf{Y}||\mathsf{Z}| d_{\mathrm{M}},$$

where d_{M} is the largest value of the function $d(x, y, z)$ over $\mathsf{X} \times \mathsf{Y} \times \mathsf{Z}$. Thus (16.38) yields for $k \geqq k_3$

$$\Pr\{d(X^k, Y^k, h(\psi(Y^k), X^k)) \leqq Ed(X, Y, Z) + \tau d_{\mathrm{M}}\} \geqq 1 - \tau. \tag{16.39}$$

The proof will be completed by applying the rate slicing corollary, Corollary 13.2. First observe that (16.37) and (16.36) yield

$$\begin{aligned} \frac{1}{k}H(\psi(Y^k)|X^k) &= \frac{1}{k}H(X^k|\psi(Y^k)) + \frac{1}{k}H(\psi(Y^k)) - H(X) \\ &\leqq H(X|U) + \tau + I(U \wedge Y) + \tau - H(X) \\ &= I(U \wedge Y) - I(U \wedge X) + 2\tau = I(U \wedge Y|X) + 2\tau. \end{aligned} \tag{16.40}$$

Consider a "hyper-DMMS" with generic variables $(\hat{X}, \hat{U})$, where $\hat{X} \triangleq X^l$, $\hat{U} \triangleq \psi(\hat{Y})$, $\hat{Y} \triangleq Y^l$ and l is so large that (16.40) and (16.39) hold for $k = l$. The latter means, with this new notation,

$$\Pr\{d(\hat{X}, \hat{Y}, h(\hat{U}, \hat{X})) \leqq Ed(X, Y, Z) + \tau d_{\mathrm{M}}\} \geqq 1 - \tau. \tag{16.41}$$

By the rate slicing corollary, it follows that for sufficiently large s there exist functions $f : \mathsf{X}^{ls} \to \mathsf{M}_\mathsf{X}$, $g : \mathsf{Y}^{ls} \to \mathsf{M}_\mathsf{Y}$, $\vartheta : \mathsf{M}_\mathsf{X} \times \mathsf{M}_\mathsf{Y} \to \mathsf{U}^{ls} \times \mathsf{X}^{ls}$ such that

$$\begin{aligned}\frac{1}{ls}\log\|f\| &\leqq H(X)+\tau,\\ \frac{1}{ls}\log\|g\| &\leqq \frac{1}{l}H(\hat{U}|\hat{X})+\tau \leqq I(U\wedge Y|X)+3\tau\end{aligned} \tag{16.42}$$

and

$$\Pr\{\vartheta(f(\hat{X}^s), g(\hat{Y}^s)) \neq (\hat{U}^s, \hat{X}^s)\} \leqq \frac{\varepsilon}{2}, \tag{16.43}$$

where $\hat{X}^s \triangleq \hat{X}_1\hat{X}_2\dots\hat{X}_s = X_1X_2\dots X_{ls}$, etc.

Let N be the number of indices $1 \leqq i \leqq s$ for which $d(\hat{X}_i, \hat{Y}_i, h(\hat{U}_i, \hat{X}_i)) > Ed(X, Y, Z) + \tau d_\mathrm{M}$. Then (16.41) implies for sufficiently large s that

$$\Pr\left\{\frac{N}{s} < 2\tau\right\} > 1 - \frac{\varepsilon}{2}. \tag{16.44}$$

Define the mapping $\varphi : \mathsf{M}_\mathsf{X} \times \mathsf{M}_\mathsf{Y} \to \mathsf{Z}^{ls}$ by $\varphi(m_1, m_2) \triangleq h(\vartheta(m_1, m_2))$. Since

$$d(\hat{X}^s, \hat{Y}^s, h(\hat{U}^s, \hat{X}^s)) = \frac{1}{s}\sum_{i=1}^{s} d(\hat{X}_i, \hat{Y}_i, h(\hat{U}_i, \hat{X}_i)),$$

from (16.43) and (16.44) one obtains

$$\Pr\{d(X^{ls}, Y^{ls}, \varphi(f(X^{ls}), g(Y^{ls})) \leqq Ed(X, Y, Z) + 3\tau d_\mathrm{M}\} > 1 - \varepsilon.$$

Recalling that τ was arbitrary, we have thus shown that for every $\delta > 0$, $k = ls$ and sufficiently large s there exist k-length block codes (f_k, g_k, φ_k) satisfying

$$\frac{1}{k}\log\|f_k\| \leqq H(X)+\delta, \quad \frac{1}{k}\log\|g_k\| \leqq I(U\wedge Y|X)+\delta$$

and

$$\Pr\{d(X^k, Y^k, \varphi_k(f_k(X^k), g_k(Y^k))) \leqq Ed(X, Y, Z) + \delta\} > 1 - \varepsilon.$$

Hence the existence of such codes for every sufficiently large k follows immediately. Recalling that $Ed(X, Y, Z) < \Delta$, this completes the proof. □

Instead of proving the corresponding (weak) converse result, we shall establish a slightly stronger theorem involving the average distortion of the code.

THEOREM 16.6 Given a source network with a distortion measure as in Theorem 16.5, for every k-length block code (f, g, φ) with average distortion

$$Ed(X^k, Y^k, \varphi(f(X^k), g(Y^k))) \leqq \Delta \tag{16.45}$$

we have

$$\frac{1}{k}\log\|g\| \geqq I(U\wedge Y|X)$$

for some pair of RVs (U, Z) satisfying

$$Z = h(U, X) \quad \text{for some mapping} \quad h : \mathsf{U} \times \mathsf{X} \to \mathsf{Z}, \tag{16.46}$$

$$U \multimap Y \multimap X, \quad |\mathsf{U}| \leqq |\mathsf{Y}| + 1, \tag{16.47}$$

$$Ed(X, Y, Z) \leqq \Delta. \tag{16.48}$$

○

→ 16.22

COROLLARY 16.6 $(H(X), R)$ is an achievable rate pair at distortion level $\Delta > 0$ for the source network of Theorems 16.5 and 16.6 iff

$$R \geqq I(U \wedge Y|X)$$

for some pair of RVs (U, Z) with the properties (16.46)–(16.48). ○

Proof Introduce the function

$$R^*(\Delta) = R^*_{XY}(\Delta) \triangleq \inf_{Ed(X,Y,Z) \leqq \Delta} I(U \wedge Y|X), \tag{16.49}$$

where (U, Z) ranges over pairs of RVs having properties (16.46) and (16.47), except possibly for the constraint on the range of U. We first show that imposing that constraint does not increase the infimum.

For any RV U satisfying $U \multimap Y \multimap X$, denote

$$\Delta(U) \triangleq \min_{h:\mathsf{U}\times\mathsf{X}\to\mathsf{Z}} Ed(X, Y, h(U, X)). \tag{16.50}$$

Clearly

$$\begin{aligned} R^*(\Delta) &= \inf_{\Delta(U)\leqq\Delta} I(U \wedge Y|X) = \inf_{\Delta(U)\leqq\Delta} [I(U \wedge Y) - I(U \wedge X)] \\ &= H(Y) - H(X) + \inf_{\Delta(U)\leqq\Delta} [H(X|U) - H(Y|U)]. \end{aligned} \tag{16.51}$$

Hence our first claim will be established by showing that the set of pairs

$$(\Delta(U), \; H(X|U) - H(Y|U)) \tag{16.52}$$

for RVs U satisfying the Markov condition in (16.47) remains the same if also the range constraint in (16.47) is imposed. This, however, is a straightforward consequence of the support lemma, Lemma 15.4. In fact, consider the following $|\mathsf{Y}| + 1$ real-valued continuous functions on $\mathcal{P}(\mathsf{Y})$:

$$f_y(P) \triangleq P(y) \quad \text{for all but one} \quad y \in \mathsf{Y},$$

$$f_{\mathsf{XY}}(P) \triangleq H(P\widehat{W}) - H(P),$$

$$f_d(P) \triangleq \min_{l:\mathsf{X}\to\mathsf{Z}} \sum_{x\in\mathsf{X}} \sum_{y\in\mathsf{Y}} P(y)\widehat{W}(x|y)d(x, y, l(x)),$$

where $\widehat{W} = P_{X|Y}$, and the minimum is for all mappings as shown.

Parallelling the proof of Lemma 15.5, let U now be any RV satisfying $U \multimap Y \multimap X$. Apply the support lemma with the distribution of U in the role of μ, i.e., considering the distribution $P_{Y|U}(\cdot|u)$ as an element of $\mathcal{P}(\mathsf{Y})$ that has μ-measure $\Pr\{U = u\}$. The corresponding averages are

$$\sum_{u \in \mathsf{U}} \Pr\{U = u\} f_y(P_{Y|U}(\cdot|u)) = \Pr\{Y = y\},$$
$$\sum_{u \in \mathsf{U}} \Pr\{U = u\} f_{\mathsf{XY}}(P_{Y|U}(\cdot|u)) = H(X|U) - H(Y|U),$$
$$\sum_{u \in \mathsf{U}} \Pr\{U = u\} f_d(P_{Y|U}(\cdot|u)) = \Delta(U).$$

By the support lemma, there exist $|\mathsf{Y}| + 1$ elements of $\mathcal{P}(\mathsf{Y})$, say P_u, $u \in \mathsf{U}_0$, and numbers $\alpha(u) \geqq 0$, $u \in \mathsf{U}_0$, of sum 1 such that

$$\sum_{u \in \mathsf{U}_0} \alpha(u) f_y(P_u) = \Pr\{Y = y\},$$
$$\sum_{u \in \mathsf{U}_0} \alpha(u) f_{\mathsf{XY}}(P_u) = H(X|U) - H(Y|U),$$
$$\sum_{u \in \mathsf{U}_0} \alpha(u) f_d(P_u) = \Delta(U).$$

Replacing the RV U by U_0 with range U_0 and distribution $\{\alpha(u),\ u \in \mathsf{U}_0\}$, and regarding P_u as the conditional distribution of Y on the condition $U_0 = u$, this establishes our first claim.

Next observe that $R^*(\Delta)$ is a continuous convex function of Δ, and in its definition the infimum is actually a minimum. On account of (16.51), this follows if we check that the set of pairs (16.52) with $U \multimap Y \multimap X$ is convex and closed. As this set does not change by imposing the range constraint on U, its closedness is obvious from the continuity of the involved quantities. Its convexity can be verified similarly to the proof of Lemma 15.6. Namely, if both U_1 and U_2 satisfy the Markov condition, then so does $U \triangleq (U_I, I)$, where I is a two-valued RV independent of $U_1 U_2 X Y$. If $\Pr(I = 1) = \lambda$, we have

$$\Delta(U) = \lambda \Delta(U_1) + (1 - \lambda) \Delta(U_2),$$

and, as in the proof of Lemma 15.6,

$$H(X|U) = \lambda H(X|U_1) + (1 - \lambda) H(X|U_2),$$
$$H(Y|U) = \lambda H(Y|U_1) + (1 - \lambda) H(Y|U_2).$$

These three relations establish the convexity.

Now the assertion of the theorem will be proved if we show that (16.45) implies

$$\frac{1}{k} \log \|g\| \geqq R^*(\Delta). \tag{16.53}$$

Writing $U \triangleq g(Y^k)$ we have

$$\begin{aligned}\frac{1}{k}\log\|g\| &\geqq \frac{1}{k}H(U) \geqq \frac{1}{k}H(U|X^k) \geqq \frac{1}{k}I(U \wedge Y^k|X^k)\\ &= H(Y|X) - \frac{1}{k}H(Y^k|X^kU).\end{aligned} \tag{16.54}$$

Here

$$\frac{1}{k}H(Y^k|X^kU) = \frac{1}{k}\sum_{i=1}^{k} H(Y_i|X^kY^{i-1}U). \tag{16.55}$$

Define new RVs by $U_i \triangleq (X_1, \dots, X_{i-1}, X_{i+1}, \dots, X_k, Y^{i-1}, U)$, $i = 1, \dots, n$. Since $U = g(Y^k)$, and $X^{i-1}X_{i+1}\cdots X_kY^k \multimap Y_i \multimap X_i$, these satisfy

$$U_i \multimap Y_i \multimap X_i. \tag{16.56}$$

With this notation, (16.55) becomes

$$\frac{1}{k}H(Y^k|X^kU) = \frac{1}{k}\sum_{i=1}^{k} H(Y_i|X_iU_i),$$

and substituting this into (16.54) we obtain

$$\frac{1}{k}\log\|g\| \geqq \frac{1}{k}\sum_{i=1}^{k} I(U_i \wedge Y_i|X_i). \tag{16.57}$$

Denote $Z^k \triangleq \varphi(f(X^k), g(Y^k))$. Since Z^k is a function of U_i and X_i, its ith component Z_i is a function of U_i and X_i. With the notation

$$\Delta_i \triangleq Ed(X_i, Y_i, Z_i), \tag{16.58}$$

it follows by the definition of $R^*(\Delta) = R^*_{XY}(\Delta) = R^*_{X_iY_i}(\Delta)$ that

$$I(U_i \wedge Y_i|X_i) \geqq R^*(\Delta_i) \quad \text{for every } i.$$

Thus, by (16.57) and the convexity of $R^*(\Delta)$

$$\frac{1}{k}\log\|g\| \geqq \frac{1}{k}\sum_{i=1}^{k} R^*(\Delta_i) \geqq R^*\left(\frac{1}{k}\sum_{i=1}^{k}\Delta_i\right). \tag{16.59}$$

On account of (16.58) and (16.45) we have

$$\frac{1}{k}\sum_{i=1}^{k}\Delta_i = Ed(X^k, Y^k, Z^k) \leqq \Delta.$$

Thus, using the monotonicity of $R^*(\Delta)$, (16.59) results in (16.53), proving the theorem.

The corollary says that $(H(X), R)$ is an achievable rate pair at distortion level $\Delta > 0$ iff $R \geqq R^*(\Delta)$. The "if-part" follows from Theorem 16.5 by the continuity of the function $R^*(\Delta)$. To prove the converse, suppose that to every $\varepsilon > 0$ and $\delta > 0$ there exists a k-length block code (f, g, φ) with

$$\Pr\{d(X^k, Y^k, \varphi(f(X^k), g(Y^k))) \leqq \Delta\} \geqq 1 - \varepsilon$$

and

$$\frac{1}{k} \log \|g\| \leqq R + \delta.$$

By the theorem, these relations imply $R + \delta \geqq R^*(\Delta + \varepsilon d_{\mathrm{M}})$. This completes the proof by continuity. □

Discussion

This chapter contains recent results, most of them presented in problems. The text serves as an illustration of the methods representative for the present state of the art. Our guiding principle in choosing the material was to concentrate on computable characterizations of capacity resp. achievable rate regions. Of course, these characterizations required bounds on the ranges of the involved auxiliary random variables. We did not attempt, however, to find the best possible bounds; for example, the bound on $|\mathsf{U}|$ in the ABC coding theorem could be easily improved by 1. When no computable charac- → 15.1
terization of the full region was available, we tried to give partial results of *conclusive* type, as, e.g., the characterization of a projection of the region in question. While the methods used often yield non-trivial inner or outer bounds to capacity resp. achievable rate regions, non-coinciding bounds of this type are considered non-conclusive results and are mentioned only exceptionally.

In the field of channel networks, almost all the known results not treated in Chapter 14 relate to broadcast channels. The capacity region of a general two-output broadcast channel (Fig. 16.1) is still unknown. Out of its three projections, only those to the (R_1, R_0) and (R_0, R_2) planes have been characterized (ABC coding theorem). In the case of comparable component channels and a few other special cases the whole capacity region has been determined (see Problems 16.4, 16.5, 16.8, 16.9 and 16.11). Recall that for broadcast channels the m- and a-capacity regions are the same (Problem 14.13). In all cases mentioned above, i.e., when a conclusive result is known, the strong converse holds for the maximum probability of error. For the average probability of error this is no longer true (Problem 14.13), but for the ABC the ε-capacity region is the same for every $0 < \varepsilon < 1/2$.

Our knowledge of source networks, though limited, is more complete. Using the general results of Chapters 13 and 15, one can determine the achievable rate region of a number of source networks. Theorem 16.4 illustrates this approach. Source networks with arbitrary fidelity criteria are beyond the scope of this method; even for the simple network of Fig. 16.5, only a projection of the achievable rate region is known (Corollary 16.6).

Problems 16.13–16.18 provide examples of the general method. The source networks in these problems, when reduced to normal form, contain one helper. Thus the product-space characterization of Theorem 13.15 leads to an entropy characterization problem. Since the corresponding set $\mathcal{F}$ has a computable characterization, so does the achievable rate region. Further, whenever the corresponding sets $\mathcal{F}$ and $\mathcal{G}$ are equal, the strong converse can be proved. Of course, for the solution of an individual source network problem, partial knowledge about the corresponding set $\mathcal{F}$ (resp. $\mathcal{G}$) often suffices, as only a part of the boundary of $\mathcal{F}$ enters the characterization of the achievable rate region. This is why the achievable rate region of certain source networks with many inputs can be determined; see Problems 16.6 and 16.31.

As a rule, source networks with more than one helper represent very hard problems. → 16.19
An exception is when no pair of helpers is connected to the same output. In particular, the problem of Fig. P16.20 appears formidable. Even in the very special case of Problem 16.20, the solution relies on a trick. Difficulties also arise concerning the source → 16.22
network of Fig. 16.5 with general fidelity criteria. Note that a general solution of this problem would contain a solution of the problem of Fig. P16.20 for the case when Z is → 16.23
an arbitrary function of the pair X, Y. Further, the problem of Fig. P16.20 for an arbitrary 3-source is a special case of a fork-network-type problem with an arbitrary fidelity criterion and the same 3-source.

Since the first edition of this book, the body of multi-user information theory has grown substantially. Some of the developments were hinted at in the Discussion of Chapter 14. It should be noted that conclusive results are still rare, and "hard problems" from the 1970s are still unresolved, including those discussed in the preceding paragraph. Regarding channel networks, the achievable rates for a general two-output broadcast channel given in Problem 16.10, or for the interference channel referred to (but not stated) in Problem 14.17, have neither been improved nor shown to be best possible.

Unlike elsewhere, in this chapter we did not make additions to the Problems to illustrate progress. Only one problem has been added (Problem 16.22). However, the new Chapter 17 is devoted to the subject of information-theoretic security, where progress has been very impressive. Whereas in the models treated so far all parties were assumed to cooperate towards optimally utilizing the available resources, models involving security have a non-cooperative aspect: a major goal is to conceal (at least part of) the exchanged information from (at least) one party. Note, however, that non-cooperative → 16.33
models may arise also outside the field of security. In the first edition, the then available information-theoretic results about secure transmission over insecure channels were included in the Problems part of this chapter; these results have now been moved to Chapter 17. ○

Problems

16.1. Show that the capacity region of a k-output broadcast channel can be obtained from the special case of such a channel network where the underlying network has only $2^k - 1$ inputs.

16.2. (a) Show that for every (X, Y, Z) such that Y resp. Z is connected with X by the channels V resp. W, the pairs (R_1, R_0) defined by

$$R_1 \triangleq \lambda I(X \wedge Y), \quad R_0 \triangleq (1-\lambda)\min[I(X \wedge Y), I(X \wedge Z)]$$

are achievable rate pairs for the ABC with component channels V and W, whatever the value of $0 \leqq \lambda \leqq 1$. Observe that the region of the rate pairs so obtained is smaller than that in Theorem 16.1, except for special cases.

Hint Apply the compound channel coding theorem (Corollary 10.10) and the time sharing principle.

(b) Show that (16.1) and (16.2) determine the same region as (16.2) and

$$R_1 \leqq I(X \wedge Y|U), \quad R_0 + R_1 \leqq I(X \wedge Y).$$

Hint Show that the new region is convex for fixed *XYZ*. (Körner and Marton (1977b).)

(c) Let S and T be independent RVs, and let $X = g(S, T)$ be a function of S and T. Further, let Y, Z be any RVs such that $ST \multimap X \multimap YZ$, $P_{Y|X} = V$, $P_{Z|X} = W$. Show that a pair of non-negative numbers (R_1, R_0) satisfying the inequalities

$$R_1 \leqq I(S \wedge Y|T), \quad R_0 \leqq I(T \wedge Y|S), \quad R_0 \leqq I(T \wedge Z),$$
$$R_1 + R_0 \leqq I(ST \wedge Y)$$

is an achievable rate pair for our ABC.

Hint Consider the channel network of Fig. P14.18A with component channels $\hat{V} : \mathsf{S} \times \mathsf{T} \to \mathsf{Y}$ and $\hat{W} : \mathsf{S} \times \mathsf{T} \to \mathsf{Z}$, where $\hat{V} \triangleq P_{Y|ST}$, $\hat{W} \triangleq P_{Z|ST}$, and apply the result of Problem 14.18. (Cover (1975a); Van der Meulen (1975).)

(d) Show that the result of (c) implies Theorem 16.1.

Hint It follows from the proof of Theorem 16.1 (see (16.3)–(16.5)) that for every quadruple of RVs (T, X, Y, Z) such that $T \multimap X \multimap YZ$ and Y resp. Z are connected with X by the respective channels V and W, and $I(T \wedge Z) \leqq I(T \wedge Y)$, a pair (R_1, R_0) with

$$R_1 \leqq I(X \wedge Y|T), \quad R_0 \leqq I(T \wedge Z)$$

is an achievable rate pair for the ABC. Further, the achievability of these pairs implies Theorem 16.1, as shown there. Hence it is sufficient to prove that these pairs can be obtained in the form of (c). For this sake represent the stochastic matrix $\hat{V} \triangleq P_{X|T}$ as the convex combination of stochastic zero–one matrices V_k as

$$\hat{V} = \sum_{k=1}^{K} \lambda_k V_k.$$

Let S be a RV independent of T with $\Pr\{S = k\} = \lambda_k$. Set $\hat{X} \triangleq f_S(T)$, where f_k is the function uniquely determined by the zero–one matrix V_k. Let $\hat{Y}$ resp. $\hat{Z}$ be any RVs connected with $\hat{X}$ by the channels V resp. W, so that $ST \multimap \hat{X} \multimap \hat{Y}\hat{Z}$. Prove that this quintuple satisfies (c). (Körner and Marton (1977b).)

16.3. (*Converse-type bound for two-output BC*) Show that if $(R_1, 0, R_2)$ is an achievable rate triple for a BC with component channels $V : \mathsf{X} \to \mathsf{Y}$, $W : \mathsf{X} \to \mathsf{Z}$, then there exist RVs U, X, Y, Z as in Theorem 16.3 and a number $0 \leqq t \leqq \min[I(U \wedge Y), I(U \wedge Z)]$ such that

$$R_1 \leqq I(X \wedge Y|U) + t, \quad R_2 \leqq I(U \wedge Z) - t.$$

Hint The proof of Theorem 16.3 literally applies. In fact, in that proof the appearance of input 0 also in the addressing at the output of channel V was exploited only in deriving the bound $R_0 + R_1 \leqq (X \wedge Y)$. (Joint result of J. Körner and K. Marton, published in Marton (1979).)

16.4. (*Degraded broadcast channel*) Consider a two-output BC such that channel W is a degraded version of channel V, see Problem 14.16, that is,

$$W(z|x) = \sum_{v \in \mathsf{Y}} V(y|x) D(z|y) \quad \text{for every } x \in \mathsf{X}, z \in \mathsf{Z},$$

for some stochastic matrix $D : \mathsf{Y} \to \mathsf{Z}$.

(a) For such a BC, prove the weak converse to Theorem 16.1 by applying Fano's inequality.

(b) Show that $(R_1, R_0, R_2) \geqq \mathbf{0}$ is an achievable rate triple for this BC iff $(R_1, R_0 + R_2)$ satisfy the inequalities

$$R_1 \leqq I(X \wedge Y|U),$$
$$R_0 + R_2 \leqq I(U \wedge Z),$$
$$R_1 + R_0 + R_2 \leqq I(X \wedge Y)$$

for some quadruple of RVs (U, X, Y, Z) as in Theorem 16.3; moreover, it suffices to consider RVs satisfying the Markov condition $U \multimap X \multimap Y \multimap Z$, and $P_{Y|X} = V$, $P_{Z|X} = D$.

Hint (a) Supposing that (R_1, R_0) is an achievable rate pair, consider a sequence of n-length block codes (f_n, φ_n, ψ_n) achieving (R_1, R_0). Let the RVs M_1 and M_0 be independent and uniformly distributed over the corresponding message sets. Put

$$X^n \triangleq f_n(M_1, M_0)$$

and let the RVs (Y^n, Z^n) satisfy the relations

$$M_1 M_0 \multimap X^n \multimap Y^n \multimap Z^n, \quad P_{Y^n|X^n} = V^n, \quad P_{Z^n|Y^n} = D^n,$$

where D is the stochastic matrix as above. We have

$$R_0 - \delta \leqq \frac{1}{n} H(M_0) = \frac{1}{n} I(M_0 \wedge Z^n) + \frac{1}{n} H(M_0|Z^n)$$

and

$$R_1 - \delta \leqq \frac{1}{n} H(M_1|M_0) = \frac{1}{n} I(M_1 \wedge Y^n|M_0) + \frac{1}{n} H(M_1|Y^n M_0)$$
$$\leqq \frac{1}{n} I(X^n \wedge Y^n|M_0) + \frac{1}{n} H(M_1|Y^n).$$

In both cases, the right-most term is converging to zero by Fano's inequality. Further

$$I(M_0 \wedge Z^n) = H(Z^n) - H(Z^n|M_0) \leqq \sum_{i=1}^{n} H(Z_i) - \sum_{i=1}^{n} H(Z_i|M_0 Z^{i-1})$$
$$\leqq \sum_{i=1}^{n} [H(Z_i) - H(Z_i|M_0 Z^{i-1} Y^{i-1})].$$

Here for every fixed value of M_0 the output Z_i of the DMC $\{D\}$ is conditionally independent of the previous outputs Z^{i-1}, if Y^{i-1} and this value of M_0 are given. Thus we have

$$H(Z_i|M_0 Z^{i-1} Y^{i-1}) = H(Z_i|M_0 Y^{i-1}),$$

whence

$$I(M_0 \wedge Z^n) \leqq \sum_{i=1}^{n} I(M_0 Y^{i-1} \wedge Z_i). \tag{*}$$

Further, using the Markov property

$$Y_i \multimap M_0 X_i \multimap Y^{i-1} Y_{i+1} \cdots Y_n X^{i-1} X_{i+1} \cdots X_n,$$

which is a consequence of Y^n and X^n being connected by a DMC, we get

$$I(X^n \wedge Y^n|M_0) = H(Y^n|M_0) - H(Y^n|M_0 X^n)$$
$$= \sum_{i=1}^{n} [H(Y_i|M_0 Y^{i-1}) - H(Y_i|M_0 X_i Y^{i-1})]$$
$$= \sum_{i=1}^{n} I(X_i \wedge Y_i|M_0 Y^{i-1}). \tag{**}$$

Introduce a RV I uniformly distributed over $\{1, 2, \ldots, n\}$ and independent of all the others. Define

$$U \triangleq M_0 Y^{I-1} I, \quad X \triangleq X_I, \quad Y \triangleq Y_I, \quad Z \triangleq Z_I.$$

Check that (*) implies

$$\frac{1}{n} I(M_0 \wedge Z^n) \leqq I(U \wedge Z|I) \leqq I(U \wedge Z),$$

while (**) means

$$\frac{1}{n} I(X^n \wedge Y^n|M_0) = I(X \wedge Y|U).$$

(b) Use Theorem 16.1 and Problem 6.16.

(Cover (1972) raised the problem and conjectured a similar result. The direct part of the coding theorem was proved by Bergmans (1973). For the case of binary symmetric V and W the weak converse is due to Wyner and Ziv (1973). Gallager (1974a) proved that the convex hull of the above region is the true capacity region, leaving open the question whether the region itself is convex, as conjectured by both Cover and Bergmans. This was proved by Ahlswede and Körner (1975). The corresponding strong converse (which follows now from Theorem 16.3 by (b)) was established by Ahlswede, Gács and Körner (1976).)

16.5. (*Broadcast channel with comparable components*) Consider a two-output BC such that the channel V is more capable than channel W in the sense of Problem 6.18.

(a) Show that the necessary and sufficient condition for the achievability of a rate triple $(R_1, R_0, R_2) \geqq \mathbf{0}$ is the same as in the degraded case; see Problem 16.4(b) (but the last assertion of Problem 16.4(b) no longer holds). *Hint* Use the ABC coding theorem, Problem 16.2(b) and Problem 6.18. (Körner and Marton (1977a) proved this when V is less noisy than W in the sense of Problem 15.11. Their result was extended to the more capable case by El Gamal (1979).)

(b) Deduce the corresponding strong converse.

Hint Use Theorem 16.3 and Problem 6.18.

16.6. (*One side-information source*)

(a) Prove the direct part of Theorem 16.4 without explicitly using Theorem 13.15 and the general solution of the entropy characterization problem.

Hint Deduce from the rate slicing corollary that for every n and function f with range X^n

$$\left(\frac{1}{n}H(f(X^n)),\quad \frac{1}{n}H(Y^n|f(X^n))\right)$$

is an achievable rate pair. Then use Corollary 15.3. (This proof is due to Ahlswede and Körner (1975).)

(b) Give a direct proof of the weak converse part of Theorem 16.4.

Hint Suppose that $(R_{\mathsf{X}}, R_{\mathsf{Y}})$ is an achievable rate pair and consider a sequence of k-length block codes (f_k, g_k, φ_k) with error probability converging to zero and

$$\overline{\lim_{k\to\infty}} \frac{1}{k}\log \|f_k\| \leqq R_{\mathsf{X}}, \qquad \overline{\lim_{k\to\infty}} \frac{1}{k}\log \|g_k\| \leqq R_{\mathsf{Y}}.$$

Denoting $U \triangleq f_k(X^k)$, $\hat{Y}^k \triangleq \varphi_k(f_k(X^k), g_k(Y^k))$, we have

$$\begin{aligned}\frac{1}{k}\log \|g_k\| &\geqq \frac{1}{k}H(g_k(Y^k)|U) \geqq \frac{1}{k}H(\hat{Y}^k|U) \geqq \frac{1}{k}I(Y^k \wedge \hat{Y}^k|U)\\ &= \frac{1}{k}H(Y^k|U) - \frac{1}{k}H(Y^k|\hat{Y}^k, U),\end{aligned}$$

where $(1/k)H(Y^k|\hat{Y}^k, U)$ tends to zero by Fano's inequality and the properties of (f_k, g_k, φ_k). Further,

$$\frac{1}{k}H(Y^k|U) = \frac{1}{k}\sum_{i=1}^{k} H(Y_i|Y^{i-1}U) \geqq \frac{1}{k}\sum_{i=1}^{k} H(Y_i|Y^{i-1}X^{i-1}U).$$

Check the Markov property $Y_i \multimap X^{i-1}U \multimap Y^{i-1}$; then the last bound gives

$$\frac{1}{k}H(Y^k|U) \geqq \frac{1}{k}\sum_{i=1}^{k} H(Y_i|X^{i-1}U).$$

On the other hand,

$$\frac{1}{k}\log\|f_k\| \geqq \frac{1}{k}H(U) \geqq \frac{1}{k}I(U \wedge X^k) = H(X) - \frac{1}{k}H(X^k|U)$$
$$= H(X) - \frac{1}{k}\sum_{i=1}^{k} H(X_i|X^{i-1}U).$$

Introduce a RV I uniformly distributed over $\{1, 2, \ldots, n\}$ and independent of all the others. Define

$$\tilde{U} \triangleq UX^{I-1}I, \quad \tilde{X} \triangleq X_I, \quad \tilde{Y} \triangleq Y_I.$$

Then $\tilde{U} \multimap \tilde{X} \multimap \tilde{Y}$ and $P_{\tilde{X}\tilde{Y}} = P_{XY}$. With these definitions the preceding two inequalities yield

$$\frac{1}{k}H(Y^k|U) \geqq H(\tilde{Y}|\tilde{U}) \quad \text{and} \quad \frac{1}{k}H(U) \geqq I(\tilde{X} \wedge \tilde{U}).$$

Finally, apply the support lemma, Lemma 15.4. (This proof is due to Gray and Wyner (1974), Wyner (1975b) and Ahlswede and Körner (1975).)

(c) Consider a DMMS with $r+1$ component sources and corresponding generic variables $X, Y^{(1)}, \ldots, Y^{(r)}$. Prove that the achievable rate region of the source network of Fig. P16.6 consists of those $(r+1)$-tuples of real numbers $(R_0, R_1, \ldots, R_r)$ which satisfy the inequalities

$$R_0 \geqq I(X \wedge U),$$
$$R_i \geqq H(Y^{(i)}|U), \quad i = 1, \ldots, r,$$

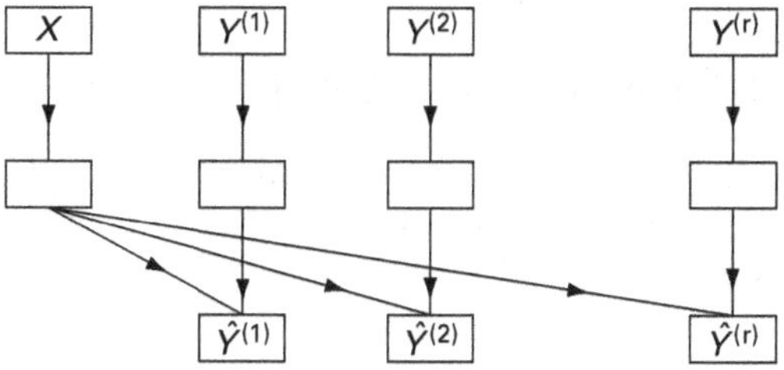

Figure P16.6

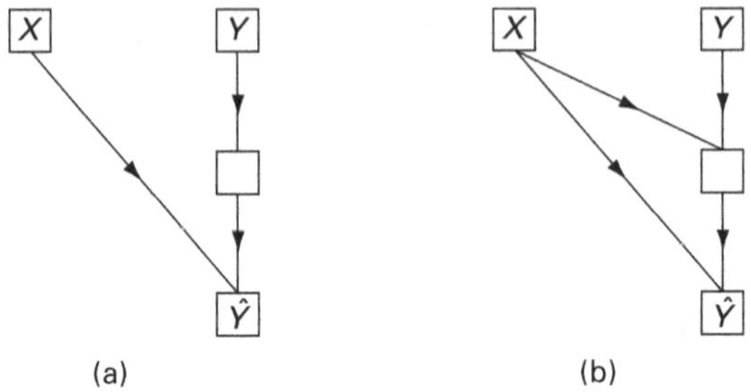

Figure P16.7

for some RV U such that $U \multimap X \multimap Y^{(1)} \ldots Y^{(r)}$ and the range of U has at most $|\mathsf{X}| + r$ elements.

Hint Proceed as in (a) and (b). (For $r \geqq 2$, the result is due to Wyner (1975b) and Ahlswede and Körner (1974, unpublished), quoted in Körner (1975).)

(d) Establish the corresponding strong converse.

Hint Bound the rates by image sizes as in the proof of Theorem 16.4. Bound the latter by entropies using Lemma 15.2. Then proceed as in (b). (This method is due to Ahlswede, Gács and Körner (1976).)

16.7. (*Limitations on rate slicing*) The remarkable phenomenon underlying the rate slicing corollary, Corollary 13.2 (which was basic for Chapter 13) means that the two source networks of Fig. P16.7 have the same interval of achievable rates. Prove that this is not the case for an arbitrary fidelity criterion.

Hint Observe that Corollary 16.6 gives the achievable rate interval of the source network of Fig. P16.7(a) with an arbitrary fidelity criterion. Consider a binary 2-source with generic variables having joint distribution given by

$$P_{XY}(0,0) = P_{XY}(1,1) = \frac{1-p}{2}, \quad P_{XY}(0,1) = P_{XY}(1,0) = \frac{p}{2}, \quad 0 < p < \frac{1}{2}$$

and the distortion measure

$$d(x, y, z) \triangleq \begin{cases} 0 & \text{if} \quad y = z \\ 1 & \text{else.} \end{cases}$$

Show that if $\Delta < p$ then for Fig. P16.7 (b) the minimum achievable rate equals $h(p) - h(\Delta)$, while for Fig. P16.7 (a) this is not an achievable rate unless $\Delta = 0$. (Wyner and Ziv (1976). For a similar phenomenon, see Problem 16.21.)

Broadcast channels (Problems 16.8–16.12)

16.8. (*Product of degraded broadcast channels*) Consider a two-output BC with component channels $V : \mathsf{X}_1 \times \mathsf{X}_2 \to \mathsf{Y}_1 \times \mathsf{Y}_2$ and $W : \mathsf{X}_1 \times \mathsf{X}_2 \to \mathsf{Z}_1 \times \mathsf{Z}_2$, where

$$V(y_1, y_2|x_1, x_2) \triangleq V_1(y_1|x_1)V_2(y_2|x_2),$$
$$W(z_1, z_2|x_1, x_2) \triangleq W_1(z_1|x_1)W_2(z_2|x_2).$$

Suppose that W_1 is a degraded version of V_1, while V_2 is a degraded version of W_2, i.e., the factor channels are degraded in opposed directions.

(a) Prove that $(R_1, R_0, R_2) \geqq \mathbf{0}$ is an achievable rate triple for this BC iff there exists an 8-tuple of RVs U_i, X_i, Y_i, Z_i $(i = 1, 2)$ such that the two quadruples (U_i, X_i, Y_i, Z_i) are independent of each other, Y_i resp. Z_i is connected with X_i by V_i resp. W_i, $U_i \multimap X_i \multimap Y_i Z_i$, $|\mathsf{U}_i| \leqq |\mathsf{X}_i| + 2$ and the following inequalities hold:

$$\begin{aligned}
R_0 &\leqq I(U_1 \wedge Y_1) + I(U_2 \wedge Y_2),\\
R_0 &\leqq I(U_1 \wedge Z_1) + I(U_2 \wedge Z_2),\\
R_0 + R_1 &\leqq I(X_1 \wedge Y_1) + I(U_2 \wedge Y_2),\\
R_0 + R_2 &\leqq I(U_1 \wedge Z_1) + I(X_2 \wedge Z_2),\\
R_0 + R_1 + R_2 &\leqq I(X_1 \wedge Y_1) + I(U_2 \wedge Y_2) + I(X_2 \wedge Z_2|U_2),\\
R_0 + R_1 + R_2 &\leqq I(U_1 \wedge Z_1) + I(X_2 \wedge Y_2) + I(X_1 \wedge Y_1|U_1).
\end{aligned}$$

(Poltyrev (1977); El Gamal (1980). Poltyrev solved the case $R_0 = 0$.)

(b) Check that this capacity region may be larger than the (Minkowski) sum of the capacity regions of the factor BCs, i.e., of those with component channels (V_i, W_i), $i = 1, 2$.

Hint The direct part follows using the construction of Theorem 16.1 with $U = U_1 U_2$.

To prove the converse, the condition of independence of the quadruples (U_i, X_i, Y_i, Z_i) may be disregarded since the joint distribution of the two quadruples does not enter the characterization. Fix some $\delta > 0$, and consider codes with $(1/n) \log |\mathsf{M}_j| > R_j - \frac{\delta}{4}$, $j = 0, 1, 2$ and error probabilities converging to zero.

Let M_0, M_1, M_2 be independent RVs each uniformly distributed over the respective message sets $\mathsf{M}_0, \mathsf{M}_1, \mathsf{M}_2$ and let X^n be the codeword corresponding to the message triple M_1, M_0, M_2. We denote by X_1^n and X_2^n the corresponding inputs of the factor channels, while Y_i^n resp. Z_i^n are the corresponding outputs of V_i^n resp. W_i^n. Then, by Fano's inequality, for n large enough

$$\begin{aligned}
nR_0 &\leqq H(M_0) + n\frac{\delta}{4} \leqq I(M_0 \wedge Y_1^n Y_2^n) + n\delta,\\
n(R_0 + R_1) &\leqq H(M_0) + H(M_1) + n\frac{\delta}{2} \leqq I(M_0 M_1 \wedge Y_1^n Y_2^n) + n\delta,\\
n(R_0 + R_1 + R_2) &\leqq H(M_0) + H(M_1) + H(M_2) + n\frac{3\delta}{4},\\
&\leqq I(M_0 M_1 \wedge Y_1^n Y_2^n) + I(M_2 \wedge Z_1^n Z_2^n) + n\delta.
\end{aligned}$$

Further bounding nR_0 resp. $n(R_0 + R_1)$, one has

$$\begin{aligned}
n(R_0 - \delta) &\leqq I(M_0 \wedge Y_2^n) \leqq I(M_0 \wedge Y_1^n | Y_2^n)\\
&\leqq I(M_0 M_1 \wedge Y_2^n) + I(M_0 M_2 Y_2^n \wedge Y_1^n),
\end{aligned}$$

$$\begin{aligned} n(R_0+R_1-\delta) &\leqq I(M_0M_1 \wedge Y_2^n) + I(M_0M_1 \wedge Y_1^nY_2^n) \\ &\leqq I(M_0M_1 \wedge Y_2^n) + I(X_1^n \wedge Y_1^n). \end{aligned}$$

Finally, the bound of $n(R_0 + R_1 + R_2)$ yields

$$\begin{aligned} n(R_0+R_1+R_2-\delta) &\leqq I(M_0M_1 \wedge Y_1^nY_2^n) + I(M_2 \wedge Z_1^nZ_2^n|M_0M_1) \\ &= I(M_0M_1 \wedge Y_2^n) + I(M_0M_1 \wedge Y_1^n|Y_2^n) + I(M_2 \wedge Z_1^n|M_0M_1) \\ &\quad + I(M_2 \wedge Z_2^n|M_0M_1Z_1^n) = I(M_0M_1Z_1^n \wedge Y_2^n) - I(Z_1^n \wedge Y_2^n|M_0M_1) \\ &\quad + I(M_2 \wedge Z_1^n|M_0M_1) + I(M_0M_1 \wedge Y_1^nY_2^n) + I(M_2 \wedge Z_2^n|M_0M_1Z_1^n) \\ &\leqq I(M_0M_1Z_1^n \wedge Y_2^n) + I(M_2 \wedge Z_1^n|M_0M_1Y_2^n) + I(M_0M_1 \wedge Y_1^n|Y_2^n) \\ &\quad + I(M_2 \wedge Z_2^n|M_0M_1Z_1^n). \end{aligned}$$

Since W_1 is a degraded version of V_1, we further obtain

$$\begin{aligned} &n(R_0+R_1+R_2-\delta) \\ &\leqq I(M_0M_1Z_1^n \wedge Y_2^n) + I(M_2 \wedge Y_1^n|M_0M_1Y_2^n) \\ &\quad + I(M_0M_1 \wedge Y_1^n|Y_2^n) + I(M_2 \wedge Z_2^n|M_0M_1Z_1^n) \\ &= I(M_0M_1Z_1^n \wedge Y_2^n) + I(M_0M_1M_2 \wedge Y_1^n|Y_2^n) + I(M_2 \wedge Z_2^n|M_0M_1Z_1^n) \\ &\leqq I(M_0M_1Z_1^n \wedge Y_2^n) + I(M_2 \wedge Z_2^n|M_0M_1Z_1^n) + I(X_1^n \wedge Y_1^n) \\ &\leqq I(M_0M_1Z_1^n \wedge Z_2^n) + I(M_2 \wedge Z_2^n|M_0M_1Z_1^n) + I(X_1^n \wedge Y_1^n), \end{aligned}$$

where the last step follows from the fact that V_2 is a degraded version of W_2. For $i = 1, 2, \ldots, n$, define

$$U_{1i} \triangleq Y_2^nM_2M_0Y_1^{i-1} \quad \text{and} \quad U_{2i} \triangleq Z_1^nM_1M_0Z_2^{i-1}.$$

Then $U_{1i} -\!\!\circ\!\!- X_{1i} -\!\!\circ\!\!- Y_{1i}Z_{1i}$, $U_{2i} -\!\!\circ\!\!- X_{2i} -\!\!\circ\!\!- Y_{2i}Z_{2i}$, and one easily sees that

$$\begin{aligned} I(M_0M_1 \wedge Y_2^n) &\leqq \sum_{i=1}^n I(U_{2i} \wedge Y_{2i}), \\ I(M_0M_2Y_2^n \wedge Y_1^n) &\leqq \sum_{i=1}^n I(U_{1i} \wedge Y_{1i}), \\ I(M_1 \wedge Y_1^n|M_0M_2Y_2^n) &\leqq \sum_{i=1}^n I(X_{1i} \wedge Y_{1i}|U_{1i}). \end{aligned}$$

Using these inequalities and their counterparts obtained by replacing (V_1, W_1) by (W_2, V_2), one sees that

$$\begin{aligned} n(R_0-\delta) &\leqq \sum_{i=1}^n [I(U_{1i} \wedge Y_{1i}) + I(U_{2i} \wedge Y_{2i})], \\ n(R_0-\delta) &\leqq \sum_{i=1}^n [I(U_{1i} \wedge Z_{1i}) + I(U_{2i} \wedge Z_{2i})], \end{aligned}$$

$$n(R_0 + R_1 - \delta) \leqq \sum_{i=1}^{n} [I(U_{2i} \wedge Y_{2i}) + I(X_{1i} \wedge Y_{1i})],$$

$$n(R_0 + R_2 - \delta) \leqq \sum_{i=1}^{n} [I(U_{1i} \wedge Z_{1i}) + I(U_{2i} \wedge Z_{2i})],$$

$$n(R_0 + R_1 + R_2 - \delta) \leqq \sum_{i=1}^{n} [I(X_{1i} \wedge Y_{1i}) + I(U_{2i} \wedge Y_{2i}) + I(X_{2i} \wedge Z_{2i}|U_{2i})],$$

$$n(R_0 + R_1 + R_2 - \delta) \leqq \sum_{i=1}^{n} [I(X_{2i} \wedge Z_{2i}) + I(U_{1i} \wedge Z_{1i}) + I(X_{1i} \wedge Y_{1i}|U_{1i})].$$

The rest is similar to the proof in the hint to Problem 16.4.

16.9. (*Sum of degraded broadcast channels*) Consider a two-output BC with component channels $V : \mathsf{X}_1 \cup \mathsf{X}_2 \to \mathsf{Y}_1 \cup \mathsf{Y}_2$ and $W : \mathsf{X}_1 \cup \mathsf{X}_2 \to \mathsf{Z}_1 \cup \mathsf{Z}_2$ such that V is the sum (see Problem 6.15) of the channels $V_i : \mathsf{X}_i \to \mathsf{Y}_i$, whereas W is the sum of $W_i : \mathsf{X}_i \to \mathsf{Z}_i$, $i = 1, 2$. Determine the capacity region of the BC (V, W) if W_1 is a degraded version of V_1, while V_2 is a degraded version of W_2. (El Gamal (1980). A solution for $R_0 = 0$ was given by Poltyrev (1979).)

Hint Set $\mathsf{X} \triangleq \mathsf{X}_1 \cup \mathsf{X}_2$, $\mathsf{Y} \triangleq \mathsf{Y}_1 \cup \mathsf{Y}_2$, $\mathsf{Z} \triangleq \mathsf{Z}_1 \cup \mathsf{Z}_2$ and define $F : \mathsf{X} \to \{1, 2\}$ by $F(x) = i$ if $x \in \mathsf{X}_i$. Prove that (R_1, R_0, R_2) is an achievable rate triple iff there exists a quadruple of RVs (U, X, Y, Z) such that Y resp. Z is connected with X by channel V resp. W, $U \multimap X \multimap YZ$, the range U of U has size $|\mathsf{U}| \leqq |\mathsf{X}| + 4$, and one has the inequalities

$$R_0 \leqq \min[I(U, F(X) \wedge Y), I(U, F(X) \wedge Z)],$$

$$R_0 + R_1 \leqq I(U, F(X) \wedge Y) + \Pr\{F(X) = 1\} I(X \wedge Y|U, F(X) = 1),$$

$$R_0 + R_2 \leqq I(U, F(X) \wedge Z) + \Pr\{F(X) = 2\} I(X \wedge Z|U, F(X) = 2),$$

$$R_0 + R_1 + R_2 \leqq \min[I(U, F(X) \wedge Y), I(U, F(X) \wedge Z)] + \Pr\{F(X) = 1\} I(X \wedge Y|U, F(X) = 1) + \Pr\{F(X) = 2\} I(X \wedge Z|U, F(X) = 2).$$

The direct part easily follows from the result of Problem 16.4.

To prove the converse part, fix $\delta > 0$ and consider codes as in Problem 16.8. Let M_1, M_0, M_2 be independent RVs each uniformly distributed over the respective message sets $\mathsf{M}_1, \mathsf{M}_0, \mathsf{M}_2$ and let X^n be the codeword corresponding to the message triple M_1, M_0, M_2. Denote by Y^n resp. Z^n the outputs of V^n resp. W^n if the input is X^n, write $F^n(X^n) \triangleq F(X_1) \dots F(X_n)$ and define the RV U_i $(i = 1, 2, \dots, n)$ as follows:

$$U_i \triangleq \begin{cases} M_0 M_1 F^n(X^n) Y^{i-1} Z^{i-1} & \text{if} \quad F(X_i) = 2 \\ M_0 M_2 F^n(X^n) Y^{i-1} Z^{i-1} & \text{if} \quad F(X_i) = 1. \end{cases}$$

Fano's inequality implies for n large enough

$$n(R_0 - \delta) \leqq I(M_0 \wedge Y^n) \leqq I(M_0, F^n(X^n) \wedge Y^n)$$
$$\leqq \sum_{i=1}^{n}[I(F(X_i) \wedge Y_i) + I(M_0 \wedge Y_i|F^n(X^n)Y^{i-1})]$$
$$\leqq \sum_{i=1}^{n}[I(F(X_i) \wedge Y_i) + I(U_i \wedge Y_i|F(X_i))]. \qquad (*)$$

Again by Fano's inequality

$$n(R_0 + R_1 - \delta) \leqq I(M_0M_1 \wedge Y^n) \leqq I(M_0M_1F^n(X^n) \wedge Y^n)$$
$$\leqq \sum_{i=1}^{n}[I(F(X_i) \wedge Y_i) + I(M_0M_1 \wedge Y_i|F^n(X^n)Y^{i-1})].$$

Here

$$I(M_0M_1 \wedge Y_i|F^n(X^n)Y^{i-1})$$
$$\leqq \Pr\{F(X_i) = 1\}I(X^nM_0M_1M_2 \wedge Y_i|F^n(X^n), F(X_i) = 1, Y^{i-1})$$
$$+ \Pr\{F(X_i) = 2\}I(M_0M_1 \wedge Y_i|F^n(X^n), F(X_i) = 2, Y^{i-1})$$
$$\leqq \Pr\{F(X_i) = 1\}[I(U_i \wedge Y_i|F(X_i) = 1) + I(X^n \wedge Y_i|F(X_i) = 1, U_i)]$$
$$+ \Pr\{F(X_i) = 2\}I(U_l \wedge Y_i|F(X_i) = 2).$$

Hence by the memoryless character of V^n

$$n(R_0 + R_1 - \delta) \leqq \sum_{i=1}^{n}[I(F(X_i) \wedge Y_i) + I(U_i \wedge Y_i|F(X_i))$$
$$+ \Pr\{F(X_i) = 1\}I(X_i \wedge Y_i|F(X_i) = 1, U_i)]. \qquad (**)$$

Finally, using Fano's inequality as before,

$$n(R_0 + R_1 + R_2 - \delta)$$
$$\leqq I(F^n(X^n)M_0M_1 \wedge Y^n) + I(M_2 \wedge Y^nZ^n|F^n(X^n)M_0M_1)$$
$$\leqq \sum_{i=1}^{n}[I(F(X_i) \wedge Y_i) + I(M_0M_1 \wedge Y_i|F^n(X^n)Y^{i-1})$$
$$+I(M_2 \wedge Y_iZ_i|F^n(X^n)M_0M_1Y^{i-1}Z^{i-1})]. \qquad (***)$$

Here, conditioning on the event $F(X_i) = 1$, one has

$$I(M_0M_1 \wedge Y_i|F^n(X^n)Y^{i-1},\ F(X_i) = 1)$$
$$+ I(M_2 \wedge Y_iZ_i|F^n(X^n)\,M_0M_1Y^{i-1}Z^{i-1},\ F(X_i) = 1)$$
$$\leqq I(X^nU_iM_1 \wedge Y_iZ_i|F(X_i) = 1) = I(U_i \wedge Y_iZ_i|F(X_i) = 1)$$
$$+ I(X^n \wedge Y_iZ_i|U_i,\ F(X_i) = 1)$$
$$\leqq I(U_i \wedge Y_i\,|F(X_i) = 1) + I(X_i \wedge Y_i|U_i,\ F(X_i) = 1),$$

where the final inequality follows from the fact that W_1 is a degraded version of V_1 and the memoryless character of the channel V_1^n. On the other hand, conditioning on $F(X_i) = 2$ one obtains

$$I(M_0M_1 \wedge Y_i|F^n(X^n)Y^{i-1},\ F(X_i) = 2) \leqq I(U_i \wedge Y_i|F(X_i) = 2)$$

and

$$\begin{aligned} &I(M_2 \wedge Y_iZ_i|F^n(X^n)\ M_0M_1\ Y^{i-1}\ Z^{i-1},\ F(X_i) = 2) \\ &\quad \leqq I(X^n \wedge Y_iZ_i|U_i,\ F(X_i) = 2) = I(X_i \wedge Z_i|U_i,\ F(X_i) = 2), \end{aligned}$$

where the preceding inequality follows from the memoryless character of W_2^n and the fact that V_2 is a degraded version of W_2.
The preceding three bounds and (***) result in

$$\begin{aligned} n(R_0 + R_1 + R_2 - \delta) \leqq \sum_{i=1}^{n} [&I(F(X_i) \wedge Y_i) + I(U_i \wedge Y_i|F(X_i)) \\ &+ \Pr\{F(X_i) = 1\} I(X_i \wedge Y_i|F(X_i) = 1, U_i) \\ &+ \Pr\{F(X_i) = 2\} I(X_i \wedge Z_i|F(X_i) = 2, U_i)]. \end{aligned}$$

This, (*) and (**) with their analogs obtained upon reversing the roles of the two component channels yield the result similar to the concluding part of the hint of Problem 16.4.

16.10. (*Some achievable rates for a general two-output BC*)

(a) Let S, T, U be independent RVs and let $X \triangleq g(S, T, U)$ be a function of this triple. Further, let Y, Z be RVs such that $STU \multimap X \multimap YZ$, $P_{Y|X} = V$, $P_{Z|X} = W$. Show that any triple of non-negative numbers $(R_1,\ R_0,\ R_2)$ satisfying the following inequalities:

$$\begin{aligned} &R_1 \leqq I(S \wedge Y|T),\ \ R_2 \leqq I(U \wedge Z|T), \\ &R_0 \leqq \min\,[I(T \wedge Y|S), I(T \wedge Z|U)],\ R_0 + R_1 \leqq I(TS \wedge Y), \\ &R_0 + R_2 \leqq I(TU \wedge Z) \end{aligned}$$

is an achievable triple for the BC with component channels $\{V\}$ and $\{W\}$. (Cover (1975a), Van der Meulen (1975).)

Hint Consider the channel network of Fig. P14.18B(a) with component channels $\hat{V} : \mathsf{S} \times \mathsf{T} \times \mathsf{U} \to \mathsf{Y}$ and $\hat{W} : \mathsf{S} \times \mathsf{T} \times \mathsf{U} \to \mathsf{Z}$, where $\hat{V} \triangleq P_{Y|STU}$, $\hat{W} \triangleq P_{Z|STU}$ and generalize Problem 16.2(b).

(b) Let (S, U, X, Y, Z) be a quintuple of RVs such that $SU \multimap X \multimap YZ$ and Y resp. Z are connected with X by the respective channels V and W. In particular, S and U need not be independent. Prove that if

$$\begin{aligned} &0 \leqq R_1 \leqq I(S \wedge Y),\ 0 \leqq R_2 \leqq I(U \wedge Z), \\ &R_1 + R_2 \leqq I(S \wedge Y) + I(U \wedge Z) - I(S \wedge U), \end{aligned}$$

then $(R_1, 0, R_2)$ is an achievable rate triple for the BC with component channels $\{V\}$ and $\{W\}$. (Marton (1979). In 2010, this is still the best achievability result known for general BCs with common rate zero, and so is its counterpart in (c) below without that constraint.)

Hint It is sufficient to prove that the triple $(R_1, 0, R_2)$ with

$$R_1 \triangleq I(S \wedge Y), \quad R_2 \triangleq I(U \wedge Z) - I(U \wedge S)$$

is an achievable rate triple. For this sake, fix some $\delta > 0$ and write

$$\begin{aligned} J &\triangleq \lfloor \exp\{n(I(S \wedge Y) - \delta)\} \rfloor, \\ K &\triangleq \lfloor \exp\{n(I(U \wedge Z) - (U \wedge S))\} \rfloor, \\ L &\triangleq \lfloor \exp\{n(I(U \wedge S)\} \rfloor . \end{aligned}$$

Using the method of random selection, choose the n-length sequences $\mathbf{s}_j \in \mathsf{S}^n$, $\mathbf{u}_{kl} \in \mathsf{U}^n$, $1 \leqq j \leqq J$, $1 \leqq k \leqq K$, $1 \leqq l \leqq L$, independently of each other according to the distributions P_S^n resp. P_U^n. For every j and k denote by $l(j,k)$ the smallest index l for which

$$(\mathbf{s}_j, \mathbf{u}_{kl}) \in \mathsf{T}^n_{[SU]}$$

if there is any such pair, and put $l(j,k) = 1$, say, otherwise. Set

$$\mathsf{B}_j \triangleq \mathsf{T}^n_{[Y|S]}(\mathbf{s}_j) - \bigcup_{j' \neq j} \mathsf{T}^n_{[Y|S]}(\mathbf{s}_{j'}),$$

$$\mathsf{C}_k \triangleq \bigcup_{l=1}^{L} \mathsf{T}^n_{[Z|U]}(\mathbf{u}_{kl}) - \bigcup_{k' \neq k} \bigcup_{l=1}^{L} \mathsf{T}^n_{[Z|U]}(\mathbf{u}_{k'l}).$$

Associate with the random collection

$$\{(\mathbf{s}_j, \mathbf{u}_{kl}) : 1 \leqq j \leqq J, \ 1 \leqq k \leqq K, \ 1 \leqq l \leqq L\}$$

an n-length block code (f, φ, ψ) for the given BC, taking for $f(j,k)$ any element of $\mathsf{T}^n_{[X|SU]}(\mathbf{s}_j, \mathbf{u}_{kl(j,k)})$, and setting $\varphi(\mathbf{y}) = j$ iff $\mathbf{y} \in \mathsf{B}_j$, and $\psi(\mathbf{z}) = k$ iff $\mathbf{z} \in \mathsf{C}_k$. Prove that the expectation of the average probability of error of this randomly selected code (f, φ, ψ) converges to zero as $n \to \infty$.

(c) Generalizing (b), show that for every sextuple of RVs (S, T, U, X, Y, Z) such that $STU \multimap X \multimap YZ$, with X, Y, Z satisfying the same conditions as in (b), every triple of non-negative numbers (R_1, R_0, R_2) satisfying the inequalities

$$\begin{aligned} &R_0 \leqq \min\{I(T \wedge Y), \ I(T \wedge Z)\}, \\ &R_0 + R_1 \leqq I(ST \wedge Y), \quad R_0 + R_2 \leqq I(UT \wedge Z), \\ &R_0 + R_1 + R_2 \leqq \min\{I(T \wedge Y), \ I(T \wedge Z)\} + I(S \wedge Y|T) \\ &\qquad\qquad + I(U \wedge Z|T) - I(U \wedge S|T) \end{aligned}$$

is an achievable rate triple for the BC of (b). (Marton (1979).)

16.11. (a) (*Deterministic BC*) Consider a two-output BC for which V and W are zero–one matrices, i.e., deterministic mappings $f : \mathsf{X} \to \mathsf{Y}$, $g : \mathsf{X} \to \mathsf{Z}$. Prove that the capacity region consists of those triples of non-negative numbers

(R_1, R_0, R_2) for which there is a pair of RVs (T, X) with $|\mathsf{T}| \leqq |\mathsf{X}| + 2$ such that for $Y \triangleq f(X)$, $Z \triangleq g(X)$

$$\begin{aligned} &R_0 \leqq \min\{I(T \wedge Y),\ I(T \wedge Z)\}, \\ &R_0 + R_1 \leqq H(Y), \quad R_0 + R_2 \leqq H(Z), \\ &R_0 + R_1 + R_2 \leqq \min\{I(T \wedge Y),\ I(T \wedge Z)\} + H(Y, Z|T). \end{aligned}$$

Hint The direct part of the result is a special case of Problem 16.10(c). The converse part follows in the usual way (set $T_i \triangleq M_0 Y^{i-1} Z^{i-1}$).
(Blackwell (1963, unpublished) asked the problem in a special case, proving that time sharing between the two components is not optimal. In that special case the capacity region was determined by Gelfand (1977). The above result was proved (with $R_0 = 0$) by Pinsker (1978) and Marton (1979).)

(b) (*Semi-deterministic BC*) More generally, consider a two-output BC for which only V is a zero–one matrix while W is arbitrary. Prove that $(R_1, 0, R_2)$ is an element of the capacity region iff there is a quadruple of RVs (U, X, Y, Z) satisfying

$$U \multimap X \multimap YZ, \quad P_{Y|X} = V, \quad P_{Z|X} = W, \quad |\mathsf{U}| \leqq |\mathsf{X}| + 2$$

for which

$$\begin{aligned} 0 \leqq R_1 \leqq H(Y), \quad 0 \leqq R_2 \leqq I(U \wedge Z) \\ R_1 + R_2 \leqq H(Y|U) + I(U \wedge Z). \end{aligned}$$

(Marton (1979); Gelfand and Pinsker (1980a) independently determined the whole capacity region.)

Hint The direct part is a special case of Problem 16.10(b) (choose $T \triangleq$ constant, $S \triangleq Y$). The converse part is a special case of Problem 16.3.

(c) (*Zero error*) Show that for a two-output BC as in (a), to every achievable rate triple (R_1, R_0, R_2) there exist codes with zero probability of error and rates converging to (R_1, R_0, R_2). (Pinsker (1978).)

Hint Use Problem 14.13.

16.12. (*ABC with input constraint*) Define the region of achievable rate pairs under input constraint (c, Γ) (see Chapter 6) for an ABC with component channels $\{V\}$, $\{W\}$, i.e., restrict attention to codes for which every codeword $f(m_1, m_0)$ satisfies the condition $c(f(m_1, m_0)) \leqq \Gamma$.
Prove that (R_1, R_0) is an ε-achievable rate pair ($\varepsilon \in (0, 1/2)$) under input constraint (c, Γ) iff there exists a quadruple of RVs (U, X, Y, Z) such that

$$U \multimap X \multimap YZ, \quad |\mathsf{U}| \leqq |\mathsf{X}| + 3, \quad Ec(X) \leqq \Gamma,$$

Y resp. Z are connected with X by the channel V resp. W, and the inequalities (16.1) and (16.2) hold.
Hint The proof is analogous to those of Theorems 16.1 and 16.3.

Source networks with three inputs and one helper (Problems 16.13–16.18)

16.13. (*Zigzag source network*) Determine the achievable rate region of the source network of Fig. P16.13(a) with an arbitrary 3-source. Prove that $(R_{\mathsf{X}}, R_{\mathsf{Y}}, R_{\mathsf{Z}})$ is an achievable rate triple iff there exist a RV U and a number t such that

$$U \multimap Y \multimap XZ, \quad |\mathsf{U}| \leqq |\mathsf{Y}| + 2, \quad 0 \leqq t \leqq \min[I(U \wedge X), I(U \wedge Z)],$$

and one has

$$R_{\mathsf{X}} \geqq H(X), \quad R_{\mathsf{Y}} \geqq H(XY) - H(X|U) - t, \quad R_{\mathsf{Z}} \geqq H(Z|U) + t.$$

(Körner and Marton (1977c).)

Hint By Lemma 13.11 the achievable rate region of this source network equals that of the normal source network with one helper of Fig. P16.13 (b). By the helpers theorem, Theorem 13.15 this achievable rate region equals the closure of the set of those vectors $(R_{\mathsf{X}}, R_{\mathsf{Y}}, R_{\mathsf{Z}})$, which for some integer n and mapping g with domain Y^n satisfy

$$R_{\mathsf{X}} \geqq H(X), \quad R_{\mathsf{Y}} \geqq \frac{1}{n} H(g(Y^n)) + \frac{1}{n} H(Y^n | g(Y^n), X^n),$$

$$R_{\mathsf{Z}} \geqq \frac{1}{n} H(Z^n | g(Y^n)).$$

Observing that

$$H(g(Y^n)) + H(Y^n | g(Y^n), X^n) = nH(X, Y) - H(X^n | g(Y^n)),$$

this and Theorem 15.20 complete the proof.

16.14. (*Zigzag source network, strong converse*) Prove that the ε-achievable rate region of a zigzag source network equals the achievable rate region for every $0 < \varepsilon < 1$. (Körner and Marton (1977c).)

Hint Fix some $\delta > 0$ and consider a code $(f_k, g_k, h_k, \varphi_k, \psi_k, \vartheta_k)$ with probability of error less than ε. Parallelling the proof of the strong converse part of Theorem 16.4, for every $\mathbf{y} \in \mathsf{Y}^k$ denote by $\mathsf{A}(\mathbf{y}) \subset \mathsf{X}^k$ resp. $\mathsf{C}(\mathbf{y}) \subset \mathsf{Z}^k$ the

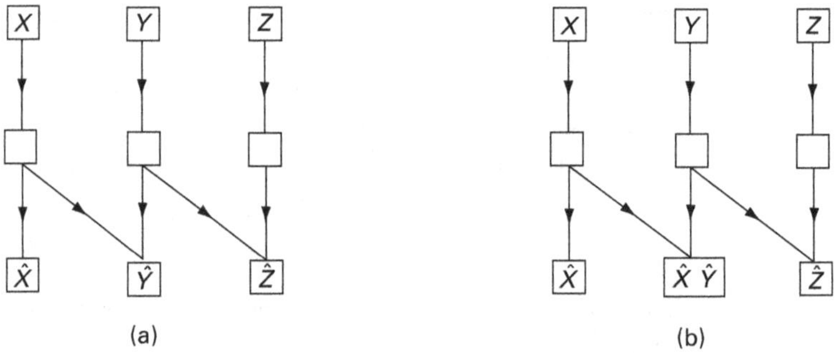

Figure P16.13 (a) Zigzag source network. (b) The corresponding NSN.

set of those elements of X^k resp. Z^k for which no error occurs, i.e., for which $\psi_k(f_k(\mathbf{x}), g_k(\mathbf{y})) = \mathbf{y}$ resp. $\vartheta_k(g_k(\mathbf{y}), h_k(\mathbf{z})) = \mathbf{z}$. By assumption,

$$\sum_{\mathbf{y}\in\mathsf{Y}^k} P_Y^k(\mathbf{y}) P_{XZ|Y}^k(\mathsf{A}(\mathbf{y}) \times \mathsf{C}(\mathbf{y})|\mathbf{y}) \geqq 1 - \varepsilon.$$

Denote by $\tilde{\mathsf{B}}$ the set of those sequences $\mathbf{y} \in \mathsf{T}_{[Y]}^k \subset \mathsf{Y}^k$ for which

$$P_{XZ|Y}^k(\mathsf{A}(\mathbf{y}) \times \mathsf{C}(\mathbf{y})|\mathbf{y}) > \frac{1-\varepsilon}{2}.$$

Then, for k large enough,

$$P_Y^k(\tilde{\mathsf{B}}) \geqq \frac{1-\varepsilon}{3}$$

and thus

$$\frac{1}{k} \log |\tilde{\mathsf{B}}| \geqq H(Y) - \delta.$$

As in the proof of Theorem 16.4, there exists a set $\mathsf{B} \subset \tilde{\mathsf{B}}$ such that g_k is constant on B and $|\mathsf{B}| \geqq |\tilde{\mathsf{B}}|/\|g_k\|$, thus

$$\frac{1}{k} \log |\mathsf{B}| \geqq H(Y) - \delta - \frac{1}{k} \log \|g_k\| .$$

Consider channels $V = P_{X|Y}$ and $W = P_{Z|Y}$. Observe that B is the codeword set of an $(n, (1-\varepsilon)/3)$-code for the DMC $\{V\}$. Hence, by Lemma 6.8, for every $\eta > 0$ and large enough k,

$$g_{V^k}(\mathsf{B}, \eta) \geqq |\mathsf{B}| \exp\{k(H(X|Y) - \delta)\},$$

yielding

$$R_{\mathsf{Y}} \geqq \frac{1}{k} \log \|g_k\| - \delta \geqq H(X, Y) - \frac{1}{k} \log g_{V^k}(\mathsf{B}, \eta) - 3\delta.$$

As in the proof of Theorem 16.4, one also has for large k

$$R_{\mathsf{Z}} \geqq \frac{1}{k} \log g_{W^k}\left(\mathsf{B}, \frac{1-\varepsilon}{2}\right) - \delta.$$

Setting $\eta \triangleq (1-\varepsilon)/2$, the preceding two inequalities and Theorem 15.10 complete the proof.

16.15. Determine the achievable rate region of the source network of Fig. P16.15(a).

(a) Prove that $(R_{\mathsf{X}}, R_{\mathsf{Y}}, R_{\mathsf{Z}})$ is an achievable rate triple iff there exist a RV U and a number t such that

$$U \multimap Y \multimap XZ, \quad |\mathsf{U}| \leqq |\mathsf{Y}| + 2, \quad 0 \leqq t \leqq \min[I(U \wedge X), I(U \wedge Z)],$$

and

$$R_{\mathsf{X}} \geqq H(X), \quad R_{\mathsf{Y}} \geqq H(X, Y, Z) - H(X, Z|U) - t, \quad R_{\mathsf{Z}} \geqq H(Z|U) + t.$$

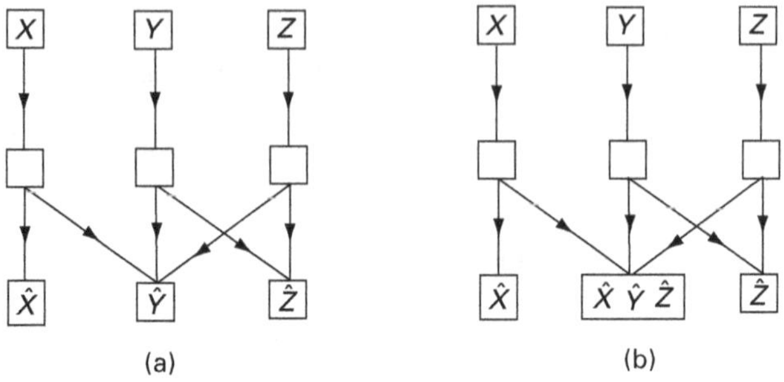

Figure P16.15

(b) Prove the corresponding strong converse.
(Körner (1975).)

Hint The corresponding NSN with one helper is shown in Fig. P16.15(b). It follows from Theorem 13.15 that the achievable rate region equals the closure of the set of those vectors $(R_{\mathsf{X}}, R_{\mathsf{Y}}, R_{\mathsf{Z}})$ which for some n and mapping g with domain Y^n satisfy the inequalities

$$R_{\mathsf{X}} \geqq H(X), \quad R_{\mathsf{Y}} \geqq \frac{1}{n} H(Y^n|X^n, Z^n, g(Y^n)) + \frac{1}{n} H(g(Y^n)),$$
$$R_{\mathsf{Z}} \geqq \frac{1}{n} H(Z^n|g(Y^n)).$$

Hence the statement of (a) follows by Theorem 15.20. The strong converse can be proved as in Problem 16.14.

(c) Without relying on any previous result, show directly that the achievable rate region of the source network of Fig. P16.15(a) has the same projection on the $(R_{\mathsf{Y}}, R_{\mathsf{Z}})$-plane as that of the zigzag source network with a 3-source having generic variables (X, Z), Y, Z.

16.16. Determine the achievable rate region of the source network of Fig. P16.16(a) with an arbitrary 3-source.

(a) Prove that $(R_{\mathsf{X}}, R_{\mathsf{Y}}, R_{\mathsf{Z}})$ is an achievable rate triple iff for some U and t as in Problem 16.15 the inequalities

$$R_{\mathsf{X}} \geqq H(X|U) + t, \; R_{\mathsf{Y}} \geqq H(X, Y) - [H(X|U) + t], \; R_{\mathsf{Z}} \geqq H(Z|U) + t$$

hold.

(b) Prove the corresponding strong converse.

Hint The given source network has the same achievable rate region as the NSN with one helper shown in Fig. P16.16 (b). Proceed as in Problem 16.15.

16.17. Determine the achievable rate region of the source network of Fig. P16.17 with an arbitrary 3-source. Prove that $(R_{\mathsf{X}}, R_{\mathsf{Y}}, R_{\mathsf{Z}})$ is an achievable rate triple iff for some U and t as in Problem 16.15

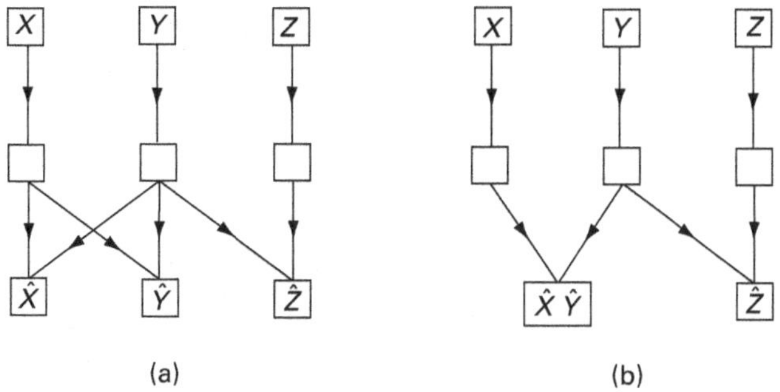

Figure P16.16

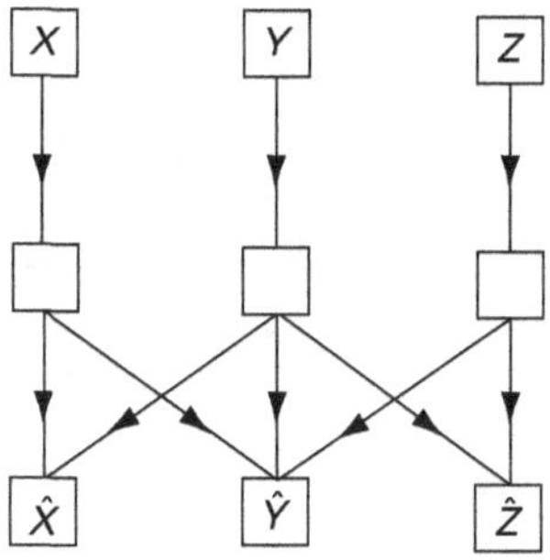

Figure P16.17

$$R_X \geqq H(X|U) + t, \quad R_Y \geqq H(X, Y, Z) - H(X, Z|U) - t,$$
$$R_Z \geqq H(Z|U) + t.$$

Hint The corresponding NSN (with one helper) is obtained by replacing $\hat{Y}$ at the middle output of Fig. P16.17 by $\hat{X}\hat{Y}\hat{Z}$. Now, by Theorem 13.15 the achievable rate region equals the closure of the set of those vectors (R_X, R_Y, R_Z) which for some n and mapping g with domain Y^n satisfy

$$R_X \geqq \frac{1}{n} H(X^n | g(Y^n)),$$
$$R_Y \geqq \frac{1}{n} H(g(Y^n)) + \frac{1}{n} H(Y^n | X^n, Z^n, g(Y^n)),$$
$$R_Z \geqq \frac{1}{n} H(Z^n | g(Y^n)).$$

Note that although Theorem 15.20 does not apply to this case, the result of Problem 15.21 (b) does.

16.18. Determine the achievable rate region and prove the corresponding strong converse for the source network of Fig. P16.18. Prove that (R_X, R_Y, R_Z) is an ε-achievable rate triple iff there exists a RV U with

$$U \multimap Y \multimap XZ, \quad |\mathsf{U}| \leqq |\mathsf{Y}| + 2$$

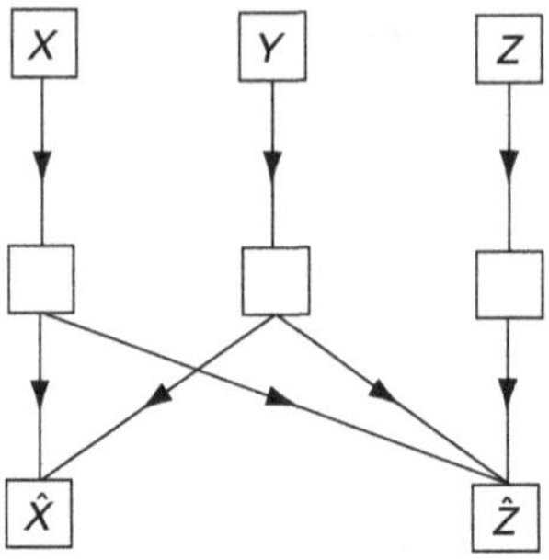

Figure P16.18

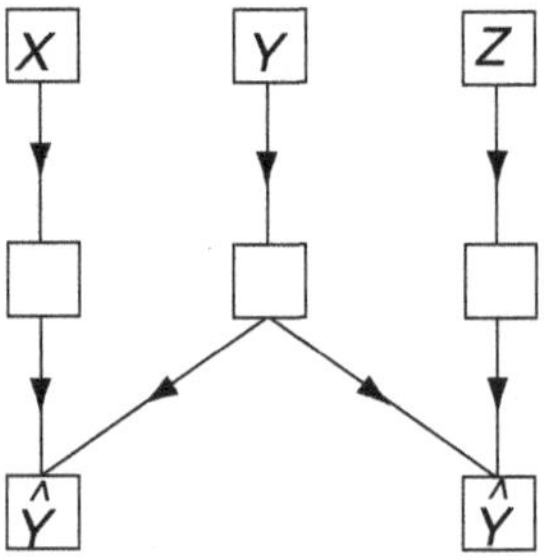

Figure P16.19

such that

$$R_X \geqq H(X|U), \quad R_Y \geqq I(U \wedge Y), \quad R_Z \geqq H(Z|U, X).$$

(Unpublished result of J. Körner and K. Marton from 1975.)
Hint The proof is analogous to the preceding ones.

Source networks with two helpers

16.19. Prove that (R_X, R_Y, R_Z) is an ε-achievable rate triple for the source network of Fig. P16.19 iff $(R_X, R_Y) \in \mathcal{R}(\langle X \rangle, Y)$ and $(R_Z, R_Y) \in \mathcal{R}(\langle Z \rangle, Y)$. (Sgarro (1977).)
Hint The asserted form of the achievable rate region is obvious from Theorem 13.15. The strong converse follows from the fact that it is valid for the source network of Theorem 16.4.

16.20. (*Binary adder*) Consider the source network of Fig. P16.20. Let X, Y, Z have range $\{0, 1\}$. Let X, Y have *symmetric joint distribution,* i.e.,

$$P_{XY}(0, 0) = P_{XY}(1, 1) = \frac{1-p}{2}, \quad P_{XY}(0, 1) = P_{XY}(1, 0) = \frac{p}{2};$$

further, suppose that

$$Z = \begin{cases} 0 & \text{if} \quad X = Y \\ 1 & \text{if} \quad X \neq Y, \end{cases}$$

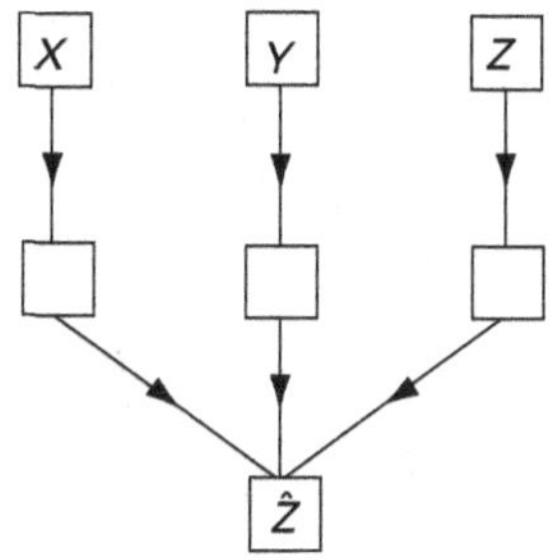

Figure P16.20

i.e., $Z = X \oplus Y$, where "$\oplus$" stands for addition mod 2.

(a) Prove that there exist codes $(f_k, g_k, h_k, \varphi_k)$, where h_k takes a single value and

$$f_k : \mathsf{X}_k \to \mathsf{M_X}, \quad g_k : Y^k \to \mathsf{M_Y}, \quad \varphi_k : \mathsf{M_X} \times \mathsf{M_Y} \to \mathsf{Z}^k,$$

such that

$$\lim_{k\to\infty} \Pr\left\{\varphi_k(f_k(X^k), g_k(Y^k)) \neq Z^k\right\} = 0$$

and

$$\lim_{k\to\infty} \frac{1}{k} \log \|f_k\| = \lim_{k\to\infty} \frac{1}{k} \log \|g_k\| = H(Z).$$

(Joint result of J. Körner and K. Marton published in Körner (1975).)

(b) Determine the achievable rate region. Show that it consists of those triples $(R_\mathsf{X}, R_\mathsf{Y}, R_\mathsf{Z})$ for which

$$R_\mathsf{X} + R_\mathsf{Z} \geqq H(Z), \quad R_\mathsf{Y} + R_\mathsf{Z} \geqq H(Z).$$

Prove the corresponding strong converse. (Körner and Marton (1979).)

Hint (a) By Problem 1.7, there exist functions $\hat{f}_k : \{0,1\}^k \to \{0,1\}^{l_k}$ and $\hat{\varphi}_k : \{0,1\}^{l_k} \to \{0,1\}^k$ such that $\hat{f}_k$ is a linear mapping, i.e.,

$$\hat{f}_k(\mathbf{z}' \oplus \mathbf{z}'') = \hat{f}_k(\mathbf{z}') \oplus \hat{f}_k(\mathbf{z}''),$$

and

$$\lim_{k\to\infty} \Pr\left\{\hat{\varphi}_k(\hat{f}_k(Z^k)) \neq Z^k\right\} = 0, \quad \lim_{k\to\infty} \frac{l_k}{k} = H(Z).$$

Check that the choice

$$f_k(\mathbf{x}) \triangleq \hat{f}_k(\mathbf{x}), \quad g_k(\mathbf{y}) \triangleq \hat{f}_k(\mathbf{y}), \quad \varphi_k(f_k(\mathbf{x}), g_k(\mathbf{y})) \triangleq \hat{\varphi}_k(f_k(\mathbf{x}) \oplus g_k(\mathbf{y}))$$

is appropriate.

(b) The direct part follows from (a) by the time sharing principle. For the converse part, consider codes $(f_k, g_k, h_k, \varphi_k)$ having error probability less than ε. Replacing g_k by the identity mapping on Y^k, one can modify the decoder so as to yield a fork network code for the same DMMS having the same probability of error. Then, by the fork network coding theorem, $R_\mathsf{X} + R_\mathsf{Z} \geqq H(XZ|Y)$. By symmetry, $R_Y + R_Z \geqq H(YZ|X)$. Note finally that $H(XZ|Y) = H(YZ|X) = H(Z)$.

16.21. (*Limitations on rate slicing*) If (R_X, R_Y, R_Z) is an achievable rate vector for the source network of Fig. P16.20 with an arbitrary 3-source, then

$$(R_X + R_Y, R_Z) \in \mathcal{R}(\langle XY\rangle, Z).$$

Is it true that every element of $\mathcal{R}(\langle XY\rangle, Z)$ is equal to $(R_X + R_Y, R_Z)$ for some achievable rate vector of the above source network?

Hint No; see Problem 16.20.

General fidelity criteria (Problems 16.22–16.24)

16.22. (*Rate-distortion with side information; proof valid also for* $\Delta = 0$)

(a) Give a proof of Theorem 16.5 that does not rely on rate slicing.

Hint To complete a proof as in the text in a direct manner, it suffices to find a function $\psi\colon \mathsf{Y}^k \to \mathsf{U}^k$ satisfying (16.38) and an encoder $f\colon \mathsf{Y}^k \to \mathsf{M}_1$ of rate close to $I(U \wedge Y|X)$ such that $\mathbf{x}$ and $f(\mathbf{y})$ determine $\psi(\mathbf{y})$ unless $(\mathbf{x}, \mathbf{y})$ is in a subset of $\mathsf{X}^k \times \mathsf{Y}^k$ of negligible P^k_{XY}-probability.

Now consider $\exp\{k[I(U \wedge Y) + \delta]\}$ sequences $\mathbf{u}_{ij} \in \mathsf{U}^k$, $i \in \mathsf{M}_1$, $j \in \mathsf{M}_2$, randomly selected from the distribution P^k_U, where

$$|\mathsf{M}_1| = \exp\{k[I(U \wedge Y|X) + 3\delta]\}, \quad |\mathsf{M}_2| = \exp\{k[I(U \wedge X) - 2\delta]\}.$$

With probability close to 1, to each $\mathbf{y} \in \mathsf{T}_{[Y]}$ there exist $i = f(\mathbf{y})$, $j = \varphi(\mathbf{y})$ such that $(\mathbf{u}_{ij}, \mathbf{y}) \in \mathsf{T}_{[UY]}$. Verify that $\psi(\mathbf{y}) \triangleq \mathbf{u}_{f(\mathbf{y})\varphi(\mathbf{y})}$ and f have the required property. For details, see Lemma 17.22(i) and its proof (where the roles of X and Y are interchanged).

(b) Show that Corollary 16.6 is valid also for $\Delta = 0$.

Hint Only the direct part needs a modified proof. For this purpose, apply the proof in the hint to part(a), replacing the condition (16.35) by $Ed(X, Y, Z) = 0$; this implies $d(\mathbf{x}, \mathbf{y}, h(\mathbf{u}, \mathbf{x})) = 0$ if $(\mathbf{x}, \mathbf{y}, h(\mathbf{u}, \mathbf{x})) \in \mathsf{T}_{[XYZ]}$.)

16.23. (*Binary adder*) Consider the source network of Fig. P16.23 with $\mathsf{X} = \mathsf{Y} \triangleq \{0, 1\}$ and the distortion measure

$$d(x, y, z) \triangleq \begin{cases} 0 & \text{if} \quad z = x \oplus y. \\ 1 & \text{otherwise.} \end{cases}$$

Here $\oplus$ denotes addition mod 2.

(a) Prove that if (X, Y) have a symmetric joint distribution as in Problem 16.20, then (R_X, R_Y) is an ε-achievable rate vector at distortion level 0 iff

$$R_X \geqq H(Z), \quad R_Y \geqq H(Z), \quad \text{where} \quad Z \triangleq X \oplus Y.$$

(Körner and Marton (1979).)

Hint see Problem 16.20.

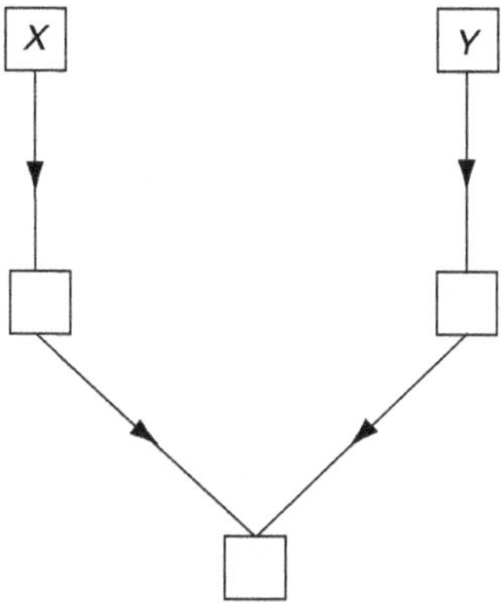

Figure P16.23

(b) Prove that for any binary X and Y with $H(Z) \geqq \min[H(X), H(Y)]$ the pair $(R_\mathsf{X}, R_\mathsf{Y})$ is an achievable rate vector at distortion level 0 iff

$$R_\mathsf{X} \geqq H(X|Y), \quad R_\mathsf{Y} \geqq H(Y|X), \quad R_\mathsf{X} + R_\mathsf{Y} \geqq H(X, Y).$$

This means that in order to reconstruct the sum-sequence $X^k \oplus Y^k$, one needs rates as high as for the reconstruction of the pair (X^k, Y^k).

(c) Prove the corresponding strong converse. (Körner (1977, unpublished).)

Hint The direct part of (b) is an obvious consequence of the fork network coding theorem, Theorem 13.5. To prove the weak converse suppose, e.g., that $H(Z) \geqq H(Y)$ and consider channels V: $\mathsf{X} \to \mathsf{Z}$ and $\mathsf{W} : \mathsf{X} \to \mathsf{Y}$ such that $V = P_{Z|X}$ and $W = P_{Y|X}$. Observe that these channels have one row in common, whereas in the other row the same distribution appears in a reversed order. Check that this implies that one channel is the degraded version of the other. Since $H(Z) \geqq H(Y)$ implies $I(X \wedge Z) \geqq I(X \wedge Y)$, conclude that W is a degraded version of V. Fix some $\delta > 0$. By Fano's inequality, for large k and any code (f_k, g_k, φ_k) with small probability of error one has

$$k(R_\mathsf{X} + \delta) \geqq H(f_k(X^k)) + H(Z^k|f_k(X^k), Y^k) = kH(X, Y) - H(Y^k|f_k(X^k)),$$
$$k(R_\mathsf{Y} + \delta) \geqq H(Z^k|f_k(X^k)) \geqq H(Y^k|f_k(X^k)),$$

where the final inequality is a consequence of

$$I(Z^k \wedge X^k|f_k(X^k)) \geqq I(Y^k \wedge X^k|f_k(X^k)),$$

which holds as W is a degraded version of V. Part (c) is proved analogously to the strong converse part of Theorem 16.4, using the degradedness of the channels as in (b).

Remark There exist binary-valued RVs X, Y with a non-symmetric joint distribution (see Problem 16.20) such that some achievable rate pair for the binary adder satisfies

$$R_\mathsf{X} + R_\mathsf{Y} < H(X, Y).$$

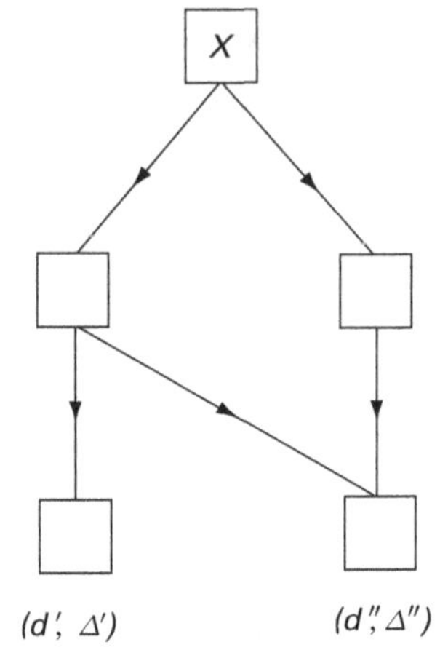

Figure P16.24

Namely, $(H(Z), H(Z))$ is an achievable rate pair by the construction of Problem 16.20.

16.24. (*Two-step source coding*)

(a) Consider a DMS with generic RV X and two ε-fidelity criteria, (d', Δ') and (d'', Δ''). Determine the ε-achievable rate region for the source network of Fig. P16.24 for this source if the fidelity criteria at the two decoders are the given ones. Prove that (R', R'') is an achievable rate pair iff there exist RVs Z' and Z'' taking values in the corresponding reproduction alphabets such that

$$R' \geqq I(X \wedge Z'), \quad R' + R'' \geqq I(X \wedge Z'Z''),$$
$$Ed'(X, Z') \leqq \Delta', \quad Ed''(X, Z'') \leqq \Delta''$$

(Gray and Wyner (1974).)

(b) Prove the corresponding strong converse.

Hint The converse part is proved as in Theorem 7.3. For the direct part, it suffices to prove that $(I(X \wedge Z'), I(X \wedge Z'|Z'))$ is an achievable rate pair, whenever $Ed'(X, Z') \leqq \Delta'$, $Ed''(X, Z'') \leqq \Delta''$. This can be done similarly to the proof of Theorem 7.3, using a channel code construction analogous to (though simpler than) that of Theorem 16.1. Define $\{V : \mathsf{Z}' \times \mathsf{Z}'' \to \mathsf{X}\}$ and $\{W : \mathsf{Z}' \to \mathsf{X}\}$ by $V = P_{X|Z'Z''}$, $W = P_{X|Z'}$. First choose a maximal code for the DMC $\{W\}$ with Z'-typical codewords $\mathbf{z}_i$. Then, for every i, consider a maximal code for $\{V\}$ with codewords which are $Z'Z''$-typical and have projection $\mathbf{z}_i$ on Z'^n. Apply the blowing up lemma to the decoding sets at the output of the DMC $\{V\}$, and make the resulting sets disjoint in an arbitrary manner. The obtained partition of X^n defines the two source codes similarly to the construction in Theorem 7.3.

16.25. (*Double Markovity*) Prove that a triple of RVs (U, X, Y) satisfies the conditions

$$U \multimap X \multimap Y, \quad X \multimap U \multimap Y \qquad (*)$$

iff there exist functions f of U and g of X such that

(i) $f(U) = g(X)$ with probability 1;

(ii) Y is conditionally independent of UX given $(f(U), g(X))$.

In particular, if X and U have an indecomposable joint distribution (see Problem 15.12) then (*) implies the independence of Y and UX.

16.26. (*Information of X about Y*) Let us denote by $A(X; Y)$ the smallest possible R for which $(R, H(Y|X)) \in \mathcal{R}(\langle X \rangle, Y)$ see Theorem 16.4. Intuitively, $A(X; Y)$ is the rate of the information contained in the X-source about the Y-source. Clearly, $I(X \wedge Y) \leqq A(X; Y) \leqq H(X)$. Prove that $A(X; Y) < H(X)$ iff there is a function $\psi(X)$ of X such that

$$X \multimap \psi(X) \multimap Y.$$

Hint Apply Theorem 16.4 and Problem 16.25.

Common information (Problems 16.27–16.30)

16.27. Let (f_k, φ_k) resp. (g_k, ψ_k) be k-to-l_k resp. k-to-m_k binary block codes for the component sources of a DMMS with generic variables X, Y see Chapter 1. Fix some $\varepsilon \in (0,1)$.

(a) Denote by N_k the (random) length of the longest coinciding beginning segment of the two binary sequences $f_k(X^k)$ and $g_k(Y^k)$. Prove that for asymptotically optimal codes, i.e., when

$$\lim_{k\to\infty} \frac{l_k}{k} = H(X), \quad \lim_{k\to\infty} \frac{m_k}{k} = H(Y),$$
$$\varliminf_{k\to\infty} \Pr\{\varphi_k(f_k(X^k)) = X^k, \psi_k(g_k(Y^k)) = Y^k\} > 1 - \varepsilon, \tag{*}$$

the *expected common length* EN_k of the two codes satisfies

$$\varlimsup_{k\to\infty} \frac{EN_k}{k} \leqq H(J),$$

where J is a common function of X and Y having maximum range. More precisely, $J = f(X) = g(Y)$, where f and g are functions of X resp. Y such that $f(X) = g(Y)$ with probability 1 and the number of values taken by f (or g) with positive probability is largest possible. (Obviously, all these RVs J have the same entropy.) Show further that there exist sequences of codes as above achieving

$$\lim_{k\to\infty} \frac{EN_k}{k} = H(J).$$

For this reason, $H(J)$ may be called the *common information* of X and Y. In particular, if X and Y have an indecomposable joint distribution then this common information is zero.

(b) Prove that for codes satisfying (*) such that the beginning segments of non-random length n_k of $f_k(X^k)$ and $g_k(Y^k)$ coincide with probability at least

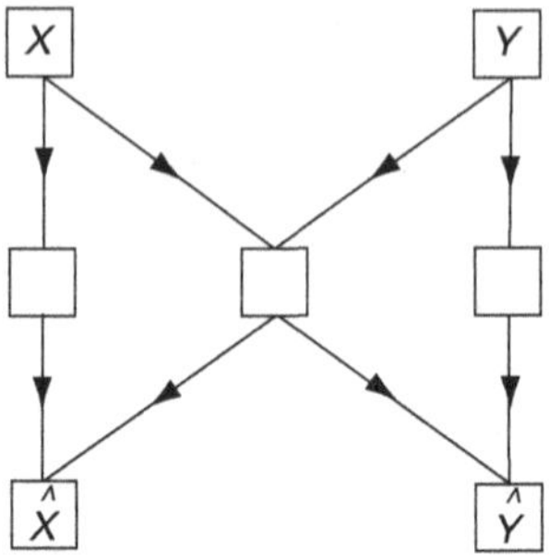

Figure P16.28

ε, one has $\overline{\lim}_{k\to\infty} \frac{n_k}{k} \leqq H(J)$; further, $\lim_{k\to\infty} \frac{n_k}{k} = H(J)$ can be achieved. (Gács and Körner (1973).)

Hint Apply Problem 15.14(a).

16.28. For two RVs X, Y, denote by $C(X \wedge Y)$ the largest R for which the rate triple $(H(X) - R, R, H(Y) - R)$ is achievable for the source network of Fig. P16.28. (If f_n, g_n, h_n are the encoders of an n-length block code for this network, then X^n is reproduced from $(f_n(X^n), g_n(X^n, Y^n))$ and Y^n from $(h_n(Y^n), g_n(X^n, Y^n))$. It is reasonable to require that $(1/n) \log \|f_n\| + (1/n) \log \|g_n\|$ be close to $H(X)$ and $(1/n) \log \|h_n\| + (1/n) \log \|g_n\|$ be close to $H(Y)$; in this case $(1/n) \log \|g_n\|$ can be interpreted as a measure of *common information* of the two sources.) Prove that $C(X \wedge Y) = H(J)$ for J as in Problem 16.27.

Hint Apply Problem 16.6. It follows that

$$C(X \wedge Y) = \max_{\substack{I(XY\wedge U)+H(X|U)=H(X)\\ I(XY\wedge U)+H(Y|U)=H(Y)}} I(XY \wedge U).$$

Consider some triple (U, X, Y) which achieves the maximum. The constraint equalities are equivalent to $U \multimap X \multimap Y$ and $U \multimap Y \multimap X$, respectively. By Problem 16.25 there exists some common function of X and Y, say J, such that U is conditionally independent of XY given J. Hence

$$C(X \wedge Y) = I(XY \wedge U) = I(XYJ \wedge U) = I(J \wedge U) \leqq H(J).$$

The opposite inequality is obvious. (Ahlswede and Körner (1974, unpublished).)

16.29. (*Wyner's common information*) Let us denote by $W(X \wedge Y)$ and call *Wyner's common information* the smallest R for which there exist R_X, R_Y such that (R_X, R, R_Y) is an achievable rate triple for the source network of Fig. P.28 and $R_X + R + R_Y = H(X, Y)$. Intuitively, $W(X \wedge Y)$ is the minimum rate of a joint code of the two component sources, which assures an essentially rate-optimal operation of the source network of Fig. P16.28. Prove that

$$W(X \wedge Y) = \min_{X \multimap U \multimap Y} I(XY \wedge U).$$

(Wyner (1975a), (1975b).)

16.30. (*Common information and mutual information*) Prove that

$$C(X \wedge Y) \leqq I(X \wedge Y) \leqq W(X \wedge Y)$$

and that in both inequalities there is equality iff the correlation of X and Y is deterministic. The latter means that there exists a common function of X and Y such that X and Y are independent given that function. (Ahlswede–Körner (1974, unpublished).)

Hint Apply Problem 16.25. Let (U, X, Y) be a triple of RVs achieving the minimum in the characterization of $W(X \wedge Y)$; see Problem 16.29. Then the condition $I(XY \wedge U) = I(X \wedge Y)$ is equivalent to $H(X, Y|U) = H(X|Y) + H(Y|X)$. Taking into account the Markov condition $X \multimap U \multimap Y$, this amounts to $H(X|Y) + H(Y|X) = H(X|U, Y) + H(Y|U, X)$. This, in turn, is equivalent to $X \multimap Y \multimap U$ and $U \multimap X \multimap Y$. The latter means that $I(XY \wedge U) = H(J)$, where J is the same as in Problem 16.27.

Miscellaneous source networks (Problems 16.31–16.33)

16.31. (*Fork network with side information*) Determine the achievable rate region of the source network of Fig. P16.31 with an $(r + 1)$-source having arbitrary generic variables $X, Y^{(1)}, \ldots, Y^{(r)}$. Show that $\mathbf{R} = (R_0, R_1, \ldots, R_r)$ is an achievable rate vector iff there exists a RV U and a number $t \geqq 0$ with

$$U \multimap X \multimap Y^{(1)} \ldots Y^{(r)}, \quad |\mathsf{U}| \leqq |\mathsf{X}| + r + 1, \quad t \leqq I(U \wedge Y^{(1)} \ldots Y^{(r)}) \quad (*)$$

such that

$$R_0 \geqq I(U \wedge X) - t, \quad \sum_{i=1}^{r} R_i \geqq H(Y^{(1)} \ldots Y^{(r)}|U) + t,$$

and for every $\mathsf{L} \subset \{1, \ldots, r\}$, denoting $Y_{\mathsf{L}} \triangleq \{Y^{(i)}\}_{i \in \mathsf{L}}$,

$$\sum_{i \in \mathsf{L}} R_i \geqq H(Y_{\mathsf{L}}|Y_{\bar{\mathsf{L}}}U) + |t - I(U \wedge Y_{\bar{\mathsf{L}}})|^+.$$

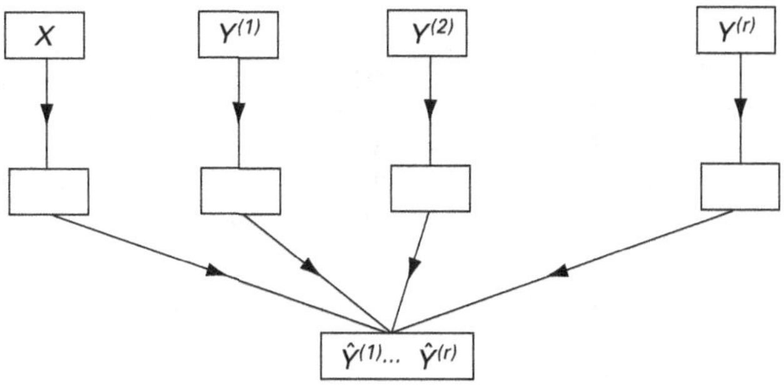

Figure P16.31

(A single-letter characterization of this achievable rate region has been given in a more complicated algebraic form by Gelfand and Pinsker (1979).)

Hint Denote by $\mathcal{R}^*$ the set of all vectors satisfying the above inequalities for some U and t as in (*). By the helpers theorem, Theorem 13.15, the achievable rate region is the closure of the set of vectors $\hat{\mathbf{R}} = (\hat{R}_0, \hat{R}_1, \ldots, \hat{R}_r)$, which for some n and function f with domain X^n satisfy the inequalities

$$\hat{R}_0 \geqq \frac{1}{n} H(f(X^n)),$$
$$\sum_{i \in \mathsf{L}} \hat{R}_i \geqq \frac{1}{n} H(Y^n_{\mathsf{L}} | Y^n_{\bar{\mathsf{L}}} f(X^n)), \quad \mathsf{L} \subset \{1, \ldots, r\}.$$

Hence the direct part, i.e., the achievability of every $\mathbf{R} \in \mathcal{R}^*$, follows by Problem 15.18. The converse part can be verified using (the converse part of the result of) Problem 15.20, setting $S \triangleq Y^{(1)} \ldots Y^{(r)}$. In this way we obtain from the preceding inequalities that to every achievable rate vector $\mathbf{R} = (R_0, R_1, \ldots, R_r)$ there exist a RV U and a number t satisfying (*) so that the asserted inequalities hold.

16.32. (*Source networks of depth* > 2)

(a) It is easily seen that for an arbitrary DMMS with two components the achievable rate region of the source network of Fig. P16.32(a) is $\mathcal{R}(\langle X \rangle, Y)$, i.e., the connection from the encoder of the X-source to that of the Y-source is of no use in an asymptotic sense. Prove that connection in the reversed direction is useful. More precisely, show that that $(R_{\mathsf{X}}, R_{\mathsf{Y}})$ is an ε-achievable rate pair for the source network of Fig. P16.32(b) iff

$$R_{\mathsf{X}} + R_{\mathsf{Y}} \geqq H(Y), \quad R_{\mathsf{Y}} \geqq H(Y|X).$$

Hint The converse part of the previous statement is obvious. To establish the direct part it is sufficient to prove that

$$R_{\mathsf{X}} \triangleq I(X \wedge Y), \quad R_{\mathsf{Y}} \triangleq H(Y|X)$$

is an achievable rate pair. Let $g : \mathsf{Y}^k \to \mathsf{M}_{\mathsf{Y}}$ define a partition of Y^k into codeword sets of (k, ε_k)-codes (with $\varepsilon_k \to 0$) for the DMC $\{V : \mathsf{Y} \to \mathsf{X}\}$

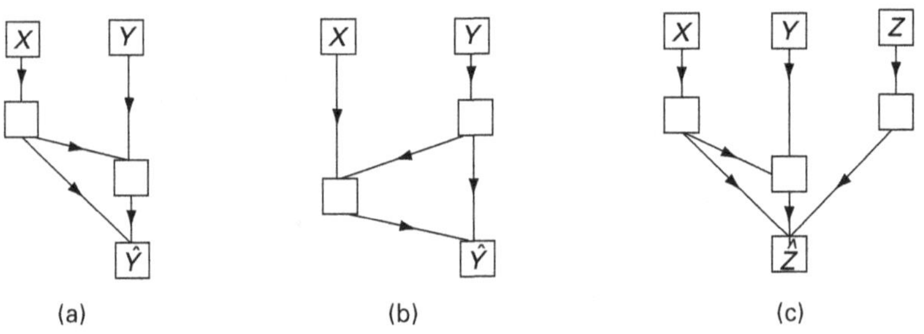

Figure P16.32

(with $V \triangleq P_{X|Y}$) as in Theorem 13.2. Put

$$f(m, x) \triangleq \varphi_m(x) \quad \text{if} \quad g(\mathbf{y}) = m,$$

where φ_m is the decoder of the mth (k, ε_k)-code.

(b) Conclude that the achievable rate region of the source network of Fig. P16.32(c) is, in general, different from that of the source network of Fig. P16.20.

Hint (a) is a special case.

16.33. (*Non-cooperative source network*) Consider the source network of Fig. P16.6 in the case $r = 2$, writing Y and Z for $Y^{(1)}$ and $Y^{(2)}$, respectively. The achievable rate region of this source network was determined in Problem 16.6(c). Now we modify the problem and assume that the encoder of the X-source is chosen with the purpose to "help" only the Y-decoder, perhaps deliberately trying to make the side information it provides possibly useless to the Z-decoder. The question is how much the latter can still benefit from this side information. In order to take maximum advantage of the side information available, the encoder and decoder of the Z-source are allowed to be of block length greater than k, say $n_k k$, where n_k is a positive integer.

(a) Define achievable rate triples (R_X, R_Y, R_Z) for the present problem by the following conditions. To every $\varepsilon > 0$, $\delta > 0$ and sufficiently large k

(i) there exists a k-length block code (f_k, g_k, φ_k) for a source network as in Fig. 16.4 having the X-source as the "side information source" and the Y-source as the "main source" such that

$$\frac{1}{k} \log \|f_k\| \leqq R_X + \delta, \qquad \frac{1}{k} \log \|g_k\| \leqq R_Y + \delta$$

and the probability of error is less than ε;

(ii) to every (f_k, g_k, φ_k) as in (i), there exists an n_k and an $n_k k$-length block code (f_k^*, h_k, ψ_k) for the source network having the X-source as the "side information source" and the Z-source as the "main source" such that f_k^* is the n_k-fold juxtaposition of f_k with itself, h_k has rate

$$\frac{1}{n_k k} \log \|h_k\| \leqq R_Z + \delta,$$

and this code yields probability of error less than ε.

Prove that (R_X, R_Y, R_Z) is an achievable rate triple in this sense iff $(R_X, R_Y) \in \mathcal{R}(\langle X \rangle, Y)$, see Theorem 16.4, and, in addition,

$$R_Z \geqq \max_{\substack{H(Y|U)+t \leqq R_Y \\ I(U \wedge X)-t \leqq R_X}} (H(Z|U) + t), \qquad (*)$$

where the maximization refers to RVs U and non-negative numbers t satisfying the conditions (15.26), (15.27) and (15.32).

Hint We have to prove that for any $(R_X, R_Y) \in \mathcal{R}(\langle X \rangle, Y)$, the triple (R_X, R_Y, R_Z) is achievable iff (*) holds. This is an easy consequence of the entropy characterization theorem, Theorem 15.20. To see this, note first that (*) can be equivalently stated as follows: for every (a, b) with $R_X \leqq H(X) - a$, $R_Y \leqq b$ we have

$$R_Z \leqq c \quad \text{whenever} \quad (a, b, c) \in \mathcal{H}^*(X; Y; Z|X); \qquad (**)$$

see (15.31).

Now, given any (f_k, g_k, φ_k) as in (i), we have – as in the hint to Problem 16.6(b) –

$$R_X + \delta \geqq \frac{1}{k} \log \|f_k\| \geqq H(X) - \frac{1}{k} H(X^k | f_k(X^k))$$

and (if ε is small enough)

$$R_Y + 2\delta \geqq \frac{1}{k} \log \|g_k\| + \delta \geqq \frac{1}{k} H(Y^k | f_k(X^k)).$$

Setting

$$a \triangleq \frac{1}{k} H(X^k | f_k(X^k)), \quad b \triangleq \frac{1}{k} H(Y^k | f_k(X^k)), \quad c \triangleq \frac{1}{k} H(Z^k | f_k(X^k)),$$

the existence of a code as in (ii) follows from (**) by Theorem 13.2 (applied to the 2-source with generic variables $f_k(X^k)$ and Z^k).

Conversely, we have to show for every $(a, b, c) \in \mathcal{H}^*(X; Y; Z|X)$ with

$$R_X + \delta \geqq H(X) - a, \quad R_Y + \delta \geqq b \qquad (***)$$

the existence of codes (f_k, g_k, φ_k) as in (i) such that if to them there are codes satisfying (ii) then necessarily $R_Z + \delta \leqq c$. By Theorem 15.20, it suffices to prove this for the case when

$$a = \frac{1}{l} H(X^l | \tilde{f}(X^l)), \quad b = \frac{1}{l} H(Y^l | \tilde{f}(X^l)), \quad c = \frac{1}{l} H(Z^l | \tilde{f}(X^l))$$

for some integer l and function $\tilde{f}$ of X^t. Applying the rate slicing corollary to the 2-source with generic variables $\tilde{f}(X^l)$ and Y^l, for sufficiently large integer multiples k of l we get codes (f_k, g_k, φ_k) with rate pairs close to $\left((1/l)H(\tilde{f}(X^l)), (1/l)H(Y^l | \tilde{f}(X^l))\right)$, i.e., satisfying (i) with (R_X, R_Y) as in (***), such that f_k is a function of the k/l-fold juxtaposition of $\tilde{f}$ with itself. Then every code as in (ii) (with this f_k) necessarily satisfies

$$R_Z + \delta \geqq \frac{1}{k} H(Z^k | f_k(X^k)) = \frac{1}{l} H(Z^l | \tilde{f}(X^l)) = c.$$

(b) In (a) we have implicitly made the usual assumption that when the encoder–decoder pair for the Z-source is chosen, the X-encoder f_k is known. Reformulate the problem without this assumption. Show that the achievable rate region remains the same.

Hint We have to prove that if (*) holds then codes as in (ii) can be constructed even without knowledge of the mapping f_k (assuming, of course, that the range of f_k is given). To this, the proof in the hint to (a) literally applies, except that instead of Theorem 13.2 we have to use its universal version, see Problem 13.6(b). (Csiszár–Körner (1976); an error in that paper is corrected here.)

Story of the results

Broadcast channels were introduced by Cover (1972). Cover conjectured the coding theorem for degraded broadcast channels and invented the idea of superposition coding, which was generalized by Bergmans (1973), who proved the direct part of that theorem. For further history of the problem (in the degraded case) see Problem 16.4. The ABC coding theorem is due to Körner and Marton (1977a).

The problem solved by Theorem 16.4 was raised by Wyner and Ziv (1973). They proved the (weak) converse part of the result for symmetrically correlated binary sources. For general 2-sources the achievable rate region was determined independently by Ahlswede and Körner (1975) and in a series of papers by Gray and Wyner (1974) and Wyner (1975a), (1975b). The strong converse is due to Ahlswede, Gács and Körner (1976). Theorems 4.5 and 4.6 reformulate a result of Wyner and Ziv (1976).

At the time of writing, only partial results are available for many important problems. Most of them are not conclusive (in the sense of the Discussion) and therefore have not been included. The interested reader may consult the survey papers of Berger (1977), Van der Meulen (1977) and Gelfand and Prelov (1978).

Addition An up-to-date survey of some basic results of multi-user information theory is Kramer (2008). Further results can be found in the lecture notes of El Gamal and Kim (2010, unpublished).

17 Information-theoretic security

Contemporary techniques of data security are primarily based on computational complexity, typically on the infeasibility of inverting certain functions using currently available mathematical techniques and computing power. Among the various protocols based on such ideas, those approved by the cryptography community appear secure enough, but mathematical and technological progress may render them insecure in the future. Indeed, past experience suggests that this is likely to happen.

Information-theoretic secrecy offers *provable security* even against an adversary with unlimited computing power. This chapter provides a glimpse into the substantial progress that has been made towards clarifying the theoretical possibilities in this direction. Practical applications are reasonably expected within a much shorter time than capacity-achieving coding techniques have followed Shannon's discovery of the noisy channel coding theorem.

Two kinds of problems will be addressed: secure transmission over insecure channels and secret key generation taking advantage of public communication. After introducing necessary concepts and tools in Section 17.1, these problems will be treated in Sections 17.2 and 17.3. Let us emphasize that the mathematical models and techniques will be similar to those in previous chapters. These models, however, are now studied from a non-cooperative aspect: a major goal is to keep (at least) one party ignorant of (at least part of) the information exchanged.

17.1 Basic concepts and tools

Perfect secrecy resp. *ε-secrecy* of a RV K from an adversary whose knowledge is represented by a RV Z, means that $I(K \wedge Z) = 0$ (stochastic independence) resp. $I(K \wedge Z) \leqq \varepsilon$. In addition to this secrecy, often the nearly uniform distribution of K on its range K is also required, and if both are satisfied K is said to represent $\log |\mathsf{K}|$ secret random bits.

The two requirements are conveniently expressed by a single one stating that the *security index* of K against Z,

$$S(K|Z) \triangleq \log |\mathsf{K}| - H(K) + I(K \wedge Z) = \log |\mathsf{K}| - H(K|Z), \tag{17.1}$$

has to be close to zero. Note that (17.1) is equivalent to

$$S(K|Z) = D(P_{KZ} \| P_0 \times P_Z), \tag{17.2}$$

where P_0 denotes the uniform distribution on K. → 17.1

The following proposition provides an immediate application of this concept; K and M are RVs with values in K, which, for convenience, is regarded as a commutative group with group operation $+$.

PROPOSITION 17.1 Suppose M is independent of (K, Z) and that an adversary who knows Z learns $K + M$. Then if $S(K|Z) < \varepsilon$, the message M remains ε-secret from this adversary. In particular, if $S(K|Z) = 0$ then M remains perfectly secret. ○

Proof

$$\begin{aligned} I(M \wedge Z, M + K) = I(M \wedge M + K|Z) &= H(M + K|Z) - H(K|MZ) \\ &\leqq \log|\mathsf{K}| - H(K|Z) = S(K|Z), \end{aligned}$$

because the independence of M and (K, Z) implies $H(K|MZ) = H(K|Z)$. □

Proposition 17.1 implies that if $S(K|Z)$ is close to zero, any one of parties *knowing the secret key* K can securely send a message M as above to the others, over an insecure channel: the sender encrypts M by adding K and sends $M + K$. The receiver(s) can decipher simply by subtracting K, while an adversary whose original knowledge was Z gains effectively no benefit from learning $M + K$.

If $S(K|Z) = 0$, this method, called *Vernam cipher* or *one time pad*, guarantees perfectly secure transmission (alas, for secure transmission of several messages, each has to be encrypted by a new key).

In this chapter, it will be investigated for various models, each involving sequences of length $n \to \infty$, what "secrecy rates" $(1/n) \log|\mathsf{K}|$ can be achieved with RVs K generated within the model, whose security index approaches zero and which are accessible to two or more parties, with probability of error approaching zero. The following extractor lemma will be a key tool, like the covering and packing lemmas in Chapters 9 and 10.

Informally, a mapping $\kappa : \mathsf{U} \to \{1, \ldots, k\}$ will be said to *extract* $\log k$ *random bits* from a PD P on the set U if the distribution $\{P(\kappa^{-1}(i)),\ i = 1, \ldots, k\}$, the κ-image of P, is nearly uniform. Near uniformity of a PD Q on a finite set K could be defined in various ways; for example, by the smallness of $\log|\mathsf{K}| - H(Q) = D(Q||P_0)$, where P_0 is the uniform distribution on K. It is most common to use variation distance, and in the following Q will be called ε-*uniform* if

$$d(Q, P_0) = \sum_{i \in \mathsf{K}} \left| Q(i) - \frac{1}{|\mathsf{K}|} \right| \leqq \varepsilon.$$

DEFINITION 17.2 A mapping $\kappa : \mathsf{U} \to \{1, \ldots, k\}$ is an ε-*extractor* for a family $\mathcal{P}$ of PDs on U if the κ-image of P is ε-uniform for each $P \in \mathcal{P}$, or equivalently if the distribution of $\kappa(U)$ is ε-uniform for each RV U with distribution $P_U \in \mathcal{P}$. ○

This term is more frequently used for *seeded extractors* that extract random bits from a (typically unknown) distribution also employing known random bits. Often, a computability requirement is also imposed as part of the definition. Here (in the text part) only deterministic extractors as defined above will be used.

The first part of the following basic lemma gives a sufficient condition for a randomly selected mapping to extract $\log k$ random bits from a PD P, with probability close to unity, even in a stronger sense of near uniformity than ε-uniformity defined above. Random selection means that the values $\kappa(u)$, $u \in \mathsf{U}$, are chosen independently, uniformly from $\{1, \ldots, k\}$. That sufficient condition is intuitively appealing but too restrictive for our purposes. The second part of the lemma, a direct consequence of the first part, relaxes that condition to one commonly satisfied in applications and still sufficient for obtaining an ε-extractor (with ε replaced by $\varepsilon + 2\eta$; see (17.4)). For a better understanding of the significance of the following lemma see also the comment after its proof.

LEMMA 17.3 (*Extractor*)

(i) For a PD P on a finite set U and $\varepsilon > 0$, if $P(u) \leqq 1/d$ for each $u \in \mathsf{U}$ then a randomly selected mapping $\kappa : \mathsf{U} \to \{1, \ldots, k\}$ satisfies

$$\left| P(\kappa^{-1}(i)) - \frac{1}{k} \right| \leqq \frac{\varepsilon}{k}, \quad i = 1, \ldots, k, \tag{17.3}$$

with probability at least

$$1 - 2ke^{-\varepsilon^2 d/2k(1+\varepsilon)}.$$

(ii) The weaker hypothesis $P(\{u : P(u) \leqq 1/d\}) \geqq 1 - \eta$ still suffices for

$$\sum_{i=1}^{k} \left| P(\kappa^{-1}(i)) - \frac{1}{k} \right| \leqq \varepsilon + 2\eta \tag{17.4}$$

to hold with probability at least

$$1 - 2ke^{-\varepsilon^2(1-\eta)d/2k(1+\varepsilon)}. \tag{17.5}$$

In particular, if each P in a family $\mathcal{P}$ of PDs on U satisfies that hypothesis then a randomly selected κ as above will be an $(\epsilon + 2\eta)$-extractor for the family $\mathcal{P}$ with probability at least

$$1 - 2k|\mathcal{P}|e^{-\varepsilon^2(1-\eta)d/2k(1+\varepsilon)}.$$ ○

Proof Fixing $i \in \{1, \ldots, k\}$, for a random mapping κ define $\chi(u) = 1$ or 0 according as $\kappa(u) = i$ or not. Then $\chi(u)$, $u \in \mathsf{U}$, are independent, identically distributed (i.i.d.) RVs with $\Pr\{\chi(u) = 1\} = 1/k$.

Chernoff bounding will be applied to

$$P(\kappa^{-1}(i)) = \sum_{u \in \mathsf{U}} P(u)\chi(u).$$

For any $\beta > 0$,

$$\Pr\left\{P(\kappa^{-1}(i)) > \frac{1+\varepsilon}{k}\right\} = \Pr\left\{\exp\left(\beta d \sum_{u \in \mathsf{U}} P(u)\chi(u)\right) > \exp\left(\beta d \frac{1+\varepsilon}{k}\right)\right\}$$
$$\leqq \exp\left(-\beta d \frac{1+\varepsilon}{k}\right) \prod_{u \in \mathsf{U}} \left\{1 + \frac{1}{k}\left[\exp(\beta d P(u)) - 1\right]\right\}. \tag{17.6}$$

Here, using the hypothesis that $P(u) \leqq 1/d$ for all $u \in \mathsf{U}$,

$$\exp(\beta d P(u)) - 1 = \sum_{j=1}^{\infty} \frac{(\beta d P(u) \ln 2)^j}{j!} < \beta d P(u) \ln 2 \cdot \left[1 + \frac{1}{2}\sum_{j=2}^{\infty} (\beta \ln 2)^{j-1}\right]$$
$$= \beta d P(u) \ln 2 \cdot (1 + \beta^*),$$
$$\beta^* \triangleq \frac{\beta \ln 2}{2(1 - \beta \ln 2)},$$

provided that $\beta \ln 2 < 1$.

Using the inequality $1 + t \ln 2 < \exp t$, it follows that the final product in (17.6) is bounded above by

$$\exp\left(\sum_{u \in \mathsf{U}} \frac{1}{k} \beta d P(u)(1 + \beta^*)\right) = \exp\left(\frac{\beta d (1 + \beta^*)}{k}\right).$$

Hence the bounding (17.6) can be continued by

$$\leqq \exp\left(-\frac{\beta d}{k}(\varepsilon - \beta^*)\right).$$

Setting $\beta = \varepsilon \log e/(1 + \varepsilon)$, when $\beta^* = \varepsilon/2$, we finally obtain

$$\Pr\left\{P(\kappa^{-1}(i)) > \frac{1+\varepsilon}{k}\right\} \leqq e^{-\frac{\varepsilon^2}{2(1+\varepsilon)} \cdot \frac{d}{k}}. \tag{17.7}$$

Similarly,

$$\Pr\left\{P(\kappa^{-1}(i)) < \frac{1-\varepsilon}{k}\right\} = \Pr\left\{\exp\left(-\beta d \sum_{u \in \mathsf{U}} P(u)\chi(u)\right) > \exp\left(-\beta d \frac{1-\varepsilon}{k}\right)\right\}$$
$$\leqq \exp\left(\beta d \frac{1-\varepsilon}{k}\right) \prod_{u \in \mathsf{U}} \left\{1 + \frac{1}{k}\left[\exp(-\beta d P(u)) - 1\right]\right\}. \tag{17.8}$$

Bounding the square bracket is even simpler than for (17.6):

$$\exp(-\beta d P(u)) - 1 = \sum_{j=1}^{\infty} \frac{(-\beta d P(u) \ln 2)^j}{j!} \leqq -\beta d P(u) \ln 2 \left[1 - \frac{1}{2}\beta d P(u) \ln 2\right]$$
$$\leqq -\beta d P(u) \ln 2 \left(1 - \frac{1}{2}\beta \ln 2\right).$$

Using this, we obtain as before that (17.8) can be continued by

$$\leqq \exp\left(-\frac{\beta d}{k}\left(\varepsilon - \frac{1}{2}\beta \ln 2\right)\right).$$

Setting $\beta = \varepsilon \log e$, this yields

$$\Pr\left\{P(\kappa^{-1}(i)) < \frac{1-\varepsilon}{k}\right\} \leqq e^{-\varepsilon^2 d/2k}. \tag{17.9}$$

Equations (17.7) and (17.9) complete the proof of the first assertion.

The second assertion is an easy consequence. Indeed, given a PD P as in this assertion, denote by P_{F} its normalized restriction to $\mathsf{F} \triangleq \{u : P(u) \leqq 1/d\}$. Thus

$$P_{\mathsf{F}}(u) = \begin{cases} \dfrac{P(u)}{P(\mathsf{F})} \leqq \dfrac{1}{d(1-\eta)} & \text{if} \quad u \in \mathsf{F} \\ 0 & \text{if} \quad u \notin \mathsf{F}. \end{cases}$$

As P_{F} satisfies the hypothesis of the first assertion with $d(1-\eta)$ in the role of d, the κ-image of P_{F} is ε-uniform with probability at least (17.5). Finally, the variation distance of the κ-image of P_{F} from that of P does not exceed the variation distance of P_{F} from P, which is equal to $2(1 - P(\mathsf{F}))$, i.e., no more than 2η. □

COMMENT The logarithm of the largest d that meets the hypothesis of part (i) is the *minentropy*, $H_\infty(P) \triangleq \min_{u \in \mathsf{U}}(-\log P(u))$, see Problem 1.15, and that of the largest d meeting the hypothesis of (ii) for a fixed P is the largest minentropy of a distribution
→ 17.3,17.4 P_{F} conditioned on a set F satisfying $P(\mathsf{F}) \geqq 1 - \eta$; clearly, the best F has the form in the proof. In typical applications of Lemma 17.3, U, $\mathcal{P}$, etc. depend on $n \to \infty$, where $|\mathsf{U}|$, $|\mathcal{P}|$ and d grow exponentially with n, and ε, η go to zero exponentially fast, with a small exponent. Then the number $\log k$ of extracted random bits may grow with almost the same rate as $\log d$, the latter having the intuitive meaning as above. In fact, Lemma 17.3 guarantees that if $\log k$ is less than $\log d$ by an arbitrarily small multiple of n then a randomly selected mapping κ meets (17.4) except with doubly exponentially
→ 17.5 small probability. ○

As the latter qualification will often occur in this chapter, we introduce the following.

CONVENTION 17.4 In assertions depending on $n \to \infty$ and on a random selection, the term *doubly exponentially surely* (d.e.s.) will mean that the total probability of such outcomes of the random selection, for which the assertion is false, is not larger than $\exp(-\exp(nc))$, if $n \geqq n_0$ (for suitable $c > 0$ and n_0). ○

Clearly, if each of N_n assertions individually hold d.e.s. (with the same c and n_0), they hold simultaneously, d.e.s., providing N_n does not grow doubly exponentially. Thus, in
→ 17.6 the setting of the above comment, Lemma 17.3 highlights the existence of ε-extractors that extract nearly $\log d$ random bits from either PD in a large class. Note that if we have an ε-extractor for the family of conditional distributions of a RV U conditioned on all possible values of another RV V (U and V typically take values in sets U, V of

exponentially growing size), the random bits extracted will be *secret* from an adversary whose knowledge is represented by V.

The next consequence of Lemma 17.3 explicitly brings out the secrecy aspect. It will be of central importance in this chapter.

LEMMA 17.5 (*Secrecy*) Let U, V be RVs with values in finite sets U, V.

(i) The hypotheses

$$P_{UV}\left(\left\{(u,v):\ P_{U|V}(u|v) \leqq \frac{1}{d}\right\}\right) \geqq 1-\eta^2, \quad \eta \leqq \frac{1}{3}, \tag{17.10}$$

$$k\ln(2k|\mathsf{V}|) < \alpha^2 d, \quad \alpha \leqq \frac{1}{6}, \tag{17.11}$$

imply the existence of a mapping $\kappa : \mathsf{U} \to \{1,\ldots,k\}$ satisfying the security bound

$$S(\kappa(U)|V) \leqq (\alpha + 2\eta)\log k + h(\alpha+\eta). \tag{17.12}$$

Indeed, a randomly selected mapping satisfies (17.12) with probability at least $1 - 2k|\mathsf{V}|e^{-\alpha^2 d/k}$.

(ii) If there exists a set $\mathsf{B} \subset \mathsf{U} \times \mathsf{V}$ such that

$$P_{UV}(u,v) < \frac{1}{\alpha|\mathsf{B}|} \quad \text{when } (u,v) \in \mathsf{B}, \tag{17.13}$$

$$P_{UV}(\mathsf{B}) \geqq 1 - (\eta^2 - \alpha^2), \tag{17.14}$$

with $\alpha \leqq 1/6$, $\eta \leqq 1/3$, then there exists a mapping $\kappa : \mathsf{U} \to \{1,\ldots,k\}$ satisfying the security bound (17.12) provided that

$$k < \min\left[\alpha^6 \min|\mathsf{B}_v|, \frac{1}{2|\mathsf{V}|}e^{1/\alpha}\right],$$

where $\mathsf{B}_v \triangleq \{u :\ (u,v) \in \mathsf{B}\}$ and the minimization is for all v in the projection of B to V. Indeed, a randomly selected mapping satisfies (17.12) with probability at least

$$1 - 2k|\mathsf{V}|e^{-\alpha^5 \min|\mathsf{B}_v|\cdot k^{-1}}. \qquad \bigcirc$$

Proof (i) Denote by G_v the set $\{u :\ P_{U|V}(u|v) \leqq 1/d\}$. Since

$$P_{UV}\left(\left\{(u,v):\ P_{U|V}(u|v) \leqq \frac{1}{d}\right\}\right) = \sum_{v\in\mathsf{V}} P_V(v) P_{U|V}(\mathsf{G}_v|v),$$

the hypothesis (17.10) implies that the set

$$\mathsf{E} \triangleq \{v :\ P_{U|V}(\mathsf{G}_v|v) < 1-\eta\}$$

has P_V-probability $P_V(\mathsf{E}) < \eta$. As each PD $P_{U|V}(\cdot|v)$, $v \notin \mathsf{E}$, satisfies the hypothesis of the second assertion of Lemma 17.3, the bound (17.4), with arbitrary $\varepsilon > 0$, holds simultaneously for each $P_{U|V}(\cdot|v)$, $v \notin \mathsf{E}$, in the role of P, at least with probability

$$1 - 2k|\mathsf{V}|e^{-\frac{\varepsilon^2(1-\eta)}{2(1+\varepsilon)}\cdot\frac{d}{k}}.$$

Set for convenience $\varepsilon = 2\alpha$, $\alpha \leqq 1/6$, then $\varepsilon^2(1-\eta)/2(1+\varepsilon) > \alpha^2$ (if $\eta \leqq 1/3$), and thus the final probability is at least $1 - 2k|\mathsf{V}|e^{-\alpha^2 d/k}$.

The variation distance bound (17.4) allows us to bound the difference of $H(\kappa(U)|\ V = v)$ from the entropy $\log k$ of the uniform distribution, e.g., by Lemma 2.7. We use here the sharper bound in Problem 3.10 (though the sharpening is insignificant for our purposes), to obtain for $v \notin \mathsf{E}$

$$\log k - H(\kappa(U)|\ V = v) \leqq \frac{\varepsilon + 2\eta}{2}\log k + h\left(\frac{\varepsilon + 2\eta}{2}\right),$$

where $\varepsilon = 2\alpha$. For $v \in \mathsf{E}$ we bound the entropy difference trivially by $\log k$. Thus, recalling the definition of the security index (17.1), the assertion (17.12) follows.

(ii) Given B satisfying (17.13), denote

$$\mathsf{B}' \triangleq \mathsf{B} \cap \{\mathsf{U} \times \mathsf{C}\}, \qquad \mathsf{C} \triangleq \left\{v :\ P_V(v) \geqq \frac{\alpha^2|\mathsf{B}_v|}{|\mathsf{B}|}\right\}.$$

Then for $(u, v) \in \mathsf{B}'$

$$P_{U|V}(u|v) = \frac{P_{UV}(u,v)}{P_V(v)} \leqq \frac{(\alpha|\mathsf{B}|)^{-1}}{\alpha^2|\mathsf{B}_v|\cdot|\mathsf{B}|^{-1}} \leqq \frac{1}{\alpha^3 \min|\mathsf{B}_v|}.$$

This implies that (17.10) holds with $d = \alpha^3 \min|\mathsf{B}_v|$, because $P_{UV}(\mathsf{B}') \geqq P_{UV}(\mathsf{B}) - P_V(\overline{\mathsf{C}})$, where $P_V(\overline{\mathsf{C}}) < \alpha^2$ due to $\sum_{v\in\mathsf{V}}|\mathsf{B}_v| = |\mathsf{B}|$.

Hence the assertion follows by (i), since the hypothesis on k implies (17.11). □

COROLLARY 17.5 For a DMMS with generic variables (X, Y)

(i) to any $\delta > 0$ there exists $\xi > 0$ such that for $k \leqq \exp\{n[H(X|Y) - \delta]\}$ a randomly selected mapping $\kappa : \mathsf{X}^n \to \{1, \dots, k\}$ gives

$$S(\kappa(X^n)|Y^n) < \exp(-\xi n), \quad \text{d.e.s.};$$

(ii) let $Z^{(n)}$ be any RV jointly distributed with (X^n, Y^n) that has at most $\exp(nr)$ possible values. Then, if $k \leqq \exp\{n[H(X|Y) - r - \delta]\}$, assertion (i) remains valid for $S(\kappa(X^n)|Y^n, Z^{(n)})$. ○

Proof (i) Apply the secrecy lemma to $(U, V) \triangleq (X^n, Y^n)$. Note that setting $d \triangleq \exp\{n[H(X|Y) - \delta/2]\}$ the hypothesis (17.10) is met with $\eta = \eta_n \to 0$ exponentially fast. Taking $\alpha = \alpha_n$, going to zero exponentially with a sufficiently small exponent, condition (17.11) holds for $k \leqq \exp\{n[H(X|Y) - \delta]\}$ if n is sufficiently large. Then (17.12) gives the assertion.

(ii) Apply the secrecy lemma to $U = X^n$, $V = (Y^n, Z^{(n)})$. Consider exceptional sets E_1 and E_2 consisting of those pairs $(\mathbf{x}, \mathbf{y})$ for which

$$P_{X^n|Y^n}(\mathbf{x}|\mathbf{y}) > \exp\left\{-n\left[H(X|Y) - \frac{\delta}{2}\right]\right\}$$

resp. of those pairs $(\mathbf{y}, \mathbf{z})$ for which

$$P_{Z^{(n)}|Y^n}(\mathbf{z}|\mathbf{y}) \leqq \zeta \exp(-nr),$$

with $\zeta > 0$ specified later. Then for $(\mathbf{x}, \mathbf{y}, \mathbf{z})$ with $(\mathbf{x}, \mathbf{y}) \notin \mathsf{E}_1$, $(\mathbf{y}, \mathbf{z}) \notin \mathsf{E}_2$,

$$P_{X^n|Y^n Z^{(n)}}(\mathbf{x}|\mathbf{y}, \mathbf{z}) \leqq \frac{P_{X^n|Y^n}(\mathbf{x}|\mathbf{y})}{P_{Z^{(n)}|Y^n}(\mathbf{z}|\mathbf{y})} \leqq \zeta^{-1} \exp\left\{-n\left[H(X|Y) - r - \frac{\delta}{2}\right]\right\}.$$

Take $\zeta = \exp(-n\delta/4)$, say, then the last bound yields (17.10) with

$$d = \exp\left\{n\left[H(X|Y) - r - \frac{3\delta}{4}\right]\right\}$$

and $\eta^2 = P_{X^n Y^n}(\mathsf{E}_1) + P_{Y^n Z^{(n)}}(\mathsf{E}_2)$. Here the first term is exponentially small as in (i), while the second term is obviously not larger than ζ (since $Z^{(n)}$ has no more than $\exp(nr)$ possible values), and is thus also exponentially small.

The proof is completed as in (i). □

In information theory, randomly selected mappings are even more frequently used for source coding (see Chapter 13) than for extracting random bits. Lemma 17.6 below states a well-known universal coding result, for reference purposes. It is interesting to compare Lemma 17.6 with Corollary 17.5: a randomly selected mapping is "good" for extracting secret random bits, resp. for source coding with side information, according to whether its rate $(1/n)\log k$ is smaller or larger than $H(X|Y)$.

LEMMA 17.6 Given finite sets X, Y, and $H > 0$, $\delta > 0$, for a randomly selected mapping $\kappa : \mathsf{X}^n \to \{1, \ldots, k\}$ with $(1/n)\log k \geqq H + \delta$ and a suitable decoder mapping $\varphi : \mathsf{Y}^n \times \{1, \ldots, k\} \to \mathsf{X}^n$ (depending on κ), it holds with probability approaching unity as $n \to \infty$ that, simultaneously for all DMMSs with generic variables X, Y satisfying $H(X|Y) \leqq H$,

$$\Pr\left\{\varphi\left(Y^n, \kappa(X^n)\right) \neq X^n\right\} \leqq \varepsilon_n,$$

where $\varepsilon_n \to 0$ exponentially rapidly. ○

The proof is omitted. See Problem 13.6 for a stronger result.

At this point, we make a digression about typical sequences. In this Chapter, unlike elsewhere in this book, typical sequences will be used with constants (arbitrarily small but) not depending on the block length n. This will make sure that non-typicality occurs with exponentially small probability; see Lemma 17.8.

For brevity, a simplified version of the terminology introduced in Definition 2.8 will be used, when this does not cause ambiguity.

CONVENTION 17.7 Given RVs U, X, Y, etc. with values in sets $\mathsf{U}, \mathsf{X}, \mathsf{Y}$, etc., the brief term *$\xi$-typical* will refer to sequences $\mathbf{u} \in \mathsf{T}^n_{[U]_\xi}$ (or $\mathbf{x} \in \mathsf{T}^n_{[X]_\xi}$, etc.), or pairs $(\mathbf{u}, \mathbf{x}) \in \mathsf{T}^n_{[UX]_\xi}$, triples $(\mathbf{u}, \mathbf{x}, \mathbf{y}) \in \mathsf{T}^n_{[UXY]_\xi}$, etc. Similarly, *joint ξ-typicality* of $\mathbf{y}$ with $\mathbf{x}$ means that the pair $(\mathbf{x}, \mathbf{y})$ is ξ-typical, and thus $\mathbf{y}$ belongs to

$$\mathsf{T}^n_{[XY]_\xi}(\mathbf{x}) \triangleq \left\{\mathbf{y} : (\mathbf{x}, \mathbf{y}) \in \mathsf{T}^n_{[XY]_\xi}\right\};$$

analogously for joint ξ-typicality of $\mathbf{y}$ with $(\mathbf{u}, \mathbf{x})$, etc. ○

The set $\mathsf{T}^n_{[XY]_\xi}(\mathbf{x})$ is an analog of, but different from, $\mathsf{T}^n_{[Y|X]_\xi}(\mathbf{x})$ in Definition 2.9. The latter will not be explicitly used in this chapter, but the implication of Lemma 2.10 that

$$\mathsf{T}^n_{[XY]_\zeta}(\mathbf{x}) \supset \mathsf{T}^n_{[Y|X]_{\zeta-\xi}}(\mathbf{x}) \quad \text{if } \mathbf{x} \in \mathsf{T}^n_{[X]_\xi},\ \xi < \zeta \tag{17.15}$$

will be referred to occasionally. From now on, the upper index n of T^n will be omitted.

Some basic properties of typical sequences are recalled in Lemma 17.8, in which U, X, Y are RVs with values in finite sets U, X, Y, and typicality constants are denoted by ξ, ζ, σ. These are small positive numbers, and τ denotes a positive number which is arbitrarily small if the corresponding typicality constant is small (τ may depend on the underlying distribution but not on the sequences $\mathbf{u}$, $\mathbf{x}$, $\mathbf{y}$ in the assertions).

LEMMA 17.8 (*Typicality*) It holds, assuming in (iii)–(iv) that n is sufficiently large:

(i)

$$P^n_X\left(\mathsf{T}_{[X]_\xi}\right) \geqq 1 - 2|\mathsf{X}|e^{-2\xi^2 n},$$

$$P^n_{X|U}\left(\mathsf{T}_{[UX]_\zeta}(\mathbf{u})|\mathbf{u}\right) \geqq 1 - 2|\mathsf{U}||\mathsf{X}|e^{-2(\zeta-\xi)^2 n} \quad \text{if } \mathbf{u} \in \mathsf{T}_{[U]_\xi},\ \xi < \zeta;$$

(ii)

$$\left|-\frac{1}{n}\log P^n_X(\mathbf{x}) - H(X)\right| < \tau \quad \text{if } \mathbf{x} \in \mathsf{T}_{[X]_\xi},$$

$$\left|-\frac{1}{n}\log P^n_{X|U}(\mathbf{x}|\mathbf{u}) - H(X|U)\right| < \tau \quad \text{if } (\mathbf{u}, \mathbf{x}) \in \mathsf{T}_{[UX]_\xi};$$

(iii)

$$\left|\frac{1}{n}\log\left|\mathsf{T}_{[X]_\xi}\right| - H(X)\right| < \tau,$$

$$\left|\frac{1}{n}\log\left|\mathsf{T}_{[UX]_\zeta}(\mathbf{u})\right| - H(X|U)\right| < \tau \quad \text{if } \mathsf{T}_{[UX]_\zeta}(\mathbf{u}) \neq \emptyset$$

(a sufficient condition for $T_{[UX]_\zeta} \neq \emptyset$ is $\mathbf{u} \in \mathsf{T}_{[U]_\xi}$ with $\xi < \zeta$);

(iv)

$$\left|\frac{1}{n}\log P^n_X\left(\mathsf{T}_{[XY]_\zeta}(\mathbf{y})\right) - I(X \wedge Y)\right| < \tau \quad \text{if } \mathsf{T}_{[XY]_\zeta}(\mathbf{y}) \neq \emptyset,$$

$$\left|\frac{1}{n}\log P^n_{X|U}\left(\mathsf{T}_{[UXY]_\sigma}(\mathbf{u}, \mathbf{y}) \,\middle|\, \mathbf{u}\right) - I(X \wedge Y|U)\right| < \tau \quad \text{if } \mathsf{T}_{[UXY]_\sigma}(\mathbf{u}, \mathbf{y}) \neq \emptyset;$$

(v) If $U \multimap X \multimap Y$ and $(\mathbf{u}, \mathbf{x}) \in \mathsf{T}_{[UX]_\xi}$, $\xi < \sigma$, then

$$P^n_{Y|X}\left(\mathsf{T}_{[UXY]_\sigma}(\mathbf{u}, \mathbf{x}) \,\middle|\, \mathbf{x}\right) \geqq 1 - 2|\mathsf{U}||\mathsf{X}||\mathsf{Y}|e^{-2(\sigma-\xi)^2 n}.$$ ○

Proof (i) holds by the Remark to Lemma 2.12, using for the second assertion (17.15) also. (ii) is obvious from the definitions. (iii) follows similarly to Lemma 2.13, or directly from (i), (ii); the sufficient condition for non-emptiness follows from (17.15). (iv) is a consequence of (ii) and (iii): for the first assertion, one needs the bound on $\left|\mathsf{T}_{[XY]_\zeta}(\mathbf{y})\right|$ obtained from (iii), with Y in the role of U; for the second assertion, the similar bound on $\left|\mathsf{T}_{[UXY]_\sigma}(\mathbf{u},\mathbf{y})\right|$ is needed, obtained from (iii) with (U, Y) in the role of U. Finally, if $U \multimap X \multimap Y$ then

$$P^n_{Y|X}\left(\mathsf{T}_{[UXY]_\sigma}(\mathbf{u},\mathbf{x}) \,\middle|\, \mathbf{x}\right) = P^n_{Y|UX}\left(\mathsf{T}_{[UXY]_\sigma}(\mathbf{u},\mathbf{x}) \,\middle|\, \mathbf{u},\mathbf{x}\right),$$

and (v) follows from the second assertion in (i), applied to (U, X) and Y in the role of U and X. □

Simple typicality properties of randomly selected codeword sets will also be needed; these are stated as corollaries of the following lemma.

LEMMA 17.9 The probability that in N independent trials an event of probability q occurs less/more than $\alpha q N$ times, according as $\alpha \lessgtr 1$, is bounded above by $e^{-c(\alpha)Nq}$, where $c(\alpha) = \alpha \ln \alpha - \alpha + 1$. ○

Proof A standard upper bound to the probability of less/more than pN occurrences (according as $p \lessgtr q$) is $\exp\left[-ND(p\|q)\right]$, where $D(p\|q)$ is a shorthand for the divergence $D((p, 1-p)\|(q, 1-q))$; see Problem 2.8. Hence it suffices to show that $D(\alpha q\|q)\ln 2 \geqq q(\alpha \ln \alpha - \alpha + 1)$.

Now,

$$f(q) \triangleq D(\alpha q\|q)\ln 2 = \alpha q \ln q - (1-\alpha q)\ln \frac{1-\alpha q}{1-q}$$

is a convex function of q, with $f(0) = 0$, $f'(0) = \alpha \ln \alpha - \alpha + 1$. □

COROLLARY 17.9A Consider $N = \exp(nR)$ sequences $\mathbf{x}_i \in \mathsf{X}^n$ independently drawn from the distribution P^n_X.

(i) To any $\xi > 0$ there exists $\rho > 0$ such that all but a fraction $\exp(-\rho n)$ of the sequences $\mathbf{x}_i$ are ξ-typical, d.e.s.
(ii) If $I(X \wedge Y) < R$, to any $\tau > 0$ there exists $\zeta > 0$ such that

$$\left|\frac{1}{n}\log\left|\{i : \mathbf{x}_i \in \mathsf{T}_{[XY]_\zeta}(\mathbf{y})\}\right| - (R - I(X \wedge Y))\right| < \tau, \quad \text{d.e.s.}$$

simultaneously for all $\mathbf{y} \in \mathsf{Y}^n$ with $\mathsf{T}_{[XY]_\zeta}(\mathbf{y}) \neq \emptyset$. ○

The hypothesis $\mathsf{T}_{[XY]_\zeta}(\mathbf{y}) \neq \emptyset$ is met for all ξ-typical $\mathbf{y}$ with $\xi < \zeta$, by (17.15).

Proof (i) The probability q_n that a particular $\mathbf{x}_i$ is not ξ-typical goes to zero exponentially as $n \to \infty$, by Lemma 17.8(i). Take $0 < \rho < R$ such that $q_n < (2/3)\exp(-n\rho)$, and consider independent events $A_i \supset \{\mathbf{x}_i \text{ not } \xi\text{-typical}\}$ of probability $q = (2/3)\exp(-n\rho)$, $i = 1, \ldots, N$. Then, if more than a fraction $\exp(-n\rho)$ of the $\mathbf{x}_i$ are not ξ-typical, more than $(3/2)Nq$ of the events A_i occur. Lemma 17.9 gives the probability of the latter an upper bound that goes to zero doubly exponentially.

(ii) By Lemma 17.8(iv), the probability $q_n(\mathbf{y})$ that a particular $\mathbf{x}_i$ is jointly ζ-typical with a given $\mathbf{y} \in \mathsf{Y}^n$ (such that $\mathsf{T}_{[XY]_\zeta}(\mathbf{y}) \neq \emptyset$) is between $\exp\{-n[I(X \wedge Y) + \tau]\}$ and $\exp\{-n[I(X \wedge Y) - \tau]\}$, where τ is arbitrarily small if ζ is. Applying Lemma 17.9

both with $\alpha = 1/2$ and $\alpha = 3/2$, say, it follows that the number of those $\mathbf{x}_i$ which are jointly ζ-typical with $\mathbf{y}$ is between $(1/2)Nq_n(\mathbf{y})$ and $(3/2)Nq_n(\mathbf{y})$, d.e.s., providing ζ is small enough to make $\tau < R - I(X \wedge Y)$. Thus the asserted bound holds d.e.s. for each $\mathbf{y} \in \mathsf{Y}^n$ with $\mathsf{T}_{[XY]_\zeta}(\mathbf{y}) \neq \emptyset$, and therefore – as there are only exponentially many $\mathbf{y}$ – also simultaneously for all of them. □

COROLLARY 17.9B Consider $N = \exp(nR)$ sequences $\mathbf{x}_i \in \mathsf{X}^n$ drawn independently from the distribution $P^n_{X|U}(\cdot|\mathbf{u})$, for given $\mathbf{u} \in \mathsf{U}^n$. The following hold uniformly with respect to the choice of $\mathbf{u}$.

(i) If $\mathbf{u}$ is ξ-typical and $\zeta > \xi$, there exists $\rho > 0$ such that all but a fraction $\exp(-n\rho)$ of the $\mathbf{x}_i$ are jointly ζ-typical with $\mathbf{u}$, d.e.s.
(ii) If $I(X \wedge Y|U) < R$, to any $\tau > 0$ there exists $\sigma > 0$ such that

$$\left| \frac{1}{n} \log \left| \left\{ i : \mathbf{x}_i \in \mathsf{T}_{[UXY]_\sigma}(\mathbf{u}, \mathbf{y}) \right\} \right| - (R - I(X \wedge Y|U)) \right| < \tau, \quad \text{d.e.s.,}$$

simultaneously for all $\mathbf{y} \in \mathsf{Y}^n$ with $\mathsf{T}_{[UXY]_\sigma}(\mathbf{u}, \mathbf{y}) \neq \emptyset$. ○

A sufficient condition for $\mathsf{T}_{[UXY]_\sigma}(\mathbf{u}, \mathbf{y}) \neq \emptyset$ is $(\mathbf{u}, \mathbf{y}) \in \mathsf{T}_{[UY]_\zeta}$, with $\zeta < \sigma$.

Proof The proof of Corollary 17.9A applies with obvious modifications. □

17.2 Secure transmission over an insecure channel

The model called *wiretap channel* involves two DMC's $\{W_1 : \mathsf{X} \to \mathsf{Y}\}$ and $\{W_2 : \mathsf{X} \to \mathsf{Z}\}$, like the broadcast channel in Chapter 16, but here a different problem is addressed. The first channel is regarded as connecting the sender to a legal receiver, while the second channel is used by an eavesdropper to wiretap the traffic of the first channel. In this scenario, in addition to making sure that the legal receiver can decode the transmitted message with small probability of error, another goal is to keep the wiretapper ignorant about it. The sender and receiver are assumed to have no other means of communication than the channel $\{W_1\}$, thus no secret key is available to them that could be used to maintain secrecy via encryption. However, the sender is allowed to use a stochastic encoder, which may be more efficient to achieve secrecy than a deterministic one.

Let the RVs M, X^n, Y^n, Z^n represent the message to be transmitted, the channel input obtained applying some stochastic encoder $F : \mathsf{M} \to \mathsf{X}^n$ to M, and the corresponding outputs of the channels $\{W_1\}$ and $\{W_2\}$. The formal meaning of the above is nothing more than the Markov condition $M \multimap X^n \multimap Y^n Z^n$ plus the conditions that Y^n and Z^n are connected with X^n via the DMC's $\{W_1\}$ and $\{W_2\}$. Hence, the (stochastic) encoder will not explicitly enter the following definition.

DEFINITION 17.10 A non-negative number R is an *achievable rate* for the wiretap channel if for each $\varepsilon > 0$, $\delta > 0$, and sufficiently large n there exist RVs $M \multimap X^n \multimap Y^n Z^n$ such that M is distributed on a set M with $\frac{1}{n} \log |\mathsf{M}| > R - \delta$, Y^n

and Z^n are connected with X^n via the channels $\{W_1\}$ and $\{W_2\}$, and with a suitable (deterministic) decoder $\varphi : \mathsf{Y}^n \to \mathsf{M}$

$$\Pr\left\{\varphi(Y^n) \neq M\right\} < \varepsilon, \quad S(M|Z^n) < \varepsilon. \tag{17.16}$$

The largest achievable rate is the *secrecy capacity* C_S. ○

REMARKS (i) It is common to require the message RV M to be uniformly distributed on M. That condition could be added to Definition 17.10 causing no change. Indeed, the second condition in (17.16) implies by (17.2) that the distribution of M is close to the uniform distribution on M in I-divergence, hence by Pinsker's inequality also in variation distance. → 3.18 Thus there exists an M' uniformly distributed on M with $\Pr\left\{M' \neq M\right\}$ small, and it can be chosen to satisfy $M' -\!\!\circ\!\!- M -\!\!\circ\!\!- X^n -\!\!\circ\!\!- Y^n Z^n$. This M' meets the second condition in (17.16) because

$$S(M'|Z^n) = I(M' \wedge Z^n) \leqq I(M \wedge Z^n) \leqq S(M|Z^n).$$

Of course, M' also satisfies the first condition in (17.16), with some $\varepsilon' > \varepsilon$, which is arbitrarily small if ε is.

(ii) A stronger, as well as a weaker, version of the above definition also deserves mention. The stronger one requires $\varepsilon = \varepsilon_n$ to go to zero exponentially; the weaker version requires, instead of $S(M|Z^n) < \varepsilon$, only $(1/n)S(M|Z^n) < \varepsilon$. Originally, the weaker version was the standard one. Our results will be the same under either version, and this applies also to the models treated later on. ○

THEOREM 17.11 The secrecy capacity of the wiretap channel is

$$C_S = \max\left[I(V \wedge Y) - I(V \wedge Z)\right],$$

for RVs $V -\!\!\circ\!\!- X -\!\!\circ\!\!- YZ$ such that $P_{Y|X} = W_1$, $P_{Z|X} = W_2$. The auxiliary RV V may be assumed to satisfy the range constraint $|\mathsf{V}| \leqq |\mathsf{X}|$. Moreover, the secrecy capacity remains the same under both the stronger and weaker versions of Definition 17.10 in → 17.8 the above Remark. ○

COROLLARY 17.11 $C_S > 0$ unless the channel W_2 is less noisy than W_1. ○

By definition (see Problem 15.11), W_2 is less noisy than W_1 if it holds for all RVs $V -\!\!\circ\!\!- X -\!\!\circ\!\!- YZ$ with $P_{Y|X} = W_1$ and $P_{Z|X} = W_2$ that $I(V \wedge Y) \leqq I(V \wedge Z)$. Note that Corollary 17.11 provides an operational characterization of the "less noisy" relation.

Proof of the direct part It suffices to prove that $I(X \wedge Y) - I(X \wedge Z)$ is an achievable rate when positive. Indeed, if that is proven and

$$R = I(V \wedge Y) - I(V \wedge Z) > 0, \quad \text{where} \quad V -\!\!\circ\!\!- X -\!\!\circ\!\!- YZ,$$

then R is an achievable rate for the auxiliary wiretap channel defined by $\widetilde{W}_1 = P_{Y|V}$, $\widetilde{W}_2 = P_{Z|V}$. Accordingly, take RVs $M -\!\!\circ\!\!- V^n -\!\!\circ\!\!- Y^n Z^n$ as in Definition 17.10 with V^n

in the role of X^n, in particular Y^n and Z^n connected with V^n by the channel $\{\widetilde{W}_1\}$ resp. $\{\widetilde{W}_2\}$. Then X^n can be defined so as to satisfy $M \multimap V^n \multimap X^n \multimap Y^n Z^n$, with Y^n and Z^n connected with X^n by channels $\{W_1\}$ and $\{W_2\}$, respectively; it follows that R is an achievable rate for the original channel as well.

Now fix $\delta > 0$ arbitrarily small and select randomly $N = \exp\{n[I(X \wedge Y) - \delta]\}$ sequences $\mathbf{x}_i \in \mathsf{X}^n$ from the distribution P_X^n. It is well known that, with probability close to unity, the resulting set $\mathsf{U} \triangleq \{\mathbf{x}_1, \dots, \mathbf{x}_N\}$ is the codeword set of a code for channel $\{W_1\}$ whose average probability of error is exponentially small.

Denote by U a RV uniformly distributed on U, and let Y^n, Z^n be the outputs of the channels $\{W_1\}$, $\{W_2\}$ corresponding to input U. Due to the decodability of U from Y^n, the achievability of $R \triangleq I(X \wedge Y) - I(X \wedge Z)$ (in the stronger sense) will follow if we prove the existence of a mapping $\kappa : \mathsf{U} \to \{1, \dots, k\}$ with $(1/n)\log k$ arbitrarily close to R such that $M = \kappa(U)$ satisfies the secrecy condition in (17.16), with $\varepsilon = \varepsilon_n \to 0$ exponentially fast.

To this, we use the secrecy lemma, Lemma 17.5 with the present U, and $V = Z^n$, applying its part (ii) to the set of all jointly ζ-typical pairs $(\mathbf{x}_i, \mathbf{z}) \in \mathsf{U} \times \mathsf{Z}^n$ in the role of B.

By the typicality lemma, Lemma 17.8, (ii) and (iii) (with Z in the role of U), this set B satisfies (17.13) with $\alpha = \exp(-2n\tau)$, where τ is arbitrarily small if ζ is. We also have to show that $P_{UV}(\overline{\mathsf{B}})$ is exponentially small. This holds d.e.s. because

$$P_U\left(\mathsf{U} - \mathsf{T}_{[X]_\xi}\right) = \frac{1}{N}\left|\{i : \mathbf{x}_i \notin \mathsf{T}_{[X]_\xi}\}\right|$$

is exponentially small by Corollary 17.9A for any $\xi > 0$, and if $\mathbf{x}_i \in \mathsf{U} \cap \mathsf{T}_{[X]_\xi}$, $\xi < \zeta$, then

$$W_2\left(\overline{\mathsf{T}}_{[XZ]_\zeta}(\mathbf{x}_i) \,\middle|\, \mathbf{x}_i\right) \leqq 2|\mathsf{X}||\mathsf{Z}|e^{-2(\zeta-\xi)^2 n}$$

by Lemma 17.8 (i).

Finally, we need a lower bound to $|\mathsf{B}_v|$, i.e., the number of those $\mathbf{x}_i \in \mathsf{U}$ which are jointly ζ-typical with a given $\mathbf{z}$ in the projection of B to Z^n. As such $\mathbf{z}$ satisfy $\mathsf{T}_{[XZ]_\zeta}(\mathbf{z}) \neq \emptyset$, Corollary 17.9A applied with Z in the role of Y yields

$$\left|\frac{1}{n}\log\left|\{i : \mathbf{x}_i \in \mathsf{T}_{[XZ]_\zeta}(\mathbf{z})\}\right| - I(X \wedge Y) - \delta - I(X \wedge Z)\right| < \tau, \quad \text{d.e.s.,}$$

where τ is arbitrarily small if ζ is. This completes the considerations needed to apply part (ii) of Lemma 17.5 to this scenario, and it follows that k there can grow with an exponential rate arbitrarily close to $R = I(X \wedge Y) - I(X \wedge Z)$, if δ and ζ are small enough. □

COMMENT It is worth emphasizing that the above proof starts with a classical code that admits reliable (but non-secret) transmission to the intended receiver, with rate larger than secrecy capacity. Then, the secrecy lemma is used to "extract secret bits" from the transmitted information, at a rate attaining the claimed secrecy capacity. This general idea underlies all the achievability proofs in this chapter. The step of extracting secret bits is commonly referred to as *privacy amplification.* ○

To prove the converse part of Theorem 17.11, and also other converses in this chapter, the following analog of the identity (15.34) will be used.

LEMMA 17.12 (*Key identity*) For arbitrary RVs U, V, and sequences of RVs Y^n, Z^n,

$$\begin{aligned} &I(V \wedge Y^n|U) - I(V \wedge Z^n|U) \\ &= \sum_{i=1}^{n} \Big[I(V \wedge Y_i|Y^{i-1}Z_{i+1}\dots Z_nU) - I(V \wedge Z_i|Y^{i-1}Z_{i+1}\dots Z_nU) \Big] \\ &= n\Big[I(\tilde{V} \wedge Y_J|\tilde{U}) - I(\tilde{V} \wedge Z_J|\tilde{U}) \Big], \end{aligned}$$

where $\tilde{U} = JUY^{J-1}Z_{J+1}\dots Z_n$, $\tilde{V} = \tilde{U}V$, and J denotes a RV uniformly distributed on $\{1, \dots, n\}$, independent of U, V, Y^n, Z^n. ○

Proof The ith term of the sum equals

$$\begin{aligned} &H(V|Y^{i-1}Z_{i+1}\dots Z_nU) - H(V|Y^iZ_{i+1}\dots Z_nU) \\ &\quad - H(V|Y^{i-1}Z_{i+1}\dots Z_nU) + H(V|Y^{i-1}Z_i\dots Z_nU). \end{aligned}$$

Summing these terms gives, after cancellations,

$$H(V|Z^nU) - H(V|Y^nU),$$

which equals the left-most side of the claimed identity. The second equality is obvious. □

Proof of the converse part of Theorem 17.11 Suppose R is an achievable rate, perhaps only in the weak sense, i.e., with condition (17.16) in Definition 17.10 relaxed, as in Remark (ii) after that definition.

Accordingly, consider RVs $M \multimap X^n \multimap Y^nZ^n$ with $(1/n)\log|\mathsf{M}| > R - \delta$ such that Y^n and Z^n are connected with X^n by the channels $\{W_1\}$ and $\{W_2\}$, and such that for some $\varphi : \mathsf{Y}^n \to \mathsf{M}$ both $\Pr\{\varphi(Y^n) \neq M\} < \varepsilon$ and

$$S(M|Z^n) = \log|\mathsf{M}| - H(M) + I(M \wedge Z^n) < n\varepsilon. \tag{17.17}$$

Since

$$H(M|Y^n) \leqq H(M|\varphi(Y^n)) < \varepsilon \log|\mathsf{M}| + 1,$$

by Fano's inequality, it follows from (17.17) that

$$(1-\varepsilon)\log|\mathsf{M}| - 1 - I(M \wedge Y^n) + I(M \wedge Z^n) < n\varepsilon,$$

i.e.,

$$(1-\varepsilon)(R-\delta) - \frac{1}{n} - \varepsilon < \frac{1}{n}\big[I(M \wedge Y^n) - I(M \wedge Z^n)\big]. \tag{17.18}$$

Using Lemma 17.12 with $U = \text{const}$, $V = M$, the right-hand side of (17.18) becomes

$$I(\tilde{V} \wedge Y_J|\tilde{U}) - I(\tilde{V} \wedge Z_J|\tilde{U}),$$

where $\tilde{U} \triangleq JY^{J-1}Z_{J+1}\dots Z_n$, $\tilde{V} \triangleq M\tilde{U}$, with J uniformly distributed on $\{1, \dots, n\}$, independent of (M, X^n, Y^n, Z^n). Renaming $\tilde{U}$, $\tilde{V}$, X_J, Y_J, Z_J as U, V, X, Y, Z, the Markov relation $U \multimap V \multimap X \multimap YZ$ is clearly satisfied, and so are the conditions $P_{Y|X} = W_1$, $P_{Z|X} = W_2$. Thus, as $\varepsilon > 0$, $\delta > 0$ can be arbitrarily small, (17.18) implies that R cannot exceed the supremum of $I(V \wedge Y|U) - I(V \wedge Z|U)$, for RVs U, V, X, Y, Z as above. Clearly, this supremum equals that of $I(V \wedge Y) - I(V \wedge Z)$ for $V \multimap X \multimap YZ$.

Finally, the assertion about the range of V follows similarly to the proof of Lemma 15.5. Consider, on the set of all PDs on X, the functions $f_x \triangleq P(x)$, for each $x \in \mathsf{X}$ but one, and $g(P) \triangleq H(PW_1) - H(PW_2)$. Given any RV V with $V \multimap X \multimap YZ$, averaging the values of these functions at the points $P_v \triangleq P_{X|V}(\cdot|v)$, with weights $P_V(v)$, $v \in \mathsf{V}$, gives the probabilities $P_X(x)$, $x \in \mathsf{X}$, resp. $H(Y|V) - H(Z|V)$. By the support lemma, Lemma 15.4, these averages can be represented as $\sum \alpha_i P_i(x)$ resp. $\sum \alpha_i g(P_i)$ for suitable PDs P_i and weights α_i, $1 \leqq i \leqq |\mathsf{X}|$. Thus V may be replaced by a new RV $\overline{V}$ with range $\{1, \dots, |\mathsf{X}|\}$ such that

$$P_{\overline{V}XYZ}(i, x, y, z) = \alpha_i P_i(x) P_{YZ|X}(y, z|x),$$

retaining the value of $H(Y|V) - H(Z|V)$, thus also of $I(V \wedge Y) - I(V \wedge Z)$. □

Next we consider a more general model called the *broadcast channel with confidential messages* (BCC). It also involves two DMCs $\{W_1\}$, $\{W_2\}$, but now both receivers are "legal," and a common message M_0 is sent to both of them. Another message M_1 is sent specifically to receiver 1, as in the ABC model in Chapter 16. Now, however, a secrecy aspect is added to that model: the message meant for receiver 1 consists of two parts, $M_1 = (M_1', M_s)$, where M_s is confidential and has to be concealed from receiver 2, while no secrecy requirement is imposed on M_1'.

Thus, the BCC model involves three message sets, M_0, M_1', M_s, and a triple of message RVs M_0, M_1', M_s uniformly distributed on $\mathsf{M}_0 \times \mathsf{M}_1' \times \mathsf{M}_s$. A perhaps stochastic encoder assigns a channel input X^n to this triple, receivers 1 and 2 observe the channel outputs Y^n resp. Z^n and have to decode (M_0, M_1', M_s) resp. M_0. In addition, the security index $S(M_s|Z^n)$ has to be small. A formal definition of the achievable rate triples (R_0, R_1', R_s) and the corresponding capacity region is left to the reader.

REMARK In the original version of this model, instead of assuming that the message intended for receiver 1 consisted of a secret and a non-secret part, this message M_1 was required to have a possibly large *equivocation rate* $(1/n)H(M_1|Z^n)$. Then, the achievable *rate-equivocation triples* (R_0, R_1, R_e) were of interest, for which (informally) common messages and messages for receiver 1 can be reliably sent at rates R_0, R_1, while the equivocation rate is at least R_e. In our model, the message for receiver 1 is $M_1 = (M_1', M_s)$, whose equivocation rate is (asymptotically) not less than the rate of the secret message M_s. It follows that whenever (R_0, R_1', R_s) is an achievable rate triple in our sense, $(R_0, R_1' + R_s, R_s)$ is an achievable rate-equivocation triple. The proof of

Theorem 17.13 will naturally imply that all achievable rate-equivocation triples arise in this manner. ○

THEOREM 17.13 The capacity region of the BCC consists of those triples of non-negative numbers (R_0, R_1', R_s) that satisfy, for some RVs $U \multimap V \multimap X \multimap YZ$ with $P_{Y|X} = W_1$, $P_{Z|X} = W_2$, the inequalities

$$\begin{aligned} R_0 &\leqq \min[I(U \wedge Y), I(U \wedge Z)], \\ R_s &\leqq I(V \wedge Y|U) - I(V \wedge Z|U), \\ R_0 + R_1' + R_s &\leqq I(V \wedge Y|U) + \min[I(U \wedge Y), I(U \wedge Z)]. \end{aligned}$$

Moreover, it may be assumed that $V = (U, V')$ and that the range sizes of U and V' are at most $|\mathsf{X}| + 3$ and $|\mathsf{X}| + 1$. ○

An equivalent characterization of the BCC capacity region is given by the above inequalities with U, V satisfying only $UV \multimap X \multimap YZ$ and the range constraints $|\mathsf{U}| \leqq |\mathsf{X}| + 3$, $|\mathsf{V}| \leqq |\mathsf{X}| + 1$. This follows from the final assertion of the theorem, renaming V' to V, because

$$I(UV' \wedge Y|U) = I(V' \wedge Y|U), \qquad I(UV' \wedge Z|U) = I(V' \wedge Z|U).$$

Proof of the converse part of Theorem 17.13 We start with the converse, for its proof is quite straightforward, using the key identity (Lemma 17.12). Actually, the stronger converse result indicated in the remark preceding Theorem 17.13 will be proved that each achievable rate-equivocation triple satisfies

$$\begin{aligned} R_0 &\leqq \min[I(U \wedge Y), I(U \wedge Z)], \\ R_e &\leqq I(V \wedge Y|U) - I(V \wedge Z|U), \\ R_0 + R_1 &\leqq I(V \wedge Y|U) + \min[I(U \wedge Y), I(U \wedge Z)], \end{aligned}$$

for some RVs $U \multimap V \multimap X \multimap YZ$ with $P_{Y|X} = W_1$, $P_{Z|X} = W_2$.

Consider RVs M_0 and M_1, independent and uniformly distributed on message sets M_0, M_1, let a (perhaps stochastic) encoder assign channel input X^n to (M_0, M_1), and let Y^n, Z^n be the corresponding outputs. Suppose that M_0 is decodable both from Y^n and from Z^n, and that M_1 is decodable from Y^n, with small probability of error.

The equivocation rate satisfies

$$\begin{aligned} \frac{1}{n}H(M_1|Z^n) &\leqq \frac{1}{n}H(M_1|Z^n M_0) + \frac{1}{n}H(M_0|Z^n) \\ &= \frac{1}{n}H(M_1|M_0) - \frac{1}{n}I(M_1 \wedge Z^n|M_0) + \frac{1}{n}H(M_0|Z^n) \\ &= \frac{1}{n}\left[I(M_1 \wedge Y^n|M_0) - I(M_1 \wedge Z^n|M_0)\right] \\ &\quad + \frac{1}{n}H(M_1|Y^n M_0) + \frac{1}{n}H(M_0|Z^n). \end{aligned}$$

The final two terms are negligible by Fano's inequality. Hence, using Lemma 17.12, the equivocation rate is bounded above, up to negligible terms, by

$$I(\tilde{V} \wedge Y_J|\tilde{U}) - I(\tilde{V} \wedge Z_J|\tilde{U}), \quad \tilde{U} = JM_0Y^{J-1}Z_{J+1}\ldots Z_n, \; \tilde{V} = \tilde{U}M_1, \tag{17.19}$$

where J, independent of M_0, M_1, X^n, Y^n, Z^n, is uniformly distributed on $\{1, \ldots, n\}$.

Clearly, these RVs $\tilde{U}$, $\tilde{V}$, X_J, Y_J, Z_J satisfy the Markov condition $\tilde{U} \multimap \tilde{V} \multimap X_J \multimap Y_J Z_J$, as well as $P_{Y_J|X_J} = W_1$, $P_{Z_J|X_J} = W_2$.

To bound the common rate R_0 note that

$$\frac{1}{n}\log|\mathsf{M}_0| = \frac{1}{n}H(M_0) = \frac{1}{n}I(M_0 \wedge Y^n) + \frac{1}{n}H(M_0|Y^n).$$

The final term is negligible by Fano's inequality, while

$$\begin{aligned}\frac{1}{n}I(M_0 \wedge Y^n) &= \frac{1}{n}\sum_{i=1}^{n} I(M_0 \wedge Y_i|Y^{i-1}) \leqq \frac{1}{n}\sum_{i=1}^{n} I(M_0 Y^{i-1} Z_{i+1}\ldots Z_n \wedge Y_i)\\ &= I(\tilde{U} \wedge Y_J|J) \leqq I(\tilde{U} \wedge Y_J).\end{aligned}$$

This and a similar argument replacing Y^n by Z^n, this time using

$$I(M_0 \wedge Z^n) = \sum_{i=1}^{n} I(M_0 \wedge Z_i|Z_{i+1}\ldots Z_n),$$

establish

$$R_0 \leqq \min\left[I(\tilde{U} \wedge Y_J), I(\tilde{U} \wedge Z_J)\right].$$

Finally,

$$\frac{1}{n}\log|\mathsf{M}_1| = \frac{1}{n}H(M_1|M_0) = \frac{1}{n}I(M_1 \wedge Y^n|M_0) + \frac{1}{n}H(M_1|Y^n M_0),$$

where the last term is again negligible. Thus, up to negligible terms, the rate sum $(1/n)\log|\mathsf{M}_1| + (1/n)\log|\mathsf{M}_0|$ is equal to

$$\begin{aligned}\frac{1}{n}I(M_1 \wedge Y^n|M_0) + \frac{1}{n}I(M_0 \wedge Y^n) &= \frac{1}{n}I(M_0M_1 \wedge Y^n)\\ &= \frac{1}{n}\sum_{i=1}^{n} I(M_0M_1 \wedge Y_i|Y^{i-1}) \leqq \frac{1}{n}\sum_{i=1}^{n} I(M_0M_1Y^{i-1}Z_{i+1}\ldots Z_n \wedge Y_i)\\ &= I(\tilde{V} \wedge Y_J|J) \leqq I(\tilde{V} \wedge Y_J) = I(\tilde{V} \wedge Y_J|\tilde{U}) + I(\tilde{U} \wedge Y_J)\end{aligned}$$

and also to

$$\begin{aligned}&\frac{1}{n}I(M_1 \wedge Y^n|M_0) + \frac{1}{n}I(M_0 \wedge Z^n)\\ &\quad = \frac{1}{n}I(M_1 \wedge Y^n|M_0) - \frac{1}{n}I(M_1 \wedge Z^n|M_0) + \frac{1}{n}I(M_0M_1 \wedge Z^n).\end{aligned}$$

Here the difference of the conditional mutual informations is equal to (17.19), while the final term is bounded as above by $I(\tilde{V} \wedge Z_J) = I(\tilde{V} \wedge Z_J|\tilde{U}) + I(\tilde{U} \wedge Z_J)$, completing the proof of the asserted bound also for $R_0 + R_1$.

To prove the final assertion of Theorem 17.13, we show the following: any U, V satisfying the Markov condition, but not the final assertion of Theorem 17.13, may be replaced by $\overline{U}$, $\overline{V}$ that satisfy also the final assertion, in such a way that this does not affect the values of

$$I(U \wedge Y), I(U \wedge Z), I(V \wedge Y|U), I(V \wedge Z|U). \tag{17.20}$$

To this end, the support lemma, Lemma 15.4 will be applied twice.

First consider the following $|\mathsf{X}| + 3$ continuous functions on the set $\mathcal{P}(\mathsf{V})$ of all PDs on V, regarding $P_{X|V}$ and hence also $\widetilde{W}_1 \triangleq P_{Y|V}$ and $\widetilde{W}_2 \triangleq P_{Z|V}$ as fixed:

$$f_x(P) \triangleq \sum_{v \in \mathsf{V}} P(v) P_{X|V}(x|v), \qquad \text{for all } x \in \mathsf{X} \text{ but one},$$

$$f_Y(P) \triangleq H(P\widetilde{W}_1), \qquad f_{Y|V}(P) \triangleq \sum_{v \in \mathsf{V}} P(v) H(\widetilde{W}_1(\cdot|v)),$$

$$f_Z(P) \triangleq H(P\widetilde{W}_2), \qquad f_{Z|V}(P) \triangleq \sum_{v \in \mathsf{V}} P(v) H(\widetilde{W}_2(\cdot|v)).$$

The probabilities $P_X(x)$ and the conditional entropies $H(Y|U)$, $H(Y|V)$, $H(Z|U)$, $H(Z|V)$ are equal to averages of values taken by these functions at points $P_u \triangleq P_{V|U}(\cdot|u)$, with weights $P_U(u)$, $u \in \mathsf{U}$. By the support lemma, these quantities can also be represented as averages of function values at $|\mathsf{X}| + 3$ points $P_i \in \mathcal{P}(\mathsf{V})$, with suitable weights α_i, $i = 1, \ldots, |\mathsf{X}| + 3$. Hence, one can define RVs $\overline{U}$ and $\tilde{V}$ with values in $\overline{\mathsf{U}} \triangleq \{1, \ldots, |\mathsf{X}| + 3\}$ and in V satisfying

$$P_{\overline{U}\tilde{V}XYZ}(i, v, x, y, z) = \alpha_i P_i(v) P_{X|V}(x|v) P_{YZ|X}(y, z|x) \tag{17.21}$$

such that replacing (U, V) by $(\overline{U}, \tilde{V})$ does not affect the corresponding conditional entropies, and hence neither the mutual informations (17.20).

In the second step, $\overline{U}$ is fixed and $\tilde{V}$ will be replaced by $\overline{V} = (\overline{U}, V')$. To this, consider $|\mathsf{X}| + 1$ continuous functions on $\mathcal{P}(\mathsf{X})$ that assign to $P \in \mathcal{P}(\mathsf{X})$ the values $P(x)$ (for all but one $x \in \mathsf{X}$), $H(PW_1)$ and $H(PW_2)$. Fixing any $i \in \overline{\mathsf{U}}$, the conditional probabilities $P_{X|\overline{U}}(x|i)$ and entropies $H(Y|\tilde{V})$, $H(Z|\tilde{V})$ are equal to averages of values of these functions at points $P_v \triangleq P_{X|V}(\cdot|v)$, with weights $P_i(v)$, $v \in \mathsf{V}$. By the support lemma, these quantities can also be represented as averages of function values at $|\mathsf{X}| + 1$ points $P_{ij} \in \mathcal{P}(\mathsf{X})$, with suitable weights β_{ij}, $j = 1, \ldots, |\mathsf{X}| + 1$. It follows that one can define a RV V' with values in $\{1, \ldots, |\mathsf{X}| + 1\}$ such that

$$P_{V'XYZ|\overline{U}=i}(j, x, y, z) = \beta_{ij} P_{ij}(x) P_{YZ|X}(y, z|x),$$

and, for each $i \in \overline{\mathsf{U}}$,

$$H(Y|\tilde{V}) = H(Y|V', \overline{U} = i), \quad H(Z|\tilde{V}) = H(Z|V', \overline{U} = i).$$

It follows that

$$I(\tilde{V} \wedge Y|\overline{U}) = I(V' \wedge Y|\overline{U}), \quad I(\tilde{V} \wedge Z|\overline{U}) = I(V' \wedge Z|\overline{U}),$$

and on replacing V' by $\overline{V} \triangleq (\overline{U}, V')$ these mutual informations remain the same. □

The proof of the direct part of Theorem 17.13 will rely upon a classical ABC code, and will be completed using the secrecy lemma, Lemma 17.5 (see the comment after the proof of Theorem 17.11). The first ingredient could well be a code as in the direct part of the ABC coding theorem (Theorem 16.1). Instead, an ABC code constructed as in the following lemma via the method of random selection will be used; this will facilitate the secrecy extraction part of the proof. Note that the lemma is not related to secrecy, and it provides an independent proof of Theorem 16.1. Although similar constructions have already appeared in the Problems sections, it will be proved in detail, for completeness.

LEMMA 17.14 Given RVs $U \multimap X \multimap YZ$ with $P_{Y|X} = W_1$, $P_{Z|X} = W_2$, select at random N_0 sequences $\mathbf{u}_i \in \mathsf{U}^n$, $i \in \mathsf{I} \triangleq \{1, \ldots, N_0\}$, from the distribution P_U^n, and then for each $i \in \mathsf{I}$ select N_1 sequences $\mathbf{x}_{ij} \in \mathsf{X}^n$, $j \in \mathsf{J} \triangleq \{1, \ldots, N_1\}$, from the distribution $P_{X|U}^n(\cdot\,|\mathbf{u}_i)$. If

$$N_0 = \exp\{n\,(\min[I(U \wedge Y), I(U \wedge Z)] - \delta)\},$$
$$N_1 = \exp\{n\,(I(X \wedge Y|U) - \delta)\}$$

and n is sufficiently large, the sequences $\mathbf{x}_{ij}$, $i \in \mathsf{I}$, $j \in \mathsf{J}$, represent with probability close to unity a good codeword set for the ABC defined by W_1, W_2 in the following sense. There exist decoders $\varphi_1 : \mathsf{Y}^n \to \mathsf{I} \times \mathsf{J}$, $\varphi_2 : \mathsf{Z}^n \to \mathsf{I}$ such that for (M_0, M_1) uniformly distributed on $\mathsf{I} \times \mathsf{J}$, and Y^n, Z^n denoting the outputs of the DMCs $\{W_1\}$, $\{W_2\}$ corresponding to input $X^n \triangleq \mathbf{x}_{M_0 M_1}$, the error probabilities

$$\bar{e}_1 \triangleq \Pr\left\{\varphi_1(Y^n) \neq (M_0, M_1)\right\},$$
$$\bar{e}_2 \triangleq \Pr\left\{\varphi_2(Z^n) \neq M_0\right\}$$

are less than $\exp(-n\rho)$, for some $\rho > 0$. ○

REMARK When the value of the expression specifying N_0 resp. N_1 is not an integer, the smallest integer larger than this value is taken. In particular, if the exponent of that expression is negative, we set N_0 resp. N_1 equal to unity. The proof below assumes that both exponents are positive; in the opposite (degenerate) case, only trivial changes are needed. ○

Proof of Lemma 17.14 We will use decoders φ_1, φ_2 defined by joint typicality as follows, with $\sigma > 0$ sufficiently small. If to $\mathbf{y} \in \mathsf{Y}^n$ there exists a unique $i \in \mathsf{I}$ with $(\mathbf{u}_i, \mathbf{y})$ jointly $\sigma|\mathsf{X}|$-typical, and also a unique $j \in \mathsf{J}$ with $(\mathbf{u}_i, \mathbf{x}_{ij}, \mathbf{y})$ jointly σ-typical, set $\varphi_1(\mathbf{y}) = (i, j)$; otherwise, declare an error. Similarly, if to $\mathbf{z} \in \mathsf{Z}^n$ there exists a unique $i \in \mathsf{I}$ with $(\mathbf{u}_i, \mathbf{z})$ jointly $\sigma|\mathsf{X}|$-typical, set $\varphi_2(\mathbf{z}) = i$; otherwise declare an error.

Denote by A and B the sets of those triples $(i, j, \mathbf{y})$ in $\mathsf{I} \times \mathsf{J} \times \mathsf{Y}^n$ resp. $(i, j, \mathbf{z})$ in $\mathsf{I} \times \mathsf{J} \times \mathsf{Z}^n$ for which $\mathbf{y}$ resp. $\mathbf{z}$ is jointly σ-typical with $(\mathbf{u}_i, \mathbf{x}_{ij})$. Of course, these sets depend on the outcome of the random selection. Our first claim is that A and B have, d.e.s., the property that $(M_0, M_1, Y^n) \notin \mathsf{A}$ or $(M_0, M_1, Z^n) \notin \mathsf{B}$ occurs with probability less than $\exp(-n\rho)$, for some constant $\rho > 0$. Indeed, the conditional probability on the condition $X^n = \mathbf{x}_{ij}$ that Y^n (or Z^n) is not jointly σ-typical with $(\mathbf{u}_i, \mathbf{x}_{ij})$ has an exponentially small upper bound when $(\mathbf{u}_i, \mathbf{x}_{ij}) \in \mathsf{T}_{[UX]_\zeta}$, $\zeta < \sigma$;

see the typicality lemma, Lemma 17.8(v). Moreover, $(\mathbf{u}_i, \mathbf{x}_{ij}) \in \mathsf{T}_{[UX]_\zeta}$ holds for all but an exponentially small fraction of the pairs $(i, j) \in \mathsf{I} \times \mathsf{J}$, d.e.s., on account of Corollaries 17.9A and 17.9B.

Having established our first claim, it suffices to consider instead of $\bar{e}_1$ and $\bar{e}_2$ the probabilities $\tilde{e}_1$ resp. $\tilde{e}_2$ that $(M_0, M_1, Y^n) \in \mathsf{A}$ and $\varphi_1(Y^n) \neq (M_0, M_1)$ resp. $(M_0, M_1, Z^n) \in \mathsf{B}$ and $\varphi_2(Z^n) \neq M_0$.

Denote by $\chi_0(\mathbf{u}, \mathbf{y})$, $\chi_1(\mathbf{u}, \mathbf{x}, \mathbf{y})$, $\chi_2(\mathbf{u}, \mathbf{z})$ the (indicator) functions equal to unity if $(\mathbf{u}, \mathbf{y}) \in \mathsf{T}_{[UY]_{\sigma|\mathsf{X}|}}$ resp. $(\mathbf{u}, \mathbf{x}, \mathbf{y}) \in \mathsf{T}_{[UXY]_\sigma}$ resp. $(\mathbf{u}, \mathbf{z}) \in \mathsf{T}_{[UZ]_{\sigma|\mathsf{X}|}}$, and zero otherwise. Then for $(M_0, M_1, Y^n) = (i, j, \mathbf{y}) \in \mathsf{A}$, the decoder φ_1 correctly decodes (M_0, M_1) iff

$$\sum_{l \neq i} \chi_0(\mathbf{u}_l, \mathbf{y}) + \sum_{m \neq j} \chi_1(\mathbf{u}_i, \mathbf{x}_{im}, \mathbf{y}) = 0$$

(using the fact that $(\mathbf{u}_i, \mathbf{x}_{ij}, \mathbf{y}) \in \mathsf{T}_{[UXY]_\sigma}$ implies $(\mathbf{u}_i, \mathbf{y}) \in \mathsf{T}_{[UY]_{\sigma|\mathsf{X}|}}$). Similarly, for $(M_0, M_1, Z^n) = (i, j, \mathbf{z}) \in \mathsf{B}$ the decoder φ_2 correctly decodes $M_0 = i$ iff

$$\sum_{l \neq i} \chi_2(\mathbf{u}_l, \mathbf{z}) = 0.$$

It follows that $\tilde{e}_1$ is bounded above by the sum over $(i, j, \mathbf{y}) \in \mathsf{A}$ of

$$\frac{1}{N_0 N_1} W_1^n(\mathbf{y}|\mathbf{x}_{ij}) \left[\sum_{l \neq i} \chi_0(\mathbf{u}_l, \mathbf{y}) + \sum_{m \neq j} \chi_1(\mathbf{u}_i, \mathbf{x}_{im}, \mathbf{y}) \right] \tag{17.22}$$

end $\tilde{e}_2$ is bounded above by the sum over $(i, j, \mathbf{z}) \in \mathsf{B}$ of

$$\frac{1}{N_0 N_1} W_2^n(\mathbf{z}|\mathbf{x}_{ij}) \sum_{l \neq i} \chi_2(\mathbf{u}_l, \mathbf{z}). \tag{17.23}$$

We are going to show that these upper bounds to $\tilde{e}_1$ and $\tilde{e}_2$ are exponentially small, perhaps not d.e.s. but at least with probability close to unity. To this, the standard technique of bounding expectation with respect to the random selection will be applied.

Consider the conditional expectation of (17.22) on the condition that $(\mathbf{u}_i, \mathbf{x}_{ij})$ are given. As $\mathbf{u}_l$ $(l \neq i)$ is independent of $(\mathbf{u}_i, \mathbf{x}_{ij})$, and $\mathbf{x}_{im}$ $(m \neq j)$ is conditionally independent of $\mathbf{x}_{ij}$ given $\mathbf{u}_i$, we obtain using the typicality lemma, Lemma 17.8 (iv) that

$$\begin{aligned}
E(\chi_0(\mathbf{u}_l, \mathbf{y})|\mathbf{u}_i, \mathbf{x}_{ij}) &= E(\chi_0(\mathbf{u}_l, \mathbf{y})) = \Pr\left\{\mathbf{u}_l \in \mathsf{T}_{[UY]_{\sigma|\mathsf{X}|}}(\mathbf{y})\right\} \\
&\leqq \exp\left\{-n\left[I(U \wedge Y) - \tau'\right]\right\}, \\
E(\chi_1(\mathbf{u}_i, \mathbf{x}_{im}, \mathbf{y})|\mathbf{u}_i, \mathbf{x}_{ij}) &= E(\chi_1(\mathbf{u}_i, \mathbf{x}_{im}, \mathbf{y})|\mathbf{u}_i) = \Pr\left\{\mathbf{x}_{im} \in \mathsf{T}_{[UXY]_\sigma}(\mathbf{u}_i, \mathbf{y})|\mathbf{u}_i\right\} \\
&\leqq \exp\{-n\left[I(X \wedge Y|U) - \tau\right]\},
\end{aligned}$$

where τ' and τ are positive numbers, arbitrary small if σ is. It follows that the conditional expectation, given $(\mathbf{u}_i, \mathbf{x}_{ij})$, of the sum of (17.22) over $\mathbf{y}$ with $(i, j, \mathbf{y}) \in \mathsf{A}$ is bounded above by

$$\frac{1}{N_0 N_1}\left[N_0 \exp\left\{-n\left[I(U \wedge Y) - \tau'\right]\right\} + N_1 \exp\left\{-n\left[I(X \wedge Y|U) - \tau\right]\right\}\right].$$

Of course, this is also an upper bound to the unconditional expectation.

Summing over $(i, j) \in \mathsf{I} \times \mathsf{J}$ yields

$$E(\tilde{e}_1) \leqq N_0 \exp\left\{-n\left[I(U \wedge Y) - \tau'\right]\right\} + N_1 \exp\left\{-n\left[I(X \wedge Y|U) - \tau\right]\right\}.$$

It follows similarly that

$$E(\tilde{e}_2) \leqq N_0 \exp\left\{-n\left[I(U \wedge Z) - \tau'\right]\right\}.$$

Recalling the choice of N_0 and N_1, this shows that $E(\tilde{e}_1)$ and $E(\tilde{e}_2)$ are exponentially small, if σ is small enough. This completes the proof of Lemma 17.14. □

Proof of the direct part of Theorem 17.13 By the same argument as in the proof of Theorem 17.11, it suffices to show that those triples (R_0, R_1', R_s) are achievable that satisfy the inequalities in the theorem with $V = X$. Note that if (R_0, R_1', R_s) is an achievable triple and $\tilde{R}_0 < R_0$, $\tilde{R}_0 + \tilde{R}_1' = R_0 + R_1'$, or if $\tilde{R}_s < R_s$ and $\tilde{R}_1' + \tilde{R}_s = R_1' + R_s$, then $(\tilde{R}_0, \tilde{R}_1', R_s)$ resp. $(R_0, \tilde{R}_1', \tilde{R}_s')$ is also achievable. Hence, it suffices to show that the triple

$$\begin{aligned} R_0 &= \min\left[I(U \wedge Y), I(U \wedge Z)\right], \\ R_1' &= I(X \wedge Z|U), \\ R_s &= I(X \wedge Y|U) - I(X \wedge Z|U) \end{aligned} \tag{17.24}$$

is achievable whenever $U \multimap X \multimap YZ$ (with $P_{Y|X} = W_1$, $P_{Z|X} = W_2$) and the difference in (17.24) is positive.

→ 17.9

We will use the construction in Lemma 17.14. As the lemma takes care of decodability, we henceforth concentrate on the secrecy aspect. Our goal is to identify M_1, uniformly distributed on J, with a pair (M_1', M_s) of independent RVs uniformly distributed on $\{1, \ldots, m\}$ resp. $\{1, \ldots, k\}$, with $mk = N_1$, such that the secrecy index $S(M_s|Z^n)$ is exponentially small and $(1/n)\log k$ is close to R_s in (17.24); then $(1/n)\log m$ will be close to R_1' by the choice of N_1.

As the key step, we show that random selection yields with probability close to unity a mapping $\kappa : \mathsf{J} \to \{1, \ldots, k\}$ such that $\kappa(M_1)$ satisfies the security requirement. Then the proof will be completed by modifying the nearly uniformly distributed RV $\kappa(M_1)$ to an exactly uniformly distributed M_s.

For technical convenience, the exponential smallness of $S(\kappa(M_1)|Z^n)$ will be established by proving the stronger assertion that $S(\kappa(M_1)|M_0, Z^n)$ is exponentially small. As in the proof of Theorem 17.11, the secrecy lemma, Lemma 17.5 will be used, this time with $U = M_1$, $V = (M_0, Z^n)$.

Take for B in part (ii) of Lemma 17.5 the set of all $(i, j, \mathbf{z}) \in \mathsf{I} \times \mathsf{J} \times \mathsf{Z}^n$ with $(\mathbf{u}_i, \mathbf{x}_{ij}, \mathbf{z})$ jointly σ-typical, denoted by B also in the proof of Lemma 17.14; as shown there, the probability of $(M_0, M_1, Z^n) \notin \mathsf{B}$ is less than $\exp(-n\rho)$, d.e.s. As in the proof of Theorem 17.11, B satisfies (17.13) with $\alpha = \exp(-2n\tau)$, where τ is arbitrary small if σ is.

Finally, Corollary 17.9B applied with Z in the role of Y yields, for $v = (i, \mathbf{z})$ in the projection of B to $\mathsf{I} \times \mathsf{Z}^n$

$$|\mathbf{B}_v| = \left|\{j : \mathbf{x}_{ij} \in \mathsf{T}_{[UXZ]_\sigma}(\mathbf{u}_i, \mathbf{z})\}\right| \geqq \exp\{n\,[I(X \wedge Y|U) - I(X \wedge Z|U) - 2\delta]\},$$

d.e.s., if σ is sufficiently small to make τ in Corollary 17.9B less than δ. Hence, it follows by Lemma 17.5 that for a randomly selected mapping $\kappa : \mathsf{J} \to \{1, \ldots, k\}$ with k growing at an exponential rate arbitrary close to $I(X \wedge Y|U) - I(X \wedge Z|U)$, the security index $S(\kappa(M_1)|M_0, Z^n)$ will be exponentially small, d.e.s.

Having established this key fact, the proof that (17.24) gives an achievable rate triple for the BCC is easily completed. By the first assertion of the extractor lemma, Lemma 17.3, the sets $\mathsf{A}_l \triangleq \kappa^{-1}(l)$, $l = 1, \ldots, k$ satisfy, d.e.s.,

$$\left|\frac{|\mathsf{A}_l|}{N_1} - \frac{1}{k}\right| \leqq \frac{\varepsilon}{k}, \qquad \varepsilon = \varepsilon_n \to 0 \text{ exponentially fast.}$$

Assuming with no loss of generality that $N_1 = mk$ for an integer m, it follows that $\mathsf{J} = \{1, \ldots, N_1\}$ can be partitioned into k sets B_l of size m such that $|\mathsf{A}_l \cap \mathsf{B}_l| \geqq (1 - \varepsilon)m$, $l = 1, \ldots, k$. This allows us to identify the RV M_1, uniformly distributed on J, with a pair (M_1', M_s), where $M_s = l$ if $M_1 = i$ with $i \in \mathsf{B}_l$, while M_1' (with values in $\{1, \ldots, m\}$) specifies the value i of M_1 within B_l.

Then M_1' and M_s are independent and uniformly distributed on $\{1, \ldots, m\}$ and $\{1, \ldots, k\}$ as required, and since M_s differs from $\kappa(M)$ only with exponentially small probability, $S(M_s|Z^n)$ is exponentially small as required. □

17.3 Secret key generation using public discussion

In the following, rather than addressing secure transmission directly, we concentrate on the problem of how two or more parties can generate a common secret key at an optimum rate. Of course, this key then may be used for secure transmission via encryption as in Proposition 17.1.

A basic concept for these considerations is *common randomness* (CR), also relevant in several other contexts not discussed here. A *secret key* (SK) is formally defined as CR secret from an adversary.

DEFINITION 17.15 A RV V with values in V is *ε-recoverable* from a RV U with values in U if there exists a function $f : \mathsf{U} \to \mathsf{V}$ such that $\Pr\{f(U) = V\} \geqq 1 - \varepsilon$. A RV K represents ε-CR for two or more parties if it is ε-recoverable from each of the RVs that represent the knowledge of these parties. An ε-SK for two or more parties, secret from an adversary whose knowledge is represented by a RV Z, is such an ε-CR K whose security index $S(K|Z)$ does not exceed ε. ○

A *source model* of generating a SK for two parties, usually referred to as Alice and Bob, involves a DMMS with generic variables (X, Y, Z). In time duration n, Alice and Bob observe the n-length source outputs X^n and Y^n, respectively, while an eavesdropper

Eve observes Z^n. Alice and Bob, in order to generate a SK secret from Eve, may use (i) their observations X^n and Y^n, (ii) the outcomes of random experiments they may perform (independently of each other and of (X^n, Y^n, Z^n)) and (iii) communication over a noiseless public channel.

A *channel model* of generating a SK for Alice and Bob involves a DMC $\{W : \mathsf{X} \to \mathsf{Y} \times \mathsf{Z}\}$. Alice selects the inputs, Bob observes the Y-outputs and Eve observes the Z-outputs. In addition, Alice and Bob are allowed to communicate over a noiseless public channel.

In both source and channel models, Eve is assumed to observe all public communication of Alice and Bob, but she is not allowed to interfere with it. It may appear counterintuitive that communication known to the adversary can be beneficial for generating a SK, but in fact it can. It will turn out that in typical source models and also in some channel models, no SK can be generated without public communication.

→ 17.11(c)

→ 17.12

We first concentrate on *unrestricted models* in which the public communication may be interactive and of unlimited rate. However, models restricting the possible communication in some way are also of interest. For example, the wiretap channel may be regarded as a channel model in which no public communication is allowed.

A *communication protocol* for a source model, where the original knowledge of Alice and Bob consists in X^n and Y^n, respectively, starts with Alice generating a RV Q_1 and Bob generating Q_2, where Q_1, Q_2 and (X^n, Y^n, Z^n) are mutually independent. Then Alice and Bob exchange public messages $F_{11}, \dots, F_{1r}$ and $F_{21}, \dots, F_{2r}$ in alternating order, where F_{1i} is any function of X^n, Q_1 and of $(F_{21}, \dots, F_{2,i-1})$ previously received from Bob, while F_{2i} is any function of Y^n, Q_2 and of $(F_{11}, \dots, F_{1i})$ previously received from Alice. The number r of exchanges may be arbitrary. After performing the protocol, Alice's knowledge is represented by (X^n, Q_1, F_2), Bob's by (Y^n, Q_2, F_1) and Eve's by (Z^n, F_1, F_2), where $F_j = (F_{j1}, \dots, F_{jr})$, $j = 1, 2$.

COMMENT The RVs Q_1, Q_2 serve for modeling *randomization*. In practice, Alice and Bob may send messages which are "randomized functions" of what they know at the actual instant. Equivalently, these may be regarded as deterministic functions if Alice's and Bob's knowledge includes the initially generated RVs Q_1 and Q_2. Note that while complex protocols as described above and, for channel models, below, have to be taken into account when proving converse results, our direct results about generating a SK will involve very simple protocols, each party sending at most one public message. In our direct results for source models, randomization (i.e., introducing RVs Q_1, Q_2) will not be needed either. ○

A *communication protocol* for a channel model, of time duration n, also starts with Alice and Bob generating independent RVs Q_1, Q_2. Then, at time instant $t = 1$, Alice selects a DMC input X_1 as a function of Q_1, and Bob and Eve observe the corresponding outputs Y_1 and Z_1. After this, Alice and Bob may exchange public messages $F_{11}^{(1)}, \dots, F_{1r_1}^{(1)}$ and $F_{21}^{(1)}, \dots, F_{2r_1}^{(1)}$ in alternating order, $F_{1i}^{(1)}$ being a function of Q_1 and of $F_{21}^{(1)}, \dots, F_{2,i-1}^{(1)}$ previously received from Bob, and $F_{2i}^{(2)}$ being a function of Y_1, Q_2 and $F_{11}^{(1)}, \dots, F_{1i}^{(1)}$ previously received from Alice. At time instant t, denote by F_1^{t-1}

and F_2^{t-1} the public message sequences $F_{11}^{(1)}, \dots, F_{1r_1}^{(1)}, \dots, F_{11}^{(t-1)}, \dots, F_{1r_{t-1}}^{(t-1)}$ and $F_{21}^{(1)}, \dots, F_{2r_1}^{(1)}, \dots, F_{21}^{(t-1)}, \dots, F_{2r_{t-1}}^{(t-1)}$ sent by Alice and Bob preceding this instant (of course, both may be empty). Then Alice selects the DMC input X_t as a function of Q_1 and F_2^{t-1}, and Bob and Eve observe the corresponding outputs Y_t, Z_t. The latter are conditionally independent of Q_1, Q_2, Y^{t-1}, Z^{t-1}, F_1^{t-1}, F_2^{t-1} on the condition that X_t is given, with conditional distribution $P_{Y_t Z_t | X_t} = W$. After this, Alice and Bob may exchange public messages $F_{11}^{(t)}, \dots, F_{1r_t}^{(t)}$ and $F_{21}^{(t)}, \dots, F_{2r_t}^{(t)}$, where $F_{1i}^{(t)}$ resp. $F_{2i}^{(t)}$ is a function of Q_1 and $(F_2^{t-1}, F_{21}^{(t)}, \dots, F_{2,i-1}^{(t)})$ resp. of Y^t, Q_2 and $(F_1^{t-1}, F_{11}^{(t)}, \dots, F_{1i}^{(t)})$. After performing the protocol, Alice's knowledge is represented by (Q_1, F_2), Bob's by (Y^n, Q_2, F_1) and Eve's by (Z^n, F_1, F_2), where $F_j = F_j^n \triangleq (F_{j1}^{(1)}, \dots, F_{jr_1}^{(1)}, \dots, F_{j1}^{(n)}, \dots, F_{jr_n}^{(n)})$, $j = 1, 2$.

DEFINITION 17.16 A non-negative number H is an *achievable SK rate* for a source or channel model if for every $\varepsilon > 0$, $\delta > 0$ and sufficiently large n there exists a communication protocol that makes some RV K with $(1/n) \log |\mathsf{K}| > H - \delta$ an ε-SK for Alice and Bob, secret from Eve. In other words, this K has to be ε-recoverable for Alice and Bob after having performed the protocol, and has to have security index

$$S(K | Z^n, F_1, F_2) \leqq \varepsilon. \tag{17.25}$$

The largest achievable SK rate is the *SK capacity* C_{SK}. ○

The formal meaning of the ε-recoverability condition is that

$$\Pr\{K = K_A\} \geqq 1 - \varepsilon, \quad \Pr\{K = K_B\} \geqq 1 - \varepsilon, \tag{17.26}$$

where K_A and K_B are functions of (X^n, Q_1, F_2) resp. (Y^n, Q_2, F_1). The notations K_A, K_B will always be used in this sense, both K_A and K_B having the same range K as K. In the case of a channel model, X^n is determined by (Q_1, F_2), thus K_A is actually a function of (Q_1, F_2).

REMARKS (i) The remarks after Definition 17.10 also apply here. In particular, the definition of achievable SK rates could be strengthened requiring $\varepsilon = \varepsilon_n \to 0$ exponentially, or weakened, relaxing the security condition (17.25) to

$$\frac{1}{n} S(K | Z^n, F_1, F_2) \leqq \varepsilon. \tag{17.27}$$

Under the weakened or strengthened definitions (but perhaps not under the original one) one may assume that K is actually a function of Alice's (or Bob's) knowledge, i.e., $K = K_A$ (or $K = K_B$). Indeed, if $\Pr\{K = K_A\} \geqq 1 - \varepsilon$ and K satisfies the conditions of the weakened/strengthened definition, then so does K_A, with a modified ε (assuming, with no loss of generality, that $(1/n) \log |\mathsf{K}| < 2H$, say). A still weaker definition of SK capacity, not requiring nearly uniform distribution of K, has also been used. For models treated in this chapter, all these definitions give the same SK capacity. → 17.1(d) → 17.10

(ii) A simpler concept than SK capacity is *CR capacity*, the largest rate at which CR can be generated in a given model without any secrecy requirements. Typically, in the → 17.11

first step of generating a SK, non-secret CR is generated, and in the second step secret bits are extracted from it (privacy amplification). In the first step it need not be best to generate CR at a maximum rate; quite frequently, "good" CR that allows the extraction of secret bits at SK capacity has a smaller rate than the CR capacity. ○

It is instructive to compare the SK capacity C_{SK} of a channel model defined by $W : \mathsf{X} \to \mathsf{Y} \times \mathsf{Z}$ with the secrecy capacity C_S of the wiretap channel defined by the components (marginals) W_1 and W_2 of W. Here, $W_1(y|x) \triangleq \sum_{z \in \mathsf{Z}} W(y, z|x)$, and W_2 is defined analogously. Clearly, a message securely transmitted over this wiretap channel represents a SK for Alice and Bob, thus $C_{SK} \geqq C_S$ always. This inequality may be strict. Note that while C_S depends on W only through the component channels W_1, W_2, this is not so for C_{SK}.

→ 17.12

THEOREM 17.17 The SK capacity of an unrestricted source model is bounded by

$$I(X \wedge Y) - \min[I(X \wedge Z), I(Y \wedge Z)] \leqq C_{SK} \leqq I(X \wedge Y|Z), \tag{17.28}$$

and that of an unrestricted channel model is bounded by

$$\max\{I(X \wedge Y) - \min[I(X \wedge Z), I(Y \wedge Z)]\} \leqq C_{SK} \leqq \max I(X \wedge Y|Z), \tag{17.29}$$

where the maximum is for RVs X, Y, Z satisfying $P_{YZ|X} = W$.

In (17.28) resp. (17.29) the upper bound is tight if X, Y, Z form a Markov chain in some order, resp. if the channel W has independent components or is physically degraded. ○

→ 17.13

A channel $W : \mathsf{X} \to \mathsf{Y} \times \mathsf{Z}$ has *independent components* if

$$W(y, z|x) = W_1(y|x)W_2(z|x),$$

and W is *physically degraded* from Y to Z if

$$W(y, z|x) = W_1(y|x)V(z|y) \quad \text{for some channel } V : \mathsf{Y} \to \mathsf{Z};$$

physical degradedness from Z to Y is defined similarly. Note that physical degradedness of W implies that one of its component channels W_1 and W_2 is a degraded version of the other (see Problem 6.16), but not conversely.

Proof of Theorem 17.17, lower bounds Given a DMMS with generic variables (X, Y, Z) such that $I(X \wedge Y) > I(X \wedge Z)$ and a small $\delta > 0$, take an encoder $f : \mathsf{X}^n \to \{1, \dots, \exp(nr)\}$, $r = H(X|Y) + \delta$, such that X^n is ε-recoverable from Y^n and $f(X^n)$; such an f exists by Lemma 17.6 (or simply by Theorem 13.2) if n is sufficiently large.

By Corollary 17.5(ii), there exists $\kappa : X^n \to \{1, \dots, k\}$ with

$$k = \exp\{n[H(X|Z) - r - \delta]\} = \exp\{n[I(X \wedge Y) - I(X \wedge Z) - 2\delta]\}$$

such that the security index $S(\kappa(X^n)|Z^n, f(X^n))$ is exponentially small. This proves that $I(X \wedge Y) - I(X \wedge Z)$ is an achievable SK rate, with the simple protocol of Alice sending Bob the message $F_1 = f(X^n)$ (Bob remaining silent). In Lemma 17.6, one can take $\varepsilon = \varepsilon_n \to 0$ exponentially fast, thus $I(X \wedge Y) - I(X \wedge Z)$ is an achievable SK rate even under the strengthened definition.

By symmetry, this completes the proof of the lower bound in (17.28).

The lower bound in (17.29) immediately follows, as channel models allow emulation of source models: Alice may choose i.i.d. repetitions of a RV X as DMC inputs, then these inputs X^n and the corresponding outputs Y^n, Z^n are i.i.d. repetitions of (X, Y, Z). Hence any achievable SK rate for a source model with generic variables (X, Y, Z) satisfying $P_{YZ|X} = W$ is also achievable for the channel model.

The assertion that $I(X \wedge Y|Z)$ is an achievable SK rate for a source model whose generic variables form a Markov chain in some order, is a consequence of the already proved result. Indeed, $I(X \wedge Y|Z)$ is equal for $X \multimap Y \multimap Z$ respectively $Y \multimap X \multimap Z$ to $I(X \wedge Y) - I(X \wedge Z)$ or $I(X \wedge Y) - I(Y \wedge Z)$, and for $X \multimap Z \multimap Y$ to zero. The corresponding assertion for channel models follows because if W has independent components or is physically degraded then $P_{YZ|X} = W$ implies that X, Y, Z form a Markov chain in some order. □

For the proof of the upper bounds in Theorem 17.17 we state two technical lemmas.

LEMMA 17.18 Let X, Y, Z and $F_1 = (F_{11}, \ldots, F_{1r})$, $F_2 = (F_{21}, \ldots, F_{2r})$ be such that, for each $1 \leqq i \leqq r$, F_{1i} is a function of X, Z and $F_{21}, \ldots, F_{2,i-1}$, while F_{2i} is a function of Y, Z and $F_{11}, \ldots, F_{1,i}$. Then

$$I(X \wedge Y|Z) \geqq I(X \wedge Y|Z F_1 F_2). \qquad \bigcirc$$

Proof

$$\begin{aligned} I(X \wedge Y|Z) &= I(X F_{11} \wedge Y|Z) \geqq I(X \wedge Y|Z F_{11}) \\ &= I(X \wedge Y F_{21}|Z F_{11}) \geqq I(X \wedge Y|Z F_{11} F_{21}). \end{aligned}$$

Repeating this argument r times, the assertion follows. □

LEMMA 17.19 If the RVs Q, S, T, X, Y, Z satisfy the Markov condition $QST \multimap X \multimap YZ$, and X is a function of (Q, T), then

$$I(Q \wedge SY|ZT) \leqq I(Q \wedge S|T) + I(X \wedge Y|Z). \qquad \bigcirc$$

Proof

$$\begin{aligned} I(Q \wedge SY|ZT) &= I(Q \wedge SYZT) - I(Q \wedge ZT) \\ &= I(Q \wedge ST) + I(Q \wedge YZ|ST) - I(Q \wedge T) - I(Q \wedge Z|T) \\ &= I(Q \wedge S|T) + H(YZ|ST) - H(YZ|QST) - H(Z|T) \\ &\quad + H(Z|QT). \end{aligned}$$

Here

$$H(YZ|ST) - H(Z|T) \leqq H(YZ|T) - H(Z|T) = H(Y|ZT) \leqq H(Y|Z),$$

and, using the hypothesis,

$$-H(YZ|QST) + H(Z|QT) = -H(YZ|X) + H(Z|X) = -H(Y|XZ). \qquad \square$$

Proof of Theorem 17.17, upper bounds Consider first the source model, and suppose that a suitable protocol with communication $F_1 = (F_{11}, \dots, F_{1r})$, $F_2 = (F_{21}, \dots, F_{2r})$ produces a RV K ε-recoverable for both Alice and Bob, and that K satisfies at least the weakened security condition (17.27).

Let K_A and K_B be functions of Alice's resp. Bob's knowledge (X^n, Q_1, F_2) resp. (Y^n, Q_2, F_1) satisfying (17.26). By Remark (i) to Definition 17.16, it may be assumed that $K = K_A$. Then, using (17.27) and Fano's inequality,

$$\begin{aligned} \log|\mathsf{K}| - n\varepsilon &\leqq H(K|Z^n F_1 F_2) \leqq H(K|K_B) + I(K \wedge K_B|Z^n F_1 F_2) \\ &\leqq \varepsilon \log|\mathsf{K}| + 1 + I(X^n Q_1 \wedge Y^n Q_2|Z^n F_1 F_2). \end{aligned} \tag{17.30}$$

By Lemma 17.18, the final term is bounded above by

$$I(X^n Q_1 \wedge Y^n Q_2|Z^n) = I(X^n \wedge Y^n|Z^n) = nI(X \wedge Y|Z).$$

As $\varepsilon > 0$ can be arbitrarily small, this proves that $I(X \wedge Y|Z)$ is an upper bound to achievable SK rates.

Turning to the channel model, consider any protocol that has been described for that case and a RV K $(= K_A)$ which is a function of Alice's knowledge (Q_1, F_2) and is ε-recoverable also for Bob. Then (17.30) holds with the right-most term replaced by $I(Q_1 \wedge Y^n Q_2|Z^n F_1 F_2)$.

Apply Lemma 17.18 to Q_1, $Y^n Q_2$, $Z^n F_1^{n-1} F_2^{n-1}$ in the role of X, Y, Z, with $(F_{11}^{(n)}, \dots, F_{1r_n}^{(n)})$ and $(F_{21}^{(n)}, \dots, F_{2r_n}^{(n)})$ in the role of F_1 and F_2. This yields

$$I(Q_1 \wedge Y^n Q_2|Z^n F_1 F_2) \leqq I(Q_1 \wedge Y^n Q_2|Z^n F_1^{n-1} F_2^{n-1}).$$

Next, apply Lemma 17.19 to $Q = Q_1$, $S = Y^{n-1} Q_2$, $T = Z^{n-1} F_1^{n-1} F_2^{n-1}$, $X = X_n$, $Y = Y_n$, $Z = Z_n$. Clearly, this choice satisfies the hypotheses, and we obtain

$$I(Q_1 \wedge Y^n Q_2|Z^n F_1^{n-1} F_2^{n-1}) \leqq I(Q_1 \wedge Y^{n-1} Q_2|Z^{n-1} F_1^{n-1} F_2^{n-1}) + I(X_n \wedge Y_n|Z_n).$$

Iterating this argument, we finally obtain that

→ 17.14

$$I(Q_1 \wedge Y^n Q_2|Z^n F_1 F_2) \leqq \sum_{i=1}^{n} I(X_i \wedge Y_i|Z_i) \leqq n \max I(X \wedge Y|Z). \qquad \square$$

In the general case, no single-letter characterization of the SK capacity of unrestricted source and channel models is known. The modified models below, however, can be fully solved via Theorem 17.17.

In the modified models, the outputs of the Z-source, respectively the channel outputs Z_t, $t = 1, \dots, n$, are known to Alice and Bob. One scenario for which such a source model is appropriate is when a centralized server, with access to Z^n as well as to all public communication, is supposed to help Alice and Bob in generating a SK about which this very server remains ignorant. Clearly, this goal is served most effectively by fully revealing Z^n. Models of this kind will be distinguished from the previous ones by calling the key generated by Alice and Bob a *private key* (PK), referring to privacy from the centralized server.

Formally, the modified source models are defined by allowing the public messages of Alice and Bob to depend functionally on Z^n. Similarly in channel models, as Alice and Bob are assumed to learn Z_t immediately at instant t, their public messages $F_{1i}^{(t)}$ and $F_{2i}^{(t)}$ may functionally depend on $Z^t = Z_1 \dots Z_t$, and Alice may select the DMC inputs X_t depending on Z^{t-1}. The formal definition of achievable PK rates and *PK capacity* is similar to Definition 17.16; of course, now the RVs K_A and K_B in (17.26) may functionally depend on Z^n.

THEOREM 17.20 The PK capacity of a modified source or channel model as above is equal to the upper bound to SK capacity in Theorem 17.17. ○

Proof The proof of the upper bounds in Theorem 17.17 applies verbatim for the modified models. To see that these bounds are now achievable, apply the lower bounds in Theorem 17.17 with Y substituted by (Y, Z). □

REMARK Theorem 17.20 provides an operational characterization of conditional mutual information. ○

Next, a one-way version of the original source model will be addressed in which the permissible communication is restricted to one message that Alice may send to Bob. We concentrate on the case when this message F is a deterministic function of X^n (allowing randomization would not lead to larger achievable SK rates). The *communication rate* $(1/n) \log \|F\|$ may be constrained or unconstrained.

In this model, an *achievable SK rate*, with *constraint* R or with no constraint, is defined as a number $H \geqq 0$ with the following property. For each $\varepsilon > 0$, $\delta > 0$ and sufficiently large n there exists a function $F = F(X^n)$, of rate less than $R + \delta$ in the case of constraint R, and a RV K ε-recoverable both from X^n and from (Y^n, F), such that

$$\frac{1}{n} \log |\mathsf{K}| > H - \delta, \quad S(K|Z^n, F) \leqq \varepsilon.$$

The largest achievable SK rate, with constraint R or with no constraint, is the *SK capacity* $C_{SK}(R)$ or $C_{SK}(\infty)$.

THEOREM 17.21 For the one-way source model,

$$C_{SK}(R) = \max [I(V \wedge Y|U) - I(V \wedge Z|U)]$$

for RVs U, V such that

$$U \multimap V \multimap X \multimap YZ, \qquad I(V \wedge X|Y) \leqq R, \tag{17.31}$$

and $C_{SK}(\infty)$ equals the maximum of the same expression subject only to the Markov condition. Moreover, it may be assumed that $V = (U, V')$, both U and V' having range of size at most $|\mathsf{X}| + 1$. ○

Similarly to Theorem 17.13, this characterization of $C_{SK}(R)$ has an equivalent form in which U, V satisfy, instead of (17.31),

$$UV \multimap X \multimap YZ, \quad I(UV \wedge X|Y) \leqq R,$$

and both have range of size at most $|\mathsf{X}| + 1$.

Proof of the converse part of Theorem 17.21 Supposing H is an SK rate achievable with constraint R (or with no constraint), consider a message $F = F(X^n)$ of Alice with $(1/n)\log \|F\| < R + \delta$ (or arbitrary if no constraint is present), and a RV K with $(1/n)\log|\mathsf{K}| > H - \delta$, which is ε-recoverable both from X^n and (Y^n, F), and satisfies at least the weakened secrecy condition (as in the previous proofs)

$$\frac{1}{n}S(K|Z^n, F) = \frac{1}{n}\log|\mathsf{K}| - \frac{1}{n}H(K|Z^n, F) < \varepsilon. \tag{17.32}$$

We may assume as before that $K = K_A$, i.e., K is a function of X^n. Its ε-recoverability by Bob implies $H(K|Y^n F) < \varepsilon \log|\mathsf{K}| + 1$, by Fano's inequality. Thus, from (17.32)

$$\frac{1}{n}\log|\mathsf{K}| < \frac{1}{n}\big[H(K|Z^n F) - H(K|Y^n F)\big] + \varepsilon \cdot \frac{1}{n}\log|\mathsf{K}| + \frac{1}{n} + \varepsilon,$$

and consequently

$$H < \frac{1}{n}\log|\mathsf{K}| + \delta < \frac{1}{1-\varepsilon} \cdot \frac{1}{n}\big[H(K|Z^n F) - H(K|Y^n F)\big] + \varepsilon',$$

where $\varepsilon' > 0$ is arbitrarily small if δ and ε are.

The final bracket is equal to $I(K \wedge Y^n|F) - I(K \wedge Z^n|F)$, to which the key identity (Lemma 17.12) applies. It follows that

$$H < \frac{1}{1-\varepsilon}[I(V \wedge Y_J|U) - I(V \wedge Z_J|U)] + \varepsilon', \tag{17.33}$$

with $U = FY^{J-1}Z_{J+1}\ldots Z_n J$, $V = KU$, where J is uniformly distributed on $\{1, \ldots, n\}$ and independent of (X^n, Y^n, Z^n). The RVs U, V, X_J, Y_J, Z_J satisfy the Markov relation $U \multimap V \multimap X_J \multimap Y_J Z_J$ because K is a function of X^n, and because the fact that (X^n, Y^n, Z^n) are i.i.d. repetitions of (X, Y, Z) implies that $X^{i-1}X_{i+1}\ldots X_n Y^{i-1} Z_{i+1}\ldots Z_n \multimap X_i \multimap Y_i Z_i$. Clearly, one can write X, Y, Z instead of X_J, Y_J, Z_J, hence (17.33) already proves that no achievable SK rate can exceed the claimed value of $C_{SK}(\infty)$.

Next, in the presence of rate constraint we also have – using Fano's inequality and Lemma 17.12 –

$$\begin{aligned} R + \delta &\geqq \frac{1}{n}\log\|F\| \geqq \frac{1}{n}H(F|Y^n) \geqq \frac{1}{n}H(FK|Y^n) - \varepsilon' \\ &= \frac{1}{n}\big[H(FK|Y^n) - H(FK|X^n)\big] - \varepsilon' \\ &= \frac{1}{n}\big[I(FK \wedge X^n) - I(FK \wedge Y^n)\big] - \varepsilon' \end{aligned}$$

$$= \frac{1}{n}\sum_{i=1}^{n}\Big[I(FK \wedge X_i|X_{i+1}\dots X_n Y^{i-1}) - I(FK \wedge Y_i|X_{i+1}\dots X_n Y^{i-1})\Big] - \varepsilon'. \tag{17.34}$$

Using the facts that (X_i, Y_i, Z_i), $i = 1, \dots, n$, are i.i.d. and both F and K are functions of X^n,

$$\begin{aligned} I(FK \wedge X_i|X_{i+1}\dots X_n Y^{i-1}) &= I(FKX_{i+1}\dots X_n Y^{i-1} \wedge X_i) \\ &= I(FKX_{i+1}\dots X_n Y^{i-1} Z_{i+1}\dots Z_n \wedge X_i) \\ &= I(FKY^{i-1}Z_{i+1}\dots Z_n \wedge X_i) + I(X_{i+1}\dots X_n \wedge X_i|FKY^{i-1}Z_{i+1}\dots Z_n). \end{aligned}$$

A similar identity holds when X_i is replaced by Y_i; moreover, the final term in the above chain of equalities is not smaller than its counterpart with Y_i replacing X_i. The latter is a consequence of the Markov relation $X_{i+1}\dots X_n \multimap FKY^{i-1}Z_{i+1}\dots Z_n X_i \multimap Y_i$, which can be verified using the properties of X_i, Y_i, Z_i, F, K as above.

Using these facts, we obtain from (17.34)

$$\begin{aligned} R + \delta + \varepsilon' &\geqq \frac{1}{n}\sum_{i=1}^{n}\Big[I(FKY^{i-1}Z_{i+1}\dots Z_n \wedge X_i) - I(FKY^{i-1}Z_{i+1}\dots Z_n \wedge Y_i)\Big] \\ &= I(V \wedge X_J|J) - I(V \wedge Y_J|J) \\ &= I(V \wedge X_J) - I(V \wedge Y_J) = I(V \wedge X_J|Y_J) \end{aligned}$$

with V and J as before. As X_J and Y_J may again be replaced by X and Y, this completes the proof of the converse, except for the final assertion.

The final assertion follows as the analogous one of Theorem 17.13. The support lemma is now applied first to $|\mathsf{X}| + 1$ continuous functions on $P(\mathsf{V})$: with the previous notation, to the $|\mathsf{X}| - 1$ functions $f_x(P)$, the function $f_Y(P) - f_Z(P)$ and the function

$$\sum_{v \in \mathsf{V}} P(v)\,[H(Y|V = v) - H(X|V = v)]\,.$$

The role of the final function is to take care of the constraint: the average of its values at the points $P_u \triangleq P_{V|U}(\cdot|u)$, with weights $P_U(u)$, is given by

$$\begin{aligned} H(Y|V) - H(X|V) &= H(Y) - I(V \wedge Y) - H(X) + I(V \wedge X) \\ &= H(Y) - H(X) + I(V \wedge X|Y). \end{aligned}$$ □

The proof of the direct part of Theorem 17.21 will use a random construction, similar to but more complex than that used in the proof of Theorem 17.13. Note that a weaker version of Theorem 17.21 allows for a much simpler proof. The basic construction is provided by the following lemma. The Remark following Lemma 17.14 also applies here. → 17.15

LEMMA 17.22 Suppose $U \multimap V \multimap X \multimap Y$, let δ be a small positive number, and n sufficiently large. Select independently $N_1 N_2$ sequences $\mathbf{u}_{ij} \in \mathsf{U}^n$ from the distribution P_U^n, where $i \in \mathsf{I} \triangleq \{1, \dots, N_1\}$, $j \in \mathsf{J} \triangleq \{1, \dots, N_2\}$,

$$N_1 = \exp\{n[I(U \wedge X|Y) + 3\delta]\}, \quad N_2 = \exp\{n[I(U \wedge Y) - 2\delta]\}.$$

Next, for each $i \in \mathsf{I}$, $j \in \mathsf{J}$, select $N_3 N_4$ random sequences $\mathbf{v}_{rs}^{ij} \in \mathsf{V}^n$, $r \in \mathsf{R} \triangleq \{1, \ldots, N_3\}$, $s \in \mathsf{S} \triangleq \{1, \ldots, N_4\}$,

$$N_3 = \exp\{n[I(V \wedge X|UY) + 3\delta]\}, \quad N_4 = \exp\{n[I(V \wedge Y|U) - 2\delta]\},$$

such that on the condition that $\mathbf{u}_{ij}$ is given, the selection of $\mathbf{v}_{rs}^{ij}$ is governed by the (conditional) distribution $P_{V|U}^n(\cdot|\mathbf{u}_{ij})$, and is (conditionally) independent of the other randomly selected sequences. Then for $0 < \zeta < \sigma$, both sufficiently small, it holds with probability close to unity that

(i) there exist functions f and φ on

$$\mathsf{T} \triangleq \left\{\mathbf{x} : \mathsf{T}_{[UX]_\zeta}(\mathbf{x}) \neq \emptyset\right\} \tag{17.35}$$

with values in I and J such that

$$(\mathbf{u}_{ij}, \mathbf{x}) \in \mathsf{T}_{[UX]_\zeta} \text{ if } \quad f(\mathbf{x}) = i, \ \varphi(\mathbf{x}) = j, \tag{17.36}$$

and for any such f and φ, extending them to functions on X^n by setting $f(x) = \varphi(x) = 0$ for $x \notin \mathsf{T}$, the RV $\varphi(X^n)$ is ε_n-recoverable from $(Y^n, f(X^n))$;

(ii) for any f and φ as in (i) there exist functions g and γ on T such that

$$(\mathbf{u}_{ij}, \mathbf{v}_{rs}^{ij}, \mathbf{x}) \in \mathsf{T}_{[UVX]_\sigma} \text{ if } \quad f(\mathbf{x}) = i, \ \varphi(\mathbf{x}) = j, \ g(\mathbf{x}) = r, \ \gamma(\mathbf{x}) = s, \tag{17.37}$$

and for any such g and γ, extending them to functions on X^n by setting $g(\mathbf{x}) = \gamma(\mathbf{x}) = 0$ for $\mathbf{x} \notin \mathsf{T}$, the RV $\gamma(X^n)$ is ε_n-recoverable from $(Y^n, f(X^n), g(X^n))$.

Here ε_n is a suitable sequence going to zero exponentially fast as $n \to \infty$. ○

Note that the set T in (17.35) contains $\mathsf{T}_{[X]_\xi}$ for any $0 < \xi < \zeta$, by (17.15), which implies by the typicality lemma, Lemma 17.8 that $P_X^n(\overline{\mathsf{T}})$ is exponentially small.

Proof of Lemma 17.22 (i) It follows from Corollary 17.9A (ii) (applied to U and X in the role of X and Y) that in the case $\mathbf{x} \in \mathsf{T}$ the number of sequences $\mathbf{u}_{ij}$ in $\mathsf{T}_{[UX]_\zeta}(\mathbf{x})$ increases exponentially, and hence is non-zero, d.e.s. (if ζ is sufficiently small). Thus, functions $f : \mathsf{T} \to \mathsf{I}$, $\varphi : \mathsf{T} \to \mathsf{J}$ satisfying (17.36) do exist, d.e.s.

We claim that if f, φ satisfy (17.36) (and $f(\mathbf{x}) = \varphi(\mathbf{x}) = 0$ if $\mathbf{x} \notin \mathsf{T}$) then the decoder $\hat{\varphi}$ defined below recovers $\varphi(X^n)$ from $(Y^n, f(X^n))$ with error probability

$$e \triangleq \sum_{(\mathbf{x},\mathbf{y}):\ \hat{\varphi}(\mathbf{y}, f(\mathbf{x})) \neq \varphi(\mathbf{x})} P_{XY}^n(\mathbf{x}, \mathbf{y}) < \varepsilon_n, \tag{17.38}$$

except for outcomes of the random selection whose probability approaches zero. Our $\hat{\varphi}$ will be a typicality decoder: if $(\mathbf{y}, i) \in \mathsf{Y}^n \times \mathsf{I}$, and a unique $j \in \mathsf{J}$ exists such that $\mathbf{u}_{ij}$ is jointly $\sigma|\mathsf{X}|$-typical with $\mathbf{y}$, set $\hat{\varphi}(\mathbf{y}, i) = j$; otherwise set $\hat{\varphi}(\mathbf{y}, i) = 0$.

Note first that it suffices to concentrate on those terms in (17.38) for which $(\mathbf{x}, \mathbf{y})$ belongs to

$$\mathsf{T}' \triangleq \left\{(\mathbf{x}, \mathbf{y}) : \mathbf{x} \in \mathsf{T}, \ (\mathbf{u}_{f(\mathbf{x})\varphi(\mathbf{x})}, \mathbf{x}, \mathbf{y}) \in \mathsf{T}_{[UXY]_\sigma}\right\}, \tag{17.39}$$

because $P^n_{XY}(\overline{\mathsf{T}}')$ has an exponentially small upper bound that depends neither on the actual $\mathbf{u}_{ij}$ nor on the choice of f and φ (as long as (17.36) holds). Indeed,

$$P^n_{XY}(\overline{\mathsf{T}}') = P^n_X(\overline{\mathsf{T}}) + \sum_{\mathbf{x}\in\mathsf{T},\ (\mathbf{x},\mathbf{y})\notin\mathsf{T}'} P^n_X(\mathbf{x}) P^n_{Y|X}\left(\overline{\mathsf{T}}_{[UXY]_\sigma}(\mathbf{u}_{f(\mathbf{x})\varphi(\mathbf{x})}, \mathbf{x})|\mathbf{x}\right),$$

where the first term is exponentially small (as noted after the statement of the lemma), and so is the second term, for it can be bounded using the typicality lemma, Lemma 17.8 (v).

When $(\mathbf{x}, \mathbf{y}) \in \mathsf{T}'$, the joint σ-typicality of $(\mathbf{u}_{f(\mathbf{x})\varphi(\mathbf{x})}, \mathbf{x}, \mathbf{y})$ implies that $\mathbf{u}_{f(\mathbf{x})\varphi(\mathbf{x})}$ is jointly $\sigma|\mathsf{X}|$-typical with $\mathbf{y}$; then $\hat{\varphi}(\mathbf{y}, f(\mathbf{x})) \neq \varphi(\mathbf{x})$ is equivalent to the existence of $m \neq \varphi(\mathbf{x})$ such that $\mathbf{u}_{f(\mathbf{x})m}$ is jointly $\sigma|\mathsf{X}|$-typical with $\mathbf{y}$. A necessary condition for the latter is the existence of $i \in \mathsf{I}$ and $j \neq m$ in J such that $(\mathbf{u}_{ij}, \mathbf{x}) \in \mathsf{T}_{[UX]_\zeta}$ and $(\mathbf{u}_{im}, \mathbf{y}) \in \mathsf{T}_{[UY]_{\sigma|\mathsf{X}|}}$. It follows that, denoting by $\chi_1(\mathbf{u}, \mathbf{x})$ and $\chi_2(\mathbf{u}, \mathbf{y})$ the (indicator) functions equal to unity if $(\mathbf{u}, \mathbf{x}) \in \mathsf{T}_{[UX]_\zeta}$ resp. $(\mathbf{u}, \mathbf{y}) \in \mathsf{T}_{[UY]_{\sigma|\mathsf{X}|}}$, and zero otherwise, the sum of terms in (17.38) with $(\mathbf{x}, \mathbf{y}) \in \mathsf{T}'$ is bounded above by

$$\sum_{(\mathbf{x},\mathbf{y})\in\mathsf{T}'} P^n_{XY}(\mathbf{x}, \mathbf{y}) \sum_{i\in\mathsf{I},\ j\in\mathsf{J},\ m\in\mathsf{J}-\{j\}} \chi_1(\mathbf{u}_{ij}, \mathbf{x})\chi_2(\mathbf{u}_{im}, \mathbf{y}). \tag{17.40}$$

This bound depends on the outcome of the random selection but not on the choice of f and φ.

We show that the bound (17.40) is exponentially small with probability approaching unity by the standard device of bounding its expectation. As the $\mathbf{u}_{ij}$ are drawn independently from the distribution P^n_U, it follows, using Lemma 17.8(iv), that

$$\begin{aligned}
E(\chi_1(\mathbf{u}_{ij}, \mathbf{x})\chi_2(\mathbf{u}_{im}, \mathbf{y})) &= \Pr\left\{\mathbf{u}_{ij} \in \mathsf{T}_{[UX]_\zeta}(\mathbf{x}),\ \mathbf{u}_{im} \in \mathsf{T}_{[UY]_{\sigma|\mathsf{X}|}}(\mathbf{y})\right\} \\
&= \Pr\left\{\mathbf{u}_{ij} \in \mathsf{T}_{[UX]_\zeta}(\mathbf{x})\right\} \Pr\left\{\mathbf{u}_{im} \in \mathsf{T}_{[UY]_{\sigma|\mathsf{X}|}}(\mathbf{y})\right\} \\
&\leqq \exp\{-n\,[I(U \wedge X) - \tau]\} \exp\left\{-n\left[I(U \wedge Y) - \tau'\right]\right\},
\end{aligned}$$

where τ, τ' are arbitrary small when σ is.

By the preceding result and the definition of N_1, N_2, the expectation of the inner sum in (17.40) – which has $N_1N_2(N_2 - 1)$ terms – is bounded above by an exponentially small quantity, providing σ is small enough to guarantee $\tau + \tau' < \delta$. Then the expectation of the sum (17.40) is bounded above by the same quantity. This implies that with probability approaching unity (exponentially fast), the sum (17.40) has an exponentially decreasing upper bound. As the probability of error (17.38) is bounded above by (17.40) plus the exponentially small contribution of the terms with $(\mathbf{x}, \mathbf{y}) \notin \mathsf{T}'$, this completes the proof of our claim about the recoverability of $\varphi(X^n)$.

(ii) As $(\mathbf{u}_{ij}, \mathbf{x}) \in \mathsf{T}_{[UX]_\zeta}$ implies $\mathsf{T}_{[UVX]_\sigma}(\mathbf{u}_{ij}, \mathbf{x}) \neq \emptyset$, it follows from Corollary 17.9B (ii) that to any such $\mathbf{u}_{ij}$ there exist, d.e.s., sequences $\mathbf{v}^{ij}_{rs} \in \mathsf{T}_{[UVX]_\sigma}(\mathbf{u}_{ij}, \mathbf{x})$, proving the existence of functions g and γ satisfying (17.37). To prove the assertion about the ε_n-recoverability of $\gamma(X^n)$, by the result of (i) it suffices to establish ε_n-recoverability of $\gamma(X^n)$ from $(Y^n, f(X^n), \varphi(X^n), g(X^n))$. The proof is similar to the previous one, hence we only sketch it.

Fixing $\vartheta > \sigma$, sufficiently small, a suitable decoder mapping $\hat{\gamma}$ is defined setting $\hat{\gamma}(\mathbf{y}, i, j, r) = s$ if a unique $s \in \mathsf{S}$ exists for which $\mathbf{y}$ is jointly $\vartheta|\mathsf{X}|$-typical with $(\mathbf{u}_{ij}, \mathbf{v}_{rs}^{ij})$ and zero otherwise. The analog of (17.38) is

$$e = \sum_{(\mathbf{x},\mathbf{y}):\ \hat{\gamma}(\mathbf{y}, f(\mathbf{x}), \varphi(\mathbf{x}), g(\mathbf{x})) \neq \gamma(\mathbf{x})} P_{XY}^n(\mathbf{x}, \mathbf{y}) < \varepsilon_n, \tag{17.41}$$

and to establish this, attention may be concentrated on the terms with $\mathbf{x} \in \mathsf{T}$ such that $(\mathbf{u}_{ij}, \mathbf{v}_{rs}^{ij}, \mathbf{x}, \mathbf{y})$ is jointly ϑ-typical for $i = f(\mathbf{x})$, $j = \varphi(\mathbf{x})$, $r = g(\mathbf{x})$, $s = \gamma(\mathbf{x})$. Denoting by T^* the set of pairs $(\mathbf{x}, \mathbf{y})$ with this property, the sum of terms in (17.41) with $(\mathbf{x}, \mathbf{y}) \in \mathsf{T}^*$ is bounded above by the following analog of (17.40):

$$\sum_{i \in \mathsf{I},\ j \in \mathsf{J}} \ \sum_{\substack{(\mathbf{x},\mathbf{y}) \in \mathsf{T}^* \\ f(\mathbf{x}) = i,\ \varphi(\mathbf{x}) = j}} P_{XY}^n(\mathbf{x}, \mathbf{y}) \sum_{r \in \mathsf{R},\ s \in \mathsf{S},\ t \in \mathsf{S} - \{s\}} \chi_1(\mathbf{u}_{ij}, \mathbf{v}_{rs}^{ij}, \mathbf{x}, \mathbf{y}) \chi_2(\mathbf{u}_{ij}, \mathbf{v}_{rt}^{ij}, \mathbf{y});$$

here χ_1 and χ_2 denote the indicator functions of the sets $\mathsf{T}_{[UVXY]_\vartheta}$ and $\mathsf{T}_{[UVY]_{\vartheta|\mathsf{X}|}}$, respectively.

Similarly as the expectation of (17.40) has been bounded, one verifies that the expectation of the final sum has an exponentially small upper bound (for the terms with given $(i, j) \in \mathsf{I} \times \mathsf{J}$, take first their conditional expectation on the condition that $\mathbf{u}_{ij} = \mathbf{u}$ is fixed). We leave to the reader to fill in the details. □

Proof of the direct part of Theorem 17.21

(i) First we show that if $U -\!\!\circ\!\!- X -\!\!\circ\!\!- YZ$ and $I(U \wedge X|Y) \leqq R$ (when a constraint on the communication rate is present) then

$$H = I(U \wedge Y) - I(U \wedge Z) \tag{17.42}$$

is an achievable SK rate, if positive. Note that – renaming U to V – this will prove the direct part of Theorem 17.21 in the special case when the maximum in the SK capacity formula is attained with $U = \text{const}$. As in previous proofs, achievability will be proved even under the "strengthened" definition.

Fix small positive numbers δ and $\xi < \zeta < \sigma$. Randomly selecting sequences $\mathbf{u}_{ij}$, $i \in \mathsf{I}$, $j \in \mathsf{J}$, as in Lemma 17.22, concentrate on such outcomes (of probability close to unity) for which $\varphi(X^n)$ is ε_n-recoverable from $(Y^n, f(X^n))$.

Let Alice send Bob the message $F = f(X^n)$; this satisfies the constraint in the definition of achievable SK rates, with 3δ instead of δ. Then $\varphi(X^n)$ will be ε_n-CR for Alice and Bob, and it suffices to show that there exists a mapping $\kappa : \mathsf{J} \to \{1, \ldots, k\}$ with $(1/n) \log k$ close to H such that the security index $S(K|Z^n, f(X^n))$ of $K \triangleq \kappa(\varphi(X^n))$ is exponentially small.

Consider the analog

$$\mathsf{T}'' \triangleq \left\{(\mathbf{x}, \mathbf{z}) :\ \mathbf{x} \in \mathsf{T},\ \ (\mathbf{u}_{f(\mathbf{x})\varphi(\mathbf{x})}, \mathbf{x}, \mathbf{z}) \in \mathsf{T}_{[UXZ]_\sigma}\right\} \tag{17.43}$$

of the set T' in (17.39); the same argument as for T' shows that $P_{XZ}^n(\overline{\mathsf{T}}'')$ is exponentially small. For technical convenience, instead of the claim that $S(K|Z^n, f(X^n))$ is exponentially small, we will prove the formally stronger claim that $S(K|Z^n, f(X^n), \chi(X^n, Z^n))$ is exponentially small, where χ denotes the indicator function of T''.

The secrecy lemma, Lemma 17.5 will be applied, giving the role of U and V to $\varphi(X^n)$ and $(f(X^n), Z^n, \chi(X^n, Z^n))$, whose joint distribution P is given by

$$P(i, j, \mathbf{z}, \nu) \triangleq \Pr\left\{f(X^n) = i,\ \varphi(X^n) = j,\ Z^n = \mathbf{z},\ \chi(X^n, Z^n) = \nu\right\}$$
$$= \sum_{\mathbf{x}:\ f(\mathbf{x})=i,\ \varphi(\mathbf{x})=j,\ \chi(\mathbf{x},\mathbf{z})=\nu} P^n_{XZ}(\mathbf{x}, \mathbf{z}), \qquad \nu \in \{0, 1\}. \tag{17.44}$$

Take for B in Lemma 17.5 the set

$$\mathsf{B} \triangleq \left\{(i, j, \mathbf{z}, 1) : (i, j) \in \mathsf{I} \times \mathsf{J},\ \mathbf{z} \in \mathsf{T}_{[Z]_\xi},\ \mathsf{T}_{[UXZ]_\sigma}(\mathbf{u}_{ij}, \mathbf{z}) \neq \emptyset\right\}. \tag{17.45}$$

This meets the condition that $\overline{\mathsf{B}}$ has exponentially small probability, as $P(\mathsf{B}) \geqq P^n_{XZ}(\mathsf{T}'') - P^n_Z(\overline{\mathsf{T}}_{[Z]_\xi})$.

Turning to the condition (17.13), note that, since $\mathsf{T}_{[UXZ]_\sigma}(\mathbf{u}_{ij}, \mathbf{z}) \neq \emptyset$ implies $(\mathbf{u}_{ij}, \mathbf{z}) \in \mathsf{T}_{[UZ]_{\sigma|\mathsf{X}|}}$,

$$|\mathsf{B}| \leqq \sum_{\mathbf{z} \in \mathsf{T}_{[Z]_\xi}} \left|\left\{(i, j) : \mathbf{u}_{ij} \in \mathsf{T}_{[UZ]_{\sigma|\mathsf{X}|}}(\mathbf{z})\right\}\right|$$
$$\leqq \exp\{n[H(Z) + \tau]\} \exp\{n[I(U \wedge X) + \delta - I(U \wedge Z) + \tau]\}, \quad \text{d.e.s.};$$

here, Corollary 17.9A (ii) has been applied to U and Z in the role of X, Z, with $R = I(U \wedge X) + \delta$. Further, (17.44) and the typicality lemma, Lemma 17.8 give, for $(i, j, \mathbf{z}, 1) \in \mathsf{B}$, the bound

$$P(i, j, \mathbf{z}, 1) \leqq \sum_{\mathbf{x} \in \mathsf{T}_{[UXZ]_\sigma}(\mathbf{u}_{ij}, \mathbf{z})} P^n_{XZ}(\mathbf{x}, \mathbf{z})$$
$$\leqq \exp\{n[H(X|UZ) + \tau]\} \exp\{-n[H(XZ) - \tau]\}.$$

The preceding two bounds establish that (17.13) holds with $\alpha = \exp\{-n(\delta + 3\tau)\}$ because

$$H(X|UZ) - H(XZ) = H(XZ|U) - H(Z|U) - H(XZ)$$
$$= -I(U \wedge XZ) - H(Z|U)$$
$$= -I(U \wedge X) - H(Z) + I(U \wedge Z).$$

In order to apply Lemma 17.5, it remains to bound the size of $\mathsf{B}_\upsilon = \{j : (i, j, \mathbf{z}, 1) \in \mathsf{B}\}$ from below when it is non-empty; note that then $\mathbf{z}$ is ξ-typical and consequently $\mathsf{T}_{[UZ]_\zeta}(\mathbf{z}) \neq \emptyset$. Using the fact that $\mathbf{u}_{ij} \in \mathsf{T}_{[UZ]_\zeta}(\mathbf{z})$ is a sufficient condition for $\mathsf{T}_{[UXZ]_\sigma}(\mathbf{u}_{ij}, \mathbf{z}) \neq \emptyset$, and then applying Corollary 17.9A (ii) with $R = I(U \wedge Y) - 2\delta$, it follows for $2\delta < I(U \wedge Y) - I(U \wedge Z)$ that

$$|\mathsf{B}_\upsilon| \geqq \left|\left\{j : \mathbf{u}_{ij} \in \mathsf{T}_{[UZ]_\zeta}(\mathbf{z})\right\}\right|$$
$$\geqq \exp\{n[I(U \wedge Y) - I(U \wedge Z) - 2\delta - \tau]\}, \quad \text{d.e.s.} \tag{17.46}$$

Hence, our claim that H in (17.42) is an achievable SK rate follows as a consequence of Lemma 17.5(ii).

(ii) To settle the general case, we have to show that if (U, V) satisfy (17.31) and

$$H = I(V \wedge Y|U) - I(V \wedge Z|U) > 0, \tag{17.47}$$

then H is an achievable SK rate with constraint R. It may be assumed that $I(U \wedge Y) < I(U \wedge Z)$, for otherwise

$$H = I(V \wedge Y) - I(V \wedge Z) - [I(U \wedge Y) - I(U \wedge Z)] \leqq I(V \wedge Y) - I(V \wedge Z)$$

is achievable by part (i).

As the proof is similar to that in (i), some details will be omitted. Concentrate on the outcomes of the random selection (of probability close to unity) for which the assertions of Lemma 17.22 hold.

Let Alice send Bob the message $F = (f(X^n), g(X^n))$, making $\gamma(X^n)$ an ε_n-CR for Alice and Bob; of course, $\varphi(X^n)$ is also an ε_n-CR, but it is recoverable to Eve due to $I(U \wedge Z) > I(U \wedge Y)$, and thus useless for generating secrecy.

The rate $(1/n) \log \|F\| = (1/n) \log(N_1 N_3)$ satisfies the constraint in the definition of achievable SK rates (with 6δ instead of δ) because

$$I(U \wedge X|Y) + I(V \wedge X|UY) = I(V \wedge X|Y).$$

It remains to prove the existence of a mapping $\kappa : \mathsf{S} \to \{1, \ldots, k\}$ with $(1/n) \log k$ close to H in (17.47) such that the security index $S(K|Z^n, F)$ of $K = \kappa(\gamma(X^n))$ is exponentially small.

As in part (i), again a formally stronger result will be proved, namely that $S(K|Z^n, f(X^n), \varphi(X^n), g(X^n), \chi(X^n, Z^n))$ is exponentially small. Here χ denotes the indicator function of the set of pairs $(\mathbf{x}, \mathbf{z})$ with $\mathbf{x} \in \mathsf{T}$, see (17.35), such that $(\mathbf{u}_{ij}, \mathbf{v}^{ij}_{rs}, \mathbf{x}, \mathbf{z})$ are jointly ϑ-typical if $i = f(\mathbf{x})$, $j = \varphi(\mathbf{x})$, $r = g(\mathbf{x})$, $s = \gamma(\mathbf{x})$. For convenience, we assume $\vartheta > \sigma|\mathsf{X}|$ rather than merely $\vartheta > \sigma$.

Again the secrecy lemma, Lemma 17.5 is applied, this time to $\gamma(X^n)$ in the role of U and $(f(X^n), \varphi(X^n), g(X^n), Z^n, \chi(X^n, Z^n))$ in the role of V. The role of B in Lemma 17.5 is played by the set B^* of those sixtuples $(i, j, r, s, \mathbf{z}, 1)$ for which $(i, j, \mathbf{z}, 1)$ belongs to the set in (17.45), and $\mathsf{T}_{[UVXZ]_\vartheta}(\mathbf{u}_{ij}, \mathbf{v}^{ij}_{rs}, \mathbf{z}) \neq \emptyset$.

The conditions of Lemma 17.5(ii) for this B^* are checked similarly as in part (i); let us concentrate on the key part, the analog of (17.46).

Fix $(i, j, r, \mathbf{z}) \in \mathsf{I} \times \mathsf{J} \times \mathsf{R} \times \mathsf{Z}^n$ such that B^*_v is non-empty when $(i, j, r, \mathbf{z}, 1)$ plays the role of v, in particular

$$\mathbf{z} \in \mathsf{T}_{[Z]_\xi}, \quad \mathsf{T}_{[UXZ]_\sigma}(\mathbf{u}_{ij}, \mathbf{z}) \neq \emptyset. \tag{17.48}$$

Fix θ with $\sigma|\mathsf{X}| < \theta < \vartheta$. Then joint θ-typicality of $(\mathbf{u}_{ij}, \mathbf{v}^{ij}_{rs}, \mathbf{z})$ is a sufficient condition for $\mathsf{T}_{[UVXZ]_\vartheta}(\mathbf{u}_{ij}, \mathbf{v}^{ij}_{rs}, \mathbf{z}) \neq \emptyset$, hence

$$|\mathsf{B}^*_v| \geqq \left|\left\{s : \mathbf{v}^{ij}_{rs} \in \mathsf{T}_{[UVZ]_\theta}(\mathbf{u}_{ij}, \mathbf{z})\right\}\right|.$$

Here $\mathsf{T}_{[UVZ]_\theta}(\mathbf{u}_{ij}, \mathbf{z}) \neq \emptyset$ because the second condition in (17.48) implies that $(\mathbf{u}_{ij}, \mathbf{z})$ are jointly $\sigma|\mathsf{X}|$-typical. Therefore, Corollary 17.9B(ii) applies (to V and Z in the role of X and Y, with $R = I(V \wedge Y|U) - 2\delta)$ and yields, for 2δ less than H in eq. (17.45), that

$$|\mathsf{B}_v^*| \geqq \exp\{n\,[I(V \wedge Y|U) - I(V \wedge Z|U) - 2\delta - \tau]\}, \quad \text{d.e.s.}$$

Here d.e.s. refers to the random choice of the sequences $\mathbf{v}_{rs}^{ij}$ when $\mathbf{u}_{ij}$ is fixed; as Corollary 17.9B holds uniformly with respect to the choice of $\mathbf{u}$ there, this result suffices for our purposes. Thus, the claim that H in (17.47) is an achievable SK rate follows as a consequence of the secrecy lemma. □

Next we discuss a *multi-terminal source model*. It involves $m \geqq 2$ parties, the ith party observing the n-length output X_i^n of the ith component source of a DMMS with generic variables $(X_1, \ldots, X_m)$.

Unrestricted public communication is allowed, in any number of rounds. The messages of either party are assumed to be noiselessly and instantenously received by all parties. The key has to be generated for a set A of parties. All parties in $\mathsf{M} \triangleq \{1, \ldots, m\}$ are assumed to cooperate towards this goal, even if A is a proper subset of M.

Recall that for unrestricted source models, even if only two parties and an eavesdropper are involved, no single-letter characterization of the SK capacity is known. The mathematical difficulty is caused by Eve having side information which is unavailable to Alice and Bob. The version of the two-party model when Eve does not posses undisclosed side information has a full solution; see Theorem 17.20.

Here, the multi-terminal generalization of the latter scenario is addressed. Thus, no component source of the given DMMS is observed by the eavesdropper alone. On the other hand, some of the m parties may be compromised: we admit a set $\mathsf{D} \subset \overline{\mathsf{A}}$ such that, for $i \in \mathsf{D}$, not only party i, but also the eavesdropper, observes X_i^n. The compromised parties $i \in \mathsf{D}$ (if any) can best contribute to the goal of generating a key for the parties in A by fully revealing X_i^n; henceforth we assume they do just that.

As in Chapter 13, we will use notation such as

$$X_{\mathsf{B}} \triangleq \{X_i : i \in \mathsf{B}\}, \qquad X_{\mathsf{B}}^n \triangleq \{X_i^n : i \in \mathsf{B}\}.$$

A *communication protocol* for the present model is similar to that in the two-party case. Each party $i \in \mathsf{D}$ (if any) reveals X_i^n immediately. The remaining parties exchange messages in r rounds. In round $t \in \{1, \ldots, r\}$, party i sends a message F_{it} which is a function of X_i^n and of the previously received messages $\left(X_{\mathsf{D}}^n, \{F_{js} : j \neq i,\ s < t\}, \{F_{jt} : j < i\}\right)$.

For this model, randomization offers no advantages; if it were allowed, the messages F_{it} could also depend on "randomizing" RVs Q_i generated by the parties $i \in \overline{\mathsf{D}}$ at the outset, independently of each other and of the source outputs.

The *total communication* of party $i \in \overline{\mathsf{D}}$ is $F_i \triangleq \{F_{it} : 1 \leqq t \leqq r\}$. The total communication of all non-compromised parties is $F_{\overline{\mathsf{D}}} \triangleq \{F_i : i \in \overline{\mathsf{D}}\}$, and the total communication of all parties is $F_{\mathsf{M}} \triangleq \left(X_{\mathsf{D}}^n, F_{\overline{\mathsf{D}}}\right)$.

Non-interactive communication means that each party sends (at most) one message, and this does not depend on any received messages.

After the communication protocol has been completed, the eavesdropper's knowledge is $F_{\mathsf{M}} = \left(X^n_{\mathsf{D}}, F_{\overline{\mathsf{D}}}\right)$. We are interested in generating CR K for the parties $i \in \mathsf{A}$ whose security index $S(K|F_{\mathsf{M}})$ is small. No assumption is made on the information available to those parties about K which are neither in A nor in D.

To emphasize that in this model all of the eavesdropper's knowledge is public, as in Theorem 17.20, we will use the term *private key* rather than secret key. Achievable PK rates, and PK capacity, are defined as in Definition 17.16, with the obvious modifications that K has to be ε-recoverable from $\left(X^n_i, F_{\mathsf{M}}\right)$ for all $i \in \mathsf{A}$, and the security requirement is $S(K|F_{\mathsf{M}}) \leqq \varepsilon$.

The PK capacity for this model is denoted by $C_{PK}(\mathsf{A}|\mathsf{D})$, or $C_{PK}(\mathsf{A})$ if $\mathsf{D} = \emptyset$.

The following concept does not involve secrecy but will turn out intrinsically related to PK capacity.

Definition 17.23 Given a DMMS with generic variables $(X_1, \ldots, X_m) = X_{\mathsf{M}}$, and a set $\mathsf{A} \subset \mathsf{M}$ of size $|\mathsf{A}| \geqq 2$, a communication protocol admits *ε-omniscience* (ε-OS) for the parties $i \in \mathsf{A}$ if X^n_{M} is ε-recoverable from either X^n_i, $i \in \mathsf{A}$, and the total communication of all parties. A non-negative number R is an *achievable OS rate* for A if for each $\varepsilon > 0$, $\delta > 0$ and sufficiently large n there exists a protocol that admits ε-OS for the parties $i \in \mathsf{A}$, with total communication of rate not exceeding $R + \delta$. When the admissible protocols are restricted, requiring all parties j in a given set $\mathsf{D} \subset \mathsf{M}$ to reveal X^n_j completely, this definition is modified: then it is the rate of the total communication of the parties in $\overline{\mathsf{D}}$ that has not to exceed $R + \delta$. The smallest achievable OS rate is denoted by $R_{OS}(\mathsf{A})$ resp. $R_{OS}(\mathsf{A}|\mathsf{D})$. ○

We will prove the intuitive result that the PK capacity $C_{PK}(\mathsf{A})$ is equal to the difference of $H(X_{\mathsf{M}})$ and the minimum achievable OS rate $R_{OS}(\mathsf{A})$. A similar result holds also for $C_{PK}(\mathsf{A}|\mathsf{D})$ with $\mathsf{D} \neq \emptyset$. These results will yield single-letter characterizations of the PK capacities, with the attractive feature that no auxiliary RVs are needed.

We first show that the smallest achievable OS rates can be characterized via constraints as in the fork network coding theorem, see Theorem 13.5 and Lemma 13.13, but not constraining rate sums over subsets of M that contain A. Specifically, we need the following sets of vectors, where $\mathcal{B}(\mathsf{A})$ and $\mathcal{B}(\mathsf{A}|\mathsf{D})$ denote the family of those non-empty subsets of M resp. $\overline{\mathsf{D}}$ that do not contain A:

$$\mathcal{R}(\mathsf{A}) \triangleq \left\{ (R_1, \ldots, R_m) : \sum_{i \in \mathsf{B}} R_i \geqq H(X_{\mathsf{B}}|X_{\overline{\mathsf{B}}}),\ \mathsf{B} \in \mathcal{B}(\mathsf{A}) \right\} \tag{17.49}$$

$$\mathcal{R}(\mathsf{A}|\mathsf{D}) \triangleq \left\{ \left\{ R_i : i \in \overline{\mathsf{D}} \right\} : \sum_{i \in \mathsf{B}} R_i \geqq H(X_{\mathsf{B}}|X_{\overline{\mathsf{B}}}),\ \mathsf{B} \in \mathcal{B}(\mathsf{A}|\mathsf{D}) \right\}. \tag{17.50}$$

Proposition 17.24 $R_{OS}(\mathsf{A})$ is equal to the minimum of $\sum_{i=1}^m R_i$ subject to $(R_1, \ldots, R_m) \in \mathcal{R}(\mathsf{A})$. Similarly, $R_{OS}(\mathsf{A}|\mathsf{D})$ equals the minimum of $\sum_{i \in \overline{\mathsf{D}}} R_i$ subject to $\left\{ R_i : i \in \overline{\mathsf{D}} \right\} \in \mathcal{R}(\mathsf{A}|\mathsf{D})$. Moreover, these minimal OS rates are achievable with non-interactive communication. ○

Proof The assertion about $R_{OS}(\mathsf{A})$ is the special case $\mathsf{D} = \emptyset$ of that about $R_{OS}(\mathsf{A}|\mathsf{D})$, thus it suffices to prove the latter assertion.

The direct part, including achievability with non-interactive communication, will be proved by establishing the following claim. Encode the component sources with $i \in \overline{\mathsf{D}}$ by randomly selected encoders F_i of rates $R_i + \delta$, where $\left\{R_i : i \in \overline{\mathsf{D}}\right\} \in \mathcal{R}(\mathsf{A}|\mathsf{D})$; then, with probability close to unity, X^n_{M} is ε-recoverable from X^n_{D}, $F_{\overline{\mathsf{D}}} \triangleq \left\{F_i(X^n_i) : i \in \overline{\mathsf{D}}\right\}$, and either X^n_j, $j \in \mathsf{A}$.

This claim follows from Lemma 13.13, if we show that $(R'_1, \ldots, R'_m)$ defined by

$$R'_i = \begin{cases} R_i & \text{if} \quad i \in \overline{\mathsf{D}} - \{j\} \\ H(X_i) & \text{if} \quad i \in \mathsf{D} \cup \{j\}, \end{cases}$$

where $j \in \mathsf{A}$ is arbitrary, satisfies the conditions

$$\sum_{i \in \mathsf{L}} R'_i \geqq H(X_{\mathsf{L}}|X_{\overline{\mathsf{L}}}), \qquad \text{for every } \mathsf{L} \subset \mathsf{M}.$$

These inequalities hold because $\mathsf{L} - (\mathsf{D} \cup \{j\})$ belongs to $\mathcal{B}(\mathsf{A}|\mathsf{D})$ (if non-empty), and therefore

$$\begin{aligned} \sum_{i \in \mathsf{L}} R'_i &= \sum_{i \in \mathsf{L} \cap \left(\overline{\mathsf{D}} - \{j\}\right)} R_i + \sum_{i \in \mathsf{L} \cap (\mathsf{D} \cup \{j\})} H(X_i) \\ &\geqq H\left(X_{\mathsf{L} \cap \left(\overline{\mathsf{D}} - \{j\}\right)} | X_{\overline{\mathsf{L}} \cup \mathsf{D} \cup \{j\}}\right) + H\left(X_{\mathsf{L} \cap (\mathsf{D} \cup \{j\})} | X_{\overline{\mathsf{L}}}\right) = H(X_{\mathsf{L}}|X_{\overline{\mathsf{L}}}). \end{aligned}$$

The converse will follow from the proof of Theorem 17.26; it also follows directly from the following technical lemma, with the choice $K = X^n_{\mathsf{M}}$. □

Note that the converse part of Proposition 17.24 is not covered by the results of Chapter 13 because the models there do not admit interactive communication.

LEMMA 17.25 Suppose a communication protocol as described before Definition 17.23 makes a RV K ε-recoverable from X^n_j and the total communication $F_{\mathsf{M}} = \left(X^n_{\mathsf{D}}, F_{\overline{\mathsf{D}}}\right)$, for every j in a given set $\mathsf{A} \subset \overline{\mathsf{D}}$. Then

$$\frac{1}{n} H(K|F_{\mathsf{M}}) = H(X_{\mathsf{M}}|X_{\mathsf{D}}) - \sum_{i \in \overline{\mathsf{D}}} R_i + \eta, \tag{17.51}$$

for some $\left\{R_i : i \in \overline{\mathsf{D}}\right\} \in \mathcal{R}(\mathsf{A}|\mathsf{D})$ and an error term $\eta \geqq 0$, which is arbitrarily small if ε is. ○

Proof Assume without any loss of generality that $\overline{\mathsf{D}} = \{1, \ldots, l\}$, $l \leqq m$, and that the communication takes r rounds: $F_{\overline{\mathsf{D}}} = (F_{11}, \ldots, F_{l1}, \ldots, F_{1r}, \ldots, F_{lr})$. Denote by F^-_{it} the sequence of messages received by party i from all parties $j \in \overline{\mathsf{D}}$ before sending the message F_{it}. Thus F_{it} is a function of $\left(X^n_i, X^n_{\mathsf{D}}, F^-_{it}\right)$ for all $i \in \overline{\mathsf{D}}$ and $1 \leqq t \leqq r$ (recall that we are considering models without randomization). Then

$$nH(X_{\mathsf{M}}|X_{\mathsf{D}}) + H(K|X^n_{\mathsf{M}}) = H(KX^n_{\mathsf{M}}|X^n_{\mathsf{D}}) = H(F_{\overline{\mathsf{D}}}KX^n_1 \dots X^n_l|X^n_{\mathsf{D}})$$
$$= \sum_{i=1}^{l}\left[\sum_{t=1}^{r} H(F_{it}|F^-_{it}X^n_{\mathsf{D}})\right] + H(K|F_{\mathsf{M}}) + \sum_{i=1}^{l} H(X^n_i|F_{\overline{\mathsf{D}}}KX^n_{\{1,\dots,i-1\}\cup\mathsf{D}}). \quad (17.52)$$

Setting

$$R_i \triangleq \frac{1}{n}\left[\sum_{t=1}^{r} H(F_{it}|F^-_{it}X^n_{\mathsf{D}}) + H(X^n_i|F_{\overline{\mathsf{D}}}KX^n_{\{1,\dots,i-1\}\cup\mathsf{D}})\right] + \delta, \quad (17.53)$$

it follows that

$$\frac{1}{n}H(K|F_{\mathsf{M}}) = H(X_{\mathsf{M}}|X_{\mathsf{D}}) - \sum_{i=1}^{l} R_i + l\delta + \frac{1}{n}H(K|X^n_{\mathsf{M}}).$$

Here the final term is negligible as K is ε-recoverable from X^n_{M}. The proof of (17.51) will be complete if we show that $(R_1, \dots, R_d)$ belongs to $\mathcal{R}(\mathsf{A}|\mathsf{D})$, for any $\delta > 0$ in (17.53) if ε is small enough.

Similarly to (17.52), for each $\mathsf{B} \in \mathcal{B}(\mathsf{A}|\mathsf{D})$,

$$nH(X_{\mathsf{B}}|X_{\overline{\mathsf{B}}}) \leqq H(F_{\overline{\mathsf{D}}}KX^n_{\mathsf{B}}|X^n_{\overline{\mathsf{B}}}) = \sum_{i=1}^{l}\left[\sum_{t=1}^{r} H(F_{it}|F^-_{it}X^n_{\overline{\mathsf{B}}})\right]$$
$$+ H(K|F_{\overline{\mathsf{D}}}X^n_{\overline{\mathsf{B}}}) + \sum_{i\in\mathsf{B}} H\left(X^n_i|F_{\overline{\mathsf{D}}}KX^n_{(\{1,\dots,i-1\}\cap\mathsf{B})\cup\overline{\mathsf{B}}}\right)$$
$$\leqq \sum_{i\in\mathsf{B}}\left[\sum_{t=1}^{r} H(F_{it}|F^-_{it}X^n_{\mathsf{D}}) + H\left(X^n_i|F_{\overline{\mathsf{D}}}KX^n_{\{1,\dots,i-1\}\cup\mathsf{D}}\right)\right] + H(K|F_{\overline{\mathsf{D}}}X^n_{\overline{\mathsf{B}}}). \quad (17.54)$$

Here the final inequality holds since $H(F_{it}|F^-_{it}X^n_{\overline{\mathsf{B}}}) = 0$ if $i \in \overline{\mathsf{B}}$ and

$$\overline{\mathsf{B}} \supset \mathsf{D}, \quad (\{1, \dots, i-1\} \cap \mathsf{B}) \cup \overline{\mathsf{B}} \supset \{1, \dots, i-1\} \cup \mathsf{D} \quad \text{if } i \in \mathsf{B}.$$

The inequalities (17.54) complete the proof, taking into account also that $(1/n)H(K|F_{\overline{\mathsf{D}}}X^n_{\overline{\mathsf{B}}})$ has an upper bound which is arbitrarily small if ε is. Indeed, by hypothesis, K is ε-recoverable from F_{D} and $X^n_{\overline{\mathsf{B}}}$ if $\mathsf{A} \cap \overline{\mathsf{B}} \neq \emptyset$, and all $\mathsf{B} \in \mathcal{B}(\mathsf{A}|\mathsf{D})$ meet the preceding condition, by definition. □

THEOREM 17.26 For the multi-terminal source model,

$$C_{PK}(\mathsf{A}) = H(X_{\mathsf{M}}) - R_{OS}(\mathsf{A}), \quad (17.55)$$
$$C_{PK}(\mathsf{A}|\mathsf{D}) = H(X_{\mathsf{M}}|X_{\mathsf{D}}) - R_{OS}(\mathsf{A}|\mathsf{D}). \quad (17.56)$$

Communication protocols achieving the OS rates suffice also to achieve the PK capacities. Moreover, the PK capacities can be achieved, with non-interactive communication, in such manner that the generated PK is a function of X^n_j, for an arbitrary fixed $j \in \mathsf{A}$. ○

→ 17.16
→ 17.17

Proof Similarly to the proof of Proposition 17.24, it suffices to address the PK capacities $C_{PK}(\mathsf{A}|\mathsf{D})$.

For the direct part, consider any communication protocol that makes X^n_{M} ε-recoverable from the total communication $F_{\mathsf{M}} = (X^n_{\mathsf{D}}, F_{\overline{\mathsf{D}}})$ and either X^n_i, $i \in \mathsf{A}$, such that the rate of $F_{\overline{\mathsf{D}}}$ is close to $R_{OS}(\mathsf{A}|\mathsf{D})$. Apply Corollary 17.5(ii), to X_{M} and X_{D} in the role of X and Y, with $Z^{(n)} = F_{\overline{\mathsf{D}}}$. Then there exists a mapping $\kappa : \mathsf{X}^n_M \to \{1, \ldots, k\}$ with $(1/n)\log k$ close to $H(X_{\mathsf{M}}|X_{\mathsf{D}}) - R_{OS}(\mathsf{A}|\mathsf{D})$ such that $K \triangleq \kappa(\mathsf{X}^n_M)$ has exponentially small security index $S(K|X^n_{\mathsf{D}}, F_{\overline{\mathsf{D}}}) = S(K|F_{\mathsf{M}})$. This proves that the right-hand side of (17.56) is an achievable PK rate, with any communication protocol that achieves $R_{OS}(\mathsf{A}|\mathsf{D})$; in particular, it is a lower bound to $C_{PK}(\mathsf{A}|\mathsf{D})$.

For the converse, suppose K is ε-recoverable from $F_{\mathsf{M}} = (X^n_{\mathsf{D}}, F_{\overline{\mathsf{D}}})$ and X^n_i, for each $i \in \mathsf{A}$, and

$$S(K|F_{\mathsf{M}}) = \log|\mathsf{K}| - H(K|F_{\mathsf{M}}) < \varepsilon n$$

(this "weakened" secrecy condition suffices for the converse). Applying (17.51) to this F_{M} yields

$$\frac{1}{n}\log|\mathsf{K}| \leqq H(X_{\mathsf{M}}|X_{\mathsf{D}}) - \sum_{i\in\overline{\mathsf{D}}} R_i + \eta + \varepsilon,$$

for some $\left\{R_i : i \in \overline{\mathsf{D}}\right\} \in \mathcal{R}(\mathsf{A}|\mathsf{D})$. This proves that

$$C_{PK}(\mathsf{A}|\mathsf{D}) \leqq H(X_{\mathsf{M}}|X_{\mathsf{D}}) - \min_{\left\{R_i : i\in\overline{\mathsf{D}}\right\}\in\mathcal{R}(\mathsf{A}|\mathsf{D})} \sum_{i\in\overline{\mathsf{D}}} R_i .$$

Note that Proposition 17.24 has not been used so far. Comparing the preceding upper bound to $C_{PK}(\mathsf{A}|\mathsf{D})$ with the previous lower bound yields

$$\min_{\left\{R_i : i\in\overline{\mathsf{D}}\right\}\in\mathcal{R}(\mathsf{A}|\mathsf{D})} \sum_{i\in\overline{\mathsf{D}}} R_i \leqq R_{OS}(\mathsf{A}|\mathsf{D}),$$

i.e., the converse assertion of Proposition 17.24. Invoking now the direct part of that proposition, the equality holds, which completes the proof of the converse part of Theorem 17.26.

To prove the final assertion of Theorem 17.26, fix n and $F_{\overline{\mathsf{D}}} = \left\{F_i(X^n_i) : i \in \overline{\mathsf{D}}\right\}$, such that X^n_{M} is ε-recoverable from $F_{\mathsf{M}} = (X^n_{\mathsf{D}}, F_{\overline{\mathsf{D}}})$ and X^n_i, for each $i \in \mathsf{A}$. Such n and $F_{\overline{\mathsf{D}}}$ exist for any $\varepsilon > 0$, with $F_{\overline{\mathsf{D}}}$ of rate R arbitrary close to $R_{OS}(\mathsf{A}|\mathsf{D})$, by Proposition 17.24. The ε-recoverability implies, by Fano's inequality, that

$$H(X^n_i|X^n_j F_{\mathsf{M}}) < \varepsilon n \log|\mathsf{X}_i| + h(\varepsilon) \quad \text{if } i \in \overline{\mathsf{D}},\ j \in \mathsf{A}. \tag{17.57}$$

Moreover, for each $j \in \mathsf{A}$,

$$\begin{aligned} H(X^n_j|F_{\mathsf{M}}) + H(X^n_{\overline{\mathsf{D}}-\{j\}}|X^n_j F_{\mathsf{M}}) &= H(X^n_{\overline{\mathsf{D}}}|F_{\mathsf{M}}) = H(X^n_{\mathsf{M}}|X^n_{\mathsf{D}} F_{\overline{\mathsf{D}}}) \\ &= H(X^n_{\mathsf{M}}|X^n_{\mathsf{D}}) - H(F_{\overline{\mathsf{D}}}|X^n_{\mathsf{D}}) \geqq n\left[H(X_{\mathsf{M}}|X_{\mathsf{D}}) - R\right]. \end{aligned}$$

If $\varepsilon > 0$ is sufficiently small and R is sufficiently close to $R_{OS}(\mathsf{A}|\mathsf{D})$, it follows, using (17.57), that $(1/n)H(X_j^n|F_{\mathsf{M}})$ has a lower bound arbitrarily close to $H(X_{\mathsf{M}}|X_{\mathsf{D}}) - R_{OS}(\mathsf{A}|\mathsf{D}) = C_{PK}(\mathsf{A}|\mathsf{D})$.

Now fix any $j \in \mathsf{A}$. Consider outputs of length nN of the given DMMS ($N \to \infty$), regarded as N-length outputs of a DMMS with generic variables X_{M}^n. Perform the given protocol on each of the N i.i.d. repetitions of X_{M}^n and complement this communication F_{M}^N, retaining non-interactivity, by a message $F_0 = F_0(X_j^{nN})$ sent by party j. This F_0 is chosen to make X_j^{nN} ε_N-recoverable from F_{M}^N, F_0 and X_i^{nN}, for each $i \in \mathsf{A} - \{j\}$, with $\varepsilon_N \to 0$ exponentially rapidly. It follows from Lemma 17.6 and (17.57) (interchanging i and j) that such an F_0 exists with $N^{-1} \log \|F_0\|$ only slightly larger than $\varepsilon n \log \max_{i \in \mathsf{A}} |\mathsf{X}_i| + h(\varepsilon)$. In particular, $(nN)^{-1} \log \|F_0\|$ is arbitrary small, if ε is.

Finally, apply Corollary 17.5(ii), to X_j^n and F_{M} in the role of X and Y, with $Z^{(N)} = F_0$. It follows that there exists a mapping $\kappa : \mathsf{X}_j^{nN} \to \{1, \ldots, k\}$ with $N^{-1} \log k$ close to $H(X_j^n|F_{\mathsf{M}}) - N^{-1} \log \|F_0\|$, and therefore $(nN)^{-1} \log k$ close to $C_{PK}(\mathsf{A}|\mathsf{D})$ such that for $K \triangleq \kappa(X_j^{nN})$ the security index $S(K|F_{\mathsf{M}}^N, F_0)$ is less than $\exp(-\xi N)$, for some $\xi > 0$. This completes the proof because one easily sees that our having concentrated on source output lengths equal to multiples of a fixed n did not cause any loss of generality. □

Theorem 17.26 provides a most satisfactory single-letter characterization of multi-
→ 17.18 terminal PK capacity (no auxiliary RVs!). To compute this PK capacity may still be hard if m is large. It is a linear programming problem, and additional insight may be
→ 17.19 expected from the duality theorem of linear programming. The following lemma gives a dual characterization of the minimum OS rates in Proposition 17.24.

Recalling the notation in (17.49) and (17.50), denote for $\mathsf{A} \subset \mathsf{M}$

$$\Lambda(\mathsf{A}) \triangleq \left\{ \lambda = \{\lambda_{\mathsf{B}} : \mathsf{B} \in \mathcal{B}(\mathsf{A})\} : \lambda_{\mathsf{B}} \geqq 0, \ \sum_{\mathsf{B}:\, i \in \mathsf{B}} \lambda_{\mathsf{B}} = 1, \ \forall i \in \mathsf{M} \right\}. \tag{17.58}$$

For disjoint subsets A and D of M, define $\Lambda(\mathsf{A}|\mathsf{D})$ similarly, with $\mathcal{B}(\mathsf{A})$ and M replaced by $\mathcal{B}(\mathsf{A}|\mathsf{D})$ and $\overline{\mathsf{D}}$, respectively.

LEMMA 17.27 The OS rates $R_{OS}(\mathsf{A})$ and $R_{OS}(\mathsf{A}|\mathsf{D})$ in Proposition 17.24 are equal to maxima for vectors λ in $\Lambda(\mathsf{A})$ or $\Lambda(\mathsf{A}|\mathsf{D})$ of sums $\sum \lambda_{\mathsf{B}} H(X_{\mathsf{B}}|X_{\overline{\mathsf{B}}})$, where B ranges over $\mathcal{B}(\mathsf{A})$ or $\mathcal{B}(\mathsf{A}|\mathsf{D})$. Moreover, there exists a maximizer λ such that the sets B with $\lambda_{\mathsf{B}} > 0$ have linearly independent incidence vectors. ○

Proof The lemma follows from the duality theorem of linear programming. Formally, it is a special case of the following version of that theorem (Schrijver (1986), Cor. 7.11, replacing **A**, **b**, **c** there by −**A**, −**b**, −**c**).

Let **A** be a $k \times m$ matrix, **b** a k-dimensional column vector and **c** an m-dimensional row vector, such that the minimum of **cx** subject to $\mathbf{Ax} \geqq \mathbf{b}$ is finite. Then this minimum
→ 17.20 equals the maximum of **yb** subject to $\mathbf{y} \geqq 0$, $\mathbf{yA} = \mathbf{c}$, and there exists a **y** attaining the maximum whose positive components correspond to linearly independent rows of **A**. □

As a counterpart of the source model considered above, we briefly address a multi-terminal channel model. The model involves a DMC with one input and multiple outputs, say $\{W : \mathsf{X}_1 \to \mathsf{X}_2 \times \cdots \times \mathsf{X}_m\}$; party 1 controls the DMC input and parties $2, \ldots, m$ observe the corresponding outputs. In addition, unlimited public communication of all parties is allowed. As in the preceding source model, there may be compromised parties (those in a set $\mathsf{D} \subset \mathsf{M}$), and all the eavesdropper's information is assumed to come from observing the public communication and the compromised parties.

As before, a PK has to be generated for the parties in a given set $\mathsf{A} \subset \mathsf{M}$, disjoint from D, and all parties including those in D are cooperating for this goal. Party 1, who controls the DMC input, may belong to either A or D or neither of them. The particular case $\mathsf{D} = \{1\}$ is suitable to model the scenario when a trusted center (in the role of Party 1) helps the parties $2, \ldots, m$ to generate a PK about which this very center remains ignorant.

A detailed description of permissible protocols for this model is omitted as it is similar to that for channel models with two legitimate users. To be specific, randomization is permitted. In the previous source model, randomization did not offer any benefits, but here the ability of party 1 to select the DMC inputs in a randomized manner is essential. Achievable PK rates and PK capacity $C_{PK}(\mathsf{A})$ resp. $C_{PK}(\mathsf{A}|\mathsf{D})$ are defined similarly as for the previous source model.

As for the two-party scenario, this channel model allows us to emulate a (multi-terminal) source model. Namely, if party 1 sends i.i.d. repetitions of a RV X_1 over the DMC, then parties $2, \ldots, m$ will observe i.i.d. repetitions of the corresponding outputs $X_2, \ldots, X_m$. Thus a DMMS with generic variables $X_{\mathsf{M}} = (X_1, \ldots, X_m)$ is emulated, where

$$P_{X_1 \ldots X_m}(x_1, \ldots, x_m) = P_{X_1}(x_1) W(x_2, \ldots, x_m | x_1). \tag{17.59}$$

This implies that the PK capacity of the channel model is bounded below by the maximum of the PK capacities of all source models with underlying distribution of form (17.59), obtainable by such emulation. By the following theorem, this bound is tight.

THEOREM 17.28 For the multi-terminal channel model

$$\begin{aligned} C_{PK}(\mathsf{A}) &= \max_{P_{X_1}} \min_{\lambda \in \Lambda(\mathsf{A})} \left[H(X_{\mathsf{M}}) - \sum_{\mathsf{B} \in \mathcal{B}(\mathsf{A})} \lambda_{\mathsf{B}} H(X_{\mathsf{B}} | X_{\overline{\mathsf{B}}}) \right] \\ &= \min_{\lambda \in \Lambda(\mathsf{A})} \max_{P_{X_1}} \left[H(X_{\mathsf{M}}) - \sum_{\mathsf{B} \in \mathcal{B}(\mathsf{A})} \lambda_{\mathsf{B}} H(X_{\mathsf{B}} | X_{\overline{\mathsf{B}}}) \right]. \end{aligned}$$

A similar result holds for $C_{PK}(\mathsf{A}|\mathsf{D})$, replacing $\Lambda(\mathsf{A})$, $\mathcal{B}(\mathsf{A})$ and $H(X_{\mathsf{M}})$ by $\Lambda(\mathsf{A}|\mathsf{D})$, $\mathcal{B}(\mathsf{A}|\mathsf{D})$ and $H(X_{\mathsf{M}}|X_{\mathsf{D}})$, respectively.

Here, the distribution of X_{M} depends on P_{X_1}, as in (17.59). ○

Proof sketch The first minimum over $\lambda \in \Lambda(\mathsf{A})$ is equal to the PK capacity of the emulated source model corresponding to a particular choice of the input RV X_1, by Theorem 17.26 and Lemma 17.27. Hence the maxmin is certainly a lower bound to $C_{PK}(\mathsf{A})$. The maxmin and minmax are equal due to the minimax theorem (Karlin (1959), Th. 1.1.5), since the expression in brackets is a linear function of λ and a continuous concave function of P_{X_1} (check this!).

The key part of the proof is to show that the maximum over P_{X_1} of the expression in brackets is an upper bound to $C_{PK}(\mathsf{A})$, for every λ in $\Lambda(\mathsf{A})$. This is done by showing that if a communication protocol involving n uses of the DMC lets the parties in A generate an ε-PK K, then $\log|\mathsf{K}|$ is (approximately) bounded above, for every $\lambda \in \Lambda(\mathsf{A})$, by

$$\sum_{t=1}^{n}\left[H(X_{\mathsf{M}t}) - \sum_{\mathsf{B}\in\mathcal{B}(\mathsf{A})} \lambda_{\mathsf{B}} H(X_{\mathsf{B}t}|X_{\overline{\mathsf{B}}t})\right],$$

where the distribution of $X_{\mathsf{M}t}$ is of the form (17.59), for each $1 \leqq t \leqq m$. As one would expect, this needs judicious manipulation of information quantities, in a similar but more complex manner than in the proof of the converse part of Theorem 17.17 for (two-party) channel models. The details are omitted. □

THEOREM 17.29 The PK capacities in Theorem 17.28 are achievable with communication such that party 1 does not send or receive public messages, while each other party sends at most one public message, equal to a function of the observed DMC outputs, after the DMC transmissions have been completed. In addition, if $1 \in \mathsf{A}$ then party 1 can generate the PK at the outset, as a uniformly distributed RV, and select the DMC inputs applying to this PK a stochastic encoder. If $1 \in \mathsf{D}$ then party 1 can chose the DMC inputs as a deterministic sequence. ○

Proof By Theorem 17.28, the PK capacity of the channel model is achievable by emulating a source model. It follows, using Theorem 17.26, that a communication protocol of the following kind is sufficent. First the DMC is used, party 1 sending i.i.d. X_1^n; then, non-interactive communication $F_{\mathsf{M}} = \{F_i(X_i^n) : i \in \mathsf{M}\}$ follows, with suitable functions F_i defined on X_i^n, $i \in \mathsf{M}$ (F_i is the identity mapping if $i \in \mathsf{D}$).

Let K be an ε-PK generated using such a protocol; assume (again using Theorem 17.26) that K is a function of X_1^n if $1 \in \mathsf{A}$, otherwise of X_j^n for some $j \in \mathsf{A}$. As K is ε-recoverable from F_{M} and X_i^n if $i \in \mathsf{A}$, there exist functions $\varphi_i(F_{\mathsf{M}}, X_i^n)$, $i \in \mathsf{A}$, such that

$$\Pr\left\{K \neq \varphi_i(F_{\mathsf{M}}, X_i^n)\right\} < \varepsilon, \quad i \in \mathsf{A}. \tag{17.60}$$

Note that

$$\Pr\left\{K \neq \varphi_i(F_{\mathsf{M}}, X_i^n)\right\} = \sum_{f\in\mathsf{F}} P_f\left(K \neq \varphi_i(F_{\mathsf{M}}, X_i^n)\right) \Pr\{F_1 = f\},$$

$$S(K|F_{\mathsf{M}}) = \sum_{f\in\mathsf{F}} S_f(K|F_{\mathsf{M}}) \Pr\{F_1 = f\},$$

where F denotes the set of possible values of F_1, and P_f, S_f denote the indicated probability resp. security index in the case for which the DMC inputs are governed by the conditional distribution $P_{X_1^n|F_1=f}$ rather than by $P_{X_1^n}$.

It follows using (17.60) and the security condition $S(K|F_{\mathsf{M}}) \leqq \varepsilon$ that there exists $f^* \in \mathsf{F}$ such that

$$\sum_{i \in \mathsf{A}} P_{f^*}\left(K \neq \varphi_i(F_{\mathsf{M}}, X_i^n)\right) + S_{f^*}(K|F_{\mathsf{M}}) < \varepsilon(|\mathsf{A}| + 1).$$

Consequently, when the (non-i.i.d.) distribution P_{f^*} governs the DMC inputs, the public communication F_{M} makes K an $\varepsilon(|\mathsf{A}| + 1)$-achievable PK. As F_1 is constant with probability 1 under P_{f^*}, it follows that the PK capacity is achievable without any public communication by party 1.

If $1 \in \mathsf{D}$ then F_1 above is the identity mapping, hence P_{f^*} is concentrated on a single sequence $\mathbf{x}^* \in \mathsf{X}^n$. If $1 \in \mathsf{A}$, when K is a function of X_1^n by assumption, it is clearly possible to generate K first and then the DMC input with distribution P_{f^*} via a stochastic encoder. The assertion that K may be taken exactly uniformly distributed follows in the standard manner; see Remark (i) after Definition 17.10. □

COMMENT The instance $1 \in \mathsf{A}$ of Theorem 17.29 can also be regarded as a result about secure transmission over a multiple-output DMC to several receivers allowed to communicate publicly with each other but not with the sender. Indeed, the theorem provides a single-letter characterization of secure-transmission capacity for this model, while demonstrating that even if the sender were admitted to participate in the public communication, PK could not be generated at a higher rate than by securely transmitting a message generated at the sender. Recall that the situation is different for the channel model in Theorem 17.17, whose SK capacity can be strictly larger than the secrecy capacity of the corresponding wiretap channel. ○

Discussion

In the literature, the information-theoretic approach has been applied to a variety of cryptographic tasks. Here, attention is concentrated on secure transmission over insecure channels, and on generating a secret key for two or more parties taking advantage of public communication. The adopted simplifying assumptions include the eavesdropper's inability to tamper with the legal parties' communication. Even within this restricted scope, the literature is rich enough not to permit comprehensive coverage. Rather, a few basic models were selected for detailed treatment, subjectively but preferring conclusive results; see the Discussion in Chapter 16.

The common approach to generating a SK is highlighted: first agree on non-secret CR, then "extract secret bits" (privacy amplification). The power of the techniques of multi-user information theory is amply demonstrated; note the remarkable efficiency of the key identity (Lemma 17.12) in proving converse results. Care has been taken to present most results with complete proofs, to enable the reader to get a working acquaintance with the mathematical techniques which are needed in related problems.

Regarding terminology, *secrecy capacity* is a generic term commonly used both in the context of secure transmission and of SK generation. In the latter context, we have preferred the terms SK and PK capacity, the latter to distinguish models of the mathematically less difficult kind in which the eavesdropper does not have information inaccessible to the legal parties. Note that in the literature the terms "secret key" and "private key" are not distinguished in this way.

In order to prove the initial (CR agreement) parts of the direct results, in the current literature random selection is the overwhelmingly preferred method. The alternative approach developed in Chapters 15 and 16 would also be adequate; see the original proofs of Theorems 17.13 and 17.21. The random constructions in Lemmas 17.14 and 17.22 illustrate frequently used techniques of multi-user information theory. This very fact has been one of our reasons for presenting proofs of Theorems 17.13 and 17.21 relying upon them. A technical advantage of this approach is that it facilitates "secrecy extraction" to obtain secrecy capacities directly in the sense of the strongest definition among several alternatives.

The secrecy extraction (or privacy amplification) parts of the proofs are based on the secrecy lemma, Lemma 17.5, a consequence of the extractor lemma, Lemma 17.3. The secret bits are extracted via a deterministic mapping which is obtained by random selection. In computer science, the theory of *randomness extractors* has a wide
→ 17.21 range of applications. This theory deals mainly with stochastic or "seeded" extractors, and regards their explicit construction a primary goal. Our approach uses deter-
→ 17.22 ministic extractors, and we do not address the construction issue. Note that good seeded extractors are also adequate (and may even have adventages) for privacy
→ 17.23 amplification.

In treating transmission with security requirements we have chosen the historically first models: the wiretap channel and its immediate generalization, the BCC. Note that a further generalization, the cognitive interference channel (see Fig. P14.18C) with confidential messages, still admits a single-letter solution. The latter model, and others treated in the literature for which few conclusive results are available, are not addressed here. See, however, the Comment preceding this Discussion.

Our treatment of generating a SK for more than two parties concentrates on PK capacities, i.e., on models where the legal parties have access to all the information available to the eavesdropper. Scenarios where the eavesdropper has side information not accessible to the legal parties can be modeled similarly; the only formal difference is in the admissible communication: the "compromised parties" $i \in \mathsf{D}$ do not communicate. It is obvious how to define formally, conforming with previous definitions, the SK capacity $C_{SK}(\mathsf{A}|\mathsf{D})$ of a multi-terminal source or channel model, for a set of parties A with secrecy from an eavesdropper having access to X^n_{D} which is *not accessible* to the parties in $\overline{\mathsf{D}}$. This means that the study of multiterminal SK capacities does not require a new mathematical framework. It may, however, require new mathematical tools, since cur-
→ 17.24 rently a single-letter characterization of $C_{SK}(\mathsf{A}|\mathsf{D})$ is known only in very special cases. Let us mention without any details that Theorem 17.26 can be extended to SK capacities, generalizing the concept of achievable omniscience rate, though a single-letter characterization of the minimum generalized OS rate remains elusive. ○

Problems

17.1. (*Properties of security index*) Verify the following relations. In (d), K_1 and K_2 have a common range K. In (e), S_{var} denotes *variational security index*, defined by replacing I-divergence by variation distance in (17.2).

(a) $S(K|Z_1Z_2) = S(K|Z_1) + I(K \wedge Z_2|Z_1)$;

(b) if $K \multimap Z_1 \multimap Z_2$ then $S(K|Z_1) = S(K|Z_1Z_2) \geqq S(K|Z_2)$;

(c) $S(K_1K_2|Z) = S(K_1|Z) + S(K_2|K_1Z)$
$\qquad = S(K_1|Z) + S(K_2|Z) + I(K_1 \wedge K_2|Z)$;

(d) $|S(K_1|Z) - S(K_2|Z)| \leqq \Pr\{K_1 \neq K_2\} \log |\mathsf{K}| + h\,(\Pr\{K_1 \neq K_2\})$.

Hint Use $|S(K_1|Z) - S(K_2|Z)| \leqq \max[H(K_1|K_2), H(K_2|K_1)]$ and Fano.

(e) $\dfrac{\log e}{2} S^2_{var}(K|Z) \leqq S(K|Z) \leqq \dfrac{1}{2} S_{var}(K|Z) \log |\mathsf{K}| + h(S_{var}(K|Z))$.

Hint For the lower bound, use the Pinsker inequality. For the upper bound, represent $S(K|Z)$ as the average of differences $\log |\mathsf{K}| - H(K|Z = z)$ with weights $\Pr\{Z = z\}$, bound these differences using Problem 3.9 and use Jensen's inequality for h.

17.2. (a) Complementing Proposition 17.1, show that a message cannot be securely encrypted with a key whose entropy is less than the entropy of the message. (Shannon (1949).)

(b) More generally, show that for any RVs K, Z and M independent of (K, Z), a necessary condition for the existence of Y ("ciphertext") such that M is a function of (Y, K) while ε-secret from (Y, Z), is $H(M) \leqq H(K|Z) + \varepsilon$.

Hint Check that for any RVs M, Y, Z, K

$$\begin{aligned} I(M \wedge Y|KZ) - I(M \wedge Y|Z) &= I(M \wedge K|YZ) - I(M \wedge K|Z) \\ &\leqq H(K|Z). \end{aligned}$$

The given hypotheses imply $I(M \wedge Y|KZ) = H(M)$ and $I(M \wedge Y|Z) \leqq \varepsilon$.

17.3. (*Goodness of a randomly selected extractor*)

(a) Show via the extractor lemma that a random mapping $\kappa : \mathsf{U} \to \{1, \ldots, k\}$ with $1 \leqq \log k \leqq H_\infty(P) - \ln H_\infty(P) - c$ satisfies (17.3) with probability at least $1 - \delta$, where c is a constant depending only on ε and δ. Show also that $\log k$ cannot significantly exceed $H_\infty(P)$ for any mapping satisfying (17.3).

Hint Show that (17.3) can hold only if $\log k \leqq H_\infty(P) + \log(1 + \varepsilon)$.

(b) Show that a randomly selected mapping $\kappa : \mathsf{U} \to \{1, \ldots, k\}$ with $1 \leqq \log k \leqq H_{\infty,\eta}(P) - \ln H_{\infty,\eta}(P) - c$ satisfies (17.4) with probability at least $1 - \delta$, where

$$H_{\infty,\eta}(P) \triangleq \max_{\mathsf{F} \subset \mathsf{U},\ P(\mathsf{F}) \geqq 1-\eta} \ \min_{u \in \mathsf{F}} (-\log P(u))$$

and c is a constant depending only on ε, η and δ. Moreover, show that if the κ-image of P is ε-uniform for some mapping κ as above, then

$$\log k \leqq H_{\infty,\eta}(P) - \log\left(1 - \frac{\varepsilon}{\eta}\right) \quad \text{if} \quad 0 < \varepsilon < \eta < 1.$$

Hint Let $\kappa : \mathsf{U} \to \{1, \ldots, k\}$ be a mapping for which the image of P is ε-uniform. Denote this image by Q, i.e., $Q(i) \triangleq P(\kappa^{-1}(i))$. Verify for

$$\mathsf{G} \triangleq \left\{ i : \left| Q(i) - \frac{1}{k} \right| \leqq \frac{\varepsilon}{\eta} Q(i) \right\}$$

that $Q(\overline{\mathsf{G}}) \leqq \eta$ and $(1 - \varepsilon/\eta)\, Q(i) \leqq 1/k$ if $i \in \mathsf{G}$. Conclude that

$$H_{\infty,\eta}(P) \geqq \min_{u \in \kappa^{-1}(\mathsf{G})} (-\log P(u)) \geqq \log k + \log\left(1 - \frac{\varepsilon}{\eta}\right).$$

(c) The *intrinsic randomness* of a RV U is the maximum number of random bits that can be extracted from U. Formally, define $IR_\varepsilon(U)$ as the maximum of $\log k$ for mappings $\kappa : \mathsf{U} \to \{1, \ldots, k\}$ for which the distribution of $\kappa(U)$ is ε-uniform. Show that one can come close to attaining $IR_\varepsilon(U)$ by random selection of κ; more exactly, taking a suitable k with $\log k$ asymptotically equal to $IR_\varepsilon(U)$ when the latter is large, for the randomly selected mapping the distribution of $\kappa(U)$ will be nearly uniform with probability close to 1, perhaps not ε-uniform but at least 4ε-uniform.

Hint Use (b) with $\eta = \frac{3}{2}\varepsilon$.

(For the concept of intrinsic randomness, see Vembu and Verdú (1995), and for non-asymptotic results relating it to so-called *smooth minentropy* see Renner and Wolf (2005). The latter concept has several variants; $H_{\infty,\eta}$ above may be regarded as one of them.)

17.4. (*Intrinsic randomness rate*) Show that for any source $X_1, X_2, \ldots$

$$\lim_{\varepsilon \to 0} \varliminf_{n \to \infty} \frac{1}{n} IR_\varepsilon(X^n) = \lim_{\eta \to 0} \varliminf_{n \to \infty} \frac{1}{n} H_{\infty,\eta}(X^n),$$

and that the right-hand side, called the *inf-entropy rate*, is equal to the entropy rate $\bar{H}$ if the source is stationary and ergodic. Moreover, in the latter case, show that

$$\lim_{n \to \infty} \frac{1}{n} IR_\varepsilon(X^n) = \bar{H} \qquad \text{for each } 0 < \varepsilon < 1.$$

(Vembu and Verdú (1995).)

Hint Use parts (b) and (c) of Problem 17.3 and Problem 4.6(c).

17.5. (a) Show that the hypothesis $P\left(\{u : P(u) \leqq 1/d\}\right) \geqq 1 - \eta$ of the extractor lemma is satisfied for $\mathsf{U} = \mathsf{X}^n$, $P = P_1 \times \cdots \times P_n$,

$$d = \exp\left[\sum_{i=1}^{n} H(P_i) - n\zeta \log |\mathsf{X}|\right], \quad \zeta = \sqrt{\frac{2 \ln 1/\eta}{n}};$$

here $|\mathsf{X}|$ has to be replaced by 3 if $|\mathsf{X}| = 2$.

Hint Chernoff bounding gives for any $\zeta > 0, t > 0$

$$P\left(\left\{u:\ P(u) \geqq \exp\left[-\sum_{i=1}^{n} H(P_i) + n\zeta \log|\mathsf{X}|\right]\right\}\right)$$
$$\leqq \exp\left[t\sum_{i=1}^{n} H(P_i) - tn\zeta \log|\mathsf{X}|\right] \prod_{i=1}^{n} \sum_{x\in\mathsf{X}} P_i^{1+t}(x).$$

Show via Taylor expansion of $P_i^t(x) = e^{t\ln P_i(x)}$ that

$$\sum_{x\in\mathsf{X}} P_i^{1+t}(x) \leqq 1 - tH(P_i)\ln 2 + \frac{t^2}{2}\sum_{x\in\mathsf{X}} P_i(x)\,[\ln P_i(x)]^2,$$

and verify that the last sum is bounded above by $(\ln|\mathsf{X}|)^2 = (\log|\mathsf{X}|\ln 2)^2$, providing $|\mathsf{X}| \geqq 3$. Then use

$$1 - tH(P_i)\ln 2 + \frac{t^2}{2}(\log|\mathsf{X}|\ln 2)^2 \leqq \exp\left[-tH(P_i) + \frac{t^2}{2}(\log|\mathsf{X}|)^2 \ln 2\right]$$

and substitute $t = (\zeta \log e)/\log|\mathsf{X}|$, with the given ζ.

(b) Extend the result of (a) to conditional probabilities for independent pairs of RVs $(X_i, Y_i), i = 1, \ldots, n$, and ζ in (a):

$$P_{X^nY^n}\left(\left\{(\mathbf{x},\mathbf{y}):\ P_{X^n|Y^n}(\mathbf{x}|\mathbf{y}) < \exp\left[-\sum_{i=1}^{n} H(X_i|Y_i) + n\zeta \log|\mathsf{X}|\right]\right\}\right)$$
$$\geqq 1-\eta,$$

again replacing $|\mathsf{X}|$ by 3 if $|\mathsf{X}| = 2$.

17.6. (a) Let $\mathcal{P}$ be a finite family of product distributions $P = P_1 \times \cdots \times P_n$ on X^n such that $\sum_{i=1}^{n} H(P_i) \geqq nH$ for each $P \in \mathcal{P}$. Show that for

$$k = \exp\left[n(H - \delta\log|\mathsf{X}|)\right],$$
$$\delta \geqq \sqrt{\frac{2\ln 3/\varepsilon}{n}} + \frac{1}{n\log|\mathsf{X}|}\left[2\log\frac{6}{\varepsilon} + \log\log\left(|\mathcal{P}||\mathsf{X}|^n\right)\right]$$

there exists a mapping $\kappa : \mathsf{X}^n \to \{1, \ldots, k\}$ which is an ε-extractor for the family $\mathcal{P}$; if $|\mathsf{X}| = 2$, it has to be replaced by 3 as in Problem 17.5. (Ahlswede and Csiszár (1998).)

Hint By Problem 17.5(a), each $P \in \mathcal{P}$ meets the hypothesis of the extractor lemma with

$$d = \exp\left[n(H - \zeta\log|\mathsf{X}|)\right], \quad \zeta = \sqrt{\frac{2\ln 1/\eta}{n}}.$$

Given any ε, set $\overline{\varepsilon} = \eta = \varepsilon/3$ and apply the extractor lemma with $\overline{\varepsilon}$ in the role of ε. It follows that on randomly selecting the mapping κ, the probability that each $P \in \mathcal{P}$ has an ε-uniform κ-image is at least $1 - 2|\mathcal{P}|ke^{-\beta d/k}$, with $\beta \triangleq \overline{\varepsilon}^2(1-\eta)/2(1+\overline{\varepsilon}) \geqq \varepsilon^2/36$. Check that this is positive for k as given.

(b) Show that even if $\mathcal{P}$ is the family of *all* product distributions on X^n with $\sum_{i=1}^{n} H(P_i) \geqq nH$, for each $\delta > 0$ and sufficiently large n there exists an ε_n-extractor with $k = \exp\left[n(H - \delta \log|\mathsf{X}|)\right]$ for the family $\mathcal{P}$, such that ε_n goes to zero exponentially fast.

Hint Check via the result of (a) that for any sub-family $\mathcal{P}_n$ of $\mathcal{P}$ with $|\mathcal{P}_n| < \exp(\exp(\xi n))$, the assertion holds if $\delta > \xi$. Then show that for any $\xi > 0$ there exists such a $\mathcal{P}_n$ exponentially dense in $\mathcal{P}$, i.e., each $P \in \mathcal{P}$ has variation distance at most ε_n from some $P' \in \mathcal{P}_n$, where $\varepsilon_n \to 0$ exponentially.

17.7. (a) In the context of Corollary 17.5, show that for $n \geqq 2$ and $k = \exp\left[n(H(X|Y) - \delta \log|\mathsf{X}|)\right]$ with

$$\delta \geqq \sqrt{\frac{4}{n}\ln\left(\frac{3n\log|\mathsf{X}|}{\varepsilon}\right)} + \frac{1}{n\log|\mathsf{X}|}\left[2\log\frac{3}{\varepsilon} + 3\log\log(|\mathsf{X}|^n|\mathsf{Y}|^n)\right],$$

there exists a mapping $\kappa : \mathsf{X}^n \to \{1, \ldots, k\}$ such that

$$S(\kappa(X^n)|Y^n) < \varepsilon + h\left(\frac{\varepsilon}{n}\right)$$

(replacing $|\mathsf{X}|$ by 3 if $|\mathsf{X}| = 2$, as before).

Hint Writing $H \triangleq H(X|Y)$, by Problem 17.5(b) the hypothesis (17.10) of the secrecy lemma is met for

$$d = \exp\left[n(H - \zeta \log|\mathsf{X}|)\right], \quad \zeta = \sqrt{\frac{4}{n}\ln\frac{1}{\eta}}.$$

Set $\eta \triangleq \varepsilon(3n\log|\mathsf{X}|)^{-1}$. Then, if $k = \exp\left[n(H - \delta\log|\mathsf{X}|)\right]$ meets the condition (17.11) with ($V = Y^n$ and) $\alpha = \eta$, the claim follows from (17.12). Check that (17.11) holds if δ is chosen as stated.

(b) Extend the result of (a) to the case when the joint distribution $P_{X^nY^n}$ can be any member of a finite family $\mathcal{P}$ of distributions of form $P_1 \times \cdots \times P_n$ on $(\mathsf{X} \times \mathsf{Y})^n$. Establish also an analog of Problem 17.6(b).

17.8. (a) Simplify Theorems 17.11 and 17.13 for the case when the channel W_1 is more capable than W_2; see Problem 6.18. Show that then $V = X$ may be taken in both theorems.

Hint Check that $U \multimap V \multimap X \multimap YZ$ implies

$$\begin{aligned} &I(V \wedge Y|U) - I(V \wedge Z|U) \\ &\quad = I(X \wedge Y|U) - I(X \wedge Z|U) - [I(X \wedge Y|V) - I(X \wedge Z|V)], \end{aligned}$$

and that the difference in the bracket is non-negative if W_1 is more capable than W_2.

(b) Show that in the general case a simplification as in (a) need not be possible.

Hint See Problem 15.11(b).

17.9. Prove the direct part of the BCC coding theorem, Theorem 17.13, relying exclusively upon techniques developed in Chapter 15, supposing that confidentiality is required in the "weak" sense that $(1/n)S(M_s|Z^n)$ has to be small (this weaker requirement causes no loss of generality; see Problem 17.10).

Hint Consider first the case of no common message, and show that for the direct part of the theorem to hold for this special case it suffices if $(0,\ R_1',\ R_s)$ is achievable whenever

$$R_1' = I(X \wedge Z|U) + t, \quad R_s = I(X \wedge Y|U) - I(X \wedge Z|U),$$

with $U,\ X,\ Y,\ Z$ as in the theorem and $0 \leqq t \leqq \min[I(U \wedge Y),\ I(U \wedge Z)]$.

To prove the latter, apply Lemma 15.18 to these RVs and t. Take the sets $\mathsf{A}^{(l)}$ and $\mathsf{A}^{(l,m)}$ in that lemma for a fixed l, denoting them briefly by A and $\mathsf{A}^{(m)}$. Thus $\mathsf{A} \subset \mathsf{T}_{[X]}$ is the codeword set of an (n, ε)-code for the DMC $\{W_1\}$, which is the disjoint union of the sets $\mathsf{A}^{(m)}$, $m = 1, \ldots, M$, each the codeword set of an (n, ε)-code for $\{W_2\}$, where, by (ii) of Lemma 15.18,

$$\left|\frac{1}{n}\log M - R_s\right| \leqq \tau.$$

Moreover, by (ii) and (iii) of Lemma 15.18

$$\left|\frac{1}{n}\log g_{W_2^n}(\mathsf{A},\ 1-\varepsilon) - \frac{1}{n}\log g_{W_2^n}(\mathsf{A}^{(m)},\ 1-\varepsilon)\right| \leqq 2\tau,$$

and by (15.107)

$$\left|\frac{1}{n}\log|\mathsf{A}^{(m)}| - R_1'\right| \leqq \tau, \quad m = 1, \ldots, M.$$

Suppose without any loss of generality that $|\mathsf{A}^{(m)}| = K$ for each m, with $(1/n)\log K$ close to R_1'. Verify the achievability of $(0, R_1', R_s)$ by taking (M_1', M_s) uniformly distributed on $\{1, \ldots, K\} \times \{1, \ldots, M\}$ and a (deterministic) encoder f that maps this product onto A, such that $f(k, m) \in \mathsf{A}^{(m)}$.

To check the smallness of $(1/n)S(M_s|Z^n) = (1/n)I(M_s \wedge Z^n)$, where Z^n is the output of the DMC $\{W_2\}$ for input $X^n \triangleq f(M_1', M_s)$, note that the distribution of Z^n and its conditional distribution on the condition $M_s = m$ are both output distributions of the DMC $\{W_2\}$, the former for an input distributed on A and the latter for an input uniformly distributed on $\mathsf{A}^{(m)}$, the codeword set of an (n, ε)-code for this channel. Invoke Lemma 15.2 and the relation stated above on the image sizes of A and $\mathsf{A}^{(m)}$ to conclude that for no m can $(1/n)H(Z^n|M_s = m)$ be significantly smaller than $(1/n)H(Z^n)$.

In the general case the achievability of $(R_0,\ R_0',\ R_s)$ given by (17.24) has to be proven. This can be done similarly, using not only the statements of Lemma 15.18, but also the underlying construction.

17.10. (*Equivalence of definitions of SK capacity*) By a familiar "weak" definition (still weaker than the "weakened" definition in Remark (i) to Definition 17.16), H is an achievable SK rate for a two-party source or channel model if for each

$\varepsilon > 0$, $\delta > 0$ and sufficiently large n there exists a permissible protocol with public communication (F_1, F_2), and a RV K, such that

(i) performing the protocol makes K an ε-CR for Alice and Bob,

(ii) $(1/n)H(K) > H - \delta$, $\quad (1/n)I(K \wedge F_1F_2Z^n) < \varepsilon$,

(iii) $(1/n)\log|\mathsf{K}|$ is bounded by a constant c not depending on ε, δ, n.

(a) Show that condition (iii) is essential for this definition to make sense. Why is (iii) not needed in Definition 17.16 and its weakened/strengthened versions?

Hint In the absence of (iii), one could take a RV K satisfying (ii) with arbitrarily large H, which is constant with probability $1 - \varepsilon$, thus trivially ε-recoverable for Alice and Bob. On the other hand, any K satisfying the conditions of either version of Definition 17.16, can be modified to satisfy also (iii), say with $c = 2H$.

(b) Show that Definition 17.16, its weakened/strengthened versions in Remark (i) to Definition 17.16, the above weak definition, and also the modification of either version replacing the security index (17.1) by the variational security index, lead to the same SK capacity for "most" DM models. Show this even for a further weakening of the "weak" definition, in which (ii) is relaxed to

$$\text{(ii}'\text{)} \quad \frac{1}{n}H(K|F_1F_2Z^n) > H - \delta.$$

(Maurer and Wolf (2000) proved for unrestricted models, and for the wiretap channel, that the "weak" definition leads to the same SK capacity as the "strong" definition they suggested, adopted here as standard. The hint below follows their proof, simplified by the availability of Corollary 17.5. Note that some DM models are not covered by this proof, see (c) below; for those, the equivalence of the different definitions of SK capacity remains an open problem.)

Hint Fix a protocol and corresponding K, K_A, K_B (see (17.26)) such that (i), (ii$'$) and (iii) hold, for fixed n, ε, δ. As in Remark (i) to Definition 17.16, one may assume $K = K_A$. Verify via Fano that $(1/n)H(K_A|K_B)$ is arbitrarily small if ε is. Repeat the protocol in N consecutive time intervals of length n, and apply Corollary 17.5(ii) to the i.i.d. repetitions K_A^N and (F_1^N, F_2^N, Z^{nN}) of K_A and (F_1, F_2, Z^n) in the role of X^n, Y^n, and to a function $g(K_A^N)$ of K_A^N in the role of $Z^{(n)}$, such that K_A^N is ε_N-recoverable from K_B^N and $g(K_A^N)$, with $\varepsilon_N \to 0$ exponentially. Lemma 17.6 guarantees the existence of such g with $(1/N)\log\|g\|$ arbitrarily close to $H(K_A|K_B)$, then $(1/nN)\log\|g\|$ is arbitrarily small if ε is. Draw the conclusion that if H is achievable in the sense of (i), (ii$'$) and (iii), it is achievable also in the sense of our "strengthened" definition, providing (F_1^N, F_2^N, g) is an admissible communication in the given model when (F_1, F_2) is (or, if the latter holds, interchanging the roles of Alice and Bob.) For the case of a channel model not admitting public communication, the same follows, letting Alice transmit $g(K_A^N)$ via the DMC, using a small fraction of the nN inputs for this purpose.

(c) Give an example of a DM model not covered by the hint to part (b).

Hint Consider a source model with Alice and Bob restricted to one public message each, which cannot rely on information received from the other party.

17.11. (*CR capacity*) A "weak" definition of achievable CR rates is similar to that of achievable SK rates in Problem 17.10, omitting the secrecy requirement in condition (ii) of Problem 17.10. As for SK rates, stronger definitions are possible: the strongest one modifies (i), requiring K to be an ε_n-CR, and replaces (iii) by $\log |\mathsf{K}| - H(K) < \varepsilon_n$, where $\varepsilon_n \to 0$ exponentially. Each definition gives rise to a corresponding definition of CR capacity, C_{CR}.

(a) Show that under either definition of CR or SK capacity in models not allowing communication by Bob, one can take K_B as a function of Y^n and of Alice's message F_1 (if any).

Hint Suppose K_B depends, in addition to Y^n and F_1, also on a randomizing Q_2, with range Q, say. As (X^n, F_1, Y^n) is independent of Q_2,

$$\begin{aligned}\Pr\left\{K_A = K_B(Y^n, F_1, Q_2)\right\} \\ = \sum_{q \in \mathsf{Q}} \Pr\left\{K_A = K_B(Y^n, F_1, q)\right\} \Pr\left\{Q_2 = q\right\}.\end{aligned}$$

Thus, K_B may be replaced by $K'_B \triangleq K_B(Y^n, F_1, q)$, for some $q \in \mathsf{Q}$.

(b) Verify the analog of Problem 17.10(b) for CR capacity.

(c) Show that for a source model that admits no public communication, C_{CR} is equal to the common information of the generic variables X, Y; see Problems 16.27–16.30. In particular, if X and Y have an indecomposable joint distribution, show that the CR capacity, let alone the SK capacity, cannot be positive if no public communication is allowed.

Hint Use Problem 16.28. Show that if H is an achievable CR rate without communication, under the weak definition, then $(H(X) - H, H, H(Y) - H)$ is an achievable rate triple for the source network referred to there. Check first, using (a), that an ε-CR K may be assumed to be a function of X^n which is ε-recoverable from Y^n. Alternatively, you may use (d) below and Problem 16.25. (Non-existence results about CR without communication when X, Y have indecomposable joint distribution, stronger than non-achievability of a positive CR rate, follow from Problem 15.14.)

(d) Find the CR capacity of a source model with communication restricted to one message by Alice, of rate R, both when she may or may not randomize. Show that if $R \leqq H(X|Y)$ then C_{CR} equals in both cases the maximum of $I(U \wedge X)$ subject to $U \multimap X \multimap Y$, $I(U \wedge X|Y) \leqq R$, and the range constraint $|\mathsf{U}| \leqq |\mathsf{X}| + 1$, while if $R \geqq H(X|Y)$ then C_{CR} equals $I(X \wedge Y) + R$ or $H(X)$ according to whether randomization is permitted or not. In particular, show that the CR capacity with no constraints on the communication rate equals $H(X)$ when randomization is not permitted and $+\infty$ when it is. (Ahlswede and Csiszár (1998). As the CR rate R is trivially achievable with communication of rate R when randomization is permitted, achievability of $I(X \wedge Y) + R$ suggests to interpret mutual information as *latent common information*, which can be recovered by communication of sufficiently large rate.)

Hint A "baby version" of the proof of Theorem 17.21 applies (which has been the origin of that proof). In particular, for the direct part, Alice sends $F = f(X^n)$ as there; this makes $\varphi(X^n)$ a CR almost independent of F, thus $(F, \varphi(X^n))$ will be CR of the required rate $I(U \wedge X)$.

17.12. (a) Show that the SK capacity of a channel model as in Theorem 17.17 may be positive when the secrecy capacity of the corresponding wiretap channel is zero. (Maurer (1993).)

Hint Consider $W : \mathsf{X} \to \mathsf{Y} \times \mathsf{Z}$ with independent components W_1, W_2, where W_1 is a degraded version of W_2 (or, at least, W_2 is less noisy than W_1).

(b) Show that the PK capacity of a channel model as in Theorem 17.20 equals the secrecy capacity of a modified version of the corresponding wiretap channel, in which the legal receiver knows the outputs of the wiretapper's channel. In particular, show that this PK capacity can be achieved without any public communication.

17.13. Show that the upper bound in (17.28) may be tight also when X, Y, Z do not form a Markov chain in any order and the SK capacity of an unrestricted source model need not be achievable with one-way communication. (Ahlswede and Csiszár (1993).)

Hint Consider two independent triples (X', Y', Z') and (X'', Y'', Z'') with $X' \multimap Y' \multimap Z'$ and $Y'' \multimap X'' \multimap Z''$. Show using Theorem 17.17 that for the DMMS with generic variables $X = (X', X'')$, $Y = (Y', Y'')$, $Z = (Z', Z'')$, Alice and Bob can generate SK $K = (K_1, K_2)$ via sending messages $F_1 = F_1(X'^n)$ and $F_2 = F_2(Y''^n)$, with $n^{-1} \log |\mathsf{K}| = n^{-1} \log |\mathsf{K}_1| + n^{-1} \log |\mathsf{K}_2|$ arbitrarily close to $I(X' \wedge Y'|Z') + I(X'' \wedge Y''|Z'') = I(X \wedge Y|Z)$. Verify that if (X', Y', Z') and (X'', Y'', Z'') are suitably chosen, X, Y, Z do not form a Markov chain in any order, and that $I(X \wedge Y|Z)$ is not an achievable SK rate with one-way communication.

17.14. (*Upper bounds to SK capacity*)

(a) Verify that the upper bound in (17.17) may be improved to the infimum of $I(X \wedge Y|V)$ for RVs V satisfying $XY \multimap Z \multimap V$. (Ahlswede and Csiszár (1993). Maurer and Wolf (1999) called this infimum *intrinsic conditional mutual information*; they denoted it by $I(X \wedge Y \downarrow Z)$ and studied its properties.)

Hint Use (17.28) and Problem 17.1(b).

(b) Show that if a source model is modified replacing Z by ZU, where U is arbitrary, the SK capacity does not decrease by more than $H(U)$. Draw the conclusion that $C_{SK} \leqq \inf_U [I(X \wedge Y \downarrow ZU) + H(U)]$. (Renner and Wolf (2003).)

Hint For an achievable SK rate H, fix F_1, F_2, K, K_A, K_B as in Definition 17.16 and the passage that follows it. Consider an auxiliary source model with generic variables K_A, K_B, $F_1 F_2 Z^n U^n$. Show via (17.28) that

$$H' \triangleq I(K_A \wedge K_B) - I(K_A \wedge F_1 F_2 Z^n U^n)$$

is an achievable SK rate for the latter, and consequently that $(1/n)H'$ is an achievable SK rate for the source model with generic variables X, Y, ZU. Verify that $(1/n)H'$ is not less than $H - H(U)$, up to an arbitrarily small error term; you may use the inequality in the hint to Problem 17.2(b), with U^n and $F_1 F_2 Z^n$ in the role of K and Z.

(c) Give an example where the bound in (b) strictly improves that in (a). (Renner and Wolf (2003).)

Hint Consider $\mathsf{X} = \mathsf{Y} = \{0, 1, 2, 3\}$, $\mathsf{Z} = \{0, 1\}$, $P_{XY}(i, j) = 1/8$ if $i \in \{0, 1\}$, $j \in \{0, 1\}$, $P_{XY}(3, 3) = P_{XY}(4, 4) = 1/4$; set $Z = X + Y \pmod 2$ or $Z = X \pmod 2$ according to whether $X \in \{0, 1\}$ or $X \in \{2, 3\}$. Check that then $I(X \wedge Y \downarrow Z) = 3/2$, while the bound in (b) is equal to unity, achieved for $U = \lceil X/2 \rceil$; this is actually the SK capacity of this model.

(d)* Show that the bound in (b) is not always tight. (Gohari and Anantharam (2010).)

17.15. (*One-way source model with randomization*)

(a) Show that the SK capacity of the one-way source model is not increased if randomization is permitted.

Hint By Problem 17.11(a), randomization by Bob cannot help. Any protocol with randomization by Alice may be regarded as a deterministic protocol for a modified DMMS, with the generic variable X replaced by (X, Q), where Q is independent of (X, Y, Z). Applying the converse proof for this modified DMMS, verify that randomization cannot make $C_{SK}(R)$ larger than a supremum as in Theorem 17.21, subject to the conditions $U \multimap V \multimap QX \multimap YZ$ and $I(V \wedge QX|Y) \leqq R$, which imply (17.31).

(b) For the case when randomization is allowed and the communication rate is unconstrained, prove the direct part of Theorem 17.21 dispensing with Lemma 17.22.

Hint Given any RVs U, V with $U \multimap V \multimap X \multimap YZ$, let Alice generate U^n, V^n, which together with X^n, Y^n, Z^n represent i.i.d. repetitions of (U, V, X, Y, Z). She keeps V^n and reveals U^n. By Theorem 17.17 applied to the DMMS with generic variables V, UY, UZ, it follows that $I(V \wedge UY) - I(V \wedge UZ) = I(V \wedge Y|U) - I(V \wedge Z|U)$ is an achievable SK rate.

17.16. By Theorem 17.26, the PK capacity of a multi-terminal source model can always be achieved by first generating CR at the largest rate possible (without randomization), via communication of rate R_{OS}.

(a) Show that communication of a smaller rate, perhaps with some parties remaining silent, may also suffice to achieve C_{PK}.

Hint For $m = 2$, $C_{PK} = I(X \wedge Y)$ is achievable with only Alice communicating, at rate $H(X|Y)$, generating CR of rate $H(X)$.

(b) Give a multi-terminal source model of positive PK capacity that does not admit generating any PK if one of the parties remains silent.

Hint See Problem 17.17.

Computation of PK capacities (Problems 17.17–17.20)

17.17. Let $X_1, \ldots, X_m$ be binary RVs such that $X_1, \ldots, X_{m-1}$ are i.i.d. $(1/2, 1/2)$, and $X_m = X_1 + \cdots + X_{m-1}$ (mod 2).

(a) Show that the joint distribution of $X_1, \ldots, X_m$ is invariant under permutations of $\mathsf{M} = \{1, \ldots, m\}$.

(b) Show for the corresponding source model that

$$C_{PK}(\mathsf{A}) = C_{PK}(\mathsf{A}|\overline{\mathsf{A}}) = \frac{1}{|\mathsf{A}| - 1}, \quad \mathsf{A} \subset \mathsf{M}, \quad |\mathsf{A}| \geqq 2,$$

and that these are achievable with a perfect PK ($\varepsilon = 0$ in the definition). (Csiszár and Narayan (2004).)

Hint It can be supposed that $\mathsf{A} = \{1, \ldots, k, m\}$, $1 \leqq k \leqq m - 1$. Take block length $n = k = |\mathsf{A}| - 1$. Let party $i \in \{1, \ldots, k\}$ reveal all but the ith component X_{ii} of X_i^n, let party m send $X_{m1} + X_{m2}, \ldots, X_{m1} + X_{mn}$, while the parties $j \in \overline{\mathsf{A}}$ (if $\mathsf{A} \neq \mathsf{M}$) reveal X_j^n. Show that this public communication yields 1 bit of a perfect PK for the parties in A, say $K = X_{11}$. This achievability result is straightforward. If you cannot find a direct proof of optimality, use Theorem 17.26.

17.18. Determine all PK capacities for three-terminal source models.

Hint Derive from Theorem 17.26 that

$$C_{PK}(\{1, 2, 3\}) = \min\left[I(X_1 \wedge X_2X_3),\ I(X_2 \wedge X_1X_3),\ I(X_3 \wedge X_1X_2),\ \frac{1}{2}I \right],$$

where

$$I \triangleq H(X_1) + H(X_2) + H(X_3) - H(X_1X_2X_3) \text{ (“multi-information”)},$$
$$C_{PK}(\{1, 2\}) = \min\left[I(X_1 \wedge X_2X_3),\ I(X_2 \wedge X_1X_3)\right],$$
$$C_{PK}(\{1, 2\}|\{3\}) = I(X_1 \wedge X_2|X_3) \qquad \text{(see Theorem 17.20)}.$$

17.19. Suppose $X_1, \ldots, X_m$ is a Markov chain.

(a) Show that $C_{PK}(\mathsf{M}) = \min_{1 \leqq i < m} I(X_i \wedge X_{i+1})$.

Hint Let $i = t$ attain the minimum. Due to

$$C_{PK}(\mathsf{M}) \leqq I(X_1 \ldots X_t \wedge X_{t+1} \ldots X_m) = I(X_t \wedge X_{t+1})$$

and Theorem 17.17, it suffices to show that $H(X_{\mathsf{M}}) - \sum_{i=1}^{m} R_i$ is equal to $I(X_t \wedge X_{t+1})$ for some $(R_1, \ldots, R_m) \in \mathcal{R}(\mathsf{M})$. Verify that a suitable choice is $R_i = H(X_i|X_{i+1})$ if $i \leqq t$ and $R_i = H(X_i|X_{i-1})$ if $i > t$.

(b) For $\mathsf{A} \subset \mathsf{M}$, denote by $i_{\min}$ and $i_{\max}$ the smallest and largest elements of A. Deduce from (a) that

$$C_{PK}(\mathsf{A}) = \min_{i_{\min} \leqq i < i_{\max}} I(X_i \wedge X_{i+1}).$$

(c) Show that

$$C_{PK}(\mathsf{A}|\mathsf{D}) = \min_{i_{\min} \leqq i < i_{\max}} I(X_i \wedge X_{i+1}|X_{\mathsf{D}}).$$

Hint The case $\mathsf{D} \cap [i_{\min}, i_{\max}] \neq \emptyset$ is trivial. Concentrate on the case when D is the complement of $[i_{\min}, i_{\max}]$, and proceed as in (a), taking $R_i = H(X_i|X_{i+1}, X_{\mathsf{D}})$, $i \leqq t$, and $R_i = H(X_i|X_{i-1}, X_{\mathsf{D}})$, $i > t$. (Csiszár and Narayan (2004).)

17.20. (*PK capacity and multi-information*)

(a) For multi-terminal source models, give an upper bound to $C_{PK}(\mathsf{A})$ in terms of multi-informations $I_{\mathsf{B}_1,\ldots,\mathsf{B}_k} \triangleq \sum_{i=1}^{k} H(X_{\mathsf{B}_k}) - H(X_{\mathsf{M}})$ corresponding to partitions $(\mathsf{B}_1, \ldots, \mathsf{B}_k)$ of $\mathsf{M} = \{1, \ldots, m\}$, such that each B_i intersects A. Consider $\lambda \in \Lambda(\mathsf{A})$, see (17.58), with $\lambda_{\mathsf{B}} = 1/(k-1)$ if $\mathsf{B} = \overline{\mathsf{B}}_i$ for some i and $\lambda_{\mathsf{B}} = 0$ otherwise, to obtain from Theorem 17.26 and Lemma 17.27 that

$$C_{PK}(\mathsf{A}) \leqq \frac{1}{k-1} I_{\mathsf{B}_1,\ldots,\mathsf{B}_k}.$$

(Csiszár and Narayan (2004).)

(b)* Show that for $\mathsf{A} = \mathsf{M}$ there exists a partition of M for which the bound in (a) is tight, but that this does not always hold otherwise. (Chan (2008, unpublished).)

Seeded extractors (Problems 17.21–17.23)

A mapping $g : \mathsf{U} \times \{0, 1\}^s \to \{0, 1\}^m$ is a *seeded ε-extractor* for a family $\mathcal{P}$ of PDs on U if for each RV U with $P_U \in \mathcal{P}$ and RV T uniformly distributed on $\{0, 1\}^s$, independent of U, the joint distribution of $g(U, T)$ and T is ε-uniform on $\{0, 1\}^{m+s}$. Informally, T represents a *seed* of length s which is used as a catalyst to extract m random bits from any U as above, while the seed is recovered eventually. (This concept was formally defined by Nisan and Zuckerman (1996) for $\mathsf{U} = \{0, 1\}^n$ and $\mathcal{P} = \{P : H_\infty(P) \geqq n\delta\}$, but relevant results were also obtained earlier. In the literature, the term often refers to this specific $\mathcal{P}$. One rationale for stochastic (seeded) extractors is that for a "very large" family $\mathcal{P}$, such as that above, deterministic extractors do not exist; see Problem 17.21(b).)

17.21. (a) Show that deterministic extractors are sufficient for extracting random bits from an individual PD: if $g : \mathsf{U} \times \{0, 1\}^s \to \{0, 1\}^m$ is a seeded ε-extractor for $\{P\}$ then, for some $t \in \{0, 1\}^s$, the mapping $g_t : \mathsf{U} \to \{0, 1\}^m$ defined by $g_t(u) \triangleq g(u, t)$ is a deterministic ε-extractor for $\{P\}$.

Hint Represent the variation distance of the distribution of $(g(T, U), T)$ from the uniform distribution on $\{0, 1\}^{m+s}$ as an average variation distance from the uniform distribution on $\{0, 1\}^m$ of the distributions of the RVs $g_t(U)$, for $t \in \{0, 1\}^s$.

(b) Let $\mathcal{P}$ be the family of those PDs P on a set U which are uniform on their support, and let the size of this support be at least $(1/2)|\mathsf{U}|$. Verify that no mapping $g : \mathcal{P} \to \{1, \ldots, k\}$ can be a (deterministic) ε-extractor (with $\varepsilon < 1$) for this $\mathcal{P}$, even if $k = 2$, although $H_{\min}(P) \geqq \log |\mathsf{U}| - 1$ for each $P \in \mathcal{P}$. (Sántha and Vazirani (1986) proved a similar but stronger result.)

Hint Consider for any $g : \mathsf{U} \to \{0, 1\}$ the conditional distributions P_0 and P_1 of a RV U uniformly distributed on U, on the conditions $g(U) = 0$ and $g(U) = 1$. Then either P_0 or P_1 belongs to $\mathcal{P}$, while for U_0 and U_1 with distribution P_0 resp. P_1, both $g(U_0)$ and $g(U_1)$ have degenerate distribution, whose variation distance from the uniform distribution on $\{0, 1\}$ is 1.

17.22. (*Leftover hash lemma*)

(a) Let $g_t : \mathsf{U} \to \{0, 1\}^m$, $t \in \mathsf{T}$, be a universal family of hash functions, i.e. (see Problem 1.8), a family of mappings such that if T is uniformly distributed on T then

$$\Pr\{g_T(u_1) = g_T(u_2)\} \leqq 2^{-m} \qquad \text{for } u_1 \neq u_2 \text{ in } \mathsf{U}.$$

Show that for any RV U with values in U, independent of T, with Rényi entropy of order 2 satisfying

$$H_2(U) \triangleq -\log \sum_{u \in \mathsf{U}} P_U^2(u) > m,$$

the joint distribution of $g_T(U)$ and T is ε-uniform on $\{0, 1\}^m \times \mathsf{T}$ with

$$\varepsilon = \exp\left(\frac{m - H_2(U)}{2}\right).$$

(Impagliazzo, Levin and Luby (1989) established a weaker version; an improvement close to that above appears in Impagliazzo and Zuckerman (1989), attributed to C. Rackoff. See also Bennett, Brassard and Robert (1988) and Bennett *et al.* (1995).)

Hint For any RV X with range X, denote by X' an independent copy of X, and let $c(X) \triangleq \Pr\{X = X'\} = \exp(-H_2(X))$. Check via the Cauchy inequality that

$$\sum_{x \in \mathsf{X}} \left| P_X(x) - \frac{1}{|\mathsf{X}|} \right| \leqq \left[|\mathsf{X}| \sum_{x \in \mathsf{X}} \left(P_X(x) - \frac{1}{|\mathsf{X}|} \right)^2 \right]^{1/2} = \left[|\mathsf{X}| c(X) - 1 \right]^{1/2}.$$

To apply this bound to $X = (g_T(U), T)$, calculate

$$c(g_T(U), T) = \Pr\left\{(g_T(U), T) = (g_{T'}(U'), T')\right\}$$

$$
\begin{aligned}
&= \Pr\{T = T'\}\Pr\{g_T(U) = g_T(U')|T = T'\} \\
&= \frac{1}{|\mathsf{T}|}\left[\Pr\{U = U'\} + \Pr\{g_T(U) = g_T(U')|U \neq U'\}\right] \\
&\leqq \frac{1}{|\mathsf{T}|}\left[c(U) + 2^{-m}\right] = \frac{1}{|\mathsf{T} \times \{0,1\}^m|}\left[\exp(m - H_2(U)) + 1\right].
\end{aligned}
$$

(b) Given a set U and $m < \log|\mathsf{U}| \leqq n$, construct $g : \mathsf{U} \times \{0,1\}^n \to \{0,1\}^m$, which is an ε-extractor with (seed length n and)

$$
\varepsilon = \exp\left(\frac{m - H_2(P)}{2}\right),
$$

for each PD P on U with $H_2(P) > m$.

Hint Assume with no loss of generality that $\mathsf{U} = \{0,1\}^n$, and regard $\{0,1\}^n$ as a finite field, with componentwise mod 2 addition and suitable multiplication. Define $g(u,t)$ as the m-bit truncation of $ut \in \{0,1\}^n$. Check that the family of mappings $g_t(u) \triangleq g(u,t)$, $t \in \{0,1\}^n$ is a universal family of hash functions and apply the result of (a).

(c) Let $\mathsf{U} = \{0,1\}^n$, $0 < H' < H < 1$. Give an explicit ε-extractor with

$$
\varepsilon = \exp\left(-\frac{H - H'}{2}n\right),
$$

seed length n and $m = \lfloor nH' \rfloor$, for the family $\mathcal{P} \triangleq \{P : H_2(P) \geqq nH\}$, and hence also for $\mathcal{P}' \triangleq \{P : H_\infty(P) \geqq nH\}$.

Hint Use $g(u,t)$ in the hint to (b), noting that $H_2(P) \geqq H_\infty(P)$.

17.23. (a) Let $g : \mathsf{U} \times \{0,1\}^s \to \{0,1\}^m$ be a seeded ε-extractor for the family $\mathcal{P} \triangleq \{P : H_\infty(P) \geqq \log d\}$, and let T be uniformly distributed on $\{0,1\}^s$. Verify that simultaneously for each pair of RVs (U,V) satisfying (17.10), and independent of T,

$$
S(g(U,T)|T,V) \leqq \left(\frac{\varepsilon}{2} + 2\eta\right)m + h\left(\frac{\varepsilon}{2} + \eta\right).
$$

Hint The argument in the proof of the secrecy lemma (with m playing the role of $\log k$) gives the claimed bound for $S(g(U,T),T|V)$. Check that the latter is equal to $S(g(U,T)|T,V)$; see Problem 17.1(c).

(b) Give a similar counterpart to Corollary 17.5.

Hint For an analog of part (i) of the corollary verify that if $g : \mathsf{X}^n \times \{0,1\}^s \to \{0,1\}^m$ is a seeded ε_n-extractor for the family of those PDs on X^n whose minentropy is bounded below by $n(H - \delta/2)$, where s is arbitrary, $m = n(H - \delta)$ and $\varepsilon_n \to 0$ exponentially, then $S(g(X^n,T)|T,Y^n)$ is exponentially small whenever $H(X|Y) \geqq H$ (by Problem 17.22, such extractors do exist). Part (ii) has a similar analog.

(c) Show that in the proof of Theorem 17.17, one could use instead of Corollary 17.5 its counterpart with seeded extractors as above.

Hint The crucial point is that Alice (if permitted to randomize) may generate the seed T and send it Bob over the public channel (in addition to her message of rate $r = H(X|Y) + \delta$ that makes X^n ε-recoverable to Bob). Then $g(X^n, T)$ will represent a SK, for revealing T does not affect secrecy: when bounding the security index of $g(X^n, T)$, Eve has been assumed to know T.

(d) The seeded extractors constructed in Problem 17.22(b) are inadequate for proving Theorems 17.11, 17.13, and 17.21 (why?). Show, however, that extractors whose seed length is a negligible fraction of n are suitable to prove these results also.
("Good" extractors have seed lengths of order $\log(n/\varepsilon)$, see, e.g., Zuckermann (1997). Their suitability for proving secrecy capacity results is demonstrated in Maurer and Wolf (2000).)

Hint In the case of Theorems 17.11, 17.13, the small seed length allows Alice to generate the seed T and send it to Bob via DMC, using up a small fraction of her transmissions; again, it does not matter if Eve learns T. In Theorem 17.21 Alice is not allowed to randomize, but she can apply a deterministic extractor to a small fraction of the source outputs she observes and send Bob the resulting (nearly) uniformly distributed T, using up only a negligible fraction of the allowed communication rate.

17.24. (*Multi-terminal SK capacity*) The definition implies, see the Discussion, that $C_{SK}(\mathsf{A}|\mathsf{D}) \leqq C_{PK}(\mathsf{A}|\mathsf{D})$. Show that the equality holds if $\mathsf{D} = \{m\}$ and
(a) (source model) $X_1, \ldots, X_m$ is a Markov chain or
(b) (channel model) for some $1 \leqq l \leqq m - 1$

$$W(x_2, \ldots x_m|x_1) = \left(\prod_{i=2}^{l} W(x_i|x_{i-1}) \right) W(x_{l+1}|x_1) \prod_{j=l+2}^{m} W(x_j|x_{j-1}).$$

For a simple formula of $C_{PK}(\mathsf{A}|\mathsf{D})$, see Problem 17.16.
(Csiszár and Narayan (2008). Previously, Csiszár and Narayan (2004) claimed equality under weaker conditions, regrettably with a flawed proof. It remains open whether the equality holds, for example, for channels with independent components, if $m > 3$.)

Hint By Theorem 17.17, the PK capacity is achievable with non-interactive communication, taking for K a function of $X^n_{i_{\min}}$ (see Problem 17.16 for notation). To show that this K is also a SK, i.e., ε-recoverable without relying on X^n_m, consider the best estimator $\hat{K}_i$ of K from (X^n_i, F_{M}), $i \in \mathsf{A}$, namely $\hat{K}_i = \arg\max \Pr\left\{K = k|X^n_i, F_{\mathsf{M}}\right\}$. Use the Markov assumption to verify that $\hat{K}_i$ is a function of X^n_i and $\{F_j : j < i\}$.

Story of the results

An information-theoretic approach to secrecy was first suggested by Shannon (1949). Proposition 17.1 is, in effect, his. The security index (17.1) was introduced in Csiszár and Narayan (2004) for a compact joint representation of two criteria that go back to Shannon.

The extractor lemma, Lemma 17.3 is taken from Ahlswede and Csiszár (1998), where it was called the coloring lemma. An inconsequential error in its original statement was pointed out by F. Matúš (personal communication). As a consequence of this basic lemma, part (i) of the secrecy lemma, Lemma 17.5 was stated in Csiszár (1996) (which, though written later, appeared earlier). It was used to prove "strong" secrecy results including a "strong" version of Theorem 17.11; for the history of "weak" and "strong" definitions, see below. Part (ii) of the secrecy lemma formalizes a proof technique used in Csiszár and Narayan (2000).

Deterministic extractors (under a different name) have been used for purposes like here at least since 1996; see the preceding paragraph. The computer science literature of extractors goes back to the 1980s. There, attention has been focused on seeded extractors for a while (see Problems 17.21–17.23), and the emergence of interest in deterministic extractors is commonly attributed to and Trevisan and Vadhan (2000).

The wiretap channel was introduced by Wyner (1975c), who first applied advanced techniques of information theory to a secrecy problem. In his model, channel W_2 was a degraded version of W_1. The general case and the more general BCC model were studied by Csiszár and Körner (1978), who proved (slightly weaker versions of) Theorems 17.11 and 17.13. The key identity (Lemma 17.12) is theirs. The achievability proof of Csiszár and Körner (1978) employed techniques presented in Chapters 15 and 16. The random coding approach used here relies upon Lemma 17.14, which represents a classical form of superposition coding. This technique of multi-user information theory dates back to Cover (1972) and Bergmans (1973).

The definition of secrecy capacities adopted here is due to Maurer (1994), who argued that the previous definition, dating back to Wyner (1975c), was too weak for cryptographic purposes. For most DM models, however, all reasonable definitions give the same secrecy capacity, as proved in essence by Maurer and Wolf (2000); see Problem 17.10 for details.

The first information theoretic results about common randomness were proved by Gács and Körner (1973) for models not admitting communication; see Problems 16.27–16.30 and 17.10(c). A systematic study of secret and non-secret CR for models admitting communication appears in Ahlswede and Csiszár (1993), (1998); CR capacity is formally defined in the latter paper. Informally, it emerged in Ahlswede and Dueck (1989b), who demonstrated its intrinsic relationship to *identification capacity*, a concept introduced by Ahlswede and Dueck (1989a), not covered here.

The fact, which appears counterintuitive at first, that public discussion may help in generating a SK, was pointed out by Bennett, Brassard and Robert (1988) and by Maurer (1993), who proved Theorem 17.17. Independently, but aware of a preliminary result of Maurer announced in 1990, Ahlswede and Csiszár (1993) also proved Theorem 17.17, as well as Theorem 17.20 and part of Theorem 17.21. The full Theorem 17.21 is due to Csiszár and Narayan (2000).

Our treatment of multi-terminal source and channel models follows Csiszár and Narayan (2004), (2008); Proposition 17.24 is due, in effect, to Wyner, Wolf and Willems (2002). General results on multi-terminal SK capacities, including that mentioned at the end of the Discussion, were obtained by Gohari and Anantharam (2010).

Another result mentioned in the Discussion concerns the cognitive interference channel with confidential messages; this is due to Liang *et al.* (2009). For a variety of models which are in the scope of this chapter, but could not be covered, the reader may consult Liang, Poor and Shamai (2009).

References

Abbreviations of frequently quoted journals and collections of papers:

AMS	*Annals of Mathematical Statistics*
BSTJ	*Bell System Technical Journal*
IC	*Information and Control*
IEEE-IT	*IEEE Transactions on Information Theory*
IRE-IT	*IRE Transactions on Information Theory*
KP	*Key Papers in the Development of Information Theory* (ed. D. Slepian), IEEE Press, New York, 1974
PCIT	*Problems of Control and Information Theory*
PPI	*Problemy Peredači Informacii* (translated into English as *Problems of Information Transmission*)
Topics in IT	*Topics in Information Theory* (eds. I. Csiszár and P. Elias). Coll. Math. Soc. J. Bolyai, No. 16. North Holland, Amsterdam, 1977
ZW	*Zeitschrift für Wahrscheinlichkeitstheorie und verwandte Gebiete*

Aczél, J. and Daróczy, Z. (1975). *On Measures of Information and Their Characterizations.* Academic Press, New York.

Aczél, J., Forte, B. and Ng, C. T. (1974). Why the Shannon and Hartley entropies are "natural." *Adv. Appl. Prob.* **6**, 131–146.

Ahlswede, R. (1968). Certain results in coding theory for compound channels. In *Proceedings of Colloquium on Information Theory*, Debrecen, 1967 (ed. A. Rényi). J. Bolyai Math. Soc., Budapest, Hungary. vol. 1, pp. 35–60.

Ahlswede, R. (1970) A note on the existence of the weak capacity for channels with arbitrarily varying channel probability functions and its relation to Shannon's zero error capacity. *AMS* **41**, 1027–1033.

Ahlswede, R. (1973a) Multi-way communication channels. In *Proceedings of 2nd International Symposium on Information Theory*, Tsahkadsor, Armenian SSR, 1971. Akadémiai Kiadó, Budapest, pp. 23–52.

Ahlswede, R. (1973b) Channel capacities for list codes. *J. Appl. Prob.* **10**, 824–836.

Ahlswede, R. (1973c) Channels with arbitrarily varying channel probability functions in the presence of noiseless feedback. *ZW* **25**, 239–252.

Ahlswede, R. (1974) The capacity region of a channel with two senders and two receivers. *Ann. Prob.* **2**, 805–814.

Ahlswede, R. (1978) Elimination of correlation in random codes for arbitrarily varying channels. *ZW* **33**, 159–175.

Ahlswede, R. (1980) A method of coding and an application to arbitrarily varying channels. *J. Combinatorics, Inf. Syst. Sci.* **5**, 10–35.

Ahlswede, R. (1986) Arbitrarily varying channels with state sequence known to the sender. *IEEE-IT* **32**, 621–629.

Ahlswede, R. (1990) Extremal properties of rate-distortion functions. *IEEE-IT* **36**, 166–171.

Ahlswede, R. (1993) The maximal error capacity of arbitrarily varying channels with constant list sizes. *IEEE-IT* **39**, 1416–1417.

Ahlswede, R. and Cai, N. (1991) Two proofs of Pinsker's conjecture concerning arbitrarily varying channels. *IEEE-IE* **37**, 1647–1649.

Ahlswede, R. and Cai, N. (1999) Arbitrarily varying multiple-access channels, part I – Ericson's symmetrizability is adequate, Gubner's conjecture is true. *IEEE-IT* **45**, 742–749.

Ahlswede, R. and Cai, N. (2000) The AVC with noiseless feedback and maximal error probability: a capacity formula with a trichotomy. In *Numbers, Information and Complexity (Bielefeld 1998)*. Kluwer, Boston MA, pp. 151–178.

Ahlswede, R., Cai, N., Li, S.-Y. R. and Yeung, R. W. (2000) Network information flow. *IEEE-IT* **46**, 1204–1216.

Ahlswede, R. and Csiszár, I. (1993) Common randomness in information theory and cryptography. Part 1, Secret sharing. *IEEE-IT* **39**, 1121–1132.

Ahlswede, R. and Csiszár, I. (1998) Common randomness in information theory and cryptography. Part 2, CR capacity. *IEEE-IT* **44**, 225–240.

Ahlswede, R. and Dueck, G. (1976) Every bad code has a good subcode: a local converse to the coding theorem. *ZW* **34**, 179–182.

Ahlswede, R. and Dueck, G. (1989a) Identification via channels. *IEEE-IT* **35**, 15–29.

Ahlswede, R. and Dueck, G. (1989b) Identification in the presence of feedback: a discovery of new capacity formulas. *IEEE-IT* **35**, 30–36.

Ahlswede, R. and Gács, P. (1976) Spreading of sets in product spaces and hypercontraction of the Markov operator. *Ann. Prob.* **4**, 925–939.

Ahlswede, R. and Gács, P. (1977) Two contributions to information theory. In: *Topics in IT*, pp. 17–40.

Ahlswede, R., Gács, P. and Körner, J. (1976) Bounds on conditional probabilities with applications in multi-user communication. *ZW* **34**, 157–177. Erratum: *ZW* **39**, 353–354.

Ahlswede, R. and Katona, G. O. H. (1977) Contributions to the geometry of Hamming spaces. *Discrete Mathematics* **17**, 1–22.

Ahlswede, R. and Körner, J. (1974) On common information and related characteristics of correlated information sources. Preprint. Presented at the *7th Prague Conference on Information Theory,* September 1974. Reprinted in *General Theory of Information Transfer and Combinatorics* (eds. R. Ahlswede *et al.*). Lecture Notes in Comp. Sci. **4123** (2006), Springer Verlag, New York, pp. 664–677.

Ahlswede, R. and Körner, J. (1975) Source coding with side information at the decoder and a converse for degraded broadcast channels. *IEEE-IT* **21**, 629–637.

Ahlswede, R. and Körner. J. (1977) On the connections between the entropies of input and output distributions of discrete memoryless channels. In *Proc. Fifth Conference on Probability Theory, Brasov 1974*. Editura Academiei Rep. Soc. Romania, Bucuresti 1977, pp. 13–23.

Ahlswede, R. and Wolfowitz, J. (1969) Correlated decoding for channels with arbitrarily varying channel probability functions. *IC* **14**, 457–473.

Ahlswede, R. and Wolfowitz, J. (1970) The capacity of a channel with arbitrarily varying cpf's and binary output alphabet. *ZW* **15**, 186–194.

Amari, S. and Nagaoka, H. (2000) *Methods of Information Geometry*. Amer. Math. Soc., Providence, RI. Japanese original: Iwanami Shoten, Tokyo, 1993.

Arimoto, S. (1972) An algorithm for computing the capacity of arbitrary discrete memoryless channels. *IEEE-IT* **18**, 14–20.

Arimoto, S. (1973) On the converse to the coding theorem for discrete memoryless channels. *IEEE-IT* **19**, 357–359.

Arimoto, S. (1976) Computation of random coding exponent functions. *IEEE-IT* **22**, 665–671.

Audenaert, K. M. R. (2007) A sharp Fannes-type inequality for the von Neumann entropy. *J. Phys. A* **40**, 8127–8136.

Augustin, U. (1978) Noisy channels. Habilitation Thesis, Universität Erlangen-Nürnberg.

Barg, A. and McGregor, A. (2005) Distance distribution of binary codes and the error probability of decoding. *IEEE-IT* **51**, 4237–4246.

Barron, A. R. (1985) *Logically Smooth Density Estimation*. PhD Thesis, Stanford University, Stanford, CA.

Bennett, C. H., Brassard, G., Crépeau, C. and Maurer, U. (1995) Generalized privacy amplification. *IEEE-IT* **41**, 1915–1923.

Bennett, C. H., Brassard, G. and Robert, J.-M. (1988) Privacy amplification by public discussion. *SIAM J. Comput.* **17**, 210–229.

Berge, C. (1962) Sur une conjecture relative au problème des codes optimaux. In: *Commun. 13ème Assemblée Gén. URSI*, Tokyo, 1962.

Berge, C. and Simonovits, M. (1974) The coloring numbers of the direct product of two hypergraphs. In *Hypergraph Seminar* (eds. C. Berge and D. Ray-Chaudhuri). Lecture Notes in Mathematics **411**. Springer, Berlin, pp. 21–33.

Berger, T. (1971) *Rate Distortion Theory: A Mathematical Basis for Data Compression*. Prentice Hall, Englewood Cliffs, NJ.

Berger, T. (1977) Multiterminal source coding. In *The Information Theory Approach to Communications* (ed. G. Longo). CISM Courses and Lecture Notes no. **229**, Springer, New York, pp. 172–231.

Bergmans, P. P. (1973) Random coding theorems for broadcast channels with degraded components. *IEEE-IT* **19**, 197–207.

Berlekamp, E. R. (1964) Block coding with noiseless feedback. Ph.D. Thesis, MIT, Cambridge, MA.

Berlin, P., Nakiboğlu, B., Rimoldi, B. and Telatar, E. (2009) A simple converse of Burnašev's reliability function. *IEEE-IT* **55**, 3074–3080.

Berrou, C., Glavieux, A. and Thitimajshima, P. (1993) Near Shannon limit error-correcting coding and decoding: turbo codes. In *Proc. IEEE Intl Conf. Commun.*, Geneva, 1993, pp. 1064–1070.

Bierbaum, M. and Wallmeier, H. M. (1979) A note on the capacity region of the multiple-access channel. *IEEE-IT* **25**, 484.

Blackwell, D. (1963) Statistics 262. Course given at the University of California, Berkeley 1963. Quoted by Van der Meulen (1977).

Blackwell, D., Breiman, L. and Thomasian, A. J. (1959) The capacity of a class of channels. *AMS* **30**, 1229–1241. (Reprinted in *KP*.)

Blackwell, D., Breiman, L. and Thomasian, A. J. (1960) The capacities of certain channel classes under random coding. *AMS* **31**, 558–567.

Blahut, R. E. (1972) Computation of channel capacity and rate-distortion functions. *IEEE-IT* **18**, 460–473.

Blahut, R. E. (1974) Hypothesis testing and information theory. *IEEE-IT* **20**, 405–417.

Blahut, R. E. (1977) Composition bounds for channel block codes. *IEEE-IT* **23**, 656–674.

Blinovsky, V., Narayan, P. and Pinsker, M. (1995) Capacity of the arbitrarily varying channel under list decoding. *PPI* **31**, 99–113.

Bloh, E. L. (1960) Optimal code construction from elementary symbols of different length. In *Problemy Peredači Informacii*, vol. 5, AN SSSR, Moscow, pp. 100–111. In Russian.

Blokhuis, A. (1993) On the Sperner capacity of the cyclic triangle. *J. Algebraic Combin.* **2**, 123–124.

Boë, J. M. (1978) Une famille remarquable de codes indécomposables. In *Automata, Languages and Programming*. Lecture Notes in Computer Science **62**, Springer, New York, pp. 105–112.

Bollobás, B. (1965) On generalized graphs. *Acta Math. Acad. Sci. Hung.* **16**, 445–452.

Boltzmann, L. (1877) Über die Beziehung zwischen dem zweiten Hauptsatze der mechanischen Wärmetheorie und der Wahrscheinlichkeitsrechnung respektive den Sätzen über das Wärmegleichgewicht. *Wien. Ber.* **76**, 373–435.

Brightwell, G., Cohen, G., Fachini, E., Fairthorne, M., Körner, J., Simonyi, G. and Tóth, Á. (2010) Permutation capacities of families of oriented infinite paths. *SIAM J. Discrete Math.* **24**, 441–456.

Burnašev, M. V. (1976) Information transmission over a discrete channel with feedback. Random transmission time. *PPI* **12**, (*4*), 10–30. In Russian.

Calderbank, Frankl, P., Graham, R. L., Li, W. and Shepp, L. (1993) The Sperner capacity of the cyclic triangle for linear and nonlinear codes. *J. Algebraic Combin.* **2**, 31–48.

Carleial, A. B. (1975) On the capacity of multi-terminal communication networks. Ph.D. Dissertation, Stanford University, Stanford, CA. Quoted by: Kramer (2003).

Carleial, A. B. (1982) Multiple-access channels with different generalized feedback signals. *IEEE-IT* **28**, 841–850.

Carter, J. L. and Wegman, M. N. (1979) Universal classes of hash functions. *J. Comput. Syst. Sci.* **18** 143–154.

Čencov, N. N. (1972) *Statistical Decision Rules and Optimal Inference.* Nauka, Moscow. In Russian. English translation: Amer. Math. Soc., Providence, RI, 1982.

Césari, Y. (1974) Sur l'application du théorème de Suschkevitch à l étude des codes rationnels complets. In *Automata, Languages and Programming* (ed. J. Loeckx). Lecture Notes in Computer Science, Springer-Verlag, pp. 342–350.

Chan, C. (2008) On the tightness of mutual dependence upperbound for secret-key capacity of multiple terminals. *arXiv*: 0805.3299v2 [cs.IT].

Chaundy, T. W. and McLeod, J. B. (1960) On a functional equation. *Edinburgh Math. Notes* **43**, 7–8.

Chernoff, H. (1952) A measure of asymptotic efficiency for tests of a hypothesis based on a sum of observations. *AMS* **23**, 493–507.

Chernoff, H. (1956) Large-sample theory: parametric case. *AMS* **27**, 1–22.

Cohen, G., Körner, J. and Simonyi, G. (1990) Zero-error capacities and very different sequences. In *Sequences, Combinatorics, Security and Transmission* (ed. R. M. Capocelli). Advanced International Workshop on Sequences, Positano, Italy, 1988. Springer, New York, pp. 144–155.

Cover, T. M. (1972) Broadcast channels. *IEEE-IT* **18**, 2–14. (Reprinted in *KP*.)

Cover, T. M. (1975a) An achievable rate region for the broadcast channel. *IEEE-IT* **21**, 399–401.

Cover, T. M. (1975b) A proof of the data compression theorem of Slepian and Wolf for ergodic sources. *IEEE-IT* **21**, 226–228.

Cover, T. M. and Leung C. S. K. (1981) An achievable rate region for the multiple-access channel with feedback. *IEEE-IT* **27**, 292–298.

Cover, T.M. and Thomas, J.A. (2006) *Elements of Information Theory*, 2nd edn. Wiley-Interscience, Hoboken, NJ.

Csiszár, I. (1967) Information-type measures of difference of probability distributions and indirect observations. *Studia Sci. Math. Hungar.* **2**, 299–318.

Csiszár, I. (1969) Simple proofs of some theorems on noiseless channels. *IC* **14**, 285–298.

Csiszár, I. (1970) On noiseless channels. *PPI* **6**, (*4*), 3–15. In Russian.

Csiszár, I. (1972) A class of measures of informativity of observation channels. *Periodica Math. Hungar.* **2**, 191–213.

Csiszár, I. (1973) On the capacity of noisy channels with arbitrary signal costs. *PCIT* **2**, 283–304.

Csiszár, I. (1974) On the computation of rate-distortion functions. *IEEE-IT* **20**, 122–124.

Csiszár, I. (1975) *I*-divergence geometry of probability distributions and minimization problems. *Ann. Prob.* **3**, 146–158.

Csiszár, I. (1980) Joint source-channel error exponent. *PCIT* **9**, 315–328.

Csiszár, I. (1982a) Linear codes for sources and source networks: error exponents, universal coding. *IEEE-IT* **28**, 585–592.

Csiszár, I. (1982b) On the error exponent of source-channel transmission with a distortion threshold. *IEEE-IT* **28**, 823–828.

Csiszár, I. (1995) Generalized cutoff rates and Rényi's information measures. *IEEE-IT* **41**, 26–34.

Csiszár, I. (1996) Almost independence and secrecy capacity. *PPI* **32**, 48–57. In Russian.

Csiszár, I. (1998) The method of types. *IEEE-IT* **44**, 2505–2523.

Csiszár, I., Katona, G. and Tusnády, G. (1969) Information sources with different cost scales and the principle of conservation of entropy. *ZW* **12**, 185–222.

Csiszár, I. and Komlós, J. (1968) On the equivalence of two models of finite-state noiseless channels from the point of view of the output. In *Proc. Coll. Inf. Th, vol. 1* Debrecen, 1967 (ed. A. Rényi) J. Bolyai Math. Soc., Budapest, Hungary, pp. 101–128.

Csiszár, I. and Körner, J. (1976) Source networks with unauthorized users. *J. Combinatorics, Inf. Syst. Sci.* **1**, 25–40.

Csiszár, I. and Körner, J. (1978) Broadcast channels with confidential messages. *IEEE-IT* **24**, 339–348.

Csiszár, I. and Körner, J. (1980) Towards a general theory of source networks. *IEEE–IT* **26**, 155–165.

Csiszár, I. and Körner, J. (1981a) Graph decomposition: a new key to coding theorems. *IEEE-IT* **27**, 6–11.

Csiszár. I. and Körner, J. (1981b) On the capacity of the arbitrarily varying channel for maximum probability of error. *ZW* **57**, 87–101.

Csiszár, I. and Körner, J. (1982) Feedback does not affect the reliability function of a DMC at rates above capacity. *IEEE-IT* **28**, 92–93.

Csiszár, I., Körner, J., Lovász, L., Marton, K. and Simonyi, G. (1990) Entropy splitting for antiblocking corners and perfect graphs. *Combinatorica* **10**, 27–40.

Csiszár, I., Körner, J. and Marton, K. (1977) A new look at the error exponent of a discrete memoryless channel. Preprint. Presented at the *IEEE International Symposium on Information Theory*, October 1977, Cornell University, Ithaca, NY.

Csiszár, I. and Longo, G. (1971) On the error exponent for source coding and for testing simple statistical hypotheses. *Studia Sci. Math. Hungar.* **6**, 181–191.

Csiszár, I. and Narayan, P. (1988) The capacity of the arbitrarily varying channel revisited: positivity, constraints. *IEEE-IT* **34**, 181–193.

Csiszár, I. and Narayan, P. (1995) Channel capacity for a given decoding metric. *IEEE-IT* **41**, 35–43.

Csiszár, I. and Narayan, P. (2000) Common randomness and secret key generation with a helper. *IEEE-IT* **46**, 344–366.

Csiszár, I. and Narayan, P. (2004) The secret key capacity for multiple terminals. *IEEE-IT* **50**, 3047–3061.

Csiszár, I. and Narayan, P. (2008) Secrecy capacities for multiterminal channel models. *IEEE-IT* **54**, 2437–2452.

Csiszár, I. and Shields, P. (2004) *Information Theory and Statistics: A Tutorial.* Now Publishers, Delft.

Daróczy, Z. (1964) Über Mittelwerte und Entropien vollständiger Wahrscheinlichkeitsverteilungen. *Acta Math. Acad. Sci. Hungar.* **15**, 203–210.

Davisson, L. D. (1966) Comments on "sequence time coding for data compression." *Proc. IEEE* **54**, 2010.

Davisson, L. D. (1973) Universal noiseless coding. *IEEE-IT* **19**, 783–796.

Davisson, L. D. and Leon-Garcia, A. (1980) A source matching approach to finding minimax codes. *IEEE-IT* **26**, 166–174.

Davisson, L. D., McEliece, R., Pursley, M. B. and Wallace, M. S. (1981) Efficient universal noiseless source codes. *IEEE-IT* **27**, 269–279.

Dobrušin, R. L. (1958) Information transmission in a channel with feedback. *Teor. Veroyatnost. i Primenen.* **34**, 367–383. (In Russian. Translation reprinted in *KP*.)

Dobrušin, R. L. (1959a) Optimal information transfer over a channel with unknown parameters. *Radiotechn. i Elektron.* **4**, 1951–1956. In Russian.

Dobrušin, R. L. (1959b) A general formulation of the basic Shannon theorem in information theory. *Usp. Mat. Nauk* **14**, (*6*), 3–104. In Russian.

Dobrušin, R. L. (1962a) Asymptotic bounds of the probability of error for the transmission of messages over a discrete memoryless channel with a symmetric transition probability matrix. *Teor. Veroyatnost. i Primenen.* **7**, 283–311. In Russian.

Dobrušin, R. L. (1962b) Asymptotic bounds of the probability of error for the transmission of messages over a memoryless channel using feedback. *Probl. Kibern.* **8**, 161–168. In Russian.

Dobrušin, R. L. (1970) Unified methods of optimal quantizing of messages. *Probl. Kibern.* **22**, 107–156. In Russian.

Dobrušin, R. L. and Stambler, S. Z. (1975) Coding theorems for classes of arbitrarily varying discrete memoryless channels. *PPI* **11**, (*2*), 3–22. In Russian.

Dobrušin, R. L. and Tsybakov, B. S. (1962) Information transmission with additional noise. *IRE-IT* **8**, 293–304. (Reprinted in *KP*.)

Dueck, G. (1978) Maximal error capacity regions are smaller than average error capacity regions for multi-user channels. *PCIT* **7**, 11–19.

Dueck, G. (1979) The capacity region of the two-way channel can exceed the inner bound. *IC* **40**, 258–266.

Dueck, G. and Körner, J. (1979) Reliability function of a discrete memoryless channel at rates above capacity. *IEEE–IT* **25**, 82–85.

Dyačkov, A. G. (1975) Upper bounds on the error probability for transmission with feedback in case of memoryless discrete channels. *PPI* **11**, (*4*), 13–28. In Russian.

Dyačkov, A. G. (1980) Bounds to average error probability for the fixed-composition-code ensemble. *PPI* **16**, (*4*), 3–8. In Russian.

Eggleston, H. G. (1958) *Convexity.* Cambridge University Press, Cambridge.

El Gamal, A. A. (1979) The capacity region of a class of broadcast channels. *IEEE-IT* **25**, 166–169.

El Gamal, A. A. (1980) The capacity of the product and sum of two inconsistently degraded broadcast channels. *PPI* **16**, (*1*), 3–23. In Russian.

El Gamal, A. A. and Kim, Y. H. (2010) *Lecture Notes on Network Information Theory.* arXiv:1001.3404 (cs.IT).

Elias, P. (1955) Coding for noisy channels. In *IRE Convention Record*, Part 4, pp. 37–46. (Reprinted in *KP*.)

Elias, P. (1957) List decoding for noisy channels. In *IRE WESCON Convention Record*, vol. 2, pp. 94–104.

Elias, P. (1958) Zero error capacity for list detection. Quarterly Progress Report no. 48, Research Laboratory of Electronics, MIT, Cambridge, MA. Quoted by: Ahlswede (1973b).

Erdős, P. (1946) On the distribution function of additive functions. *Ann. of Math.* **47**, 1–20.

Erdős, P., Füredi, Z., Hajnal, A., Komjáth, P., Rödl, V. and Seress, Á. (1986) Coloring graphs with locally few colors, *Discrete Math.* **59**, 21–34.

Ericson, T. (1985) Exponential error bounds for random codes in the arbitrarily varying channel. *IEEE-IT* **31**, 42–48.

Erokhin, V. D. (1958) ε-entropy of discrete random objects. *Teor. Veroyatnost. i Primenen.* **3**, 103–107. In Russian.

Faddeev, D. K. (1956) On the notion of entropy of a finite probability scheme. *Usp. Mat. Nauk* **11**, 227–231. In Russian.

Falk, F. (1970) Inequalities of J. W. Gibbs. *Am. J. Phys.* **38**, 858–869.

Fannes, M. (1973) A continuity property of the entropy density for spin lattice systems. *Commun. Math. Phys.* **31**, 291–294.

Fano, R. M. (1952) Class notes for transmission of information. Course 6.574, MIT, Cambridge, MA.

Fano, R. M. (1961) *Transmission of Information, A Statistical Theory of Communications.* Wiley, New York.

Fedotov, A., Harremoës, P. and Topsøe, F. (2003) Refinements of Pinsker's inequality. *IEEE-IT* **49**, 1491–1498.

Feinstein, A. (1954) A new basic theorem of information theory. *IRE-IT* **4**, 2–22. (Reprinted in *KP*.)

Fekete, M. (1923) Über die Verteilung der Wurzeln bei gewissen algebraischen Gleichungen mit ganzzähligen Koeffizienten. *Math. Z.* **17**, 228-249.

Feller, W. (1966) *An Introduction to Probability Theory and its Applications, vol. 2*, 2nd edn. Wiley, New York.

Feller, W. (1968) *An Introduction to Probability Theory and its Applications, vol. 1*, 3rd edn. Wiley, New York.

Fenchel, W. (1929) Über die Krümmung und Windung geschlossener Raumkurven. *Math. Ann.* **101**, 239–252.

Fisher, R. A. (1925) Theory of statistical estimation. *Proc. Camb. Phil. Soc.* **22**, 700–725.

Fitingof, B. M. (1966) Coding in the case of unknown and changing message statistics. *PPI* **2**, (*2*), 3–11. In Russian.

Forney, G. D. (1968) Exponential error bounds for erasure, list and decision feedback schemes. *IEEE-IT* **14**, 206–220. (Reprinted in *KP*.)

Forte, B. (1975) Why Shannon's entropy. In *Conv. Inform. Teor. Ist. Naz. Alta Mat.* Roma 1973, *Symposia Mathematica* **15**. Academic Press, New York, pp. 137–152.

Fredman, M. and Komlós, J. (1984) On the size of separating systems and perfect hash functions, *SIAM J. Algebraic Discr. Meth.* **5**, 61–68.

Fujishige, S. (1978) Polymatroidal dependence structure of a set of random variables. *IC* **39**, 55–72.

Fulkerson, D. R. (1973) On the perfect graph theorem. In *Mathematical Programming* (eds. T. C. Hu and S. M. Robinson), Academic Press, New York, pp. 69–76.

Gaarder, N. T. and Wolf, J. K. (1975) The capacity region of a multiple access discrete memoryless channel can increase with feedback. *IEEE-IT* **21**, 100–102.

Gabidulin, E. M. (1967) Bounds for the probability of decoding error when using linear codes over memoryless channels. *PPI* **3**, (*2*), 55–62. In Russian.

Gács, P. and Körner, J. (1973) Common information is far less than mutual information. *PCIT* **2**, 149–162.

Gallager, R. G. (1963) *Low Density Parity Check Codes.* MIT Press, Cambridge, MA.

Gallager, R. G. (1964) Information theory. In *Mathematics of Physics and Chemistry, vol. 2* (eds. H. Margenau and G. M. Murphy). Van Nostrand, Princeton, NJ, chap. 4.

Gallager, R. G. (1965) A simple derivation of the coding theorem and some applications. *IEEE-IT* **11**, 3–18. (Reprinted in *KP*.)

Gallager, R. G. (1968) *Information Theory and Reliable Communication.* Wiley, New York.

Gallager, R. G. (1973) The random coding bound is tight for the average code. *IEEE-IT* **19**, 244–246.

Gallager, R. G. (1974a) Capacity and coding for degraded broadcast channels. *PPI* **10**, (*1*), 3–14. In Russian.

Gallager, R. G. (1974b) MIT Lecture Notes on Universal Coding. Quoted in Editor's Note to a Correspondence by B. Ryabko, *IEEE-IT* **27**, 781.

Gallager, R. G. (1976) Source coding with side information and universal coding. Preprint. Presented at the *IEEE International Symposium on IT*, June 1976, Ronneby, Sweden.

Galluccio, A., Gargano, L., Körner, J. and Simonyi, G. (1994) Different capacities of digraphs. *Graph. Combinator.* **10**, 105–121.

Gargano, L., Körner, J. and Vaccaro, U. (1992) Qualitative independence and Sperner problems for directed graphs. *J. Comb. Theory, Ser. A.* **61**, 173–192.

Gargano, L., Körner, J. and Vaccaro, U. (1993) Sperner capacities. *Graph. Combinator.* **9**, 31–46.

Gargano, L., Körner, J. and Vaccaro, U. (1994) Capacities: from information theory to extremal set theory. *J. Comb. Theory, Ser. A.* **68**, 296–315.

Gelfand, S. I. (1977) The capacity of a broadcast channel. *PPI* **13**, (*3*), 106–108. In Russian.

Gelfand, S. I. and Pinsker, M. S. (1979) Source coding with incomplete side information. *PPI* **15**, (*2*), 45–57. In Russian.

Gelfand, S. I. and Pinsker, M. S. (1980a) Capacity of a broadcast channel with one deterministic component. *PPI* **16**, (*1*), 24–34. In Russian.

Gelfand, S. I. and Pinsker, M. S. (1980b) Coding for channels with random parameters. *PCIT* **9**, 19–31.

Gelfand, S. I. and Prelov, V. V. (1978) Multiple user communication. *Itogi Nauki i Tekhniki, Teoriya Veroyatn*, vol. 15, Moscow, pp. 123–162.

Gerrish, A. M. (1963) Estimation of information rates. Ph.D. Thesis, Dept. Electrical Engineering, Yale University, New Haven, CT. Quoted by Berger (1971).

Gibbs, J. W. (1902) *Elementary Principles in Statistical Mechanics.* Yale University Press, New Haven, CT.

Gilbert, E. N. (1952) A comparison of signalling alphabets. *BSTJ* **31**, 504–522. (Reprinted in *KP*.)

Gilbert, E. N.–Moore, E. F. (1959) Variable length binary encodings. *BSTJ* **38**, 933–967.

Gohari, A. A. and Anantharam, V. (2010) Information-theoretic key agreement of multiple terminals I–II. *IEEE-IT* **56**, 3973–3996, 3997–4010.

Goppa, V. D. (1975) Nonprobabilistic mutual information without memory. *PCIT* **4**, 97–102.

Gray, R. M. and Wyner, A. D. (1974) Source coding for a simple network. *BSTJ* **58**, 1681–1721.

Greco, G. (1998) Capacities of graphs and 2-matchings. *Discrete Math.* **186**, 135–143.

Gubner, J. (1990) On the deterministic-code capacity of the multiple-access arbitrarily varying channel. *IEEE-IT* **36**, 262–275.

Haemers, W. (1979) On some problems of Lovász concerning the Shannon capacity of a graph. *IEEE-IT* **25**, 231–232.

Hajnal, A. and Surányi, J. (1958) Über die Auflösung von Graphen in vollständige Teilgraphen. *Ann. Univ. Sci. Budapest. Sectio Math.* **1**, 113–121.

Hamming, R. V. (1950) Error detecting and error correcting codes. *BSTJ* **29**, 147–160.

Han, T. S. (1979) The capacity region of general multiple-access channel with certain correlated sources. *IC* **40**, 37–60.

Han, T. S. (1980) Slepian–Wolf–Cover theorem for a network of channels. IC **47**, 67–83.

Han, T. S. (1981) A uniqueness of Shannon's information distance and related non-negativity problems. *J. Comb. Inform. Syst. Sci.* **6**, 320–331.

Han, T. S. (2003) *Information Spectrum Methods in Information Theory.* Springer, New York.

Han, T. S. and Kobayashi, K. (1980) A unified achievable rate region for a general class of multiterminal source coding systems. *IEEE-IT* **26**, 277–288.

Han, T. S. and Kobayashi, K. (1981) A new achievable rate region for the interference channel. *IEEE-IT* **27**, 49–60.

Hardy, G. H., Littlewood, J. E. and Pólya, G. (1934) *Inequalities.* Cambridge University Press, Cambridge.

Haroutunian, E. A. (1968) Estimates of the error exponent for the semi-continuous memoryless channel. *PPI* **4**, (*4*), 37–48. In Russian.

Haroutunian, E. A. (1977) A lower bound of the probability of error for channels with feedback. *PPI* **3**, (*2*), 36–44. In Russian.

Harper, L. H. (1966) Optimal numberings and isoperimetric problems on graphs. *J. Comb. Theory* **1**, 385–394.

Hartley, R. V. L. (1928) Transmission of information. *BSTJ* **7**, 535.

Hoeffding, W. (1963) Probability inequalities for sums of bounded random variables. *J. Am. Stat. Assoc.* **58**, 13–30.

Hoeffding, W. (1965) Asymptotically optimal tests for multinomial distributions. *AMS* **36**, 369–400.

Horibe, Y. (1973) A note on entropy metrics. *IC* **22**, 403–404.

Horstein, M. (1963) Sequential transmission using noiseless feedback. *IEEE-IT* **9**, 136–143.

Hughes, B. (1997) The smallest list for the arbitrarily varying channel. *IEEE-IT* **43**, 803–815.

Hu Guo Ding (1962) On the amount of information. *Teor. Veroyatnost. i Primenen.* **4**, 447–455. In Russian.

Huffman, D. A (1952) A method for the construction of minimum redundancy codes. *Proc. IRE* **40**, 1098–1101. (Reprinted in *KP*.)

Impagliazzo, R., Levin, L. and Luby, M. (1989) Pseudo-random generation from one-way functions. *Proc. 21th ACM Symp. Theory of Computing*, pp. 12–24.

Impagliazzo, R. and Zuckerman, D. (1989) How to recycle random bits. In *30th IEEE Symp. Found. Computer Sci.*, pp. 248-253.

Jahn, J. H. (1981) Coding of arbitrarily varying multiuser channels. *IEEE-IT* **27**, 212–226.

Jeffreys, H. (1946) An invariant form for the prior probability in estimation problems. *Proc. Roy. Soc. (London) Ser. A*, **186**, 453–461.

Jelinek, F. (1967) Evaluation of distortion-rate functions for low distortions. *Proc. IEEE* **55**, 2067–2068.

Jelinek, F. (1968a) Evaluation of expurgated bound exponents. *IEEE-IT* **14**, 501–505.

Jelinek, F. (1968b) *Probabilistic Information Theory.* McGraw Hill, New York.

Jelinek, F. and Schneider, K. (1972) On variable-length-to-block coding. *IEEE-IT* **18**, 765–774.

Jiang, J., Xin, Y. and Garg, H. K. (2008) Interference channels with common information. *IEEE-IT* **54** 171–187.

Karamata, J. (1932) Sur une inégalité relative aux fonctions convexes. *Publ. Math. Univ. Belgrade* **1**, 145–148.

Karlin, S. (1959) *Mathematical Methods and Theory in Game, Programming and Economics.* Addison-Wesley, Reading, MA.

Karmažin, M. A. (1964) Solution of a problem of Shannon. *Probl. Kibern.* **11**, 263–266. In Russian.

Karush, J. (1961) A simple proof of an inequality of McMillan. *IRE-IT* **7**, 118. (Reprinted in *KP.*)

Katona, G. O. H. (2004) Strong qualitative independence. *Discrete Appl. Math.* **1**, 87–95.

Katona, G. O. H. and Nemetz, T. O. H. (1976) Huffman codes and self information. *IEEE-IT* **22**, 337–340.

Kawabata, T. and Yeung, R. (1992) The structure of the I-measure of a Markov chain. *IEEE-IT* **38**, 1146–1149.

Kemperman, J. H. B. (1969) On the optimum rate of transmitting information. In *Probability and Information Theory*. Lecture Notes in Mathematics **89**. Springer, New York, pp. 126–169.

Kiefer, J. and Wolfowitz, J. (1962) Channels with arbitrarily varying channel probability functions. *IC* **5**, 44–54.

Kolmogorov, A. N. (1958) A new invariant for transitive dynamical systems. *Dokl. AN SSSR* **119**, 861–864. In Russian.

Korn, I. (1968) On the lower bound of zero-error capacity. *IEEE-IT* **14**, 509–510.

Körner, J. (1973a) Coding of an information source having ambiguous alphabet and the entropy of graphs. In *Proc. 6th Prague Conference on Inf. Theory*. Academia, Prague, pp. 411–425.

Körner, J. (1973b) Coding of finite sources with sum-type distortion. Preprint.

Körner, J. (1973c) An extension of the class of perfect graphs. *Studia Sci. Math. Hungar.* **8**, 405–409.

Körner, J. (1975) Some methods in multi-user communication: a tutorial survey. In *Information Theory, New Trends and Open Problems* (ed. G. Longo). CISM Courses and Lectures no. 219, Springer, Wien. pp. 173–224.

Körner, J. (1977) On a simple source network. Paper presented at the *IEEE Int. Symp. on IT*, October 1977, Cornell University Ithaca, NY.

Körner, J. (1984) OPEC or a basic problem in source networks. *IEEE-IT* **30**, 68–77.

Körner, J. (1986) Fredman–Komlós bounds and information theory, *SIAM J. Algebraic Discrete Meth.* **7**, 560–570.

Körner, J. and Malvenuto, C. (2006) Pairwise colliding permutations and the capacity of infinite graphs. *SIAM J. Discrete Math.* **20**, 203–212.

Körner, J. and Marton, K. (1977a) The comparison of two noisy channels. *Topics in IT*, pp. 411–423.

Körner, J. and Marton, K. (1977b) General broadcast channels with degraded message sets. *IEEE–IT* **23**, 60–64.

Körner, J. and Marton, K. (1977c) Images of a set via two channels and their role in multi-user communication. *IEEE-IT* **23**, 751–761.

Körner, J. and Marton, K. (1979) How to encode the modulo 2 sum of two binary sources. *IEEE-IT* **25**, 219–221.

Körner, J. and Marton, K. (1988a) New bounds for perfect hashing via information theory. *Eur. J. Combinatorics* **9**, 523–530.

Körner, J. and Marton, K. (1988b) Graphs that split entropies. *SIAM J. Discrete Math.* **1**, 71–79.

Körner, J. and Orlitsky, A. (1998) Zero-error information theory. *IEEE-IT* **44**, 2207–2229.

Körner, J., Pilotto, C. and Simonyi, G. (2005) Local chromatic number and Sperner capacity. *J. Comb. Theory Ser. B* **95**, 101–117.

Körner, J. and Simonyi, G. (1992) A Sperner-type theorem and qualitative independence. *J. Comb. Theory Ser. A.* **59**, 90–103.

Körner, J., Simonyi, G. and Sinaimeri, G. (2009) On types of growth for graph–different permutations, *J. Comb. Theory Ser. A* **116**, 713–723.

Körner, J., Simonyi, G. and Tuza, Zs. (1992) Perfect couples of graphs. *Combinatorica* **12**, 179–192.

Koševel, V. N. (1977) On a problem of separate coding of two dependent sources. *PPI* **13**, (*1*), 26–32. In Russian.

Kraft, L. G. (1949) A device for quantizing, grouping and coding amplitude modulated pulses. MS Thesis, Dept. of Electrical Engineering, MIT, Cambridge, MA.

Kramer, G. (2003) Capacity results for the discrete memoryless network. *IEEE-IT* **46**, 4–21.

Kramer, G. (2008) *Topics in Multi-User Information Theory.* Now Publishers, Delft.

Krause, R. M. (1962) Channels which transmit letters of inequal duration. *IC* **5**, 13–24.

Kričevskiĭ, R. E. (1968) The relation between redundancy of coding and the reliability of information about a source. *PPI* **4**, (*3*), 48–57. In Russian.

Kričevskiĭ, R. E. (1970) *Lectures on Information Theory.* Novosibirsk State University. In Russian.

Kričevskiĭ, R. E. and Trofimov, V. K. (1977) Optimal coding of unknown and inaccurately known sources. *Topics in IT*, pp. 425–430.

Kullback, S. (1959) *Information Theory and Statistics.* Wiley, New York.

Kullback, S. (1967) A lower bound for discrimination in terms of variation. *IEEE-IT* **13**, 126–127.

Kullback, S. and Leibler, R. A. (1951) On information and sufficiency. *AMS* **22**, 79–86.

Kuznetsov, A. V. and Tsybakov, B. S. (1974) Coding in a memory with defective cells. *PPI* **10**, 52–60. In Russian.

Ledoux, M. (2001) *The Concentration of Measure Phenomenon.* Amer. Math. Soc., Providence, RI.

Liang, Y., Poor, V. and Shamai, S. (2009) *Information Theoretic Security.* Now Publishers, Delft.

Liang, Y., Somekh-Baruch, A., Poor, V., Shamai, S. and Verdú, S. (2009) Capacity of cognitive interference channels with and without secrecy. *IEEE-IT* **55**, 604–619.

Liao, H. J. (1972) Multiple access channels. Ph.D. Thesis, Dept. of Electrical Engineering, University of Hawaii, Honolulu. Quoted by Slepian and Wolf (1973b).

Lind, D. and Marcus, B. (1995) *An Introduction to Symbolic Dynamics and Coding.* Cambridge University Press, Cambridge.

Linder, T., Lugosi, G. and Zeger, K. (1995) Fixed-rate universal lossy source coding and rates of convergence for memoryless sources. *IEEE-IT* **41**, 665–676.

Ljubič, Yu. I. (1962) Remark on the capacity of the noiseless communication channels. *Usp. Mat. Nauk* **17**, 191–198. In Russian.

Longo, G. and Sgarro, A. (1979) The source coding theorem revisited: a combinatorial approach. *IEEE-IT* **25**, 544–548.

Lovász, L. (1972) Normal hypergraphs and the perfect graph conjecture. *Discrete Math.* **2**, 253–267.

Lovász, L. (1975) On the ratio of optimal integral and fractional covers. *Discrete Math.* **13**, 383–390.

Lovász, L. (1979) On the Shannon capacity of a graph. *IEEE–IT* **25**, 1–7.

Lubell, D. (1966) A short proof of Sperner's lemma *J. Comb. Theory* **1**, 299

Lynch, T. J. (1966) Sequence time coding for data compression. *Proc. IEEE* **54**, 1490–1491.

McEliece, R. J. and Omura, J. K. (1977) An improved upper bound on the block coding error exponent for binary-input discrete memoryless channels. *IEEE-IT* **23**, 611–613.

McEliece, R. J. and Posner, E. C. (1971) Hide and seek, data storage, and entropy. *AMS* **42**, 1706–1716.

McEliece, R. J., Rodemich, E. R., Rumsey, H. Jr. and Welch, L. R. (1977) New upper bounds on the rate of a code via the Delsarte–MacWilliams inequalities. *IEEE-IT* **23**, 157–166.

McMillan, B. (1956) Two inequalities implied by unique decipherability. *IRE-IT* **2**, 115–116.

Margulis, G. A. (1974) Probabilistic characteristics of graphs with large connectivity. *PPI* **10**, (2), 101–108. In Russian.

Marton, K. (1974) Error exponent for source coding with a fidelity criterion. *IEEE-IT* **20**, 197–199.

Marton, K. (1979) A coding theorem for the discrete memoryless broadcast channel. *IEEE-IT* **25**, 306–311.

Marton, K. (1986) A simple proof of the blowing-up lemma. *IEEE-IT* **32**, 445–446.

Massey, J. L. (1974) On the fractional weight of distinct binary n-tuples. *IEEE-IT* **20**, 131.

Matúš, F. (2007) Infinitely many information inequalities. *Proc. ISIT 2007*, Nice, France, pp. 41–44.

Maurer, U. M. (1993) Secret key agreement by public discussion from common information. *IEEE-IT* **39**, 733–742.

Maurer, U. M. (1994) The strong secret key of discrete random triples. In *Communication and Cryptography – Two Sides of One Tapestry* (eds. R. E. Blahut, J. Costello, U. Maurer and T. Mittelholzer). Kluwer, Dordrecht, pp. 271–285.

Maurer, U. M. and Wolf, S. (1999) Unconditionally secure key agreement and the intrinsic conditional information. *IEEE-IT* **45**, 499–514.

Maurer, U. M. and Wolf, S. (2000) Information-theoretic key agreement: from weak to strong secrecy for free. In *Advances in Cryptology – Eurocrypt 2000.* Lecture Notes in Computer Science **1807**. Springer, Berlin, pp. 351–368.

Meshalkin, L. D. (1963) A generalization of Sperner's theorem on the number of subsets of a finite set. *Teor. Veroyatn.i Primenen.* **8**, 219–220. In Russian.

Muroga, S. (1953) On the capacity of a discrete channel. *J. Phys. Soc. Japan* **8**, 484–494.

Nayak, I. J. and Rose, K. (2005) Graph capacities and zero-error transmission over compound channels. *IEEE-IT* **51**, 4374–4378.

Nemetz, T. and Simon. J. (1977) Self-information and optimal codes. *Topics in IT*, pp. 457–468.

Neuhoff, D. L., Gray, R. M. and Davisson, L. D. (1975) Fixed rate universal block source coding with a fidelity criterion. *IEEE-IT* **21**, 511–524.

Ng, C. T. (1974) Representation for measures of information with the branching property. *IC* **25**, 45–56.

Nisan, N. and Zuckerman, D. (1996) Randomness is linear in space. *J. Comput. Syst. Sci.* **52**, 43–52.

Omura, J. K. (1974) Expurgated bounds, Bhattacharyya distance and rate distoriton functions. *IC* **24**, 358–383.

Omura, J. K. (1975) A lower bounding method for channel and source coding probabilities. *IC* **27**, 148–177.

Oohama, Y. and Han, T.S. (1994) Universal coding for the Slepian–Wolf data compression system and the strong converse. *IEEE-IT* **40**, 1908–1919.

Ornstein, D. S. (1970) Bernoulli shifts with the same entropy are isomorphic. *Adv. Math.* **4**, 337–352.

Ornstein, D. S. (1973) *Ergodic Theory, Randomness and Dynamical Systems.* Yale University Press, New Haven, CT.

Pasco, R. (1976) *Source Coding Algorithms for Fast Data Compression.* Ph.D. Thesis, Stanford University, Stanford, CA. Quoted by Cover (2006).

Pilc, R. (1968) The transmission distortion of a source, as a function of the encoding block-length. *BSTJ* **47**, 827–885.

Pinsker, M. S. (1960) *Information and Information Stability of Random Variables and Processes.* Problemy Peredači Informacii **7**, AN SSSR, Moscow. English translation: Holden-Day, San Francisco, CA, 1964.

Pinsker, M. S. (1978) Capacity region of noiseless broadcast channels. *PPI* **14**, (2), 28–32. In Russian.

Pinsker, M. S. and Ševerdyaev, A. Yu. (1970) Zero error capacity with erasure. *PPI* **6**, (*1*), 20–24. In Russian.

Plotkin, M. (1960) Binary codes with specified minimum distance. *IRE-IT* **6**, 445–450.

Poljak, S. and Rödl, V. (1980) Orthogonal partitions and coverings of graphs. *Czech Math. J.* **30**, 475–485.

Poljak, S., Pultr. A. and Rödl, V. (1983) On qualitatively independent partitions and related problems. *Discrete Appl. Math.* **6**, 193–205.

Poljak, S. and Tuza, Zs. (1989) On the maximum number of qualitatively independent partitions. *J. Comb. Theory Ser. A.* **51**, 111–116.

Poltyrev, G. S. (1977) Carrying capacity for parallel broadcast channels with degraded components. *PPI* **13**, (2), 23–35. In Russian.

Poltyrev, G. S. (1979) Capacity for a sum of broadcast channels. *PPI* **15**, (2), 40–44. In Russian.

Rajski, C. (1961) A metric space of discrete probability distributions. *IC* 4, 371–377.

Renner, R. and Wolf, S. (2003) New bounds in secret-key agreement: the gap between formation and secrecy extraction. In *Advances in Cryptology – Eurocrypt 2003.* Lecture Notes in Computer Science **2656**. Springer, New York, pp. 562–577.

Renner, R. and Wolf, S. (2005) Simple and tight bounds for information reconciliation and privacy amplification. In *Advances in Cryptology – Asiacrypt 2005.* Lecture Notes in Computer Science **3788**. Springer, New York, pp. 199–216.

Rényi, A. (1961) On measures of entropy and information. In *Proc. 4th Berkeley Symp. Math. Statist. Prob.*, vol. 1. University of California Press, Berkeley, pp. 547–561.

Rényi, A. (1965) On the foundations of information theory. *Rev. Inst. Int. Stat.* **33**, 1–14.

Rényi, A. (1970) *Foundations of Probability.* Holden-Day, San Francisco, CA.

Reza, F M. (1961) *An Introduction to Information Theory.* McGraw-Hill, New York.

Richardson, T. and Urbanke, R. (2008) *Modern Coding Theory.* Cambridge University Press, Cambridge.

Rissanen, J. (1976) Generalized Kraft inequality and arithmetic coding. *IBM J. Res. Dev.* **20**, 198–203.

Rissanen, J. (1989) *Stochastic Complexity in Statistical Inquiry.* World Scientific, Singapore.

Ryabko, B. (1979) Encoding a source with unknown but ordered probabilities. *PPI* **15**, 71–77. In Russian.

Sali, A. and Simonyi, G. (1999) Orientations of self-complementary graphs and the relation of Sperner and Shannon capacities. *Eur. J. Combin.* **20**, 93–99.

Sanov, I. N. (1957) On the probability of large deviations of random variables. *Mat. Sbornik* **42**, 11–44. In Russian.

Sántha, M. and Vazirani, U. (1986) Generating quasi-randomness from semi-random sources. *J. Comp. Syst. Sci.* **33**, 75–87.

Sardinas, A. A. and Patterson, G. W. (1953) A necessary and sufficient condition for the unique decomposition of coded messages. *IRE Convention Record*, Part 8, pp. 104–108.

Sauer, N. (1972) On the density of families of sets. *J. Comb. Theory Ser.* A, **13**, 145–147.

Schmetterer, L. (1974) *Introduction to Mathematical Statistics*. Springer, Berlin.

Schrijver, A. (1986) *Theory of Linear and Integer Programming.* Wiley, New York.

Schützenberger, M. P. (1954) Contribution aux applications statistiques de la théorie de l'information. *Publ. Inst. Statist. Univ. Paris* **3**, 3–117.

Schützenberger, M. P. (1967) On synchronizing prefix codes. *IC* **11**, 396–401.

Schützenberger, M. P. and Marcus. R. S. (1959) Full decodable codeword sets. *IRE-IT* **5**, 12–15.

Sgarro, A. (1977) Source coding with side information at several decoders. *IEEE-IT* **23**, 179–182.

Shannon, C. E. (1948) A mathematical theory of communication. *BSTJ* **27**, 379–423, 623–656. (Reprinted in *KP*.)

Shannon, C. E. (1949) Communication theory of secrecy systems. *BSTJ* **28**, 656–715.

Shannon, C. E. (1953) The lattice theory of information. *Trans. IRE Prof. Group Inform. Theory* **1**, 105–107.

Shannon, C. E. (1956) The zero error capacity of a noisy channel. *IRE-IT* **2**, 8–19. (Reprinted in *KP*.)

Shannon, C. E. (1957a) Certain results in coding theory for noisy channels. *IC* **1**, 6–25. (Reprinted in *KP*.)

Shannon, C. E. (1957b) Geometrische Deutung einiger Ergebnisse bei der Berechnung der Kanalkapazität. *Nachrichtentechn. Zeitschr.* **10**, 1–8.

Shannon. C. E. (1958a) A note on a partial ordering for communication channels. *IC* **1**, 390–398. (Reprinted in *KP*.)

Shannon. C. E. (1958b) Channels with side information at the transmitter. *IBM J. Res. Dev.* **2**, 289–293. (Reprinted in *KP*.)

Shannon, C. E. (1959) Coding theorems for a discrete source with a fidelity criterion. In *IRE Nat. Convention Record*, Part 4, pp. 142–163. (Reprinted in *KP*.)

Shannon, C. E. (1961) Two-way communication channels. In *Proc. 4th Berkeley Symp. Math. Statist. Prob.*, vol. 1. University of California Press, Berkeley, pp. 611–644. (Reprinted in *KP*.)

Shannon, C. E., Gallager, R. G. and Berlekamp, E. R. (1967) Lower bounds to error probability for coding in discrete memoryless channels I–II. *IC* **10**, 65–103, 522–552. (Reprinted in *KP*.)

Shelah, S. (1972) A combinatorial problem: stability and order for models and theories in infinitary languages. *Pacific J. Math.* **41**, 247–261.

Shor, P. W. (1985) A counterexample to the triangle conjecture. *J. Comb. Theory Ser. A.* **38**, 110–112.

Simonyi, G. (2003) On Witsenhausen's zero-error rate for multiple sources. *IEEE-IT* **49**, 3258–3261.

Sinaĭ, J. G. (1959) On the notion of entropy of a dynamical system. *Dokl. Akad. Nauk SSSR* **124**, 708–711. In Russian.

Slepian, D. and Wolf., J. K. (1973a) Noiseless coding of correlated information sources. *IEEE-IT* **19**, 471–480. (Reprinted in *KP*.)

Slepian, D. and Wolf., J. K. (1973b) A coding theorem for multiple access channels with correlated sources. *BSTJ* **52**, 1037–1076.

Smorodinsky, M. (1968) The capacity of a general noiseless channel and its connection with Hausdorff dimension. *Proc. Am. Math. Soc.* **19**, 1247–1254.

Sobel, M. (1960) Group testing to classify efficiently all defectives in a binomial sample. In *Information and Decision Processes* (ed. R. E. Machol). McGraw Hill, New York, pp. 127–161.

Sperner, E. (1928) Ein Satz über Untermengen einer endlichen Menge. *Math. Z.* **27**, 544–548.

Stambler, S. Z. (1975) Shannon theorems for a full class of channels with state known at the output. *PPI* **14**, (*4*), 3–12. In Russian.

Stiglitz, I. G. (1966) Coding for a class of unknown channels. *IEEE-IT* **12**, 189–195.

Strassen, V. (1964) Asymptotische Abschätzungen in Shannon's Informationstheorie. In *Transactions of the Third Prague Conference on Information Theory, 1962*. Academia, Prague, pp. 689–723.

Szekeres, Gy. and Turán, P. (1937) An extremum problem in the theory of determinants. *Matematikai és Természettudományi Értesitő* **56**, 796–804. In Hungarian.

Talagrand, M. (1995) Concentration of measure and isoperimetric inequalities in product spaces. *Inst. Hautes Études Sci. Math.* **81**, 73–205.

Thomasian, A. J. (1961) Error bounds for continuous channels. In *Fourth London Symposium on Information Theory* (ed. C. Cherry). Butterworths, Washington, D.C., pp. 46–60.

Topsøe, F. (1967) An information theoretical identity and a problem involving capacity. *Studia Sci. Math. Hungar.* **2**, 291–292.

Topsøe, F. (1972) A new proof of a result concerning computation of capacity for a discrete channel. *ZW* **22**, 166–168.

Trevisan, L. and Vadhan, S.P. (2000) Extracting randomness from samplable distributions. In *41st IEEE Symp. Found. Computer. Sci.*, pp. 32–42.

Tunstall, B. P. (1968) Synthesis of noiseless compression codes. Ph.D. Thesis, Georgia Institute of Technology. Quoted in Jelinek and Schneider (1972).

Ulrey, M. L. (1975) The capacity region of a channel with *s* senders and *r* receivers. *IC* **29**, 185–203.

Urbanke, R. and Li, Q. (1988) The zero-error capacity region of the 2-user synchronous BAC is strictly smaller than its Shannon capacity region. In *Proc. IEEE IT Workshop, Killarney, Ireland*, p. 61.

Van der Meulen, E. C. (1975) Random coding theorems for the general discrete memoryless broadcast channel. *IEEE-IT* **21**, 180–190.

Van der Meulen, E. C. (1977) A survey of multi-way channels in information theory: 1961–1976. *IEEE-IT* **23**, 1–37.

Vapnik, V. N. and Červonenkis, A. Ya. (1971) On the uniform convergence of relative frequencies of events to their probabilities. *Teor. Veroyatnost. i Primenen.* **16**, 264–280. In Russian.

Vembu, S. and Verdú, S. (1995) Generating random bits from an arbitrary source: fundamental limits. *IEEE-IT* **41** 1322–1332.

Verdú, S. (1990) On channel capacity per unit cost. *IEEE-IT* **36**, 1019–1030.

Verdú, S. and Han, T. S. (1994) A general formula for channel capacity. *IEEE-IT* **40**, 1147–1157.

Weiss, L. (1960) On the strong converse of the coding theorem for symmetric channels without memory. *Quart. Appl. Math.* **18**, 209–214.

Willems, F. (1982) The feedback capacity region of a class of discrete memoryless multiple access channels. *IEEE-IT* **28**, 93–95.

Willems, F. and Van der Meulen, E. (1983) Partial feedback for the discrete memoryless multiple access channel. *IEEE-IT* **29**, 287–290.

Willems, F., Shtarkov, Yu. and Tjalkens, J. (1995) The context-tree weighting method: basic properties. *IEEE-IT* **41**, 653–664.

Witsenhausen, H. S. (1974) Entropy inequalities for discrete channels. *IEEE-IT* **20**, 610–616.

Witsenhausen, H. S. (1975) On sequences of pairs of dependent random variables. *SIAM J. Appl. Math.* **28**, 100–113.

Witsenhausen, H. S. (1976) The zero-error side information problem and chromatic numbers. *IEEE-IT* **22**, 592–593.

Wolf, J. K. (1974) Data reduction for multiple correlated sources. In *Proc. of the Fifth Colloquium on Microwave Communication*, Budapest, 1974, pp. 287–295.

Wolfowitz, J. (1957) The coding of messages subject to chance errors. *Illinois J. Math.* **1**, 591–606. (Reprinted in *KP*.)

Wolfowitz, J. (1960) Simultaneous channels. *Arch. Rat. Mech. Anal.* **4**, 371–386.

Wolfowitz, J. (1963) On channels without capacity. *IC* **6**, 49–54.

Wolfowitz, J. (1964) *Coding Theorems of Information Theory*, 2nd edn. Springer, Berlin.

Wolfowitz, J. (1966) Approximation with a fidelity criterion. In *Proc. 5th Berkeley Symp. Math. Statist. Prob.*, vol. 1. University of California Press, Berkeley, pp. 565–573.

Wyner, A. D. (1974) Recent results in the Shannon theory. *IEEE-IT* **20**, 2–10.

Wyner, A. D. (1975a) The common information of two dependent random variables. *IEEE-IT* **21**, 163–179.

Wyner, A. D. (1975b) On source coding with side information at the decoder. *IEEE-IT* **21**, 294–300.

Wyner, A. D. (1975c) The wire-tap channel. *BSTJ* **54**, 1355–1387.

Wyner, A. D., Wolf, J. K. and Willems, F. M. J. (2002) Communicating via a processing satellite. *IEEE-IT* **48**, 1243–1249.

Wyner, A. D. and Ziv, J. (1973) A theorem on the entropy of certain binary sequences and applications I–II. *IEEE-IT* **19**, 769–778.

Wyner, A. D. and Ziv, J. (1976) The rate-distortion function for source coding with side information at the decoder. *IEEE-IT* **22**, 1–11.

Yamamoto, K. (1954) Logarithmic order of free distributive lattices. *J. Math. Soc. Japan* **6**, 343–353.

Yamamoto, H. and Itoh, K. (1979) Asymptotic performance of a modified Schalkwijk–Barron scheme for channels with noiseless feedback. *IEEE-IT* **25**, 729–733.

Yeung, R.W. (1991) A new outlook on Shannon's information measures. *IEEE-IT* **37**, 466–477.

Yeung, R.W. (2008) *Information Theory and Network Coding*. Springer.

Zhang, Z. (2007) Estimating mutual information via Kolmogorov distance. *IEEE-IT* **53**, 3280–3283.

Zhang, Z., Yang, E. and Wei, V. K. (1997) The redundant source coding with a fidelity criterion. *IEEE-IT* **43**, 71–91.

Zhang, Z. and Yeung, R.W. (1997) A non-Shannon-type conditional inequality of information quantities. *IEEE-IT* **43**, 1982–1986.

Zhang, Z. and Yeung, R. W. (1998) On characterization of entropy functions via information inequalities. *IEEE-IT* **44**, 1440–1452.

Zigangirov, K. Sh. (1970) Upper bounds on the error probability for channels with feedback. *PPI* **6**, (*2*), 87–92. In Russian.

Zigangirov, K. Sh. (1978) Optimum zero rate transmission through binary symmetric channel with feedback. *PCIT* **7**, 183–198.

Zuckerman, D. (1997) Randomness-optimal oblivious sampling. *Random Struct. Algor.* **11**, 345–367.

Name index

Index of symbols and abbreviations

For the basic notation and symbols used throughout the book see pp. xii–xiii. This list includes notation introduced in the text and used repeatedly without reference. The page numbers in the right-hand column show the number of the page where first used.

Subject index

Printed in the United States
by Baker & Taylor Publisher Services